# ENCYCLOPEDIA OF PHYSICS

EDITED BY

S. FLÜGGE

VOLUME XIX

## ELECTRICAL CONDUCTIVITY I

WITH 208 FIGURES

SPRINGER-VERLAG
BERLIN · GÖTTINGEN · HEIDELBERG
1956

# HANDBUCH DER PHYSIK

HERAUSGEGEBEN VON

S. FLÜGGE

BAND XIX

# ELEKTRISCHE LEITUNGSPHÄNOMENE I

MIT 208 FIGUREN

SPRINGER-VERLAG
BERLIN · GÖTTINGEN · HEIDELBERG
1956

SOFTCOVER REPRINT OF THE HARDCOVER 1ST EDITION 1956

ISBN-13: 978-3-642-62041-6 e-ISBN-13: 978-3-642-62039-3
DOI: 10.1007/978-3-642-62039-3

DRUCK DER UNIVERSITÄTSDRUCKEREI H. STÜRTZ AG., WÜRZBURG

# Contents.

# The Electronic Structure of Solids.

By

J. C. SLATER.

With 60 Figures.

## I. The periodic potential problem: General properties.

**1. The self-consistent method for solids.** In principle, we should be able to find out everything about a crystal by solving SCHRÖDINGER's equation for the problem consisting of the nuclei and electrons forming the crystal. In most of our work the nuclei will be assumed held fixed, in the positions characteristic of a perfect crystal at the absolute zero of temperature, and we shall interpret our task as the investigation of the electronic wave functions and energy levels, and their bearing on the electric, magnetic, mechanical, and other properties of the crystals. The most useful method for handling this problem has proved to be HARTREE's [*84*][1] method of the self-consistent field. As is well known, HARTREE assumes each electron in an atom to move in a field produced electrostatically by the nucleus, and the charge densities of all other electrons, averaged over their wave functions. The condition of self-consistency implies that the wave function of each electron is that resulting from SCHRÖDINGER's equation for a one-electron problem arising from the electrostatic field so determined.

Three sorts of information are given by the self-consistent field method. First, we find the charge distributions in the atoms. Secondly, the one-electron energy parameters of the various electrons prove to represent rather accurately the ionization energies of the corresponding electrons of the atom, giving x-ray and optical term values. Thirdly, from the one-electron wave functions, one can construct a many-electron wave function, in determinantal form, which is assumed to form a good approximation to a solution of SCHRÖDINGER's equation for the many-electron problem. The expectation value of the energy of this many-electron wave function forms a good approximation to the energy required to form the atom from its nucleus and separated electrons.

One can apply the variation principle of quantum mechanics, and can vary the one-electron wave functions, in such a way as to minimize the expectation value of the energy; in this way one gets the best possible approximation, in the form of a single determinant, to a solution of the problem of the motion of the electrons in the atom. It was shown by FOCK [*71*] and the present writer [*214*] that this variation leads essentially to HARTREE's procedure. The slight modification arising from exchange, leading to the so-called HARTREE-FOCK equations, will be discussed in Sect. 17. It was demonstrated by KOOPMANS [*136*] that there is theoretical justification for the fact, noted empirically by HARTREE, that the one-electron energy parameters agree with the ionization energies. The

[1] References throughout are to the Bibliography at the end of this article. Many references to papers, by HARTREE and others, using this method for atomic structure, are given in LANDOLT-BÖRNSTEIN, Zahlenwerte und Funktionen aus Physik, Chemie, Astronomie. (Berlin, Springer 1950), sixth edition, Vol. I, p. 276.

method of the self-consistent field has been extended by many other authors, notably by BRILLOUIN [*34*], [*35*], [*36*], [*38*], [*39*] whose treatment makes close connection with problems of solids.

We shall return in Chapter III to the detailed problem of finding the potential in which an electron moves, given the wave functions. For a perfect crystal, the result is what we should expect: the potential is periodic, repeating its value at all equivalent points in the crystal. Near each atom, the potential is similar to that of an electron moving in that atom. According to the HARTREE-FOCK equations, the potential is slightly different for each electron, though it still maintains the same symmetry; but in preliminary work we may disregard this fact, and treat the problem of an electron moving in a single periodic potential having the symmetry of the crystal. It is this problem which we shall handle in the next few sections.

As with atoms, there are three sorts of information which we can get out of the results. First, we find the charge and current density. Secondly, we have the one-electron energies, measuring the energies required to remove the various electrons. It is this information which leads us to the theory of energy bands. Thirdly, we can find the total energy of the crystal, and as we study this as it varies with varying positions of the nuclei, we have information about elastic properties and cohesive energy.

The HARTREE-FOCK method is only an approximation, and for some purposes it is not adequate, though for many problems, such as electrical conductivity, it is satisfactory. In the later chapters of this article, we shall take up some higher approximations, related to the HEITLER-LONDON method, and the configuration interaction methods used in molecular theory, and shall see that these methods are particularly necessary in such problems as magnetism.

**2. Crystalline properties of elements and simple compounds.** Before entering on the mathematical study of the motion of electrons in a periodic potential, let us consider the structure of some of the crystals we shall meet, so as to know what periodic potentials we must deal with. Instead of covering the whole vast subject of solids in general, we shall limit ourselves to a few types of substances of particular importance, on which a fair amount of theoretical work has been done. In particular, we shall consider the typical metals, having the body- or face-centered cubic or hexagonal close-packed structures; those elements, like diamond, silicon, and germanium, which are important insulators or semiconductors; and a number of compounds of simple type, including ionic crystals of the alkali halide type, and crystals having the zinc-blende structure, intermediate in their properties between ionic crystals and semiconductors. It is well for us to remember what a minute fraction of the whole field of solids these simple substances form; it will be important in future work on solids to consider more complicated substances, both theoretically and experimentally.

We start our survey by considering the chemical nature of the elements, and their crystal structures. We notice in the first place that a very large majority of the chemical elements are metals. Obviously non-metallic are the inert gases, the halogens, and the light elements hydrogen, nitrogen, and oxygen. Of the remaining ones, we start each group in the periodic table with alkali and alkaline earth metals; in the later groups, we add $3d$, $4d$, or $5d$ electrons, and in the rare earths $4f$ electrons, all these transition elements being distinctly metallic. Then we close the group by completing the groups of $p$ electrons. As the shells of $p$ electrons become filled, the properties become continuously less metallic. Thus, most of the elements with two $p$ electrons, namely C, Si, Ge, and Sn, have diamond structures, and are insulators or semiconductors, though Pb is metallic. The elements with three $p$ electrons, N, P, As, Sb, Bi, start as definitely non-metallic, but As, Sb, and Bi have rather metallic properties, though they are unusual metals. Elements with four $p$ electrons, namely O, S, Se, Te, are all non-metallic, as are the halogens, with five $p$ electrons.

Among the typical metallic elements, three crystal structures preponderate: the two close-packed structures, namely the face-centered cubic and hexagonal close-packed structures, and the body-centered cubic. There are exceptions; for instance Mn, which crystallizes in several complicated structures; but these cases are the exceptions rather than the rule. By the time we reach the elements with unfilled $p$ shells, however, we meet quite different structures. These structures suggest covalent bonding: each atom tends to have a number of nearest neighbors related to its valence. Thus Ga has a peculiar structure with five near neighbors; the elements C, Si, Ge, and one modification of Sn have the diamond structure, each atom having four nearest neighbors, corresponding to four single covalent bonds; C also crystallizes in the graphite structure, forming hexagonal rings as in benzene, corresponding to a partial double bond; the elements P, As, Sb, Bi, have structures in which each atom has three nearest neighbors; S, Se, Te, crystallize in structures with rings or chains of atoms, each atom having two nearest neighbors. Finally the halogens form diatomic molecules in their crystals, each atom having only one nearest neighbor. None of these peculiar structures, except the diamond and graphite structures, have simplicity or high symmetry, and the only ones which we shall consider in this article in detail are those two.

The impression we get from this survey is that the definitely metallic elements try to surround themselves by many neighbors equally tightly bound. Thus, in the close-packed structures, each atom has twelve nearest neighbors, and in the body-centered cubic, eight neighbors. As a broad generalization, the elements with these structures tend to be ductile, suggesting that the atoms can be disarranged without great increase of energy, as if each atom were rather indifferent as to the precise number and position of its neighbors. On the contrary, in the covalent type of element, with the uncompleted $p$ shell, brittleness is very common, as if each atom were bound to its restricted number of neighbors by bonds which were definitely determined in number and direction. In these elements with uncompleted $p$ shells, there are just enough electrons in the outer shell to form covalent or electron-pair bonds with the nearest neighbor atoms. On the other hand, in the typical metals, there are not nearly enough outer electrons for this purpose. Thus, in an alkali metal, with its one outer electron per atom and its eight nearest neighbors in the body-centered cubic structure, we have only one-eighth as many electrons as would be required to form electron-pair bonds between each pair of atoms. In a very broad way, these facts account for the main features of the behavior of the elements. We shall find that elements forming electron-pair bonds have energy bands which are filled, with energy gaps above them, the typical situation for an insulator or semiconductor. The reason why we have intermediate cases between insulators, semiconductors, and metals among the elements with unfilled $p$ shells is simply that the gaps tend to get narrower as we go to heavier elements, and the bands eventually overlap. On the other hand, in the typical metals, the bands are only partially filled, leading to the conducting properties characteristic of these elements.

In spite of this distinction between the typical metals and the elements held by covalent binding, nevertheless there is a fundamental similarity between them, in the mechanism by which they are held together mechanically. We shall find that in every case the mechanism of mechanical binding is essentially the same: a concentration of electronic charge in the region between atoms which are bonded together, holding the crystal together by electrostatic attraction for the neighboring nuclei. This is the mechanism of covalent binding, and it is found in the typical metals as well as in the elements forming electron-pair bonds. This concentration of charge arises from the overlapping of the wave functions of the outer electrons of the atoms, and it is a remarkable fact that the distance between nearest neighboring atoms, in the elements of both typically metallic form and those forming electron-pair bonds, is very approximately twice the distance out to the radius of maximum charge density in the wave function of the atoms. That is, the atoms approach until their wave functions overlap at a point of maximum density. This fact is illustrated in Fig. 1, where we compare the interatomic distance in the elements from K to Br with twice the radius of maximum charge density, as determined from the self-consistent field calculations. Similar striking correlations are found in other periods in the periodic table. The observed atomic radii, or atomic volumes, in other words, merely reflect the sizes of the atoms. And the fact that all these elements lie on essentially the same curve shows the similarity of the mechanism responsible for their binding.

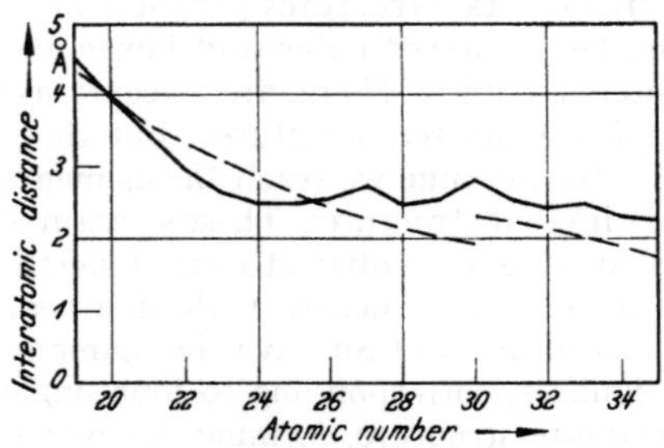


Fig. 1. Full line, observed interatomic distances in elements from K [*19*] to Br [*35*]. Dotted line, twice radius of maximum density of wave function of outermost shell of atom, estimated from self-consistent field calculations.

In this sketch of the nature of the elements, and their crystal structures, we see that the binding forces are sometimes covalent, and sometimes the metallic bond, which is not completely different in its nature. Compounds of the alkali halide type, however, show an entirely different sort of binding, the ionic binding. We shall consider the sodium chloride structure, which most of the alkali halides have; it is well known that we can get a good understanding of the properties of such crystals by assuming that the atoms are ionized, so that such a crystal as NaCl is really composed of $Na^+$ and $Cl^-$ ions, held together by COULOMB forces, and prevented from collapsing by the repulsions between the remaining closed shells. In such a case, of course, it is not the radius of the outermost shell in the neutral atom which determines the size of the crystal, but rather the radii of the two ions. Here, since the electronic shells of electrons are repelling rather than attracting each other, we expect the interatomic distances in the crystal to be a good deal bigger than the sum of the radii of the ions, computed out to the points of maximum density. This is illustrated in Fig. 2, where we give the sum of ionic radii, as computed in this way, for some of the alkali halides, and compare them with the observed interatomic distances. We see in this case that the observed interatomic distances are from two to three times the sum of the radii

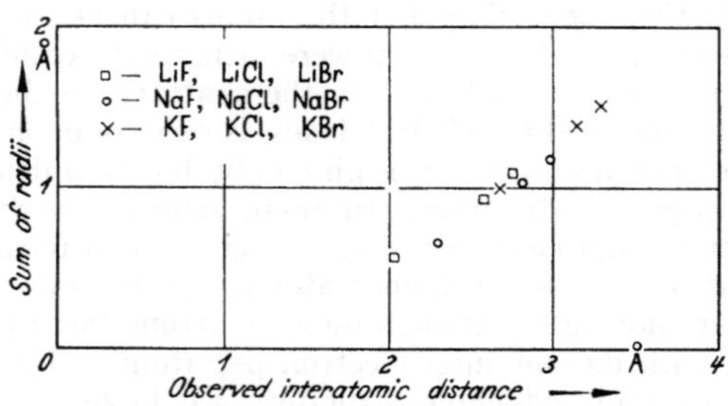


Fig. 2. Abscissa, observed interatomic distance in various alkali halides. Ordinates, sum of radii of maximum density of wave functions of outermost shells of ions, estimated from self-consistent field calculations.

of maximum charge density, showing that large repulsion between closed shells arises when the charge densities overlap only slightly, at much larger distances than that at which the maximum of the charge densities in the two ions would overlap.

An interesting class of crystals, intermediate between the ionic and the covalent type, is formed of those having the zinc-blende structure, and we shall take these up, as well as those having the sodium chloride structure. The zinc-blende structure is similar to the diamond structure, but the two atoms in the unit cell are now different, rather than alike as in diamond. We can understand the nature of this structure by starting with germanium, which has the diamond structure, and then going to GaAs, in which gallium has one electron less than germanium, arsenic one more. This substance has the zinc-blende structure, and we can build up a picture of it as a covalent structure like germanium, in which the arsenic atom has donated one electron to the gallium, so that each one has the same outer electrons which germanium has, permitting the formation of four covalent bonds in the tetrahedral directions, which is characteristic of this structure.

We can go further in this direction, with the compound ZnSe, in which zinc has two electrons fewer than germanium, selenium has two more. This is similar to ZnS, the sulfur atom having the same number of outer electrons as selenium; ZnS is in fact zinc-blende, for which the structure is named. Proceeding still further, we have CuBr, in which copper has three electrons fewer than germanium, bromine has three more. But by the time we reach this crystal, it is rather obvious that we are not going to have the bromine donating three electrons to the copper, to give each of them the same number of outer electrons as germanium, since this would lead to the highly charged ions $Cu^{---}$ and $Br^{+++}$. We could much more plausibly assume that in this case the copper atom was in the positively charged ionic form $Cu^{+}$, and bromine the negative ion $Br^{-}$, as in ordinary ionic compounds. Once we notice this, we begin to be sceptical of the simple covalent picture of all the crystals having the zinc-blende structure. We shall find later, in Sect. 37, that it is possible to give a description of such substances preserving the concept of the covalent bond, and yet allowing the charges to be arranged in such a way that the atoms are approximately electrically neutral, rather than being highly ionized, and this is almost certainly closer to the truth than either of the extreme types of structure suggested above.

The compounds which we have mentioned include a number of the technically most important, as well as the most interesting from the point of view of the nature of the chemical bond. As we have mentioned earlier, they are the ones which will be discussed in this article, but we do not wish to suggest, by excluding others, that there are not many other types of substances of great practical and theoretical interest and importance. We shall now proceed to a formal study of some of the simple properties of crystals, and of the solutions of SCHRÖDINGER's equation in the type of periodic potential found in them, and in particular in crystals of the structures which occur in the elements and compounds which we have just been describing.

**3. Crystal periodicity and WIGNER-SEITZ cells.** The characteristic of a crystal is its periodicity in space. That is, we can set up three vectors $\boldsymbol{a}_1$, $\boldsymbol{a}_2$, $\boldsymbol{a}_3$, such that if we perform a translation by a vector $m_1 \boldsymbol{a}_1 + m_2 \boldsymbol{a}_2 + m_3 \boldsymbol{a}_3$, where the $m$'s are arbitrary integers, we go from one point of the crystal to another point of identical properties. These vectors are not uniquely fixed; if we start with three such vectors and set up three new ones which are linear combinations of the original ones with integral coefficients, the new vectors will have the same properties. A parallelopiped whose three sides are $\boldsymbol{a}_1$, $\boldsymbol{a}_2$, $\boldsymbol{a}_3$ is called a unit cell; it is not uniquely fixed either. The properties of a unit cell are that such cells can be stacked together to fill space and the properties of the crystal at corresponding points of any two unit cells are identical.

If we start with a unit cell which is larger than necessary we can subdivide it into the smallest possible cells, called primitive unit cells, which still retain the two fundamental properties which we have just mentioned. For instance, in the face-centered and body-centered cubic structures, we find that we can choose primitive unit cells each containing only one atom. On the other hand, the cubic cells often used to describe these structures contain four and two atoms respectively. This is illustrated in Fig. 3, where we show the cubic cells, and one

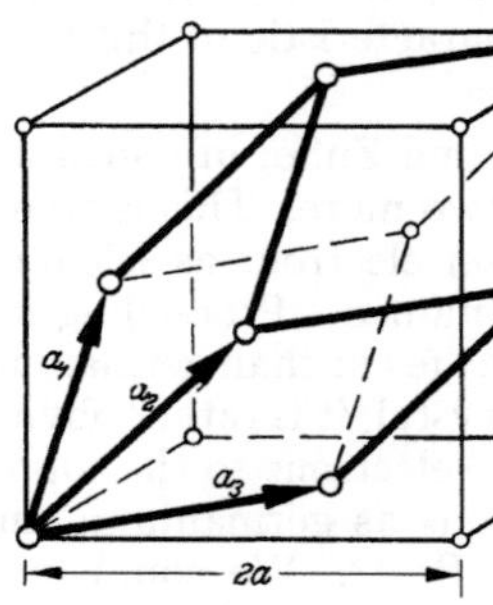


a

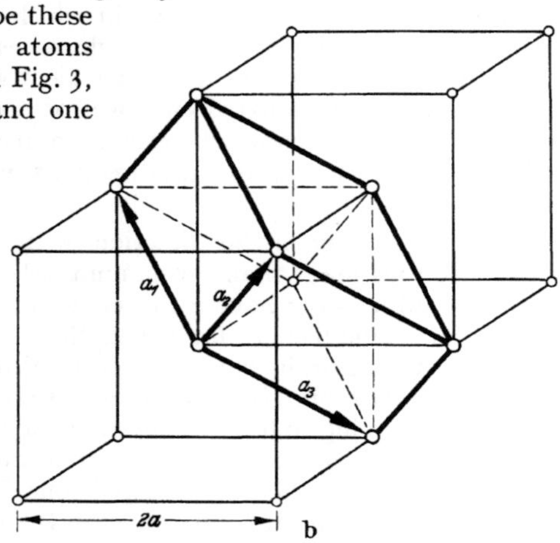


b

**Fig. 3a and b. Non-primitive and primitive unit cells for (a) the face-centered cubic structure; (b) the body-centered cubic structure. In each case the non-primitive cell is the cube of side $2a$. The primitive unit cells are shown by heavier lines.**

form of a primitive unit cell, for both structures. If the side of the cube shown in Fig. 3 is $2a$, in each case, then the vectors $\boldsymbol{a}_1$, $\boldsymbol{a}_2$, $\boldsymbol{a}_3$ are given by

$$\boldsymbol{a}_1 = a(\boldsymbol{j}+\boldsymbol{k}), \quad \boldsymbol{a}_2 = a(\boldsymbol{k}+\boldsymbol{i}), \quad \boldsymbol{a}_3 = a(\boldsymbol{i}+\boldsymbol{j}) \tag{3.1}$$

for the face-centered case, where $\boldsymbol{i}$, $\boldsymbol{j}$, $\boldsymbol{k}$ are the three unit vectors, and by

$$\boldsymbol{a}_1 = a(-\boldsymbol{i}+\boldsymbol{j}+\boldsymbol{k}), \quad \boldsymbol{a}_2 = a(\boldsymbol{i}-\boldsymbol{j}+\boldsymbol{k}), \quad \boldsymbol{a}_3 = a(\boldsymbol{i}+\boldsymbol{j}-\boldsymbol{k}) \tag{3.2}$$

for the body-centered case. We verify that the volume of the unit cell, $\boldsymbol{a}_1 \cdot (\boldsymbol{a}_2 \times \boldsymbol{a}_3)$, equals one fourth the volume of the cube $(2a)^3$ for the face-centered case, one-half the volume for the body-centered case.

There is one obvious objection to the use of these primitive unit cells: they do not show the symmetry of the crystal. Most crystals have not only a periodicity in space, but also certain symmetry properties under operations of rotation, inversion, and combinations of these operations and translations. Thus, the face-centered and body-centered cubic structures show all the symmetry properties of the cubic point group. If we locate the center of a cube at one of the atoms, any one of the 48 operations which will carry the cube into itself will also carry the crystal into itself. This symmetry is shown by the cubic cells shown in Fig. 3, but not by the smaller primitive unit cells we have chosen. For this reason, crystallographers often prefer the larger cubic cells. We shall find important reasons, however, for taking our unit cells as small as possible. We naturally ask if we cannot choose primitive unit cells which preserve the symmetry of the cube.

The way to do this, for such simple structures, was demonstrated by WIGNER and SEITZ [255]. They carried out the following construction. They set up the line joining an atom to each nearest or second-nearest neighbor, and constructed the plane which was the perpendicular bisector of each such line. These planes enclosed a solid surrounding the atom, which we may call the WIGNER-SEITZ

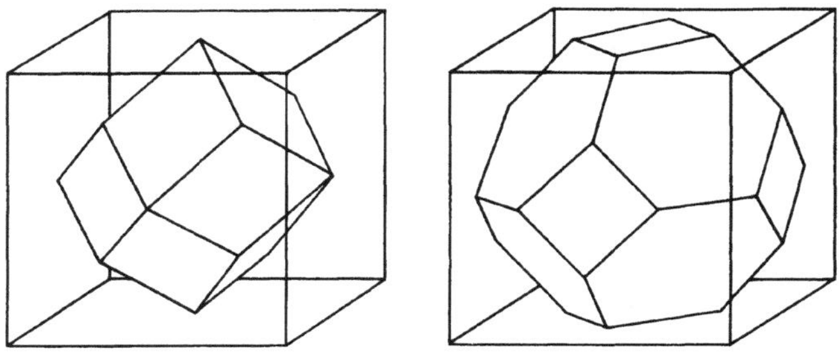

a b

Fig. 4a and b. WIGNER-SEITZ cells for (a) face-centered cubic structure, (b) body-centered cubic structure, enclosed in non-primitive unit cells of side $2\boldsymbol{a}$.

cell. Such cells, for the face-centered and body-centered structures, are shown in Fig. 4. Each of these cells has the same volume as the primitive unit cells which we have just set up, and in fact we can make up the WIGNER-SEITZ cell by taking a primitive unit cell, cutting it up into several segments by cutting along various planes, and making translations of some of these segments by appropriate translation vectors $m_1\boldsymbol{a}_1 + m_2\boldsymbol{a}_2 + m_3\boldsymbol{a}_3$. The WIGNER-SEITZ cells for these structures have the two fundamental properties of unit cells: they can be stacked together

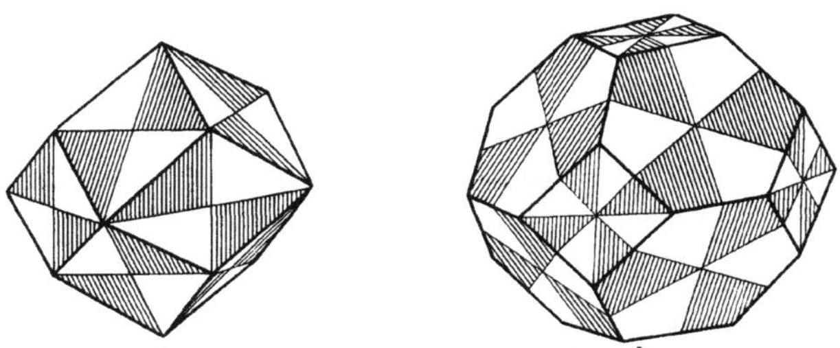

a b

Fig. 5a and b. WIGNER-SEITZ cells for face-centered and body-centered cubic structures, showing the 48 equivalent segments into which each can be subdivided. Equivalent points in each segment of the surface can be carried into each other by one of the 48 symmetry operations of the cubic point group. Alternate segments are shaded and unshaded.

to fill all space, and the crystal has identical properties at equivalent points of each cell. Hence we may use them as unit cells, in a slightly generalized sense. But they have the additional virtue that they are carried into themselves by all operations of the cubic point group, so that each of them can be subdivided into 48 segments, as shown in Fig. 5, such that the crystal will have identical properties at equivalent points of each of these segments.

In the hexagonal structure, and the diamond structure, the smallest unit cell which one can choose has two atoms per unit cell. In these structures, we can also carry out the WIGNER-SEITZ construction, and in this case we are led to WIGNER-SEITZ cells which are half as large as the unit cells. Two WIGNER-SEITZ cells stack together, with different orientations, to form a cell of complex shape which can be considered a unit cell, if we choose. In the hexagonal case, these WIGNER-SEITZ cells show the three-fold rotation axis characteristic of the hexagonal structure, and in the diamond case the tetrahedral symmetry characteristic of the diamond structure. These WIGNER-SEITZ cells are illustrated in Fig. 6. In these cases the symmetry of the unit cell is not completely exhibited by the WIGNER-SEITZ cell. Thus, in diamond, the WIGNER-SEITZ cell of tetrahedral symmetry has 24 symmetry operations in its point group, illustrated by the 24 equivalent points shown in the cell of Fig. 6. But the diamond structure has 48 symmetry operations: the 24 points shown in Fig. 6 are equivalent to 24 points in the other WIGNER-SEITZ cell which must be combined with the one shown to form the unit cell.

The structures we have been mentioning so far are structures in which elements crystallize, and in which all atoms are essentially in equivalent positions, so that

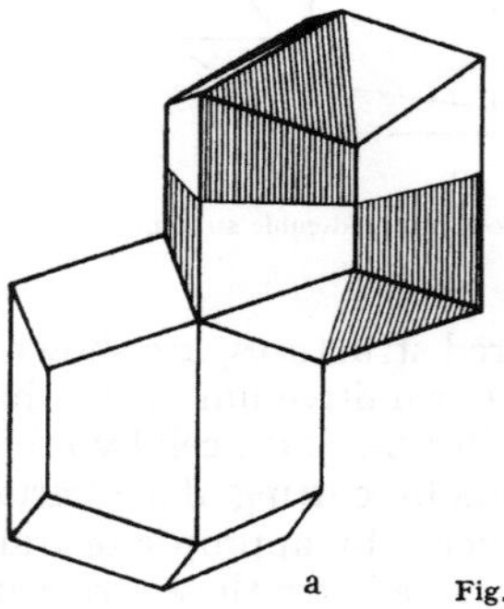
a

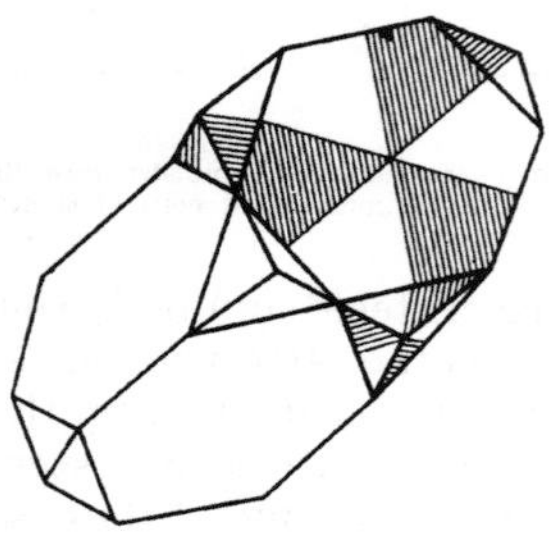
b

Fig. 6.

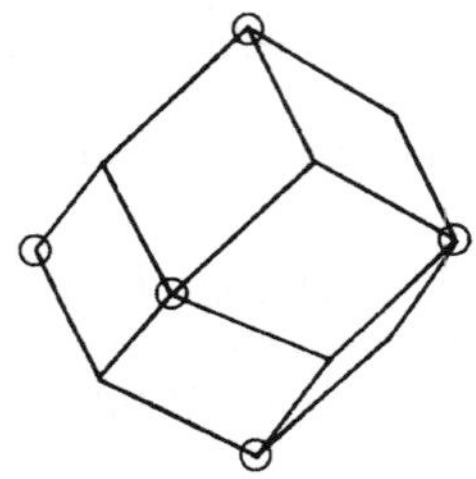

Fig. 7.

Fig. 6a and b. WIGNER-SEITZ cells for (a) the hexagonal close-packed structure; (b) the diamond structure. In each case two WIGNER-SEITZ cells are shown, surrounding the two atoms in the unit cell. In one of the cells, equivalent segments are indicated by presence or absence of shading.

Fig. 7. WIGNER-SEITZ cell for face-centered cubic structure, as adapted for NaCl structure: one type of atom at center of cell, the other type at six corners.

the introduction of the WIGNER-SEITZ cells, each holding a single atom, is natural. In a compound, however, the situation is quite different, and the unit cell must in every case hold more than one atom. The sodium chloride structure is built crystallographically from a face-centered cubic lattice, and has two atoms, one of each type, in the unit cell. The crystallographers ordinarily use a cubic cell, containing four atoms of each type. However, if we wish to use a primitive unit cell, we again have the problem of symmetry which we discussed above in connection with the face-centered cubic structure. If we choose to use the WIGNER-SEITZ cell shown in Fig. 4, it will now have to contain two atoms, one of each type. We may, if we choose, place the atom of one type at the center of the cell, and place atoms of the other type at six corners of the cell, as shown in Fig. 7. Each of these atoms will be shared between six cells. For most purposes, however, there is no particular merit in this type of description of the sodium chloride structure.

The zinc-blende structure, as we have already mentioned, is like the diamond structure, but with two different atoms in the unit cell. If we choose to use the concept of WIGNER-SEITZ cells, we may still do so as with diamond, introducing a cell which holds just one atom, so that two of them, with different orientations, are needed to build up a unit cell. The symmetry situation is now different, however; since the two cells are no longer equivalent to each other, we find that there are only 24 symmetry operations in the zinc-blende structure, just as for the point group of the regular tetrahedron.

The WIGNER-SEITZ cells, which we have taken up in this discussion, were introduced on account of their convenience in the cellular method for approximating the solution of SCHRÖDINGER's equation in a crystal, which we shall discuss in Sect. 10. We shall see there that each WIGNER-SEITZ cell is near enough to a sphere so that we can approximate the SCHRÖDINGER problem within this cell as a spherically symmetrical problem. This method is more useful for a typical metal, in which the environment of an atom is really close to being spherically symmetrical, than it is in such a case as diamond or zinc-blende, where the departure from spherical symmetry is considerable. Quite aside from this value of the WIGNER-SEITZ method in the solution of SCHRÖDINGER's equation, however, the use of these cells is convenient in visualizing the symmetry relations and environment of an atom in those very few simple crystal structures which we have mentioned; it is not a general method, adaptable to any very large groups of crystals.

**4. The reciprocal lattice, and BRILLOUIN zones.** In many cases it is desirable to form a FOURIER expansion of a function which is periodic, repeating itself at equivalent points of each unit cell. This is conveniently done by introducing a reciprocal lattice. One starts with the primitive translation vectors $\boldsymbol{a}_1, \boldsymbol{a}_2, \boldsymbol{a}_3$, determining the unit cell, and sets up three new vectors $\boldsymbol{b}_1, \boldsymbol{b}_2, \boldsymbol{b}_3$, defined by the conditions

$$\boldsymbol{a}_i \cdot \boldsymbol{b}_j = \delta_{ij} = 1 \quad \text{if} \quad i = j, \qquad 0 \quad \text{if} \quad i \neq j. \tag{4.1}$$

This is equivalent to the conditions

$$\boldsymbol{b}_1 = \frac{\boldsymbol{a}_2 \times \boldsymbol{a}_3}{\boldsymbol{a}_1 \cdot (\boldsymbol{a}_2 \times \boldsymbol{a}_3)}, \qquad \boldsymbol{b}_2 = \frac{\boldsymbol{a}_3 \times \boldsymbol{a}_1}{\boldsymbol{a}_1 \cdot (\boldsymbol{a}_2 \times \boldsymbol{a}_3)}, \qquad \boldsymbol{b}_3 = \frac{\boldsymbol{a}_1 \times \boldsymbol{a}_2}{\boldsymbol{a}_1 \cdot (\boldsymbol{a}_2 \times \boldsymbol{a}_3)}. \tag{4.2}$$

On account of the nature of these relations, the vectors $\boldsymbol{b}$ are called reciprocal to the $\boldsymbol{a}$'s. Now let us set up a sinusoidal function of position,

$$\exp 2\pi i (h_1 \boldsymbol{b}_1 + h_2 \boldsymbol{b}_2 + h_3 \boldsymbol{b}_3) \cdot \boldsymbol{r},$$

where $h_1, h_2, h_3$ are arbitrary integers, and $\boldsymbol{r}$ is the radius vector, of components $x, y, z$. If we make a translation $m_1\boldsymbol{a}_1 + m_2\boldsymbol{a}_2 + m_3\boldsymbol{a}_3$, this will multiply the exponential function written above by the factor

$$\exp 2\pi i (h_1 \boldsymbol{b}_1 + h_2 \boldsymbol{b}_2 + h_3 \boldsymbol{b}_3) \cdot (m_1 \boldsymbol{a}_1 + m_2 \boldsymbol{a}_2 + m_3 \boldsymbol{a}_3).$$

On account of the relations given in Eq. (4.1), this factor is unity. In other words, such an exponential function will repeat at equivalent points of each unit cell in the crystal. One can show that these are the only exponential functions which have this property, so that a general FOURIER expansion of a periodic function with the periodicity of the lattice is given by a linear combination of such functions. That is, a periodic function $F(\boldsymbol{r})$ can be written in the form

$$F(\boldsymbol{r}) = \Sigma (h_1 h_2 h_3)\, F(h_1 h_2 h_3) \exp 2\pi i (h_1 \boldsymbol{b}_1 + h_2 \boldsymbol{b}_2 + h_3 \boldsymbol{b}_3) \cdot \boldsymbol{r} \tag{4.3}$$

where the $F(h_1 h_2 h_3)$'s are expansions coefficients.

The crystallographers have found it convenient to set up a reciprocal space, in which the end-points of the vectors $h_1 \boldsymbol{b}_1 + h_2 \boldsymbol{b}_2 + h_3 \boldsymbol{b}_3$ form a lattice of points, which they call the reciprocal lattice[1]. Associated with each lattice point we have a set of integers $h_1, h_2, h_3$, and we can visualize one expansion coefficient as being associated with each point of the reciprocal lattice. The physicists have

[1] Detailed treatment in vol. VII, part 1 (articles by H. JAGODZINSKI and G. LEIBFRIED) and in vol. XXXII (article by J. BOUMAN).

found it rather more convenient to introduce a reciprocal space, differing in scale from this, in either of two different ways. In the first place, the physicists often define the propagation vectors $\boldsymbol{K}_i = 2\pi(h_1\boldsymbol{b}_1 + h_2\boldsymbol{b}_2 + h_3\boldsymbol{b}_3)$, where the index $i$ stands for the three integers $h_1, h_2, h_3$. These propagation vectors point along the wave normals of the plane waves represented by the exponential functions, which can now be written in the more convenient form $\exp i\boldsymbol{K}_i\cdot\boldsymbol{r}$, and the magnitudes of the propagation vectors are equal to $2\pi$ divided by the wavelength of the wave. The end-points of the vectors $\boldsymbol{K}_i$ form a lattice of points in the $\boldsymbol{K}_i$ space, differing in scale from the reciprocal space by the factor $2\pi$. Again, a wavelength $\lambda$ can be associated with a momentum by DE BROGLIE's relation that the momentum vector points in the direction of the wave normal, and has a magnitude $h/\lambda$, where $h$ is PLANCK's constant. We can then set up vectors

$$\boldsymbol{P}_i = \hbar\boldsymbol{K}_i \tag{4.4}$$

where $\hbar = h/2\pi$, and can set up a momentum space in which we have a lattice of points determined by the ends of the vectors $\boldsymbol{P}_i$, which again will be identical with the reciprocal space except for a scale factor.

It is interesting to consider the reciprocal lattice, for the face-centered and body-centered cubic lattices. If we apply the definition of Eq. (4.2) to the vectors $\boldsymbol{a}_1, \boldsymbol{a}_2, \boldsymbol{a}_3$ defined in Eq. (3.1) and (3.2) for these two cases, we find that

$$\boldsymbol{b}_1 = \frac{1}{2a}(-\boldsymbol{i}+\boldsymbol{j}+\boldsymbol{k}),\quad \boldsymbol{b}_2 = \frac{1}{2a}(\boldsymbol{i}-\boldsymbol{j}+\boldsymbol{k}),\quad \boldsymbol{b}_3 = \frac{1}{2a}(\boldsymbol{i}+\boldsymbol{j}-\boldsymbol{k}) \tag{4.5}$$

for the face-centered lattice, and

$$\boldsymbol{b}_1 = \frac{1}{2a}(\boldsymbol{j}+\boldsymbol{k}),\quad \boldsymbol{b}_2 = \frac{1}{2a}(\boldsymbol{k}+\boldsymbol{i}),\quad \boldsymbol{b}_3 = \frac{1}{2a}(\boldsymbol{i}+\boldsymbol{j}) \tag{4.6}$$

for the body-centered lattice. In other words, the lattice of points in the reciprocal space for the face-centered lattice is like the set of atomic positions in the body-centered structure in ordinary space, and vice versa. We can set up unit cells in reciprocal space, such that one point of the reciprocal lattice is found in an equivalent position of each unit cell. Just as in ordinary space, we can introduce the equivalent of WIGNER-SEITZ cells, and in this case the central cell, surrounding the origin, is identical with the central BRILLOUIN zone, introduced by BRILLOUIN [*31*], [*32*], [*33*], [*37*] in the early days of the study of the solution of SCHRÖDINGER'S equation for a periodic potential; we may define the central BRILLOUIN zone as being a solid of the same volume as the primitive unit cell in reciprocal space, but of such a shape as to remain invariant under as many as possible of the symmetry operations of the crystal. We see that these BRILLOUIN zones are given by Fig. 4, provided we make two changes: interchange the face-centered and body-centered cubic structures, and change the side of the cube from $2a$, in ordinary space, to $1/a$ in reciprocal space, or $2\pi/a$ in $\boldsymbol{K}$-space, or $h/a$ in momentum space. The volume of the BRILLOUIN zone in reciprocal space is in either case equal to the reciprocal of the volume of the WIGNER-SEITZ cell in ordinary space.

For the hexagonal and the diamond or zinc-blende structure, we define the BRILLOUIN zones, not by starting with the WIGNER-SEITZ cells containing one atom per cell, but by starting with real unit cells containing two atoms per cell. The hexagonal structure consists of layers of close-packed spheres, with distances $a$ between centers, arranged in alternating array, the distance between every second layer being $c$. This is illustrated in Fig. 8a, where we have shown one layer, and part of the second; the third is like the first. If $c/a = \sqrt{8/3} = 1.63$, the spheres are closely packed, but larger $c/a$ ratios are common among some of the metals. A primitive unit cell, containing two atoms, may be set up as in Fig. 8b, the translation vectors being given by $\boldsymbol{a}_1 = a\boldsymbol{i}$, $\boldsymbol{a}_2 = a\left(\frac{\boldsymbol{i}}{2} + \frac{\boldsymbol{j}\sqrt{3}}{2}\right)$, $\boldsymbol{a}_3 = c\boldsymbol{k}$. By analogy with the WIGNER-SEITZ method, however, we can construct a more

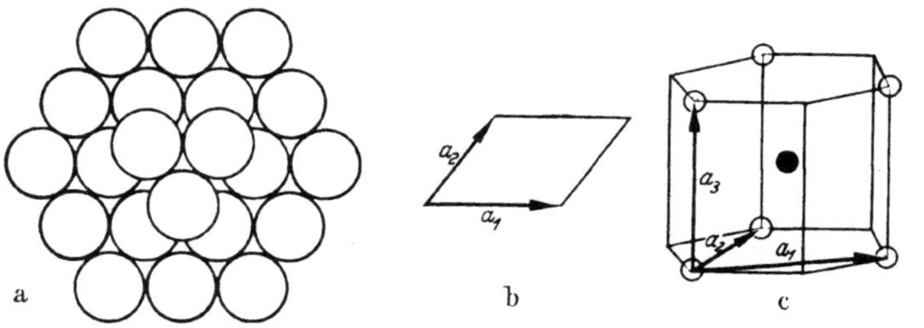


Fig. 8a–c. (a) One layer of atoms, and part of another, in the hexagonal close-packed structure; (b) a primitive unit cell, projected on the basal plane; (c) a unit cell in the form of a hexagonal prism.

symmetrical cell of the same volume. We do this by setting up planes which are the perpendicular bisectors of lines joining a central atom to the nearest neighboring atoms in its same plane. Then we have the hexagonal cell shown in Fig. 8c. This contains one atom at the center, and atoms at six out of the twelve corners, each of these being shared between six unit cells, so that one out

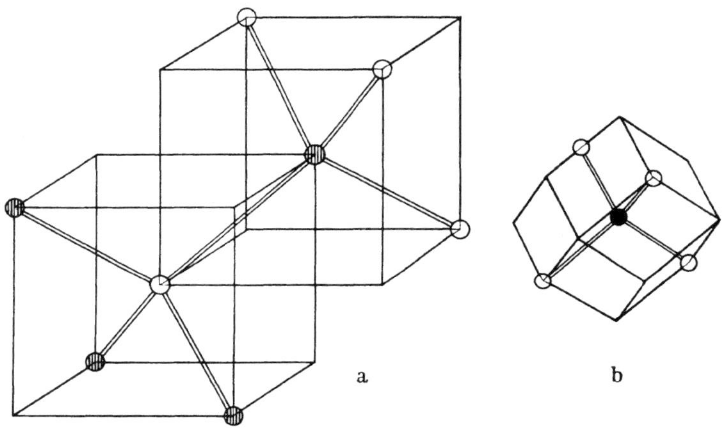


Fig. 9a and b. (a) The diamond structure. Shaded atoms lie on one face-centered cubic lattice, unshaded on the other interpenetrating lattice. The cubes shown have side $a$, half the side of the non-primitive cell shown in Fig. 3a. (b) WIGNER-SEITZ cell for face-centered cubic structure, as adapted for the diamond structure; one atom at center of the cell, and atoms at four corners as shown, each shared between four cells.

of these six atoms belongs to the cell in question. We now find that the reciprocal vectors form a lattice of points similar to the array of atoms, but rotated through 90°, so that we can choose a hexagonal BRILLOUIN zone, equal in volume to the primitive unit cell in reciprocal space. The height of this zone along the $c$ axis is $1/c$, and the volume of the zone is the reciprocal of the volume of the unit cell in ordinary space.

The diamond structure is built up from two interpenetrating face-centered cubic lattices of atoms, one displaced with respect to the other by the vector $(a/2)(\boldsymbol{i}+\boldsymbol{j}+\boldsymbol{k})$. This way of regarding it is shown in Fig. 9a. We can then use the same unit cell or WIGNER-SEITZ cell as for the face-centered structure. In Fig. 9b we show how the atoms are arranged if we use the WIGNER-SEITZ cell: there is one atom at the center, and four atoms at four of the eight corners pointing in 111 directions, each shared between four cells. Since the structure is built up from a face-centered cubic lattice, we can use the same BRILLOUIN zone as for the face-centered cubic structure.

There has been a good deal of discussion in the literature of further BRILLOUIN zones outside the central ones, which are the only ones we have so far discussed. We shall consider this subject later in Sect. 8.

**5. BLOCH's theorem and the expansion of the wave function in plane waves.** Profound results concerning the solutions of SCHRÖDINGER's equation flow from the periodicity of the potential energy, and hence of the HAMILTONian function. If we have an operator which commutes with the HAMILTONian $H$, then we can simultaneously diagonalize the HAMILTONian and the operator. If more than one operator commutes with the HAMILTONian and they all commute with each other, we can simultaneously diagonalize them all. The operators involved in adding the vectors $m_1\boldsymbol{a}_1 + m_2\boldsymbol{a}_2 + m_3\boldsymbol{a}_3$ to the radius vector of components $x, y, z$ all commute with the HAMILTONian, since they leave it unchanged, and with each other, since successive translations can be carried out in any order. Thus, we can set up solutions of SCHRÖDINGER's equation which diagonalize all these operators; that is, solutions such that when we increase the radius vector by $m_1\boldsymbol{a}_1 + m_2\boldsymbol{a}_2 + m_3\boldsymbol{a}_3$, we multiply the solution by a constant.

If the original and transformed wave functions are both to be normalized, the absolute value of this constant must be unity. In this case we can find a real vector $\boldsymbol{k}$, such that the constant by which the wave function is multiplied when we make the translation $\boldsymbol{a}_1$ is $\exp i\boldsymbol{k}\cdot\boldsymbol{a}_1$; if we make a translation $\boldsymbol{a}_2$, the constant is $\exp i\boldsymbol{k}\cdot\boldsymbol{a}_2$; and if we make a translation $\boldsymbol{a}_3$, the constant is $\exp i\boldsymbol{k}\cdot\boldsymbol{a}_3$. Then if we make the translation $\boldsymbol{R}_j = m_1\boldsymbol{a}_1 + m_2\boldsymbol{a}_2 + m_3\boldsymbol{a}_3$, where the index $j$ indicates the set of integers $m_1, m_2, m_3$, we multiply the wave function by the factor $\exp i\boldsymbol{k}\cdot\boldsymbol{R}_j$. In this case, if the wave function is $u(\boldsymbol{r})$, then the quantity $u(\boldsymbol{r})\exp -i\boldsymbol{k}\cdot\boldsymbol{r}$ must be unchanged when we make the translation $\boldsymbol{R}_j$, or must be a periodic function of $r$, repeating its value in each unit cell. We then have the result

$$u(\boldsymbol{r}) = w(\boldsymbol{r})\exp i\boldsymbol{k}\cdot\boldsymbol{r} \tag{5.1}$$

where $w(\boldsymbol{r})$ is a periodic function of $\boldsymbol{r}$, such that

$$w(\boldsymbol{r} + \boldsymbol{R}_j) = w(\boldsymbol{r}) \tag{5.2}$$

This very important result is BLOCH's theorem, and was suggested by him [*21*] in one of the earliest papers on the solution of SCHRÖDINGER's equation in a periodic potential. The wave function obviously has a resemblance to a plane wave of propagation constant $\boldsymbol{k}$. From $\boldsymbol{k}$ we can set up a quantity $\boldsymbol{k}\hbar$ of the dimensions of a momentum; we shall call it a pseudomomentum. We have assumed $\boldsymbol{k}$ to be real; if it is complex, which is mathematically allowed, the wave will be damped, and we usually do not use such solutions of SCHRÖDINGER's equation in problems of solids, except in cases involving solutions near surfaces of discontinuity.

The function $w(\boldsymbol{r})$, like any periodic function of position with the periodicity of the crystal, can be expanded in FOURIER series, following the method of Sect. 4. Thus, we can write

$$w(\boldsymbol{r}) = \Sigma(h_1 h_2 h_3)\, F(h_1 h_2 h_3)\exp i\boldsymbol{K}_i\cdot\boldsymbol{r} \tag{5.3}$$

where as before $\boldsymbol{K}_i = 2\pi(h_1\boldsymbol{b}_1 + h_2\boldsymbol{b}_2 + h_3\boldsymbol{b}_3)$. We then see that $u(\boldsymbol{r})$ can be written as an expansion in terms of plane waves of form $\exp i(\boldsymbol{k}+\boldsymbol{K}_i)\cdot\boldsymbol{r}$. It is convenient to denote the coefficient of one such plane wave in the expansion as $v(\boldsymbol{k}+\boldsymbol{K}_i)$. Then we have

$$u(\boldsymbol{r}) = \Sigma(\boldsymbol{K}_i)\, v(\boldsymbol{k}+\boldsymbol{K}_i)\exp i(\boldsymbol{k}+\boldsymbol{K}_i)\cdot\boldsymbol{r}. \tag{5.4}$$

In this expression, we have exhibited a solution of SCHRÖDINGER's equation in a periodic potential as an expansion in terms of plane waves. The propagation vectors $\boldsymbol{k}+\boldsymbol{K}_i$ of these waves have the property that they occur at equivalent points of every unit cell of the reciprocal space. We may, if we choose, associate each coefficient $v(\boldsymbol{k}+\boldsymbol{K}_i)$ with its appropriate point in $\boldsymbol{K}$-space.

It is customary in discussing solid-state problems to use what are called periodic boundary conditions. We fix our attention on a block of material consisting of the parallelopiped whose dimensions are $N_1$ unit cells along the direction of $\boldsymbol{a}_1$, $N_2$ unit cells along $\boldsymbol{a}_2$, $N_3$ unit cells along $\boldsymbol{a}_3$; the total number of unit cells in this block, which is $N_1 N_2 N_3$, will be called $N$, and its volume is $N\Omega$, if $\Omega$ is the volume of a unit cell. Then we assume that we are dealing with a wave function which repeats periodically from one block of this size to an adjacent identical, block. If the block is large enough, no difficulty arises from imposing this artificial condition of periodicity. This implies that each plane wave $\exp i(\boldsymbol{k}+\boldsymbol{K}_i)\cdot\boldsymbol{r}$ is unchanged if $\boldsymbol{r}$ is increased by the amount $N_1\boldsymbol{a}_1 + N_2\boldsymbol{a}_2 + N_3\boldsymbol{a}_3$. Hence we have

$$\left(\frac{\boldsymbol{k}}{2\pi} + h_1\boldsymbol{b}_1 + h_2\boldsymbol{b}_2 + h_3\boldsymbol{b}_3\right)\cdot(N_1\boldsymbol{a}_1 + N_2\boldsymbol{a}_2 + N_3\boldsymbol{a}_3) = \text{integer} \tag{5.5}$$

where we have rewritten $\boldsymbol{K}_i$ in terms of the $\boldsymbol{b}$'s. Using the relations between the $\boldsymbol{a}$'s and $\boldsymbol{b}$'s, this reduces to $(\boldsymbol{k}/2\pi)\cdot(N_1\boldsymbol{a}_1 + N_2\boldsymbol{a}_2 + N_3\boldsymbol{a}_3) = \text{integer}$. Let us express $\boldsymbol{k}/2\pi$ as a linear combination of the vectors $\boldsymbol{b}_1, \boldsymbol{b}_2, \boldsymbol{b}_3$, so that $\boldsymbol{k}/2\pi = h_1'\boldsymbol{b}_1 + h_2'\boldsymbol{b}_2 + h_3'\boldsymbol{b}_3$, where $h_1', h_2', h_3'$ are non-integral values, unlike $h_1, h_2, h_3$. Then we see that $h_1' = \text{integer}/N_1$, $h_2' = \text{integer}/N_2$, $h_3' = \text{integer}/N_3$. In other words, the $\boldsymbol{k}$ vectors lie on a lattice, similar to the reciprocal lattice, but more closely spaced, so that there are $N_1$ independent points along the $\boldsymbol{b}_1$ direction, $N_2$ along the $\boldsymbol{b}_2$ direction, $N_3$ along the $\boldsymbol{b}_3$ direction, in a unit cell in the reciprocal space, or in the central BRILLOUIN zone. That is, there are as many independent allowed $\boldsymbol{k}$ vectors in the central BRILLOUIN zone as there are unit cells in our fundamental block of unit cells. One can then prove simply that the plane waves associated with different allowed values of $\boldsymbol{k}+\boldsymbol{K}_i$ are orthogonal to each other, when integrated over the block of $N$ cells.

Let us assume that $u(\boldsymbol{r})$ is to be normalized over our fundamental block of cells. The functions $(N\Omega)^{-\frac{1}{2}}\exp i(\boldsymbol{k}+\boldsymbol{K}_i)\cdot\boldsymbol{r}$ will also be normalized over the block of cells, and as we have just seen they are orthogonal, for different $\boldsymbol{k}+\boldsymbol{K}_i$ values. Thus Eq. (5.4) represents an expansion of our wave function $u(\boldsymbol{r})$ in terms of a set of orthonormal functions, the expansion coefficients being $v(\boldsymbol{k}+\boldsymbol{K}_i)(N\Omega)^{\frac{1}{2}}$. These quantities have a straightforward physical meaning. If the wave function is a single plane wave, we know that we have a flow of electrons, all of the same momentum, given by $\hbar$ times the propagation vector. Then the square of the expansion coefficient of a given plane wave gives the probability that, if the particle is moving according to the wave function $u(\boldsymbol{r})$, it is found in one of the plane waves. That is, it gives the probability that the momentum should have a specified value. The quantities $(N\Omega)^{\frac{1}{2}}v(\boldsymbol{k}+\boldsymbol{K}_i)$, then, form what is called the momentum eigenfunction. The square of such a quantity, for any value of momentum $(\boldsymbol{k}+\boldsymbol{K}_i)\hbar$, gives the probability of finding the electron with that momentum, just as the square of the magnitude of the ordinary wave function $u(\boldsymbol{r})$ gives probability of finding the electron with the coordinate $\boldsymbol{r}$. The probability of finding the electron with a given momentum is only different from zero for the discrete momenta $(\boldsymbol{k}+\boldsymbol{K}_i)\hbar$.

In Eq. (5.4), we have written the wave function as an expansion in terms of plane waves, the coefficients forming the momentum eigenfunction. We can now find the equations determining the $v$'s by substituting $u(\boldsymbol{r})$ from Eq. (5.4) into SCHRÖDINGER's equation $Hu=Eu$, where $H=-\hbar^2\nabla^2/2m+V$, $V$ being the periodic potential. On account of its periodic nature, we can expand $V$ in a FOURIER series.

$$V(\boldsymbol{r})=\Sigma(\boldsymbol{K}_j)\,W(\boldsymbol{K}_j)\exp i\,\boldsymbol{K}_j\cdot\boldsymbol{r}. \tag{5.6}$$

We then substitute the function of Eq. (5.4) into SCHRÖDINGER'S equation, multiply by the complex conjugate of one of the exponential functions $\exp\, i\,(\boldsymbol{k}+\boldsymbol{K}_i)\cdot\boldsymbol{r}$, integrate over the block of cells, and find that the quantities $v$ satisfy the equations

$$\frac{\hbar^2}{2m}(\boldsymbol{k}+\boldsymbol{K}_j)^2\,v_n(\boldsymbol{k}+\boldsymbol{K}_j)+\Sigma(\boldsymbol{K}_j')\,W(-\boldsymbol{K}_j')\,v_n(\boldsymbol{k}+\boldsymbol{K}_j+\boldsymbol{K}_j')=E_{\boldsymbol{k}n}\,v_n(\boldsymbol{k}+\boldsymbol{K}_j). \tag{5.7}$$

Here we have given subscripts $\boldsymbol{k}$ and $n$ to $E_{\boldsymbol{k}n}$, and subscript $n$ to the $v$'s, to indicate that we are dealing with the $n$th eigenfunction of the problem associated with a particular value of $\boldsymbol{k}$.

In Eq. (5.7), we have the equivalent of SCHRÖDINGER'S equation for the momentum eigenfunctions. It takes the familiar form of a set of simultaneous algebraic equations for the expansion coefficients of the wave function $u(\boldsymbol{r})$ in terms of plane waves. For a given $\boldsymbol{k}$, we shall have an infinite set of solutions, denoted by the index $n$, each one with its energy $E_{\boldsymbol{k}n}$. From these facts we can draw one very important conclusion. The solutions depend on the lattice of points $\boldsymbol{k}+\boldsymbol{K}_j$, where $\boldsymbol{K}_j$ runs over all values of the reciprocal lattice. If we consider a new $\boldsymbol{k}$, given by $\boldsymbol{k}'=\boldsymbol{k}+\boldsymbol{K}_i$, where $\boldsymbol{K}_i$ is one of the vectors of the reciprocal lattice, then the lattice of points $\boldsymbol{k}'+\boldsymbol{K}_j$ comprises the same points as the lattice $\boldsymbol{k}+\boldsymbol{K}_j$. Hence the solutions of Eq. (5.7) for this new $\boldsymbol{k}'$ must be identical with those of the original $\boldsymbol{k}$, only labelled differently, and $E_{\boldsymbol{k}'n}$ must equal $E_{\boldsymbol{k}n}$. In other words, the energy $E_{\boldsymbol{k}n}$ repeats itself at equivalent points in each unit cell of the reciprocal lattice. Since the allowed values of $\boldsymbol{k}$ are very closely spaced in the reciprocal lattice, we may regard $E_{\boldsymbol{k}n}$ as a continuous function of $\boldsymbol{k}$, which then varies periodically with $\boldsymbol{k}$, having the periodicity of the reciprocal lattice.

Such a set of energy levels as a function of $\boldsymbol{k}$ is an energy band. All distinct wave functions and energy levels will be found if $\boldsymbol{k}$ is allowed to range through the central unit cell of momentum space, or the central BRILLOUIN zone. Such a $\boldsymbol{k}$ vector, in the central zone, is called a reduced propagation vector, and it is generally, though not always, most convenient to use it, restricting $\boldsymbol{k}$ to the central zone. Since the number of distinct $\boldsymbol{k}$ values in the central zone is just equal to the number of unit cells in our block of cells, we see that an energy band has enough states so that if each state in it is filled with one electron of each spin, as allowed by the PAULI principle, there will be just two electrons, one of each spin, per unit cell arising from this energy band. We shall find, in Sect. 6, that in general, except for special symmetry points in the BRILLOUIN zone, there is no degeneracy between the energy bands; that is, for a general value $\boldsymbol{k}$, the $E_{\boldsymbol{k}n}$'s are all distinct.

**6. The effect of crystal symmetry on the wave functions and energy bands; general properties.** Crystals possess many elements of symmetry in addition to the translation which we have been discussing in the three preceding sections. The effect of these symmetries on the nature of the energy bands is best discussed by the group theory; we shall now give a short account of the results of such study. Important work on the subject has been done by BETHE [*17*], SEITZ [*199*], HUND [*109*], [*110*], BOUCKAERT, SMOLUCHOWSKI, and WIGNER [*29*], HERRING [*92*], [*93*], [*96*], VON DER LAGE and BETHE [*249*], BELL [*13*], and ANTONCIK and TRLIFAJ [*7*]. We shall discuss the body-centered and face-centered cubic, the diamond and zinc-blende, and the hexagonal close-packed structures, as mentioned in Sect. 2.

In a body-centered or face-centered structure, the symmetry operations which will carry the crystal into coincidence with itself are of two sorts. First, there are the translations by vectors $\boldsymbol{R}_i = m_1 \boldsymbol{a}_1 + m_2 \boldsymbol{a}_2 + m_3 \boldsymbol{a}_3$, which we have already considered. If we use periodic boundary conditions, so that a translation through $N_1 \boldsymbol{a}_1$ or $N_2 \boldsymbol{a}_2$ or $N_3 \boldsymbol{a}_3$ is regarded as equivalent to the identity, there are just $N_1 N_2 N_3 = N$ independent translational operations. We have seen that these operations all commute with each other and with the HAMILTONian, and that we can therefore choose wave functions which are transformed into multiples of themselves by these operations, the factors by which they are multiplied being the BLOCH factors $\exp i\boldsymbol{k} \cdot \boldsymbol{R}_i$, where $\boldsymbol{k}$ is one of the $N$ propagation vectors satisfying Eq. (5.5). On account of this property, the wave functions can be written in the form of Eq. (5.1) or Eq. (5.4), which are equivalent to each other.

The other symmetry operations in the body-centered and face-centered cubic structures are the operations of the cubic point group. If we imagine a cube, or a WIGNER-SEITZ cell, to be located with its center at one of the atoms, and its axes along the axes of the crystal, there are 48 operations which will transform the cube or cell into itself: twenty-four rotations, and twenty-four operations consisting of an inversion and a rotation. These operations can be visualized from Fig. 5, in which we drew a WIGNER-SEITZ cell, showing the 48 equivalent points. The operations are those which transform one of these points into any other point (including itself). If we now locate such a cube or WIGNER-SEITZ cell at one of the atoms of the crystal, and perform any of the 48 operations on it, the crystal will be left unchanged. Furthermore, the combined operation formed from one of these 48 operations, followed by a translation $\boldsymbol{R}_i$, will leave the crystal unchanged. We can state this condition more precisely in the following way. Start with a point with vector position $\boldsymbol{r}$ in the crystal, measured from one of the atoms as an origin. Introduce the vector $\boldsymbol{r}'$, which is a vector of the same length as $\boldsymbol{r}$, but pointing to one of the 48 points equivalent to $\boldsymbol{r}$. Add one of the $N$ translational vectors $\boldsymbol{R}_i$. Then if $V(\boldsymbol{r})$ is the potential energy at point $\boldsymbol{r}$, the condition is that

$$V(\boldsymbol{r}' + \boldsymbol{R}_j) = V(\boldsymbol{r}). \tag{6.1}$$

There are $48N$ such operations, and they can be proved to form a group, called the space group. The 48 operations of rotation and inversion, without regard to translation, form what is called the point group.

The situation with the diamond, zinc-blende, and hexagonal structures is similar, but somewhat more complicated on account of the additional operations connected with having two atoms per unit cell. In the diamond and zinc-blende structures, in addition to the translations, we can locate a tetrahedron at one of the atoms, and any one of the 24 operations which will carry the tetrahedron into itself will also carry the crystal into itself. These 24 operations are illustrated by the 24 points shown in Fig. 6. They are the only symmetry operations, aside from the lattice translations, in the zinc-blende structure, but in diamond there are other operations which will carry one of the atoms in the unit cell into the other atom in the unit cell. Such an operation consists of an inversion, followed by a non-primitive translation carrying the center of the original atom into the other atom. By combining the 24 operations of the tetrahedral group, with the identity or the operation of inversion-translation, and the $N$ translational operations, we again find $48N$ operations in the diamond structure, where as before $N$ is the number of unit cells. As with the cubic structures, these operations again form a group. The point group in this case consists of the 48 operations of rotation and inversion, taking no account of the non-primitive translations.

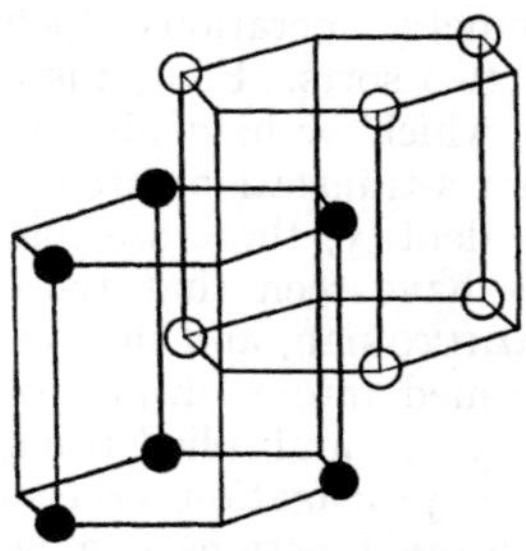

Fig. 10. Two interpenetrating lattices of atoms forming the hexagonal close-packed structure.

Finally, for the hexagonal structure, we can place a prism whose base is an equilateral triangle with its center at one of the atoms, and any one of the 12 operations which will carry this prism into itself will carry the crystal into itself. But here again, as with the diamond structure, we have an additional operation which will carry the environment of one of the atoms in the unit cell into the environment of the other. In this case, as seen in Fig. 10, this operation consists of a rotation of 60° about the hexagonal axis, followed by a translation carrying the center of one of the atoms into the center of the other. In this case, when we combine these operations with the translations, we find $24N$ symmetry operations.

We can now draw conclusions from the fact that the potential energy is invariant under a group of symmetry operations, in each of these cases. The HAMILTONian operator commutes with each operator of the group, though these operators do not all commute with each other (the translations and rotations do not commute with each other, for example). In such a case, one can prove that all the solutions of SCHRÖDINGER's equation must transform according to one of the irreducible representations of the space group. This means the following. Let us start with a wave function $\psi(\boldsymbol{r})$ which is a solution of SCHRÖDINGER's equation corresponding to a definite propagation constant $\boldsymbol{k}$, and perform the various operations of the group on the coordinates. That is, for the cubic structures, we change the vector position $\boldsymbol{r}$ to $\boldsymbol{r}'+\boldsymbol{R}_i$, as in Eq. (6.1), and for the diamond and hexagonal structures we perform similar operations, which in these cases can also include the non-primitive translations plus inversions or rotations. We set up the function $\psi(\boldsymbol{r}'+\boldsymbol{R}_i)$, or the corresponding thing for the diamond and hexagonal structures. The $48N$ or $24N$ functions, respectively, which we obtain in this way, will not all be linearly independent. There will be, in fact, only $n$ linearly independent functions which can be found in this way, where $n$ is some number less than $48N$ (or $24N$). We now find that each of these $n$ functions will transform into a linear combination of the same $n$ functions under any one of the operations of the group. We then say that the $n$ functions so found form a basis for an $n$-dimensional irreducible representation of the space group. The great importance of this situation arises from the fact that all of these $n$ functions, since they arise from each other only by symmetry operations which leave the HAMILTONian invariant, must correspond to the same energy, or must be degenerate with each other.

To investigate the dimensionality of the representations, let us start with a wave function $\psi_k(\boldsymbol{r})$, corresponding to a general value of the propagation vector $\boldsymbol{k}$; that is, a value having no symmetry properties (it is not $\boldsymbol{k}=0$, or along the $x$ or $y$ or $z$ axes, or along the direction $x=y$, etc.). Then we let any one of the symmetry operations of the group operate on this function. We can prove that the result is a multiple of one of 48 (or 24, for the hexagonal case) similar functions corresponding to different $\boldsymbol{k}$ vectors which are equivalent to each other, the multiple depending on the BLOCH factor $\exp i\boldsymbol{k}\cdot\boldsymbol{R}_i$ associated with the translational operation. By equivalent $\boldsymbol{k}$ vectors, we mean in the case of the cubic or diamond structures vectors related to each other like the vectors pointing from the origin to the 48 equivalent points in Fig. 5, and in the hexagonal structure vectors pointing to such points on the face of a hexagon as are shown in Fig. 11. Such a set of $\boldsymbol{k}$-vectors is called by BOUCKAERT, SMOLUCHOWSKI, and WIGNER a star, and we see that each $\boldsymbol{k}$-vector of a star is associated with a wave function of the same energy, so that for a general $\boldsymbol{k}$-value we have 48-dimensional irreducible representations of the space group for the cubic and diamond structures, and 24-dimensional for the hexagonal structure.

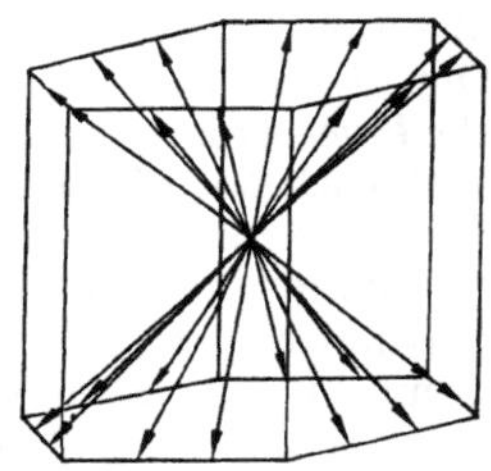

Fig. 11. A star of $\boldsymbol{k}$-vectors for a problem of hexagonal symmetry.

We then have the result that the energy, as a function of $\boldsymbol{k}$, transforms into itself under each of the 48 (or 24) operations interchanging the $\boldsymbol{k}$-vectors of the star. In other words, $E(\boldsymbol{k})$ shows a much higher type of symmetry than appeared from our discussion of Sect. 5. There we showed that $E(\boldsymbol{k})$ was a periodic function of $\boldsymbol{k}$, repeating in each unit cell of reciprocal space. Here we have shown further that each unit cell is subdivided into 48 (or 24) equivalent regions, such that the function $E(\boldsymbol{k})$ in any one of these regions repeats in each of the others, the regions being derived from each other by suitable rotations and inversions. It is easy to set up the rules relating the wave functions corresponding to the various $\boldsymbol{k}$-vectors of a star. We may start with Eq. (5.4), expressing the wave function as a FOURIER expansion, and can show that the momentum eigenfunctions $v(\boldsymbol{k}+\boldsymbol{K}_i)$, or FOURIER coefficients of the various functions, when regarded as functions of position in $\boldsymbol{k}$-space, transform into each other in appropriate ways under the action of the symmetry operations.

**7. The effect of crystal symmetry: Special symmetry points.** In the preceding section we have sketched the situation arising for a general propagation vector $\boldsymbol{k}$. More interesting predictions of the group theory arise for special values of $\boldsymbol{k}$. Such values come at the center of the BRILLOUIN zone, along such axes as the (001), (110), (111) in the cubic system, and similar axes in the hexagonal case, at the intersections of these axes with the boundaries of the BRILLOUIN zone, and at certain other points, lines, and planes in $\boldsymbol{k}$-space. We shall not give the details of the results, which are found in the literature quoted above in Sect. 6, but shall indicate the general nature of the situation.

The characteristic of the special values of $\boldsymbol{k}$ is that the star of $\boldsymbol{k}$-vectors has less than its normal number of independent vectors. Thus, for $\boldsymbol{k}=0$, there is only one vector, of zero length. For a vector along the (001) direction, in a cubic crystal, the $\boldsymbol{k}$-vector forms one of six equivalent vectors, vectors of equal length along the $\pm x$, $\pm y$, $\pm z$ axes. Similar situations arise at the other symmetry points. In such a case, out of all the symmetry operations made up by combining rotations, inversions, and non-primitive translations, as discussed in Sect. 6 (48 for the cubic and diamond structures, 24 for the hexagonal structure), some will transform a given function into a function with a different $\boldsymbol{k}$-vector, but others will transform it into a function with the same $\boldsymbol{k}$. Those operations which leave the $\boldsymbol{k}$-vector invariant form a sub-group of the complete group, and are called the group of the $\boldsymbol{k}$-vector. In the case of a general $\boldsymbol{k}$-vector, the group of the $\boldsymbol{k}$-vector consists only of the identity, but for the special vectors there are more such operations.

We now consider the irreducible representations of the space group connected with a special $\boldsymbol{k}$-vector. To find the dimensionality of the irreducible representations, we must consider the group of the $\boldsymbol{k}$-vector, and the representations of that group. For each representation of that group, we can build up a representation of the spacegroup, its dimensionality being the product of the dimensionality of the representation of the group of the $\boldsymbol{k}$-vector, and of the number of equivalent $\boldsymbol{k}$-vectors in the star. Let us illustrate this by two examples.

First we take the case of a cubic crystal, with its $\boldsymbol{k}$-vector along the (001) direction. As stated above, we have a star consisting of six equivalent $\boldsymbol{k}$-vectors. Out of the 48 operations of the cubic point group, there are eight which leave the $\boldsymbol{k}$-vector invariant, and eight which transform the function into other functions with each one of the other $\boldsymbol{k}$-vectors. The eight operations leaving the $\boldsymbol{k}$-vector invariant form the group of the vector, and are just the operations which will transform a square into itself. It is shown in group theory that there are four one-dimensional and one two-dimensional irreducible representations for this group. Let us ask how we can describe the resulting wave functions. The wave function in the neighborhood of each atom will behave like an atomic function, but will be multiplied by a factor $\exp i k_z Z_i$, where $Z_i$ is the $z$ component of the translation vector from one atom to another, when we go from one atom to a neighboring atom. We can then describe the functions most simply by stating the symmetry of the atomic functions which resemble them near each atom.

There is first a function, usually denoted as $\Delta_1$, which can be made up of a linear combination of an $s$ function, a $p$ function $p_z$, a $d$ function behaving like $3z^2-r^2$, as well as functions of higher azimuthal quantum numbers, on each atom. The common characteristic of all these is that they all change into themselves under all the operations which transform a square in the $xy$ plane into itself. We then have a six-dimensional representation of the space group, for which six basis functions are functions of the type just described, but with propagation vectors along $\pm x$, $\pm y$, $\pm z$ with appropriate atomic functions in each case. Under each of the symmetry operations, one such function will transform either into a multiple of itself, or a multiple of one of the other five functions.

There are three other one-dimensional representations of the group of the $\boldsymbol{k}$-vector, each leading to a six-dimensional representation of the space group. The one which is usually denoted by $\Delta_2$ is composed of atomic functions having the symmetry of the $d$ function behaving like $x^2-y^2$ on each atom. The function denoted by $\Delta_2'$ is composed of atomic functions of the symmetry of $d$ functions behaving like $xy$ on each atom, and $\Delta_1'$ is composed of atomic $g$ functions behaving like $xy(x^2-y^2)$. The two-dimensional representation of the group of the $\boldsymbol{k}$-vector, denoted by $\Delta_5$, consists of two functions which may be of type $p_x$ and $p_y$, or of $d$ type like the functions $xz$ and $yz$. Under such an operation as a rotation of 90° about the $z$ axis, the function $p_x$ would transform into $p_y$, and $p_y$ would transform into $-p_x$. On the other hand, other operations would transform one of these functions into functions with the $\boldsymbol{k}$-vector along another of the six possible directions. This two-dimensional representation leads, then, to a twelve-dimensional representation of the space group.

This two-dimensional representation $\Delta_5$ of the group of the $\boldsymbol{k}$-vector along (001) leads to a situation quite different from what we meet for a general $\boldsymbol{k}$-vector. For here we have two functions, of $p_x$ and $p_y$-like symmetry, with the same propagation vector along the $z$ axis, and degenerate with each other. In other words, if we have such a state, we necessarily have two degenerate states with the same $\boldsymbol{k}$-vector. For a general $\boldsymbol{k}$-vector, on the contrary, only one function among the basis functions of an irreducible representation corresponds to the same $\boldsymbol{k}$-vector; the 48 different functions correspond to 48 different $\boldsymbol{k}$'s. Thus there is no reason arising from symmetry why two or more energy levels associated with a general $\boldsymbol{k}$-vector should be equal. This is the basis for our earlier statement that in general, the energies of different energy bands are all separated, at a general point in $\boldsymbol{k}$-space. However, for $\boldsymbol{k}$ along (001), we have just found a two-dimensional representation. In such a case, two energy bands are said to stick together along the (001) direction, necessarily having the same energy. If we depart from the (001) direction in any arbitrary direction, however, the energy bands will in general separate, and interesting studies have been made of the way in which this comes about, for instance by SHOCKLEY [*207*] and the present writer [*227*].

As a second example, we take the case $\boldsymbol{k}=0$ in a cubic crystal. For $\boldsymbol{k}=0$, the wave function must repeat periodically at each atom of the crystal. The group of the $\boldsymbol{k}$-vector is now the same as the group of 48 operations carrying a cube into itself. In this case there are ten irreducible representations, of which four are one-dimensional, two are two-dimensional, and four are three-dimensional; every solution of the wave equation for $\boldsymbol{k}=0$ must belong to one of these representations. We shall describe the representations which are most often met in practice. The representation denoted by $\Gamma_1$ is what we should have with atomic $s$ functions on each atom; it is a one-dimensional representation. The three-dimensional representation $\Gamma_{15}$ is like what we should have with atomic $p_x$ or $p_y$ or $p_z$ wave functions on each atom. Since these three functions belong to the same representation, we see that an atomic $p$-state is not split at $\boldsymbol{k}=0$. The representation $\Gamma_{12}$ consists of two $d$-like functions behaving like the polynomials $x^2-y^2$ and $3z^2-r^2$, which transform into linear combinations of each other under each of the operations of the group, and $\Gamma'_{25}$ consists of three $d$-like functions containing polynomials $xy$, $yz$, $zx$. The fact that $\Gamma_{12}$ and $\Gamma'_{25}$ are different representations shows that in general a five-fold degenerate $d$ state of an atom will be split into a two-fold and a three-fold degenerate state in a cubic crystal at $\boldsymbol{k}=0$. Similar remarks can be made for the other irreducible representations, which are formed from atomic wave functions of higher azimuthal quantum number.

We note that at $\boldsymbol{k}=0$, the representation $\Gamma_{15}$ is a three-dimensional one with functions behaving like $p_x$, $p_y$, $p_z$. As we depart from $\boldsymbol{k}=0$ along the (001) direction, we have already seen that there we have representations formed from atomic $p$ functions. Thus the representation $\Delta_1$, as we have seen, can be made up of a mixture of atomic $p_z$ with atomic $s$ and $d$ functions, while the two-dimensional representation $\Delta_5$ is made up of atomic functions of type $p_x$ and $p_y$, which may be mixed with $d$-like functions $xz$ and $yz$. We see several things from this simple example. As we go away from $\boldsymbol{k}=0$ along (001), the three-fold degenerate state $\Gamma_{15}$ becomes split into a non-degenerate state $\Delta_1$, and a two-fold degenerate state $\Delta_5$. Furthermore, however, these states can change their character as $\boldsymbol{k}$ increases, partially taking on the behavior of $s$ or $d$ atomic functions. Conversely, a state of type $\Delta_1$ could emerge from an $s$-like state of type $\Gamma_1$ at $\boldsymbol{k}=0$, or from a $d$-like state of type $\Gamma_{12}$. Such relations, stating which states at a point like $\boldsymbol{k}=0$ can join definite states along a line such as (001), are called compatibility relations, and have been worked out in detail for the various structures of crystals.

These few examples will indicate the general nature of the results obtained at symmetry points in the Brillouin-zone; space does not permit of the complete classification of the symmetry relations. We shall see further examples of the symmetry types in later sections, where we examine in detail the calculations which have been made of energy bands of various crystals.

In our discussions we have been disregarding one feature which can actually be of considerable importance: the spin-orbit coupling. This can have the effect of removing part of the degeneracy at special symmetry points; for instance, of splitting a three-fold degenerate state into a two-fold and a non-degenerate state. This question has been exhaustively considered by Elliott [*57*], [*58*], and the reader is referred to his papers for further information.

## II. The periodic potential problem: Methods and results.

**8. The free-electron approximation.** In Chapter I, we have been considering the general properties of solutions of the periodic potential problem, representing the one-electron wave functions of electrons in a crystal, determined by the method of the self-consistent field. We have derived results arising from the symmetry properties of the crystal, both its translational and rotational symmetry. These results must be true quite generally, irrespective of the particular method used for solving the periodic potential problem. To get useful results for individual crystals, however, we need something much more detailed than this: we wish to have as good approximations as we can get to the exact solutions of Schrödinger's equation for a potential representing as well as possible that actually found in the crystal. In the present chapter, we shall take up some of the methods which have been used in practice for getting approximate solutions of the periodic potential problem, together with the results which have been obtained by these methods for specific substances. The present section will deal with the free-electron approximation, one of the two approximations introduced in the earliest days of the problem.

In Sect. 5, we have seen that the wave function can be expanded in a series of plane waves, given in Eq. (5.4), and we have exhibited in Eq. (5.7) the algebraic equations which must be solved to find the coefficients of the expansion, and the energy levels. These equations were set up in some of the very earliest papers on the periodic potential: papers by Bethe [*16*] and by Morse [*166*] on electron scattering in a crystal, one by Peierls [*184*], [*185*] on conductivity and other properties, and several papers by Brillouin [*31*], [*32*], [*33*], [*37*] dealing with the general problem of the motion of electrons in a periodic potential. These writers all examined the form which the equations take in the limit where the electrons are almost free, or where the potential energy is almost constant, so that the Fourier components of expansion of the potential energy, which we have called $W(\boldsymbol{K}_j)$, are all small. We shall see later that this free-electron limit does not agree very well with the actual situation found in metals, but nevertheless the treatment of the free-electron case has had wide significance in the further treatment of the problem, and is interesting in many ways.

Since Eq. (5.7) takes the form of the well-known equations of quantum mechanics leading to perturbation theory, one can use perturbation methods here. If we disregard $W(\boldsymbol{K}_j)$ altogether, except for the diagonal term $W(0)$, which measures the average potential energy, then Eq. (5.7) has a solution representing a single plane wave. That is, we can assume that only one term in Eq. (5.4) is different from zero, say that for $\boldsymbol{K}_i = \boldsymbol{K}_0$, leading to $u(\boldsymbol{r}) = v(\boldsymbol{k}+\boldsymbol{K}_0)\exp i(\boldsymbol{k}+\boldsymbol{K}_0)\cdot\boldsymbol{r}$, all coefficients $v(\boldsymbol{k}+\boldsymbol{K}_i)$ equalling zero if $\boldsymbol{K}_i \neq \boldsymbol{K}_0$. When we substitute this in Eq. (5.7), we find a solution, with energy equal to $(\hbar^2/2m)(\boldsymbol{k}+\boldsymbol{K}_0)^2 + W(0)$, simply the kinetic energy of the plane wave plus the average potential energy. We now ask what is the effect of introducing small FOURIER components $W(\boldsymbol{K}_j)$, looking only for the second order of perturbations. We find in the usual way that the perturbed wave function is

$$u(\boldsymbol{r}) = v(\boldsymbol{k}+\boldsymbol{K}_0)\exp i(\boldsymbol{k}+\boldsymbol{K}_0)\cdot\boldsymbol{r}\left\{1 + \frac{\sum\limits_{\boldsymbol{K}_j \neq 0} W(\boldsymbol{K}_j)\exp i\,\boldsymbol{K}_j\cdot\boldsymbol{r}}{(\hbar^2/2m)\left[(\boldsymbol{k}+\boldsymbol{K}_0)^2 - (\boldsymbol{k}+\boldsymbol{K}_0+\boldsymbol{K}_j)^2\right]}\right\} \tag{8.1}$$

and the corresponding energy is

$$E = \frac{\hbar^2}{2m}(\boldsymbol{k}+\boldsymbol{K}_0)^2 + W(0) + \sum_{\boldsymbol{K}_j \neq 0} \frac{|W(\boldsymbol{K}_j)|^2}{(\hbar^2/2m)\left[(\boldsymbol{k}+\boldsymbol{K}_0)^2 - (\boldsymbol{k}+\boldsymbol{K}_0+\boldsymbol{K}_j)^2\right]}. \tag{8.2}$$

Since the energy and wave function are only slightly different from those corresponding to the unperturbed case with the propagation vector $\boldsymbol{k}+\boldsymbol{K}_0$, it is most convenient in this case to denote energy and wave function by using $\boldsymbol{k}+\boldsymbol{K}_0$ as equivalent to a quantum number, rather than by using the reduced propagation vector $\boldsymbol{k}$ (assumed to lie in the central BRILLOUIN zone).

The perturbation treatment indicated in Eqs. (8.1) and (8.2) is satisfactory, provided we are far from degeneracy; that is, provided none of the denominators approach zero. If two or more plane waves have approximately the same energy, however, so that we find one or more $\boldsymbol{K}_j$'s for which $(\boldsymbol{k}+\boldsymbol{K}_0)^2$ is approximately equal to $(\boldsymbol{k}+\boldsymbol{K}_0+\boldsymbol{K}_j)^2$, then we must handle the problem like a degenerate system. That is, we must first set up a problem derived from Eq. (5.7), in which we include only those $v$'s associated with the degenerate waves, and must solve this problem, to get suitable linear combinations of the waves which diagonalize the HAMILTONian. Then we can use these linear combinations as a starting point of a second-order perturbation problem involving all the other states. The simplest case is that in which only two waves are approximately degenerate with each other. Then our secular problem reduces to the very simple form

$$\left.\begin{aligned}
&\frac{\hbar^2}{2m}(\boldsymbol{k}+\boldsymbol{K}_0)^2\, v(\boldsymbol{k}+\boldsymbol{K}_0) + W(0)\, v(\boldsymbol{k}+\boldsymbol{K}_0) + W(-\boldsymbol{K}_j)\, v(\boldsymbol{k}+\boldsymbol{K}_0+\boldsymbol{K}_j)\\
&\qquad\qquad = E\, v(\boldsymbol{k}+\boldsymbol{K}_0),\\
&\frac{\hbar^2}{2m}(\boldsymbol{k}+\boldsymbol{K}_0+\boldsymbol{K}_j)^2\, v(\boldsymbol{k}+\boldsymbol{K}_0+\boldsymbol{K}_j) + W(0)\, v(\boldsymbol{k}+\boldsymbol{K}_0+\boldsymbol{K}_j) + W(\boldsymbol{K}_j)\, v(\boldsymbol{k}+\boldsymbol{K}_0)\\
&\qquad\qquad = E\, v(\boldsymbol{k}+\boldsymbol{K}_0+\boldsymbol{K}_j).
\end{aligned}\right\} \tag{8.3}$$

These lead in the usual way to a secular equation

$$\begin{vmatrix} \frac{\hbar^2}{2m}(\boldsymbol{k}+\boldsymbol{K}_0)^2 + W(0) - E & W(-\boldsymbol{K}_j) \\ W(\boldsymbol{K}_j) & \frac{\hbar^2}{2m}(\boldsymbol{k}+\boldsymbol{K}_0+\boldsymbol{K}_j)^2 + W(0) - E \end{vmatrix} = 0 \tag{8.4}$$

which can at once be solved. At exact degeneracy, the solutions will be the sum and difference of the two waves, and the energy values will be the diagonal energy value of either wave, plus or minus $|W(\boldsymbol{K}_j)|$.

This case of degeneracy of two waves is a very interesting one, and was shown by the various writers whom we have mentioned to be equivalent to

BRAGG reflection. Thus, in Fig. 12, we show the two vectors $\boldsymbol{k}+\boldsymbol{K}_0$ and $\boldsymbol{k}+\boldsymbol{K}_0+\boldsymbol{K}_j$, the propagation vectors of the two interacting waves, which must have the same length on account of the degeneracy. The planes of atoms described by the vector $\boldsymbol{K}_j$ in the reciprocal lattice have a spacing $d$ given by $|\boldsymbol{K}_j|=2\pi/d$, and the vector $\boldsymbol{K}_j$ is normal to the planes, so that one of the planes will be in the direction shown in the figure, such that the incident wave vector makes a glancing angle of incidence $\vartheta$ with it. The angle $\vartheta$ is obviously given by the condition

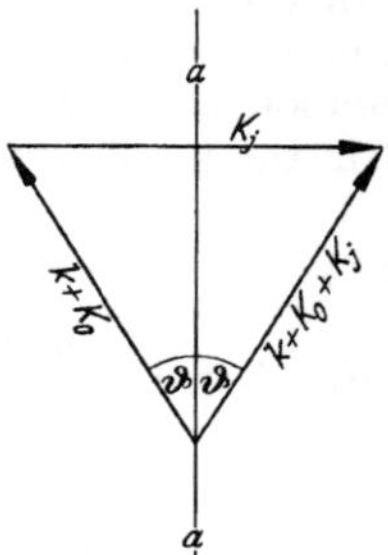


Fig. 12. Diagram to illustrate BRAGG's law. Scattering planes of atoms are parallel to line $aa$.

$$\sin\vartheta=\frac{|\boldsymbol{K}_j|}{2\,|\boldsymbol{k}+\boldsymbol{K}_0|}=\frac{\lambda}{2d} \tag{8.5}$$

since $|\boldsymbol{k}+\boldsymbol{K}_0|=2\pi/\lambda$. But this is BRAGG's law for scattering of the electron wave by the atomic planes. We see, then, that two waves will hardly interact with each other at all unless they are related by BRAGG's law, but in that case there will be strong reflection.

From Fig. 12 we see that any propagation vector terminating on a plane in the reciprocal space which is the perpendicular bisector of a vector $\boldsymbol{K}_j$ will be degenerate with another wave. For propagation vectors in the neighborhood of such a plane, we must get the energy by solution of a secular equation like Eq. (8.4). In Fig. 13 we show the nature of the resulting solution for the energy; we have taken the case where the vector $\boldsymbol{k}+\boldsymbol{K}_0$ is parallel to $\boldsymbol{K}_j$, but of variable magnitude, and have plotted the energy as a function of the magnitude of the vector. There are two ways in which we can regard the situation. First, we may consider the line drawn heavy in the figure, representing the branch of the curve which most nearly agrees with the value $(\hbar^2/2m)\,(\boldsymbol{k}+\boldsymbol{K}_0)^2+W(0)$, the energy of a single plane wave. When we consider it in this way, we find a break in the curve, across the plane in question, of magnitude $2|W(\boldsymbol{K}_j)|$. This is usually called an energy gap. But secondly, we may take advantage of the periodicity of the energy in reciprocal space, which has been pointed out in Sect. 5, and draw the energy curve displaced by amounts $\pm\boldsymbol{K}_j$. Then we see that we have two branches of the curve, or two energy surfaces, one reaching a maximum and the other a minimum at the plane perpendicular to $\boldsymbol{K}_j$.

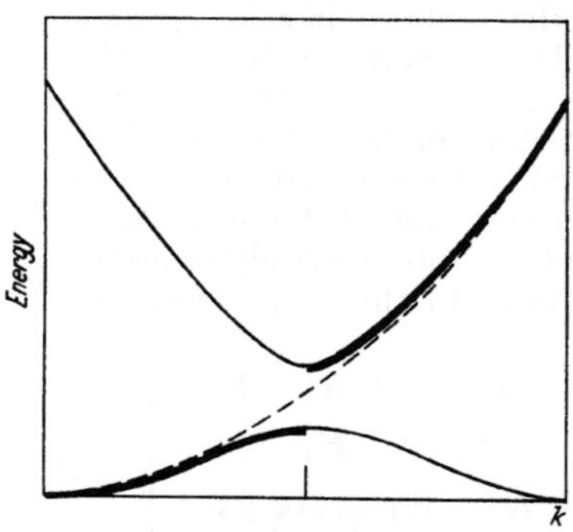


Fig. 13. Energy as function of $\boldsymbol{k}$, showing break in curve, or energy gap, at value of $\boldsymbol{k}$ given by BRAGG's law. Parts of curve drawn heavy most nearly agree with free-electron parabola, which is shown by dotted line.

BRILLOUIN [*31*], [*32*], [*33*], [*37*] has made much use of the volumes in the reciprocal space bounded by the surfaces which are the perpendicular bisectors of all vectors $\boldsymbol{K}_j$. These volumes are ordinarily called BRILLOUIN zones, and the smallest of them is the one which we have discussed in Sect. 4. We see that if we choose to plot the energy as a function of $k$ as in the heavy line of Fig. 13, the energy will be approximately that of a plane wave, but will show an energy gap across each surface of a BRILLOUIN zone. On the other hand, if we use a reduced propagation vector, plotting all energy surfaces in the central BRILLOUIN zone, we shall not have energy gaps as such, but shall merely have energy surfaces of complicated forms. Since we shall see later that the free-electron approximation is not very good for actual metals, and therefore Fig. 13 is not a very good representation of the energy in the neighborhood of one of the reflecting planes, these BRILLOUIN zones are of only limited usefulness, in contrast to the central zone, which is of fundamental significance, since it is equivalent to the unit cell in the reciprocal space.

We have considered the case where two plane waves are degenerate with each other; but in many cases a much larger number of waves may be degenerate with each other. These cases arise at the symmetry points in reciprocal space which were discussed in Sect. 7. For example, let the reduced vector $\boldsymbol{k}$ be equal to zero. We take a plane wave with propagation vector $\boldsymbol{K}_i$, where $\boldsymbol{K}_i$ does not have any special symmetry properties (it is not along the $x$ or $y$ or $z$ axis, or the line $x=y=z$, etc.). Then there will be 48 equivalent vectors $\boldsymbol{K}_i$ in the reciprocal lattice all of equal length with $\boldsymbol{K}_i$, if we are dealing with a cubic crystal, so that the energies of all 48 plane waves will be identical. Our problem of degeneracy will then consist of 48 functions, and this would be a difficult problem to solve if it were not for the use of symmetry.

Our results of Sect. 7, however, show that the wave functions must correspond to one of the various irreducible representations of the cubic group, in this case. It is not hard by inspection to set up linear combinations of the 48 plane waves which have these symmetry properties. We shall find automatically that there are no non-diagonal matrix components of energy between such wave functions of different symmetry types. In other words, by use of symmetry, we have solved the secular equation at the point $\boldsymbol{k}=0$, corresponding to degeneracy. As mentioned in Sect. 7, we shall find in this case one-, two-, and three-dimensional representations of the cubic group. In other words, we shall find that the energy levels are split, on account of the non-diagonal matrix components $W(\boldsymbol{K}_j)$, into a number of different sets, each one not more than three-fold degenerate, rather than having the 48-fold degeneracy which we should encounter at $\boldsymbol{k}=0$ in the absence of the periodic potential. It is not hard to start with these wave functions of the proper symmetry properties, and investigate the behavior of the energy bands as we depart slightly from $\boldsymbol{k}=0$. We have to handle only a two- or three-fold degeneracy, finding non-diagonal matrix components between the degenerate states which vanish at $\boldsymbol{k}=0$, but appear as $\boldsymbol{k}$ departs from zero. By proceeding along these lines, one can investigate the behavior of the energy bands in the neighborhood of the various symmetry points considered in Sect. 7, on the basis of the free-electron approximation. Such a general treatment, however, was not given in the references to the free-electron method cited above.

We have now outlined the general procedure for using the free-electron approximation. Its results are consistent with the conclusions regarding symmetry derived in Sect. 7. However, it forms a very poor approximation to the actual situation in a solid, for the reason that the actual effect of the periodic potential is much too great to handle by perturbation methods. The early workers in the field were misled by the analogy between their problem of electron diffraction, and the similar case of x-ray diffraction: the dynamical theory of x-ray diffraction[1] is formally identical with the theory which we have outlined in this section, and for the x-ray case it is entirely satisfactory. The point is that the quantities similar to $W(\boldsymbol{K}_j)$, in the x-ray case, are extremely small, so that the energy gaps arising from them are almost negligibly small; on the other hand, when the

[1] A good treatment of this theory, with additional references, is given in R. W. JAMES, The Optical Principles of the Diffraction of X-Rays (London, G. Bell and Sons, Ltd., 1948), Chapters II, VIII.

magnitudes of the Fourier components of the actual potential are computed in the Schrödinger case, they prove to be of the order of magnitude of several electron-volts, and correspond to large effects.

We can investigate the effect of such large components rigorously in a simple one-dimensional problem, that in which the potential is a sinusoidal function of position. This leads to the Mathieu equation, which can be solved exactly, and its application to the present problem has been discussed by many writers, including Morse [*166*], Brillouin [*31*], the present writer [*227*], and others. This is the case in which only one of the coefficients $W(\boldsymbol{K}_j)$ is different from zero, but where that coefficient can be as large as we please. The result of study of this soluble case is that when $W(\boldsymbol{K}_j)$ is very small, energy gaps are introduced, just as we have described; but as it increases, these gaps become wider and wider, until finally the energy bands between the gaps become narrower and narrower, and narrow bands split off from the bottom of the spectrum of energy bands, as shown in Fig. 14. This is a situation analogous to what we have in a real solid: the narrow bands correspond to the x-ray energy levels, while the continuum above them, with narrow gaps, corresponds to the outer energy bands. When we now investigate the exact solutions of this Mathieu case, we find that once $W(\boldsymbol{K}_j)$ is great enough so that several narrow bands have split off from the bottom of the continuum, it becomes completely incorrect to represent the solution approximately by one, or a few, plane waves, even for the energy levels in the continuum. The series of Eq. (5.4) becomes very slowly convergent, and perturbation methods such as we have been using in this section become completely incorrect. We must therefore conclude that the type of free-electron calculation used in this section is not adequate as a starting point for a study of real crystals.

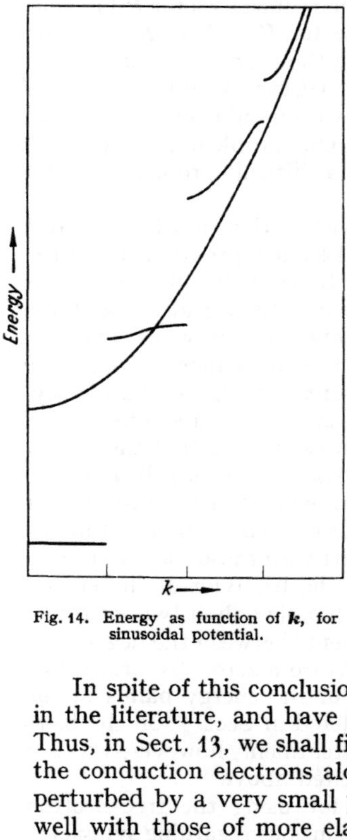

Fig. 14. Energy as function of $\boldsymbol{k}$, for sinusoidal potential.

In spite of this conclusion, the methods of this section have been widely used in the literature, and have found surprisingly good agreement with experiment. Thus, in Sect. 13, we shall find that for an alkali metal like sodium, one can treat the conduction electrons alone as if they formed a gas of almost free electrons, perturbed by a very small periodic potential, and the results agree surprisingly well with those of more elaborate methods of calculation. The success of this method has led to widespread belief that the free-electron approximation is valid, and forms the basis of such treatments as Sommerfeld's [*233*] free-electron treatment of conductivity, which have been very fruitful. The reason for the success of the method seems to be the following. We shall see in Sect. 11 that quite different types of approximation to the solution of Schrödinger's equation for a periodic potential, the orthogonalized and augmented plane wave methods, which are capable of giving very good approximations, show a close superficial resemblance to the free-electron method. They lead to secular problems very similar to those discussed in this section. It thus seems that the apparent success of the free-electron method should not be regarded as a verification of the correctness of the postulates of that method, but rather as justifying these other types of approximation.

**9. The BLOCH or tight-binding approximation.** BLOCH [21], in his fundamental paper on the periodic potential problem, proposed an approximation for the wave functions which in many ways is entirely satisfactory, particularly for the lower occupied energy bands. This approximation expresses a wave function for the crystal as a linear combination of atomic orbitals, which would be abbreviated LCAO by those who work in the theory of molecular orbitals. Let $\varphi(\boldsymbol{r}-\boldsymbol{R}_i)$ be an atomic wave function centered about an atom in the unit cell displaced from the unit cell at the origin by the vector $\boldsymbol{R}_i$, which equals $m_1\boldsymbol{a}_1+m_2\boldsymbol{a}_2+m_3\boldsymbol{a}_3$. BLOCH then set up the combination

$$b_{\boldsymbol{k}}(\boldsymbol{r})=\sum_{\boldsymbol{R}_i}\exp i\,\boldsymbol{k}\cdot\boldsymbol{R}_i\,\varphi(\boldsymbol{r}-\boldsymbol{R}_i). \tag{9.1}$$

This is a linear combination of atomic orbitals $\varphi(\boldsymbol{r}-\boldsymbol{R}_i)$ located in equivalent positions in all unit cells of the crystal, with the coefficients $\exp i\,\boldsymbol{k}\cdot\boldsymbol{R}_i$. One can demonstrate very simply that this function possesses the fundamental property of periodicity, as described in Eqs. (5.1) and (5.2): the function is multiplied by the factor $\exp i\,\boldsymbol{k}\cdot\boldsymbol{R}_j$ when $\boldsymbol{r}$ is increased by $\boldsymbol{R}_j$. Furthermore, for the tightly bound electrons of the atom, $b_{\boldsymbol{k}}(\boldsymbol{r})$ forms a very accurate approximation to a solution of SCHRÖDINGER's equation for the periodic potential problem. The reason is obvious: near the atom on which $\varphi$ is located the potential in the periodic potential problem is very close to the atomic potential, so that the atomic wave function is a good approximation to a solution. Furthermore, for the tightly bound electrons, the $\varphi$'s located in different unit cells will hardly overlap at all, so that near each atom $b_{\boldsymbol{k}}(\boldsymbol{r})$ will reduce almost exactly to a single atomic wave function on that atom.

BLOCH assumed that a single function of the type given in Eq. (9.1) formed a satisfactory approximation, and calculated the expectation value of the energy for this function. We can easily follow his calculation. We let the HAMILTONian function $H$ operate on $b_{\boldsymbol{k}}(\boldsymbol{r})$, multiply by $b_{\boldsymbol{k}}^*(\boldsymbol{r})$, and integrate over our fundamental block of cells. In doing so, we meet integrals of the nature of $\int\varphi^*(\boldsymbol{r}-\boldsymbol{R}_i)\times H\,\varphi(\boldsymbol{r}-\boldsymbol{R}_j)\,dv$. On account of the periodicity of $H$, such an integral depends only on the distance $\boldsymbol{R}_i-\boldsymbol{R}_j$ between the unit cells in question. If we let $\boldsymbol{R}=0$ for the central unit cell, we may then define

$$H(\boldsymbol{R}_i)=\int\varphi^*(\boldsymbol{r})\,H\,\varphi(\boldsymbol{r}-\boldsymbol{R}_i)\,dv. \tag{9.2}$$

Then we find that

$$\int b_{\boldsymbol{k}}^*(\boldsymbol{r})\,H\,b_{\boldsymbol{k}}(\boldsymbol{r})\,dv=N\sum_{\boldsymbol{R}_i}\exp i\boldsymbol{k}\cdot\boldsymbol{R}_i\,H(\boldsymbol{R}_i) \tag{9.3}$$

where $N$ as before is the number of unit cells in the block of cells over which we integrate. The expression of Eq. (9.3) is not, however, the expectation value of energy, because the wave function in Eq. (9.1) is not normalized. If we define

$$S(\boldsymbol{R}_i)=\int\varphi^*(\boldsymbol{r})\,\varphi(\boldsymbol{r}-\boldsymbol{R}_i)\,dv \tag{9.4}$$

as the overlap integral between an orbital in the central unit cell, and an identical orbital in the unit cell at $\boldsymbol{R}_i$, we find that

$$\int b_{\boldsymbol{k}}^*(\boldsymbol{r})\,b_{\boldsymbol{k}}(\boldsymbol{r})\,dv=N\sum_{\boldsymbol{R}_i}\exp i\,\boldsymbol{k}\cdot\boldsymbol{R}_i\,S(\boldsymbol{R}_i). \tag{9.5}$$

Using this result, we may normalize $b_{\boldsymbol{k}}(\boldsymbol{r})$, and find that the expectation value of the energy is given by

$$(H)_{av}=\frac{\sum\limits_{\boldsymbol{R}_i}\exp i\,\boldsymbol{k}\cdot\boldsymbol{R}_i\,H(\boldsymbol{R}_i)}{\sum\limits_{\boldsymbol{R}_j}\exp i\,\boldsymbol{k}\cdot\boldsymbol{R}_j\,S(\boldsymbol{R}_j)}. \tag{9.6}$$

The expression of Eq. (9.6) has a very simple form, which expresses a number of facts of importance. In the first place, the energy shows the periodicity in the reciprocal space which we have demonstrated in Sect. 5. This is proved at once: if $\boldsymbol{k}$ is increased by one of the vectors $\boldsymbol{K}_s$ of the reciprocal lattice, each of the exponentials in the numerator or denominator of Eq. (9.6) will be unchanged, on account of the relation $\boldsymbol{R}_i \cdot \boldsymbol{K}_s = 2\pi \times$ integer, which follows from Eq. (4.1). Secondly, if the atomic orbitals $\varphi$ in different unit cells hardly overlap at all, the energy will be almost independent of $\boldsymbol{k}$, or the energy band will hardly be split at all. This is obvious because for non-overlapping orbitals, the quantities $H(\boldsymbol{R}_i)$ and $S(\boldsymbol{R}_i)$ of Eqs. (9.2) and (9.4) must be vanishingly small for $\boldsymbol{R}_i \neq 0$, since the product of the wave functions in the two unit cells appears in the integrand, and consequently the expression $(H)_{av}$ reduces to a single term, $H(0)/S(0)$, independent of $\boldsymbol{k}$. If we assume the atomic orbitals to be normalized, $S(0) = 1$, and $H(0)$ is the expectation value of the energy for an unmodified atomic orbital.

BLOCH tacitly assumed that his atomic orbitals were $s$ orbitals, and showed that the expression of Eq. (9.6) led to a minimum energy at $\boldsymbol{k} = 0$. It was some time before the case of degenerate atomic orbitals was properly handled by the tight-binding method. It is clear that we can set up a BLOCH sum, using Eq. (9.1), for each of the atomic orbitals arising from the various atoms in the unit cell. We should expect that we could get a better approximation to a solution of SCHRÖDINGER's equation by using not a single BLOCH sum, but a linear combination of a number of such functions, with appropriate coefficients. If we are not dealing with approximate degeneracy, it is not imperative to use such linear combinations, but if several BLOCH sums are exactly or almost degenerate, it is necessary to do so. By methods similar to those used in deriving Eq. (9.6), it is easy to set up the formula for the non-diagonal matrix component of the HAMILTONian $H$ between two BLOCH sums. Such components are zero, by symmetry, between BLOCH sums of different $\boldsymbol{k}$ values (we assume throughout this section that we are dealing with reduced $\boldsymbol{k}$-vectors, in the central BRILLOUIN zone), but we must make a linear combination of the various BLOCH sums associated with the same $\boldsymbol{k}$-vector, and different atomic orbitals. It has been found in the last few years that we get results of qualitative correctness for all the occupied energy bands in the crystal, if we make such linear combinations of the various occupied atomic orbitals in the atoms, and minimize the energy to determine the coefficients.

For a general $\boldsymbol{k}$-vector, the BLOCH sums formed from all types of atomic orbitals must be combined to make up the approximate wave function just described, so that one must solve a secular equation of high order to carry out the calculation. For a $\boldsymbol{k}$-vector showing the type of symmetry discussed in Sect. 7, however, the order of the secular equation is reduced, for we need to use only a restricted set of atomic orbitals to set up a wave function transforming according to one of the irreducible representations of the group of the $\boldsymbol{k}$-vector. Thus, in Sect. 7, we have used as an example the propagation along the (001) direction in a cubic crystal, and have set up five irreducible representations, denoted by $\Delta_1$, $\Delta_2$, $\Delta_1'$, $\Delta_2'$, $\Delta_5$, which represent the only types of symmetry which are allowed along this particular direction. Furthermore, we have described the behavior of the wave function near an atom in terms of the atomic functions which go to produce a function of the proper symmetry. In the case of the tight-binding method, we then find that only the BLOCH sums formed from those particular atomic functions will be concerned in setting up the wave function. For instance, for the state $\Delta_1$, referring to Sect. 7, we see that we shall have a combination of BLOCH sums formed from atomic $s$ functions, from $p_z$, and from the $d$ function varying as $3z^2 - r^2$; these are all the types of atomic functions of azimuthal quantum number up to two which we should meet in this particular case. Similar rules allow the tight-binding method to be applied in the special symmetry directions in a fairly simple way.

Comparatively few calculations have been made by the method we have just sketched, in spite of the fact that it seems to be one of the most promising methods for finding the wave functions of the occupied energy bands. One of the first uses of the method was by JONES, MOTT, and SKINNER [*127*] in 1934, in which they treated the quantities $H(\boldsymbol{R}_i)$ as disposable parameters. Recent calculations have been those of WALLACE [*250*] on graphite, PARMENTER [*175*] on Li, and of FLETCHER and WOHLFARTH [*69*], [*70*] on Ni. Part of the difficulty of the method arises because the integrals $H(\boldsymbol{R}_i)$ are hard to calculate. They involve what the workers in the field of molecular structure call three-center integrals: in Eq. (9.2), $\varphi(\boldsymbol{r})$ is centered around one atom, $\varphi(\boldsymbol{r}-\boldsymbol{R}_i)$ around another, and the potential energy term in $H$ will involve potentials of spherical form centered around third atoms. Until recently, no practical method has been available for computing such integrals, but very recent developments in molecular theory, using digital computers, make such integrals possible to work with. It is to be anticipated that future quantitative work with this method will be much commoner than in the past. Lacking quantitative values of the integrals, it has been suggested by the writer and KOSTER [*230*] that this method could well be used as an interpolation method, in case precise values of the energy were available at certain points in the BRILLOUIN zone, using these known values to determine semi-empirical values of the parameters such as $H(\boldsymbol{R}_i)$ concerned in the calculation.

Though the tight-binding method is fairly satisfactory for the occupied energy bands, it is of small use for higher, excited energy bands, as has been pointed out particularly by PARMENTER [*175*]. The reason is that, in a metal at least, the excited atomic orbitals from one unit cell are so extended in space that they mostly are actually located in neighboring unit cells. That this is so is natural from our comments in Sect. 2, to the effect that the interatomic distances in metals are such that the outer occupied orbitals of adjacent atoms have their maximum densities overlapping each other. The excited orbitals are much larger, and shows much greater overlapping. When now we make a BLOCH sum of these excited orbitals, their atomic character is almost entirely lost, and they take on the behavior of plane waves, rather independent of which atomic orbital we start with. Other methods must be used for the excited energy bands.

Since the BLOCH functions form good approximations to solutions of SCHRÖDINGER'S equation, for the bound states, it is very informing to express them as linear combinations of plane waves, according to Eq. (5.4), so as to investigate the actual behavior of the momentum eigenfunctions for these states; a discussion of this problem has been given by the writer [*219*]. It is very easy to show that for a single BLOCH sum, such as is given in Eq. (9.1), the result is as follows, provided we can disregard the overlapping of the atomic orbitals. In the first place, we make a FOURIER resolution of the atomic function $\varphi(\boldsymbol{r})$. This can be expressed in the form of Eq. (5.4), except that the quantity $v(\boldsymbol{k}+\boldsymbol{K}_i)$ will now be a continuous function of the variable $\boldsymbol{k}+\boldsymbol{K}_i$, there being no periodicity in the problem. Then we pick out the values of this continuous function given by the lattice of points $v(\boldsymbol{k}+\boldsymbol{K}_i)$, where $\boldsymbol{k}$ is fixed, $\boldsymbol{K}_i$ goes through the reciprocal lattice. The discrete quantities so found represent the momentum eigenfunctions for the BLOCH sum. We can then at once investigate the convergence of the sum of Eq. (5.4) by finding the FOURIER resolution of the atomic orbital.

The results of such an analysis, determining the momentum eigenfunction of an isolated atom, are well known; for the hydrogen atom, the problem was solved many years ago by PODOLSKY and PAULING [*188*], and the results for hydrogen are typical of those for other atoms. The momentum eigenfunction extends through a volume of momentum space which is inversely proportional to the volume of ordinary space occupied by the ordinary eigenfunction. More precisely, if the ordinary eigenfunction is large through a fraction $f$ of the unit cell, the momentum eigenfunction will be large in $1/f$ unit cells in momentum space. For one of the inner electrons of the atom, where $f$ is small, the momentum eigenfunction will then extend through many unit cells in momentum space, showing that the expression of Eq. (5.4) converges very slowly. For an outer electron, for which $f$ is of the order of magnitude of unity, the convergence is faster. However, as we shall show later, even in this case there must be significant contributions to the momentum eigenfunction from many unit cells of momentum space. Thus we verify our statements of Sect. 8 regarding the slow convergence of the expansion of the wave function of a crystal in plane waves.

**10. The cellular method.** We have now discussed the general properties of the solutions of SCHRÖDINGER's equation in the type of periodic potential found in a crystal. We have found one type of approximate solution, the free-electron approximation which is hardly applicable directly to any cases of real crystals, and another approximation, the tight-binding or BLOCH method, which works well for the bound energy bands in crystals. We shall next pass on to other approximate methods of solution which have given promise of leading to more accurate or more informing wave functions. First we take up the cellular method first proposed by WIGNER and SEITZ [*255*].

WIGNER and SEITZ wished to consider the wave function at the bottom of the conduction band in sodium, the wave function which would be expressed in the tight-binding approximation as the sum of the $3s$ atomic orbitals on all atoms of the crystal (the factor $\exp i\boldsymbol{k}\cdot\boldsymbol{R}_i$ being unity for all atoms, since $\boldsymbol{k}=0$). They set up the WIGNER-SEITZ cells for the body-centered cubic structure. They assumed that the potential function within each cell was the same as that used for the valence electron in a self-consistent problem of a sodium atom; this is approximately the potential of the $Na^+$ ion, though they actually used a semi-empirical potential proposed for this problem by PROKOFJEW [*190*]. They then noted that the SCHRÖDINGER equation within the cell was spherically symmetrical, so that one should be able to solve it just as in an atom, only with different boundary conditions. Instead of going asymptotically to zero at infinite distance, the wave function within one cell should join smoothly onto the wave function in the next cell.

Actual boundary conditions over the surface of the polyhedral cell are difficult to apply, and WIGNER and SEITZ substituted for this a set of boundary conditions over a sphere of the same volume as the cell. Since the wave function must be periodic from one cell to another, for $\boldsymbol{k}=0$, they assumed that the wave function within the cell should have $s$-like symmetry, being independent of angle, and that its slope should vanish on the surface of the sphere. They solved SCHRÖDINGER's equation for the PROKOFJEW potential, subject to this boundary condition, for a variety of internuclear distances, and found the energy of the bottom of the conduction band, as a function of internuclear distance. Instead of trying to apply the same method for other $\boldsymbol{k}$ values, they estimated the energy of the other states in the band by adding a free-electron kinetic energy term $\hbar^2 k^2/2m$ to the energy at the center of the band; in later work by the same method, $m$ is replaced by an effective mass.

It was then suggested by the present writer [*218*] that the method could be extended to handle other $\boldsymbol{k}$ values. For a general $\boldsymbol{k}$ value, the wave function would not be spherically symmetric within the WIGNER-SEITZ cell. Rather, an expansion was assumed, the wave function being written within the cell as a linear combination of solutions of SCHRÖDINGER's equation for a given energy parameter, but for different values of $l$ and $m$, the azimuthal and magnetic quantum numbers. For an arbitrary energy, solutions exist for each $l$ and $m$ value which are regular at the origin, though they would not have the correct boundary conditions at infinity for an isolated atom. The coefficients in the linear combination of wave functions were determined by applying joining conditions to approximate a function continuous from cell to cell. These conditions were applied in the following way. Let us consider the center of one

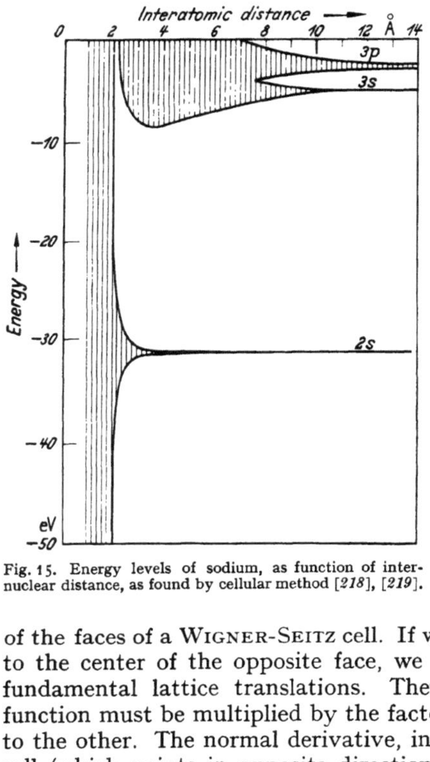


Fig. 15. Energy levels of sodium, as function of internuclear distance, as found by cellular method [218], [219].

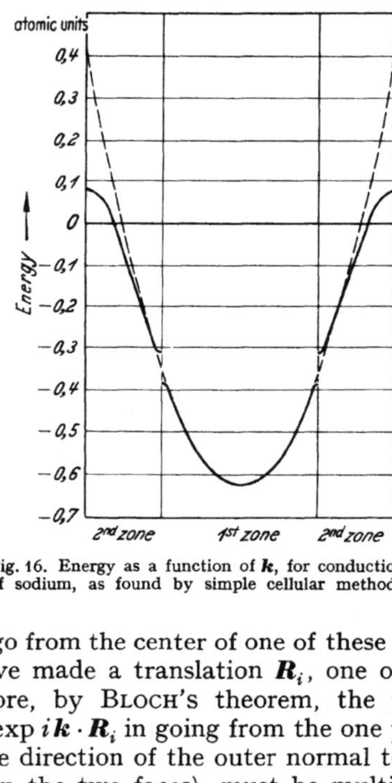


Fig. 16. Energy as a function of $\boldsymbol{k}$, for conduction band of sodium, as found by simple cellular method [218].

of the faces of a WIGNER-SEITZ cell. If we go from the center of one of these faces to the center of the opposite face, we have made a translation $\boldsymbol{R}_i$, one of the fundamental lattice translations. Therefore, by BLOCH's theorem, the wave function must be multiplied by the factor $\exp i\boldsymbol{k}\cdot\boldsymbol{R}_i$ in going from the one point to the other. The normal derivative, in the direction of the outer normal to the cell (which points in opposite directions on the two faces), must be multiplied by the factor $-\exp i\boldsymbol{k}\cdot\boldsymbol{R}_i$. The writer determined the linear combination by demanding this type of continuity of the function and its normal derivative, at the centers of all the faces of the WIGNER-SEITZ cell. This obviously did not ensure continuity all over the surface of the cell, but for a first approximation it seemed satisfactory.

In order to apply the method, one uses a suitably selected set of values of $l$ and $m$, or of spherical harmonic dependences of the wave functions, the number of independent functions being related to the number of faces of the WIGNER-SEITZ cell. When this is done, it is found that the conditions can only be imposed if the energy is suitably chosen as a function of $\boldsymbol{k}$. In other words, this method leads to a calculation of the energy bands. It can be easily applied to finding the width of the bands as a function of internuclear distance. In Fig. 15, we show the energy bands of sodium as determined in this way. We also show in Fig. 16 the curve of $E$ as a function of $\boldsymbol{k}$ determined for sodium, and note how closely it approaches the parabola characteristic of free electrons. Finally we show in Fig. 17 the wave function of sodium, for a given $\boldsymbol{k}$-vector in the conduction band. This curve shows that the wave function is extraordinarily similar to a sinusoidal wave, or plane wave, in the region between the atoms, though near each atom it takes on the character of an atomic wave function.

The cellular method in this simple form was applied, starting in the 1930's, to a considerable number of crystals, both metallic and non-metallic. Work was carried out by HUND and MROWKA [*110*], [*111*], [*112*], KIMBALL [*131*], KRUTTER [*143*], MILLMAN [*160*], SHOCKLEY [*204*], [*205*], CHODOROW and MANNING [*156*], [*44*], MANNING and KRUTTER [*157*], MANNING [*155*], GREENE and MANNING [*82*], EWING and SEITZ [*62*], TIBBS [*241*], [*242*], MULLANEY [*173*], VON DER LAGE and BETHE [*249*], and MORITA and HORIE [*165*]. The results undoubtedly were of qualitative value, but increasing doubt has been throw on the quantitative accuray of this method. The first serious doubt arose from a paper of SHOCKLEY [*206*], in which he carried out a so-called empty-lattice test. That is, he tried to solve the SCHRÖDINGER equation for a free electron by the cellular method. Here of course we know that the solution is a plane wave, and the cellular method gave a rather poor approximation to this wave function and its energy. Clearly the difficulty with the method lay not in principle, but in the fact that continuity was obtained only at the centers of the faces of the WIGNER-SEITZ cell, and consequently only a few terms were retained in the expansion of the solution in terms of spherical harmonics.

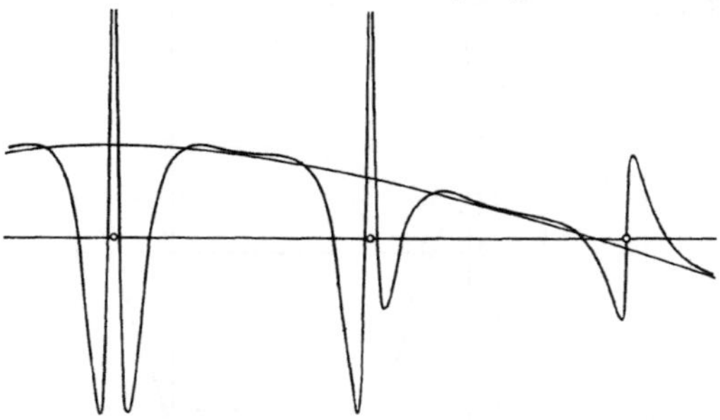

Fig. 17. Electronic wave function, real part, for points on a line in 111 direction in sodium, as found by simple-cellular method [*218*].

To avoid these difficulties, the method has recently been extended and improved by VON DER LAGE and BETHE [*249*], by HOWARTH and JONES [*103*], [*105*], by SCHIFF [*197*], and by BELL, HUM, PINCHERLE, SCIAMA, and WOODWARD [*15*]. These writers have used a linear combination of many more spherical harmonics, and have fitted the boundary conditions at many more points than in the earlier work. In doing this, they have considered special symmetry points in the BRILLOUIN zone, and those particular combinations of spherical harmonics which were appropriate to fit the symmetry behavior of the wave function at each of these points. Their work is believed to be among the most accurate of any of the various methods which we now have for solving the periodic potential problem. They have tested their method by use of the empty-lattice test, and it shows much better agreement than with the earlier applications of the cellular method. It is interesting to notice that work on copper, by this method, is in rather good agreement with the results of FLETCHER and WOHLFARTH on copper by the tight-binding method [*69*], [*70*], but disagrees badly with the earlier work of KRUTTER [*143*] on copper by the earlier cellular method. This throws serious doubt on all the earlier applications of the cellular method, but not on the method as refined in this recent work.

**11. The orthogonalized plane wave and augmented plane wave methods.** We have mentioned that the tight-binding approximation is entirely unsuited for finding the excited energy bands, and the cellular method also is not very good for them, for as we go to higher and higher bands, more and more spherical harmonics will be needed to fit the wave functions. A suggestion arises from the form of the wave function for the conduction band in sodium, shown in Fig. 17, as to how to set up wave functions for the excited bands. We notice that this function is very much like a plane wave in the region between atoms, but like an atomic function near each nucleus. One may guess that this situation will also hold for more highly excited states, and this gives a hint as to how to set up approximate wave functions. It also shows us why the free-electron approximation will not be good, even for excited states, to which it has sometimes been applied: to expand the part of the wave function near the various nuclei in plane waves, in such a wave function as shown in Fig. 17, requires just as slowly-convergent a series of plane waves as would be required to expand a tightly-bound atomic wave function. A superposition of a few plane waves can never show anything like the proper behavior, near the nuclei.

Several methods of expanding the wave function have been derived from this general view of the excited wave functions. One of these is the orthogonalized plane wave method, first suggested by HERRING [*95*], and shortly after that applied to beryllium by HERRING and HILL [*98*], and recently discussed by CALLAWAY [*40*]. HERRING noted that the inner part of a wave function such as that shown in Fig. 17 is similar to what could be built up out of the wave functions of the inner electrons of the atom, in this case out of sodium $1s$, $2s$, and $2p$ wave functions. Thus he suggested writing the wave function as the sum of a plane wave, and a linear combination of the BLOCH sums formed from these inner wave functions. He noted that the coefficients of this linear combination could be obtained in a very simple way. We know that the various solutions of SCHRÖDINGER'S equation must be orthogonal to each other. This suggested using such a linear combination that the resultant wave function would be orthogonal to the BLOCH sums formed from all inner wave functions, a suggestion which led to the name of orthogonalized plane wave method for this procedure.

We should suspect that a single orthogonalized plane wave would form a rather good approximation to a solution of SCHRÖDINGER'S equation. By making a linear combination of a number of different orthogonalized plane waves corresponding to propagation constants $\boldsymbol{k}+\boldsymbol{K}_i$, we should have a problem analogous to that of the free-electron approximation, already discussed in Sect. 9, but we should hope that the series would converge much more rapidly. We have just stated that the reason for slow convergence of the free-electron approximation is that it takes many plane waves to approximate properly to the wave function near the nuclei; but a single orthogonalized plane wave does this rather well. The method has recently been applied to the problems of diamond and germanium by HERMAN and CALLAWAY [*87*], [*88*], [*91*], and they find that in fact the series of orthogonalized plane waves converges fast enough so that one can handle the secular problem of interaction of these waves, leading to a set of equations similar to Eq. (5.7), though with different values of the matrix components of energy, by means of a digital computer. This method has not been applied to excited energy levels, above the lowest unoccupied band in semiconductors, but it begins to approach a method which could be used for these levels.

A somewhat different method, which seems to converge even faster, has been proposed by the writer and SAFFREN [*196*], [*229*]. This method, which has been called the augmented plane wave method, has some of the aspects both of the orthogonalized plane wave method and of the cellular method. It starts by assuming a special form of potential function, though this restriction is removed later: it is assumed that the potential is spherically symmetrical within certain spherical regions surrounding the various atoms, and is constant in the region between these spheres. Ordinarily in a metal the spheres would be chosen to touch. Then one takes a wave function which is a plane wave, $\exp i\,\boldsymbol{k}\cdot\boldsymbol{r}$ in the region between the spheres, and which is a solution of SCHRÖDINGER'S equation for the spherical potential, corresponding to an energy parameter $E$, inside each of the spheres, joining continuously at the boundaries of the spheres. This joining can be carried out as follows.

It is well known that a plane wave $\exp i\,\boldsymbol{k}\cdot\boldsymbol{r}$ can be expanded in a series of spherical harmonics of angle, and spherical BESSEL functions of $r$, in the form

$$\exp i\,\boldsymbol{k}\cdot\boldsymbol{r} = \sum_l (2l+1)\, i^l\, P_l(\cos\vartheta)\, j_l(k\,r). \tag{11.1}$$

Here the axis of the spherical coordinates is taken to be along the propagation vector. If we set $r$ equal to the radius of the sphere surrounding an atom at the origin, we have expanded the plane wave in spherical harmonics of the angle, over the surface of the sphere. We now find solutions of SCHRÖDINGER'S equation, for energy $E$, and each $l$ value, regular at the origin, inside the sphere. We make a linear combination of these solutions, the coefficients being determined so as to join continuously onto the plane wave, as expressed in the form of Eq. (11.1). Though the function is continuous, the first derivative in general will not be.

Next we compute the expectation value of the energy for the wave function consisting of the plane wave, joined in this way to spherical solutions within each atom of the crystal. There will be contributions from the interiors of the spheres, and from the region outside, each of which can easily be computed. In addition, there will be a contribution from the surface of each sphere; for the surface discontinuity of slope corresponds to an infinite value of the LAPLACian, or kinetic energy, on the surface, which integrates to a finite contribution over the surface of each sphere. Having computed the expectation value of the energy, as a function of the energy parameter $E$, we then vary $E$ to make the energy stationary, making use of the variation method to get the best possible solution of this type. When we do this, we find the interesting fact that the value of $E$ which makes the expectation value of energy stationary is in fact equal to this expectation value itself. In other words, our augmented plane wave, consisting of a plane wave outside the spheres, spherical solutions within the spheres, has an expectation value of energy such that the function is an exact solution of SCHRÖDINGER'S equation within each atomic sphere, corresponding to this energy value. It will not be an exact solution outside the spheres, for the resulting $E$ is not in general equal to $E_0$, the energy of a plane-wave solution of propagation constant $\boldsymbol{k}$. But the error in expectation value of energy arising from this fact is compensated by an equal and opposite contribution arising from the surface integrals.

It has been found practicable to carry through the calculation using this method, and the first step is to find the energy $E$ of the augmented plane wave, as a function of $\boldsymbol{k}$, or of $E_0$. The result obtained is similar to that shown in Fig. 18. For each $\boldsymbol{k}$-value, we have an infinite number of energy values $E$. The lower ones correspond almost exactly to atomic values, for the inner shells. These wave functions are large within the atomic spheres, almost precisely zero at the boundaries of the spheres, and join onto a plane wave of very small amplitude, whose only function is to impose the proper phase, so that the wave function acts just like a BLOCH sum. A single such augmented plane wave forms a very good approximation to a bound wave function. There are no matrix components of energy between two such wave functions of the same $\boldsymbol{k}$ value, and the wave functions are automatically orthogonal to each other. so that we see that for the bound states we have excellent approximations in this way.

For higher energies, corresponding to the outer occupied states in the crystal, and excited energy bands, we see that one branch of the curve of $E$ vs. $E_0$ tends to approach the straight line $E=E_0$, though the curve has of course an infinite number of branches. If $E=E_0$, the single augmented plane wave is an exact solution of SCHRÖDINGER's equation outside as well as inside the spheres; its only error comes on the surface of the spheres, where we still do not have to have a continuous slope, but the average error arising from this source, averaged over the surface of the sphere, vanishes. In other words, the best solutions to choose are those which lie close to this line $E=E_0$. It is obvious that these solutions have energies approximately equal to those of a plane wave. Hence for these higher energies, our functions form by themselves approximate solutions of SCHRÖDINGER's equation, equalling plane waves between the atoms, but joining onto spherical functions within the atoms.

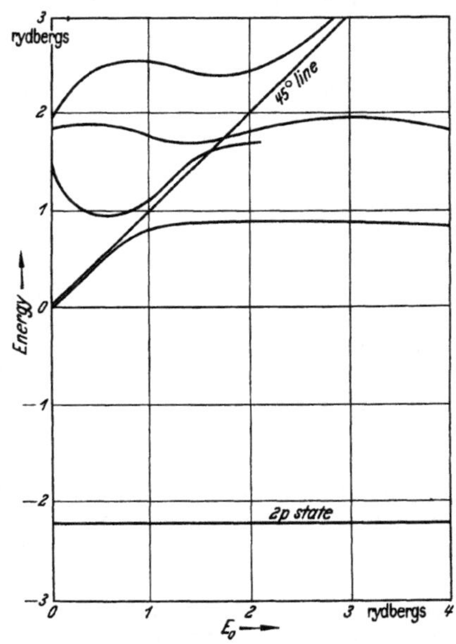


Fig. 18. Energy of augmented plane wave for sodium, as function of $E_0$, energy of plane-wave function, as computed by SAFFREN. The branches of the curve which approach the 45° line are those for which the energy is most nearly equal to that of the plane wave.

The procedure then is to take these augmented plane waves as the starting point of a secular problem like that met in the free-electron method, or in the orthogonalized plane wave method. That is, we set up a number of augmented plane waves, with propagation vectors $\boldsymbol{k}+\boldsymbol{K}_i$, where $\boldsymbol{k}$ is a fixed value, assume that the solution can be expressed as a linear combination of such waves, and determine the coefficients from a secular equation. This procedure has been carried through by HOWARTH [*104*] for copper. He finds that the convergence is extremely rapid, more so than in the orthogonalized plane wave method, suggesting that only a very few augmented plane waves must be combined to give an acceptable approximation to a solution of SCHRÖDINGER's equation. Furthermore, he finds that the results check very well with the ones which he had already obtained for copper, using the same potential, by the cellular method [*103*]. In other words, it now seems possible to solve the same problem by different methods, and obtain concordant results. Further calculations are underway using the augmented plane wave method, for a number of other metals, and the indications seem to be that this method will prove to be the most rapidly convergent and accurate method for solving the periodic potential problem. An earlier version of the method [*221*] has been applied to the case of beryllium by MATYAS [*158*].

We have mentioned the fact that the limitation on the potential function which we have described, namely that it be spherically symmetrical within atomic spheres, constant outside, can be removed. This can be done as follows. The augmented plane waves are set up using the special potential just described, but when we set up the matrix components of energy, in preparation for solving the secular problem of interaction of a number of augmented plane waves, we use the correct potential function, both in diagonal and non-diagonal matrix components. We may then assume that the resulting energy values and wave functions will correct rather accurately for the actual potential. This procedure has not yet been carried through in an actual case.

**12. Other approaches to the periodic potential problem.** Several other schemes have been suggested for solving the periodic potential problem, using the special form of potential just described in connection with the augmented plane wave method. Thus, KORRINGA [*138*] has described a method which has close analogies to the dynamical theory of x-ray diffraction. KORRINGA starts by considering the problem of scattering of a plane wave by a single spherical atom. The method of doing this is familiar, and it depends on expansion of the solution within the sphere in functions of $r$ times spherical harmonics, joining them onto spherical outgoing waves plus the incoming plane wave at the surface of the sphere. This joining can be done exactly, unlike the situation encountered in the augmented plane wave method, where we tried to join a plane wave outside the sphere to a solution of SCHRÖDINGER's equation inside the sphere, without an outgoing wave. Such a joining of course cannot be exact, and this was why we were left with a discontinuity of slope at the surface of the sphere.

Having discussed the scattering of a wave by a single spherical atom, KORRINGA then imposed the following condition: that the wave impinging on each atom be in fact the sum of the spherical waves emitted from all other atoms of the crystal. We then have a self-consistent problem: the spherical waves emitted by one atom must be just such as to be produced by the wave actually incident on that atom, but at the same time this incident wave must be the sum of the waves scattered by all other atoms, and furthermore each atom must scatter just the same sort of wave as the one which we are considering, only of course with its phase properly modulated according to the BLOCH formula. This condition is just like that used in the study of the scattering of electromagnetic radiation by EWALD [*59*], [*60*], [*61*] in his treatment of dispersion and of x-ray diffraction. EWALD worked out a great deal of mathematical technique for handling the sums of waves scattered by atoms on a crystal lattice, and KORRINGA showed how this technique could be applied to the periodic potential problem.

KORRINGA's method has been applied to the case of niobium by TREFFTZ [*243*]. A similar method has been proposed by KOHN, ROSTOKER, and SILVERMAN [*133*], [*135*], [*213*], and by DANK and CALLEN [*49*], and has been applied to lithium. MORSE, in unpublished work, is carrying through a similar method. There is nothing in principle which makes these methods unsatisfactory, though the mathematical difficulties are great. It is to be anticipated that before long, they will also be available for getting accurate solutions for the periodic potential problem, and they may well be as good as the augmented plane wave method, or better. It seems a reasonable prediction that, within a very few years, when the methods now being developed are completely in operation, using the facilities of modern digital computers, the periodic potential problem of the type met in a crystal will be capable of solution, to a high degree of accuracy.

**13. Results of energy-band calculations: body-centered cubic structure.** In the preceding sections, we have outlined many approaches to the problem of the solution of SCHRÖDINGER's equation in a periodic potential. Unfortunately, the most promising methods have only been worked out recently, and few results have yet been obtained on actual crystals. By putting together the results of these methods with earlier and less satisfactory ones, however, we can get a fairly reliable estimate of the nature of the energy bands in actual substances. We shall try to make such an estimate in this section and the following ones. In the first place, for the inner electrons of the atom—electrons with a binding energy of more than forty or fifty electron-volts—the energy bands are hardly split at all, so that a tight-binding approximation is satisfactory. We shall not discuss these levels further. The really interesting ones are the higher occupied levels, and lower excited levels, which we shall now take up. We consider in succession the body-centered cubic, face-centered cubic and sodium chloride, diamond and zinc-blende, hexagonal close-packed, and other structures.

The metals crystallizing in the body-centered cubic structure are Li, Na, K, Rb, Cs, Ba, V, Cr, Fe, Mo, Ta, W. By far the most extensive study has been devoted to sodium, where the leading papers are by WIGNER and SEITZ [*255*], the present author [*218*], VON DER LAGE and BETHE [*249*], and HOWARTH and JONES [*105*], all by the cellular method; there is also very recent unpublished work by SAFFREN using the augmented plane wave method. Preliminary results on this metal have already been mentioned in Sect. 10. Let us, however, go much more into detail regarding the nature of the energy as a function of $\boldsymbol{k}$, as determined by the more recent work. Before describing the results, we should discuss the form in which they are usually determined and presented.

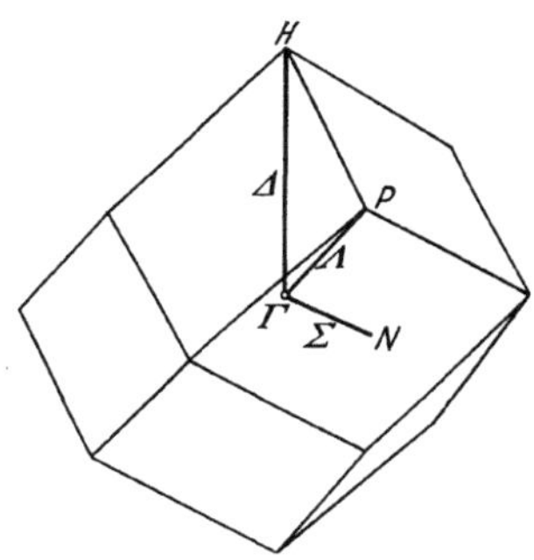


Fig. 19. Central BRILLOUIN zone of the body-centered cubic structure, showing the notation of BOUCKAERT, SMOLUCHOWSKI, and WIGNER [*29*] for symmetry points and lines.

Ideally, we should like to know the energy as a function of the three components of $\boldsymbol{k}$. This is ordinarily more than can be computed. Most of the calculations are limited to finding the energy at certain symmetry points, or along certain symmetry lines (or, in a few cases, in certain symmetry planes) in $\boldsymbol{k}$ space. The notation of BOUCKAERT, SMOLUCHOWSKI, and WIGNER [*29*] is generally used to describe these points or lines. In Fig. 19 we show the central BRILLOUIN zone for the body-centered cubic structure, with the notation of BOUCKAERT, SMOLUCHOWSKI, and WIGNER. That is, the center of the zone is denoted as $\Gamma$; the line along the (001) direction as $\Delta$; the line along the (110) direction as $\Sigma$; the line along the (111) direction as $\Lambda$; and the extremities of these three lines, at the surfaces of the BRILLOUIN zones, as H, N, and P respectively. Each type of symmetry at one of these points or along one of these lines is indicated by certain subscripts, of which we have given a few examples in Sect. 7. Then the commonest way to present the results of the calculation is to draw the energy as a function of the magnitude of $k$, for propagation vectors along these three symmetry directions (001), (110), and (111), labelling each state with the symbol indicating its symmetry type.

In Fig. 20, we give curves of this type for sodium, for the 100 direction, as determined by combining the results of VON DER LAGE and BETHE and of HOWARTH and JONES. There are some small uncertainties in the results concerning the relative order of the energies at some of the symmetry points but the various writers are in agreement regarding the general nature of the curves. The remarkable thing about this case of sodium is the close approach of the results to the predictions of a free-electron approximation. In Fig. 20 we show parabolic energy curves derived from the free-electron method with neglect of the non-diagonal matrix components of energy, $W(\boldsymbol{K}_j)$.

It is remarkable how closely these free-electron energy levels agree quantitatively with those determined by the more elaborate methods. As we have stated in Sect. 8, the reason for this is not to be sought in the correctness of the postulates of the free-electron method, but rather in the fact that the augmented plane wave method, which is very accurate, is superficially very similar to the free-electron method. In Fig. 18 we saw that the energy of an augmented plane wave, for sodium, is very approximately equal to the energy of the corresponding free-electron wave. Furthermore, since these augmented plane waves form rather accurate solutions of SCHRÖDINGER's equation by themselves, the non-diagonal matrix components between them are small, as if we were dealing with almost free electrons. The general form of the non-diagonal matrix components is similar in the two methods. Hence we have a superficial resemblance between the correct solutions, and the predictions of the free-electron theory, as far as the energy is concerned. The wave functions, however, are very different, as was indicated in Sect. 10, and in Fig. 17. As we saw there, the wave functions act much like free electrons between atoms, but like atomic wave functions within the atoms. The reason why the free-electron theory is so good is that these wave functions within the atoms, for sodium, can be joined very smoothly onto plane waves outside the atoms. This situation holds to a lesser extent for the other alkali metals, and in some other cases, but is far from a general situation throughout the periodic table.

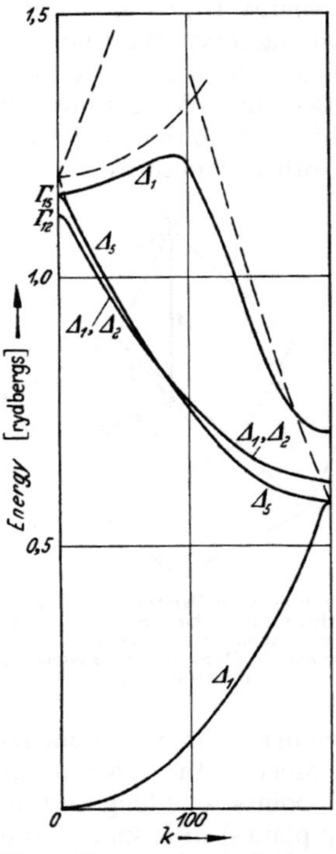


Fig. 20. Energy as function of $\boldsymbol{k}$, in 100 direction, for sodium, as determined from calculations of VON DER LAGE and BETHE [249], and HOWARTH and JONES [105], compared with the free-electron values. The values of VON DER LAGE and BETHE are used at the center of the BRILLOUIN zone, those of HOWARTH and JONES at the extremity. The energies have not been calculated except at the end points, so that the lines drawn are interpolated. Dashed lines represent the free-electron approximation; the lowest two free-electron lines coincide with the lowest $\Delta_1$ and $\Delta_5$ states.

We have discussed sodium, the case which has been most carefully worked out among the body-centered crystals. The other alkali metals behave in a rather similar way, though with differences in detail. Calculations have been made for Li by SEITZ [200], MILLMAN [160], BARDEEN [9], [10], SILVERMAN and KOHN [212], [213], PARMENTER [175], and SCHIFF [197]; for potassium by GORIN [81] and GOMBAS [80], the latter paper dealing with all the alkalies, but without detailed band calculations; for cesium by STERNHEIMER [235]. The other interesting cases of body-centered structure come in the transition elements, such as V, Cr, Fe, etc. Here we have no entirely satisfactory calculations, though a number of the elements have been discussed. The case which has received most attention is body-centered iron (the ferromagnetic modification), in which work by the older cellular method was carried out by MANNING [155], a very simplified tight-binding calculation has been made by the present writer and KOSTER [230], and a very recent calculation by the orthogonalized plane wave method has been made by CALLAWAY [41]; work involving the augmented plane wave method is under way. The interesting feature in this case is of course the behavior of the $d$ electrons.

In Fig. 21 we show the behavior of the $d$ bands along the (001) direction as found by the simplified tight-binding calculation of the writer and KOSTER. The results of CALLAWAY for this direction are qualitatively very similar, though quantitatively slightly different. Crossing this band is that derived from the atomic $4s$ electrons, which is presumably quite free-electron-like. This band will have the symmetry types $\Gamma_1$ and $\Delta_1$, and hence can interact with the $\Delta_1$ state arising from the $\Gamma_{12}$ $d$-like state. When this interaction is taken into account, the result will be something like that shown in Fig. 21b. In Fig. 21c we show the curves for the same situation found by MANNING by the older cellular method. We see that the results, though differing in detail from Fig. 21b, still are in general agreement. The way in which the $3d$ band cuts into the middle of the $4s$ band, which we see in these figures, is typical of all transition group elements. We find somewhat similar, though more complicated, results in the other directions of propagation. The calculations by the various methods do not agree so well in the other directions, and it is not clear at present whether any of them really represent very accurately the true solution of the periodic potential problem which has been assumed. The only feature which seems quite certain is the presence of the narrow $3d$ band. It is this which is responsible for the existence of ferromagnetism, and we shall point out later, in Sect. 39, the way in which ferromagnetism splits the bands into two components, one for electrons of each spin, thus greatly modifying the results of this calculation.

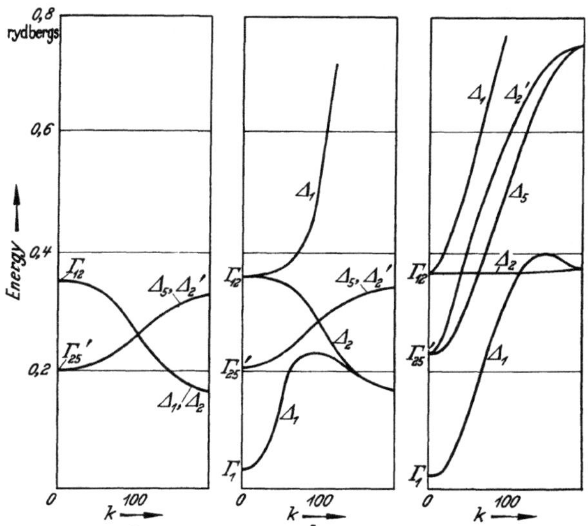


Fig. 21 a–c. Energy as function of $\boldsymbol{k}$, along the 100 direction, in body-centered iron. (a) Simplified LCAO treatment of $d$-bands, as given by SLATER and KOSTER [230]; (b) qualitative nature of modification of (a) caused by superposition of band arising from $4s$ electron; (c) results of MANNING [155] by simple cellular method.

The only other body-centered transition element for which extensive calculations have been made is tungsten, which MANNING and CHODOROW [156] discussed by the cellular method. The results were very complicated, and since we have already pointed out that this method is open to question in any case, we shall not reproduce the results obtained for this element. They indicate, as we should suppose, a rather narrow band arising essentially from $5d$ electrons, superposed on a broader band of $6s$-type.

**14. Results of energy-band calculations: face-centered cubic and sodium chloride structures.** The principal face-centered cubic metals are Al, Ca, Fe (the non-magnetic modification), Co (the non-magnetic modification), Ni, Cu, Sr, Rh, Pd, Ag, La (one modification), Ir, Pt, Au, Tl (one modification) and Pb. Among these, Ca, Fe, and Cu were discussed on the basis of the older form of the cellular method, Ca by MANNING and KRUTTER [157], Fe by GREENE and MANNING [82], and Cu by KRUTTER [143]; preliminary results on Cu by an earlier form of the augmented plane wave method were reported by CHODOROW [43]. TIBBS [241] carried out calculations on Cu and Ag. More recently, FLETCHER and WOHLFARTH [69], [70] have considered Ni on the basis of the tight-binding method, and HOWARTH [103], [104] has handled Cu by the cellular and the augmented plane wave methods. The interest in the transition group elements, here as with the body-centered structure, lies principally in the behavior of the $3d$ electrons. In Ca, we have another feature not met in the body-centered metals we discussed in Sect. 13: this element has two valence electrons, which would just fill up a $4s$-band, provided it were separated from other bands. It is then of interest, since Ca is a conductor, to see how the $4s$-band overlaps with the higher bands, so that we do not have an insulator.

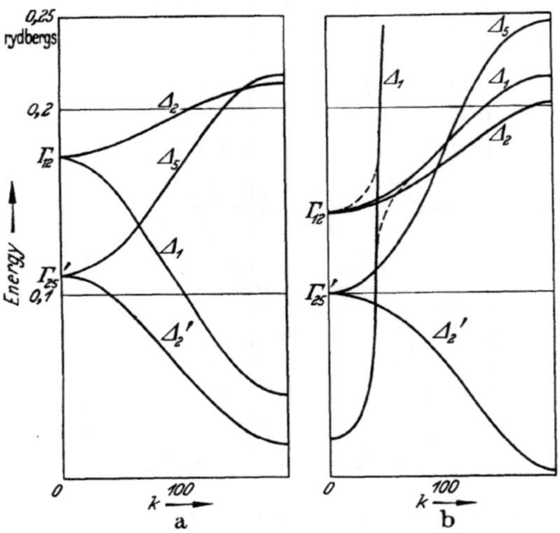


Fig. 22.

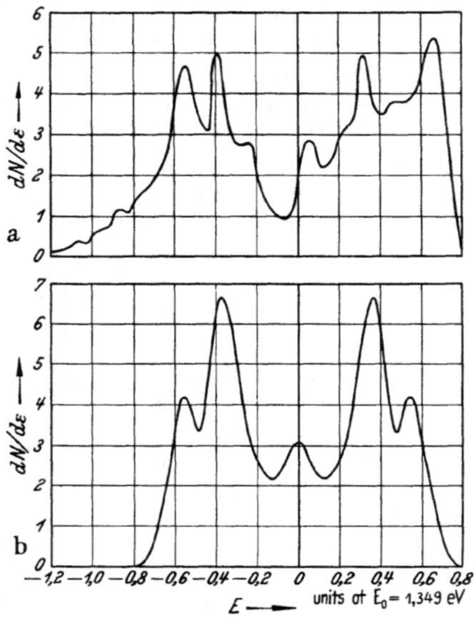


Fig. 23.

Fig. 22a and b. (a) Energy as a function of $\boldsymbol{k}$, along the 100 direction, in nickel. Calculation of FLETCHER and WOHLFARTH [69], by tight-binding method. (b) Calculation of HOWARTH [103] for copper using cellular method, and interpolation method of SLATER and KOSTER [230].

Fig. 23a and b. Density of states curve for $3d$-bands in a transition metal, with (a) face-centered cubic structure, (b) body-centered cubic structure, as calculated by KOSTER [140] by the tight-binding method.

In Fig. 22, we show the $3d$-bands of Ni, as determined by FLETCHER and WOHLFARTH by the tight-binding method, and those of Cu, as found by HOWARTH by the cellular method. The results of the two methods in general agree closely; the fact that one is for Ni, the other for Cu, would not be expected to make any qualitative difference. The small discrepancies between the results for the two metals probably come from the different methods of calculation, not from real differences. In both cases, we expect a $4s$-like band to overlap the $3d$-band, and this is indicated in HOWARTH's calculation. Here, as with the body-centered case as illustrated in Fig. 21, there will be interference between the two bands where they cross, but this will not seriously modify either band. The interference has not been calculated, and its nature is indicated qualitatively by the dotted lines in Fig. 22.

We see that the details of the $3d$-bands in the face-centered structure differ from those in the body-centered case, but the general results are very similar. In each case, we have a narrow $d$-band, with a high density of states. In Fig. 23

we show a curve of density of states of the $3d$-band for Ni, calculated by Koster [*140*] from the results of Fletcher and Wohlfarth. In this case, where we are working with the tight-binding method, it is not hard to carry out calculations at arbitrary points of the Brillouin zone, as well as for special symmetry points, and the calculations on which Fig. 23 is based were made at a large number of points in the Brillouin zone. For comparison, we also show in Fig. 23 the results of a similar calculation for the body-centered structure, such as is found in Fe. Neither of these curves is of great reliability, and we do not possess any really good calculation of the density of states in the $3d$-band; but it is interesting to note that, though the details of the energy bands are different, still the density of states curves for both types of structure have a good deal of similarity, leading to definite minima of density somewhere near the middle of the band.

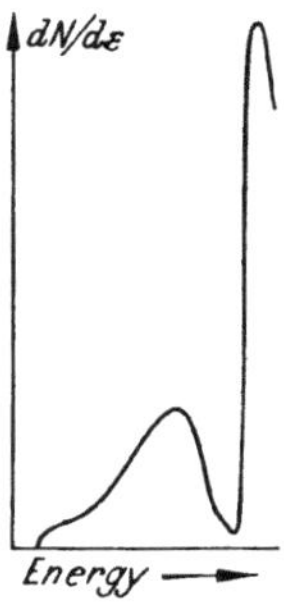


Fig. 24. Density of states curve for Ca, as calculated by Manning and Krutter [*157*].

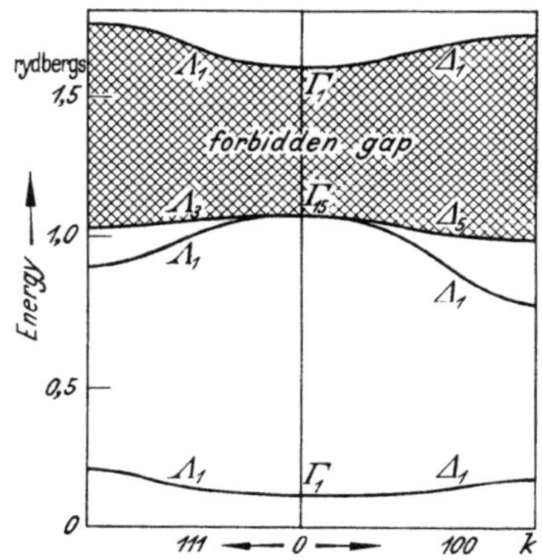


Fig. 25. Energy as a function of $\boldsymbol{k}$, along 100 and 111 directions, for LiF, as found by the cellular method by Ewing and Seitz [*62*].

We have mentioned that the case of Ca is interesting, because the two electrons per atom very nearly fill a band. The only calculation we have in this case is that of Manning and Krutter by the older cellular method, which is probably not very reliable. The density of states curve which they obtain is shown in Fig. 24. The fact that we have very nearly two independent bands, but with very slight overlapping, is obvious from this curve. The available electrons fill the bands just above the minimum in the density of states curve. However, the possibility remains that a better calculation might lead to rather different results.

The sodium chloride structure, in which most of the alkali halides, and a good many other substances, crystallize, is face-centered cubic, as we mentioned in Sect. 3, but with two atoms in the unit cell. As far as the classification of the energy bands is concerned, it is the same as for the face-centered cubic structure found in the metals. Energy-band calculations for LiF and LiH were made by the cellular method by Ewing and Seitz [*62*], and for NaCl by the same method by Shockley [*204*]; Tibbs [*242*] also discussed NaCl. These are all old calculations; there do not seem to be any very recent calculations for the alkali halides. On the oxides and sulfides of divalent metals, we have the calculation of Morita and Horie [*165*] on BaO, using the cellular method; and the elaborate calculation of Bell, Hum, Pincherle, Sciama, and Woodward [*15*] on PbS, using the modern and accurate version of the cellular method As interesting a case as any is at the same time the simplest, LiF, as worked out by Ewing and Seitz. In Fig. 25 we show their results for energy bands, in the 100 and

111 directions. The non-degenerate band, arising from $s$-like wave functions, and the higher three-fold degenerate one coming from $p$-like wave functions, are very clear. These bands are all filled in the crystal, the lowest empty band being of $s$-like nature. In Fig. 26 we give a plot showing the nature of the wave function for $k=0$, as we go along a line from a F to a Li nucleus. The functions shown by (a) and (b) are the occupied bands, that shown by (c) is the excited band. It is interesting to see how the $s$-like band has considerable contributions on both the F and Li atoms, while the $p$-like one is concentrated entirely on the F, though both give some little density in the region midway between the

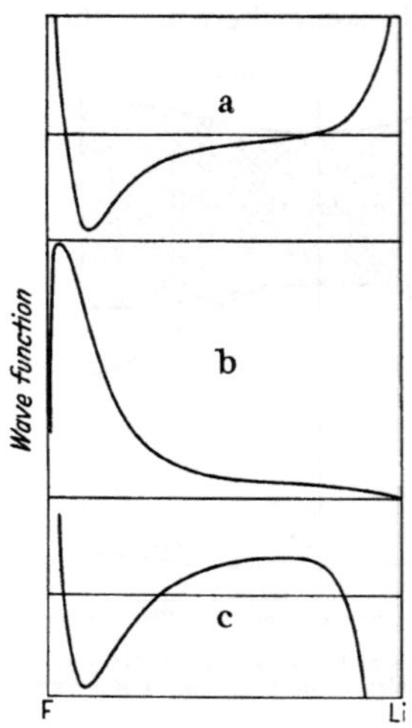


Fig. 26. Wave function for LiF, along line joining Li and F atoms. Functions $a$ and $b$ represent occupied bands, $c$ the lowest excited band. Calculated by EWING and SEITZ [62] by cellular method.

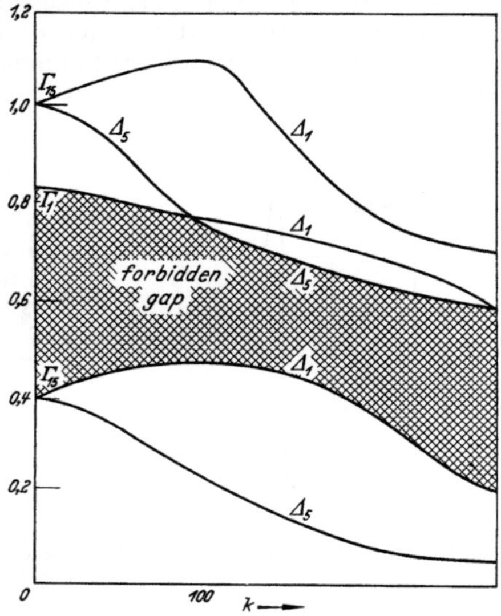


Fig. 27. Energy as function of $\mathbf{k}$, along 100 direction, for PbS, as determined by cellular method by BELL, HUM, PINCHERLE, SCIAMA, and WOODWARD [15].

two. In Fig. 27, we show the energy bands of PbS, as calculated by BELL *et al.* They are interesting as showing the great complication of energy bands which we are likely to find in a crystal composed of heavy atoms. One cannot be sure that the details of the curves are right, but all present indications are that the real energy bands can be at least as complicated as these.

**15. Results of energy-band calculations: diamond and zinc-blende structures.** No true metals crystallize in the diamond structure, but the semiconductors Si, Ge, and Sn (gray tin) have this structure, as well as diamond itself. Early calculations of diamond by the cellular method were made by KIMBALL [*131*] and by HUND and MROWKA [*111*], [*112*] independently, and calculations of silicon by the same method by MULLANEY [*173*]. More recent work on silicon has been done by HOLMES [*99*], YAMAKA and SUGITA [*263*], and BELL, HENSMAN, JENKINS, and PINCHERLE [*14*], using successively more elaborate forms of the cellular method. Recently HERMAN and CALLAWAY [*87*], [*88*], [*89*], [*91*] have made calculations using the orthogonalized plane wave method on diamond and germanium, and WOODRUFF [*262*] has used the same method for silicon. Work using the tight-binding approximation has been carried out for diamond by MORITA [*163*], by HALL [*83*], and by the writer and KOSTER [*230*]. From all of this work we can get a fairly complete picture of the nature of the energy bands in crystals of this type, though the finer details are still not known with certainty.

In Fig. 28 we show energy bands for diamond, for propagation along the (001) and the (111) directions, as calculated by an approximate tight-binding method by the writer and KOSTER, using this method as an interpolation procedure between values at the center and boundaries of the BRILLOUIN zone found by HERMAN by the orthogonalized plane wave method. The BRILLOUIN zone, as we remember, has the same form as for the face-centered cubic crystal, but the symmetry types have different significance. We shall describe the symmetries at the center of the BRILLOUIN zone, denoted by $\Gamma$, along the (001) direction, denoted by $\Delta$, and along the (111) direction, denoted by $\Lambda$.

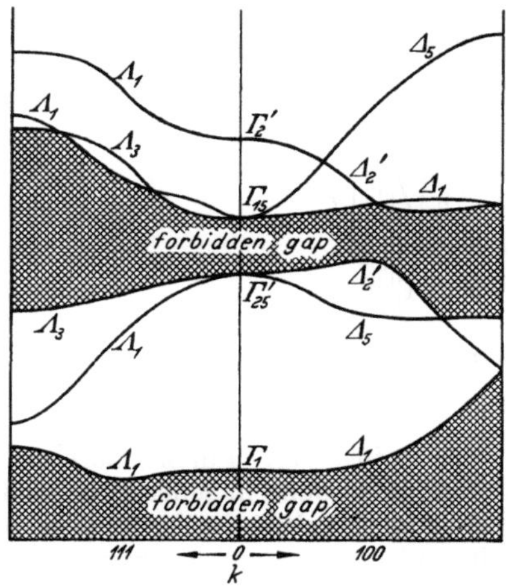


Fig. 28. Energy as a function of $\boldsymbol{k}$, along 100 and 111 directions, for diamond. Computed by tight-binding method by SLATER and KOSTER [230], interpolating between points computed by orthogonalized plane wave method by HERMAN [87].

The wave function of type $\Gamma_1$ can be considered to be made up of a BLOCH sum of atomic $s$ functions on all atoms of the crystal, all with the same coefficients. $\Gamma_2'$, like $\Gamma_1$, is made of atomic $s$ functions, but here the signs on the two atoms in the unit cell are opposite, so that the wave function is of an antibonding rather than a bonding type. The functions $\Gamma_{25}'$ and $\Gamma_{15}$ are triply degenerate, and can be considered to be made of $p_x$, $p_y$, or $p_z$ functions on the atoms. As with $\Gamma_1$ and $\Gamma_2'$, the difference between the two types of function comes in the signs of the atomic functions. In $\Gamma_{25}'$ the functions on the two atoms in the unit cell have opposite signs, so that their wave functions, being of the $p$-type, add in the region of overlapping between the two atoms, leading to a bonding type of interaction, while in $\Gamma_{15}$ the signs are the same, there are nodes of the wave function between the neighboring atoms, and we have antibonding.

For propagation along the (001) or $z$ direction, the function of symmetry $\Delta_1$ is made up on one set of atoms of a BLOCH sum of $s$ and $p_z$ functions (also, more generally, of $d$ functions behaving like $xy$ or like $3z^2-r^2$). On the other set of atoms, we have a similar BLOCH sum, the phases being determined by a BLOCH factor $\exp i\,\boldsymbol{k}\cdot\boldsymbol{R}$, where $\boldsymbol{R}$ is the radius vector to the atom in question (rather than merely to the corner of unit cell). The $\boldsymbol{k}$ in this expression is the reduced $\boldsymbol{k}$-vector. The function of symmetry $\Delta_2'$ is made up of the same functions on the atoms, but with an additional change of sign between the two atoms in the unit cell. We note the interesting degeneracy between the states of these two types at the edge of the unit cell. This arises because the ratio of the wave functions on the two atoms in unit cell is in one case $i$, in the other case $-i$, at this point, so that one wave function is the complex conjugate of the other, and they are necessarily degenerate. The function of symmetry $\Delta_5$ is two-fold degenerate, one of the functions being a linear combination of $p$ functions like $p_x+p_y$ on one set of atoms with a similar combination of $p$ functions on the other set, while the other function is an identical combination of functions $p_x-p_y$ on the two sets of atoms. For propagation along the (111) direction, $\Lambda_1$ is a combination of an $s$ wave function and $p$'s pointing along the 111 direction, while $\Lambda_3$ is a two-fold degenerate set of functions, made up from two types of $p$ functions at right angles to the (111) direction, and to each other.

The interesting fact to notice about the energy bands, as given in Fig. 28, is that they are separated into two groups, with a gap between them. The lower group of bands accomodates eight electrons per unit cell, or four per atom, which is just the number of available electrons, so that we have an insulator. Furthermore, from the nature of the wave functions, the filled bands correspond in general to bonding electrons, the empty ones to antibonding electrons, resulting in the binding energy of diamond. The physical properties of this substance, then, are correlated properly with the nature of the energy bands.

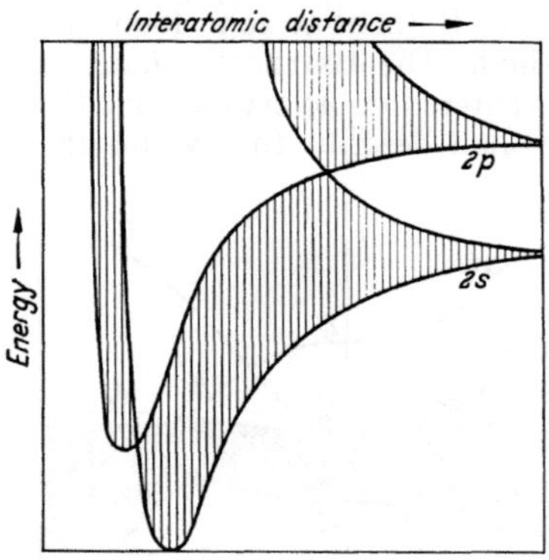


Fig. 29. Energy bands of diamond as a function of interatomic distance, as given by KIMBALL [*131*] using the simple cellular method.

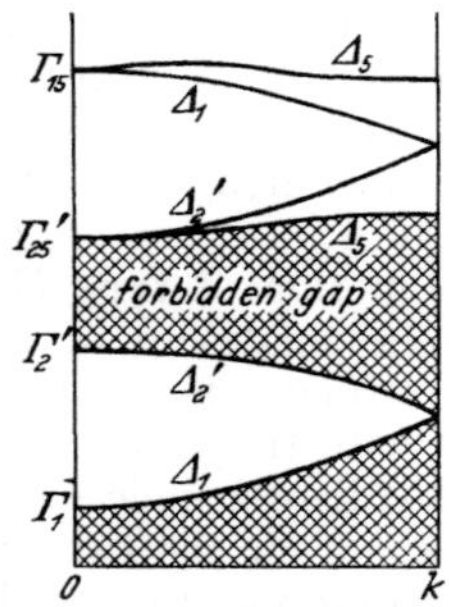


Fig. 30. Energy as a function of $\boldsymbol{k}$, along the 100 direction, for diamond, at large internuclear distance (schematic).

One very interesting feature of the energy bands of diamond is best brought out by the earlier cellular calculations of KIMBALL and of HUND and MROWKA, where they have calculated the bands as functions of interatomic distance. Such a curve is shown in Fig. 29. The detailed nature of the energy bands as given by these earlier calculations is somewhat different from the more accurate and more recent calculation given in Fig. 28, but is similar enough so that we can use the same qualitative description. The point shown in Fig. 29 is that for large internuclear distances, one band goes to the $2s$ state of carbon, the other to the $2p$ states. In this limit, the band going to the $2s$ state of course is capable of accommodating two electrons per atom, that going to can $2p$ accommodate 6 electrons, and with 4 electrons per atom the lower band would be filled, the upper one partly filled, and we should have metallic properties. At smaller internuclear distances, however, there is a crossover, and the lower band can now accomodate four electrons per atom, leaving a gap above, so that we have an insulator. This is the situation which we really have in diamond. Furthermore, the gap width would be expected to increase under pressure.

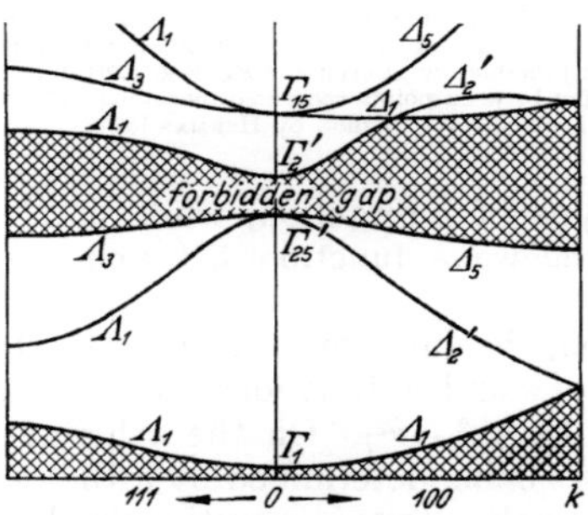


Fig. 31. Energy as a function of $\boldsymbol{k}$, along the 100 and 111 directions, for germanium, as found by HERMAN [*88*] by the orthogonalized plane wave method.

In terms of the energy bands as shown in Fig. 28, this change occurs in the following way. At large internuclear distances, the states $\Gamma_1$ and $\Gamma_2'$ will lie below $\Gamma'_{25}$ and $\Gamma_{15}$, since the atomic $2s$ state from which $\Gamma_1$ and $\Gamma'_2$ are formed will lie below the atomic $2p$ state from which $\Gamma'_{25}$ and $\Gamma_{15}$ are formed, and the bands are only slightly broadened. Then at large distances there will be a band of character $\Delta_1$ along the (001) direction, changing to $\Delta'_2$ at the edge of the band, and joining the state $\Gamma'_2$ at the center of the zone, as indicated schematically in Fig. 30. This band would hold two electrons per atom. This will lie entirely

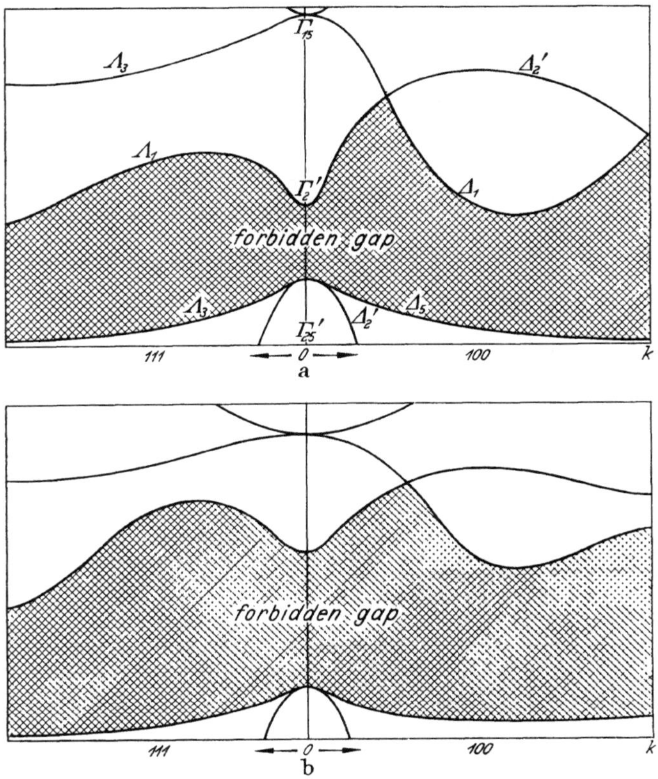


Fig. 32a and b. Energy as a function of $\boldsymbol{k}$, along the 100 and 111 directions for (a) Ge and (b) GaAs, as estimated by HERMAN [90], partly from theory, partly from experimental evidence on cyclotron resonance, properties of alloys, etc.

below the higher band, of $p$-character, shown in Fig. 30. As the bands broaden, the level $\Gamma'_2$ will be pushed above $\Gamma'_{25}$, and the levels will join as indicated in Fig. 28, so that we shall get the situation actually found in diamond.

The case of diamond, an insulator, is similar to the semiconductors silicon and germanium, and tin, which take on more and more metallic properties. In Fig. 31 we show the latest results of HERMAN on germanium, showing differences in detail from diamond, but still the same general sort of situation. In the semiconductors germanium and silicon, experimental information of many types regarding the energy bands is now becoming available, and it is likely that the structure of their energy bands will shortly be known better than that of any other elements.

As for crystals with the zinc-blende structure, the nature of the energy bands is in general like that of the diamond structure, but with minor differences at the points of degeneracy in the BRILLOUIN zone. These differences have been discussed by PARMENTER [177], and DRESSELHAUS (unpublished) has taken up the effect on this degeneracy of spin-orbit coupling. HERMAN [90] has made some speculations on the probable form of the energy bands in a number of crystals of this type, showing that they are not markedly different from those of germanium. Thus, in Fig. 32, we show HERMAN's estimate of the form of the energy bands of GaAs as compared with Ge. These curves indicate an estimate arrived at partly from the calculation of energy bands shown in Fig. 31, partly from experimental evidence on cyclotron resonance and other sorts, and they probably give very reliable indications of the actual form of the curves. So far, no direct calculations of energy bands for crystals of the zinc-blende structure have been reported, though it is understood that several such calculations are in progress.

**16. Results of energy-band calculations: hexagonal and other structures.** The metals crystallizing in the hexagonal structure are Be, Mg, Ti, Co, Zn, Zr, Ru, Cd, La, Hf, Os, and Tl. Calculations for Be were made by HERRING and HILL [98],

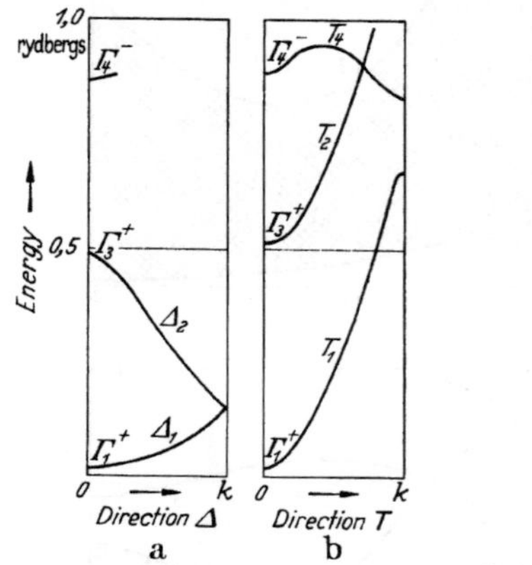


Fig. 33a and b. Energy as a function of $\boldsymbol{k}$, for Be, as found by HERRING and HILL [98] by the orthogonalized plane wave method. (a) Along the hexagonal axis; (b) along a direction in the basal plane parallel to an edge of the hexagon.

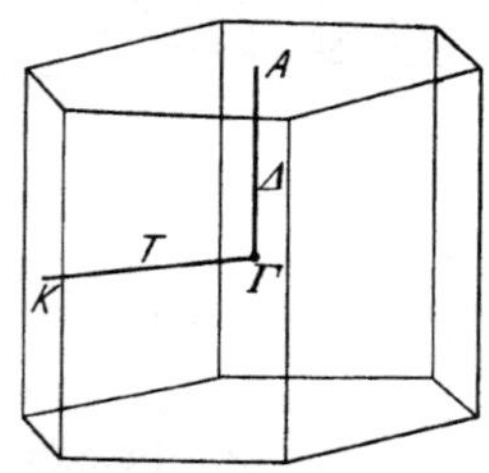


Fig. 34 Central BRILLOUIN zone for the hexagonal structure, showing notation of HERRING and HILL [98] for symmetry directions.

in the first application of the orthogonalized plane wave method. MATYAS [158] has used an older form of the augmented plane wave method for this substance, and TRLIFAJ [244] has carried out calculations for Mg using similar methods. JONES [124] has given discussions of alloys of Mg and of Zn, basing his arguments on the free-electron approximation. As yet no results have appeared on the transition-group elements having the hexagonal structure, and we can discuss only the behavior of the divalent metals, of which Be is the simplest example, and has been most carefully worked out. The interest in this substance is, of course, similar to that which we have already mentioned in Ca, which has the face-centered cubic structure: at large internuclear distance, the *s* electrons would form a band just capable of holding the valence electrons, and the substance would be an insulator. At the observed internuclear distance, this band must become mixed with the bands formed from the *p* electrons, so that we have a metal. The situation is thus in a sense the opposite of that just discussed for the diamond structure, where the substance would be a conductor at large distances, but is an insulator or semiconductor at the observed inter nuclear distance. We shall wish to investigate the way in which this mixing of the bands occurs.

The results of HERRING and HILL for Be are shown in Fig. 33. The symmetry situation here is of course entirely different from what we have met so far in the cubic crystals, and we must describe it before the results become clear. In Fig. 34 we show the central BRILLOUIN zone for the hexagonal structure, and indicate the notations used for the various symmetry points and directions by HERRING and HILL. At the center of the zone, the point $\Gamma$, we see examples of three types of symmetry. Of these, $\Gamma_1^+$ is like a BLOCH sum of $s$ functions on all atoms. $\Gamma_4^-$ is a sum of $s$ functions on the atoms of one of the two interpenetrating lattices from which the crystal is formed, with $s$ functions with a negative coefficient on the other lattice. That is, it corresponds to an antibonding combination of $s$ functions, while $\Gamma_1^+$ is bonding. The state $\Gamma_3^+$ is a bonding combination of $p_z$-like functions on the atoms of the crystal; this means that the $p_z$ functions on one atom of the unit cell have the opposite sign to the functions on the other atom. Other types of function, such as those formed from $p_x$ and $p_y$ atomic functions, have higher energy than those shown in Fig. 33.

As we go along the direction $\Delta$, parallel to the hexagonal axis, the function $\Delta_1$ is made up of a combination of $s$ and $p_z$ functions, the amplitudes and phases being determined by a BLOCH factor $\exp i\,\boldsymbol{k}\cdot\boldsymbol{R}$, where as in the diamond structure for $\Delta_1$ we are to determine the phase of each atom in the unit cell by using the appropriate vector $\boldsymbol{R}$ to that atom. The function $\Delta_2$, like $\Delta_2'$ in the diamond case, differs from $\Delta_1$ in having the negative sign on the function for the second atom in the unit cell, and as in diamond, we have a degeneracy at the boundary of unit cell, point $A$, where one function is the complex conjugate of the other. As a matter of fact, in the hexagonal structure, it can be shown that a degeneracy of this type occurs at every point of the top and bottom faces of the central BRILLOUIN zone. In the case of Be, as shown in Fig. 33, the character of the wave function in the lowest band changes gradually from a completely $s$-like nature, at the bottom of the band, where the symmetry is $\Gamma_1^+$, to a completely $p$-like nature at the top of the band. On account of the signs of the wave functions on the two atoms of the unit cell, we have bonding wave functions in both cases. On the other hand, at large internuclear distances, the level $\Gamma_4^-$, which lies above $\Gamma_3^+$ at the observed internuclear distance, will fall below it, and the lowest band will go from $\Gamma_1^+$ at the bottom to $\Gamma_4^-$ at the top, having $s$-like character throughout. This is the way in which the band disengages itself from the $p$-band at large distances, and in which the material becomes an insulator, since then there are just enough electrons to fill the lowest band.

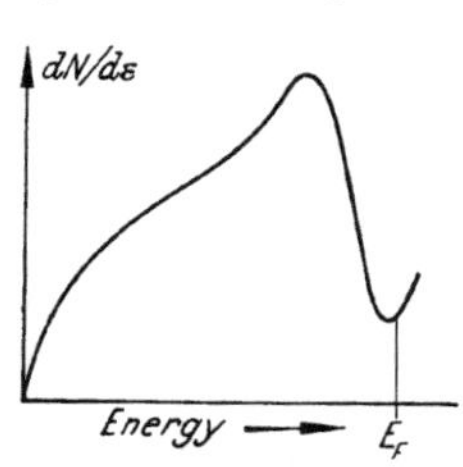


Fig. 35. Density of states for Be, as determined by HERRING and HILL [98] by the orthogonalized plane wave method.

In Fig. 33 we have also indicated the energy as a function of $k$ along the direction $T$, lying in the basal plane of the hexagon. Here the lowest state $T_1$ consists of a mixture of $s$ atomic functions, and $p_x$ functions (if the propagation is along $x$). The state $T_2$, arising from $\Gamma_3^+$, is made of $p_z$-like functions, mixed with $d$ functions of type $xz$, and the highest state $T_4$, arising from $\Gamma_4^-$, is made of $s$ and $p_x$ functions, as is $T_1$, but with a difference of sign from one atom in the unit cell to the other, so as to lead to antibonding rather than bonding. The behavior in several other directions is given by HERRING and HILL, with results qualitatively similar to what we have shown.

We have already pointed out the particularly interesting feature of this structure: the way in which the $s$-like and $p$-like bands, which would be separated at large internuclear distance, have become mixed with each other, so that this substance is a conductor rather than an insulator. This is shown also in Fig. 35, in which the density of states is given, derived from the energy bands which we have discussed. There is a sharp dip in the density of states, coming approximately at the FERMI level, but no actual gap.

The results which have been obtained for Mg and Zn, in the references cited above, though not as complete as those of HERRING and HILL for Be, give qualitatively similar results. We have stated that JONES and others have carried out discussions of these crystals by the free-electron approximation, and have applied these discussions to the properties of alloys of these elements, using methods which we shall discuss in Sect. 28. The general nature of the results obtained by the free-electron method shows surprising similarity to what is found by the more exact methods, and the reason is presumably what we have already mentioned in connection with sodium: the free-electron method shows superficial similarity to the orthogonalized or augmented plane wave methods, which are quite accurate.

One feature of the structure, found in the accurate calculations for Be and Mg, and reproduced by the free-electron calculation, is the fact that the FERMI level comes an extremely small distance below the level $\Gamma_4^-$. Thus a slight increase in the number of electrons per atom would result in the appearance of filled levels near the center of the BRILLOUIN zone, which are empty in the actual crystal. Now the general interpretation of alloy structure, which we shall discuss in Sect. 28, is that if atoms with a higher valence than Mg or Be are added to the crystal, such for instance as Al, the general structure of the energy bands is not altered, but we merely have such a change as would be produced by a small increase in the number of conduction electrons. In such a case, in Mg, a very small number of such atoms incorporated in the lattice will suffice to start filling the band originating at the level $\Gamma_4^-$. There is a good deal of experimental evidence that such a situation actually occurs, thus verifying the rather fine details of the energy-band structure which we have discussed.

We have indicated, in Sect. 2, that a number of elements, principally in the third and fifth column of the periodic table, have peculiar crystal structures, indicating the existence of covalent binding. In these cases, we should expect to find a situation reminding us of the diamond structure, or of the hexagonal structure which we have just discussed. In the diamond case, we have just enough electrons to fill certain bands, leaving a small gap above the filled bands and leading to an insulator, while in the hexagonal case the bands overlap slightly, but still to a large extent we are filling bands having wave functions of a bonding character, leaving antibonding types of wave functions unoccupied. The characteristic which we expect of the energy bands in a covalently bonded crystal is just what we find in diamond: certain energy bands will have bonding character and will be occupied in the crystal, while others will have antibonding character and will be unoccupied. As a result of the low symmetry of the structures, we may expect the bands to overlap slightly, and hence lead to metals, but nevertheless with properties very similar to those of an insulator bound by covalent bonds.

The structure of this type which has received most attention is that of Bi, where JONES [*121*] has carried out calculations by the free-electron approximation. Here, as before, we may be doubtful about the validity of this approximation, but its similarity to the augmented plane wave method may give more validity to the calculations than we should expect otherwise. Studies of Bi have also been made by MORITA [*164*]. JONES finds for Bi that we have a situation just of the type which we have sketched: there is a band, almost filled with electrons, which overlaps very slightly a higher almost empty band. The peculiar physical properties of Bi presumably come from this special type of band structure. Here, as with Mg, alloying with very small amounts of impurities of different valence from Bi can make very striking changes in the character of the substance, which can be correlated with the assumed structure of the energy bands.

Certain alloys, which we shall mention in Sect. 28, also have structures of complicated nature, and their energy bands have been discussed [*120*], [*122*], [*183*] by the free-electron approximation. Here again, as with Mg and Bi, very interesting conclusions can be drawn regarding the change of properties with changing concentration. In all of these cases, we must repeat that the sucess of the free-electron method does not rest on any sound theoretical foundation. More accurate calculations, for instance by the augmented plane wave method, would be highly desirable to put the study of these alloys on a firmer foundation. It does not seem out of the question that such studies might be made at some future time.

The one additional element whose band structure has been rather extensively studied is carbon in the form of graphite. Studies have been made by LENNARD-JONES [*147*], COULSON [*46*], WALLACE [*250*], and BRENNAN [*30*], that of WALLACE being the most complete. This is a hexagonal structure, and hence has similarities to the hexagonal close-packed structure which we have been discussing. However, the important physical fact regarding graphite is that it is a layer structure, composed of layers consisting of hexagonal rings of carbons, essentially similar to benzene rings, successive layers being quite far apart. Thus one can get a good approximation to the graphite energy bands by considering only the two-dimensional problem of a single sheet or layer. Here one finds a situation somewhat similar to diamond, in that there are certain lower, occupied bands, with upper vacant ones. The situation is special, however, in that there is no gap between these bands. The density of states goes down linearly as we go into the conduction band. Hence the substance is a conductor, of the metallic type, and yet of a peculiar sort, in that the density of states is zero just at the FERMI level.

## III. The self-consistent field.

**17. The HARTREE-FOCK equations.** In Sect. 1, we have given a very brief account of the arguments underlying the use of the self-consistent field method for solids. Now we wish to examine the basis of this method more carefully. The reason for postponing this discussion until now is that one must have a fairly accurate idea of the type of wave functions concerned in a self-consistent field calculation in order to calculate the field which in turn is needed for finding the wave functions themselves.

The HARTREE-FOCK equations provide the most straight-forward way of setting up a self-consistent field. We have indicated in Sect. 1 the procedure for finding them. We set up one-electron wave functions $u_i(x, \sigma)$, where $x$ symbolizes the space coordinates, $\sigma$ the spin, the latter appearing in such a form that $u_i$ can be written as a product of a function of coordinates, times either the PAULI function $\alpha$ or $\beta$ of spin, $\alpha$ implying that the component of spin along a fixed axis is $\hbar/2$, $\beta$ implying that it is $-\hbar/2$. We assume as many one-electron functions as there are electrons in the problem. We assume that these functions are orthogonal and normalized. We then form the determinantal function

$$\psi = (N!)^{-\frac{1}{2}} \sum_P (-1)^P P u_1(x_1 \sigma_1) \ldots u_N(x_N \sigma_N). \tag{17.1}$$

Here the sum is over all the permutations $P$ of the indices $1 \ldots N$ of the one-electron functions (or alternatively of the coordinates and spins of the electrons), and $(-1)^P$ symbolizes 1 if the order of the permutation is even, $-1$ if it is odd. It is well known that this determinantal function possesses the necessary antisymmetry to satisfy the PAULI exclusion principle.

Let $H$ be the HAMILTONian operator of the $N$-electron problem. We may write this in the form

$$H = H_0 + \sum_i f_i + \sum_{\text{pairs } ij,\ i \neq j} g_{ij}. \tag{17.2}$$

Here $H_0$ represents that part of the HAMILTONian independent of the electron coordinates; that is, the potential energies of interactions of the nuclei. The quantity $f_i$ equals that part of the HAMILTONian depending on the coordinates of the $i$-th electron alone. That is, it is the LAPLACian or kinetic energy of the $i$-th electron, its potential energy in the field of all nuclei and any external field, and the spin-orbit interaction of the $i$-th electron (which we shall disregard in most of our treatment, as representing ordinarily a small correction). The quantity $g_{ij}$ is the term in the HAMILTONian depending on the $i$-th and $j$-th electrons; that is, it is the COULOMB repulsion of these two electrons, plus interactions between the spin or orbit of one electron and the spin or orbit of the other (which we shall likewise disregard).

Now we find the expectation value of the HAMILTONian $H$ with respect to the wave function $\psi$, and vary the $u_i$'s to make the expectation value stationary. According to the RITZ variation principle, this procedure will give us that determinantal function $\psi$ which best approximates a solution of SCHRÖDINGER'S equation. In particular, we seek the ground state, and this is found by making the expectation value of the energy a minimum. The equations resulting for the $u_i$'s are the HARTREE-FOCK equations, which can be written in the form

$$\left.\begin{aligned} &f_1 u_i(x_1) + \left[\int \sum_j u_j^*(x_2)\, u_j(x_2)\, g_{12}\, dx_2\right] u_i(x_1) - \\ &\quad - \sum_{\substack{j \\ \text{spin } j = \text{spin } i}} \left[\int u_j^*(x_2)\, u_i(x_2)\, g_{12}\, dx_2\right] u_j(x_1) = \varepsilon_i u_i(x_1). \end{aligned}\right\} \tag{17.3}$$

These equations have solutions $u_i$ of two different sorts. First, there are the $N$ one-electron wave functions which are occupied in the ground state of the system being considered. The summations over $j$ in Eq. (17.3) are taken over these occupied orbitals. In the equations, the $u$'s represent merely the space parts of the one-electron wave functions, and the second summation, described by $\sum_{\substack{j \\ \text{spin } j = \text{spin } i}}$, is extended only over those wave functions whose spins are the same type as the spin of the $i$-th function. In addition to these $N$ occupied orbitals, however, the equations have an infinite number of other solutions, corresponding to excited orbitals not occupied in the ground state of the system.

One can easily prove, directly from Eq. (17.3), that the one-electron wave functions $u_i(x, \sigma)$ are orthogonal to each other, where now we consider the infinite set of solutions, corresponding both to the occupied and unoccupied orbitals. It is to be noted in the proof that the orthogonality of two one-electron wave functions of the same spin arises from the orthogonality of their space-like parts, but that of two electrons of opposite spin arises from their spin eigenfunctions, and their space-like parts do not have to be orthogonal to each other.

The whole set of $u_i$'s forms a complete set of one-electron functions, and determinantal wave functions made by assigning $N$ electrons to $N$ arbitrarily chosen $u_i$'s form a complete set of $N$-electron antisymmetric wave functions. We must remember that an exact solution of SCHRÖDINGER'S equation cannot be written in the form of a single determinantal function. To get a better approximation, we must make a linear combination of determinantal functions of the type just considered. This is the process called *configuration interaction*, and will be taken up in Sect. 30 and in Chapter VII. For many purposes, however, a single determinant is fairly satisfactory, and it is only that approximation which we shall deal with in the present chapter.

The quantity $\varepsilon_i$, appearing in Eq. (17.3), is an eigenvalue called the one-electron energy. We recall KOOPMANS' theorem, discussed in Sect. 1: the one-electron energy $\varepsilon_i$ of an occupied orbital represents a good approximation to the negative of the energy required to remove the $i$-th electron from the system. The correspondence between these one-electron energies and the ionization potentials of atoms was noted empirically by HARTREE in the first calculations of self-consistent fields for atoms, and in fact it formed the basis of the physical thinking which led to the method of the self-consistent field in the very early days of wave mechanics.

KOOPMANS' proof involved the following steps. First he computed the expectation value of the energy of the whole system, using the HAMILTONian of Eq. (17.2) and the determinantal wave function of Eq. (17.1), made up out of the $N$ one-electron functions satisfying the HARTREE-FOCK equations (17.3), and occupied in the ground state of the system. Then he set up an approximate wave function for the ion, by making a determinant out of $N-1$ one-electron wave functions, using merely the original set of functions, but omitting the appropriate one. He found the expectation value of the energy for this determinantal wave function for the ion, and subtracted from it the expectation value of the energy of the whole system. The difference proves to be $-\varepsilon_i$, as we stated before. It is clear that this procedure should give a slightly too great value for the ionization energy, for the wave function of the ion is not determined as accurately as that of the original system, since we have not solved a separate variation problem for the ion, as we did for the original system. Thus the energy determined for the ion is likely to be further above the true value than that determined for the original system, or the computed ionization energy should be too great. This is generally found in atomic calculations, but the one-electron energies as determined by KOOPMANS' theorem prove to be surprisingly accurate.

In problems having symmetry relations such as we find with crystals, we are likely to find degeneracy among the one-electron energies $\varepsilon_i$. Orbitals $u_i$ associated with the same one-electron energy may be considered together, by analogy with the atomic case, in which we speak, for instance, of all $p$ electrons of the same principal quantum number as belonging to the same shell, even though their magnetic quantum numbers are different. Specially simple cases arise when all orbitals of a degenerate set are occupied in the ground state of the crystal. In such a case we speak of a closed shell of electrons. If all of the $N$ electrons are in closed shells, analogous to the case of an inert gas atom in atomic theory, one can then prove an important theorem: in such a case, the $u_i$'s all transform according to the irreducible representations of the group of operations leaving the nuclear framework of the crystal invariant. We are then led in turn to degeneracy arising from the irreducible representations, as we have met them in Chapter II.

From the discussion of that chapter we can understand just what a closed shell consists of. In a cubic crystal, for instance, it can consist of the 48 wave functions associated with a star of $\boldsymbol{k}$-vectors of the same length, but pointing in equivalent directions; or of the smaller number of wave functions associated with a star of $\boldsymbol{k}$-vectors along certain special symmetry directions. If all of these 48 (or other number) of wave functions are occupied, we have a closed shell, for purposes of the present theorem. We see, then, that these results about the symmetry of the wave functions and their degeneracy by no means imply that we are dealing with an insulator, in which some complete energy bands are filled, others empty. They are far less restrictive, and in fact in practically all cases of crystals, we can assume that we are dealing with closed shells in the present sense, so that the $u_i$'s have the symmetry properties of the irreducible representations of the crystal symmetry group. Thus we justify, from the HARTREE-FOCK equation, the type of symmetry analysis which formed the main basis of Chapter II.

**18. Physical meaning of the HARTREE-FOCK equations.** In Sect. 17 we have discussed the HARTREE-FOCK equations mathematically, but we have not shown why they can be regarded as leading to a self-consistent field. Let us now examine this question, and try to understand the physical content of the equations.

In Eq. (17.3), if only the first summation over $j$ were present, and the second were missing, the equations would have a very simple physical meaning. The first term $f_1 u_i(x_1)$ represents the one-electron kinetic energy operator, and potential energy in the field of the nuclei, acting on $u_i(x_1)$. In the second term, the quantity $\int \sum_j u_j^*(x_2)\, u_j(x_2)\, g_{12}\, dx_2$ is the potential energy, at point $x_1$, of the charge distribution of $N$ electrons distributed in all occupied orbitals. Hence the term involving this summation would be this potential energy, regarded as an operator, acting on $u_i(x_1)$. If the second summation over $j$ were missing, then, we should have the one-electron SCHRÖDINGER equation of an electron moving in the field of the nuclei, and of all electrons, including itself, averaged over the occupied wave functions. This is precisely the assumption of the naive self-consistent field picture, except for one thing: we expect that the electron will not act electrostatically on itself, and from this we infer that the second summation must somehow correct the first summation, in such a way as to remove the interaction of the electron with itself.

In the original HARTREE self-consistent field scheme, the equations were essentially those of Eq. (17.3), except that in place of the second summation, one had merely the $i$-th term taken from that summation. That is, in place of it, one had $[-\int u_i^*(x_2)\, u_i(x_2)\, g_{12}\, dx_2]\, u_i(x_1)$. This represents just the electrostatic interaction of an electron in the wave function $u_i$, at the point $x_1$, operating on the wave function $u_i(x_1)$, so that this term had just the effect of removing the term corresponding to $j = i$ in the first summation, leaving us with the interaction of all other electrons on the one in question.

This simple HARTREE assumption is fairly satisfactory for an individual atom, and the differences between it and the HARTREE-FOCK scheme are small and unimportant. We can easily see, however, that for a crystal the HARTREE scheme is completely untenable. To see this, let us set up a crystal consisting of rather widely spaced atoms, but with a regular crystalline arrangement. Our intuition tells us that the self-consistent field problem for each of these atoms should be essentially equivalent to that of an atom at infinite distance from its neighbors. In the HARTREE scheme for an isolated atom, the $u_i$'s would be atomic wave functions concentrated around that atom, so that the term correcting for the interaction of an electron with itself would be that arising from an electron concentrated on the atom. The one-electron problem, then, is that of an electron moving in the field of a singly ionized positive ion.

In a crystal, however, the $u_i$'s are BLOCH-type wave functions, extending through the crystal. Only an exceedingly minute fraction of the charge density is located on each atom. The HARTREE correction term for the interaction of an electron with itself, using this wave function, is then the interaction with a charge whose magnitude is one electronic charge, distributed uniformly throughout the whole crystal. This correction is entirely negligible. In other words, the HARTREE equations for this problem lead in this limit of widely separated atoms to the equations for an electron moving in the field of an uncharged atom, rather than of a singly ionized positive ion, which is quite wrong.

The HARTREE-FOCK equations correct this shortcoming of the HARTREE method, and lead to equations of the same type, whether we are dealing with isolated atoms, or with atoms rather widely separated in a crystal. To understand this statement, we must first investigate more closely the meaning of these equations, following lines used by WIGNER and SEITZ [255]. To do this, it is interesting, as the writer [225] has pointed out, to rewrite the second summation of Eq. (17.3) in the form

$$-\sum_{\substack{j\\ \text{spin}\, j = \text{spin}\, i}} \left[\int u_j^*(x_2)\, u_i(x_2)\, g_{12}\, dx_2\right] u_j(x_1) = -\left[\int \varrho(x_2, x_1)\, g_{12}\, dx_2\right] u_i(x_1) \tag{18.1}$$

where

$$\varrho(x_2, x_1) = \frac{\sum\limits_{\substack{j\\ \text{spin}\, j = \text{spin}\, i}} u_j^*(x_2)\, u_i(x_2)\, u_i^*(x_1)\, u_j(x_1)}{u_i^*(x_1)\, u_i(x_1)}. \tag{18.2}$$

In Eq. (18.1) we exhibit the term as the potential energy,. at position $x_1$, of a charge distribution $\varrho(x_2, x_1)$, located at position $x_2$, and depending parametrically on the position $x_1$, this charge distribution being defined in Eq. (18.2). We may call this charge distribution the exchange charge, since it arises from exchange terms, and we may interpret the HARTREE-FOCK equations as being the ordinary SCHRÖDINGER equations of an electron moving in the field of the nuclei, and of all electrons, diminished by the exchange charge of Eq. (18.2). To investigate the exchange correction, we must then study this exchange charge.

We can prove two simple theorems about the exchange charge. In the first place, we can find its total amount, by integrating over $dx_2$. On account of the orthonormal properties of the $u$'s, we find that $\int u_j^*(x_2)\, u_i(x_2)\, dx_2 = 0$ unless $j = i$. In case $u_i$ represents one of the orbitals occupied in the ground state of the system, the term $j = i$ will be included in the summation over $j$, it alone will give something different from zero, and inspection of Eq. (18.2) shows that it will lead to $\int \varrho(x_2, x_1)\, dx_2 = 1$, independent of $x_1$. In other words, in this case, the total exchange charge is one electronic unit of charge, so that the total charge, diminished by the exchange charge, will be a charge of $N-1$ electrons, as it should be. In other words, according to the HARTREE-FOCK equations, an occupied orbital is to be found from a SCHRÖDINGER equation with a potential arising from the nuclei, and $N-1$ other electrons. On the other hand, if $u_i$ represents one of the excited orbitals, the term $j = i$ will not be present in the summation of Eq. (18.2), and the total exchange charge will be zero, so that the electron will move in a potential arising from the nuclei and $N$ electrons.

The second theorem which we can easily prove relates to the value of the exchange charge density when $x_2 = x_1$. In this case, we see from Eq. (18.2) that

$$\varrho(x_1, x_1) = \sum_{\substack{j\\ \text{spin}\, j = \text{spin}\, i}} u_j^*(x_1)\, u_j(x_1). \tag{18.3}$$

In other words, the exchange charge density when $x_2 = x_1$ equals the whole charge density of occupied states of the same spin as $u_i$.

From these two theorems, we can obtain a good picture of the nature of the charge density, equal to the charge density of all electrons diminished by the exchange charge density, which exerts an electrostatic potential acting on the one-electron wave function of an occupied state, according to the HARTREE-FOCK equations. This is made up of two parts, one coming from the electrons of the same spin as the electron in question, the other part from the electrons of the opposite spin. As far as electrons of the opposite spin are concerned, there is no exchange charge density, and the total charge of electrons of opposite spin exerts an electrostatic effect on the electron in question. As for electrons of the same spin, however, the electrostatic field is exerted by a charge whose total amount is one unit less than the total number of electrons of this spin, if we are dealing with an occupied orbital. The exchange charge is removed from a location depending parametrically on $x_1$.

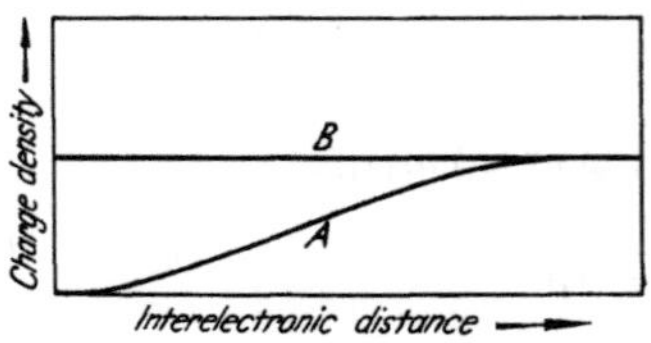


Fig. 36.

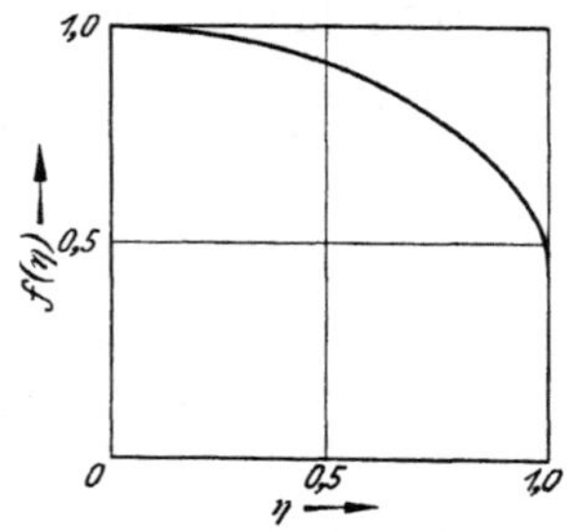


Fig. 37.

Fig. 36A and B. Density of charge near an electron, plotted against interelectronic distance, for (A) another electron of the same spin; (B) another electron of opposite spin.

Fig. 37. $F(\eta) = \frac{1}{2} + \frac{1-\eta^2}{4\eta} \log \frac{1+\eta}{1-\eta}$ as function of $\eta$. The quantity $\eta$ is the ratio of the momentum of the electron to the momentum at the top of the FERMI distribution. $F(\eta)$ is proportional to the exchange correction for a free electron, as calculated by DIRAC [53] and BLOCH [22].

That is, it is as if the electron whose wave function we are finding cleared out charge density from a hole surrounding it, often called the exchange charge hole. The total amount of this removed charge is one unit, and at the center of the hole, or when the coordinate $x_2$ equals $x_1$, all charge density is removed.

The qualitative nature of this situation is shown in Fig. 36, where we show the effective densities of charge of the same and of opposite spin to the electron in question, for a special and simple case, that of perfectly free electrons, moving in a field-free space (which would be produced if we had a continuous distribution of positive charge density, just enough to balance the electronic charge density). One can work out this case exactly, replacing the summations of Eq. (18.2) by integrations, in a way discussed by BLOCH [22] and by WIGNER and SEITZ [255]. We see from Fig. 36 that in this case, the exchange hole is not greatly different from what we should have if we merely removed all negative charge from a hole just large enough to contain one electron, with a density of negative charge equal to that found for electrons of the same spin as the electron concerned. The radius of the exchange hole would then be proportional to the $-\frac{1}{3}$ power of the electron density.

If we are dealing, not with free electrons, but with electrons in widely spaced atoms in a crystalline array, then one can use BLOCH sums for the functions $u_i$ in Eq. (18.2). In that case, one can show easily that the exchange hole for the crystal is identical with that for an isolated atom, provided the atomic wave functions do not overlap. In other words, when the electron of coordinates $x_1$, whose wave function we are finding, is found in a particular atom, it finds itself in the field of that atom lacking one electron, or forming a positive ion, while all the other atoms of the crystal are neutral. Thus we have overcome the difficulty met with the HARTREE method in this problem We can see how this difficulty is overcome, by considering Eq (18.3). When the $u$'s are BLOCH sums, the exchange charge density $\varrho(x_1, x_1)$ consists of a very large number of contributions, each very small. The contribution arising from the term $j=i$, which is the only one considered by HARTREE, is completely negligible. The sum, however, is the same quantity which we should find for a single atom, and is a correction of large amount in finding the final potential.

One characteristic of the HARTREE-FOCK method is that we have a different potential to use for each one-electron wave function $u_i$. For instance, in the free-electron case which we have illustrated in Fig. 36, the exchange hole, and the potential arising from it, are different, depending on the momentum of the orbital $u_i$, which is assumed in this limiting case to be a plane wave. In Fig. 37 we give the correction to the potential energy arising from the exchange correction, in this case, as calculated by DIRAC [*53*] and by BLOCH [*22*]; on account of the free-electron character of this example, the potential energy is independent of position, and depends only on the $\boldsymbol{k}$ of the electron. We notice that the correction is larger for the bottom of the energy band than for the highest occupied orbitals; this is because the exchange charge density is somewhat more concentrated if $\boldsymbol{k}=0$ than it is for larger $\boldsymbol{k}$'s. There has been a good deal of discussion in the literature, for example by BRILLOUIN [*38*], BARDEEN [*8*], WIGNER [*253*], [*254*], KOPPE [*137*], WOHLFARTH [*259*], RAIMES [*195*], and JONES [*126*], of the meaning of this dependence of the potential energy, and hence the one-electron energy $\varepsilon_i$, on $\boldsymbol{k}$, and the net conclusion has been that for the free-electron case this dependence does not correspond to any physical situation actually met with metals. The discussion of this point is best carried on in connection with the study of electron correlation carried out by BOHM and PINES [*23*], [*24*], [*25*], [*186*], [*187*], discussed in Sect. 32, and hence we postpone discussion of this point until that section.

**19. Simplified methods of handling the self-consistent field.** The HARTREE-FOCK method, discussed in the two preceding sections, forms in principle the best method for handling the self-consistent field. However, in practice, it is so difficult that it has never been carried through in full detail for a solid. The reason for the difficulty comes in the summations over $j$ in Eq. (17.3); these properly should be summations over all the occupied orbitals, which means over wave functions with almost infinitely many values of the propagation constant $\boldsymbol{k}$. Most of the calculations which have been actually carried out so far have been made by greatly simplified methods, bearing no very close resemblance to the HARTREE-FOCK method.

Perhaps the simplest of these methods was that used by WIGNER and SEITZ [*255*], and following them by other writers, for sodium. They started with a field for the sodium atom, which had been determined by PROKOFJEW [*190*] semi-empirically; that is, PROKOFJEW had tried to find a field such that the one-electron energies of the wave functions in it agreed as closely as possible with the observed term values in the sodium atom. This actually comes close to a HARTREE-FOCK potential for the valence electron. WIGNER and SEITZ then assumed that the field to use for the energy-band calculation for sodium was the same as the PROKOFJEW field, within each WIGNER-SEITZ cell. This would be what we should get from the HARTREE-FOCK method, if we made the following simplifications. First, we assume that the exchange hole to be used for a conduction electron simply is equivalent to the charge of that conduction electron itself within the WIGNER-SEITZ cell in which the electron is moving, and does not extend into neighboring cells. That is, we assume that the field acting on the conduction electron is the field of the positive ion nearest which it is located, plus neutral atoms on all other atomic sites. Secondly, we assume that each of the WIGNER-SEITZ cells can be replaced by a sphere, so that the neutral atoms at other atomic sites will produce no field, and the field arising from the ion nearest the conduction electron will be spherical. Finally, we replace this field produced by the ion by the PROKOFJEW approximation. For the case of sodium, these assumptions are probably well justified.

The particular feature of sodium which makes it easy to handle is that the ions are small compared to the whole crystal, and over most of the volume of the crystal it is intuitively clear that each electron will move in the field only of the nearest ion and of the neighboring atoms, as is assumed in the simple treatment just described. For other metals, however, such as a divalent metal, a conduction electron in a given cell will move in the field of the other atoms, the ion core in its own cell, but also the other conduction electron in its own cell. Approximations to such potentials have been set up in the cases of the energy-band calculations which have been carried out, all forming first approximations to a HARTREE-FOCK potential, but departing from it in more or less unknown ways. In almost all of these calculations, a single potential energy function has been used for all the wave functions, or at least for all the wave functions in a band, not depending on the $\boldsymbol{k}$ of the wave function concerned.

The writer [*225*] has suggested a simplified method of handling the exchange charge, which is qualitatively correct, and quantitatively good enough to be of use, though it is far from precise. It consists of using for the exchange charge that value which would be found from the free-electron approximation, averaged over the $\boldsymbol{k}$'s of the electrons, and using the actual density of electronic charge at the point $x_1$ where the potential is being computed, as if these electrons formed part of an infinite sea of electrons of constant density, We recall from Sect. 18 that the radius of the exchange hole comes out inversely proportional to the cube root of the density of electrons. The exchange potential energy is then inversely proportional to the radius of the hole, or is directly proportional to the cube root of the electron density. This exchange correction gives moderately good results for isolated atoms, and it has been used for crystalline calculations by HERMAN and CALLAWAY [*91*] for germanium, by CALLAWAY [*41*] for iron, and WOODRUFF [*262*] for silicon. It has the advantage that it is easy to apply, since it involves only finding the local density of electronic charge, which must be computed in any case for finding the self-consistent field. Though far from satisfactory, it represents about the limit of complication which workers have yet been willing to undertake, in attempting to solve the self-consistent field problem for solids.

An even more simplified potential is that derived from the THOMAS-FERMI method [*64*], [*240*]. We shall not go into that method in detail, since it is familiar from its use in discussing atomic structure. The simplest form of the method takes no account of exchange. At a given point of space, it assumes that the potential energy acting on an electron equals that arising from the nuclei, and from all electrons including the one being considered. It then determines the charge density at a given point of space by assuming that this is the same as the electron density in a free-electron gas satisfying the FERMI statistics, in a region where the potential energy equals the local value at the given point of space. That is, in momentum space, electrons are filled up to a sphere whose size corresponds to the FERMI level. These assumptions lead to a self-consistent problem, the FERMI level being determined so as to make the system as a whole electrically neutral. The method has been extended by DIRAC [53] to include exchange, by using as the exchange hole the value calculated for a free-electron gas in a region of constant potential energy equal to the local potential energy.

The THOMAS-FERMI method has been applied to the cellular model of a crystal by the writer and KRUTTER [231]. Other writers have improved the method, including GOMBÁS [79], whose book on the subject contains very complete descriptions of the method, and many references, and it is capable of giving potentials which show the qualitatively correct behavior[1]. However, as usual with this method, it really does not take proper account of the periodic nature of the elements. It does not give the sharp break found between the inert gases and the alkali metals, for instance. Hence, though the results obtained by this method are interesting, they are not as significant as those found by direct solution of SCHRÖDINGER'S equation for the energy bands, in an assumed potential which is as close as possible to a HARTREE-FOCK potential. For metals at very high pressure and temperature, however, the method is very appropriate, as was shown by FEYNMAN, METROPOLIS, and TELLER [67].

## IV. One-electron theory of properties of solids.

**20. The FERMI statistics.** FERMI [63] in the early days of quantum mechanics set up the statistical laws governing a gas consisting of independent particles moving in an arbitrary force field, and subject to the PAULI exclusion principle. He showed that, at a temperature $T$, the probability of finding a given stationary state occupied was given by the FERMI function

$$f(\varepsilon) = \frac{1}{e^{\frac{\varepsilon - \varepsilon_F}{kT}} - 1} \tag{20.1}$$

where $\varepsilon$ is the energy level of the stationary state of the electron which is being considered, and $\varepsilon_F$ is a constant, the FERMI energy, which is determined by the condition that the total number of particles in all stationary states should equal the number known to be present in the system. The energy levels $\varepsilon$ are assumed to be determined by SCHRÖDINGER'S equation in terms of the assumed potential energy function.

SOMMERFELD [233], [234] and his students then followed by considering a free-electron model of a metal, finding the energy levels from the quantum theory of the perfect gas, and applying the FERMI statistics to the problem. By combining these assumptions with simple treatments of the scattering of electrons by thermal oscillations, HOUSTON [100], [101] was able to set up a theory of electrical conduction in metals which showed many resemblances to experiment. Various writers, including A. H. WILSON [257], then pointed out that the energy band theory would make quite different predictions from the free electron theory, in case there were just enough electrons to fill certain energy bands, leaving an empty gap above the filled levels. In fact, one would be led in that case to an insulator, or to a semiconductor, depending on the size of the energy gap. By applying the method of the FERMI statistics to a problem with energy bands of this sort, very satisfying theories of the behavior of insulators and semiconductors have been set up, for example by SHOCKLEY and his associates [208], and these theories are at the basis of the very broad development of knowledge of the electrical properties of solids which has come about both

[1] See also GOMBÁS' article in vol. XXXVI of this Encyclopedia.

theoretically and experimentally in recent years. In the present chapter we shall examine some of the predictions of this type of theory. Since we here overlap some of the other articles of this Encyclopedia, dealing with such problems as conductivity[1], semiconductors[2], etc., our treatment will be brief, and our object is only see how the type of theory which we have been outlining in the preceding chapter leads to applications of the type which we have just mentioned.

In the present section, we shall consider the validity of the method of the FERMI statistics, which underlies all of the theory we have just described; we must emphasize at the outset that the FERMI statistics, and everything based on it, is only an approximation to the exact treatment of the problem. The model which FERMI used is analogous to the assumptions of the self-consistent field, and it is only within the framework of a theory depending on such approximations that the FERMI statistics forms a valid starting point for the study of a solid. Let us first analyze the assumptions which go into the derivation of the FERMI statistics. We may state these in terms of three postulates:

1. The fundamental postulate of the quantum statistics, or of the canonical ensemble: the probability of finding a many-electron system in a state of energy $E$ in thermal equilibrium at temperature $T$ is proportional to $\exp(-E/kT)$.

2. The postulate of the existence of one-electron stationary states, satisfying the exclusion principle: we can approximate an actual system of many electrons by assuming that each electron moves in a stationary state of a one-electron system, but no more than one electron can move in the same stationary state.

3. The postulate of the additivity of one-electron energies: the total energy $E$ of a many-electron system can be approximated by the sum of one-electron energies of the electrons in the various occupied stationary states, so that

$$E = \sum_i \varepsilon_i + \text{const} \tag{20.2}$$

where the summation extends over the one-electron energies $\varepsilon_i$ of the occupied one-electron states.

Of these three postulates, the first is fundamental to quantum statistics, and may be assumed to be true in general. The second is much more restrictive. The general statement of the exclusion principle is in terms of the antisymmetry of the wave function with respect to permutation of the electron coordinates and spins. We cannot translate this into language prohibiting two electrons from being in the same stationary state, unless we are approximating the wave function in a form built up out of one-electron wave functions. This is never better than an approximation, and of course it is the approximation underlying the use of the HARTREE-FOCK method. The third postulate is far more specialized than the other two. It is by no means true, even with the HARTREE-FOCK approximation, that the total energy of the system can be written as a sum of one-electron energies of the occupied orbitals, plus a constant. We see, in other words, that the use of the FERMI statistics as a basis for study of the properties of solids must be viewed with suspicion, unless we are dealing with a problem in which there are special reasons why the three postulates should be satisfied. Practical workers in the field of electronics, who base the whole treatment of their science on FERMI statistics, should realise these limitations of the method.

The tacit assumption underlying most of the use of the energy-band theory and its treatment by means of the FERMI statistics is that we can not only use

[1] Cf. e.g. the article of H. JONES in this volume, and several articles in vol. XIV and XV of this Encyclopedia.

[2] Cf. vol. XX of this Encyclopedia.

postulate (3), but that also the one-electron energies $\varepsilon_i$ appearing in Eq. (20.2) are the same as the one-electron energies derived from a self-consistent field calculation; for instance, they equal the one-electron energy parameter $\varepsilon_i$ of the HARTREE-FOCK equation, Eq. (17.3), or appearing in substitute equations using a single potential function in place of the more complicated potential of the HARTREE-FOCK equations. Let us ask whether this assumption is likely to be correct to a sufficiently high degree of approximation in the cases we encounter in practice.

It is obvious from simple examples that the assumption is not right in general. For an atom, for instance, the one-electron energies are equal, approximately, by KOOPMANS' theorem, to the negative of the energies required to remove any one of the electrons from the neutral atom. To find the total energy of the atom, however, referred to a zero in which all the electrons are removed to infinity, we must remove one electron after another. The first one is removed from the neutral atom, the second from a singly charged ion, the next from a doubly charged ion, and so on. The sum of these ionization potentials of the successively more highly charged ions is much greater numerically than the sum of the ionization potentials of the neutral atom. One can easily show, by considering the HARTREE-FOCK equations, that the difference between the two sums equals the total repulsive interaction energy of all electrons in the system. Thus, for an atom, the sum of one-electron energies is by no means equal to the total energy of the atom, and the difference is not a constant, as Eq. (20.2) requires, but depends on the quantum numbers of the individual electrons.

However, in the cases to which the FERMI statistics is ordinarily applied, we do not meet any such problem of multiple ionization. In the problem of the highly ionized atom, the potential field in which one electron moves is vitally affected by what the other electrons are doing. In a solid, under ordinary circumstances, the departures from the ground state are so small that these potential fields are only affected to a negligible extent by the excitation. We never deal with problems in which electrons are removed from inner shells, except in cases of x-ray excitation. With thermal excitation in a metal at ordinary temperatures, we very slightly raise the energies of a very small fraction of the conduction electrons, as we shall see in the next section, and in a semiconductor or insulator we make a very much smaller change in the electronic distribution, since only very few electrons can be raised from the valence to the conduction band at ordinary temperatures. If we use a simple intuitive point of view, we may say that any given electron is most likely to find its immediate environment completely unaffected by the thermal excitation, and if it finds any change it will be small. Hence one may conclude that we are almost always dealing with a situation so close to the ground state that the self-consistent field is affected to a negligible extent by the excitation, and assumption (3) is obeyed, for all practical purposes. This intuitive conclusion can be made quantitative in terms of the HARTREE-FOCK equations.

There are cases, however, where there are serious deviations from the validity of assumption (3), and these cases must be handled separately. They come, as we should expect, when the interaction of two or more excited electrons (or holes) becomes important, for then the field acting on one of the electrons is seriously affected by the fact that the other one is excited. One such case comes in the study of impurity atoms in semiconductors. If we have an impurity atom which has an atomic number differing by several units from that of the atoms of the semiconductor, it is possible for the impurity atom to exist in several different stages of ionization. It is clear that here, as with isolated atoms, we must modify the self-consistent field to take account of the degree of ionization. Again, we have the case of the exciton, met with insulators. This case arises when we have simultaneously an electron excited to the conduction band, and a hole left in the valence band, and when furthermore the electron and hole are bound to each other, executing orbits around each other like the electron and proton in a hydrogen atom. Obviously here it is the interaction of the electron and hole which is of importance. As another example, problems of ferromagnetism involve the cooperative interaction of magnetic electrons on account of their exchange effect. Here the exchange potential energy acting on one electron depends strongly on the amount of magnetization; we cannot consider the exchange hole, and hence the one-electron energy, of one electron as being independent of the states which the other electrons are in.

Cases of the type which we have just mentioned are typical examples of problems in which we may not use the FERMI statistics in its naive form. They have in common the fact that it is the interaction of excited electrons which is important. We must be on our guard in all such problems, and must not be surprised to find that the FERMI statistics, and simple energy-band theory, do not furnish a complete basis for the treatment. In every such case, however, we can find a satisfactory substitute for the FERMI statistics from the fundamental postulates of quantum statistics. We must first find the energy levels of the problem; then we compute the partition function, $\sum \exp(-E/kT)$, where $E$ is the energy level of the whole system, and where we sum over all possible energy levels. Finally we use the fundamental relation of statistical mechanics.

$$\exp(-A/kT) = \sum \exp(-E/kT) \tag{20.3}$$

where $A$ is the HELMHOLTZ free energy, from which the equation of state and specific heat can be determined by differentiation. If the energy levels $E$ are determined according to the postulates required for FERMI statistics, we automatically find that Eq. (20.3) leads to the results of FERMI statistics; but it gives us a straightforward method of procedure in the cases where FERMI statistics does not apply.

**21. Electronic specific heat.** For calculation of the electronic specific heat, the FERMI statistics presumably has as much validity as the HARTREE-FOCK method, provided the one-electron energies are properly computed. The calculation of the specific heat is then a simple problem in statistics, which can be carried out in terms of the curve giving the number of one-electron stationary states of the unit amount of material, per unit energy range. This quantity is usually denoted as $dN/d\varepsilon$. One finds that the heat capacity of unit amount of material can be expanded in power series in the temperature, and the first term, proportional to temperature, was found by SOMMERFELD [*233*] to be

$$C_V = \left(\frac{dN}{d\varepsilon}\right)_0 \frac{\pi^2}{3} k^2 T \tag{21.1}$$

where $(dN/d\varepsilon)_0$ represents the value of $dN/d\varepsilon$ at the energy corresponding to the FERMI energy at the absolute zero of temperature. From the form of Eq. (21.1) it is obvious that a term of this type exists only for metals, not for semiconductors or insulators, since in the latter case the FERMI level lies within an energy gap.

The proportionality of the electronic specific heat to the temperature can be understood from a simple qualitative argument. In Fig. 38 we show the curve of $dN/d\varepsilon$ *vs.* $\varepsilon$, and two curves showing the product of this and the FERMI function, for two different temperatures, showing the number of electrons actually found with the various energy levels, at the two temperatures. It is clear that the effect of raising the temperature is to take electrons from an energy range of the order of magnitude of $kT$, originally lying below the FERMI level, and to raise them above the FERMI level, increasing their average by an amount proportional to $kT$. Hence the increase of total energy in going from the absolute zero to temperature $T$ is of the order of magnitude of $kT(dN/d\varepsilon)_0$, the number of electrons affected, multiplied by $kT$, the increase of energy per electron, or is proportional to $(dN/d\varepsilon)_0 k^2 T^2$. The electronic heat capacity is proportional to the derivative of this with respect to temperature, or to $2(dN/d\varepsilon)_0 k^2 T$, agreeing with Eq. (21.1) except in the numerical factor, which should be $\pi^2/3 = 3.3$ rather than 2. It is the fact that only a number of electrons equal approximately to $kT(dN/d\varepsilon)_0$ are affected in energy by the thermal excitation, and that even they change their energy by only a small fraction, which led us in Sect. 20 to the conclusion that thermal excitation would make a negligible change in the self-consistent field, so that the assumptions of the FERMI statistics were correct in such a case.

Table 1. *Values of $dN/d\varepsilon$ at* FERMI *level, determined from low temperature specific heat.*

| Element[1] | $dN/d\varepsilon$ (electrons per atom per ev) | |
|---|---|---|
| | Experiment | Theory (free electron) |
| Mg | 0.58 | 0.45 |
| Ti | 1.48 | |
| V | 2.5 | |
| Cr | 0.66 | |
| Fe | 2.15 | |
| Ni | 3.1 | |
| Cu | 0.32 | 0.21 |
| Zn | 0.32 | 0.32 |

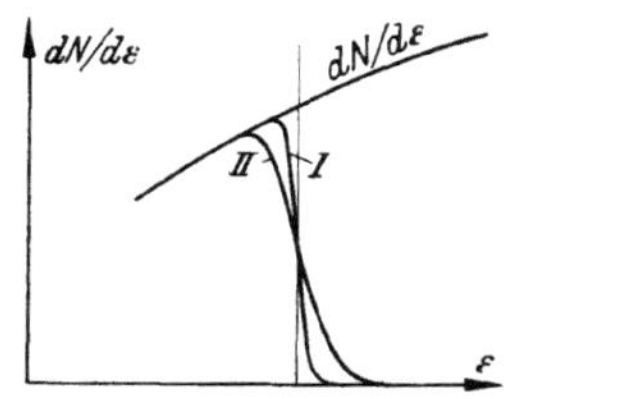


Fig. 38. Curve of $dN/d\varepsilon$, as function of energy $\varepsilon$, and two curves showing product of $dN/d\varepsilon$ and FERMI function (schematic). Curve I corresponds to a low temperature, II to a high temperature. The range of energy over which these curves are falling from the $dN/d\varepsilon$ curve to zero is of the order of magnitude of $kT$.

It is clear that Eq. (21.1) allows us to make a very direct experimental determination of $(dN/d\varepsilon)_0$. The electronic heat capacity is small compared to the contribution to the heat capacity arising from lattice vibrations, except at very low temperatures, where the lattice contribution decreases as $T^3$, leaving the electronic contribution to preponderate. Measurements of the low-temperature electronic heat capacity have been made for a number of metals. In Table 1 we give values of the quantity $(dN/d\varepsilon)_0$ determined from these measurements. The first striking observation is the high value obtained for the metals of the transition groups of the periodic table. This verifies the existence of the narrow $3d$, $4d$, $5d$ energy bands in these elements, with coorespondingly high values of $dN/d\varepsilon$ at the FERMI level. MOTT [*167*], [*169*] was one of the first to call attention to this important experimental check on the general concepts of energy-band theory.

We can go further than this, however, in comparing the experimental results of Table 1 with theory. Many writers, starting with SOMMERFELD and his co-workers, have assumed that the distribution $dN/d\varepsilon$ for the electrons of real metals can be determined by the free-electron approximation. For the alkali metals, we have seen in Sect. 13 that this assumption is well justified by more

[1] Values for Mg, Ti, V, Cr, Cu listed by S. A. FRIEDBERG, I. ESTERMANN and J. E. GOLDMAN: Phys. Rev. **85**, 375 (1952).
For Fe, W. H. KEESOM and B. KURRELMEYER: Physica, Haag **6**, 364 (1939). — Ni, W. H. KEESOM and C. W. CLARK: Physica, Haag **2**, 513 (1935). — Zn, W. H. KEESOM and J. A. KOK: Physica, Haag **1**, 770 (1934).

accurate calculations. In spite of its approximate nature, it still proves to give a fairly satisfactory value of $dN/d\varepsilon$ for the metals with two or three valence electrons (the alkaline earths, and the trivalent elements like Al). We assume in these cases that we are dealing with a free electron gas having one, two, or three electrons in the volume of an atom. Even for the $4s$-$4p$ bands of elements like Cu and Zn, containing completed bands of $3d$ electrons, we can find fairly accurate values of $dN/d\varepsilon$ for the $4s$-$4p$ electrons by assuming a free-electron gas with one or two electrons per atom respectively. To check these statements, we have calculated values of $(dN/d\varepsilon)_0$ for these elements, on the free-electron approximation, and include these values in the second column of Table 1. It is clear that the agreement is good as far as orders of magnitude are concerned,

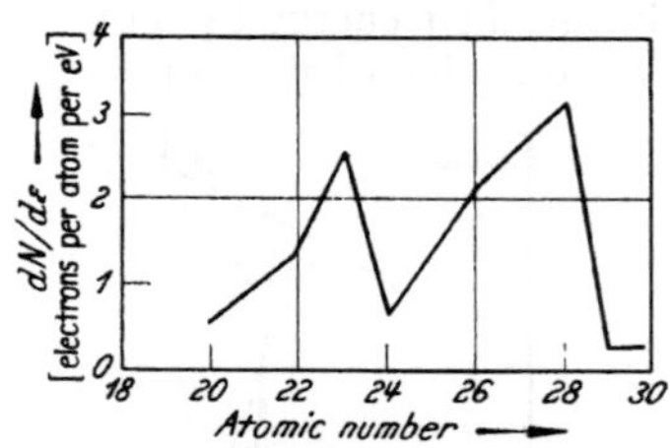


Fig. 39. Density of states $dN/d\varepsilon$ at Fermi level, in electrons per ev per atom, for elements in the $3d$ transition series, determined from electronic specific heat.

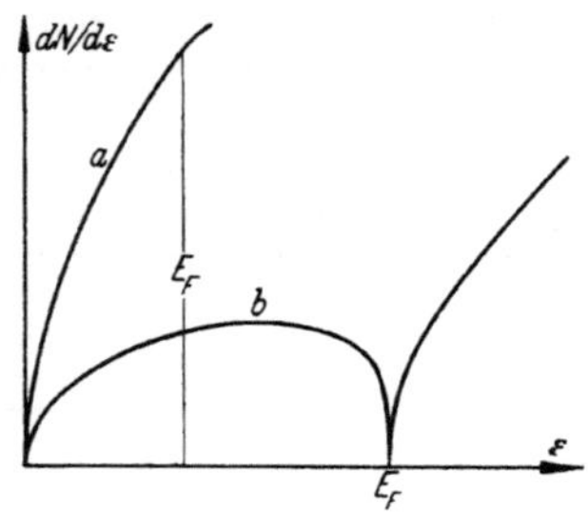


Fig. 40a and b. Density of states $dN/d\varepsilon$ for metallic sodium (a) according to the free-electron theory, (b) according to the Hartree-Fock theory. Fermi energies are shown in each case. (Curves from Raimes [*195*]).

verifying the fact that the free-electron approximation is not bad for these elements. To avoid misunderstanding, we repeat what was said in Chapt. II, namely that this does not at all justify the assumptions of the free-electron method, but merely indicates that there is a great formal resemblance between this treatment and the much more valid methods of the augmented or orthogonalized plane waves.

For the transition elements, having unfilled $3d$, $4d$, or $5d$ bands, we cannot make a simple comparison with theory of the type just discussed, for the theory is less simple and less accurate. However, one can work backward from the observations, to draw conclusions about the shape of the $d$-band. If we assume that the band has about the same value of $dN/d\varepsilon$ for the various elements in the $3d$ transitions group, but is merely filled to different heights in the different elements, we can assume that the experimental values of $dN/d\varepsilon$ for the different elements all fit on the same curve. Thus we plot in Fig. 39 the values of $dN/d\varepsilon$ for the elements in the $3d$ group, as given from experimental values in Table 1. The resemblance of this experimental curve to the theoretical curves of Fig. 23 is striking. Not only the general shape, but the height of the curve, is also reasonable. Thus the maximum height comes at about 3 electrons/ev. If we take a curve for $dN/d\varepsilon$ like that of Fig. 23, with a total width of about 5 ev, as seems to be deduced from experiments on soft x-ray spectra, the height of the highest peak comes out to be of approximately this order of magnitude. From this evidence, it seems that the theory of electronic specific heat based on the Fermi statistics and on the energy band theory is well justified by experiment.

There is one matter connected with the electronic specific heat which we should mention at this point. In Sect. 18, we pointed out that the Hartree-Fock equations for free electrons would suggest that the exchange correction should depend on the $\boldsymbol{k}$ of the electron in question, with a corresponding dependence of exchange potential on $\boldsymbol{k}$. When this is taken into account, it leads to a curve of $dN/d\varepsilon$ as given in Fig. 40, a figure given by Raimes [*195*]. For comparison, we give the curve as computed by the elementary method, which gives $dN/d\varepsilon$ proportional to $\varepsilon^{\frac{1}{2}}$. We observe the striking fact that, using the HARTREE-FOCK corrections, $dN/d\varepsilon$ goes down to zero at the FERMI level. If this were the case, Eq. (21.1) would show that there should be no electronic specific heat, contrary to the observations collected in Table 1. Experiment thus speaks against the reality of these exchange effects, and justifies the simple free-electron theory. As we have mentioned in Sect. 18, discussion of this point is best postponed until Sect. 32, in which we shall show that the theory of BOHM and PINES substantially removes the dependence of exchange energy on $\boldsymbol{k}$, and hence indicates that the simple theory is more nearly correct than the sophisticated HARTREE-FOCK theory.

**22. Relation of energy-band theory to electrical conductivity.** The subject of electrical conductivity is taken up in an extensive contribution to this Encyclopedia, the article by JONES on Theory of Electrical Conductivity in Metals, in this volume, and consequently we need do no more than sketch the relation between our discussion of the electron theory of solids, and conductivity theory; the reader is referred to JONES' article for further mathematical details. Several ingredients go into treatment of conductivity according to wave mechanics. The ordinary discussions start by considering a wave packet of electrons, all corresponding to approximately the same $\boldsymbol{k}$-value, and rather concentrated in space; though on account of the uncertainty principle, if the $\boldsymbol{k}$-values concerned in the wave packet fill only a small fraction of a BRILLOUIN zone, the wave packet must occupy a correspondingly large number of unit cells in the crystal. One finds the contribution of the electrons in this packet to the current. One then investigates the action of this packet under the combined action of an accelerating electric field, and of scattering resulting from lattice vibrations, and from irregularities in the lattice. As in any problem in kinetic theory, one sets up conditions for a steady state, in which wave packets are accelerated by the field, slowed down by scattering, resulting in a current proportional to the field, and leading to a calculation of the resistance.

The first theorem which one needs in order to carry through this program is the one stating the contribution to the current arising from a wave packet. Let us assume that we have a one-electron problem, in which the wave function is a wave packet, made up of a linear combination of wave functions all chosen from a single energy-band, with neighboring $\boldsymbol{k}$-vectors, the coefficients being set up so that the packet is moderately concentrated in space. Let the one-electron energy in this energy band as a function of the propagation constant $\boldsymbol{k}$ be $\varepsilon(\boldsymbol{k})$. Then one can prove that the average value, or expectation value, of the velocity of the electron, averaged over this wave packet, is given by

$$(v_x)_{\rm av} = \frac{1}{\hbar}\,\frac{\partial\varepsilon}{\partial k_x}, \quad (v_y)_{\rm av} = \frac{1}{\hbar}\,\frac{\partial\varepsilon}{\partial k_y}, \quad (v_z)_{\rm av} = \frac{1}{\hbar}\,\frac{\partial\varepsilon}{\partial k_z}. \tag{22.1}$$

Since $\hbar k$ represents the quantity of the dimensions of a momentum (often called the pseudomomentum) associated with the propagation vector $\boldsymbol{k}$, this means that a component of velocity is given by the derivative of $\varepsilon$ with respect to the appropriate component of the pseudomomentum.

Eq. (22.1) can be proved in several ways. One proof is a simple one using the formula for the group velocity of a wave packet. For any type of wave propagation, it is well known that the group velocity equals the partial derivative of the frequency with respect to the wave number, or reciprocal of the wave length. In this case, the frequency of the SCHRÖDINGER wave is given by $\varepsilon/h$, and the reciprocal of the wave length is $k/2\pi$, from which Eq. (22.1) follows at once.

For a more formal proof, one can start with the equation for the expectation value of the velocity of any wave function, as found from wave mechanics. This equation is well known to be

$$(v_x)_{\text{av}} = \frac{\hbar}{2mi} \int \left[ u^*(x, y, z) \frac{\partial u}{\partial x} - u(x, y, z) \frac{\partial u^*}{\partial x} \right] dx\, dy\, dz \tag{22.2}$$

with similar equations for the other components, provided there is no vector potential; the extension to the case where there is a vector potential is easily carried out. One expresses $u$ as a linear combination of BLOCH-like functions all corresponding to nearby values of $\boldsymbol{k}$, and all solutions of a periodic potential problem. It is then a straightforward mathematical procedure to show that the expression on the right side of Eq. (22.2) is equal to the value given by Eq. (22.1). It is understood that the derivatives on the right side of Eq. (22.1) are to be computed for the values of $k_x$, $k_y$, $k_z$ corresponding to the averages over the wave packet.

The next theorem required for discussing conductivity relates to the motion of a wave packet under the action of an external applied force, superposed on the periodic potential. Let there be an added non-periodic potential $V(x, y, z)$, which can represent the potential energy of an electron in an applied electric field. We note that, in order to preserve our periodic boundary conditions, this function must be chosen to be a periodic function of position, with the same periodicity as the boundary conditions; but this periodic volume is assumed to be large compared to the wave packet, so that it is only the behavior of $V$ in the neighborhood of the wave packet which is of physical significance, and we may choose $V$ in that neighborhood to agree with the actual potential, allowing this to repeat periodically in each periodic volume. Then we can prove that under the action of this applied potential, the expectation value of the $\boldsymbol{k}$ of the wave packet will change with time, according to the equations

$$\frac{d}{dt}(\hbar k_x)_{\text{av}} = -\frac{\partial V}{\partial x}, \qquad \frac{d}{dt}(\hbar k_y)_{\text{av}} = -\frac{\partial V}{\partial y}, \qquad \frac{d}{dt}(\hbar k_z)_{\text{av}} = -\frac{\partial V}{\partial z}. \tag{22.3}$$

That is, the time rate of change of the pseudomomentum $\hbar \boldsymbol{k}$ equals the force, as computed from the applied potential $V$. Here, similarly to the situation in Eq. (22.2), the quantities $\partial V/\partial x$, etc., are to be computed at points of space equal to the average or expectation values of $x$, $y$, $z$ for the wave packet.

The proof usually given for Eq. (22.3) is a simple one arising from the fact that the law of conservation of energy must be satisfied in quantum mechanics as in classical mechanics. The expectation value of the energy of the wave packet will be the sum of $\varepsilon(\boldsymbol{k})$, computed for the $\boldsymbol{k}$ which is the average over the wave packet, and $V(x, y, z)$, where $x, y, z$ are the coordinates of the center of the wave packet. As time goes on, both the coordinates and the values of $\boldsymbol{k}$ for the center of the wave packet will change, and the energy will change for both these reasons. On account of conservation, the total rate of change must be zero. Thus we have the relation

$$\left. \begin{aligned} \frac{d\,(\text{energy})}{dt} = 0 = {} & \frac{\partial V}{\partial x}(v_x)_{\text{av}} + \frac{\partial V}{\partial y}(v_y)_{\text{av}} + \frac{\partial V}{\partial z}(v_z)_{\text{av}} + \\ & + \frac{\partial \varepsilon}{\partial k_x}\frac{d(k_x)_{\text{av}}}{dt} + \frac{\partial \varepsilon}{\partial k_y}\frac{d(k_y)_{\text{av}}}{dt} + \frac{\partial \varepsilon}{\partial k_z}\frac{d(k_z)_{\text{av}}}{dt}. \end{aligned} \right\} \tag{22.4}$$

If we now assume Eq. (22.1), we see that Eq. (22.3) is the only assumption which we can make which automatically satisfies Eq. (22.4). It is interesting to note the resemblance of Eqs. (22.1) and (22.3) to classical HAMILTONian equations of a problem in which the HAMILTONian was $\varepsilon(\boldsymbol{p}) + V(x, y, z)$, where $\boldsymbol{p} = \hbar \boldsymbol{k}$. We see that Eq. (22.1) is then equivalent to the HAMILTONian equations $dq/dt = \partial H/\partial p$, and Eq. (22.3) to $dp/dt = -\partial H/\partial q$. With this identification, Eq. (22.4) then becomes equivalent to the standard proof of conservation of energy, from HAMILTON's equation.

Though this demonstration of Eq. (22.3) is simple, a more elaborate and formal proof shows that in some cases it requires modification. We have been assuming that the wave packet can be expanded in terms of one-electron wave functions arising from a single energy band. However, ZENER [*264*], HOUSTON [*102*], and many other authors, including JAMES [*116*] and FEUER [*65*], have shown that in some cases wave functions in more than one band must be used, indicating that there is a chance that the electron will jump from one band to another. One type of case in which this occurs is illustrated in Fig. 41. Here we have shown the periodic potential, plus the superposed perturbing potential $V$. We have indicated the existence of two energy bands, with a gap between; what we are plotting is the sum of the ordinary energy $\varepsilon(\boldsymbol{k})$ (we show merely the

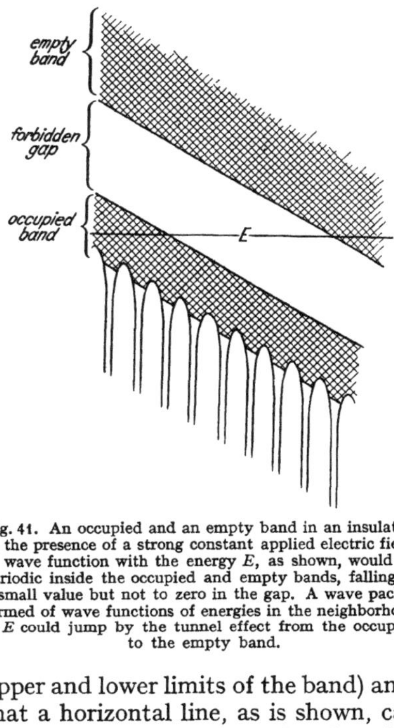


Fig. 41. An occupied and an empty band in an insulator, in the presence of a strong constant applied electric field. A wave function with the energy $E$, as shown, would be periodic inside the occupied and empty bands, falling to a small value but not to zero in the gap. A wave packet formed of wave functions of energies in the neighborhood of $E$ could jump by the tunnel effect from the occupied to the empty band.

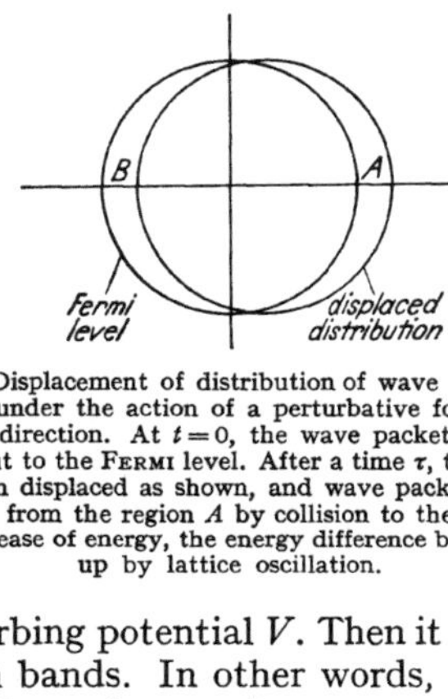


Fig. 42. Displacement of distribution of wave packets in $\boldsymbol{k}$-space, under the action of a perturbative force acting in the $x$ direction. At $t=0$, the wave packets will just extend out to the FERMI level. After a time $\tau$, they would have been displaced as shown, and wave packets in this case pass from the region $A$ by collision to the region $B$, with decrease of energy, the energy difference being taken up by lattice oscillation.

upper and lower limits of the band) and the perturbing potential $V$. Then it is clear that a horizontal line, as is shown, can cut both bands. In other words, a wave packet can jump from one band to the other with no change of energy, penetrating the region of space where its energy would lie in the gap by a tunnel effect. Calculation shows that to have appreciable probabilities of such transitions, the local field must be of the order of magnitude of $10^6$ to $10^7$ volts/cm.

Another case in which transitions from one band to another are of importance arises when we have more than one band with about the same energy, as in the transition metals, where the $3d$ and $4s$ bands overlap each other. In such a case, a wave packet formed from the $4s$-like band can be scattered into empty levels in the $3d$-band. MOTT [*167*] has called attention to such processes, and has suggested that the reason why the transition metals have considerably higher resistance than the others is that these scattering processes are important.

From our two theorems of Eqs. (22.1) and (22.3), we can now form a detailed picture of what goes on when electrons are accelerated in a conductor by an external field. We start with an equilibrium distribution of electrons, filling states in a $\boldsymbol{k}$-space up to a FERMI energy, as shown in Fig. 42, or with the occupation decreasing rapidly in the neighborhood of the FERMI level if the temperature is above the absolute zero. Normally we should build up the wave function as a determinantal combination of the occupied BLOCH orbitals. However, for the present purpose, we make linear combinations of these orbitals, in the form of wave packets of the type we have described, each extending over only a small range in **k**-space, and over a correspondingly large volume of ordinary space, but still small enough so that the perturbing potential $V$ can be regarded as approximately constant over the packet. Then we can represent the wave function by a determinantal combination of these wave packets.

At an initial instant we apply our perturbing field. Immediately the wave packets commence to move in **k**-space, with a velocity given by Eq. (22.3). Since all wave packets will move with the same velocity, the distribution of occupied states in Fig. 42 can be considered to move bodily in the direction of the external field, so that after a short time interval $\tau$ it will have moved to a new location, as indicated in Fig. 42. It is clear that some of the wave packets will have increased their energies $\varepsilon(\boldsymbol{k})$, some will have decreased them; conservation is achieved by the fact that they are simultaneously changing their perturbative energy $V$. After the time $\tau$, however, we see from Fig. 42 that the energy could be decreased if electrons represented by wave packets in the region $A$ had transitions to the region $B$. Such transitions are produced by interaction with thermal vibrations or irregularities in the lattice, in a way which we have not so far described.

The transitions just mentioned cannot be properly treated without discussing the interaction between electronic motion and lattice vibrations, a topic lying outside the scope of this article. However, we may state the general nature of the results. One can have a simultaneous transition, in which energy $\varepsilon(\boldsymbol{k})$ and **k**-vector of the electronic motion change, and in which compensating changes in energy and **k**-vector occur in the lattice vibrations. We are dealing in this case with the energy $h\nu$ of the elastic vibration, and the **k**-value $2\pi/\lambda$ associated with its propagation vector. On account of the fact that the velocity of sound is so small, the energy change $h\nu = hv/\lambda$ associated with such a transition, where $v$ is the velocity of sound, is very small compared to the energy changes met in the electronic energy. Hence in such a simultaneous transition of electron and lattice, the electron's energy is very nearly conserved, only a small amount being delivered to, or abstracted from, the lattice, though the changes in **k** are large. On the average, energy will be given up to the lattice, resulting in heating it up, and the momentum of the electron will be reversed, as in going from the region $A$ to $B$ in Fig. 42. A transition of this type can only occur to an unoccupied electronic level, on account of the exclusion principle.

We have a picture, then, of the wave packets marching in a procession in the direction of the external force up to the region $A$ where they lie above the FERMI level, and from which they can find regions $B$ of lower energy which are unoccupied by electrons; and then of the packets in $A$ undergoing scattering processes which return them to $B$, so that they must start their march again. It is clear that a steady state is possible under these circumstances, the wave packets moving beyond the FERMI level for an average time $\tau$, which is the mean free time after which they are likely to have a scattering process. The average current is then proportional to the applied field, according to Eq. (22.3), and to the mean free time $\tau$. It is in this way that OHM's law is derived. Most of the labor involved in discussing electrical conductivity comes from studying the nature of this steady state, the resulting distribution of occupied levels in the **k**-space, and the consequent current, as computed by use of Eq. (22.1). As we have mentioned, this is a large subject which is handled elsewhere in this Encyclopedia, and consequently we leave the question at this point.

**23. Relation of energy-band theory to semiconductors.** The properties of semiconductors are being treated elsewhere in this Encyclopedia[1], and as with electrical conductivity, which formed the subject of the preceding section, we can be brief in our discussion of the subject. The fundamental point connected with semiconductors is of course the fact that if a solid has certain energy bands entirely occupied, others entirely empty, with no partly-filled bands, it can carry no current. This fact arises from Eq. (22.1). If we have a filled band, this means that the wave packets which can be constructed from its wave functions completely fill the first BRILLOUIN zone. The total current carried by all electrons is found by summing or integrating the current carried by the various wave packets. That is, it involves an integral of the quantity whose $x$ component is $\partial \varepsilon / \partial (\hbar k_x)$ over the BRILLOUIN zone. On account of the periodicity of $\varepsilon$ in $\boldsymbol{k}$-space, this integral necessarily vanishes.

From the FERMI statistics, we easily find that with an insulator which carries no net charge, the FERMI level will lie within the gap between the highest occupied band (the valence band) and the next higher band (the conduction band). As the temperature rises, electrons will begin to be excited from the highest levels of the valence band to the lowest levels of the conduction band. In these circumstances, the substance becomes an intrinsic semiconductor, and we are led to study the conductivity of a conduction band containing a few electrons near its bottom energy level, and a valence band with a few holes, or vacant levels, near its top energy level.

Most actual semiconductors contain impurity atoms. In such cases, the number of electrons in the conduction band, or of holes in the valence band, can be much greater than for an intrinsic semiconductor, and can vary in quite a different way with temperature. If the electrons in the conduction band preponderate, we speak of an $n$-type semiconductor, while if the holes in the valence band preponderate, it is a $p$-type semiconductor. We shall not go into the details of the predictions of the FERMI statistics in these cases, or into the way in which current is carried by the electrons and holes; these are the main topics of discussion in a treatment of semiconductors. There are only two subjects connected with the theory which tie in so closely with our main approach that we shall discuss them here: the nature of the energy levels at the top of the valence band and the bottom of the conduction band, and hence the nature of the current carried by electrons and holes; and the energy levels surrounding an impurity atom in a semiconductor, with their consequences of $n$- or $p$-type semiconductors. We shall postpone the discussion of energy levels of impurity atoms until Chapter V. In the present section, however, we shall treat the nature of the energy bands concerned in the conductivity of semiconductors.

There are two possibilities for the functional dependence of the energy on $\boldsymbol{k}$, at the bottom or top of an energy band: either it is an analytic function which can be expanded in power series in $\boldsymbol{k}$, or it is not. The former case is the commoner, and we shall discuss it first. It is always the case if the energy band is not degenerate at the bottom or top. The latter situation occurs in cases of degeneracy, but it is rather uncommon for these to lie at the bottom or top of a band.

Let the energy $\varepsilon(\boldsymbol{k})$ have a minimum at $\boldsymbol{k} = \boldsymbol{k}_0$. Then it can be expanded in power series in $k_x - k_{x0}$, $k_y - k_{y0}$, $k_z - k_{z0}$, the series having no linear terms.

[1] By O. MADELUNG in vol. XX.

By a rotation of axes we can remove the cross terms. That is, if we let $\xi, \eta, \zeta$, be three suitably chosen orthogonal linear combinations of the quantities $k_x - k_{x0}$, $k_y - k_{y0}$, $k_z - k_{z0}$, we can write the energy near its minimum in the form

$$\varepsilon(\xi, \eta, \zeta) = \varepsilon_0 + \frac{\hbar^2 \xi^2}{2m_1} + \frac{\hbar^2 \eta^2}{2m_2} + \frac{\hbar^2 \zeta^2}{2m_3} + \text{higher terms} \tag{23.1}$$

where $m_1, m_2, m_3$ are positive constants. In such a case, Eq. (22.1), which holds equally well in the rotated coordinates, leads to

$$(v'_x)_{\text{av}} = \frac{\partial \varepsilon}{\partial(\hbar\xi)} = \frac{\hbar\xi}{m_1}, \qquad (v'_y)_{\text{av}} = \frac{\hbar\eta}{m_2}, \qquad (v'_z)_{\text{av}} = \frac{\hbar\zeta}{m_3}, \tag{23.2a}$$

That is, the velocity components $(v'_x)_{\text{av}}$, etc., with respect to the rotated axes, are related to the pseudomomentum components $\hbar\xi$, etc., in a way which suggests that $m_1, m_2, m_3$ should be treated as effective masses related to the three directions. We can now use Eq. (22.3). We find

$$\frac{d(m_1 v'_x)_{\text{av}}}{dt} = -\frac{\partial V}{\partial x'}, \qquad \frac{d(m_2 v'_y)_{\text{av}}}{dt} = -\frac{\partial V}{\partial y'}, \qquad \frac{d(m_3 v'_z)_{\text{av}}}{dt} = -\frac{\partial V}{\partial z'} \tag{23.2b}$$

where $x', y', z'$ refer to the rotated coordinates. That is, the equations of motion are like those of classical particles, except that we must use three different masses for acceleration in the three directions.

We now note that there are two cases. If we are dealing with the bottom of a conduction band, the three masses will all be positive, and there is nothing remarkable about the situation. However, if we are dealing with the top of a valence band, the three masses must all be negative, as we see from the fact that there we are expanding the energy about a maximum instead of a minimum, so that the coefficients in Eq. (23.1) must all be negative. We may instead, if we prefer, adopt in this case the convention that the masses are to be defined as positive, so that in place of Eq. (23.1) we have in this case

$$\varepsilon(\xi, \eta, \zeta) = \varepsilon_0 - \frac{\hbar^2 \xi^2}{2m_1} - \frac{\hbar^2 \eta^2}{2m_2} - \frac{\hbar^2 \zeta^2}{2m_3}. \tag{23.3}$$

Then in place of Eq. (23.2) we have

$$\frac{d(m_1 v'_x)_{\text{av}}}{dt} = \frac{\partial V}{\partial x'}, \qquad \text{etc.} \tag{23.4}$$

That is, we have the equations of motion of a classical particle with masses $m_1, m_2, m_3$, but acted on by a force which is the negative of that acting on an electron, or equal to that which would act on a positive electron at the same point.

This latter interpretation is more convenient in the case of the semiconductor, for the reason that at the top of a band, such as a valence band, we are ordinarily concerned with holes (that is, full bands lacking a single electron) rather than electrons. If we make up a wave function for a filled band lacking one electron in a determinantal form, using wave-packet wave functions of the type we have described, one of these wave-packet functions will be empty. Under the action of the field, all wave packets will be accelerated formally in the same way, whether they are occupied or not, with the result that the empty wave packet will be accelerated according to Eq. (23.4). The resulting current carried by the whole electronic system, however, will be the negative of that which would have been carried by the unoccupied wave packet. The reason is that a filled band carries no current, as we have pointed out, so that a band lacking one electron will carry just the negative of the current which that electron would have carried. Hence as far as the current is concerned, a hole carries the same current as a positive electron occupying the same wave packet. Thus the interpretation of a hole as a positive charge carrier is consistent both with the equation of acceleration, Eq. (23.4), and with the current carried, which makes it a very useful way to regard the hole.

We have now discussed the case in which the energy can be expanded in power series about its maximum or minimum values. Such cases are very important in practice. For instance, in germanium, it is known from experiments on cyclotron resonance by LAX, KIP, and their associates [*52*], [*55*] that the minimum energy in the conduction band comes at eight equivalent points in the BRILLOUIN zone, along the eight 111 directions. One has one effective mass for the direction along the 111 axis, and another mass for the two directions at right angles to 111, which by symmetry must be equivalent. Effective masses have been determined by a variety of experimental methods, and we are now quite sure of the general form of the energy bands in this case. Analogous results have been obtained for silicon, by the same writers [*50*], [*51*].

On the other hand, as we have mentioned earlier, if the energy maximum or minimum lies at a point of degeneracy, the energy cannot be expanded in power series in the momentum. Near a point of two-fold degeneracy, the energy as a function of the momentum is given approximately by the roots of a quadratic equation, and near a point of three-fold degeneracy by the roots of a cubic. In such cases the energy is proportional to the square of the magnitude of the momentum (using the point of degeneracy as the origin), but the dependence on angles cannot be expressed in terms of a quadratic expression in the three components of momentum. For instance, near a point of two-fold degeneracy, in a corresponding two-dimensional problem, we can show that the energy is given by

$$\varepsilon = k^2 \left[A \pm (B^2 \cos^2 2\vartheta + C^2 \sin^2 2\vartheta)^{\frac{1}{2}}\right] \tag{23.5}$$

where $k$ represents the magnitude of the pseudomomentum, measured from the position of the minimum, $\vartheta$ is the angle between $\boldsymbol{k}$ and one of the two principal axes, and $A$, $B$, and $C$ are constants. This cannot be expressed in the form of Eq. (23.1), using $k \cos\vartheta$ and $k \sin\vartheta$ in place of $\xi$ and $\eta$. In such a case, the simple concept of effective mass, which is based on Eq. (23.1), cannot be applied, and more complicated methods must be used.

In the neighborhood of points of degeneracy, the spin-orbit interaction, which we have disregarded so far, can become of importance, for it can lead to a splitting of the levels, and a removal of part of the degeneracy. We shall not discuss this problem, but call the attention of the reader to the fact that it has been discussed in detail in the literature, particularly by ELLIOTT [*57*], [*58*], and that it occurs in such an important case as the top of the valence band in germanium.

We have now mentioned the problems arising in studying the acceleration of wave packets in a semiconductor. As we stated earlier, this is one of the two aspects in which our study of the electronic structure of solids meets the problems of semiconductor theory. The other aspect relates to the energy levels introduced into the crystal by the presence of impurity atoms. These impurity energy levels will form one of the principal topics of the next chapter.

## V. Impurity atoms and alloys.

**24. Impurity atoms, and the effective mass approximation.** There are many important problems, both in the study of semiconductors and of metals, in which we are interested in a crystal containing certain types of imperfections. These imperfections can take the form of missing atoms, of interstitial atoms, of impurity atoms replacing atoms normally present in the crystal, of dislocations, and of many other types; a good survey of all these types of imperfections, and their importance, has been given by SEITZ [*201*]. In all such problems, we are required to solve SCHRÖDINGER'S equation, not for a periodic potential, but for a potential which is periodic except for a localized perturbation. The perturbative potential, of course, must be found by a self-consistent field method from perturbed wave functions. There are general methods of approach to all such problems, capable of giving results of any desired degree of accuracy, which we shall now take up, describing the application of the results to problems of semiconductors, metals, and alloys.

The first and simplest method of handling such problems is suggested by the analysis of Sect. 23. There we showed that a wave packet composed of wave functions in an energy band whose energy could be expressed as a quadratic function of the momentum, as in Eq. (23.1) or (23.3), moved under the action of an external perturbative field like a classical particle, but with different effective masses, $m_1$, $m_2$, $m_3$, along the three principal axes. We remember that according to quantum mechanics, a wave packet moves according to the classical equations of motion of a particle, provided the wave function follows the SCHRÖDINGER equation derived from the classical HAMILTONian. Now the classical HAMILTONian which would lead to the motion of the wave packet which we have described is

$$H = \frac{p_x^2}{2m_1} + \frac{p_y^2}{2m_2} + \frac{p_z^2}{2m_3} + V \tag{24.1}$$

where we have chosen the $x$, $y$, $z$ axes as the three principal axes of the ellipsoidal energy surfaces, and where $V$ is the perturbative potential. One suspects intuitively, then, that we can convert this into a quantum theoretical treatment by using the SCHRÖDINGER equation

$$-\hbar^2\left(\frac{\partial^2/\partial x^2}{2m_1} + \frac{\partial^2/\partial y^2}{2m_2} + \frac{\partial^2/\partial z^2}{2m_3}\right)\psi + V\psi = E\psi. \tag{24.2}$$

We should suspect that the eigenvalues of this equation would in some way represent the energy values in the neighborhood of the impurity atom, provided $V$ were the perturbative potential energy (that is, the difference between the potential energy of an electron in the perturbed crystal and in an unperturbed crystal). Furthermore, we should suspect that the eigenfunctions $\psi$ would have some relation to the correct wave functions for the perturbed problem.

A simple and very satisfactory application of this intuitive method has been made to the energy levels surrounding an impurity atom in a semiconductor. Let us consider, for instance, a germanium crystal in which one of the atoms is replaced by an arsenic atom, whose atomic number is one unit greater. The perturbative potential is what we should have by adding a unit positive charge at the position of the arsenic atom. This produces a COULOMB potential energy in the crystal, given by $-e^2/\varkappa r$, where $\varkappa$ is the dielectric constant of the material. One then has in Eq. (24.2) a hydrogen-like problem. In the earliest work by this method, before the effective masses of semiconductors were known accurately, it was assumed that $m_1$, $m_2$, $m_3$ were all equal, so that the problem was precisely hydrogen-like, except for the presence of the dielectric constant, and the fact that the effective mass was different from the real mass. One would then deduce the existence of discrete energy levels of negative energy (that is, below the bottom of the conduction band), the energy being given by the RYDBERG formula, as suitably modified for the presence of the dielectric constant and the effective mass. These modifications result in energy levels which, instead of being at $-13.6$ ev, as in hydrogen, come at only a few hundredths of an electron volt below the bottom of the conduction band, and this qualitatively agrees with what is found experimentally for the donor levels in an $n$-type semiconductor.

More accurate calculations are now being made, using Eq. (24.2), and using our present accurate knowledge of the effective masses in semiconductors. It is now known that the three effective masses are not equal. Hence, Eq. (24.2) is not really hydrogen-like, and in fact it cannot be solved analytically with a Coulomb potential. However, Kohn and Luttinger [*134*] and independently Kleiner [*132*] have used variation methods to get approximate solutions of the equation, and the energy levels so found prove to agree surprisingly well with those found experimentally, from optical absorptions in semiconductors containing impurity atoms. Thus, in Table 2, we give computed and experimental values of the energy levels of a number of the lowest energies found in silicon with arsenic impurity atoms, as computed by Kohn and Luttinger. It is clear that this method forms a good approximation to the actual facts.

Table 2. *Energy levels of As impurity in Si crystal, from* W. Kohn [Phys. Rev. **98**, 1856 (1955)], *and other references mentioned there. Zero of energy is bottom of the continuum.*

| Quantum number | Theoretical energy (ev) | Experimental energy (ev) |
|---|---|---|
| $1s$ | $-0.029$ | $-0.0533$ |
| $2p,\ m=0$ | $-0.0109$ | $-0.0110$ |
| $2p,\ m=\pm 1$ | $-0.0059$ | $-0.0061$ |
| $3p,\ m=0$ | $-0.0057$ | |
| $3p,\ m=\pm 1$ | $-0.0029$ | $-0.0027$ |
| $4p,\ m=\pm 1$ | | $-0.0012$ |

The poorest agreement comes for the lowest state, which is analogous to the $1s$ state of hydrogen. The reason for this is clear. The wave functions predicted by Eq. (24.2) are widely extended in space, on account of the high dielectric constant and small effective mass. Both of these effects tend to increase the physical size of the wave function and to decrease its energy numerically. Hence the electron in the bound state spends most of its time far from the impurity atom, and it should be a good approximation to replace its behavior in traveling through the crystal by that of a free electron with effective mass coefficients $m_1, m_2, m_3$. However, in the $1s$-like state, the electron spends a considerable fraction of its time near the impurity atom, where this approximation is not good. Kohn and Luttinger have considered the modification arising from this fact, and show that it is a reasonable explanation of the discrepancy for this lowest state.

Most of the further development of the theory of energy levels in impurity atoms is devoted to giving regorous derivations of the method leading to Eq. (24.2), which we have introduced only intuitively, and to estimating the limit of validity of that method, and to setting up more powerful methods which can be applied even in cases where the effective mass method is not appropriate. The mathematical basis for this further development was laid by Wannier [*251*], in a paper describing the structure of the exciton. The exciton is a name applied to a combination of an electron in a conduction band of a semiconductor, and a hole in the valence band, in the special case where they are close enough together so that their Coulomb attraction binds them together in a stable structure. Wannier treated this two-particle problem by his general method of analysis, which we shall take up in the next section. On account of the complication of that problem, his method seemed somewhat involved, but it has been shown by the present writer [*223*], Adams [*1*], and others, that the same method can be applied to the simpler case of a single electron or hole moving in a localized perturbative field such as we find in an impurity atom, and that this method leads to a more fundamental understanding of the problem of impurity levels. We shall now take up this analysis.

**25. WANNIER's method for the impurity problem.** WANNIER started his consideration by thinking of the tight-binding approximation. In Eq. (9.1) we have seen that BLOCH approximated the wave function of an electron in a given energy band by a linear combination of atomic orbitals

$$b_{\boldsymbol{k}}(\boldsymbol{r}) = \sum_{\boldsymbol{R}_i} \exp(i\,\boldsymbol{k}\cdot\boldsymbol{R}_i)\,\varphi(\boldsymbol{r}-\boldsymbol{R}_i) \tag{25.1}$$

where $\varphi(\boldsymbol{r}-\boldsymbol{R}_i)$ represents an atomic function localized on the atom at vector position $\boldsymbol{R}_i$. For a plane-wave-like solution, the coefficients of these various atomic functions are the exponentials $\exp(i\boldsymbol{k}\cdot\boldsymbol{R}_i)$. It seems plausible that in the presence of a perturbation, we can still approximate the wave function by a linear combination of the atomic orbitals $\varphi(\boldsymbol{r}-\boldsymbol{R}_i)$, but now with coefficients which are no longer the simple exponentials which we find for the unperturbed crystal.

WANNIER went further than this, however: he asked whether one cannot set up functions similar to the atomic orbitals, but such that the function formed from them by the method described in Eq. (25.1) would be the precise solution of the periodic potential problem, rather than the approximation given by Eq. (25.1). He showed that this could be done, the resulting functions $a(\boldsymbol{r}-\boldsymbol{R}_i)$ being the WANNIER functions. Having derived them., he then set up the linear combination of WANNIER functions,

$$\psi = \sum_{\boldsymbol{R}_i} U(\boldsymbol{R}_i)\,a(\boldsymbol{r}-\boldsymbol{R}_i) \tag{25.2}$$

in analogy to Eq. (25.1), and found the relations which the coefficients $U(\boldsymbol{R}_i)$ must satisfy, in order that the function $\psi$ may form the best approximation possible, of the given form, to a solution of SCHRÖDINGER'S equation for the perturbed problem. The resulting equations for $U(\boldsymbol{R}_i)$ can be put in a very interesting form similar to Eq. (24.2), and it is this derivation which we shall give in the present section.

First we shall define the WANNIER functions, and demonstrate their properties. We start with the $N$ functions $u_{\boldsymbol{k}}(\boldsymbol{r})$, where $N$ is the number of unit cells in the repeating volume of the crystal, and where our $N$ functions correspond to $N$ different $\boldsymbol{k}$ values, just filling the central BRILLOUIN zone. Let us try to make a linear combination of these $N$ functions which resembles an atomic function as closely as possible. That is, it should be as concentrated a wave packet as we can form. In our discussion of wave packets in Sect. 22, we pointed out that if we construct a wave packet by using only a restricted range of $\boldsymbol{k}$ values, it will fill many unit cells in the crystal. However, if we use all $\boldsymbol{k}$ values in the BRILLOUIN zone, we can construct a wave packet which is largely confined to a single unit cell, and hence resembles an atomic function. For heuristic purposes, let us suppose that all the functions $u_{\boldsymbol{k}}(\boldsymbol{r})$ resemble each other, and are in phase with each other, in one unit cell, say that for $\boldsymbol{R}_i=0$. As we go away from this cell, of course the various functions vary sinusoidally with position, with propagation constants $\boldsymbol{k}$. Now let us superpose the functions $u_{\boldsymbol{k}}$, all with the same coefficients. Then in the central cell the functions will reinforce each other, building up something resembling an atomic function, while as we go away from this cell, they will get out of phase with each other, and will rapidly cancel. In other words, the sum $\sum_{\boldsymbol{k}} u_{\boldsymbol{k}}(\boldsymbol{r})$ represents a function concentrated in the central cell. By a simple extension of this argument, the sum $\sum_{\boldsymbol{k}} \exp(-i\boldsymbol{k}\cdot\boldsymbol{R}_j)\, u_{\boldsymbol{k}}(\boldsymbol{r})$ represents an identical function centered in the cell at position $\boldsymbol{R}_j$.

The functions we have just described, when properly normalized, are the WANNIER functions. If the functions $u_{\boldsymbol{k}}$ are normalized over the volume of $N$ unit cells, then we can show that the function

$$a(\boldsymbol{r}-\boldsymbol{R}_j) = N^{-\frac{1}{2}} \sum_{\boldsymbol{k}} \exp(-i\boldsymbol{k}\cdot\boldsymbol{R}_j)\, u_{\boldsymbol{k}}(\boldsymbol{r}) \tag{25.3}$$

is also normalized, and in fact represents the WANNIER function localized in the cell at $\boldsymbol{R}_j$. These functions, for different values of $\boldsymbol{R}_j$, can be shown to be orthogonal to each other. Since the $N$ functions $a(\boldsymbol{r}-\boldsymbol{R}_j)$ are determined from the $N$ functions $u_{\boldsymbol{k}}(\boldsymbol{r})$ by a unitary transformation, it is obvious that we can write $u_{\boldsymbol{k}}(\boldsymbol{r})$ as a linear combination of the $a$'s. We find, in fact, that

$$u_{\boldsymbol{k}}(\boldsymbol{r}) = N^{-\frac{1}{2}} \sum_{\boldsymbol{R}_i} \exp(i\boldsymbol{k}\cdot\boldsymbol{R}_i)\, a(\boldsymbol{r}-\boldsymbol{R}_i) \tag{25.4}$$

which shows an obvious similarity to Eq. (25.1), so that by means of the WANNIER functions we have achieved the aim of being able to write an exact solution of the periodic potential problem as a linear combination, of the form of a BLOCH sum, of atomic types of functions.

WANNIER investigated the dependence of the function $a(\boldsymbol{r})$ on coordinates, in some simple cases, and showed that in fact it was fairly concentrated, though it has appreciable values in several unit cells, rather than being entirely confined to a single unit cell. We note in this connection the limitation which we have just imposed for heuristic purposes, assuming that the functions $u_{\boldsymbol{k}}$ all resembled each other in one unit cell. We note that nothing in the solution of SCHRÖDINGER's equation for a periodic potential determines the phase of the function $u_{\boldsymbol{k}}$; the phase can change in an arbitrary manner from one $\boldsymbol{k}$ value to another. Such an arbitrary phase would not affect the formal proofs of the properties of WANNIER functions, as expressed in Eqs. (25.3) and (25.4). However, different choices of phase would result in WANNIER functions of different appearance, and that choice of phases which would make the $u_{\boldsymbol{k}}$'s resemble each other most closely in one unit cell would result in the most concentrated WANNIER functions. This matter of the lack of uniqueness of WANNIER functions has not been exhaustively discussed, though the writer [*227*] has commented on it, but we shall assume that the choice of phases is made in such a way as to have as concentrated WANNIER functions as possible. Even in such cases, however, we find that the amplitudes of the WANNIER functions fall off only inversely with the distance from the unit cell where the function is localized, rather than exponentially as with an atomic wave function.

Now that we have set up the WANNIER functions, we can assume that the wave function $\psi$ for a perturbed problem can be approximated as a linear combination

$$\psi = \sum_{\boldsymbol{R}_i} U(\boldsymbol{R}_i)\, a(\boldsymbol{r}-\boldsymbol{R}_i) \tag{25.5}$$

as given in Eq. (25.2), and can find the equations determining the $U$'s. Let the unperturbed HAMILTONian of the periodic potential problem be $H_0$, and let the perturbative term, as before, be $V$. We can now determine the $U$'s so that the combination of functions given by Eq. (25.5) will minimize the expectation value of the HAMILTONian $H_0 + V$. The way of doing this is familiar from quantum mechanics. We must first find the matrix components of the HAMILTONian with respect to the wave functions $\boldsymbol{a}(\boldsymbol{r}-\boldsymbol{R}_i)$. If these are called $H_{ij}$, we can then set up equations

$$\sum_{\boldsymbol{R}_j} H_{ij}\, U(\boldsymbol{R}_j) = E\, U(\boldsymbol{R}_i) \tag{25.6}$$

where $E$ is the energy. These equations can, in principle, be solved in the usual way for the quantities $U(\boldsymbol{R}_i)$ and the energy; in order that there be solutions, the determinant of coefficients, $|H_{ij} - E\,\delta_{ij}|$, must vanish, giving a secular equation for the energy. We shall later consider practical methods of solving these equations.

To carry out our process, we must find matrix components $H_{ij}$ of the HAMILTONian. As far as the perturbative potential $V$ is concerned, the matrix components $V_{ij}$ will clearly be large only when both the WANNIER functions $a(\boldsymbol{r}-\boldsymbol{R}_i)$ and $a(\boldsymbol{r}-\boldsymbol{R}_j)$ are large in the part of space where the perturbative potential is large. In fact, if the perturbation is large only in one unit cell, the largest such matrix component $V_{ij}$ will be that for which $\boldsymbol{R}_i$ and $\boldsymbol{R}_j$ both refer to the unit cell in which the perturbation is located. We shall make some use later of the special case in which only this one matrix component of $V$ need be considered, the others all being negligible. If on the contrary the potential $V$ varies only very slowly from point to point, as it will do if it stretches over many unit cells, then in the matrix component

$$V_{ij} = \int a^*(\boldsymbol{r}-\boldsymbol{R}_i)\, V(\boldsymbol{r})\, a(\boldsymbol{r}-\boldsymbol{R}_j)\, dv \tag{25.7}$$

we may assume as a first approximation that $V$ is constant all over the region of space where $a^*(\boldsymbol{r}-\boldsymbol{R}_i)\, a(\boldsymbol{r}-\boldsymbol{R}_j)$ is appreciable, and hence may take it out of the integral sign. Then on account of the orthogonality and normalization of the $a$'s, we see that $V_{ij}$ will be diagonal, and the diagonal matrix component $V_{ii}$ will equal the value of $V(\boldsymbol{r})$ in the cell located at $\boldsymbol{R}_i$. This approximation will also be used in later work.

The matrix components of the unperturbed HAMILTONian $H_0$ with respect to the WANNIER functions have interesting and simple values. We can get at these by setting up the energy $\varepsilon(\boldsymbol{k})$ of the wave function $u_{\boldsymbol{k}}(\boldsymbol{r})$, which is an energy value for the periodic potential problem, or is the expectation value of $H_0$ for the function given in Eq. (25.4). This is

$$\left.\begin{aligned} \varepsilon(\boldsymbol{k}) &= \int u_{\boldsymbol{k}}^*(\boldsymbol{r})\, H_0\, u_{\boldsymbol{k}}(\boldsymbol{r})\, dv \\ &= N^{-1} \sum_{\boldsymbol{R}_i, \boldsymbol{R}_j} \exp\left(i\boldsymbol{k}\cdot(\boldsymbol{R}_j - \boldsymbol{R}_i)\right) (H_0)_{ij} \end{aligned}\right\} \tag{25.8}$$

where $(H_0)_{ij}$ is the matrix component of $H_0$, given by

$$(H_0)_{ij} = \int a^*(\boldsymbol{r}-\boldsymbol{R}_i)\, H_0\, a(\boldsymbol{r}-\boldsymbol{R}_j)\, dv. \tag{25.9}$$

On account of the periodicity of $H_0$, and the way in which the $a$'s differ from each other only by a lattice translation, it is obvious that $(H_0)_{ij}$ depends only on $\boldsymbol{R}_i - \boldsymbol{R}_j$, the distance between the two cells concerned, not on the absolute values of $\boldsymbol{R}_i$ and $\boldsymbol{R}_j$ separately. Let us denote $(H_0)_{ij}$ by

$$(H_0)_{ij} = \mathcal{E}(\boldsymbol{R}_i - \boldsymbol{R}_j) \tag{25.10}$$

where we have one value $\mathcal{E}$ for each lattice position in the crystal. Then we can rewrite Eq. (25.8) in the form

$$\varepsilon(\boldsymbol{k}) = \sum_{\boldsymbol{R}_l} \exp(-i\boldsymbol{k}\cdot\boldsymbol{R}_l)\, \mathcal{E}(\boldsymbol{R}_l). \tag{25.11}$$

In Eq. (25.11) we express $\varepsilon(\boldsymbol{k})$ as a periodic function in $\boldsymbol{k}$ space, by means of a FOURIER series of the terms $\exp(-i\boldsymbol{k}\cdot\boldsymbol{R}_l)$, the quantities $\mathcal{E}(\boldsymbol{R}_l)$ being the FOURIER coefficients. This result, taken together with Eq. (25.10), shows that we can find the matrix components of $H_0$ between WANNIER functions by making a FOURIER expansion of the energy of the energy band, as a function of $\boldsymbol{k}$; or alternatively, if we know the matrix components of $H_0$ between WANNIER functions, we can express the energy of the energy band in the form of Eq. (25.11).

We can now rewrite Eq. (25.6) in terms of the values which we have found for the matrix components of $H_0$ and $V$. We have

$$\sum_{\boldsymbol{R}_j}[\mathcal{E}(\boldsymbol{R}_i-\boldsymbol{R}_j)+V_{ij}]\,U(\boldsymbol{R}_j)=E\,U(\boldsymbol{R}_i). \tag{25.12}$$

In Eq. (25.12) we have the set of equations which must be solved, to find the eigenvalues and eigenfunctions of the problem, in so far as the functions can be expanded in a series of unperturbed functions related to a single energy band. We shall later discuss the more general case where we require unperturbed functions coming from all energy bands to get a satisfactory expansion. When we now consider Eq. (25.12), consisting of a very large number of simultaneous equations for the $U$'s, we might feel that it would be hopeless to solve these equations in a practical way. However, we shall now show how WANNIER was able, in certain important cases, to reduce these equations to the form of a differential equation similar to Eq. (24.2).

To lead to WANNIER's method, we can start by noting that though the $U$'s are a set of coefficients, nevertheless we might consider $U(\boldsymbol{r})$ as a continuous function of a variable $\boldsymbol{r}$, such that $U(\boldsymbol{R}_i)$ is the value which this function takes on when $\boldsymbol{r}=\boldsymbol{R}_i$. For instance, in the unperturbed case, where $U(\boldsymbol{R}_i)=\exp(i\boldsymbol{k}\cdot\boldsymbol{R}_i)$, clearly the continuous function $U(\boldsymbol{r})$ is simply $\exp(i\boldsymbol{k}\cdot\boldsymbol{r})$. We might hope that $U(\boldsymbol{r})$ could be determined by a differential equation. It proves to be helpful in setting up this differential equation to consider the one-electron energy $\varepsilon(\boldsymbol{k})$, and to convert this into an operator. We introduce pseudomomentum components $\boldsymbol{p}$ by the equation $\boldsymbol{p}=\hbar k$, and then replace $p_x, p_y, p_z$ by differential operations $-i\hbar\,\partial/\partial x$, $-i\hbar\,\partial/\partial y$, $-i\hbar\,\partial/\partial z$. That is, we replace $k_x, k_y, k_z$ by $-i\,\partial/\partial x$, $-i\,\partial/\partial y$, $-i\,\partial/\partial z$. If we do this in the function $\varepsilon(\boldsymbol{k})$, we have an operator, which can operate on a function of $x, y, z$.

Let us then allow this operator to operate on a continuous function $U(\boldsymbol{r})$. To evaluate the result, we express $\varepsilon(\boldsymbol{k})$ as a FOURIER expansion according to Eq. (25.11). We must then find the effect of an operator $\exp-i\boldsymbol{k}\cdot\boldsymbol{R}_l$, where $\boldsymbol{k}$ is the operator $-i\nabla$, operating on the function $U(\boldsymbol{r})$. By expanding the exponential in power series, and comparing with the formula for TAYLOR's expansion, we find at once that the result is

$$\exp(-\nabla\cdot\boldsymbol{R}_l)\,U(\boldsymbol{r})=U(\boldsymbol{r}-\boldsymbol{R}_l). \tag{25.13}$$

When we now consider Eq. (25.11), we find that

$$\varepsilon(\boldsymbol{k})\,U(\boldsymbol{r})=\sum_{\boldsymbol{R}_l}\mathcal{E}(\boldsymbol{R}_l)\,U(\boldsymbol{r}-\boldsymbol{R}_l) \tag{25.14}$$

where now $\varepsilon(\boldsymbol{k})$ is an operator. We now see, by comparing Eq. (25.14) with (25.12), that the first terms of the latter equation, $\sum_{\boldsymbol{R}_j}\mathcal{E}(\boldsymbol{R}_i-\boldsymbol{R}_j)\,U(\boldsymbol{R}_j)$, can be rewritten

$$\sum_{\boldsymbol{R}_j}\mathcal{E}(\boldsymbol{R}_i-\boldsymbol{R}_j)\,U(\boldsymbol{R}_j)=\varepsilon(\boldsymbol{k})\,U(\boldsymbol{R}_i) \tag{25.15}$$

where again $\varepsilon(\boldsymbol{k})$ is an operator. [To see the relation, let $\boldsymbol{r}$ in Eq. (25.14) be replaced by $\boldsymbol{R}_i$, and let $\boldsymbol{R}_l$ be replaced by $\boldsymbol{R}_i - \boldsymbol{R}_j$]. It is plain, then, that by use of these relations, we can write Eq. (25.12) in an operator form.

The only case in which WANNIER'S complete method applies is that in which $V_{ij}$ is diagonal, and can be written as a function $V_{ii}$, which can be replaced by a function $V(\boldsymbol{r})$. In the preceding discussion, we have seen that this will be the case if the potential varies slowly with position. Then we can replace Eq. (25.12) by the equation

$$\varepsilon(\boldsymbol{k})\, U(\boldsymbol{r}) + V(\boldsymbol{r})\, U(\boldsymbol{r}) = E\, U(\boldsymbol{r}) \tag{25.16}$$

where again $\varepsilon(\boldsymbol{k})$ stands really for the operator $\varepsilon(-i\nabla)$. In Eq. (25.16) we have WANNIER'S formulation of the differential equation method of solving the difference equations represented by Eq. (25.12). The formulation up to this point is exact, to the same approximation to which Eq. (25.12) is exact: that is, provided we can really expand the excited wave function in series of the wave functions of one band only.

Now we can simplify the problem, in case the energy $\varepsilon(\boldsymbol{k})$ can be expanded in power series, as in Eq. (23.1), so that we can use the effective mass approximation. That is, if we can write

$$\varepsilon(\boldsymbol{k}) = \varepsilon_0 + \frac{\hbar^2 k_x^2}{2m_1} + \frac{\hbar^2 k_y^2}{2m_2} + \frac{\hbar^2 k_z^2}{2m_3} \tag{25.17}$$

to a sufficiently good approximation, Eq. (25.16) reduces to

$$-\hbar^2\left(\frac{\partial^2/\partial x^2}{2m_1} + \frac{\partial^2/\partial y^2}{2m_2} + \frac{\partial^2/\partial z^2}{2m_3}\right) U(\boldsymbol{r}) + \left(\varepsilon_0 + V(\boldsymbol{r})\right) U(\boldsymbol{r}) = E\, U(\boldsymbol{r}) \tag{25.18}$$

which is identical with Eq. (24.2), which we arrived at intuitively. [In Eq. (24.2) the zero of energy was taken to be the bottom of the energy band, equivalent to setting $\varepsilon_0$ equal to zero.] We thus see the way in which WANNIER'S argument allows us to verify this intuitive discussion which we gave at the beginning of the section, and to see how it is to be extended when the effective mass approximation is no longer adequate. In the next section we shall discuss the limitations of this method, extensions to remove the limitations, and more powerful methods for solving the difference equations, Eq. (25.12), in case the equivalent differential Eq. (25.18) is not adequate.

**26. Extension of WANNIER'S method.** We may describe the results of the preceding section in the following way. In a periodic potential perturbed by certain non-periodic forces, the wave function resembles a BLOCH sum of atomic functions (more accurately, WANNIER functions), but the coefficients, instead of being the values of a plane wave $\exp(i\boldsymbol{k}\cdot\boldsymbol{r})$ at the positions of the atoms, are given by the values of a function $U(\boldsymbol{r})$ which satisfies the difference equations Eq. (25.12), which can be approximated in suitable cases by differential equations like Eq. (25.18). We may now ask two questions: how good are the approximations, and what sort of results for $U(\boldsymbol{r})$ do we find in actual problems?

In Sect. 24, we have seen that for impurities in a semiconductor, the solutions of Eq. (25.18) prove to agree quite well with experiment. Let us first inquire why we should expect this. We have stated that the wave functions forming the solutions of Eq. (25.18) are widely extended in space, stretching over many unit cells, which means that the function $U(r)$ varies slowly with position. We could expand this function, not in terms of WANNIER functions, but in terms of BLOCH functions $u_k$. On account of the slow variation with position, the major terms in this expansion come from the BLOCH functions of long wavelength, or those lying near the minimum of the energy $\varepsilon(\boldsymbol{k})$. Near this minimum, the expansion of Eq. (25.17) is justified, and consequently we expect Eq. (25.18) to hold. We see, however, that if the wave function were much more concentrated, we should need large contributions from all the BLOCH functions in the energy band to describe the wave function. In other words, the behavior of $\varepsilon(\boldsymbol{k})$ high above the minimum of the band would be important, and in this case in general we could not expect the quadratic approximation of Eq. (25.17) to be satisfactory. Consequently for more concentrated wave functions, which as we shall see can come from potentials concentrated on a single atomic site, the method of Eq. (25.18), using a differential equation for $U(r)$, is not adequate. We shall now show how quite a different method of solution can be used for such concentrated wave functions, allowing us to get a more general view of the type of wave functions and energy levels to be found for crystals containing impurity atoms.

The method which we shall use, suggested by the writer and KOSTER [*139*], [*141*] works most simply for the case where the matrix components $V_{ij}$ of the perturbative potential are all zero, except the one for which both $i$ and $j$ refer to the central unit cell, a special case which we mentioned in our discussion in Sect. 25. For this particular problem, it proves to be more convenient to expand the perturbed wave function, not in terms of the WANNIER functions, but in terms of the original functions $u_{\boldsymbol{k}}$, as described in Eqs. (25.3) and (25.4). That is, we let

$$\psi = \sum_{\boldsymbol{k}} F(\boldsymbol{k})\, u_{\boldsymbol{k}}(\boldsymbol{r}) \tag{26.1}$$

where the coefficients $F(\boldsymbol{k})$ of the $N$ functions $u_{\boldsymbol{k}}$ are to be determined, so as to minimize the expectation value of the HAMILTONian. For this purpose, we need the matrix components of the HAMILTONian $H_0 + V$ between the functions $u_{\boldsymbol{k}}$. As far as $H_0$ is concerned, the HAMILTONian is diagonal, the diagonal component being simply $\varepsilon(\boldsymbol{k})$. The matrix components of $V$ can be found by using the expression of Eq. (25.4) for $u_{\boldsymbol{k}}$, and taking advantage of the fact that only one of the matrix components of $V$ with respect to the WANNIER functions, as expressed in Eq. (25.7), is different from zero, namely that for which $i=j=0$; we may denote that component by $V_0$. We then find that all matrix components of $V$ between two functions $u_{\boldsymbol{k}}$ and $u'_{\boldsymbol{k}}$ are equal to each other, each being $V_0/N$. It is this simplifying feature which makes a solution of this problem possible.

With this information about the matrix components of the HAMILTONian between functions $u_{\boldsymbol{k}}$, we can set up the equations determining the $F(\boldsymbol{k})$'s. These equations are

$$\varepsilon(\boldsymbol{k})\, F(\boldsymbol{k}) + \sum_{\boldsymbol{k}'} (V_0/N)\, F(\boldsymbol{k}') = E\, F(\boldsymbol{k}) \tag{26.2}$$

where the summation over $\boldsymbol{k}'$ extends over all $N$ values in the BRILLOUIN zone. We can immediately solve these equations by inspection, finding

$$F(\boldsymbol{k}) = \frac{A}{E - \varepsilon(\boldsymbol{k})}, \qquad A = \sum_{\boldsymbol{k}'} (V_0/N)\, F(\boldsymbol{k}'). \tag{26.3}$$

From Eq. (26.3), we see that the coefficient of a wave function $u_{\boldsymbol{k}}$, in the expansion of the perturbed wave function, is proportional to $1/[E-\varepsilon(\boldsymbol{k})]$, decreasing as the energy of the wave function $u_{\boldsymbol{k}}$ departs from the eigenvalue. To find the eigenvalue $E$, we substitute for $F(\boldsymbol{k}')$ from the first of Eqs. (26.3) in the second equation, obtaining

$$V_0/N = \left\{\sum_{\boldsymbol{k}'} \frac{1}{E - \varepsilon(\boldsymbol{k})}\right\}^{-1}. \tag{26.4}$$

In Eq. (26.4), we have a relation for determining the eigenvalues $E$ for a given value of $V_0$. The equation can be more conveniently solved in the reverse order, finding $V_0$ as a function of $E$. We need to known merely $\varepsilon(\boldsymbol{k})$, the spectrum of eigenvalues of the uperturbed problem. The summation over $\boldsymbol{k}'$ extends over the $N$ discrete values of $\boldsymbol{k}'$. We now find in actual cases that the relation between $E$ and $V_0$ resembles that given in Fig. 43. This figure is given for a three-dimensional crystal with simple form for energy as a function of $\boldsymbol{k}$. We see that any vertical line (fixed value of $V_0$) cuts the curve in $N$ points, giving the $N$ eigenvalues of the problem. If $V_0$ numerically is less than a certain critical value, these values are close to the $N$ eigenvalues of the unperturbed problem. For $V_0$ greater than this critical value, however, one eigenvalue separates from the band, forming a discrete state. This discrete state separates from the top or bottom of the band, depending on the sign of $V_0$.

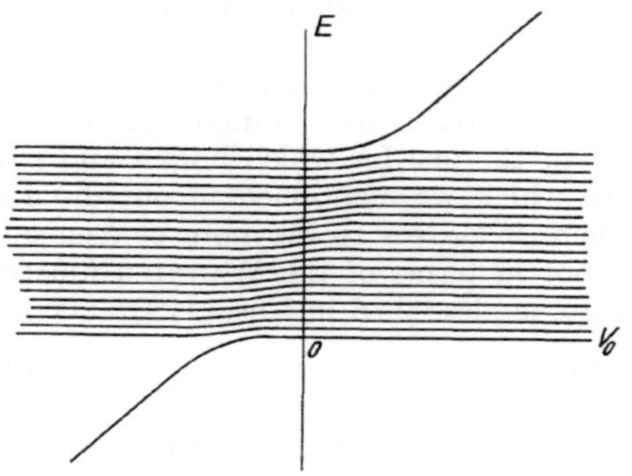


Fig. 43.

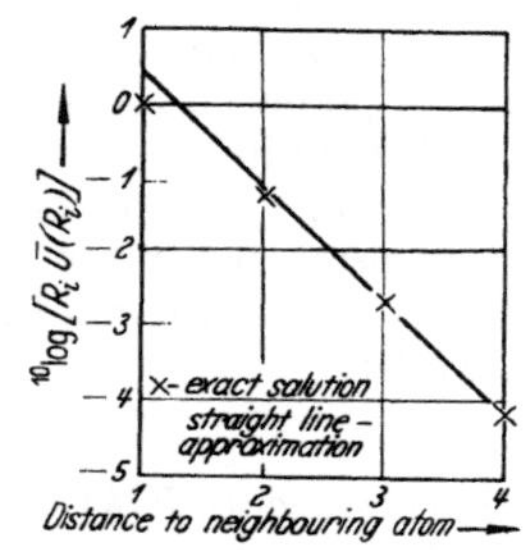


Fig. 44.

Fig. 43. Energy levels as a function of perturbing potential $V_0$, for an energy band with a perturbation at one atom, showing how a discrete level splits off from the band at a finite value of $V_0$, following KOSTER and SLATER [141].

Fig. 44. Coefficients $U(\boldsymbol{R}_i)$ of WANNIER functions in the exact solution of the perturbed potential problem, as compared with the solution $\exp(-a r)/r$ of the corresponding differential equation. Quantity plotted is $\log[R_i\,U(R_i)]$, for $\boldsymbol{R}_i$ along 100 direction (in simple cubic crystal), which would be linear function of $R_i$ for the exponential approximation. Points represent values for the exact solution, straight line the best exponential fit. Point for central atom cannot be plotted in this way. Computed [141] for case where discrete level is separated from the band by $\frac{1}{8}$ width of band.

One can investigate the nature of the wave functions of the discrete state and the remaining states in the continuum, so as to understand what is going on. For orientation, let us ask what would be the corresponding problem if we were dealing with the differential equation approach to the problem. We could obtain a similar case by using Eq. (25.18), and by assuming that $V(r)$ was a square-well potential; that is, that $V(r)$ was zero outside a small sphere of radius $R$, and had a constant value, positive or negative, inside this sphere. The case of negative potential energy is the one which splits a discrete state off the bottom of a band, in which case one must use Eq. (25.18), while with positive $V_0$ the level is split off the top of a band, like a valence band, and one must use negative effective masses in Eq. (25.18). Now one remembers that it is a standard problem in quantum mechanics to investigate the solutions of SCHRÖDINGER's equation for a square-well potential of this type. One finds very simply that there are no bound states unless the well achieves a certain size, as determined jointly by its radius and the value of $V$ within it. For wells larger than this, a bound state appears, and its energy departs more and more from the bottom of the continuum, the greater the magnitude of the well. One finds, furthermore, that the wave function outside the well falls off as $\exp(-a r)/r$, where $a$ is a constant which is greater, the greater the energy separation between the discrete state and the continuum. That is, the wave function falls off more rapidly as we leave the potential well, the greater the magnitude of the well.

One can now show, as in the references above, that the exact solution of the periodic potential problem with the perturbation at a single atom shows these same properties. In Fig. 43 we have seen how the discrete level splits off with increasing strength of the perturbation. One can investigate the wave function as well, and it is found that the expression $\exp(-a r)/r$, found by the differential equation method, works quite well qualitatively for the exact solution. This is illustrated in Fig. 44, taken from [*141*], in which we plot the dependence of the correct wave function on distance, as compared with the solution of the differential equation. The discrepancy is large only at the origin, where we should hardly except the differential equation to form a satisfactory approximation.

The effect of the perturbation on the wave functions of the continuum is interesting. As the magnitude of the perturbation $V_0$ increases from zero, the lower wave functions in the band (assuming that $V_0$ is negative, so that the discrete level will split from the bottom of the band) start to become concentrated near the perturbing atom, while the wave functions of the upper levels start to avoid this atom. Once the discrete state splits off, the situation becomes modified, in that all the wave functions of the continuum tend to avoid the perturbing atom, while the discrete wave function alone is large on it.

The case we have described is the simplest one which can be handled by the general approach which we have sketched in this section. However, the method goes far beyond what we have indicated. In the first place, it is by no means always true that we can get a good approximation to the perturbed wave function, by making a linear combination only of the unperturbed wave functions in a single energy band. In general, we should use all unperturbed energy bands in our linear combination. This is particularly important if we are dealing with a case of almost or exactly degenerate bands. For instance, if there should be degeneracy at the energy minimum, so that as we have seen it is impossible to use the effective mass approximation, it is also clearly impossible to get a good expansion of the perturbed wave function in terms of only one of the various bands which become degenerate at the energy minimum. The general method has been set up for handling such a case. Then again, it is a very artificial assumption to disregard all components of the perturbative energy except that between the two WANNIER functions at the origin. We can also take a more general case into account, in which a number of these components must be considered. We then find that the energy, instead of being given by Eq. (26.4), which can be solved directly, is to be found from a secular equation with a number of rows and columns equal to the product of the number of non-vanishing matrix components $V_{ij}$, and the number of energy bands which must be assumed. Examples have been worked out of this more general case.

On the basis of the general discussion which we have just sketched, it is possible to understand the relation between the case of the COULOMB potential, which we have taken up in Sect. 24, and the potential perturbation at a single atom, which we have just been discussing. If one has only a perturbation at a single atom, only one discrete level will split off the continuum. If there are perturbations at more atomic sites, more discrete levels will split off, but it requires a larger perturbative potential to split off the wave functions of higher quantum number. The case of the COULOMB potential is one in which there are perturbations on all atoms of the crystal, and in principle, we have an infinite number of discrete levels. However, most of these have energies which are so close to the continuum that they cannot be distinguished in practice, and they form in fact a part of the continuum.

The difficulty with the hydrogenic solution for the COULOMB potential comes, as we have suggested in Sect. 24, at the origin. Really the wave function of the 1 s-like solution consists of a linear combination of WANNIER functions on all atoms of the crystal, the largest coefficient coming for the WANNIER function at the origin, where the impurity atom is located. Now if we are dealing with an energy level only very slightly below the continuum, as described in Sect. 24, the wave function is widely extended, and the contribution of the WANNIER function at the origin is small, even though it is larger than those from WANNIER functions at other atoms.

We must realize, however, that the differential equation method is not applicable right at the origin. A COULOMB potential goes infinite at $r=0$, though it is obvious that the matrix component of the COULOMB potential between two WANNIER functions at the origin will not be infinite. This matrix component will not involve the dielectric constant, as will the matrix components corresponding to other lattice sites, for the dielectric effect really results from polarizations of the various atoms, and it will not be felt when are considering the interaction of two electrons on the central, impurity atom. Hence in the difference equation, we shall in fact have a large term $V_{ii}$ coming from the central cell, which cannot be approximated at all by means of the COULOMB approximation. If this term becomes large enough, as it would if the atomic number of the impurity atom were different enough from that of the atom which it replaces, we should tend to have a discrete level splitting off widely from the continuum. This would prove to be a modification of the 1 s-like hydrogenic solution, but with an energy so far from the continuum, and a wave function so concentrated around the impurity atom, that the ordinary hydrogenic solution would be completely inapplicable. In other words, we expect that as the impurity atom becomes more and more different from the atom of the crystal, the discrete state, instead of being only slightly below the continuum, will depart widely from it, and will show a very non-hydrogenic wave function. As we shall mention in the next section, there seems to be experimental evidence for this sort of situation.

**27. The self-consistent field in impurity problems: semi-conductors.** In the preceding sections, we have been discussing the problem of finding solutions of SCHRÖDINGER'S equation for a periodic potential modified by a localized perturbation. We have found that in some cases discrete levels are split off from the bottom or top of the energy bands, as a result of such perturbations. We must remember, however, that we are merely engaged in one aspect of a solution of the problem of the self-consistent field, and that the perturbative potential must in fact be determined from the charge distributions of the electrons on the perturbed atom, as well as on the remaining atoms of the crystal. In the present section, we shall consider qualitatively the nature of this self-consistent-field problem.

First we consider a semiconductor. Let us, for instance, think of a germanium crystal with an arsenic atom in place of one of the germanium atoms. We are interested in the wave functions of an electron in the conduction band; for electrical neutrality, there will be one electron in the conduction band for each arsenic atom. We must consider, however, the lower, occupied bands as well, in order to investigate the self-consistent field. The charge difference between germanium and arsenic brings about a change in energy level which is very large compared to the width of the energy band, for all the inner wave functions of the atom, lying below the valence band. From the theory we have been outlining in the preceding sections, the wave functions of the germanium crystal will become modified so as to avoid entirely the impurity atom, and there will be separate wave functions localized around this atom, obviously with energies and wave functions almost precisely as in the isolated arsenic atom, corresponding to all the inner levels. These states will all be filled, resulting in filled inner shells for all germanium atoms, and for the arsenic.

For the valence band, which is fairly wide, the perturbation arising from the extra nuclear charge on the arsenic atom will probably not be great enough to split a discrete level off from the band. Rather, we shall have a modification of the wave functions, such that the lower wave functions of the band will tend to concentrate nearer the arsenic, the higher ones will tend to avoid it. However, all these states will be filled in any case. And in case we are justified in building up these wave functions as linear combinations of the WANNIER functions of the unperturbed valence band, the total charge density in these states will be just the same that we should have had in the germanium crystal. The reason is that we have a certain number of WANNIER functions (actually four per unit cell, on account of the four-fold nature of the valence band in the diamond structure), we are making just the same number of linear combinations of them, by means of a unitary transformation, and then we are letting all of these transformed wave functions be occupied by electrons. It is a well known theorem that the charge distribution so resulting is the same that we should have had if the original wave functions had been occupied by one electron each.

We are left, then, with the same number of electrons on each arsenic atom which we should have on a germanium atom. We must then consider the field in which the one conduction electron moves. It seems intuitively obvious that it will move in a periodic potential, supplemented by the one extra positive charge located on the arsenic. We can get justification for this view by considering the HARTREE-FOCK equations. In Sect. 18, we showed that these equations indicated that a given electron moved in the field of the nuclei, all electrons including itself, minus the field of an exchange charge density, given by Eq. (18.2) This exchange charge will be located largely on the arsenic atom, so that the electric field itself is the field of an arsenic ion with single positive charge.

We are justified in this case, then, in using the type of analysis set up in preceding sections. The electron in the conduction band will have hydrogen-like discrete states with energies slightly below the conduction band, and a continuum of energies in the conduction band, corresponding to wave functions in which it moves through the crystal, though tending to avoid the impurity atom. At the absolute zero of temperature the electron will be bound in the lowest discrete state, a 1s-like state. At higher temperatures, however, it is likely to be excited into one of the higher discrete states, or into the continuum. It is this excitation of the electron which results in a free conduction electron for each donor atom, in the case of a semiconductor.

As another illustration, let us take an impurity atom with smaller atomic number than germanium; say gallium whose nuclear charge is one unit less. Here the situation is very similar to the case of arsenic, just discussed, except that when we come to the valence band, there will be one missing electron or hole for electrical neutrality. It is most convenient in this case to solve the self-consistent problem for the negative ion, in which this hole is filled with an electron, so that we have a filled valence band. Then we use KOOPMANS' theorem to find the energy used in removing one electron to form the neutral crystal. With the filled valence band, there will be the same number of electrons on a gallium atom as on germanium, so that its net electric field will be that of a singly negative ion, repelling neighboring electrons. This repulsive field will split off hydrogenlike discrete levels from the top of the valence band.

The ground state of the neutral crystal will then be one in which a single electron is removed from the topmost level of the valence band; that is, from the 1 *s*-like hydrogenic wave function. Removal of this electron will leave the neighborhood of the impurity atom electrically neutral. The impurity atom itself will be negatively charged, but the hole will be circulating around it at a radius of a relatively small number of atomic distances, keeping electrical neutrality. At higher temperatures, an electron can be excited from one of the continuum states of the valence band to the 1 *s*-like state, leaving the impurity atom negatively charged as before, but with the hole circulating throughout the crystal, as a free particle. This is the mechanism by which acceptor impurity atoms lead to holes in the top of the valence band.

If the atomic number of the impurity atom differs from that of the semiconductor by more than one unit, the situation can be more complicated. For instance, let us consider a copper impurity atom, whose atomic number is three units less than that of germanium. We may consider the ground state of the neutral crystal to be that in which three holes circulate around the copper ion bearing a negative charge of three units, so that it contains as many electrons as a germanium atom. Removal of a single hole to a state in the continuum results in an excited state of the crystal, leading to an acceptor state of the ordinary sort. But in this case, it is found that a second hole can also be removed, leaving the copper atom and its environment not as a singly negatively charged ion ($Cu^{---}$ surrounded by two holes) but as a doubly negatively charged ion ($Cu^{---}$ surrounded by only one hole). Experimental evidence exists, given by DUNLAP [*56*], suggsting that while it requires only a few hundredths of an electron volt to remove the first hole, it requires about 0.3 ev to remove the second, in line with our earlier remarks indicating that successive ionizations of an atom require more and more energy.

These larger ionization energies do not fall in line with the hydrogenic approximation to the discrete energy levels near the impurity atom. The reason presumably is that in such a case, the perturbative potential energy is too great to use this approximation. Near the nucleus of the impurity atom, the perturbative potential will be that of a three-fold negative charge. This is presumably a great enough perturbation to result in wave functions much more concentrated around the impurity than for the hydrogenic case, with energy levels much more displaced from the valence band. Rather than describing the various stages of ionization as in the preceding paragraph, it is really more accurate to proceed in the following way. The wave functions arising from the 4*s* and 4*p* atomic wave functions of germanium and copper, in the germanium crystal with copper impurity atom, will be divided into two classes: energy-band wave functions, with energies like those of an unperturbed germanium crystal, concentrated only on the germanium atoms, enough in number so that if they are all occupied, the germanium atoms are all neutral; and discrete wave functions concentrated almost completely around the copper atom and in its immediate neighborhood. These discrete functions will have an energy level (corresponding to a 4*s* atomic function) of such a height that one electron will be located in such a function, in the ground state of the system, rendering the copper atom electrically neutral. It requires only a small amount of energy to remove an electron from the top of the germanium valence band and attach it to the copper atom, forming a single negative ion, with both 4*s* wave functions of copper filled. It requires considerably more energy to remove a second electron from the valence band and attach it to the copper, forming a double negative ion, with both $4s$ and one $4p$ wave functions of copper filled.

**28. The self-consistent field in impurity problems: metals and alloys.** We have now given a sketch of the situation regarding impurity atoms in semiconductors, and can pass on to the case of metals. Here, as with semiconductors, there is evidence that if the atomic number of the impurity atom differs by only one or perhaps two units from that of the material composing the crystal, the wave functions and energy levels will not be greatly distorted, while if the atomic number differs by three or more units there will be a large distortion of energy levels and wave functions. The evidence in the case of metals comes not from the action of impurity atoms as donor or acceptor atoms, but rather from the properties of substitutional alloys. It is clear that as we replace a few atoms of a crystal by impurity atoms, we are really forming a substitutional solid solution, and properties of the alloy will give us information about the nature of the energy levels and wave functions.

Hume-Rothery [*106*] pointed out many years ago that there were classes of alloys in which the ranges of composition for stability seemed to depend directly on the average number of electrons per atom, and only indirectly on the relative number of atoms of various types. Jones [*120*], [*121*], [*122*] interpreted this situation in terms of energy bands in the following way. He assumed that the energy bands of such an alloy did not depend on which of the two atoms was present at any particular lattice site, but that the number of electrons in the band alone determined the properties. For instance, the brass structures, made up of copper and zinc, show such behavior, and one assumes that each copper atom contributes one electron, each zinc atom two electrons, to the energy band. Particular stability seems to occur, for the structures in question, when there are a definite number of electrons per atom in the band.

Other cases which seem to point in the same direction come from a study of the saturation moment of ferromagnetic alloys. In Sect. 41 we shall take up this question in more detail, but we may state the general situation. In a number of systems of alloys of atoms with atomic numbers differing by one or two units, such as Ni—Cu, Co—Ni, and Co—Cu, the saturation magnetization interpolates smoothly between the values characteristic of the pure elements. This can be interpreted as if the alloy had a saturation magnetization characteristic of an energy band which was filled up to the point which would be found by assigning the electrons of both types of atoms to the same energy band. Both these cases, then, indicate that in some cases the energy bands of an element are not greatly changed by alloying it with atoms of nuclear charge differing by one or two units. On the other hand, if the alloying atom has a nuclear charge differing by three or more units, the magnetic properties behave quite differently, and can be interpreted in terms of energy levels split off from the energy band by the presence of the impurity atoms.

These phenomena are just what we should expect from the discussion of Sect. 26, as has been pointed out by Friedel [*72*], [*73*], using the Thomas-Fermi method. In the first place, we note that the situation regarding the perturbative potential surrounding an impurity atom will be quite different in a metal from what it is in a semiconductor. Mott [*168*], using a simple method based on the Thomas-Fermi equations, has shown that the conduction electrons of a metal will completely shield the Coulomb potential arising from an impurity atom, at distances much greater than an interatomic spacing. Hence we have the case of a perturbation which is confined to a single atomic site, which we took up and solved completely in Sect. 26. We saw there that for a small perturbation, no discrete states will be split off from the energy band. The energy levels will be almost precisely unaffected by the perturbation, though the wave functions will be considerably modified. This situation is what seems to exist when the impurity atom has a charge differing by one or two units from the rest of the material. The modification of the wave function will always be of such a nature that the impurity atom tends to be electrically neutral, though it may not be precisely so.

For greater differences of atomic number, however, the perturbation appears to be great enough so that discrete states are pushed out of the energy band, either above or below it. Here the discrete states will tend to be occupied or vacant, in such a way as again to result approximately in electrical neutrality. For instance, in a case of a Cr impurity atom in a Ni crystal, the Cr has an atomic number four units less than Ni. Around each Cr atom, there will be localized wave functions whose energy levels will be pushed above the $3d$ band of Ni, since Cr, with a smaller atomic number, will act like a negatively charged impurity, or a positive perturbative potential energy. These discrete energy levels will presumably lie high enough so that several of them will be unoccupied in the ground state of the crystal, leaving each Cr atom on the average with four less electrons than a Ni atom. In such a case, the energy band will lose as many states per Cr atom as there are discrete Cr energy levels arising from the band, and the occupancy of the energy band must be calculated from the remaining Ni atoms, with the modified number of available energy levels. This type of behavior seems to be in agreement with the magnetic properties of such alloys as Ni-Cr, as we shall see in Sect. 41.

**29. The effect of random impurities on energy bands.** There have been a number of studies, among others by JAMES and GINZBARG [*118*], LANDAUER and HELLAND [*145*], and PARMENTER [*176*], of the effect of a large number of randomly arranged impurity atoms on energy bands. If we think of a semiconductor, we have seen in preceding sections that there can be discrete wave functions surrounding an impurity atom, either of the hydrogenic type, or with a more concentrated wave function if the perturbation is greater. If now we have many impurity atoms of the same type, it is possible for their wave functions to overlap, and for their discrete energy levels to broaden into an energy band. If the impurity atoms are arranged regularly on some type of superlattice, this will be a regular band, but if they are arranged at random, it requires further study to find the type of resulting wave functions and energy levels.

An alloy is, of course, a random distribution of atoms of one type in a crystal of another (unless it is an ordered intermetallic compound). Hence, in the preceding section, we have really been considering a problem of the present type. We have seen that if the two types of atoms are very similar, there will not be any discrete energy levels at all, and the energy bands will not be disturbed by the perturbations. In a semiconductor, this will not be the case under any circumstances, since there will always be a hydrogenic wave function with a slight splitting of discrete levels, even in the presence of a small perturbation. Thus in any case in a semiconductor, and for larger perturbations in a metal, we may expect new distributions of energy levels to appear in previously forbidden energy regions, provided we have enough impurity atoms.

To study such energy levels, the writers referred to above have made use of a one-dimensional model. One can take two different types of energy wells, representing two different types of atoms, and arrange them at random along a line. SCHRÖDINGER's equation can always be solved for such a one-dimensional model, and this allows one to study the distribution of energy levels, and compute the curve of $dN/d\varepsilon$, the number of energy levels per unit energy range. For a one-dimensional problem, an impurity atom always results in at least one discrete state, no matter how small its perturbation is, so that there is no analogy in the one-dimensional case to the situation we met in three dimensions, where a small perturbation makes no change whatever in the energy bands.

The results derived by the one-dimensional studies just described are very complicated, and cannot be easily analyzed; the reader is referred to the original papers for more details. As one would expect, if a single impurity atom would have a discrete level within a forbidden gap of the crystal, then if we have many impurity atoms we shall have a corresponding number of discrete levels within the gap. If the impurity atoms are so far apart that their wave functions do not overlap, and are identical, these discrete levels will all coincide in energy, but if the wave functions overlap, the levels will become perturbed, broadening into a more or less continuous distribution of levels filling part of the gap. The situation is different, depending on how the energy bands of the pure crystals formed from either type of atom compare with each other. Since the one-dimensional models which have been used differ so widely from actual three-dimensional crystals, it is hard to draw definite conclusions regarding the energy bands of real crystals containing many impurities, but it is clear that this is an important problem which in principle can be handled by the methods of the present chapter.

## VI. Crystalline binding energy, correlation, and elastic properties.

**30. The binding energy of a crystal.** In most of the preceding sections, our main interest has been in the behavior of single electrons in solids. We have concentrated on the self-consistent field method, and the way in which it leads to one-electron wave functions and one-electron energy levels. For most problems in the optical and electrical properties of solids, we are interested in the interaction of the individual electrons of the solid with external electromagnetic fields, and such a one-electron treatment is just what is most appropriate. Even in problems involving the energy of the crystal as a whole, as for instance in problems of electronic specific heat, we have been able to treat separately the various electrons, simply on account of the fact that so few electrons are excited at ordinary temperatures that their interactions can be disregarded.

In the present chapter, on the contrary, and in the two following chapters, we shall be considering properties of the solid in which the interaction of electrons is of fundamental importance, and in which we are concerned with the total energy of the crystal, as a function of various parameters. The present chapter will deal with elastic energy: the total energy of the crystal, as a function of volume and other geometrical parameters. Our attention will be focussed on the part of the energy dependent on the electronic motion, rather than on the vibrations of the atoms, since the latter topic is handled elsewhere in this encyclopedia. Thus we may consider our present chapter to deal largely with elastic behavior at the absolute zero of temperature, the nuclear vibrations being neglected, though we shall supplement this by considering the type of elastic vibrations which the crystal can perform, since these arise on account of elastic forces resulting from the electronic structure of the solid. We shall, however, omit discussion of the relation of these elastic vibrations to the thermal properties of the crystal.

Our task may be viewed in a simple way: we wish to find the total energy of a crystal (electronic kinetic energy, electrostatic energy of interaction between all charged particles, both electrons and nuclei) as a function of the nuclear positions, considering that the nuclei are held fixed. In principle, we know just how to do this. We set up the electronic wave functions, and find the expectation value of the HAMILTONian operator over the wave function. We have used such methods, in the special case where the wave function can be written as a determinantal combination of one-electron wave functions, in our discussion of the HARTREE-FOCK method in Chapter III. Unfortunately, this calculation, though simple in principle, is in practice so difficult that it has hardly been carried through for any actual crystal. Consequently one is forced to make various approximations to make the calculations possible.

As a first step in the direction of investigating these approximations, it is well to take a general look at the problem of the calculation of the energy of a crystal from its electronic wave function; by doing so, we can recognize certain terms in the energy which, though hard to calculate rigorously, still have a sufficiently simple meaning so that we can approximate them in plausible ways. In the present section, we give such a general discussion.

Let the electronic wave function be $\psi(x_1 \dots x_N)$. Here $x_1 \dots x_N$ symbolize the coordinates, and spins, of the $N$ electrons. The HAMILTONian, as given in Eq. (17.2), is $H_0 + \sum_i f_i + \sum_{\text{pairs } ij,\ i \neq j} g_{ij}$, where we remember that $f_i$ depends on the coordinate (and spin) of the $i$-th electron only, and $g_{ij}$ on the coordinates (and spins) of the $i$-th and $j$-th. On account of the antisymmetry of the wave function, the expectation value of any one of the operators $f_i$ will be the same, and similarly that of any one of the $g_{ij}$'s. Hence we have

$$(\text{Energy})_{\text{av}} = H_0 + N \int \psi^* f_1 \psi \, dv + \frac{N(N-1)}{2} \int \psi^* g_{12} \psi \, dv \qquad (30.1)$$

where the integration over $dv$ includes summation over spins. The term $N \int \psi^* f_1 \psi \, dv$ represents the kinetic energy of the electrons, and their potential energy in the field of the nuclei. Let us consider the latter quantity, which arises from the part of $f_1$ representing the potential energy of the first electron in the field of the nuclei. Since that quantity depends on the coordinates of the first electron only, we may first integrate $\psi^* \psi$ over the coordinates of all electrons except the first. The result will be a function of the coordinates of the first electron only, and will represent the probability of finding that electron in a volume element $dv_1$. Since all electrons have the same charge distribution, the electronic charge of all electrons is $-Ne \int \psi^* \psi \, dv_2 \dots dv_N$. Hence we see that the term of Eq. (30.1) coming from the interaction of electrons and nuclei is simply the product of the electronic charge density, and the nuclear potential, integrated over all space. That is, this term is to be computed by classical electrostatics, and will be found correctly from an approximate wave function, to the extent that the wave function leads to the correct charge density of electrons.

More interesting is the last term of Eq. (30.1), coming from the COULOMB interaction between electrons. It would be interesting to put this term in a form $\frac{1}{2} \int \varrho \, \Phi \, dv$, where $\varrho$ is the electronic charge density, $\Phi$ the electrostatic potential arising from the electrons, and the integration is over the coordinates of a single electron; for this would be the expression according to electrostatics for the energy of interaction of the electronic charge distribution with itself. The writer has shown how to do this [*219*], [*225*]. The electronic charge density, as we have just seen, equals $-Ne \int \psi^* \psi \, dv_2 \dots dv_N$. This is written as a function of the coordinates of electron 1. The last term of Eq. (30.1) can then be put in the desired form, if we assume that the potential $\Phi$ can be written in the form

$$\Phi = \frac{-(N-1)\,e\int \psi^*\,\psi\,(1/r_{12})\,dv_2\ldots dv_N}{\int \psi^*\,\psi\,dv_2\ldots dv_N}. \tag{30.2}$$

We can regard this as the potential, at the point $x_1$, arising from $(N-1)$ electrons, distributed in space with coordinates $x_2$ according to the density function

$$\varrho'(x_2) = \frac{-(N-1)\,e\int \psi^*\,\psi\,dv_3\ldots dv_N}{\int \psi^*\,\psi\,dv_2\ldots dv_N}. \tag{30.3}$$

We verify at once, by integration over $x_2$, that the charge density $\varrho'$ does in fact integrate to the charge of $N-1$ electrons.

We see, then, that the potential $\Phi$ acting on the electrons is not that arising from the actual charge density

$$\varrho(x_1) = -N\,e\int \psi^*\,\psi\,dv_2\ldots dv_N \tag{30.4}$$

which integrates to $N$ electrons, but that it arises instead from the modified charge density of Eq. (30.3), integrating to $N-1$ electrons. It is obvious that some such modification of classical electrostatics must be present, on account of the simple fact that an electron does not act electrostatically on itself. We must now investigate the difference between the actual charge density $\varrho(x_1)$, and the effective charge density $\varrho'(x_2)$ of electrons acting on the first electron. By analogy with the discussion of Chapter III, we should expect that $\varrho'$ would be equal to $\varrho$, but diminished by the charge of a single electron removed from a hole localized in the neighborhood of the point $x_1$ where the electron is located which is being acted on by the charge distribution $\varrho'$. We can easily see that this is the case.

We note first that $\varrho'(x_2)$ depends parametrically on $x_1$; this expresses the fact that the other electrons rearrange themselves, according to the position of the first electron of coordinates $x_1$. When $x_2$ equals $x_1$, we can make definite statements about $\varrho'$. Equality of $x_2$ and $x_1$ implies not only that the two electrons are at the same point of space, but also that they have the same spin. Under these circumstances, on account of antisymmetry, the charge density given by Eq. (30.3) is zero. In other words, at the position of electron 1, no charge density of other electrons will be found corresponding to the same spin as electron 1. The situation is qualitatively similar to that of Chapter III, where an electron was acted on by all electrons, minus an exchange hole, consisting of electrons of the same spin as the electron in question, and having just such a charge density at the position of electron 1 as to cancel completely the density of charge corresponding to that spin at point $x_1$.

We find, in fact, that if we approximate the wave function of the whole system of electrons by a single determinant of one-electron orbitals, the charge density $\varrho(x_1)$ is the sum of the charge densities of all one-electron orbitals, and the effective charge $\varrho'$ acting on the first electron equals the total charge, minus the charge density of the exchange distribution of charge defined by Eq. (18.2). In other words, the electrostatic energy of interaction between the electrons is equal to the energy of the charge distribution of all electrons, acted on by a charge distribution given by all electrons minus the exchange hole, provided we can use a determinantal approximation to the wave function.

We are now in position to see why this single determinant does not provide a very good approximation to the actual wave function of the electrons in a solid. The effect of the exchange hole is to remove electrons of the same spin as the electron in question from its neighborhood, and this effect results in a decrease of electrostatic potential energy. But we can be quite sure that there is another effect, which such a determinantal wave function is not able to describe: we can be quite sure that electrons of the opposite spin to the one in question will also tend to avoid it, not on account of the exclusion principle, but merely on account of electrostatic repulsion. Such an effect is found in the very few two-electron problems which have been solved accurately in quantum mechanics, such as HYLLERAAS' solution [*114*] of the helium atom, and JAMES and COOLIDGE'S solution [*117*] of the hydrogen molecule. In both of these accurate wave functions, even though the two electrons have opposite spin, it is necessary to include in the wave function terms in $r_{12}$, the distance between the electrons, of such a nature that the electrons are not found in each other's neighborhood as often as they would be if they moved independently. This effect, by which an electron tends to keep electrons of opposite spin out of its neighborhood, is called correlation, and its effect on the behavior of a solid is very important.

We have just seen that a single determinantal wave function cannot include correlation. The reason is straight-forward. The self-consistent field method assumes that each electron moves independently of the others, and there is no mechanism in it for one electron to keep another out of its way, aside from the automatic effect arising from the exclusion principle, which only affects electrons of the same spin. On the other hand, as we shall see later, the effects of correlation are quantitatively important. Hence it is very important to correct the single determinantal wave functions, if we wish to get accurate values for the electronic energy. Several possible methods suggest themselves for making such a correction.

The first of these methods is that of configuration interaction, which will be discussed more in detail in the next chapter. By making a linear combination of a sufficiently large number of determinantal wave functions formed from different one-electron orbitals, one can approach the true wave function as closely as one pleases, and consequently configuration interaction, the superposition of such determinantal wave functions, is one way of handling correlation. We shall see that this method is quite practicable in one case, that of solids held together by covalent bonds, in which correlation is extremely important; and it points the way in another case, that of ferromagnetism. However, for such a simple case as an alkali metal, in which the electronic wave functions have many analogies to plane waves, it is not particularly suitable.

There are two approximate methods for handling correlation, which have been suggested for the problem of free electrons, a problem in many ways analogous to that of the alkali metals; one of these was suggested by WIGNER and SEITZ, the other by BOHM and PINES. We shall take up these two methods in the two following sections. In both of these methods, it is assumed that the simple determinantal method is adequate for the calculation of the electronic kinetic energy, the potential energy in the field of the nuclei, and the exchange correction; only the correlation correction, which is entirely absent in the determinantal method, is handled by other methods. We shall now go on to a brief discussion of these two methods of handling correlation, and of some of the results which have been obtained by their use.

**31. Correlation energy, WIGNER-SEITZ treatment.** WIGNER and SEITZ [*255*] originated the method which bears their name for finding one-electron orbitals for alkali metals, and at the same time carried out calculations of the total energy of the crystal; their principal aim was to find the cohesive energy, the difference between the energy of the crystal at its normal interatomic distance, and the energy of the separated atoms. They first used the cellular method, which we have described in Sect. 10, to find the energy of the lowest orbital in the conduction band for sodium. They made an estimate, by a method which we have not gone into, of the effective mass, which proved to be very close to the free electron mass. They then assumed that the conduction electrons could be replaced by a free-electron gas of appropriate density. From the one-electron energies so determined, they could find the average kinetic energy of the electrons, and their average potential energy in the field of the nuclei, for these quantities appear in the one-electron energies. They used a free-electron calculation of the exchange energy, such as we have mentioned in Sect. 18. This left only the correlation energy which had to be estimated, in order to get a complete calculation of the energy of the crystal. They were concerned only with the conduction electrons and their interaction with the atomic core, since the electrons of the core lie so far inside the atom that they are not affected, in sodium, by the condensation of the atoms to form a crystal, and their terms in the energy will subtract out when one finds the cohesive energy.

It then seemed reasonable to WIGNER and SEITZ to assume that as far as the correlation energy was concerned, their problem was equivalent to that of a real free-electron gas, a gas of electrons moving in a region filled uniformly with positive charge, of just such a density as to neutralize the electrons, resulting in no net electric field. Their problem was to estimate what one might call the correlation hole: the charge distribution of electrons of opposite spin to the one in question, removed from its immediate neighborhood, and of course redistributed at greater distance from the electron.

To discuss the motion of the electrons, WIGNER and SEITZ assumed that the wave function of electrons of plus spin could be computed as if they moved in the field of the positive charge (which they assumed to be uniformly distributed), of the other electrons of plus spin (which they handled by the determinantal method, making use of an exchange hole), and of the electrons of minus spin; and they approximated by assuming that the electrons of minus spin were fixed in a certain arrangement, so that the wave function of the electrons of plus spin would contain the coordinates of the electrons of minus spin as parameters. At large internuclear distances, they assumed that the electrons of minus spin would tend to be arranged on a regular lattice, which they took to be body-centered cubic. They could then approximate the solution of the motion of the electrons of plus spin in the field of the uniform positive charge, the uniform charge of electrons of plus spin, and the lattice-like arrangement of the electrons of minus spin.

By proceeding in this way, using the perturbation theory to estimate the wave function of the electrons of plus spin in the field just described, they arrived at an expression for the correlation energy. This expression, in a form later derived by WIGNER [*253*], [*254*], was

$$\text{Correlation energy} = -\frac{0.58 \text{ Rydbergs}}{r_s + 5.21}. \tag{31.1}$$

Here $r_s$ is the radius of a sphere containing one atom, expressed in BOHR units, and the quantity 5.21 is expressed in BOHR units. The correlation energy given in Eq. (31.1) is the amount per atom, computed for an alkali metal, for which the calculations are designed. By combining the correlation energy as given by Eq. (31.1) with the other terms in the energy, WIGNER and SEITZ obtained surprisingly accurate values for the cohesive energies of the alkalies. Many calculations have been made using this method, including those of BARDEEN [*8*], [*9*], [*10*] on the alkali metals, HERRING [*94*], [*97*] on Li, RAIMES [*191*] to [*194*] on a number of metals, and DONOVAN [54] on Be. From their success one may conclude that Eq. (31.1) is well justified by its agreement with experiment. Nevertheless, one must admit that its derivation leaves a great deal to be desired. The assumption that the positive ions are uniformly distributed is of course not at all true; the wave function cannot really be treated properly by handling the electrons of plus and minus spin in quite different ways; perturbation methods were used in the derivation, in a way that can be questioned. It is probably better to regard the calculation of WIGNER and SEITZ as a lucky guess, more accurate than its derivation suggests.

**32. BOHM and PINES' theory of long-range correlation.** Another treatment of electron correlation, very different in its approach, and yet leading to final results not unlike that of WIGNER and SEITZ, has been given by BOHM and PINES [23], [24], [25], [186], [187]; additional work along the same lines has been done by WOLFF [261] and ADAMS [2], [3]. They started by thinking of the classical behavior of a plasma; that is, of a region in a gas discharge containing positive ions, and electrons, of such densities as to remain electrically neutral. It is well known that, if the positive ions remain fixed, the electrons as a whole can sustain sinusoidal oscillations of a definite frequency, called the plasma frequency. This frequency can be derived by the most elementary methods. Thus, let us assume that there is a longitudinal wave, like a sound wave, arising from the motion of electrons. For instance, let the displacement of an electron along the $x$ direction be given by the real part of the expression

$$x = x_0\, e^{i(kx-\omega t)}. \tag{32.1}$$

This will result in a sinusoidal charge density, which can be determined from the equation of continuity. If there are $N$ electrons per unit volume, each of charge $-e$, the flux will be $-Ne\dot{x} = -Ne(-i\omega)\, x_0 e^{i(kx-\omega t)}$ along the $x$ axis. The divergence of the flux will equal the negative time derivative of the charge density, so that we have

$$\frac{d\varrho}{dt} = -\operatorname{div}(-Ne\dot{x}) = (-ik)(-Ne)(-i\omega)\, x_0\, e^{i(kx-\omega t)} \tag{32.2}$$

from which the charge density is

$$\varrho = ikNe\, x_0\, e^{i(kx-\omega t)}. \tag{32.3}$$

By POISSON's equation, this charge density will result in an electric field given by $\operatorname{div} E = 4\pi\varrho$ (if we use GAUSSian units, and assume a unit dielectric constant). From this, and Eq. (32.3), we have

$$E = 4\pi Ne\, x_0\, e^{i(kx-\omega t)}. \tag{32.4}$$

But the force on an electron must be $-eE$, and this, by NEWTON's law, must equal $m\, d^2x/dt^2$, where $m$ is the electronic mass. Hence we have

$$4\pi Ne^2 x_0 = m\omega^2 x_0, \qquad \omega = (4\pi Ne^2/m)^{\frac{1}{2}}. \tag{32.5}$$

This frequency of plasma oscillation, we note, is independent of the propagation constant $k$ of the wave, though a more elaborate treatment results in a slight dependence of $\omega$ on $k$.

BOHM and PINES suggested that a similar situation should hold for the electrons of a metal. It seemed plausible, then, somehow to analyze the motion of the electrons into a set of plane waves of the type just described, and to describe their motion in terms of these plane waves. They showed how to introduce quantum-theoretical variables describing the amplitudes of the plane waves. Naturally, the sinusoidal vibrations which we have just been describing will be quantized, and it becomes important to know whether they will be excited at ordinary temperatures. When we substitute for $N$ in Eq. (32.5) the number of conduction electrons per unit volume in a metal, we find that the plasma frequencies correspond to energies of the order of ten electron volts. Hence we must conclude that these oscillations will not be excited at all at ordinary temperatures.

This lack of excitation of the plasma modes of oscillation has a definite meaning regarding electron correlation. If there were no correlation, there would be fluctuations of electron density, which could be expressed in terms of the oscillations which we have just set up. If there were no excitation of these oscillations at all, there could be no density fluctuation. The only fluctuation allowed, according to the method of BOHM and PINES, is that arising from the zero-point excitation of the modes of oscillation, and this proves to be much smaller than we should find from a theory of a free-electron gas neglecting correlation. We have, then, a plausible argument for discarding the type of charge fluctuation which would arise from a simple theory.

The situation proves to be not quite as simple, however, as this sketch would suggest. BOHM and PINES have shown that there is good reason to use the type of analysis just sketched for long waves, with $k$ less than a critical value $k_c$. For shorter wave lengths, however, they show that a different method of treatment must be used. They were led to this idea of a critical wave length from the classical plasma. In that case, it is found that there is a critical distance, called the DEBYE length, such that in case the product $k\lambda_D$, where $\lambda_D$ is the DEBYE length, is small compared to unity, one can have plasma oscillations of the type described, but waves for which the product is large compared to unity, or for which the wave length is small compared to the DEBYE length, cannot exist. They were led, by analogy with this argument, to the idea of a critical $k_c$, with a related critical distance inversely proportional to it, measuring the shortest possible wave length of plasma oscillations.

One has, then, a number of coordinates representing the amplitudes of the sinusoidal oscillations of propagation constant less than $k_c$. The number of such coordinates proves to be rather small compared to the total number of coordinates required to describe the electrons, both because only long wave lengths are represented, and because one has only longitudinal waves. We must then have many more electronic coordinates left over, but not as many as $3N$, if $N$ is the number of electrons. Alternatively, if we choose to use $3N$ electronic coordinates, there will be constraints imposed on them, equivalent to stating that the vibrational modes are not excited. When the nature of these constraints is examined, it appears from the work of BOHM and PINES that we can describe the result in simple language. It is as if the electrons were free to move according to interaction forces, but these forces, rather than being COULOMB forces of the usual type, have a screened COULOMB potential of the form

$$\text{Potential energy} = \frac{e^2}{r} \exp(-a r) \tag{32.6}$$

where $r$ is the distance between the electrons, and where $a$ is a constant, of the order of magnitude of $k_c$.

Up to this point, the quantity $k_c$ is left arbitrary. BOHM and PINES have determined this quantity, however, in a very plausible way, by writing the energy of the electrons as a function of $k_c$, and choosing $k_c$ to minimize the energy, hoping thereby to get the best approximate solution of SCHRÖDINGER's equation. When this is done, the quantity $1/a$, the distance in which the exponential in Eq. (32.6) falls to $1/e$ of its value, proves to be of the order of magnitude of interatomic distances.

The physical interpretation of these results is simple, and probably correct, whether one agrees with the analytical details of the treatment of BOHM and PINES or not. We may describe them as follows. If one of the conduction electrons in a metal becomes displaced, it would tend to create an electric field at neighboring points. But the other conduction electrons are so free to move around that they immediately proceed to shield the rest of the metal from the effect of the displaced electron, so that the field arising from its displacement is only felt at distances of the order of magnitude of $1/a$ from the electron, falling

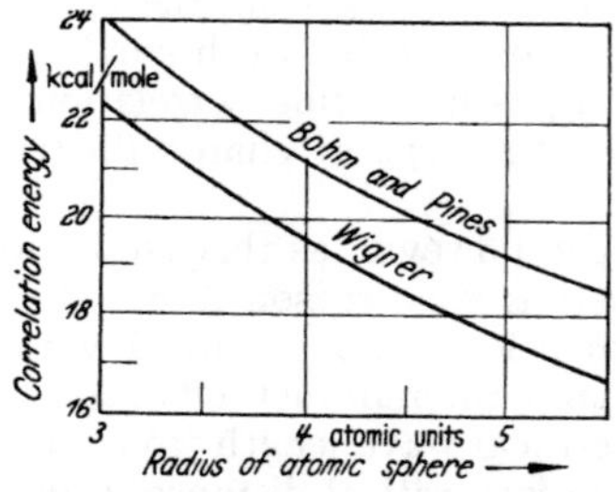


Fig. 45. Correlation energy as function of radius of atomic sphere, comparing calculation of BOHM and PINES [186] with that of WIGNER.

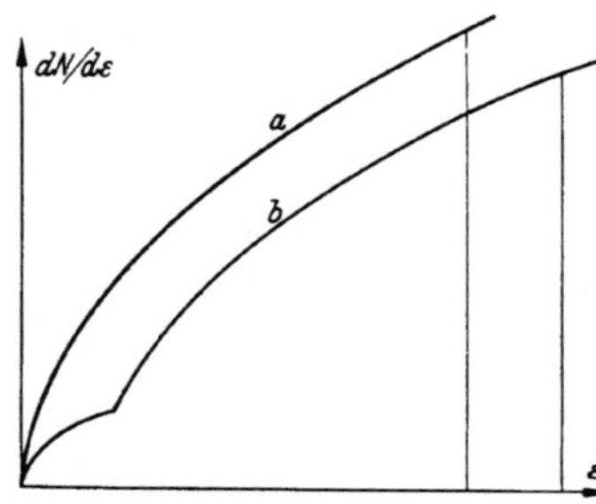


Fig. 46. Density of states $dN/d\varepsilon$ for metallic sodium, (a) according to the free-electron theory; (b) according to the theory of BOHM and PINES. FERMI energies are shown in each case. (Curves from RAIMES [195].)

off exponentially, rather than in the slow manner given by COULOMB's law. This shielding means that the effect of the motion of any electron on the distant neighbors is almost entirely cancelled, and there is no tendency for the electrons to become disturbed, and to set up charge fluctuations. This model is seen to be fundamentally similar to that which MOTT [168] used in showing that an ion in a metal would be shielded, by motion of the electrons, with a shielding length of interatomic dimensions.

From the expression of Eq. (32.6) for the screened COULOMB potential, one can derive the charge density of screening electrons. Since the field of an electron is completely screened at a distance large compared to $1/a$ it is obvious by electrostatics that it must be surrounded by a hole whose charge is just one electronic charge. This must be a combination of the exchange and correlation holes; part of the displaced charge has the same spin as the electron in question, part has opposite spin. By proceeding in this way, BOHM and PINES have been able to deduce the part of the hole arising from correlation, and consequently the correlation energy. The resulting expression cannot be put in a simple analytic expression like that of WIGNER and SEITZ as given in Eq. (31.1). However PINES [186] has given a graphical comparison of the two corrections as a function of $r_s$ which we show in Fig. 45. It is clear that the agreement is qualitatively very good and quantitatively not bad. One concludes from this agreement that it is likely that both calculations, that of WIGNER and SEITZ and that of BOHM and PINES, form tolerably good approximations to the truth, even though one may question the postulates underlying both derivations.

One particularly satisfactory result of the calculations of BOHM and PINES is the way in which they remove the paradox connected with the specific heat of free electrons as computed by the HARTREE-FOCK method, to which we have referred in Sects. 18 and 21. We saw there that the exchange hole for free electrons, as computed straightforwardly, differs decidedly depending on whether the electron in question is at the bottom or the top of the FERMI distribution, and this affects the distribution of energy levels $dN/d\varepsilon$, in a way which seems to bear no resemblance to the experimental facts. PINES points out that this difference in exchange hole arises mostly from electronic interactions at fairly large distances. When the COULOMB potential is replaced by the screened COULOMB potential of Eq. (32.6) the effect almost completely disappears. Thus, we give in Fig. 46 the curve computed by RAIMES [*195*] showing $dN/d\varepsilon$ as found by the simple free electron theory, and also by the method of BOHM and PINES. It is obvious that the BOHM and PINES method agrees rather closely with the free electron theory, giving justification for our feeling that the use of a different potential for conduction electrons of different $\boldsymbol{k}$ values is not physically justified. This result, in other words, furnishes a valuable piece of evidence that the simple picture of electrons moving in a periodic potential really agrees with surprising accuracy with the facts of electronic motion in metals.

**33. Calculation of elastic constants**[1]. In the preceding sections of this chapter, we have sketched methods by which the total energy of a crystal can be calculated, as a function of the nuclear positions. If the crystal is kept geometrically similar to itself, but the atoms are moved closer together or farther apart, and we calculate the energy for different internuclear distances, we have the material for finding the compressibility, for this depends on the second derivative of energy with respect to the lattice spacing. In the work we have described earlier, such calculations of compressibility have been made, and agreement has been obtained with experiment as far as order of magnitude is concerned, though the agreement is far from perfect.

A good deal of effort has been expended as well in computing values for the other elastic constants, the principal work being by FUCHS [*76*], [*77*], [*78*], JONES [*123*], [*124*], [*125*], and WOHLFARTH [*260*]. If we wish, for instance, to find a shear constant, it is clear that we must find the energy of a sheared crystal, as a function of the amount of shear; here again, the appropriate elastic constant will be determined by the second derivative of the energy with respect to the strain. We shall not go into the details of any of these calculations, but can indicate their general nature. Calculations for the alkali metals are based on a free-electron assumption regarding the conduction electrons. If now the crystal is sheared, without change of volume, a number of terms in the energy will be unchanged. Thus the FERMI energy of a free-electron gas depends only on its density, as do the exchange and correlation corrections. The only term in the energy, in fact, which proves to depend on shear, is that arising from the electrostatic interactions of the ions.

In an unsheared cubic crystal, we may approximately replace each WIGNER-SEITZ cell by a sphere, containing a net charge of zero, with sperical symmetry. Such a spherical charge distribution exerts no forces at distant points. Once the crystal is sheared, however, this simple situation no longer exists. The cell occupied by each atom becomes elongated, and will now exert a field at a distant point. The corrections to the electrostatic energy arising from this effect are best calculated by the EWALD method, used for evaluating electrostatic energies in crystals. In this case, we have a crystal which on account of shear is slightly distorted from cubic symmetry. By the EWALD method we can find

[1] See G. LEIBFRIED's article in vol. VII, part 1, of this Encyclopedia for more details.

the electrostatic energy of the distorted lattice of positive ions, immersed in a uniform sea of negative charge. Such calculations have been made in the work cited, and have been shown to lead to the major terms in the calculation of the shear constants of the crystals.

In some cases more complicated than the alkali metals, interesting additional features have been shown to be important in calculating elastic constants. Thus, copper is similar in many ways to an alkali metal, the conduction electrons being much like free electrons. It differs greatly, however, in that the interactions between the inner cores of the atoms are much more important than in the alkalies. The lattice spacing of copper is much less than in the alkalies, and the inner core, containing its $3d$ electrons, is much larger. FUCHS in his treatment of the copper crystal first proceeded by a standard WIGNER-SEITZ method, finding that he got a fairly accurate value of the internuclear distance and a moderately good cohesive energy in this way, treating the interactions of the cores as if they were merely point charges. However, he found much too great a compressibility in this way; experimentally, the atoms were much harder than this simple calculation would suggest. To remedy this, he treated separately the repulsive interaction between the cores resulting from the exclusion principle, as if they were inert-gas atoms. This repulsion increases very rapidly as the internuclear distance decreases. By taking account of this repulsion between inner cores, he greatly improved his calculation of elastic constants.

For the alkali metals and copper, it is fairly well justified to treat the wave functions of the conduction electrons as if they were free electrons. This is by no means the case for some other crystals, however, in which the band structure is very important. Such an example is found in magnesium, where JONES [*124*] has found that the shear constant is strongly influenced by the band structure. We remember that in Mg the FERMI level comes very close to certain symmetry points in the BRILLOUIN zone, as was pointed out in Sect. 16.

JONES found that this situation was affected strikingly by shear, with the result that the shearing constant is very sensitive to the details of the band structure and the hexagonal structure of the crystal.

These examples will suffice to indicate the type of problem which has been attacked in the study of the elastic constants of metals. The details of the calculations would hardly be of interest except to one concerned with the relative sizes of the various terms in the energy expression. One may only say that, considering the crude nature of the approximations which must be made in computing the total energy of a crystal, it is surprising that the calculations of elastic constants have been as successful as they have.

For non-metals, the situation is quite different. For ionic crystals, the well-known BORN-MADELUNG theory was used to discuss the compressibility of crystals long before the quantum theory appeared, making the assumptions that ions acted on each other by COULOMB attractions, and in addition by terms representing the repulsions of their closed shells, which were sometimes expressed by $r^{-n}$, sometimes by $e^{-ar}$. Since such assumptions are discussed elsewhere in this Encyclopedia, we shall not take them up here. Naturally one would wish for a more accurate discussion of such problems on the basis of quantum mechanics, and such a discussion, for the alkali halides, has been given by LÖWDIN [*153*].

LÖWDIN set up a wave function for an alkali halide crystal by using the tight-binding approximation, which we have described in Sect. 9. He then made a straightforward calculation of the expectation value of the HAMILTONian for this wave function, as a function of the internuclear distance. This is a particularly fortunate case for such a calculation, for each ion is really a closed-shell structure, for which a single determinantal wave function is a very good approximation. The calculation of the energy was a major effort, since the various overlap and many-center integrals met in the tight-binding method were really calculated, not neglected or guessed at as has so often been done in similar calculations. The resulting curve of energy versus internuclear distance proved to be in very good agreement with experiment. A comparsion is shown in Fig. 47, in which LÖWDIN plots the calculated pressure-volume relations for NaCl, and compares with experimental data of BRIDGMAN. This calculation is perhaps the most straight-forward application of the quantum mechanics to an equation of state which we have, and the agreement with experiment is striking. LÖWDIN has also made calculations on sodium [*154*], by a somewhat similar method, though his results in that case may be criticized on account of his neglect of correlation.

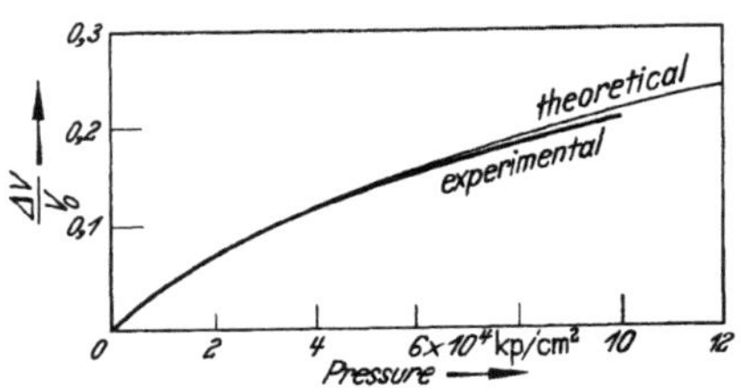


Fig. 47. Change of volume versus pressure, for NaCl, as computed by LÖWDIN (Thesis, Uppsala, 1948), compared with experimental measurement of BRIDGMAN.

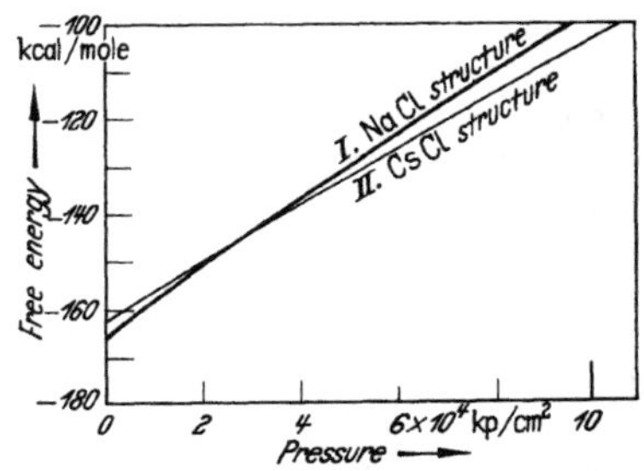


Fig. 48. Free energy of KCl as function of pressure, as computed by LÖWDIN (Thesis, Uppsala, 1948). Curve I, NaCl structure; II, CsCl structure. The observed transition pressure is $2.0 \times 10^4$ kp/cm.². The stable phase has the lower free energy.

One problem related to that of elastic constants, which has been discussed to a considerable extent, is that of the relative stability of different crystalline forms of the same substance. It is familiar experimental fact that a great many solids exist in different crystalline phases in different parts of the pressure-temperature range. It is a natural desire to calculate the free energy of the various phases, to show which one should be most stable, and if possible to predict the transitions. All efforts to do this have encountered a fundamental difficulty: the energy differences between different phases are extremely small compared to the binding energies of the various phases, and if we have difficulty in getting even approximately correct binding energies, it is very dubious whether calculations of the energy differences between different phases have enough accuracy to be of any significance. It is at least encouraging that the calculations check the experimental fact that the different phases have very nearly the same energy; it is this fact which is responsible for the widespread phenomenon of polymorphism. But it is hard to do better than this simple result.

Nevertheless, in a number of cases calculations have been made which really seem to point the way to a theoretical understanding of polymorphism. A good example is the calculation which LÖWDIN [*153*] made, by the methods just outlined, on CsCl and similar crystals. It is known that CsCl crystallizes under ordinary circumstances in a different structure from the ordinary NaCl structure; its structure is what we should find if we placed ions of one type at the lattice points of a simple cubic lattice, with ions of the other type at the body centered positions. Some of the alkali halides, such as RbCl, show transitions from the NaCl to the CsCl structure under pressure. LÖWDIN, using the methods just outlined, has calculated the free energy of crystals of both these types, and has tried to estimate where the transition should come. In Fig. 48 we show his results, compared with experiment; we see that at least qualitatively the prediction is correct.

**34. Elastic vibrations of crystals.** The forces concerned in the elastic vibrations of crystals arise from the electronic wave functions, and hence fall within the scope of this article, though the general problem of elastic vibrations is handled in Vol. VII, Part 1, of this Encyclopedia. We are not interested in the simple approximation used by DEBYE for discussing specific heat, but rather in the more detailed calculations of BORN, BLACKMAN, and many others [*18*], [*19*], [*20*], [*27*], [*68*], [*129*], [*130*], [*161*], [*162*], taking the detailed structure of the crystal into account. Recent work, which we shall mention, permits one to go further with both the experimental and the theoretical study of these vibrations than was possible several years ago.

In the first place, let us examine the microscopic elastic constants of a crystal. We start by assuming that the crystal is at zero temperature; we disregard the zero-point oscillation, and we assume that it is under a uniform state of stress (which may be zero presure), which results in a certain volume, and a correponding shape and size for the unit cell. Under these circumstances, we assume each atom to be in its equilibrium position in the unit cell, the position corresponding to lowest electronic energy. Then we subject one of the atoms in one unit cell to an infinitesimal displacement along an arbitrary direction. We shall find that each atom in the crystal is then acted on by a force, proportional to the displacement of the atom in question. The $x$, $y$, and $z$ components of the force on each such atom, per unit displacement of the displaced atom along the $x$, $y$, and $z$ directions, are all force constants of the crystal, which are independent, except for symmetry relations. These force constants can be found by computing the electronic energy as a function of small displacements of the various atoms; the force constants in question are second derivatives of the energy with respect to the coordinates of the two atoms in question.

In principle, there are an infinite number of such elastic constants. In practice, the force exerted by one displaced atom on another decreases very rapidly as the two atoms are separated, so that only a few of these force constants have appreciable magnitude. We shall later examine practicable ways of calculating them from wave mechanics. We immediately note one fact, however: there are many more microscopic elastic constants of appreciable magnitude than there are macroscopic elastic constants. Thus, we shall quote a case of a copper crystal, in which nine independent microscopic elastic constants have been measured, in contrast to the three independent macroscopic elastic constants of a cubic crystal. One can show that the macroscopic constants are linear combinations of the microscopic constants, so that the macroscopic constants can be deduced from the microscopic ones, but not vice versa. Neglect of this fact has led to many unwarranted simplifications of the problem of elastic vibrations, in an effort to deduce the microscopic constants from macroscopic information.

Once one has set up the energy of the distorted crystal, as a quadratic function of the infinitesimal displacements of the atoms, the constants being these microscopic force constants, one can then discuss the elastic vibrations by well-known methods of classical mechanics. If there are $n$ atoms in the unit cell, and $N$ unit cells in the repeating region of the crystal, there will be $3nN$ normal modes of oscillation. These may be taken to be plane waves, with propagation constants $\boldsymbol{k}$ which are just the same as the propagation constants of the SCHRÖDINGER waves which we have discussed in Chapter II. These $\boldsymbol{k}$ vectors can be represented as points inside the central BRILLOUIN zone, just as for the electronic case, and as in that case, there will be $N$ independent $\boldsymbol{k}$ vectors. For each $\boldsymbol{k}$, there will then be $3n$ independent normal modes. For a general point of $\boldsymbol{k}$ space, these will all have different frequencies, though for points of special symmetry, there may be degeneracy between them. For instance, for a cubic monatomic crystal, for which $n=1$, then in case the propagation vector is along the $x$ axis, there will be one mode which is longitudinal, representing a vibration along $x$, while there will be two transverse modes, with vibration along $y$ and $z$ respectively, degenerate with each other. For a general propagation constant, however, this degeneracy is removed, and the distinction between longitudinal and transverse vibrations is no longer definite.

In general, three modes will have zero frequency for zero propagation constant, and these are the so-called acoustical modes, which are always present, even for the simplest type of crystal, which has only one atom per unit cell. If we have more than one atom per unit cell, there are more modes. The situation is obviously similar to that of energy bands in the electronic problem; we have different frequency bands, so to speak, each one consisting of a single frequency for each $\boldsymbol{k}$ value, so that, on account of the very large number of $\boldsymbol{k}$ values, the frequency is practically a continuous function of $\boldsymbol{k}$. The most familiar case, beyond the very simple one of the monatomic crystal, is that with two atoms per unit cell; in this case, in addition to the three acoustical modes, there will be three more, lying at higher frequency. These additional modes are ordinarily called the optical branch of the vibration spectrum, since the frequencies of all such modes lie in the infrared part of the spectrum.

It is not a hard task to solve for the frequencies of each mode, for arbitrary $\boldsymbol{k}$ value, in terms of the microscopic elastic constants; it demands solving a straightforward secular equation with $3n$ rows and columns, which with modern computing methods is not difficult. We then see that, in principle, it should be possible to work backward, and from measurement of the frequency of a considerable number of normal vibrations, to find the microscopic elastic constants. It has been possible to do this, in practice, from measurements of the thermal diffuse scattering of x-rays[1]. Such experiments have been made for only a few monatomic crystals, of which copper is an example; work on the method has been carried out by CURIEN [*48*], JOYNSON [*128*], JACOBSEN [*115*], and others, based on the work of BORN and BEGBIE [*11*], [*12*], [*26*].

The method of this experiment is simple in principle. Let us assume that our crystal is heated to a temperature high enough so that each normal mode is excited to its equipartition value. Each such mode represents a sinusoidal variation of strain through the crystal, and as such, it is capable of scattering x-rays, according to BRAGG's law. For simplicity of description, let us think of an experiment such as is used to find the LAUE spots in a crystal. We pass continuous x-rays through the crystal, along a symmetry direction. For the LAUE spots, each crystal plane will pick out x-rays of an appropriate wave length, so that they can be reflected by the plane in question, satisfying BRAGG's law $\lambda = 2d\sin\vartheta$, where $\lambda$ is the wave length, $d$ the spacing of the planes, $\vartheta$ the glancing angle of incidence. From the angle of the diffracted spot, we find $\vartheta$, so that we find the angles of the various planes in the crystal. If we had used monochromatic rather than continuous x-rays, we should not ordinarily have had the correct wave length to satisfy BRAGG's law for any of the crystal planes, and no scattering would have occurred.

For the temperature diffuse scattering, we are not interested in scattering by the lattice planes, but rather by the sinusoidal distributions of strain set up

---

[1] Cf. vol. XXXII of this Encyclopedia.

by the elastic vibrations. We proceed as in the preceding paragraph, now using monochromatic x-rays, so that in general no LAUE spots will be present. However, for a given angle $\vartheta$ and the given wave length $\lambda$ of monochromatic x-rays, there will be a vibrational wave whose wave length, which we can identify with $d$, satisfies BRAGG's law. In other words, there will be scattered intensity at each angle $\vartheta$ and this is the temperature diffuse scattering. The scattering at a given angle is produced by waves whose wave length is $d$, traveling in a direction determined by $\vartheta$, and the intensity of scattering is proportional to the intensity of the vibrational wave. In other words, measurement of the temperature diffuse scattering allows us to find the intensity of the vibrational wave of a given propagation constant.

Let us forget for a moment that even in a monatomic crystal there are three modes with each propagation constant, so that the scattering from each of these modes will be scattered at the same BRAGG angle. If it were not for that, we should have the means of determining the vibrational spectrum of the crystal. For by equipartition we know that the energy of each vibrational mode is $kT$. We know the amplitude or intensity of this wave from the measurement. But the relation between the amplitude and the energy of a linear oscillator involves its frequency. Hence from the measurement of scattered intensity we can compute the frequency, as a function of propagation constant.

The difficulty remains that there are three modes, at least, with each propagation constant. This difficulty can be overcome in practice by using particular directions of scattering, and particular polarizations of the x-rays, such that only one out of the three modes will be excited. This method has been used [*115*] to find the frequencies of a considerable number of modes of vibration in the crystal of Cu, and in some other cases. One can then take the theory of the elastic vibrations, in terms of the microscopic elastic constants, and assume that all but a few of these constants are zero. Thus, in the case of Cu, it was assumed that the only non-vanishing constants related to the force exerted by a displaced atom on a nearest, second-nearest, or third-nearest neighboring atom. This resulted in nine independent elastic constants. Enough elastic vibrations were observed to make it possible to solve for these nine elastic constants. In this sort of way, one can experimentally find the microscopic elastic constants which we have described above. The first elementary test to give to the results is to see if the macroscopic elastic constants computed from them agree with experiment; this check proves to be adequately satisfied, in the cases that have been investigated.

It becomes, then, of interest to find if these microscopic elastic constants can be computed from the wave function of the electrons in the crystal. The following method has been used by H. WHITE (unpublished) in a preliminary attempt to do so. Use was made in this calculation of a theorem of FEYNMAN which is very useful in computing the forces on nuclei. FEYNMAN [*66*] showed that if one computes the expectation value of the force on a nucleus, in a problem in which the nuclei of all atoms are held fixed, this expectation value (which is simply the force exerted electrostatistically on the nucleus by all other nuclei, and by the charge distribution of the electrons, distributed according to the SCHRÖDINGER wave function) equals the force which would be found by an alternative method, namely by computing the expectation value of the energy for various positions of the nucleus in question, and by differentiating this energy with respect to the nuclear position. The great advantage of FEYNMAN's theorem is that instead of having to find wave functions for many nuclear positions, and having to find derivatives of this energy, which is usually determined only by numerical methods, one can simply use a single wave function, and use the charge distribution arising from this wave function to find the force.

The way in which this theorem can be used to find the microscopic force constants is the following. One assumes (for a monatomic crystal like Cu) that a single nucleus is displaced, say along the $x$ axis, by an infinitesimal amount. Then one finds the change in the electronic wave function as a result of this distortion. When one has found this distorted wave function, one then proceeds by electrostatics to find the force exerted on each other nucleus in the crystal. In the absence of the distortion, the force would be zero, so that one is interested only in the force exerted electrostatically by the change in the charge distribution. Furthermore, since one is interested only in forces proportional to the displacement of the atom, first-order perturbation theory is rigorously adequate for the calculation. In the calculation mentioned, WHITE has used the methods of Chapter V to find the perturbed wave functions arising from the displacement of an atom, treating the displacement like the sort of perturbation of the potential discussed in Chapter V. He has then computed the force constants, and compared them with those observed by the x-ray diffuse scattering.

The agreement between theory and experiment, in this case, is fairly good, but not strikingly so; we have discussed the method more for its general interest, and the probability that it will later be used in more cases, rather than because it has been very successfully used so far. Part of the complication in copper is the same thing mentioned in Sect. 33: the repulsions between the inner shells in copper are quite important, comparable with the importance of the elastic effects arising from the conduction electrons. WHITE assumed that the inner shells acted on each other according to the results of elastic calculations, and treated the conduction electrons as free electrons, as in elastic theory, using the methods of Chapter V only to find the modification of the charge of the conduction electrons when one nucleus was displaced. It is rather hard to disentangle the two effects in an evaluation of his results. It appears, however, that this method is capable of yielding useful results for microscopic elastic constants.

One interesting feature connected with these calculations is its relation to the problem of correlation energy, discussed in Sects. 31 and 32. In those sections, we have seen that if an electron is displaced in a conduction band of a metal, we may expect other conduction electrons to rearrange themselves to shield the effect of the displacement of the electron from parts of the crystal more distant than the order of an interatomic spacing. Now if the conduction electrons can shield the effect of the displacement of an electron, they can also shield the effect of the displacement of a positive ion. It is this shielding effect which WHITE has investigated directly, by the methods of Chapter V. He has in fact found such shielding, but the shielding charge extends out to several interatomic distances, so that the field is not perfectly shielded at distances of one or two atomic distances. This is in agreement with the experimental observations from diffuse scattering of x-rays. Thus, in Table 3, we give the elastic constants observed for copper. We see that while the constants related to second and third nearest neighbors are considerably smaller than those related to nearest neighbors, they are not negligible. Now by FEYNMAN's theorem, these elastic constants related to second and third nearest neighbors must arise on account of the unshielded part of the field of the displaced nucleus on these second and third nearest neighbors. We have, then, very direct experimental evidence bearing on the type of shielding postulated in BOHM and PINES' shielded COULOMB

Table 3. *Atomic force constants of Cu crystal, following* JACOBSEN [*115*]. *To describe a force constant, we assume the atom at the origin is displaced along the x axis. We then give the x, y, z coordinates of the atom on which the force acts, and the direction of the appropriate force component on that atom. Other force constants are determined by symmetry from those given.*

| Force constant notation | | Force constant experimental ($10^4$ dynes/cm.²) | Coordinates of atom x | y | z | Direction of force |
|---|---|---|---|---|---|---|
| Nearest neighbors dist. 1.41$a$ | $\alpha_1$ | 0.48 | 0 | $a$ | $a$ | $x$ |
| | $\beta_1$ | 0.87 | $a$ | $a$ | 0 | $x$ |
| | $\gamma_1$ | 1.25 | $a$ | $a$ | 0 | $y$ |
| Second nearest neighbors, dist. $2a$ | $\alpha_2$ | 0.35 | $2a$ | 0 | 0 | $x$ |
| | $\beta_2$ | −0.072 | 0 | $2a$ | 0 | $y$ |
| Third nearest neighbors, dist. 2.45$a$ | $\alpha_3$ | 0.09 | $2a$ | $a$ | $a$ | $x$ |
| | $\beta_3$ | −0.022 | $a$ | $a$ | $2a$ | $x$ |
| | $\gamma_3$ | −0.015 | $a$ | $a$ | $2a$ | $y$ |
| | $\varepsilon_3$ | 0.06 | $2a$ | $a$ | $a$ | $y$ |

forces, as described in Eq. (32.6), and we see that to agree with these experiments, the range of the shielded COULOMB forces must extend out to at least the second and third nearest neighbors.

## VII. Configuration interaction and the covalent bond.

**35. Configuration interaction in molecules: the hydrogen molecule.** In Sect. 30, we stated that one of the ways of handling correlation was by means of configuration interaction; that is, by building up a wave function which must be expressed as a linear combination of determinantal wave functions, rather than as a single such determinant. We shall take up this possibility in the present chapter. We shall see that it has a close bearing on other problems, and in particular on the nature of the covalent bond. It is commonly said that there are a number of different types of bonds which can be present in solids: ionic bonds, covalent bonds, metallic bonds. Closer examination, however, convinces us that there are no very clear-cut distinctions between these types of bonding. A good deal of light can be thrown on their nature, and on the way in which one type can merge into another, by study of configuration interaction. These problems can be understood more easily from a study of isolated molecules than from solids, and for that reason we devote the present and the next section to a discussion of configuration interaction in molecules, pointing out its relation to correlation on the one hand, to the nature of the chemical bond on the other.

We may well start our discussion with the most familiar case, that of the hydrogen molecule, generally regarded as a typical example of a covalent bond. Here it is of course well known that there are two approximation methods of great value, the method of molecular orbitals, and the HEITLER-LONDON method. The first of these is the prototype of the method using a single determinant, the second of the method of configuration interaction. The relations between these two methods, and various modifications of them, have been discussed by many writers, including the present writer [*215*], [*224*], WEINBAUM [*252*], COULSON [*45*], COULSON and FISCHER [*47*], CALLEN [*42*], and MUELLER and EYRING [*172*]. Let us start by considering the method of molecular orbitals.

We know that we can set up a single determinant which fairly well represents the ground state of the hydrogen molecule, by setting up molecular orbitals as linear combinations of atomic orbitals, and assigning both electrons to the molecular orbital of lowest energy, one with each spin. Thus, if $a(x)$ symbolizes a wave function of an electron in the ground state of a hydrogen atom located at the position of one of the atoms in the molecule, $b(x)$ a similar wave function about the other atom, then the symmetric combination $a(x)+b(x)$ forms an approximation to a symmetric molecular orbital for the molecule, while $a(x)-b(x)$ forms an approximation to an antisymmetric orbital. If we tried to solve the HARTREE-FOCK equation for the molecule, we should ordinarily assume that the one-electron orbital occupied in the ground state was symmetric on reflection in the midplane separating the atoms, and hence that it could be approximated by the symmetric function $a(x)+b(x)$. If we let one electron with $+$-spin, and another with $-$-spin, both occupy this orbital, and construct a determinantal wave function from these two spin-orbital functions, the resulting determinant would form an approximation to the wave function of the ground state. One can do better by replacing the symmetric function $a(x)+b(x)$ by a more accurate symmetric function, attempting thereby to minimize the energy of the determinantal function. COULSON [45] has found a good approximation to such a symmetric orbital, and thereby has come as close as is practicable to solving the HARTREE-FOCK equations for the ground state of the molecule.

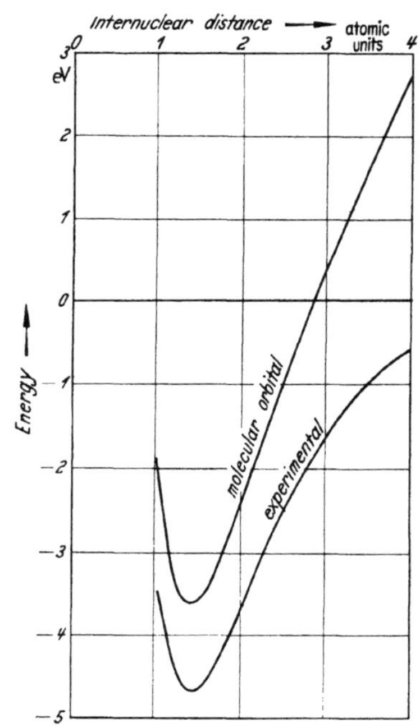


Fig. 49. Energy of best molecular orbital function for the hydrogen molecule, compared with the experimental curve. The molecular orbital function is set up from that given by COULSON, Trans. Farad. Soc. 33, 1479 (1937), modified to take account of COULSON'S more accurate calculation in Ref. 45. Experimental curve is the MORSE curve quoted by JAMES and COOLIDGE [117].

Such a molecular orbital wave function for the hydrogen molecule does not form a very good approximation. Its energy, as a function of internuclear distance, is shown in Fig. 49, where it is compared with the true energy curve. The energy lies too high, both at the equilibrium distance, and at infinite internuclear distance. The reason for the error is presumably the disregarding of correlation energy, which is implied in the HARTREE-FOCK method. The two electrons have opposite spin, and consequently the single determinantal wave function has nothing in it which keeps the electrons out of each other's way. Even at infinite internuclear distance, each electron will have equal probabilities of being found on each atom, irrespective of the behavior of the other, so that there is a fifty percent chance that both electrons will be found on one atom, leaving a positive and a negative ion, and only fifty percent that the electrons will be in their correct positions, one being found on each atom. We naturally ask, is there any way in which we can correct this fault?

The method of HEITLER and LONDON [86] takes care of this correlation correction at infinite internuclear distance, though it is less and less adequate in making the correction as the distance decreases. We remember that the HEITLER-LONDON method writes the wave function, as far as coordinates are concerned, in the form $a(x_1)\,b(x_2)+b(x_1)\,a(x_2)$, where $a$ and $b$ are as before, and $x_1$ and $x_2$ are the coordinates of the two electrons. At infinite internuclear distance, this wave function vanishes unless either electron 1 is one atom $a$, 2 on $b$, or vice versa; in other words, the state formed from positive and negative ions is eliminated, and the wave function reduces to that of two isolated atoms, so that its energy is entirely correct in this limit, and the correlation correction is exactly correct. At smaller distances, the energy is not nearly as good. This is partly a result, however, of the fact that we should not really use the functions $a$ and $b$, representing the isolated atoms, for finite distance. Various writers mentioned above have attempted to improve this situation by making the functions $a$ and $b$ depend parametrically on internuclear distance, varying them to minimize the energy, subject only to the condition that one function be found from the other by reflection in the mid-plane, so as to preserve the same type of symmetry as in the HEITLER-LONDON method.

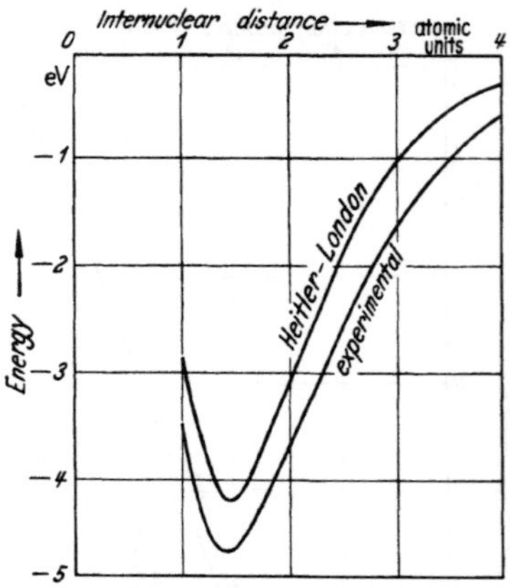


Fig. 50. Energy of best HEITLER-LONDON type wave function for the hydrogen molecule, compared with the experimental value. The HEITLER-LONDON curve is made up from the curve given by COULSON, Trans. Farad. Soc. 33, 1479 (1937), modified to give approximately the minimum energy found by CALLEN [42] from a configuration interaction of two configurations, or by MUELLER and EYRING [172] from their semilocalized orbitals; both these methods can be interpreted in terms of generalized HEITLER-LONDON functions.

The best function which has been found in this way gives an energy as a function of internuclear distance shown in Fig. 50, where again we compare with the true energy. This is a composite curve, made up of results of COULSON, CALLEN, and MUELLER and EYRING. It should not be regarded as the strict result of calculation, but the writer believes it to be close to the best energy curve which can be obtained in this way by methods so far used. It is obvious that by using a wave function of this generalized HEITLER-LONDON type, we have greatly improved the agreement with experiment, as compared to the molecular-orbital wave function shown in Fig. 49.

The process of building up a wave function of this type is a simple form of configuration interaction. To see that the wave function is to be expressed as a linear combination of determinantal wave functions, we write it in the (unnormalized) form

$$\left.\begin{aligned}&\begin{vmatrix} a(x_1)\,\alpha(s_1) & a(x_2)\,\alpha(s_2) \\ b(x_1)\,\beta(s_1) & b(x_2)\,\beta(s_2)\end{vmatrix} + \begin{vmatrix} b(x_1)\,\alpha(s_1) & b(x_2)\,\alpha(s_2) \\ a(x_1)\,\beta(s_1) & a(x_2)\,\beta(s_2)\end{vmatrix}\\ &\quad = [a(x_1)\,b(x_2) + b(x_1)\,a(x_2)]\,[\alpha(s_1)\,\beta(s_2) - \beta(s_1)\,\alpha(s_2)]\end{aligned}\right\} \tag{35.1}$$

where $\alpha$ and $\beta$ represent the PAULI spin functions, the $x$'s are space coordinates, and the $s$'s are spin coordinates. We see that the dependence of the wave function on coordinates is that already discussed, while the dependence on spin is that characteristic of a singlet state.

The method of exhibiting the generalized HEITLER-LONDON wave function in terms of configuration interaction, as shown in Eq. (35.1), is not the only way in which this can be done, and in fact it is not always the most informing way to do it. We can equally well write the function as a linear combination of the ground-state molecular-orbital wave function, and another determinantal function. Let us write the symmetric molecular orbital from which the ground-state molecular-orbital determinantal function is formed as $a(x)+b(x)$, and the lowest antisymmetric molecular orbital as $a(x)-b(x)$. If we are using the simple linear combination of atomic orbital method for setting up the molecular orbitals,

$a$ and $b$ will be wave functions of hydrogen atoms, as before; more generally, however, we can write any symmetric and antisymmetric molecular orbitals in the forms $a \pm b$, if $a$ and $b$ are properly defined.

Let us then set up determinantal wave functions arising from placing one electron of each spin in the symmetric (or alternatively the antisymmetric) molecular orbital. These two wave functions are

$$\left.\begin{aligned} &\begin{vmatrix} [a(x_1) \pm b(x_1)]\,\alpha(s_1) & [a(x_2) \pm b(x_2)]\,\alpha(s_2) \\ [a(x_1) \pm b(x_1)]\,\beta(s_1) & [a(x_2) \pm b(x_2)]\,\beta(s_2) \end{vmatrix} \\ &\qquad = \{[a(x_1)\,a(x_2) + b(x_1)\,b(x_2)] \pm [a(x_1)\,b(x_2) + b(x_1)\,a(x_2)]\} \times \\ &\qquad\quad \times \{\alpha(s_1)\,\beta(s_2) - \beta(s_1)\,\alpha(s_2)\}. \end{aligned}\right\} \tag{35.2}$$

That is, these two functions are linear combinations of the HEITLER-LONDON function, as given in Eq. (35.1), and the ionic-type function whose dependence on coordinates is given by $a(x_1)\,a(x_2) + b(x_1)\,b(x_2)$. By making linear combinations of these two functions, then, representing a molecular-orbital configuration in which both electrons are in a symmetric molecular orbital, and one in which they are both in an antisymmetric orbital, we can reproduce the HEITLER-LONDON type of function as given in Eq. (35.1). By varying the symmetric, and the antisymmetric, molecular orbitals in such a way as to minimize the energy of this linear combination, we can achieve a wave function identical with that whose energy was given in Fig. 50. This, in fact, is the way that curve was actually computed.

In making this comparison of the HEITLER-LONDON and the molecular orbital methods, we should emphasize one point, which has not been realized until recently, but which has been brought out by the work of COULSON and FISCHER [*47*], and of HURLEY, LENNARD-JONES, and POPLE [*113*]. This is the fact that if the functions $a$ and $b$ in the HEITLER-LONDON function of Eq. (35.1) are allowed to be subject to arbitrary variation, as we have assumed, then this HEITLER-LONDON type wave function, which we may describe as a generalized valence-bond function, has a great deal of flexibility, and in particular enough flexibility so that it can include the molecular-orbital method as a special case. We have implied that the functions $a$ and $b$ were different from each other; but if we take the special case where they are identical with each other, and hence each is a symmetric function on reflection in the mid-plane between the atoms (since we have stated that $a$ goes into $b$ on such a reflection), then the function (35.1) becomes simply a single determinant in which both electrons are assigned, one with each spin, to a single symmetric molecular orbital. In other words, this special case of the function of Eq. (35.1) is the ordinary molecular-orbital ground-state wave function. Since the general function (35.1) is more general than this, in that $a$ and $b$ are allowed to be different from each other, it is obvious from the variation principle of quantum mechanics that we can always obtain a lower energy, and consequently a better wave function, by using the function (35.1) than by a single determinantal wave function. This is just another way of saying that the use of configuration interaction between the two wave functions of Eq. (35.2), namely the molecular orbital functions formed from symmetric and from antisymmetric molecular orbitals, will always lower the energy with respect to the single determinantal function formed from the symmetric molecular orbital.

We see, then, that a very considerable improvement in the wave function found by the determinantal or molecular-orbital method can be achieved by a configuration interaction with one additional configuration, in a way suggested by the HEITLER-LONDON solution. Various writers, including particularly LENNARD-JONES and POPLE [149], [150], [152], have discussed the type of correlation which we can achieve between the electrons by this method. If, as we have generally been assuming, the function $a$ is located more on one atom, $b$ more on the other, then the HEITLER-LONDON type of function given by Eq. (35.1) has the effect of tending to keep one electron more on one atom, the other more on the other. This is a very effective type of correlation correction when the atoms are widely separated, but gets ineffective as the two atoms coalesce. By adding further configurations, other types of correlation can be achieved: a so-called in-out correlation, in which one electron stays nearer to the nuclei while the other one is at larger distances; an angular correlation, by which the two electrons tend to stay on opposite sides of the nuclei. In principle, by superposing enough configurations in this way, we can make our correlation correction as good as we please, and can approach the true wave function. In practice, the process has been found to converge very slowly, and the best wave functions for hydrogen obtained in this way have not had an energy very much better than that shown in Fig. 50, obtained from a single interacting configuration aside from the ground state of the HARTREE-FOCK method.

From the example of the hydrogen molecule, which we have just been discussing, we can draw several interesting conclusions, though we have not stated all the evidence on which they are based. In the first place, the covalent bond can be described either in terms of the molecular-orbital or the HEITLER-LONDON language, though the latter, when regarded in the generalized form which we have used, will always give a better wave function. The wave functions obtained by the two methods do not, as a matter of fact, differ very much as far as the electronic charge distribution is concerned. The great difference between them comes in the treatment of correlation, which is much better handled by the HEITLER-LONDON than by the molecular-orbital method. The feature of the system responsible for the binding of the two atoms into a molecule is not primarily the presence of the two electrons with opposite spins. It is rather a concentration of electronic charge between the nuclei, which is described almost equally well by either of the methods. We have seen in Sect. 34 that FEYNMAN's theorem allows us to find the force between two atoms in terms of ordinary electrostatics from the electronic charge distribution; in this case, the force of attraction between the atoms, at distances larger than the equilibrium distance, comes from the fact that either nucleus is more attracted by this concentration of electrons between the nuclei, than it is repelled by the other nucleus. Even though the molecular-orbital method gives a fairly correct account of the total charge density, however, it fails completely to take account of the correlation. But we see from Fig. 49 that this correction is in fact less important at the equilibrium distance than at infinite distance; the energy of the single molecular-orbital state is not as bad at the equilibrium distance as at infinite distance.

Many points encountered in general discussions of molecular binding are found in the hydrogen molecule, but many others are missed in this example, for two reasons: the hydrogen molecule is a specialized one in that the two atoms are alike, and obviously it is also a special case because it is diatomic, and we do not encounter the interactions between different bonds which are important in a polyatomic molecule or in a crystal. To illustrate these points, we shall discuss in the next section two other examples, one a diatomic molecule with unlike atoms, the other a polyatomic molecule.

**36. Configuration interaction in molecules, continued.** We next wish, in contrast to the case of hydrogen, to discuss a diatomic molecule with unlike atoms, and one of the simplest such molecules is LiH. Let us first ask how this would be treated by the HARTREE-FOCK method. We should expect that the most tightly bound orbital found in the self-consistent field method would be practically like a $1s$ atomic orbital around the Li atom, and that this would be occupied by two electrons in the molecule. Next, we should expect that there would be a bonding orbital linking the Li and H atoms, with charge distribution piling up between the atoms, which would likewise be occupied by two electrons in the molecule. In other words, we expect to find a picture somewhat resembling that of a covalent bond, as treated by the molecular orbital method.

As the internuclear distance increases, however, the behavior of this HARTREE-FOCK solution will be quite different from that found in hydrogen, on account of the difference of the two atoms. The bonding orbital will become more and more concentrated on one of the two atoms as the internuclear distance increases, and a little consideration shows that it will become concentrated on the hydrogen atom. Hence in the limit of infinite internuclear distance the single determinantal wave function resulting from the HARTREE-FOCK method will approach that describing a $Li^+$ ion with two $1s$ electrons, and a $H^-$ ion with two $1s$ electrons. This has an energy a good deal higher than the true ground state, which of course is a neutral Li atom and a neutral H. The expectation value of the energy of the HARTREE-FOCK wave function, as a function of internuclear distance, will then decrease rapidly with decreasing distance on account of the COULOMB attraction between the ions.

We see, in other words, that the HARTREE-FOCK method, when applied to the LiH molecule without configuration interaction, will lead to a wave function which is of purely ionic type at large internuclear distance, describing a $Li^+$ and a $H^-$ ion, but which gradually becomes more covalent as the distance decreases. This furnishes a good example of the statement made in the preceding section, that there is no clear distinction between ionic and covalent bonds. Since the true wave function of the ground state at large distances is not ionic, we see that we must carry out a configuration interaction to describe this properly; it cannot be done by means of a single determinant. As with the hydrogen molecule, this configuration interaction can be set up as a HEITLER-LONDON calculation, or as an interaction between the HARTREE-FOCK ground state and a HARTREE-FOCK excited state. Let us describe how this is to be done.

In the first place, we can set up for the two outer electrons a HEITLER-LONDON type of wave function similar to that of Eq. (35.1). If we give sufficient freedom of variation to the functions $a$ and $b$ such a wave function can be very satisfactory. There no longer is symmetry between the atoms, so that there is no requirement that $a$ and $b$ be mirror images of each other in the mid-plane between the atoms. As in the preceding section, we can write the HARTREE-FOCK solution in this form, by making $a$ and $b$ identical with each other, but no longer symmetric functions; in fact, as the internuclear distance increased, they would approach atomic functions concentrated solely on the hydrogen. But by making $a$ and $b$ different from each other, we gain a freedom which allows us to decrease the energy, and in fact to approach the correct wave function at infinite distance, where now $a$ can represent a $2s$ wave function on the Li atom, $b$ a $1s$ on the H. A calculation is not available showing the result of varying $a$ and $b$ as freely as possible, but it is to be expected that the wave function set up in this way in the form of Eq. (35.1) would be decidedly good. Or alternatively, we can combine the HARTREE-FOCK ground state with another determinantal function in which both electrons are placed in antibonding wave functions (which now, on account of the lack of symmetry of the molecule, are not antisymmetric). Such a combination of determinantal wave functions can represent rather well the results of the HEITLER-LONDON function; though here, in contrast to the case of the hydrogen molecule, we need also to include determinants in which one electron is in the symmetric, one in the antisymmetric, orbital in order to get complete agreement of the two methods.

From this sketch of the problem of LiH, we see that the new situation which arises in a molecule with two unlike atoms is that the HARTREE-FOCK function will, in general, approach the wave function of a completely ionic state at infinite internuclear distance, though it can well become more nearly covalent as the distance decreases. By configuration interaction, we can do two things: we can remove the ionic character at large distances, and can improve the description of correlation at smaller distances. We see here, as with hydrogen, that neglect of correlation by the HARTREE-FOCK wave function at large distances takes the form of describing the molecule as if it had separated into ions rather than neutral atoms, resulting in two electrons being found too close together, on one of the atoms at one of the extremities of the bond, rather than separated on the two atoms as they should be.

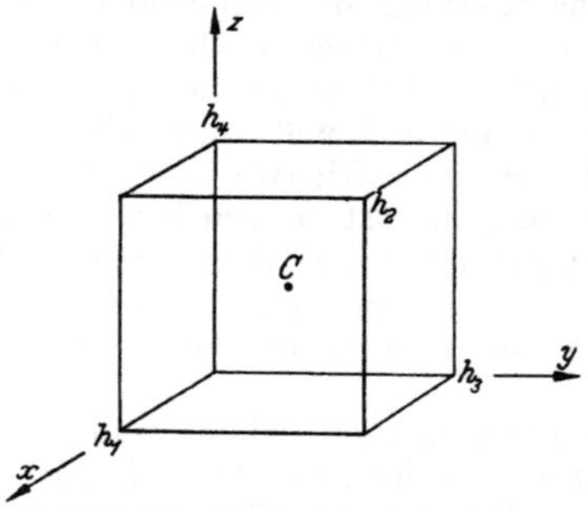


Fig. 51. Positions of the atoms in the methane molecule. $C, h_1, h_2, h_3, h_4$ represent the centers of the carbon and hydrogen atoms. In the text, $h_1 \ldots h_4$ are used to represent hydrogenic wave functions centered on the respective atoms.

As an example of a polyatomic molecule, we can take methane. The HARTREE-FOCK solution will result in a lowest orbital which is practically the carbon $1s$, and which will be occupied by two electrons. Above that, in energy, symmetry shows that there will be a non-degenerate bonding orbital, and somewhat higher in energy a three-fold degenerate bonding orbital, both having lower energies than any antibonding orbitals. The non-degenerate bonding orbital has the same type of symmetry as an $s$ state on the carbon; if we build it up by making linear combinations of atomic orbitals, it will be a linear combination of the carbon $2s$ orbital, and of the hydrogen $1s$ orbitals, the latter all having equal coefficients, so as to maintain the symmetry. To have a bonding orbital, there will be electronic charge piled up between the carbon and each of the hydrogens. As for the three-fold degenerate bonding orbitals, they will have the symmetry of $p$ functions. In Fig. 51 we show the structure of the methane molecule. The three orbitals will have the symmetries of $p_x$, $p_y$, $p_z$ atomic functions respectively, being odd respectively in $x$, $y$, and $z$. For the $p_x$-like function, for instance, we shall have something with the symmetry of a linear combination of a carbon $2p_x$ function, and a combination $h_1+h_2-h_3-h_4$ of the hydrogen orbitals on the four atoms shown. Here again, since we are dealing with bonding orbitals, there will be concentrations of charge between the carbon and the hydrogens.

In the ground state of the methane molecule, according to the HARTREE-FOCK method, each of these four bonding orbitals will be occupied by two electrons, one of each spin. It is interesting to inquire what is the relation of this picture to the familiar one of directed orbitals, in which we assume four bonding orbitals, extending from the carbon to each of the four hydrogens, each containing two electrons, one of each spin. It is well known that by making four suitable linear combinations of the non-degenerate $s$-like bonding molecular orbital, and of the three $p$-like orbitals, we can form four directed orbitals of the desired type, orthogonal to each other. It is also well known that the determinantal wave function formed from any new set of one-electron orbitals obtained from the original occupied orbitals by a unitary transformation, is identical with the original determinant. Hence we may set up a determinantal wave function in which the eight outer electrons of the molecule are assigned to these four directed bonding orbitals, two to each, and this wave function will be identical with the one found directly by the HARTREE-FOCK method. These directed bonding orbitals are what have been called equivalent orbitals, by LENNARD-JONES [*148*], [*151*].

The one-electron energy parameters, however, will be quite different in this method of representation from what they are in the ordinary HARTREE-FOCK description. In stating the HARTREE-FOCK equations, in Eq. (17.3), we stated that a certain operator, operating on the one-electron function $u_i(x_1)$, equalled $\varepsilon_i u_i(x_1)$. But if we introduce a new set of one-electron functions, determined from the occupied $u_i$'s by a unitary transformation, then the HARTREE-FOCK operator operating on one of these functions will equal a linear combination of the functions. In other words, the one-electron energy is in general a matrix, and in Eq. (17.3) we have the special case where this matrix is diagonalized. It is only in the diagonalized form that KOOPMANS' theorem applies, and that we can identify the one-electron energy with the ionization energy. We conclude, then, that if we set up the equivalent orbitals for methane, and allow the self-consistent HARTREE-FOCK operator to operate on one of the directed orbitals, the result will be a linear combination of energy parameters, multiplied by the four possible directed orbitals. This situation is what we always encounter when we try to use such localized solutions of the HARTREE-FOCK equations.

This situation which we have met with the equivalent orbitals is similar to that which we have met, in Sect. 25, in setting up the WANNIER functions. They, like our equivalent orbitals, are linear combinations of BLOCH functions, which are the solutions of the HARTREE-FOCK equations for a solid. And we have seen, in Eq. (25.9), that the one-electron HAMILTONIAN, or HARTREE-FOCK operator, has non-diagonal matrix components between different WANNIER functions. It is just these non-diagonal matrix components which are responsible for the splitting of the energy bands, as we see from Eq. (25.11). Similarly here, we have non-diagonal matrix components of one-electron energy between the equivalent orbitals, and it is these components which split the energies into that of the $2s$-like non-degenerate state of methane, and the $2p$-like three-fold degenerate state, which in fact have energies differing by a good many volts.

We see, then, that we may introduce equivalent orbitals in the methane problem, that the determinantal wave function made up from them is identical with that made up from the real solutions of the HARTREE-FOCK equations for the problem, but that there are non-diagonal matrix components of one-electron energy between the different equivalent orbitals. This situation is typical of what we shall later find in dealing with solids, if we try to introduce localized one-electron orbitals localized on the various bonds. We may now ask how we can set up configuration interaction in methane. It is certainly necessary to do so, as we can see clearly from the limiting case of infinite internuclear distance. In this limit, the $s$-like and $p$-like molecular orbitals which are occupied in the HARTREE-FOCK solution for the ground state of the molecule reduce to linear combinations of $1s$ wave functions on the hydrogen atoms. Hence this HARTREE-FOCK solution reduces to a description of a carbon ion with only two electrons, and hence with four positive charges, plus four hydrogen ions each with two negative charges. Configuration interaction is necessary in order to get the correct ground state composed of neutral atoms.

The configuration interaction required to reach the real ground state of the system of a carbon and four hydrogen atoms at infinite separation is extremely complicated. Simple configuration interactions can give wave functions reducing to neutral atoms at infinite separation, but with the carbon atom in an excited state rather than in its $2s^2\,2p^2\;{}^3P$ ground state. However, these simple configuration interactions give us some idea of how to proceed to improve the HARTREE-FOCK wave function for methane. We can set up such simple configuration interactions by handling the four bonds separately. This process can be most easily handled by a generalization of the HEITLER-LONDON method. We can set up a wave function which resembles a product of four functions of the type of Eq. (35.1), one for each bond, assigning two electrons to each of the bonds (and the remaining two to the carbon $1s$ wave functions), then permuting the electronic indices and forming a suitable antisymmetric wave function in this way. This is the ordinary valence-bond wave function, discussed by the present writer [*217*], but generalized so that the orbitals $a$ and $b$ are allowed arbitrary variation, as in our earlier discussion of the generalized HEITLER-LONDON method. The resulting wave function can be easily shown to be a linear combination of 16 determinantal wave functions.

This generalized valence-bond function then can be shown to lead to similar properties to what we have found in the case of a single valence bond. By suitable choice of the functions $a$ and $b$, we can get the special case of the molecular-orbital or HARTREE-FOCK solution. However, if we generalize these functions, we can set up a wave function in which each of the valence bonds separately is handled just as in a diatomic molecule. If we arrange it so that the functions $a$ and $b$ associated with any of the bonds are orthogonal to the corresponding $a$ and $b$ associated with any of the others, it is not difficult to set up an expression for the energy of the whole molecule, as shown by HURLEY, LENNARD-JONES, and POPLE [*113*], and there are no difficult terms associated with interactions between the bonds. The wave function goes, at infinite internuclear distance, to neutral hydrogen atoms, and a neutral carbon, though the latter is not in its ground state, but instead, as we have mentioned, in a state which has been called by VAN VLECK [*245*] the valence state, a combination of a number of excited states. In other words, we may expect that this wave function, though far from perfect, gives a fairly good account of the correlation effects which are disregarded in the simple HARTREE-FOCK solution.

As with the simpler molecules, we can also write this generalized valence-bond state in terms of a configuration interaction between the HARTREE-FOCK ground-state solution, and various other determinantal wave functions. These other determinants can be regarded as those in which in a given bond, instead of having both electrons in the bonding directed orbital, one or both of them is excited to a corresponding antibonding directed orbital, directed along the same axis. We can show that inclusion of such excited configurations will give us the same generality as the valence-bond state, though not more. It is easy to see the sort of excited state which must be used in addition, to get a more general configuration interaction: we must have configurations in which, instead of always having just two electrons associated with each bond, we may have three, four, or one, or zero electrons associated with some bonds. Configurations of this type prove to be necessary to describe properly the state of the separated atoms in their lowest energy levels. However, near the equilibrium distance in methane, it seems highly unlikely that these other configurations, which we might consider to have ionized bonds, would enter with any large coefficients, and we may expect that the simpler configuration interaction which we have been considering will represent a fairly satisfactory wave function.

As a result of the discussion of the present section, as well as of the preceding one, we can see that it is possible to generalize the treatment of configuration interaction met in partially covalent bonding, in such a way that it can be carried out in more complicated cases. We start with a molecular-orbital or HARTREE-FOCK solution. From the one-electron orbitals found by that method, and occupied in the determinantal wave function representing the ground state, we set up orthogonal linear combinations which are concentrated in the various bonds, rather than being spread through the molecule or crystal. In addition to these bonding orbitals, we set up excited antibonding orbitals, concentrated in the same regions of space as the bonding orbitals, and orthogonal to the bonding orbitals and to the other antibonding orbitals. We then set up excited determinantal wave functions in which one or both of the electrons in a bond are excited from the bonding to the antibonding orbital. By carrying out a configuration interaction between the ground state and these various excited wave functions, we get a tolerably good description of the real wave function, with a fairly good correction for correlation.

For a complicated system, this configuration interaction may be very hard to carry out, in practice. It is probably true, however, that if the original HARTREE-FOCK solution is good enough, and we are working in the neighborhood of the equilibrium internuclear distance, we can proceed by perturbation methods, and get the contributions of these excited orbitals to the total wave function by second-order perturbations. In such a case, the contribution of any one of the excited configurations to the energy will be the square of the matrix component of the HAMILTONian between the ground state and the excited state, divided by the difference between the diagonal matrix components of the energy. Each such excited configuration will depress the energy, and the more they do so, the better will be the approximation. Now the only excited configurations which have non-vanishing matrix components with the ground state were shown by BRILLOUIN [*34*] to be those in which just two electrons are excited from the orbitals which they occupied in the ground state. In other words, we need only consider those excited configurations in which the two electrons of a single bond are excited from the bonding to the antibonding orbital associated with that bond. The larger the non-diagonal matrix component of energy, the more effective will be this configuration interaction in depressing the energy. The non-diagonal matrix component of energy is an ordinary exchange integral, of the nature of

$$\int u^*(x_1)\, v^*(x_2)\, (1/r_{12})\, v(x_1)\, u(x_2)\, dv_1\, dv_2 \tag{36.1}$$

where $u$ is the bonding, $v$ the antibonding orbital. This will have its maximum value if $u$ and $v$ are concentrated as much as possible in the same region of space. This is the real reason why one must use the concentrated, directed bonding orbitals located in the regions between pairs of atoms, for the $u$'s. For the single HARTREE-FOCK wave function, the determinantal wave function is identical whether it is constructed of molecular orbitals or BLOCH functions extending over the whole molecule or crystal, or of more concentrated wave functions like equivalent orbitals or WANNIER functions, constructed as linear combinations of the occupied molecular orbitals. But the configuration interaction required to give a good account of the correlation effect, and to accomplish the same thing which is done in hydrogen by the HEITLER-LONDON method, requires the use of these concentrated bonding and antibonding orbitals.

**37. The generalized valence-bond method, applied to covalently bonded crystals.** In the preceding section, we have been led to a general method of handling configuration interaction in molecules or crystals, capable of treating the correlation effect moderately satisfactorily, and of generalizing the HEITLER-LONDON method of handling the covalent bond to the case of crystals or complicated molecules. In the present section we shall describe the application of this method to crystals such as diamond which are held by definitely covalent bonds, and crystals such as those forming the zinc-blende structure, whose bonds are part way between covalent and ionic. In the next section, we shall take up the case of metals.

For generality, we may as well discuss the zinc-blende structure; we recall that this is like the diamond structure, except that the two atoms in the unit cell are of different types. We can then treat the diamond structure as the special case where the two atoms are identical. A characteristic of the crystals having the zinc-blende structure is that there are always four electrons per atom, on an average, outside closed shells, just as in diamond, silicon, or germanium. Thus, for instance, one can have a crystal of GaAs, in which Ga has one electron less than Ge, As has one more; or ZnSe, where Zn has two less, Se two more; or CuBr, where Cu has three less, Br three more; in each case the number of electrons per unit cell is the same as in Ge. As we have mentioned in Sect. 2, we can think of these as intermediate cases between two possible types of structure. In ZnSe, for instance we can imagine the Zn to acquire two extra electrons, the Se to lose two, so that each will have four outer electrons, like Ge. and for them then to form bonds just as in Ge. On the other hand, we can also imagine Zn to lose two electrons, Se to gain two, so that Zn would form an ion $Zn^{++}$, with a closed-shell configuration like argon, and Se would form $Se^{--}$, with a configuration like krypton, then forming a purely ionic crystal. The true situation is undoubtedly between these, corresponding much more closely to neutral atoms. One of our tasks in the present section is to understand how the electronic wave function is to be expressed in such an intermediate case between covalent and ionic binding.

We have described the energy bands of such structures in Sect. 15. We have found that there are four energy bands corresponding to bonding types of wave functions, then above them an energy gap, with excited energy bands above the gap. The bonding energy bands have just enough states to accomodate eight electrons per unit cell, or four per atom, so that these are the occupied or valence bands, the higher ones being the unoccupied or conduction bands. In other words, a HARTREE-FOCK or energy-band theory accounts properly for the occupancy of the energy levels, and the fact that these materials are insulators or semiconductors.

Next we can consider the wave functions of the valence band. In diamond or silicon or germanium, these wave functions would correspond to equal amounts of charge on each atom of the unit cell, as is obvious from the fact that the two atoms of the unit cell are alike. In the zinc-blende structure, however, we should find that the valence band corresponded to having more total charge on the non-metal like As, Se, or Br, less on the metallic element like Ga, Zn, or Cu. On the other hand the conduction band would be reversed, with more charge on the metal, less on the non-metal. With the filled valence band, the metallic atom will thus have less charge than in Ge, but certainly would have some charge, so that it would be intermediate between the two limiting cases we have described, namely a germanium-like structure, or a completely ionic structure. No self-consistent calculation has been made of any such crystal, accurate enough to tell just how much charge lies on each atom, but it seems entirely likely that the atoms are nearly neutral electrically.

To proceed by analogy with the preceding section, we should now set up equivalent orbitals, linear combinations of the BLOCH functions of the valence band, corresponding to bonding orbitals between an atom and one of its four nearest neighbors. This can be done; a discussion of diamond along these lines has been given by HALL [*83*]. The one-electron functions have a resemblance to WANNIER functions, except that they are made from the BLOCH functions of the four valence bands, rather than from a single band. For diamond, silicon, or germanium, these bonding orbitals will not only be somewhat localized between neighboring pairs of atoms, but also they will be symmetric in the midplane between the atoms. For the zinc-blende structure, they will not be symmetric, and will correspond to having considerably more of the charge on the non-metallic than on the metallic atom, as we have seen in the preceding paragraph. We can likewise set up an antibonding orbital associated with each bond, for example by making similar linear combinations of the BLOCH functions in the four lowest conduction bands.

We can now set up a configuration interaction, just as in previous cases, between the ground state of the HARTREE-FOCK method, and excited states in which one or both of the electrons from a given bond are lifted from the bonding to the antibonding orbital. Just as in the general discussion which we have given previously, this configuration interaction will have the effect of describing the system in a HEITLER-LONDON-like way, and will introduce enough correlation effect to make the atoms electrically neutral at infinite internuclear distance. This gives a straightforward way of describing the partially covalent bonds in such a structure. A calculation of this type has been made by SCHMID [*198*] for diamond, with satisfactory results, though the binding energy was not calculated accurately enough to know whether it was quantitatively correct or not. One particularly interesting result came from the calculation of SCHMID: the molecular orbital solution, or the single HARTREE-FOCK determinant, formed a rather good approximation at the observed equilibrium distance. The coefficient of the excited state which had to be mixed with it was rather small, and the configuration interaction made a rather unimportant decrease in the energy. In other words, it seems that in this case that part of the correlation energy which can be handled in this way is not very large, so that an energy-band calculation seems decidedly good, for diamond. No corresponding calculation has been made for a crystal with the zinc-blende structure, but there is no reason to expect a very different result in that case.

**38. The relation of the metallic and covalent bond.** We now have enough background, from our discussion of the ionic and the covalent bonds in the preceding sections, to consider the metallic bond. Essentially the same methods of treating configuration interaction can be used for metals as for the diamond and zinc-blende structures, but a little consideration will show us that configuration interaction in metals is less important than in the semiconductors, so that the energy-band theory is a better first approximation. We wish to carry out this discussion in the light of such arguments as PAULING'S [*179*] to [*182*], to the effect that the metallic bond is in effect an unsaturated covalent type of bond.

PAULING has repeatedly emphasized that the metallic bond is very similar to a covalent bond, but without enough electrons to furnish an electron pair between adjacent atoms, such as we have in the diamond and zinc-blende structures. With this point of view there can hardly be any argument. He has further implied, however, that on account of the covalent nature of this metallic bond, it should be handled by methods similar to the HEITLER-LONDON method, and for that reason he considers the energy-band method inappropriate for discussing such metallic properties as the strength of the metallic bond. With this latter point of view we cannot agree, and the arguments of the present chapter allow us to see just how it is that the metallic bond can resemble strongly a covalent bond, and yet the energy-band treatment is well justified.

We have emphasized throughout the chapter that a covalent bond has as its principal characteristic a concentration of electronic charge between the atoms being bonded. This concentration can be described fairly well by the energy-band or HARTREE-FOCK method. The principal shortcoming of that method is its lack of consideration for correlation, and this shortcoming can be made up to a considerable extent by using the HEITLER-LONDON method. We have seen, however, that that method is equivalent to a configuration interaction, in which we start with the HARTREE-FOCK wave function, and make a combination of this with excited wave functions in which a pair of electrons is shifted from a bonding orbital to a corresponding antibonding orbital. In a metal, we may use the same scheme. We start with the ordinary energy-band or HARTREE-FOCK solution. This forms an acceptable first approximation, and really does quite well in describing the covalent nature of the metallic bond. In as simple a case as the alkalies, for instance, it is well known that the wave functions toward the bottom of the conduction band, which are the ones occupied in the ground state of the metal, have large values in the regions intermediate between atoms, and consequently correspond to a considerable concentration of charge in this intermediate region, which is the main characteristic of the covalent bond.

We can next try to improve this wave function by configuration interaction. Let us see how the analogue of the method used for the diamond or zinc-blende structure would be carried out for the metal. We must first make linear combinations of the occupied orbitals in the HARTREE-FOCK scheme, to get concentrated functions, located as nearly as possible in the bonds. But this cannot be done as satisfactorily for a metal like an alkali, as for a substance like diamond. In diamond, we could use the four occupied valence bands for making the linear combinations. As a result, we could find four concentrated orbitals per unit cell the equivalent bonding orbitals, each of which would hold two electrons, one of each spin, in the ground state. Then we set up an equal number of antibonding orbitals, equally concentrated, and set up excited configurations in which the two electrons of a bond were displaced to the antibonding orbital.

Now in an alkali metal, we have only half of a single energy band occupied in the ground state of the HARTREE-FOCK solution (aside, of course, from the inner electrons of the atom). By making linear combinations of these occupied BLOCH functions, we could make only one concentrated wave function for each pair of unit cells. But each unit cell should hold four bonds (that is, four lines joining an atom to a nearest neighbor), so that this means that we can construct only one concentrated wave function, capable of holding two electrons, for each eight bonds. This is just another way of bringing out PAULING's point, that in an alkali there are only $\frac{1}{8}$ enough electrons to form complete covalent electron-pair bonds between each pair of neighboring atoms. These wave functions, in other words, are much less concentrated than are the equivalent bonding orbitals in the diamond structure.

We can still proceed formally as with the diamond structure, setting up an antibonding concentrated wave function corresponding to each of these bonding wave functions; we could construct them from the unoccupied upper half of the lowest energy band. And we could still proceed with the configuration interaction, following the pattern used for the diamond structure. But we have seen, in Eq. (36.1), that the depression of energy as a result of the configuration interaction in each bond is larger, the more concentrated the wave function. With the very diffuse wave function which is the best we can do in an alkali metal, we should hardly expect this configuration interaction to do very much good. In other words, we expect configuration interaction to be much less important in a metal than in a covalent crystal; and this is another way of saying that the energy-band, or HARTREE-FOCK, method should be much closer to the truth for a metal than for a covalent crystal. In other words, though we feel, as PAULING does, that there is a good analogy between the metallic and the covalent bond, we do not agree with him that the energy-band method is not appropriate for handling problems of metallic cohesion.

We realize, of course, from the results of the preceding chapter, that a correlation correction is needed in such a case as an alkali metal. However, the discussion of the present chapter should convince the reader that its correct description is a more complicated problem than the free-electron description of WIGNER and SEITZ, and of BOHM and PINES, would indicate. If we take an alkali metal, and increase the internuclear distance to infinity, an energy-band wave function will go to a state in which the individual alkali atoms are found as positive or negative ions as often as they are as neutral atoms, just as in the case of the hydrogen molecule. The correction of this error represents the correlation correction which must be made at inifinite distance. The general method of doing it is clear from the discussion of the present chapter, though the exact way to do so is still a great problem. Certainly there will be a large correlation correction to the HARTREE-FOCK solution at infinite internuclear distance, rather than having the correction go to zero, as WIGNER and SEITZ's correction of Eq. (31.1) would suggest. It cannot be pretended that a proper treatment of correlation in metallic wave functions yet exists; but it seems clear that such a treatment, when it is worked out, will bear at least as much resemblance to the configuration interaction arguments of the present chapter as to the free-electron pictures of WIGNER and SEITZ, BOHM and PINES, presented in the preceding chapter.

## VIII. Magnetic properties of solids.

**39. The band theory of ferromagnetism.** In the preceding parts of this article we have been tacitly assuming that the solids we were discussing were non-magnetic; we had equal numbers of electrons in states with plus and minus spins. It is obvious, however, that many important materials show magnetic properties. The technically important metals iron, cobalt, and nickel, and many alloys and compounds containing them and other elements, are ferromagnetic. Many alloys and compounds of the transition groups of elements are antiferromagnetic or paramagnetic. In all of these cases, magnetic effects are not only interesting in themselves, but they arise from features in the electronic wave functions which are of great importance for other properties as well. Since magnetism as such is being handled elsewhere in this Encyclopedia[1], we shall not emphasize the experimental or phenomenological side of the subject. However, we cannot give an adequate treatment of the electronic structure of solids without including a treatment of magnetism.

Historically, the treatment of ferromagnetism has shown a close parallel to that of cohesive energy. The pioneer suggestion of HEISENBERG [*85*] that ferromagnetism was a phenomenon arising from the operation of the exclusion

[1] See vol. XVIII.

principle, and its relation to electron spin, followed the lines of HEITLER and LONDON'S [86] treatment of the hydrogen molecule. For a good many years, inadequate attempts were made to generalize these methods to more complicated systems than the hydrogen molecule, inadequate because they failed to recognize the very great difficulty of carrying out configuration interactions in many-electron systems, with the use of non-orthogonal one-electron orbitals. The difficulties are largely avoided in the type of discussion of configuration interaction which we have given in the preceding chapter, though we have not emphasized this fact; it has only recently been realized, and this largely by those working in the field of molecular structure, how these difficulties can be overcome As we have indicated, the general method is to start with the molecular-orbital or HARTREE-FOCK solution as a first approximation, and to treat the modifications of this method necessary to secure a HEITLER-LONDON-like result as a perturbation of the HARTREE-FOCK solution, to be treated by configuration interaction. In our discussion of ferromagnetism, we shall follow the same general outline. Thus, in the present section, we shall see how far we can go in the theory of ferromagnetism using a simple energy-band theory. Then we shall investigate to see what the effect of configuration interaction is on the problem, and we shall find, as with non-ferromagnetic materials, that the main effect is to introduce a treatment of correlation, which is disregarded in the simple energy-band method. This correlation correction is of considerable, though not decisive, importance, so that one can get a qualitatively though not quantitatively correct understanding of ferromagnetism from elementary band theory.

Ferromagnetism arises, according to the band theory, because in certain cases a HARTREE-FOCK determinantal wave function with unbalanced spin has a lower energy than the determinantal function with balanced spin, or there is a ferromagnetic state of lower energy than the non-magnetized state. Of course, when we speak in this way, we are thinking of a single ferromagnetic domain; we shall not take up the problem of interaction of domains with different orientation in the present article. Let us see how it can come about that this result follows from the HARTREE-FOCK method.

The total energy of a determinantal wave function equals

$$\left.\begin{aligned}\sum_i \int u_i^*(x_1)\, f_1\, u_i(x_1)\, dv_1 + \sum_{\text{pairs } i,j} \int u_i^*(x_1)\, u_j^*(x_2)\, g_{12}\, u_i(x_1)\, u_j(x_2)\, dv_1\, dv_2 - \\ - \sum_{\substack{\text{pairs } i,j;\\ \text{spin } i = \text{spin } j}} \int u_i^*(x_1)\, u_j^*(x_2)\, g_{12}\, u_j(x_1)\, u_i(x_2)\, dv_1\, dv_2.\end{aligned}\right\} \quad (39.1)$$

Here the first term is the sum of the kinetic energies, and potential energies in the field of the nuclei, of all electrons. The second is the sum of all COULOMB interactions between pairs of electrons. The last is the sum of all exchange interactions between electrons with parallel spin. We now ask how this total energy will be changed if we take an electron whose spin is pointing opposite to the net magnetization, and reverse the spin, thereby increasing the magnetization. If the energy increases in the process, there will be no ferromagnetism for the energy must increase to magnetize the crystal; on the contrary, if the energy decreases, the crystal will spontaneously magnetize.

Changes of two sorts will occur if the spin of an electron reverses in this way. In the first place, there will be more terms in the summation representing the exchange effect, for the electron whose spin was reversed will find more other electrons with spin parallel to its own after reversal than before. Thus as far as the exchange effect is concerned, the energy will always decrease on magnetization. On the other hand, the orbital part of the wave function will ordinarily change in such a way as to increase the other terms in the energy of Eq. (39.1), in particular the kinetic energy, and the potential energy in the field of the nuclei and all electrons. The reason for this is simple. As we can see from Fig. 52, where we show one-electron energies of electrons of both spins, it is clear that in the process of magnetization we remove electrons of minus spin from a state of low one-electron energy, and on account of the FERMI principle, we must increase their one-electron energy when we give them a plus spin, since we are assuming that there are already more electrons of plus than of minus spin.

We can now see from this simple illustration what are the factors determining whether the substance will be ferromagnetic or not. If the energy band is wide — that is, if the density of states $dN/d\varepsilon$ is small—then it will require a large increase in one-electron energy to raise electrons from the highest occupied state of minus spin and to put them in the lowest empty state of plus spin. If the band is narrow, or $dN/d\varepsilon$ is large, the increase in one-electron energy will be correspondingly small. For an exchange interaction of a given size, then, ferromagnetism will be favored by narrow energy bands. Now the magnitude of the exchange interaction proves not to be very sensitive to the type of material we are dealing with. Thus we conclude that the primary factor in determining whether a given material is ferromagnetic or not is the width of the energy bands. When we examine the question, we find, as the writer has pointed out [*216*], that the $3d$ bands in the known ferromagnetic elements Fe, Co, and Ni, are probably the narrowest partly occupied energy bands which we find among the elements, and this is consistent with the fact that these are the elements showing ferromagnetism most conspicuously.

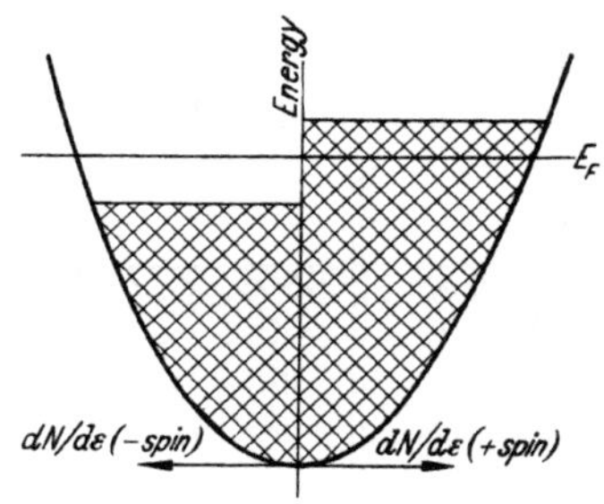


Fig. 52. Diagram to illustrate ferromagnetism. Density of states $dN/d\varepsilon$ is shown horizontally, as indicated, for electrons of both spins (the free-electron case, where $dN/d\varepsilon$ is proportional to $\varepsilon^{\frac{1}{2}}$, is shown, but the argument will hold for any $dN/d\varepsilon$). In the unmagnetized state, levels of both spins will be filled to a common FERMI level $E_F$. Diagram illustrates a case where some states of —-spin have been emptied, the electrons going into states of +-spin, to produce a magnetic moment, and increase the FERMI energy.

In Fig. 52, we have tacitly assumed that electrons of both spins have the same one-electron energies, and we have ended up with a ferromagnetic state in which the one-electron energy of the electrons of plus spin (if there is positive magnetization) is greater than that of minus spin. There is another equivalent, but perhaps better, way to look at the same question. If we compute the one-electron energy of a HARTREE-FOCK one-electron function, using Eq. (17.3), we find that it equals the average kinetic energy, the average potential energy in the field of the nuclei, the average potential energy in the field of all electrons, including itself, minus the average potential energy in the field of the exchange charge This latter term is different, depending on the spin of the electron in question, though we shall show later that it is approximately independent of the propagation constant of the electron. If the electron's spin points in the same direction as the magnetization, the exchange correction will be numerically greater, leading to a lower one-electron energy, than if its spin points in the opposite direction. Properly, then, the picture of one-electron energies should not be drawn as in Fig. 52, but rather as in Fig. 53, the whole set of energies of electrons of plus spin being displaced downward with respect to those of minus spin. When we handle the problem in this way, we can use an ordinary FERMI statistical approach: the FERMI level will be at the same height for electrons of both spins, and equilibrium will demand that the one-electron energies of the topmost electrons of both spins should be the same. This argument is really no different from the other one, but the figure is drawn differently.

This is one of the cases, as we pointed out in Sect. 20, where we must be on our guard in the application of the FERMI statistics. The reason is that the effective potential energy in which an electron moves depends on what the other electrons are doing. In this case, it depends on the total magnetization. In Eq. (17.3), we see that in a case of magnetism, the effective potential energy will be different for an electron of plus or of minus spin. For the electron of plus spin (assuming positive magnetization) the exchange correction will result in a lower potential energy than for an electron of minus spin. This difference is what is indicated in Fig. 53. But this difference in potential energy depends on the magnetization, and to a first approximation is proportional to it. Hence we must use a self-consistent field, in this magnetic problem, which is not only different for electrons of the two spins, but which depends on the magnetization. This is the reason why ferromagnetism is often referred to as a cooperative phenomenon; the potential in which a given electron moves depends in an essential way on what the other electrons are doing.

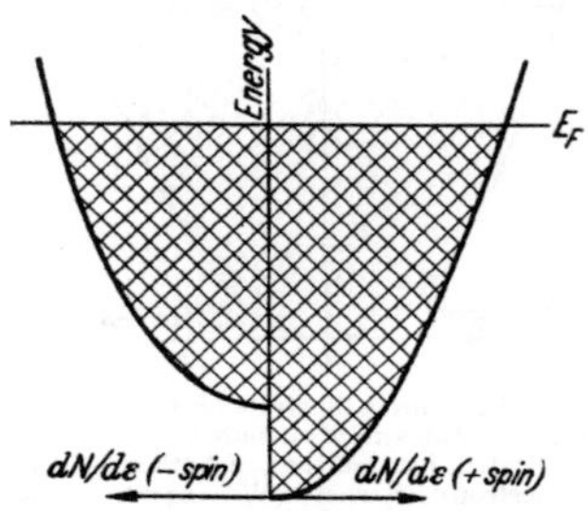


Fig. 53. Same case shown in Fig. 52, but drawn so that band of electrons of +-spin is displaced downward with respect to that of −-spin, on account of exchange, and FERMI levels are equalized.

To proceed further with our discussion of the energy-band explanation of ferromagnetism, we must consider the various terms going into the energy expression of Eq. (39.1), so that we can describe how to carry out the actual calculations of energy change on magnetization. The only terms which depend on magnetization are those in which we are dealing with an outer electron in the conduction band, which is able to reverse its spin. For such an electron, the first two terms of Eq. (39.1) give the kinetic energy of the electron, its potential energy in the field of the nuclei, and its kinetic energy in the field of inner electrons; just the same terms, in other words, which enter into the one-electron energy. Thus a calculation of energy bands, disregarding the exchange effects, will provide these terms in the energy. We are left with the exchange terms coming from pairs of electrons in the conduction band. The nature of these exchange terms, and in particular the way in which they vary with internuclear distance, is important for the theory of ferromagnetism.

An exchange integral, of form $\int u_i^*(x_1)\, u_j^*(x_2)\, g_{12}\, u_j(x_1)\, u_i(x_2)\, dv_1\, dv_2$, between two energy-band wave functions, has very different form, depending on whether we are dealing with free electrons, or with rather tightly bound electrons such as we have in the $3d$ shell in the transition elements. For two free electrons with propagation vectors $\boldsymbol{k}_i$ and $\boldsymbol{k}_j$, the integral was computed by BLOCH [22], and is found to be

$$I = \frac{4\pi e^2}{|\boldsymbol{k}_i - \boldsymbol{k}_j|^2 N \Omega} \tag{39.2}$$

where $e$ is the electronic charge, $\Omega$ the volume of the unit cell. We find, in other words, that the integral decreases to zero as the internuclear distance becomes infinite. This is natural. The charge density in this limit becomes more and more spread out, and the COULOMB interaction vanishes. This exchange integral was used by BLOCH in a study of the ferromagnetism of free electrons, which formed the first treatment of ferromagnetism by the band method.

On the other hand, if the energy-band functions represent rather tightly bound electrons, the situation is very different. In such a case we may use a tight-binding approximation for the wave functions, writing them in the form

$$u_i(\boldsymbol{r}) = N^{-\frac{1}{2}} \sum_n \exp(i\,\boldsymbol{k}_i \cdot \boldsymbol{R}_n)\, a(\boldsymbol{r} - \boldsymbol{R}_n) \tag{39.3}$$

where the summation is over the $N$ atoms in the periodic volume of the crystal (we are assuming one atom per unit cell, for this simple problem), and $a(\boldsymbol{r} - \boldsymbol{R}_n)$ is the atomic orbital centered on the lattice point at vector position $\boldsymbol{R}_n$. The wave function defined by Eq. (39.3) is not exactly normalized, since the $a$'s will not be exactly orthogonal to each other, but it will be approximately normalized [Eq. (39.3) would exactly express the wave function if the $a$'s were Wannier functions rather than atomic orbitals.] Then the exchange integral becomes

$$\left.\begin{aligned} I = N^{-2} \sum_{n,m,q,p} &\exp i\left[\boldsymbol{k}_i \cdot (\boldsymbol{R}_q - \boldsymbol{R}_n) + \boldsymbol{k}_j \cdot (\boldsymbol{R}_p - \boldsymbol{R}_m)\right] \times \\ &\times \int a^*(\boldsymbol{r}_1 - \boldsymbol{R}_n)\, a^*(\boldsymbol{r}_2 - \boldsymbol{R}_m)\, g_{12}\, a(\boldsymbol{r}_1 - \boldsymbol{R}_p)\, a(\boldsymbol{r}_2 - \boldsymbol{R}_q)\, dv_1\, dv_2. \end{aligned}\right\} \tag{39.4}$$

The integral in Eq. (39.4) is in general a four-center integral, between a charge density $a^*(\boldsymbol{r}_1 - \boldsymbol{R}_n)\, a(\boldsymbol{r}_1 - \boldsymbol{R}_p)$ shared between the $n$th and $p$th lattice sites, and a density $a^*(\boldsymbol{r}_2 - \boldsymbol{R}_m)\, a(\boldsymbol{r}_2 - \boldsymbol{R}_q)$ shared between the $m$th and $q$th sites. By far the largest such term will come when $n$, $m$, $p$, and $q$ all refer to the same atomic site; then the integral will be an intra-atomic integral, the sort encountered in the theory of atomic multiplets. If we consider only this largest term in Eq. (39.4), to investigate its general behavior, we find that we have

$$I = N^{-1} \int a^*(\boldsymbol{r}_1)\, a^*(\boldsymbol{r}_2)\, g_{12}\, a(\boldsymbol{r}_1)\, a(\boldsymbol{r}_2)\, dv_1\, dv_2 \tag{39.5}$$

where we note that we have $N$ identical terms, coming when $n$, which equals $m$, $p$, and $q$, refers to any one of the lattice sites; we have written the integral in terms of the orbitals on the lattice site at the origin.

We now observe two very interesting features of the approximate expression of Eq. (39.5) for the exchange integral in the case of tightly bound electrons: first, it is independent of the propagation constants of the electrons; secondly, it is independent of internuclear distance, since it arises from interactions entirely on a single atom. Neither of these statements will be exactly correct, if we consider in addition the terms of Eq. (39.4) for which $n$, $m$, $p$, and $q$ are not all identical, but nevertheless these correction terms, depending on the $\boldsymbol{k}$'s and on internuclear distance, are relatively small. We see, then, the completely different behavior of the exchange integral for bound electrons from that found by Bloch for free electrons, given in Eq. (39.2).

For the actual problem of ferromagnetism, this case of tight binding is much closer to the truth than the free-electron case of Eq. (39.2). There are then two consequences of these properties of the exchange integral $I$. In the first place, we were justified earlier when we assumed that the exchange terms would be approximately independent of the propagation constant, so that the effect of exchange would be merely to displace the energies of the states of one spin relative to those of the other. But secondly, there is a remarkable consequence if we consider the limiting case of large internuclear distances. In this limit, the exchange integral will stay finite, approaching in the limit the value of Eq. (39.5). On the other hand, the energy bands will always get narrower and narrower, approaching zero width in the limiting case of infinite internuclear distance. Consequently we should conclude that at sufficiently large internuclear distance, ferromagnetism would occur with any substance: the reduction of energy of the ferromagnetic state on account of the exchange effect would remain finite out to infinite distances, while in this limit it would require no one-electron energy to reverse the spins of the electrons. This conclusion, and in particular the corollary that there should be a finite difference of energy between the magnetized and non-magnetized states at infinite distance on account of the finite exchange effect, certainly seem incorrect. We shall find in the next section that they are in fact incorrect, and that they must be corrected by considering correlation. When we examine the effect of correlation, we shall find that, even though this makes a profound difference in the results for very large internuclear distance, still it seems likely that it leaves the general explanation of ferromagnetism provided by the band theory unaltered.

**40. Correlation and ferromagnetism.** At the end of the preceding section, we have pointed out that if we take the energy-band theory of ferromagnetism seriously, we should conclude that in every case where we have a partially filled energy band, the system would become ferromagnetic at infinite internuclear distance, the magnetized state having an energy lower than the non-magnetized by a finite amount. This is an incorrect result, since certainly if the atoms are an infinite distance apart, the energy cannot depend on how the spins are oriented in the various atoms. The writer [*228*] has pointed out how configuration interaction, and the correlation effect arising from it, resolve this difficulty.

Here again, as in our study of the relation of correlation energy to the binding energy of a crystal in Chapter VII, we can get a great deal of insight into the problem by considering simple molecules, starting again with hydrogen. If we consider the hydrogen molecule from the molecular orbital point of view, there are two molecular orbitals, one symmetric and one antisymmetric, which arise from the $1s$ state of atomic hydrogen. These orbitals, taken together, form the analogue of an energy band. If they were filled, they could hold four electrons in the complete molecule, one of each spin in each of the orbitals, but of course the molecule contains only two electrons. In the ground state of the molecule, as set up by the HARTREE-FOCK or molecular orbital method, we have two electrons in the lower, symmetric molecular orbital, so that we have an analogue of a half-filled band. We have seen in Sect. 35 that the determinantal wave function set up in this way behaves improperly at infinite internuclear distance. The reason is that it allows the existence of positive and negative ions in fifty percent of the cases, so that its energy is too high, the average of the correct energy of the two neutral atoms, and the higher energy of a positive and a negative ion. By making a configuration interaction between this state, and the other determinantal wave function in which both electrons are in the upper, antisymmetric orbital, we secure a final wave function whose energy goes to the correct value at infinite separation, and which in fact corresponds to a generalized HEITLER-LONDON wave function.

This situation is shown in Fig. 54. Here we show the diagonal energy of the molecular-orbital ground state, going to too high an energy at infinity. We also show the diagonal energy of the other determinantal wave function, in which both electrons are in the upper molecular orbital. The configuration interaction then spreads the energies of the resulting states apart, as shown in Fig. 54, the amount of spreading increasing at large internuclear distances. This spreading is a result of a non-diagonal matrix component of energy between the two states essentially an exchange term, but having a form much like that of Eq. (39.5), in that it reduces to an atomic term at infinite internuclear distance. The result of the configuration interaction, of course, is to introduce a correlation effect, as we indicated in Sect. 35, resulting at infinite distance in neutral atoms, rather than the charge fluctuation described by the existence of positive and negative ions.

The state of the hydrogen molecule which we have been discussing is a singlet state, corresponding to having the electronic spins oppositely directed. We remember, however, that we can form a triplet state from the same molecular orbitals. This state arises when we have one electron in the symmetric, one in the antisymmetric, molecular orbital, both with the same spin. The energy of the triplet state is also shown in Fig. 54; it is the well-known repulsive state of HEITLER and LONDON. We observe from Fig. 54 that the energy of this triplet state, like that of the correct singlet state, goes to the energy of two neutral hydrogen atoms at infinite separation; in this limit, it makes no difference whether the spins of the two electrons are parallel or not. However, if we forget the configuration interaction or correlation effect, and use merely the diagonal energy of the single determinantal wave function representing the molecular orbital or HARTREE-FOCK ground state, then we find that that diagonal energy of the singlet state rises above the energy of the triplet at infinite separation.

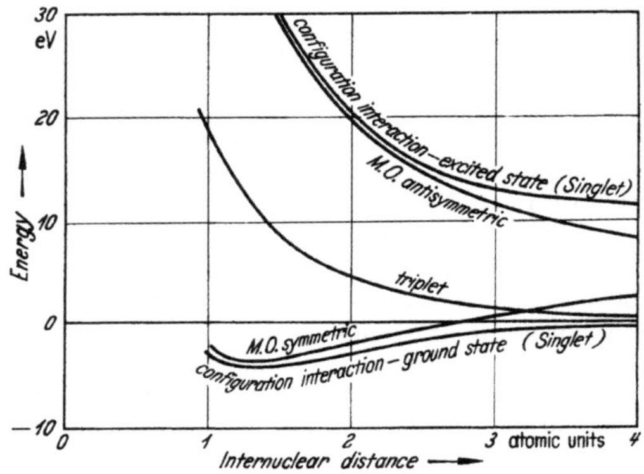


Fig. 54. States in the hydrogen molecule. State labelled "m.o. symmetric" represents energy of lowest molecular orbital state with both electrons in symmetric molecular orbitals, as in Fig. 49. That labelled "m.o. antisymmetric" is the energy of the lowest molecular orbital state with both electrons in antisymmetric molecular orbitals. (The calculation of this antisymmetric state is less accurate than for the symmetric state.) Configuration interaction spreads the states apart, as shown, the ground state resulting from this calculation being identical with that of Fig. 50. These states arising from configuration interaction are singlets. Also shown is the triplet state arising from the configuration in which one electron is in a symmetric, one in an antisymmetric, molecular orbital.

We may now draw analogies between the hydrogen case and the problem of ferromagnetism, and explain in this way the paradox which we mentioned at the beginning of this section. The singlet ground state of the hydrogen molecule is analogous to the non-magnetized state of a metal, while the triplet state is analogous to the magnetized state. According to the molecular-orbital method, the energy of the triplet state is higher than that of the singlet because we must raise one of the electrons from the symmetric to the antisymmetric molecular orbital, but it is lower on account of an exchange integral. The increase of energy in going from the symmetric to the antisymmetric molecular orbital is analogous to the increase of one-electron energy of an electron when a crystal is magnetized, as illustrated in Fig. 52. In this case, as with the metal, the exclusion principle requires that the electron be raised to a higher one-electron state if its spin is reversed, for it finds the lower one-electron state already occupied by an electron with the same spin. This increase of energy, however, vanishes at infinite internuclear distance. The decrease of energy on account of the exchange effect, on the contrary, is finite, and remains finite to infinite internuclear distance, in the molecule, as it is in the crystal. Thus, if we consider only the diagonal energy of the molecular-orbital ground state for the molecule, we should conclude that the magnetized state (the triplet) had a lower energy at infinite separation than the unmagnetized state (the singlet). It is only by configuration interaction, introducing the correlation effect which makes each atom electrically neutral at large distances, that this incorrect situation is remedied, and the energies of the magnetized and unmagnetized states are made identical at infinite distance.

It is clear that the same situation will hold for the crystal. Introduction of a correlation correction, by use of configuration interaction, is bound to lower the energy of the unmagnetized state more than that of the magnetized state. The reason is that the correlation correction affects pairs of electrons with opposite spin, lowering their energy by modifying the wave function so as to keep them out of each other's way. To a first approximation, correlation does not affect the energy of pairs of electrons with the same spin. This is why the energy of the triplet state in hydrogen is not affected by the configuration interaction. In the crystal, the magnetized state had a lower energy than the non-magnetized state, at infinite internuclear distance. A correctly carried out configuration interaction will lower the energy of the non-magnetized state more than that of the magnetized state, and in fact by just enough more so that after the correction is applied, the two states will have the same energy at infinite separation.

In this way, we see that correlation has removed the paradox which we found in our treatment of ferromagnetism by the energy band method. The question now arises, has not the correlation correction done too much? In hydrogen, it has brought the singlet state below the triplet at all internuclear distances. Will it not do the same thing in a crystal, eliminating entirely the possibility of ferromagnetism? Of course, we know that this cannot be the case, since ferromagnetism exists, but it is comforting to find that we have a diatomic molecule which can serve as a model for a ferromagnetic material, namely the oxygen molecule $O_2$. Study of this molecule can show the special situation which must arise to have a ferromagnetic solid.

When one studies the oxygen molecule by the method of molecular orbitals, one finds, as LENNARD-JONES showed [*146*], that the highest occupied orbital is a so-called $\pi$ orbital. That is, it corresponds to a wave function with one unit of orbital angular momentum about the axis of the molecule. This orbital is degenerate, for the orbital angular momentum can be pointing in either direction. Hence we really have two orbitals, capable of holding four electrons, and yet only two electrons are present in these orbitals in the ground state of the molecule. There are then several possibilities: we can have both electrons, one of each spin, in one of the orbitals, or one in each of the orbitals. In the latter case, the two electrons can have opposite spins or parallel spins, leading to a singlet or a triplet. It is found, both experimentally and by calculations of configuration interaction, that the lowest states come when one electron is in each orbital, and that of the two possibilities, the triplet lies lower, resulting in the paramagnetic nature of the oxygen molecule in its ground state.

Let us ask why the triplet state lies lower, in this case. The real reason is the degeneracy of the two $\pi$ states. Since one electron is in each of these, in either the singlet or triplet state, there is no increase in one-electron energy when the two electrons set their spins parallel to each other. Putting the situation in very simple language, if the electrons have the same spin, they are kept apart by the exclusion principle, while if they have opposite spin, they are kept apart by correlation; the exclusion principle is more effective, reducing the energy more for the triplet than for the singlet state. In either case, they have space enough (the two available orbitals) to get out of each other's way. And in either case, when we take account of configuration interaction properly, at infinite distance one electron will be on one atom, the other on the other, so that there will be no energy difference between the singlet and triplet states. This is shown in Fig. 55, in which we show an approximately calculated curve for the energy difference between the singlet and triplet energy, as a function of internuclear distance, as given by MECKLER [159]. The difference is zero at infinite distance, it increases as the distance decreases, the triplet lying lower, but then at still smaller distances there is indication of a change of sign, the triplet lying higher.

This example of oxygen suggests that in addition to a narrow energy band, there must be an additional requirement for the existence of ferromagnetism: a degenerate energy band, in the sense that it is derived from a degenerate atomic state. This conclusion was verified by a discussion given by KOSTER, STATZ, and the present writer [232], of the following problem. One started with a filled energy band, then removed two electrons. The resulting holes can have their spins parallel, giving a triplet state, or opposite, giving a singlet. The question was asked, which of these states lies lower? It was found that if the band is not degenerate, the singlet will lie lower, while if it is degenerate, there can easily be circumstances in which the triplet will lie lower. The reasons were essentially as in the oxygen molecule, but they came from typical energy-band arguments. It is, in fact, possible in this case to carry through the problem of configuration interaction, or of correlation, rather rigorously, using the general methods of Chapter V, expanding the wave functions in WANNIER functions, and treating the COULOMB interactions as perturbations.

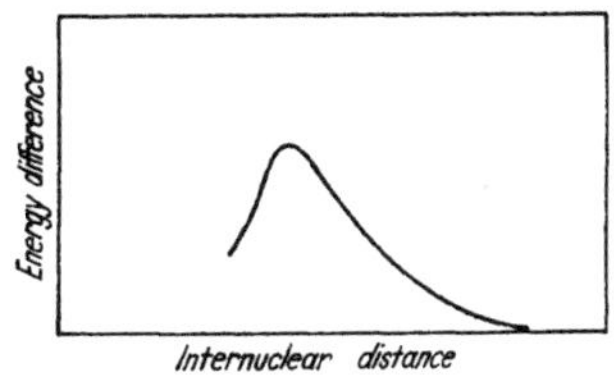


Fig. 55. Energy difference between magnetic and non-magnetic state, for oxygen molecule, from calculation of MECKLER [159], [228].

As a result of this argument, and of further study of configuration interaction, one comes to the following picture of the essential features resulting in ferromagnetism, and of the relation of correlation, and configuration interaction, to the problem. The ferromagnetic elements have $3d$-bands which are considerably more than half filled, so that it is more convenient to speak of the holes in them rather than the electrons. Suppose for example that on the average there is one hole per atom, out of the ten electrons per atom which the $d$-band could accomodate. At a finite internuclear distance, the holes will be somewhat free to circulate from atom to atom, and sometimes two or more holes will be found on the same atom. These holes will tend to have their spins line up parallel to each other. The exchange energy tending to this result is the same energy that causes atoms to have their states of maximum multiplicity lying lower: electrons keep out of each other's way better if they have parallel spin than if they have antiparallel spin, since exchange is more effective than correlation in keeping them apart. Hence when two or more holes in their wanderings are found on the same atom, their spins will line up, and the energy resulting in this is an intra-atomic integral, of the same sort found in an isolated atom. It is clearly only in the presence of a band arising from a degenerate atomic state that this can happen, since only then can two holes lie on the same atom, with approximately the same energy.

As the atoms get further and further apart, this behavior will die out, not on account of any lack of desire for the holes to set themselves with parallel spin, but simply because the correlation correction prevents the migration of holes from one atom to another, and the existence of a different number of holes on each atom from what would be required to make it electrically neutral. The energy difference between the magnetized and non-magnetized state, then, depends on an intra-atomic integral, but also on the probability of finding fluctuations in charge, the latter quantity vanishing at infinite internuclear distance.

We may now ask whether this more accurate picture of the ferromagnetic interaction in metals seriously affects the validity of the energy-band picture in actual cases. The probability is that it does not. In Fig. 56 we show schematically the energy of a magnetized and of a non-magnetized state of a metal, as computed by the energy-band theory without correlation. We then give also a curve showing how the curve for the non-magnetized state would be modified by configuration interaction or a correlation correction. From the appearance of the curve, it is very plausible that at the actual internuclear distance the correlation correction has not modified the results of the simple energy-band theory very seriously. In any case, the effect of correlation will be to diminish the effect of exchange. But if ferromagnetism actually exists in a given case, this is a good indication that the correlation correction amounts only to a relatively small fraction of the exchange term, so that if we calculate the latter from Eq. (39.5), and perhaps apply an assumed correction to it to reduce it on account of the correlation effect, we should get a fairly good calculation of the energy difference between the magnetized and non-magnetized states.

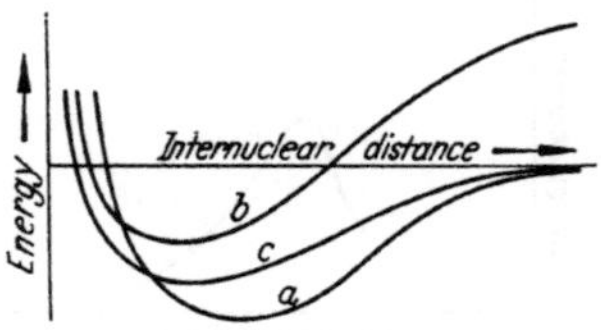


Fig. 56a—c. Energy of (*a*) magnetized state, (*b*) unmagnetized state without correlation, (*c*) unmagnetized state with correlation, as function of internuclear distance (schematic), for a magnetic material.

**41. Ferromagnetism in actual materials.** In the two preceding sections we have examined the general principles underlying ferromagnetism. Now let us see how these principles work out in actual ferromagnetic materials, principally the metals iron, cobalt, and nickel, and a series of alloys of these and neighboring elements. There are two pieces of experimental evidence which give some immediate information about the nature of their ferromagnetism. In the first place, the gyromagnetic ratio found for these cases is very close to 2, indicating that the magnetism arises from spins, not from orbital angular momenta. This justifies the general form of discussion which we have been giving in the preceding sections. Secondly, the saturation magnetization leads to non-integral numbers of BOHR magnetons: about 0.6 for nickel, 1.6 for cobalt, and 2.2 for iron. This suggests that it is the electrons in the $d$-band which are producing the ferromagnetism, and that overlying the $d$-band there is a band formed from $4s$ and $4p$-like wave functions, which holds a non-integral number of electrons, leaving a non-integral number of holes in the $d$-band. This is evidence in favor of the general band approach which we have been using. Specifically, in nickel and cobalt it seems, as MOTT has shown [*167*], as if there were about 0.6 electrons per atom in the $4s$-$4p$ band, leaving correspondingly 0.6 holes per atom in the $3d$-band in nickel, 1.6 in cobalt, explaining the observed saturation moment provided the holes have their spins lined up in the case of saturation. The case of iron, as we shall see later, is somewhat more complicated.

The calculated energy bands for the transition elements, as discussed in Chapter II, are consistent with these assumptions. One can go even further, and can attempt to calculate the various quantities which go into the theory of ferromagnetism, so as to see if the general point of view which we have been describing is quantitatively reasonable. Calculations along the lines suggested in Sect. 39 were made by the present writer [*220*] for nickel, with some extrapolation to cobalt and iron. More recently, a calculation for iron has been made by CALLAWAY [*41*], using recently-determined energy bands found by the orthogonalized plane wave method. The general results of both these calculations are that numerically the theory is entirely reasonable. Energy-band calculations were used for the one-electron energies, and the exchange integrals were estimated from known atomic integrals, using Eq. (39.5). The problem of finding the exchange integrals is really more complicated than was indicated in Sect. 39, since on account of the degeneracy of the $3d$-band there are a number of different exchange integrals between atomic orbitals, and a suitable average of these must be used. The general result of the calculations of the writer and of CALLAWAY is that the exchange terms are enough greater than the one-electron terms, for the known ferromagnetic elements, to account for the lower energy of the magnetized state, even if the exchange integrals were diminished by an appreciable amount on account of the correlation effect. The energy difference between magnetized and non-magnetized states of course is closely related to the CURIE temperature, and this comes out of the right order of magnitude, though quantitative accuracy can certainly not be claimed for the calculations. One has, in other words, a reasonable basis for a theory of ferromagnetism in the methods of the present chapter.

Table 4. *Relative probabilities of finding atoms with 2, 3, 4, 5 vacancies in Ni, and contribution of each case to magnetic energy, as estimated by* FRIEDEL *[74]*.

| *n* | Relative probability | Contribution to energy |
|---|---|---|
| 2 | 0.124 | 0.021 |
| 3 | 0.0235 | 0.012 |
| 4 | 0.0035 | 0.004 |
| 5 | 0.0004 | 0.001 |

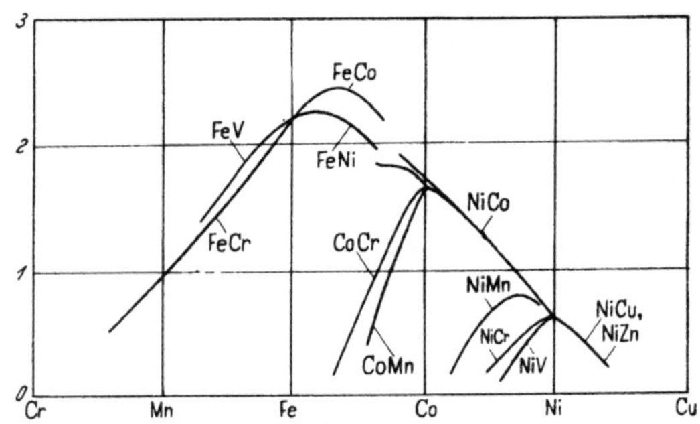


Fig. 57. Atomic moments of the ferromagnetic elements and their alloys [after J. E. GOLDMAN, Rev. Mod. Phys. 25, 108 (1953)].

A recent paper by FRIEDEL *[74]* has carried the arguments somewhat further, in that he has attempted to analyze the interactions between the holes in the $3d$-band, in terms of the probability of finding atoms with one, two, three ... holes, and the corresponding interaction energies. This is along the line of the discussion of Sect. 40, in which we have shown that the ferromagnetic tendency arises from the simultaneous presence of at least two holes in the $3d$-shell of an atom. FRIEDEL gives, in Table 4, the estimated probability of occurrence of different numbers of holes in the $3d$-shell of a Ni atom in the metal, and the contribution of each of these cases to the energy difference between magnetized and non-magnetized states. Great accuracy cannot be claimed for these calculations, but they presumably indicate the right general type of behavior. VAN VLECK *[247]* earlier had called attention to the value of making such an analysis.

Further light on the nature of ferromagnetic metals is provided by the ferromagnetic alloys. In many cases, neighboring ferromagnetic metals show wide ranges of solid solutions, which are themselves ferromagnetic. It was pointed out by PAULING *[179]* and by the present writer *[222]* that if we plot the saturation magnetization against composition, as in Fig. 57, we observe a very significant regularity of behavior. For most, though not all, of the alloys, the saturation magnetization lies on one of two straight lines. Of these, the one to the right, sloping down to zero at a point between nickel and copper, is easier to understand. This line, including the alloys of iron, cobalt, nickel, and copper with each other, behaves just as if the $3d$-bands had an approximately constant form through this series of elements, and were filled up to a point determined by the average number of electrons per atom. This is what we should find if we had the situation described in Sect. 28, indicating that the perturbations of the potential on account of the different types of atoms are not great enough to distort the bands greatly.

The other straight line, sloping downward on the left side of the diagram, from iron to a point between manganese and chromium, must have a different interpretation. It would be simple to interpret if the $3d$-band held about five electrons in all, rather than ten. Then the maximum magnetization would come when the band was half filled, between iron and cobalt, and the saturation moment would decrease either as the band filled, the case where we were proceeding toward nickel and copper, or when it was emptied, proceeding toward manganese and chromium. PAULING has postulated a theory of ferromagnetism which assumes that the $3d$-band is in fact divided in some such way as this, into a lower band which is concerned only with binding, and always has equal numbers of electrons with each spin, and an upper band which can have unbalanced spins, and hence can lead to ferromagnetism. There seems to be nothing in our discussion of the $3d$-band, in Chapter II, however, which gives any support to this hypothesis. All the calculations which have been made of this band indicate that the five sub-bands which make it up, derived from the five-fold degenerate $3d$ atomic level, are mixed up in a very complicated way, and leave no energy gap which would separate them into two bands holding about five electrons each. Consequently it seems very unlikely that PAULING's hypothesis can be justified.

Our calculations of the energy levels in the $3d$-band in Chapter II do, however, show a feature which presumably leads to the same result which would be secured by PAULING's hypothesis. We have found, in both the face-centered and body-centered structures, shown in Fig. 23, that the distribution curves, $dN/d\varepsilon$, show strong minima in density near the middles of the bands. If this dip is deep enough, it would have the same effect as a real gap between two half-bands, such as PAULING postulated. Let us suppose that in a given substance, the upper part of this band, with energies above the dip, contained only positive spins or empty states, while the lower half had all its states filled with electrons of both spins. Then we may inquire whether it would reduce the energy to remove an electron of negative spin from the levels just below the dip, and reverse its spin, raising it to the lowest unoccupied level of positive spin. From our discussion of Sect. 39, we know that the energy would be raised in this process on account of the increase of one-electron energy, lowered on account of the exchange effect, and the net result is made up of these two opposing tendencies. On account of the small density of states at the energy dip, the one-electron energy in this case would have to be increased a good deal to make this reversal of spin, so that most probably the total energy would increase. In other words, there would be a tendency for the electrons in states above the dip to line up their spins, but this would probably not extend to those electrons below the dip, so that for practical purposes, the dip in the distribution curve would act like PAULING's postulated gap between two half-bands, and would lead to the same sort of curve of saturation magnetization which we have found experimentally in Fig. 57.

PAULING has emphasized the role of the $d$ electrons in adding to the strength of binding, in the transition elements. All indices of bond strength, such as heat of vaporization, melting point, mechanical strength, and so on, rise to a maximum in the middle of the transition groups of elements, and fall at the end of the group. This is what is to be expected from the nature of the $d$-bands, and does not in any sense demand the splitting of the bands into two parts which PAULING has suggested. If we examine the nature of the wave functions in the band, we find that the lower states in the band correspond to bonding orbitals, in the sense that there is concentration of electronic charge in the regions between atoms; the upper states correspond to antibonding orbitals, with deficiency of charge in the regions between atoms. We thus expect that as we go through the transition elements, we shall find a maximum binding about at chromium, molybdenum, and tungsten, where the band is filled about half full, more or less up to the dip in energy distribution, and that the binding energy should then fall as we go to the further elements, where the added electrons are of an antibonding nature. This is qualitatively in agreement with the facts. It is unfortunate, however, that no really good calculation of metallic cohesion has been carried out on the basis of energy band theory, for one of the very tightly bound transition metals, to verify the quantitative correctness of this picture.

We have now discussed the two straight lines making up the main part of the curve of saturation magnetization, Fig. 57, but we have not taken up the special cases like the alloys of cobalt and nickel on the one hand with vanadium, chromium, and manganese on the other. The saturation magnetization of these alloys falls on entirely different curves from the other cases. We notice at once that the difference in atomic number between the two metals being alloyed together is here considerably greater than in the case of the metals whose points lie on the main curve. One suspects, then, that for these cases the alloyed elements lead to discrete energy levels separated from the energy bands of the material with which they are alloyed, while in the case of nearby elements, the energy bands are not disturbed. We have seen in Sect. 28 that this is what is to be expected.

Let us see if this hypothesis is capable of accounting directly for what is going on in these cases. Consider, for example, the case of a little chromium alloyed to nickel, which we discussed in Sect. 28. In that section, we saw that it was plausible that for each Cr atom added to the crystal, there would be several discrete levels pushed above the band. Suppose, for the sake of illustration, that there were three such discrete levels, each capable of holding two electrons. As a result of these discrete levels, the energy band will have lost enough levels to accomodate six electrons. The Cr atom has four electrons less than a Ni atom, for electrical neutrality. Hence we see that the number of levels in the band will have decreased two units more than the number of electrons, which means that the bands will fill up more than if the Cr had not replaced a Ni atom. In fact, there will be two fewer holes in the crystal, so that the saturation magnetization should decrease by two units. This is something like what is observed, the addition of Cr to the Ni resulting in a rapid decrease of saturation magnetization. It seems, then, that we are able to understand in a general way why an alloying atom whose atomic number is different by a considerable amount from the atom with which it is alloyed can behave quite differently from a neighboring atom.

**42. Antiferromagnetism.** Certain materials, principally inorganic compounds, of which MnO is an example, show antiferromagnetism; that is, at low temperature, alternating ions (in this case $Mn^{++}$) have magnetic moments pointing in opposite directions, so that though there is no magnetic moment as a whole, there is magnetism on an atomic scale. As the temperature is raised, the specific heat goes through an anomaly at a point called the Néel point, above which the material acts like a paramagnetic substance. The existence of such materials was first deduced from thermal data, and Néel [*174*] set up the postulate that they showed antiferromagnetism, and introduced a phenomenological theory to describe them. Still other materials, of which $Fe_3O_4$, and the ferrites, are examples, show what is called ferrimagnetism. Here, as in the antiferromagnetic substances, alternate ions have magnetic moments pointing in opposite directions, but those pointing one way have a greater magnetic moment than those in the other direction, so that there is a net magnetic moment, and in some ways these materials resemble ordinary ferromagnetic materials. The understanding of all these materials rested on hypotheses, until the method of neutron diffraction, in the hands of SHULL and others [*209*], [*210*], [*211*], made it possible to observe the individual atomic magnetic moments. Such observations have now given us very detailed information about a good many antiferromagnetic and ferrimagnetic substances, and have completely verified the general interpretation of their properties given principally by NÉEL.

It is obvious that the explanation of these phenomena must be closely analogous to that of ferromagnetism; in both cases the exclusion principle, and exchange effects, must be at the basis of the effect. NÉEL's theory is founded on interatomic exchange integrals of the type HEISENBERG introduced into the theory of ferromagnetism. We have not used such methods in our discussion of ferromagnetism, for these integrals, like the exchange integrals of HEITLER and LONDON, are based on methods of handling the many-electron problem which cannot be rigorously justified, on account of the difficulties of orthogonality, as the writer has pointed out [*224*], [*228*]. Consequently, as with ferromagnetism, we should like to see how the phenomena of antiferromagnetism and ferrimagnetism can fit in with our present approach, starting with the energy-band theory, and modifying it by configuration interaction to take account of correlation. We must admit at the outset that this type of explanation has not gone as far with these phenomena as with ferromagnetism.

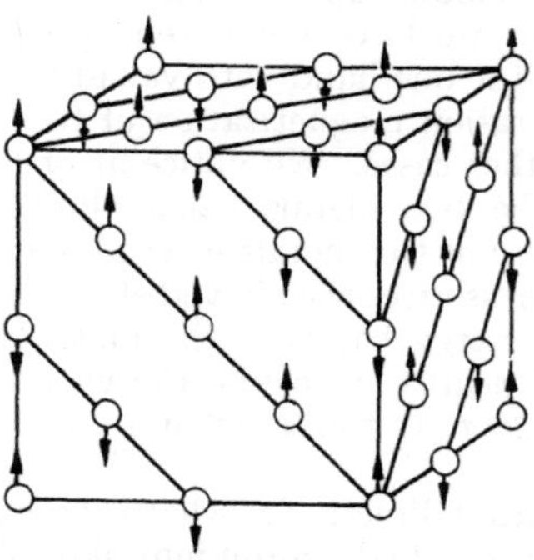

Fig. 58. Direction of spins of successive planes of $Mn^{++}$ ions in MnO crystal, following SHULL, STRAUSER, and WOLLAN [*210*]

Let us think, to be specific, of MnO. This crystal has the sodium chloride structure, and presumably can be thought of as an ionic crystal, composed of $Mn^{++}$ and $O^{--}$ ions. We note that the $Mn^{++}$ ion in its isolated state has five electrons outside of the argon shell, and has as its ground state the $^6S$ state of $3d^5$, in which each of the $3d$ orbitals contains one electron, all of parallel spin. Neutron diffraction methods indicate that the ion in the crystal has this same structure; the total magnetization of an ion, which can be determined by neutron methods, is 5 BOHR magnetons, as it should be. As for the arrangement of the spins, one finds that all the ions in one 111 plane of the crystal have their spins oriented in one way, those in the next plane in the opposite direction, and so on. Any explanation of the orientation in terms of interactions between neighboring $Mn^{++}$ ions, tending to line adjacent ions either parallel or antiparallel, is ruled out by the fact that each ion has six neighboring $Mn^{++}$ ions in its own plane with spins parallel to its own, and six equally close neighbors, three in the plane to one side and three to the other, as shown in Fig. 58, whose spins are opposite.

One can first ask what information the energy-band theory can give us about the problem, without considering configuration interaction. As in the problem of ferromagnetism, we shall have two different sets of energy bands, one for electrons of +-spin, one for electrons of −-spin. The effective potential energy in which an electron of, for instance, +-spin moves will be lower in the $Mn^{++}$ ions of +-spin than in those of −-spin, on account of the exchange effect.

Consequently, as the writer has pointed out [*226*], the periodic potential problem will have a unit cell twice as great as one ordinarily assumes, the two types of $Mn^{++}$ ions being effectively different centers of force. This will have a profound effect on the energy bands. For the electrons of +-spin, for instance, the calculation will be as if there were two quite different sorts of $Mn^{++}$ ions in the (doubled) unit cell, so that there will be two sets of energy bands, each five-fold degenerate, one with wave functions concentrated on the $Mn^{++}$ ions which are known to have +-spin, the others on the ions with —-spin. Since the effective potential energy is higher for the second type of ion, by the amount of the exchange energy, these second energy levels, concentrated on the ions with —-spin, will lie much higher than those on the ions with +-spin. The $Mn^{++}$ ions are far enough apart so that the bands are not much broadened by overlapping. We then have two quite separated sets of energy bands, each capable of holding five electrons of +-spin per (doubled) unit cell. These energy bands and wave functions are shown schematically in Fig. 59. Similarly for the electrons of —-spin we have two energy bands, with just the same energies as for +-spin, but in each case with the wave function concentrated on the opposite type of atom.

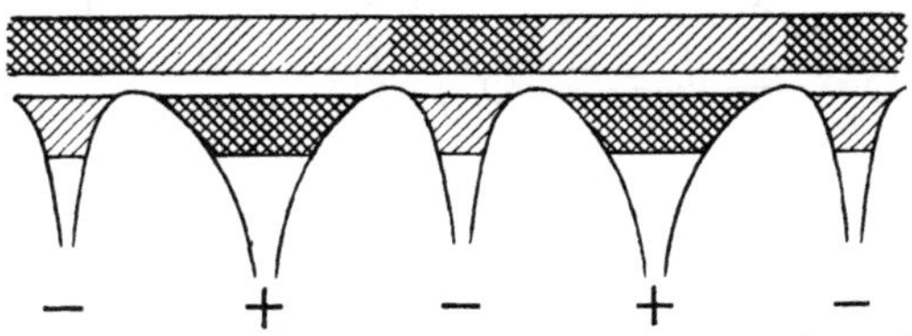


Fig. 59. Energy bands for an electron of +-spin in a MnO crystal (schematic). We show the potential energy through a series of $Mn^{++}$ ions with +- and —-spin, the potential well for an electron of +-spin being deeper in the ions of +-spin. The $3d$-band is split into two sub-bands, as shown, of which only the lower one is occupied. The wave function in each band is largest in the region indicated by cross-shading, smaller in the region with single shading.

Since we have five electrons per atom, or ten per unit cell, the electrons will fill the lower energy bands, leaving the upper ones empty. In other words, assuming these bands, the electrons of +-spin will be concentrated on half the atoms, those of —-spin on the other half, just as we are supposing. Hence, energy bands of this type form a self-consistent description of the antiferromagnetic state. But we really have not learned very much about the problem from this description of the bands. The electrons will all occupy the lower bands, lower on account of the exchange energy; but this is just the exchange effect present in the individual ions, leading to a lower energy when the spins are all parallel in that ion. Our argument has not answered at all, for instance, the question of why the electrons prefer to arrange themselves antiferromagnetically rather than ferromagnetically. To answer this question, we must proceed quite differently, and more intuitively, more along the lines of configuration interaction.

We notice, in the first place, that the nearest neighbors of a $Mn^{++}$ ion are six $O^{--}$ ions, not the other $Mn^{++}$ ions, which are considerably further away. There is good evidence that it is these oxygen ions which are the essential feature in the existence of antiferromagnetism, in such a crystal. There are several ways of describing the action of the oxygen, but the following one is perhaps the simplest. Consider a $Mn^{++}$ ion with +-spins on its $3d$ electrons. On account of the exclusion principle, the electrons of this ion will repel the electrons of +-spin in the neighboring $O^{--}$ ions, while there will not be so much repulsion of the electrons of —-spin in the oxygens; the latter repulsion would occur only on account of the correlation effect, not on account of exchange. We remember that it is the repulsion of the $Mn^{++}$ electrons for the electrons of the oxygen, on account of exchange, which is responsible for the rigidity of the crystal, preventing it from collapsing under the action of the COULOMB attractions between oppositely charged ions.

This repulsion is going to tend to deform the outer electronic shell of the oxygen ions. One notices that each oxygen has six neighboring $Mn^{++}$ ions, of which three, on one side, have plus spin, and the other three, on the other side, have minus spin, as shown in Fig. 60. There will then be a tendency for the electrons of +-spin in the oxygen to be pushed away from the $Mn^{++}$ ions with +-spin. This is possible if the $Mn^{++}$ ions on the other side of the oxygen have −-spin, as in the antiferromagnetic arrangement. On the other hand, if they had a ferromagnetic arrangement, with +-spin on both sets of $Mn^{++}$ ions, the oxygen electrons with +-spin would be pushed from both sides, and would have to be compressed, involving a much greater increase of energy than in the antiferromagnetic case. It is presumably this difference in the behavior of the oxygen ions which makes the antiferromagnetic state more stable than the ferromagnetic one, for these crystals.

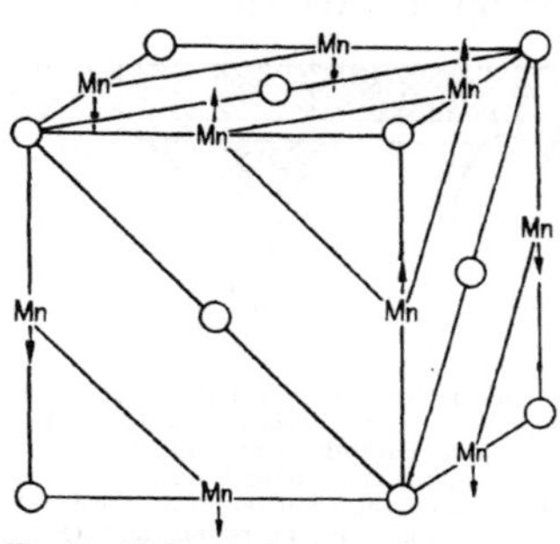

Fig. 60. MnO crystal, showing $O^{--}$-ion as well as $Mn^{++}$; arrows indicate directions of spins on $Mn^{++}$.

It is not simple to work out these ideas quantitatively. ANDERSON [4], [5], [6], following a method due to KRAMERS [142], has used a scheme called superexchange to attempt to discuss the problem. In this method, use is made of certain excited states of the oxygen electrons, and a configuration interaction is set up involving these excited states. Such a discussion might be correct if it were carried through with adequate attention to problems of orthogonality and such matters, but the discussion which ANDERSON has given cannot be considered rigorous. As a general matter, however, we see that we must describe the distortion of the oxygen electrons by the exchange interaction with the electrons of the $Mn^{++}$ on one side of it, and then in turn must consider the interaction of these distorted oxygen wave functions with the $Mn^{++}$ ion on the other side. This involves two successive exchange effects (hence the name superexchange), and to describe the modified or distorted oxygen wave functions, we must mix in excited wave functions, carrying out a configuration interaction. The general method of approach is then seen to be reasonable.

In order to get an example simple enough to carry through rigorously, PRATT [189] has discussed three atoms in a chain, the center one containing a closed shell analogous to the oxygen ion (PRATT used a closed shell of two 1$s$ electrons, one of each spin). Each of the end atoms contains a single electron (the atom is supposed to be hydrogen-like), one with a +-spin and one with a −-spin to represent the antiferromagnetic state, or both with +-spins to represent the ferromagnetic state. This problem is simple enough so that PRATT could carry through the complete configuration interaction necessary to distinguish between the two cases, and he found in fact that the antiferromagnetic state, with opposite spins in the end atoms, had a lower energy than the ferromagnetic state. The calculation was very difficult, but showed definitely that the simpler type of calculation made by ANDERSON and others was inadequate, and it has not yet been extended to more complicated cases. It does seem, however, to justify the picture of the effect of the oxygen ions which we have presented, and to show that they are the essential features in the antiferromagnetism of such crystals as MnO. Similar explanations hold for a variety of inorganic antiferromagnetic crystals, and for the ferrites as well; here again there is an oxygen ion between two sets of magnetized ions, and the ions on the two sides of the oxygen tend to set themselves with their spins opposite to each other.

The sketch which we have given of the explanation of antiferromagnetism in inorganic crystals is not very quantitative, but it probably expresses the physical facts better than the type of phenomenological theory of NÉEL, based on exchange integrals between pairs of ions. It would be desirable to investigate the relation between the two types of theories. A beginning has been made in this direction, but much further work remains to be done.

In addition to the antiferromagnetic compounds, there is some rather inconclusive evidence [*211*] that some metals and alloys are antiferromagnetic. With them, of course an explanation in the terms just given is impossible, since they do not contain oxygen ions. The writer and KOSTER [*230*] have discussed the case of chromium, which seems to have a small amount of antiferromagnetism on the basis of the energy band theory, showing that in this case the energy bands associated with electrons of +- and —-spin would not be completely separated from each other, as we have described in MnO, but rather would be only slightly modified, an energy gap appearing in the middle of the band. It is possible in this way to get an explanation of the probable behavior of chromium. However, even the existence of its antiferromagnetism is still somewhat in doubt, and the whole question of the antiferromagnetism of metals must probably wait for further experimental clarification before the theory can be developed completely.

It is desirable, however, to comment at this point on the suggestions made at one time by ZENER [*265*], [*266*], [*267*], [*268*] concerning possible antiferromagnetism of some of the metals. ZENER felt that in some of the body-centered cubic metals, such as vanadium and chromium, the atoms would exist with large magnetic moments, as the $Mn^{++}$ ion has in MnO, and that these moments would be arranged antiferromagnetically, those with one spin being located at the corners of the cubes, those with the other spin at the centers. Experiment [*211*] shows conclusively that this hypothesis is not correct; vanadium shows no magnetic moments at all, and chromium, if it shows some, has a magnetic moment very much smaller than ZENER postulated. The energy-band theory shows why this should be so, and points out the error which ZENER made in his analysis of the problem. ZENER realized that there was an exchange energy which decreased when the electrons of an atom set their spins parallel to each other. He estimated this effect from the corresponding effect in an atom, and this is in accordance with Eq. (39.5). He forgot, however, that the exchange effect is opposed by the effect of the breadth of the energy bands, as we emphasized in Sect. 39. According to our interpretation, in vanadium the energy bands are so broad that it requires more energy to set the spins of two electrons parallel to each other, on account of the increase of one-electron energy of the electron whose spin is reversed, than one gains in exchange energy. The state of lowest energy, then, is that in which each of the lowest energy-band states is filled with two electrons, one of each spin. There is no magnetic moment at all on the individual atoms. This is just what the neutron diffraction shows. The evidence is overwhelming that the magnetic moments which the atoms would possess if they were isolated from each other are completely removed when they are combined into the metallic crystal, and this is very easy to explain according to the energy-band theory.

**43. Paramagnetism.** A number of different types of paramagnetism are known. There is, in the first place, the type found in crystals containing rather widely separated magnetic ions, usually of the iron group or the rare earth group. This type of paramagnetism can be handled by simple methods, developed particularly by VAN VLECK and his students, in a large literature of which we shall list only review articles [246], [248]. We can treat each magnetic ion as being independent of the others, affected only by its immediate environment, and an external magnetic field. The conventional theory, based on a crystalline field, is too simplified, but a more accurate theory, based on the various energy levels and wave functions of a crystal containing a magnetic ion which can exist in various orientations, should not be hard to set up, and would lead to results qualitatively like those of the crystalline field theory. We shall not discuss this problem further, though it would carry one into interesting consideration of the wave functions in the neighborhood of an isolated ion in the crystal, which could be treated much like an impurity ion.

A second type of paramagnetism is that usually known as spin paramagnetism; it was treated by PAULI [178] for a free-electron model of a metal, in the early days of quantum mechanics. It is easy to describe this phenomenon in terms of the energy-band theory of a non-ferromagnetic metal. In the presence of an external magnetic field, the energy of the electrons whose spins are lined up opposite to the external field will be increased, and the energy of those lined up parallel to the field will be decreased, by an amount proportional to the external field. To attain equilibrium, enough electrons must reverse their spins so that the states of each spin will be filled up to a common FERMI level. That is, a number of electrons equal to $\frac{1}{2}\, dN/d\varepsilon\, \mu H$, where $dN/d\varepsilon$ is the density of states for all electrons, $\mu$ is the magnetic moment of the electron spin, and $H$ is the external field, must reverse their spins, resulting in a net magnetization of $\mu^2\, dN/d\varepsilon\, H$, proportional to the field, and to the density of states. This is the spin paramagnetism, and it proves to give a good account of the paramagnetic behavior of non-magnetic metals. The problem is complicated by the existence of diamagnetism, a magnetic moment of opposite sign, partly compensating the paramagnetism. We shall not go into the theory of diamagnetism, treated by LANDAU [144], but it can be shown that for free electrons, there is a diamagnetic effect equal to $\frac{1}{3}$ of the paramagnetism, so that this diamagnetic correction must be taken into account to get a reliable comparison of theory and experiment.

The remaining important types of paramagnetism are those found in ferromagnetic, antiferromagnetic, and ferrimagnetic materials, at temperatures above the CURIE or NÉEL points. Simple phenomenological theories of ferromagnetism can of course be set up, based on the WEISS theory, and these lead immediately to the conclusion that the material will become paramagnetic above the CURIE point, the magnetic susceptibility obeying CURIE's law, or being proportional to $1/(T-T_c)$, where $T_c$ is the CURIE temperature. However, this simple theory is hardly adequate for our present purposes, since the energy-band theory of ferromagnetism is rather different in its mathematical details from the WEISS theory. One can set up the temperature dependence of ferromagnetism on the basis of the energy-band theory; this has been done in particular by STONER and his school [236] to [239]. One is then led to a result similar to that found by the WEISS theory. These discussions of STONER are still, however, rather far from the methods advocated in the present article, since they are based on assumed behaviors of exchange integrals which are not strictly derived from the theory as modified by the correlation effect.

The writer believes that satisfactory theories of the temperature dependence of ferromagnetism, antiferromagnetism, and ferrimagnetism, together with the high-temperature paramagnetism to which they lead, do not exist at present. Many interesting questions are not even touched on in present theory; for instance, above the CURIE point, where long-range magnetic order ceases to exist in a ferromagnetic substance, does short-range order persist? In other words, does one have ferromagnetic domains whose dimensions are only a few atoms across, persisting at temperatures above the CURIE point? The analogy with the known behavior of order-disorder transformations in solids would suggest that one has such a situation. Furthermore, the methods of neutron diffraction are here going further than the theories, and WILKINSON and SHULL *[256]* find that such a short-range order does in fact continue in iron, for several hundred degrees above the CURIE point. It seems likely that in the near future the pace will be set in the study of magnetism by the experimental results of neutron diffraction, and further theory, based on the fundamental theory of solids rather than on the inadequate phenomenological basis of the WEISS theory, and of HEISENBERG exchange integrals, is very much needed.

## Bibliography.

Note: This bibliography includes not only the references mentioned in the text, but also a few useful books and review articles, not specifically referred to in the text.


*[1]* ADAMS, E. N. II: Motion of an electron in a perturbed periodic potential. Phys. Rev. **85**, 41 (1952); **86**, 427 (L) (1952).

*[2]* ADAMS, E. N. II: Plasma oscillations in metals. Phys. Rev. **98**, 947 (1955).

*[3]* ADAMS, E. N. II: Auxiliary conditions in the BOHM-PINES theory of the electron gas. Phys. Rev. **98**, 1130 (L) (1955).

*[4]* ANDERSON, P. W.: Antiferromagnetism. Theory of superexchange interaction. Phys. Rev. **79**, 350 (1950).

*[5]* ANDERSON, P. W.: Generalizations of the WEISS molecular field theory of antiferromagnetism. Phys. Rev. **79**, 705 (1950).

*[6]* ANDERSON, P. W.: Approximate theory of the antiferromagnetic ground state. Phys. Rev. **86**, 694 (1952).

*[7]* ANTONCIK, E., and M. TRLIFAJ: A note on the group analysis of the wave function of valency electrons in a crystal. Czech. J. Phys. **1**, 97 (1952).

*[8]* BARDEEN, J.: Electron exchange in the theory of metals. Phys. Rev. **50**, 1098 (L) (1936)

*[9]* BARDEEN, J.: An improved calculation of the energies of metallic Li and Na. J. Chem. Phys. **6**, 367 (1938).

*[10]* BARDEEN, J.: Compressibilities of the alkali metals. J. Chem. Phys. **6**, 372 (1938).

*[11]* BEGBIE, G. H.: Thermal scattering of x-rays by crystals II. The thermal scattering of the face-centered cubic and the close-packed hexagonal lattices. Proc. Roy. Soc. Lond., Ser. A **188**, 189 (1947).

*[12]* BEGBIE, G. H., and M. BORN: Thermal scattering of x-rays by crystals. Dynamical formulation. Proc. Roy. Soc. Lond., Ser. A **188**, 179 (1947).

*[13]* BELL, D. G.: Group theory and crystal lattices. Rev. Mod. Phys. **26**, 311 (1954).

*[14]* BELL, D. G., R. HENSMAN, D. P. JENKINS and L. PINCHERLE: A note on the band structure of silicon. Proc. Phys. Soc. Lond. A **67**, 562 (1954).

*[15]* BELL, D. G., D. M. HUM, L. PINCHERLE, D. W. SCIAMA, and P. M. WOODWARD: The electronic band structure of PbS. Proc. Roy. Soc. Lond., Ser. A **217**, 71 (1953).

*[16]* BETHE, H.: Theorie der Beugung von Elektronen an Kristallen. Ann. Phys., Lpz. **87**, 55 (1928).

*[17]* BETHE, H.: Termaufspaltung in Kristallen. Ann. Phys., Lpz. **3**, 133 (1929).

*[18]* BLACKMAN, M.: Contributions to the theory of the specific heat of crystals. I. Lattice theory and continuum theory. Proc. Roy. Soc. Lond., Ser. A **148**, 365 (1935). — II. On the vibrational spectrum of cubical lattices and its application to the specific heat of crystals. Proc. Roy. Soc. Lond., Ser. A **148**, 384 (1935).

*[19]* BLACKMAN, M.: On the vibrational spectrum of a three-dimensional lattice. Proc. Roy. Soc. Lond., Ser. A **159**, 416 (1937).

*[20]* BLACKMAN, M.: The theory of the specific heat of solids. Rep. Progr. Phys. **8**, 11 (1941).

*[21]* BLOCH, F.: Über die Quantenmechanik der Elektronen in Kristallgittern. Z. Physik **52**, 555 (1928).

*[22]* BLOCH, F.: Zur Theorie des Ferromagnetismus. Z. Physik **61**, 209 (1930).

*[23]* BOHM, D., and D. PINES: Screening of electronic interactions in a metal. Phys. Rev. **80**, 903 (1950).

*[24]* BOHM, D., and D. PINES: Collective description of electron interactions. I. Magnetic interactions. Phys. Rev. **82**, 625 (1951).

[25] BOHM, D., and D. PINES: Collective description of electron interactions. III. COULOMB interactions in a degenerate electron gas. Phys. Rev. **92**, 609 (1953).
[26] BORN, M.: Theoretical investigation on the relation between crystal dynamics and x-ray scattering. Rep. Progr. Phys. **9**, 294 (1942).
[27] BORN, M., and K. HUANG: Dynamical Theory of Crystal Lattices, Oxford, 1954. A general description of the theory of crystal vibrations and their relation to the physical properties of solids.
[28] BORN, M., and T. VON KARMAN: Über Schwingungen in Raumgittern. Phys. Z. **13**, 297 (1912).
[29] BOUCKAERT, L. P., R. SMOLUCHOWSKI and E. WIGNER: Theory of BRILLOUIN zones and symmetry properties of wave functions in crystals. Phys. Rev. **50**, 58 (1936).
[30] BRENNAN, R. O.: The interlayer binding in graphite. J. Chem. Phys. **20**, 40 (1952).
[31] BRILLOUIN, L.: Les électrons libres dans les métaux et la rôle des réflexions de BRAGG. C. R. Acad. Sci. Paris **191**, 198 (1930). — J. Phys. Radium **1**, 377 (1930).
[32] BRILLOUIN, L.: Les électrons dans les métaux et le classement des ondes de DE BROGLIE correspondantes. C. R. Acad. Sci. Paris **191**, 292 (1930).
[33] BRILLOUIN, L.: Quantenstatistik. Berlin: Springer 1931. A valuable summary both of quantum statistics, and of the theory of BRILLOUIN zones.
[34] BRILLOUIN, L.: Les problèmes de perturbation et les champs self-consistents. J. Phys. Radium **3**, 374 (1932).
[35] BRILLOUIN, L.: La méthode du champ self-consistent, Actualités scientifiques, Nr. 71. Paris: Hermann 1933.
[36] BRILLOUIN, L.: Champs self-consistent et électrons métalliques. III. J. Phys. Radium **4**, 1 (1933).
[37] BRILLOUIN, L.: Les électrons dans les métaux du point de vue ondulatoire, Actualités scientifiques, Nr. 88. Paris: Hermann 1934.
[38] BRILLOUIN, L.: Les champs self-consistents de HARTREE et de FOCK, Actualités scientifiques, Nr. 159. Paris: Hermann 1934.
[39] BRILLOUIN, L.: Les bases de la théorie électronique des métaux et la méthode des champs self-consistents. Helv. phys. Acta **7**, Suppl. 2, 33 (1934).
[40] CALLAWAY, J.: Orthogonalized plane wave method. Phys. Rev. **97**, 933 (1955).
[41] CALLAWAY, J.: Electronic energy bands in iron. Phys. Rev. **99**, 500 (1955).
[42] CALLEN, E.: Configuration interaction applied to the hydrogen molecule. J. Chem. Phys. **23**, 360 (1955).
[43] CHODOROW, M. I.: The band structure of metallic copper. Phys. Rev. **55**, 675 (A) (1939).
[44] CHODOROW, M. I., and M. F. MANNING: Energy bands in the body-centered lattice. Phys. Rev. **52**, 731 (1937).
[45] COULSON, C. A.: Self-consistent field for molecular hydrogen. Proc. Cambridge Phil. Soc. **34**, 204 (1938).
[46] COULSON, C. A.: Energy bands in graphite. Nature, Lond. **159**, 265 (1947).
[47] COULSON, C. A., and I. FISCHER: Notes on the molecular orbital treatment of the hydrogen molecule. Phil. Mag. **40**, 386 (1949).
[48] CURIEN, H.: Thermal scattering of x-rays by single crystals of α-Fe and dynamics of the body-centered cubic lattice. Acta crystallogr. **5**, 392 (1952).
[49] DANK, M., and H. B. CALLEN: Calculation of energy bands in solids by the integral iteration method. Phys. Rev. **86**, 622 (A) (1952).
[50] DEXTER, R. N., and B. LAX: Effective masses of holes in silicon. Phys. Rev. **96**, 223 (L) (1955).
[51] DEXTER, R. N., B. LAX, A. F. KIP and G. DRESSELHAUS: Effective masses of electrons in silicon. Phys. Rev. **96**, 222 (L) (1955).
[52] DEXTER, R. N., H. J. ZEIGER and B. LAX: Anisotropy of cyclotron resonance of holes in germanium. Phys. Rev. **95**, 557 (L) (1954).
[53] DIRAC, P. A. M.: Note on exchange phenomena in the THOMAS atom. Proc. Cambridge Phil. Soc. **26**, 376 (1930).
[54] DONOVAN, B.: A new calculation of some properties of metallic Be. Phil Mag. **43**, 868 (1952).
[55] DRESSELHAUS, G., A. F. KIP and C. KITTEL: Observation of cyclotron resonance in germanium crystals. Phys. Rev. **92**, 827 (L) (1953).
[56] DUNLAP jr., W. C.: Gold as an acceptor in germanium. Phys. Rev. **97**, 614 (1955).
[57] ELLIOTT, R. J.: Theory of the effect of spin-orbit coupling on magnetic resonance in some semiconductors. Phys. Rev. **96**, 266 (1954).
[58] ELLIOTT, R. J.: Spin-orbit coupling in band theory-character tables for some "double" space groups. Phys. Rev. **96**, 280 (1954).
[59] EWALD, P. P.: Zur Begründung der Kristalloptik. I. Theorie der Dispersion. Ann. Phys., Lpz. **49**, 1 (1916).

[60] Ewald, P. P.: Zur Begründung der Kristalloptik. II. Theorie der Reflexion und Brechung. Ann. Phys., Lpz. **49**, 117 (1916).
[61] Ewald, P. P.: Zur Begründung der Kristalloptik. III. Die Kristalloptik der Röntgenstrahlen. Ann. Phys., Lpz. **54**, 519 (1917).
[62] Ewing, D. H., and F. Seitz: On the electronic constitution of crystals: LiF and LiH. Phys. Rev. **50**, 760 (1936).
[63] Fermi, E.: Zur Quantelung des idealen einatomigen Gases. Z. Physik **36**, 902 (1926).
[64] Fermi, E.: Eine statistische Methode zur Bestimmung einiger Eigenschaften des Atoms und ihrer Anwendung auf die Theorie des periodischen Systems der Elemente. Z. Physik **48**, 73 (1928).
[65] Feuer, P.: Electronic states in crystals under large overall perturbations. Phys. Rev. **88**, 92 (1952).
[66] Feynman, R. P.: Forces in molecules. Phys. Rev. **56**, 340 (1939).
[67] Feynman, R. P., N. Metropolis and E. Teller: Equations of state of elements based on a generalized Fermi-Thomas theory. Phys. Rev. **75**, 1561 (1949).
[68] Fine, P. C.: Normal modes of vibration of a body-centered cubic lattice. Phys. Rev. **56**, 355 (1939).
[69] Fletcher, G. C.: Density of states curve for the $3d$ electron in Ni. Proc. Phys. Soc. Lond. A **65**, 192 (1952).
[70] Fletcher, G. C., and E. P. Wohlfarth: Calculation of the density of states curve for the $3d$ electrons in nickel. Phil. Mag. **42**, 106 (1951).
[71] Fock, V.: Näherungsmethode zur Lösung des quantenmechanischen Mehrkörperproblems. Z. Physik **61**, 209 (1930).
[72] Friedel, J.: The distribution of electrons around impurities in monovalent metals. Phil. Mag. **43**, 153 (1952).
[73] Friedel, J.: Electronic structure of primary solutions in metals. Phil. Mag. Suppl. **3**, 446 (1954).
[74] Friedel, J.: Sur l'origine du ferromagnetisme dans les métaux de transition. J. Phys. Radium **16**, 829 (1955).
[75] Fröhlich, H.: Elektronentheorie der Metalle. Berlin: Springer 1936. A useful reference, with good treatments of many aspects of metallic structure and properties.
[76] Fuchs, K.: A quantum mechanical investigation of the cohesive forces of metallic copper. Proc. Roy. Soc. Lond., Ser. A **151**, 585 (1935).
[77] Fuchs, K.: A quantum mechanical calculation of the elastic constants of monovalent metals. Proc. Roy. Soc. Lond., Ser. A **153**, 622 (1936).
[78] Fuchs, K.: The elastic constants and the specific heat of the alkali metals. Proc. Roy. Soc. Lond., Ser. A **157**, 444 (1936).
[79] Gombás, P.: Die statistische Theorie des Atoms und ihre Anwendungen. Vienna: Springer 1949. A very useful treatment of the Fermi-Thomas method, and all other aspects of the statistical theory, including many references to this method of treating atoms and solids.
[80] Gombás, P.: Calculation of position and width of energy bands of Na, K, Rb, and Cs. Z. Physik **113**, 150 (1939).
[81] Gorin, E.: The theoretical constitution of metallic potassium. Phys. Z. Sowjet. **9**, 328 (1936).
[82] Greene, J. B., and M. F. Manning: Electronic energy bands in face-centered iron. Phys. Rev. **63**, 203 (1943).
[83] Hall, G. G.: The electronic structure of diamond. Phil. Mag. **43**, 338 (1952). — Phys. Rev. **90** 317 (L) (1953).
[84] Hartree, D. R.: The wave mechanics of an atom with a non-coulomb central field. Proc. Cambridge Phil. Soc. **24**. 89, 111 (1928).
[85] Heisenberg, W.: Zur Theorie des Ferromagnetismus. Z. Physik **49**, 619 (1928).
[86] Heitler, W., and F. London: Wechselwirkung neutraler Atome und homöopolare Bindung nach der Quantenmechanik. Z. Physik **44**, 455 (1927).
[87] Herman, F.: Electronic structure of the diamond crystal. Phys. Rev. **88**, 1210 (1952).
[88] Herman, F.: Some recent developments in the calculation of crystal energy bands—new results for the germanium crystal. Physica, Haag **20**, 801 (1954).
[89] Herman, F.: Speculations on the energy band structure of Ge-Si alloys. Phys. Rev. **95**, 847 (L) (1954).
[90] Herman, F.: Speculations on the energy band structure of zinc-blende-type crystals. J. Electronics, Lond. **1**, 103 (1955).
[91] Herman, F., and J. Callaway: Electronic structure of the germanium crystal. Phys. Rev. **89**, 518 (L) (1953).
[92] Herring, C.: Effect of time-reversal symmetry on energy bands of crystals. Phys. **52**, 361 (1937).

[93] Herring, C.: Accidental degeneracy in the energy bands of crystals. Phys. Rev. **52**, 365 (1937).
[94] Herring, C.: Compressibility of lithium. Phys. Rev. **55**, 598 (A) (1939).
[95] Herring, C.: New method for calculating wave functions in crystals. Phys. Rev. **57**, 1169 (1940).
[96] Herring, C.: Character tables for two space groups. J. Franklin Inst. **223**, 525 (1942).
[97] Herring, C.: Correlation energy and the heat of sublimation of lithium. Phys. Rev. **82**, 282 (L) (1951).
[98] Herring, C., and A. G. Hill: The theoretical constitution of metallic beryllium. Phys. Rev. **58**, 132 (1940).
[99] Holmes, D. K.: An application of the cellular method to silicon. Phys. Rev. **87**, 782 (1952).
[100] Houston, W. V.: Elektrische Leitfähigkeit auf Grund der Wellenmechanik. Z. Physik **48**, 449 (1928).
[101] Houston, W. V.: Temperature dependence of electrical conductivity. Phys. Rev. **34**, 279 (1929).
[102] Houston, W. V.: Acceleration of electrons in a crystal lattice. Phys. Rev. **57**, 184 (1940).
[103] Howarth, D. J.: Electronic eigenvalues of copper. Proc. Roy. Soc. Lond., Ser. A **220**, 513 (1953).
[104] Howarth, D. J.: Application of the augmented plane wave method to copper. Phys. Rev. **99**, 469 (1955).
[105] Howarth, D. J., and H. Jones: The cellular method of determining electronic wave functions and eigenvalues in crystals, with applications to sodium. Proc. Phys. Soc. Lond. A **65**, 355 (1952).
[106] Hume-Rothery, W.: The Metallic State. Oxford 1931.
[107] Hume-Rothery, W.: The Structure of Metals and Alloys. London 1936.
[108] Hume-Rothery, W.: Electrons, Atoms, Metals and Alloys. London 1955. These successively more up-to-date versions of the experimental side of the structure of metals make close connections with theory, and furnish a good introduction to the study of metals.
[109] Hund, F.: Theorie der Bewegung der Elektronen in nichtmetallischen Kristallgittern. Z. techn. Phys. **16**, 331 (1935). — Phys. Z. **36**, 725 (1935).
[110] Hund, F.: Zustände der Elektronen in Kristallgittern. Z. techn. Phys. **16**, 494 (1935). — Phys. Z. **36**, 888 (1935).
[111] Hund, F., u. B. Mrowka: Über die Zustände der Elektronen in einem Kristallgitter, insbesondere beim Diamant. Sächs. Akad. Wiss., Leipzig **87**, 185 (1935).
[112] Hund, F., u. B. Mrowka: II. Energiebänder in einfachen Gittern. Sächs. Akad. Wiss., Leipzig **87**, 325 (1935).
[113] Hurley, A. C., J. Lennard-Jones and J. A. Pople: The molecular orbital theory of chemical valency. XVI. A theory of paired electrons in polyatomic molecules. Proc. Roy. Soc. Lond., Ser. A **220**, 446 (1953).
[114] Hylleraas, E. A.: Über den Grundzustand des Heliumatoms. Z. Physik **48**, 469 (1928).
[115] Jacobsen, E. H.: Elastic spectrum of copper from temperature-diffuse scattering of x-rays. Phys. Rev. **97**, 654 (1955).
[116] James, H. M.: Electronic states in perturbed periodic systems. Phys. Rev. **76**, 1611 (1949).
[117] James, H. M., and A. S. Coolidge: Ground state of the hydrogen molecule. J. Chem. Phys. **1**, 825 (1933).
[118] James, H. M., and A. S. Ginzbarg: Band structure in disordered alloys and impurity semiconductors. J. Phys. Chem. **57**, 840 (1953).
[119] Jenkins, D. P.: The electronic band structure of silicon. Physica, Haag **20**, 967 (1954).
[120] Jones, H.: The theory of alloys in the $\gamma$-phase. Proc. Roy. Soc. Lond., Ser. A **144**, 225 (1934).
[121] Jones, H.: Applications of the Bloch theory to the study of alloys and of the properties of bismuth. Proc. Roy. Soc. Lond., Ser. A **147**, 396 (1934).
[122] Jones, H.: Application de la théorie électronique des métaux à l'étude des alliages. Helv. phys. Acta **7**, Suppl. 2, 84 (1934).
[123] Jones, H.: Structural and elastic properties of metals. Physica, Haag **15**, 13 (1949).
[124] Jones, H.: The effect of electron concentration on the lattice spacings in magnesium solid solutions. Phil. Mag. **41**, 663 (1950).
[125] Jones, H.: A calculation of the elastic shear constants of $\beta$-brass. Phil. Mag. **43** 105 (1952).

[126] JONES, H.: Soft x-ray emission bands in metals. Phys. Rev. **94**, 1072 (L) (1954).
[127] JONES, H., N. F. MOTT and H. W. B. SKINNER: Theory of the form of the x-ray emission bands of mètals. Phys. Rev. **45**, 379 (1934).
[128] JOYNSON, R. E.: Elastic spectrum of zinc from the temperature scattering of x-rays. Phys. Rev. **94**, 851 (1954).
[129] KELLERMAN, E. W.: Theory of vibrations of NaCl lattice. Phil. Trans. Roy. Soc. Lond., Ser. A **238**, 513 (1940).
[130] KELLERMAN. E. W.: On the specific heat of the sodium chloride crystal. Proc. Roy. Soc. Lond., Ser. A **178**, 17 (1941).
[131] KIMBALL, G. E.: Electronic structure of diamond. J. Chem. Phys. **3**, 560 (1935).
[132] KLEINER, W.: Excited donor levels in silicon. Phys. Rev. **97**, 1722 (L) (1955).
[133] KOHN, W.: Variational methods for periodic lattices. Phys. Rev. **87**, 472 (1952).
[134] KOHN, W., and J. M. LUTTINGER: Theory of donor states in silicon. Phys. Rev. **98**, 915 (1955).
[135] KOHN, W., and N. ROSTOCKER: Solution of SCHRÖDINGER equation in periodic lattices with an application to metallic lithium. Phys. Rev. **94**, 1111 (1954).
[136] KOOPMANS, T.: Über die Zuordnung von Wellenfunktionen und Eigenwerten zu den einzelnen Elektronen eines Atoms. Physica, Haag **1**, 104 (1933).
[137] KOPPE, H.: Der Einfluß der Austauschenergie auf die spezifische Wärme des Elektronengases. Z. Naturforsch. **2**a, 429 (1947).
[138] KORRINGA, J.: On the calculation of the energy of a BLOCH wave in a metal. Physica, Haag **13**, 392 (1947).
[139] KOSTER, G. F.: Theory of scattering in solids. Phys. Rev. **95**, 1436 (1954).
[140] KOSTER, G. F.: Density of states curve for nickel. Phys. Rev. **98**, 901 (1955).
[141] KOSTER, G. F., and J. C. SLATER: Wave functions for impurity levels. Phys. Rev. **94**, 139 (L); **95**, 1167; **96**, 1208 (1954).
[142] KRAMERS, H. A.: L'interaction entre les atomes magnétogenes dans un cristal paramagnetique. Physica, Haag **1**, 182 (1934).
[143] KRUTTER, H. M.: Energy bands in copper. Phys. Rev. **48**, 664 (1935).
[144] LANDAU, L.: Diamagnetismus der Metalle. Z. Physik **64**, 629 (1930).
[145] LANDAUER, R., and J. C. HELLAND: Electronic structure of disordered one-dimensional chains. J. Chem. Phys. **22**, 1655 (1954).
[146] LENNARD-JONES, J. E.: The electronic structure of some diatomic molecules. Trans. Faraday Soc. **25**, 668 (1929).
[147] LENNARD-JONES, J. E.: Discussion on graphite. Trans. Faraday Soc. **30**, 58 (1934).
[148] LENNARD-JONES. J. E.: The molecular orbital theory of chemical valency. II. Equivalent orbitals in molecules of known symmetry. Proc. Roy. Soc. Lond., Ser. A **198**, 14 (1949).
[149] LENNARD-JONES, J. E.: Spatial correlation of electrons in molecules. J. Chem. Phys. **20**, 1024 (1952).
[150] LENNARD-JONES, J. E.: Spatial correlation of electrons in atoms and molecules. II. Proc. Nat. Acad. Sci. **38**, 496 (1952).
[151] LENNARD-JONES, J. E., and J. A. POPLE: The molecular orbital theory of chemical valency. IV. The significance of equivalent orbitals. Proc. Roy. Soc. Lond., Ser. A **202**, 166 (1950).
[152] LENNARD-JONES, J. E., and J. A. POPLE: The spatial correlation of electrons in atoms and molecules. I. Helium and similar two-electron systems in their ground states. Phil. Mag. **43**, 581 (1952).
[153] LÖWDIN, P.-O.: A quantum mechanical calculation of the cohesive energy, the interatomic distance and the elastic constants of some ionic crystals. Ark. Mat., Astronom. Fysik A **35**, Nr. 9, Nr. 30 (1947).
[154] LÖWDIN, P.-O.: On the quantum-mechanical calculation of the cohesive energy of molecules and crystals. Part I. A general formula for the ground state. J. Chem. Phys. **19**, 1570 (1951). Part II. Treatment of the alkali metals with numerical application to sodium. J. Chem. Phys. **19**, 1579 (1951).
[155] MANNING, M. F.: Electronic energy bands in body-centered iron. Phys. Rev. **63**, 190 (1943).
[156] MANNING, M. F., and M. I. CHODOROW: Electronic energy bands in metallic tungsten. Phys. Rev. **56**, 787 (1939).
[157] MANNING, M. F., and H. M. KRUTTER: Electronic energy bands in metallic calcium. Phys. Rev. **51**, 761 (1937).
[158] MATYAS, Z.: A new method for calculating the energy levels in solids. Czech. J. Phys. **1**, 3 (1952).
[159] MECKLER, A.: Electronic energy levels of molecular oxygen. J. Chem. Phys. **21**, 1750 (1953).

[160] MILLMAN, J.: Electronic energy bands in metallic lithium. Phys. Rev. **47**, 286 (1935).
[161] MONTROLL, E. W.: Frequency spectrum of crystalline solids. J. Chem. Phys. **10**, 218 (1942).
[162] MONTROLL, E. W.: Dynamics of a square lattice. J. Chem. Phys. **15**, 575 (1947).
[163] MORITA, A.: Electronic structure of diamond-lattice type crystals. Sci. Rep. Tôhoku Univ. **33**, 92 (1949).
[164] MORITA, A.: The electronic structure of bismuth crystal. Sci. Rep. Tôhoku Univ. **33**, 144 (1949).
[165] MORITA, A., and C. HORIE: Electronic structure of BaO crystal. Sci. Rep. Tôhoku Univ. **36**, 259 (1952).
[166] MORSE, P. M.: The quantum mechanics of electrons in crystals. Phys. Rev. **35**, 1310 (1930).
[167] MOTT, N. F.: A discussion of the transition metals on the basis of quantum mechanics. Proc. Phys. Soc. Lond. **47**, 571 (1935).
[168] MOTT, N. F.: The electrical resistance of dilute solid solutions. Proc. Cambridge Phil. Soc. **32**, 281 (1936).
[169] MOTT, N. F.: The basis of the electronic theory of metals, with special reference to the transition metals. Proc. Phys. Soc. Lond. A **62**, 416 (1949).
[170] MOTT, N. F., and R. W. GURNEY: Electronic Processes in Ionic Crystals, 2d edit. Oxford 1948. An excellent treatment of the electronic energy levels, electrical conductivity, optical properties, etc. of insulators and semiconductors.
[171] MOTT, N. F., and H. JONES: Properties of Metals and Alloys. Oxford 1936. A standard treatment of the theory of metals, particularly strong on the theory of BRILLOUIN zones, electrical and magnetic properties, etc.
[172] MUELLER, C. R., and H. EYRING: Semilocalized orbitals. I. The hydrogen molecule. J Chem. Phys. **19**, 1495 (1951).
[173] MULLANEY, J. F.: Optical properties and electronic structure of solid silicon. Phys. Rev. **66**, 326 (1944).
[174] NÉEL, L.: Propriétés magnétiques des ferrites; ferrimagnétisme et antiferromagnétisme. Ann. Phys., Paris **3**, 13, 137 (1948).
[175] PARMENTER, R. H.: Electronic energy bands in crystals. Phys. Rev. **86**, 552 (1952).
[176] PARMENTER, R. H.: Energy levels of a disordered alloy. Phys. Rev. **97**, 587 (1955).
[177] PARMENTER, R. H.: Symmetry properties of the energy bands of the zinc-blende structure. Phys. Rev. **10** 573 (1955).
[178] PAULI jr., W.: Über Gasentartung und Paramagnetismus. Z. Physik **41**, 81 (1927).
[179] PAULING, L.: Nature of the interatomic forces in metals. Phys. Rev. **54**, 899 (1938).
[180] PAULING, L.: The metallic state. Nature, Lond. **161**, 1019 (1948).
[181] PAULING, L.: The resonating valence-bond theory of metals. Physica, Haag **15**, 23 (1949).
[182] PAULING, L.: A resonating valence-bond theory of metals and intermetallic compounds. Proc. Roy. Soc. Lond., Ser. A **196**, 343 (1949).
[183] PAULING, L., and F. J. EWING: Ratio of valence electrons to atoms in metals and intermetallic compounds. Rev. Mod. Phys. **20**, 112 (1948).
[184] PEIERLS, R.: Zur Theorie der elektrischen und thermischen Leitfähigkeit der Metalle. Ann. Phys., Lpz. **4**, 121 (1930).
[185] PEIERLS, R.: Elektronentheorie der Metalle. Ergebn. exakt. Naturw. **21**, 264 (1932).
[186] PINES, D.: Collective description of electron interactions. IV. Electron interactions in metals. Phys. Rev. **92**, 626 (1953).
[187] PINES, D., and D. BOHM: Collective description of electron interactions. II. Collective *vs.* individual particle aspects of the interactions. Phys. Rev. **85**, 338 (1952).
[188] PODOLSKY, B., and L. PAULING: Momentum distribution in hydrogen-like atoms. Phys. Rev. **34**, 109 (1929).
[189] PRATT jr., G. W.: Antiferromagnetism. Phys. Rev. **97**, 926 (1955).
[190] PROKOFJEW, W.: Berechnung der Zahlen der Dispersionszentren des Natriums. Z. Physik **58**, 255 (1929).
[191] RAIMES, S.: The cohesive energy of metallic magnesium. Phil. Mag. **41**, 568 (1950).
[192] RAIMES, S.: A calculation of the cohesive energies and pressure-volume relations of the divalent metals. Phil. Mag. **43**, 327 (1952).
[193] RAIMES, S.: The compressibility of metallic aluminium. Proc. Phys. Soc. Lond. A **66**, 949 (1953).
[194] RAIMES, S.: Correlation energy in metals and the cohesive energy of metallic sodium. Proc. Phys. Soc. Lond. A **67**, 52 (1954).
[195] RAIMES, S.: Energy band shapes and band widths in metals. Phil. Mag. **45**, 727 (1954).
[196] SAFFREN, M. M., and J. C. SLATER: Augmented plane wave method for the periodic potential problem. II. Phys. Rev. **92**, 1126 (1953).

[*197*] Schiff, B.: A calculation of the eigenvalues of electronic states in metallic lithium by the cellular method. Proc. Phys. Soc. Lond. A **67**, 2 (1954).
[*198*] Schmid, L. A.: Calculating of the cohesive energy of diamond. Phys. Rev. **92**, 1373 (1953).
[*199*] Seitz, F.: A matrix-algebraic development of the crystallographic groups. I. Z. Kristallogr. **88**, 433 (1934). II. **90**, 289 (1935). III. **91**, 336 (1935). IV. **94**, 100 (1936).
[*200*] Seitz, F.: Theoretical constitution of metallic lithium. Phys. Rev. **47**, 400 (1935).
[*201*] Seitz, F.: Imperfections in Nearly Perfect Crystals. New York: Wiley 1952. A good review article in a volume including also discussions of various aspects of crystal imperfections, by various authors.
[*202*] Seitz, F.: Modern Theory of Solids. New York: Mc Graw-Hill Book Co. 1940. A standard very complete text on the theory of solids, representing very excellently the state of the subject at the time of writing.
[*203*] Seitz, F., and R. P. Johnson: Modern Theory of Solids. J. Appl. Phys. **8**, 84, 186, 246 (1937). A useful review of the subject, on a rather elementary level.
[*204*] Shockley, W.: Electronic energy bands in sodium chloride. Phys. Rev. **50**, 754 (1936).
[*205*] Shockley, W.: Energy bands for the face-centered lattice. Phys. Rev. **51**, 129 (1937).
[*206*] Shockley, W.: The empty lattice test of the cellular method in solids. Phys. Rev. **52**, 866 (1937).
[*207*] Shockley, W.: Energy band structure in semiconductors. Phys. Rev. **78**, 173 (L) (1950).
[*208*] Shockley, W.: Electrons and Holes in Semiconductors. New York: D. van Nostrand Co. 1950. An excellent treatment of the electronic structure of semiconductors, including electrical conductivity, rectifying behavior, etc.
[*209*] Shull, C. G., and J. S. Smart: Detection of antiferromagnetism by neutron diffraction. Phys. Rev. **76**, 1256 (L) (1949).
[*210*] Shull, C. G., W. A. Strauser and E. O. Wollan: Neutron diffraction by paramagnetic and antiferromagnetic substances. Phys. Rev. **83**, 333 (1951).
[*211*] Shull, C. G., and M. K. Wilkinson: Neutron diffraction studies on various transitions elements. Rev. Mod. Phys. **25**, 100 (1953).
[*212*] Silverman, R. A.: Fermi energy of metallic lithium. Phys. Rev. **85**, 227 (1952).
[*213*] Silverman, R. A., and W. Kohn: On the cohesive energy of metallic lithium. Phys. Rev. **80**, 912 (1950).
[*214*] Slater, J. C.: Note on Hartree's method. Phys. Rev. **35**, 210 (1930).
[*215*] Slater, J. C.: Cohesion in monovalent metals. Phys. Rev. **35**, 509 (1930).
[*216*] Slater, J. C.: Atomic shielding constants. Phys. Rev. **36**, 57 (1930).
[*217*] Slater, J. C.: Molecular energy levels and valence bonds. Phys. Rev. **38**, 1109 (1931).
[*218*] Slater, J. C.: Electronic energy bands in metals. Phys. Rev. **45**, 794 (1934).
[*219*] Slater, J. C.: The electronic structure of metals. Rev. Mod. Phys. **6**, 209 (1934).
[*220*] Slater, J. C.: The ferromagnetism of nickel. Phys. Rev. **49**, 537 (1936). — II. Temperature effects. Phys. Rev. **49**, 931 (1936).
[*221*] Slater, J. C.: Wave functions in a periodic potential. Phys. Rev. **51**, 846 (1937).
[*222*] Slater, J. C.: Electronic structure of alloys. J. Appl. Phys. **8**, 385 (1937).
[*223*] Slater, J. C.: Electrons in perturbed periodic lattices. Phys. Rev. **76**, 1592 (1949).
[*224*] Slater, J. C.: Note on orthogonal atomic orbitals. J. Chem. Phys. **19**, 220 (1951).
[*225*] Slater, J. C.: A simplification of the Hartree-Fock method. Phys. Rev. **81**, 385 (1951).
[*226*] Slater, J. C.: Magnetic effects and the Hartree-Fock equation. Phys. Rev. **82**, 538 (1951).
[*227*] Slater, J. C.: A soluble problem in energy bands. Phys. Rev. **87**, 807 (1952).
[*228*] Slater, J. C.: Ferromagnetism and the band theory. Rev. Mod. Phys. **25**, 199 (1953).
[*229*] Slater, J. C.: Augmented plane wave method for the periodic potential problem. Phys. Rev. **92**, 603 (1953).
[*230*] Slater, J. C., and G. F. Koster: Simplified LCAO method for the periodic potential problem. Phys. Rev. **94**, 1498 (1954).
[*231*] Slater, J. C., and H. M. Krutter: The Thomas-Fermi method for metals. Phys. Rev. **47**, 555 (1935).
[*232*] Slater, J. C., H. Statz and G. F. Koster: A two-electron example of ferromagnetism. Phys. Rev. **91**, 1323 (1953).
[*233*] Sommerfeld, A., W. V. Houston u. C. Eckart: Zur Elektronentheorie der Metalle auf Grund der Fermischen Statistik. Z. Physik **47**, 1 (1928).
[*234*] Sommerfeld, A., u. H. Bethe: Elektronentheorie der Metalle. In Handbuch der Physik, 2. Aufl., Bd. 24, Teil 2. 1933.
[*235*] Sternheimer, R.: On the compressibility of metallic cesium. Phys. Rev. **78**, 235 (1950).

[236] STONER, E. C.: Collective electron ferromagnetism. Proc. Roy. Soc. Lond., Ser. A **165**, 372 (1938).
[237] STONER, E. C.: II. Energy and specific heat. Proc. Roy. Soc. Lond., Ser. A **169**, 339 (1939).
[238] STONER, E. C.: Ferromagnetism. Rep. Progr. Phys. **11**, 43 (1948).
[239] STONER, E. C.: Collective electron ferromagnetism in metals and alloys. J. Phys. Radium **12**, 373 (1951).
[240] THOMAS, L. H.: The calculation of atomic fields. Proc. Cambridge Phil. Soc. **23**, 542 (1927).
[241] TIBBS, S. C.: Electronic energy bands in metallic copper and silver. Proc. Cambridge Phil. Soc. **34**, 89 (1938).
[242] TIBBS, S. C.: Electronic energy levels in NaCl. Trans. Faraday Soc. **35**, 1471 (1939).
[243] TREFFTZ, E.: Bestimmung der Eigenwerte einer elliptischen Differentialgleichung mit der Randbedingung der Periodizität (SCHRÖDINGER-Gleichung in Metallen). Z. angew. Math. Mech. **34**, 303 (1954).
[244] TRLIFAJ, M.: The electron theory of metallic magnesium. Czech. J. Phys. **1**, 110 (1952).
[245] VLECK, J. H. VAN: On the theory of the structure of $CH_4$ and related molecules. J. Chem. Phys. **1**, 177, 219 (1934); **2**, 20 (1934).
[246] VLECK, J. H. VAN: L'état actual de la théorie du paramagnétisme. Magnétisme, Paris **3**, 103 (1940).
[247] VLECK, J. H. VAN: Models of exchange coupling in ferromagnetic media. Rev. Mod. Phys. **25**, 220 (1953).
[248] VLECK, J. H. VAN: The Theory of Electric and Magnetic Susceptibilities. Oxford 1932. A standard and very useful treatment of the state of the theory of magnetism at the time of writing.
[249] LAGE, F. V. DER, and H. BETHE: Method for obtaining electronic eigenfunctions and eigenvalues in solids with an application to sodium. Phys. Rev. **71**, 612 (1947).
[250] WALLACE, P. R.: Band theory of graphite. Phys. Rev. **71**, 622 (1947).
[251] WANNIER, G.: Structure of electronic excitation levels in insulating crystals. Phys. Rev. **52**, 191 (1937).
[252] WEINBAUM, S.: Normal state of the hydrogen molecule. J. Chem. Phys. **1**, 593 (1933).
[253] WIGNER, E.: On the interaction of electrons in metals. Phys. Rev. **46**, 1002 (1934).
[254] WIGNER, E.: Effects of the electron interaction on the energy levels of electrons in metals. Trans. Faraday Soc. **34**, 678 (1938).
[255] WIGNER, E., and F. SEITZ: On the constitution of metallic sodium. Phys. Rev. **43**, 804 (1933); II. **46**, 509 (1934).
[256] WILKINSON, M. K., and C. G. SHULL: Neutron diffraction observations on Fe at high temperatures. Phys. Rev. **94**, 1439 (A) (1954).
[257] WILSON, A. H.: The theory of electronic semi-conductors. Proc. Roy. Soc. Lond., Ser. A **133**, 458 (1931); II. **134**, 277 (1931).
[258] WILSON, A. H.: Theory of Metals, 2d edit. Cambridge 1953. A good treatment particularly of the theory of electrical conductivity and related questions.
[259] WOHLFARTH, E. P.: The influence of exchange and correlation force on the specific heat of free electrons in metals. Phil. Mag. **41**, 534 (1950).
[260] WOHLFARTH, E. P.: Electrostatic contribution to the elastic constants of solids with the diamond structure. Phil. Mag. **43**, 474 (1952).
[261] WOLFF, P. A.: Theory of plasma waves in metals. Phys. Rev. **92**, 18 (1953).
[262] WOODRUFF, T. O.: Solution of the HARTREE-FOCK-SLATER equations for silicon crystal by the method of orthogonalized plane waves. Phys. Rev. **98**, 1741 (1955).
[263] YAMAKA, E., and T. SUGITA: Energy band structure in silicon crystal. Phys. Rev. **90**, 992 (L) (1953).
[264] ZENER, C.: A theory of the electrical breakdown of solid dielectrics. Proc. Roy. Soc. Lond., Ser. A **145**, 529 (1934).
[265] ZENER, C.: Interaction between the *d* shells in the transition metals. Phys. Rev. **81**, 440 (1951).
[266] ZENER, C.: II. Ferromagnetic compounds of manganese with the perovskite structure. Phys. Rev. **82**, 403 (1951).
[267] ZENER, C.: III. Calculation of the WEISS factors in Fe, Co, and Ni. Phys. Rev. **83**, 299 (1951).
[268] ZENER, C.: IV. Intrinsic antiferromagnetic character of iron. Phys. Rev. **85**, 324 (1952).

# Metallic Conductivity, Experimental Part.

By

A. N. GERRITSEN.

With 57 Figures.

## I. General remarks on the conductivity of metals.

**1. Introduction.** The subject of metallic conductivity was discussed so thoroughly from the experimental standpoint by GRÜNEISEN [*G 1*], [*G 2*] and by MEISSNER [*M 1*] that apart from modifications resulting from new experimental results, their contributions are still to be considered up to date as regards the basic principles. The present author has therefore considered it his main task to discuss concisely the results of recent experiments on electric conductivity at normal and at high temperatures. In order to meet the editoral demand for a monograph on the subject a great deal from older methods and results has been included. The present paper does not claim to be exhaustive in the choice of experiments for discussion. Owing to the regrettable inacessibility of a good deal of American and Russian published matter, the relevant information had to be taken from abstracts.

In the present paper, after introductory remarks and definitions in the first chapter, a survey is given of the methods of measurement of conductivity (Chap. II). The subsequent chapters give a discussion of metallic conductivity as related to physical parameters such as electron configuration and crystal structure (Chap. III), temperature (Chap. IV), magnetic fields (Chap. V), pressure (Chap. VI), crystal defects (Chap. VII) and alloying (Chap. VIII).

Discussion of the theoretical problems of the subject has been avoided throughout the paper, when necessary, reference was made to the following article of H. JONES [*J 1*] in preference to the original theoretical papers.

**2. Definitions for isotropic metals.** The conductance $S$ of a specimen is defined according to OHM's law by the relation:

$$S = I/V \tag{2.1}$$

that is, as the ratio between the current $I$ through the specimen and the voltage $V$ across its terminals. From the experimentally proved relation between the dimensions of a simple specimen and its conductance, a specific conductance or conductivity $\sigma$ is defined by the relation:

$$\sigma = \frac{l}{O} S, \tag{2.2}$$

where $l$ and $O$ are respectively the length and area of cross section of a cylindrical or edged specimen, these being the simplest forms for measurements of conductivity.

As a consequence of (2.1) and (2.2) the resistance $R$ of a specimen and its specific resistance or resistivity $\varrho$ are defined as:

$$R = V/I \tag{2.3}$$

and

$$\varrho = \frac{O}{l} R. \tag{2.4}$$

If the current, potential difference and length be measured in amperes, volts and meters, respectively, the resistance is expressed in ohms, the resistivity in ohm meters, the conductance in ohm$^{-1}$ [1], the conductivity in ohm$^{-1}$ meters$^{-1}$.

The *temperature coefficient of resistance* has been defined from the experimentally found linear relationship between resistance and temperature in the temperature region between ice point and boiling point of water. A more general definition follows for any temperature region where $R$ changes proportionately with temperature, by:

$$a_{R,T} = \frac{R_{T_2} - R_{T_1}}{T_2 - T_1} \quad \text{(ohms degree}^{-1}\text{)} \tag{2.5}$$

for $T_1 \leq T \leq T_2$.

Since this coefficient is not a specific number, a relative coefficient $\alpha_{R,T}$ is defined which demonstrates better the properties of the substance under investigation

$$\alpha_{R,T} = \frac{1}{R_{T_1}} \frac{R_{T_2} - R_{T_1}}{T_2 - T_1} \quad \text{(degree}^{-1}\text{)}. \tag{2.6}$$

Considering the thermal linear expansion of the specimen (coefficient $\gamma_T$, at normal pressure for the temperature interval $T_1 \leq T \leq T_2$), the temperature coefficient for the resistivity follows from:

$$\alpha_{R,T} = \alpha_{\varrho,T}(1 - \gamma_T \Delta T) - \gamma_T. \tag{2.7}$$

where

$$\alpha_{\varrho,T} = \frac{1}{\varrho_{T_1}} \frac{\varrho_{T_2} - \varrho_{T_1}}{\Delta T} \tag{2.8}$$

is the temperature coefficient of the resistivity in the temperature region $\Delta T = T_2 - T_1$.

Since

$$\alpha_{R,T}\, \gamma_T \approx \alpha_{\varrho,T}\, \gamma_T$$

(2.7) becomes

$$\alpha_{\varrho,T} = \alpha_{R,T} + \gamma_T (1 + \alpha_{R,T} \Delta T). \tag{2.9}$$

For $\Delta T$ of the order of 100 degrees at room temperature, the second term with the coefficient $\gamma_T$ may be neglected for most metals. Using the KELVIN scale with $0° \text{ K} \equiv -273.15° \text{ C}$ the temperature coefficient (2.8) for any small temperature region $\Delta T$ above or below a certain reference temperature $T$ is given by:

$$\alpha_{\varrho,T} = \frac{1}{\varrho_T} \frac{d\varrho}{dT}. \tag{2.10}$$

In the case of a linear relation between $\varrho$ and $T$, (2.10) corresponds with (2.8), as a reference temperature for the room temperature region the melting point of ice at 273.15° K, has been chosen. The definitions given so far are based on the experimentally proved relations between current, voltage, dimensions, and temperature for isotropic or quasi-isotropic conducting materials. For anisotropic crystals the relations are not so simple and a better justified, albeit more formal, discussion of the conductivity in connection with crystal parameters has to be given.

[1] In technical papers the unit "mho" (inversion of the word "ohm") is used for conductance.

**3. Definitions for anisotropic metal crystals.** We choose [*V 1*], [*B 2*] the current density $\boldsymbol{i}$ and the electric field $\boldsymbol{E}$ with their components $i_m$ and $E_n$ along arbitrary chosen mutually perpendicular axes. Then a general linear relation between $\boldsymbol{i}$ and $\boldsymbol{E}$ is:

$$i_m = \sum_n \sigma_{mn} E_n \quad (m \text{ and } n = 1, 2, 3) \tag{3.1}$$

in which the $\sigma_{mn}$ are conductivity coefficients.

Since $\sigma$ is a symmetric tensor, it follows from energy considerations, that $\sigma_{mn} = \sigma_{nm}$.

Now it depends on the crystal configuration how complicated the expression for the conductivity tensor is. For a cubic crystal the result is simple. In this case, the planes of symmetry are perpendicular to each of the crystal axes, which means that a relation between $\boldsymbol{i}$ and $\boldsymbol{E}$ will be the same whether any one of the axes is reversed in direction. Thus, $\sigma_{mn} = \sigma_{nm} = 0$. Furthermore the cubic axes are equivalent, a cyclic change from 1 to 2, 2 to 3, and 3 to 1 will leave an $\boldsymbol{i}(\boldsymbol{E})$-relation invariant. This means that $\sigma_{11} = \sigma_{22} = \sigma_{33}$ or, the conductivity is isotropic: $i = \sigma E$. With $i = I/O$, $E = V/l$ it follows that for an isotropic conductor the relation (3.1) is Ohm's law (2.1) when the definition (2.2) is applied.

For anisotropic conductors (the hexagonal, the trigonal and tetragonal crystals) it follows from their crystallographic symmetry that $\sigma_{mn} = 0$ for $m \neq n$ and that in general $\sigma_{mn} \neq 0$ for $m = n$. In the description of experimental results the conductivities $\sigma_{11}$, $\sigma_{22}$, $\sigma_{33}$ are written either as $\sigma_{\parallel}$, $\sigma_{\perp 1}$ and $\sigma_{\perp 2}$ for the conductivities parallel, or perpendicular to the principal axes, or the $\sigma$-symbols are indexed with the symbols for the axes. In most cases (for example the hexagonal cadmium or the rhombic bismuth) $\sigma_{\perp 1} = \sigma_{\perp 2}$, for a tetragonal case (gallium) three different $\sigma$-values have been found[1].

If an electric field $\boldsymbol{E}$ is applied parallel to the length of a cylindrical rod whose principal axis (3-axis) makes an angle $\varphi$ to its length, then it follows that:

$$\boldsymbol{i} = (\sigma_{\perp 1} \sin^2\varphi \cos^2\psi + \sigma_{\perp 2} \sin^2\varphi \sin^2\psi + \sigma_{\parallel} \cos^2\varphi)\, \boldsymbol{E}$$

where $\psi$ is the angle between the plane through the 3-axis and $\boldsymbol{E}$, and that through the 1 and 3-axes.

For the most general case of $\sigma_{\perp 1} = \sigma_{\perp 2}$, this becomes:

$$\boldsymbol{i} = (\sigma_{\perp} \sin^2\varphi + \sigma_{\parallel} \cos^2\varphi)\, \boldsymbol{E}$$

or, with $\sigma_\varphi = \boldsymbol{i}/\boldsymbol{E}$:

$$\sigma_\varphi = (\sigma_{\perp} \sin^2\varphi + \sigma_{\parallel} \cos^2\varphi). \tag{3.2}$$

If $\sigma_\perp = \sigma_\parallel$ this reduces to $\sigma_\varphi = \sigma_\perp = \sigma_\parallel$, or, the cubic metal is an isotropic conductor.

By interchanging the places of $i_m$ and $E_n$ in (3.1), a $\varrho_{mn}$ is defined for which, in general, $\varrho_{mn} \neq 1/\sigma_{mn}$. In the case, however, of the common metals, where only the diagonal elements of $\sigma$ exist, it is true that for each crystallographic direction $i_m = \sigma_{mn} E_m$ and $\sigma_{nm} = 1/\varrho_{mn}$ $(m = 1, 2, 3)$.

Then, under the condition $\sigma_{\perp 1} = \sigma_{\perp 2}$, it follows from (3.2) that, though

$$\sigma_\perp = \frac{1}{\varrho_\perp} \quad \text{and} \quad \sigma_\parallel = \frac{1}{\varrho_\parallel}, \quad \text{still} \quad \sigma_\varphi \neq \frac{1}{\varrho_\varphi} \tag{3.3}$$

unless $\varphi = 0$ or $\varphi = \pi/2$. It is obvious [cf. (2.8)], that the anisotropic conductivity will lead to anisotropic temperature coefficients of resistivity.

[1] Cf. Sect. 8, ref. [*174a*].

Most conductivity experiments have been carried out with quasi-isotropic material. Then the metal consists of an aggregate of microscopic crystals, with a random orientation of their crystallographic axes. The resistivity of a cubic single crystal is the same as that of a quasi-isotropic polycrystalline cubic specimen of the same metal. A difference due to an extra resistance at the crystalline boundaries,—where large internal strains may occur—, may be expected, especially at low temperatures. For anisotropic crystals the resistivity $\bar{\varrho}$ of polycrystalline samples can be found from the axial resistivities of a single crystal of the same material by:

$$\frac{1}{\bar{\varrho}} = \frac{1}{3}\left(\frac{1}{\varrho_{\parallel}} + \frac{2}{\varrho_{\perp}}\right). \tag{3.4}$$

This relation due to VOIGT is in good agreement with experimental results, though it has been criticised [23]. It is generally applied for a calculation of $\bar{\varrho}$ from $\varrho_{\parallel}$ and $\varrho_{\perp}$.

The question whether there exists a limit of validity of OHM's law has been the scope of some experiments. Apart from obvious cases as semi- or superconductors, the law holds for small (normal) and medium current densities. For large densities ($\sim 10^{10}$ amp/m.$^2$) which can be obtained at point contacts or in thin films, there have been found indications [18a] of deviation from (2.1). One should be careful in interpreting the results since the galvanomagnetic effects, caused by the large magnetic field of the current in the conductor, influence the measured data.

**4. Experiments with metals.** In experiments with metals, extreme care should be taken during the preparation of the specimen. It should not be forgotten, that apart from factors which can be easily controlled during the preparation of the metal (such as for single crystals the crystal orientation) or during the experiment (for example the temperature of the specimen), there are a number of circumstances which influence the results and which are not so easily controlled by the experimenter. When a metal is solidified after the refining process, it has already acquired certain characteristics as to its physical properties, which for example, may depend on the amount and quality of contamination, the number and types of dislocations.

Unless extensive precautions are taken, an increase of the contamination is surely to be expected during the handling of the metal (for example, while drawing into wires, welding or soldering of contact leads). Even under the most strictly followed precautions and identical treatments, electric properties in particular may differ for two wires made from the same parts of a pure melt. Such effects as found by investigation of the influence of dislocations and other faults on the physical properties of the metal may enter unobserved into particular results. These are the factors, which make it sometimes hazardous to draw definite conclusions from experimental results in metalphysics.

## II. Methods of measurement of electrical resistance.

**5. Introduction.** The determination of the resistance of a current-carrying conductor is generally an application of OHM's law, the quantity actually measured being a potential difference. This is true for direct current (d.c.) resistance as well as for alternating current (a.c.) resistance (impedance). The methods applied can be distinguished as *the direct and the comparison methods*. In the former, the resistance is determined simply as the ratio between voltage and current. In the latter method the same current flows through the known and the unknown resistance in series, so that the unknown resistance can be found from the ratio between the voltages across the two resistances.

The voltage is determined either from direct readings of a voltmeter (direct-reading method), or by compensating the unknown voltage by a known controllable voltage, using the voltmeter as a zero instrument (compensation method).

The above methods require the attachment of current and potential leads to the specimen. This is not desirable when contamination of extremely pure material should be avoided; or it may not be possible to weld or solder to the specimen under consideration. In these cases an inductive method of measurement is advisable.

In the following the sensitivity of a method for indicating the smallest measurable part $\delta R_x$ of $R_x$, defined as

$$S_x = \frac{R_x}{\delta R_x} \tag{5.1}$$

will be discussed. The value of the relative error $e_x$ in the determination of $R_x$ will be given for some cases:

$$e_x = \frac{\Delta R_x}{R_x} \times 100\,\%. \tag{5.2}$$

Both numbers are given unless otherwise stated, for the case of most favourable conditions for the determination of $R_x$.

**6. Direct current methods.** *α) Direct reading methods.* The simplest method, namely putting $R_x$ in series with an ammeter and connecting a voltmeter across its terminals does not call for much discussion. The sensitivity and accuracy of the method are determined by the sensitivities, indicated on the instruments.

Since

$$S_x = \frac{R_x}{\delta R_x} = \frac{V_x}{\delta V_x} = \frac{a_{\mathrm{V}}}{\delta a_{\mathrm{V}}},$$

where $a_{\mathrm{v}}$ is the deflection of the voltmeter in scale units, $S_x = 100n$ for $a_{\mathrm{v}} = 100$ and $\delta a_{\mathrm{v}} = 1/n$ of a scale division, $1/n$ being the smallest deflection that can be estimated. For most volt- and amperemeters $n \approx 5$, and thus $S_x \approx 500$; for precision instruments with a mirror scale and a very fine pointer $n$ may be as large as 20, and thus $S_x \approx 2000$. For the latter type of meters the relative error $e = 0.1\,\%$.

The measurement of a number of unknown resistances can be rendered fast and simple by putting them in series with a known resistance and by connecting the voltmeter successively across the resistances. The simplicity of the method fails when the current through the voltmeter is comparable with that through the resistor.

A constancy of the measuring current, better than the possible value of $e$, has to be provided for. This method is preferable for a check on a number of resistances, when high accuracy is not needed. Another simple method is the substitutional method. There, the unknown resistance is put into the circuit, the current being readjusted to its original value by a variable resistor[1] in the circuit. The change in the variable resistor equals the unknown resistance.

*β) Compensation methods.* When the unknown voltage is compensated by an adjustable, known voltage, the difference between the known and unknown voltages decreases during the procedure of adjustment. An instrument, sensitive for extremely small voltage differences will enable the compensation to be made

[1] By resistor we indicate any system of resistance boxes, rheostats and potentiometers which may be used for controlling a current in a circuit.

to voltage differences which are many orders of magnitude smaller than the voltages themselves. Consequently, when maximum compensation has been reached, the system can be made very sensitive for any change in the balancing. For example, with a voltage sensitive galvanometer, voltage differences of the order of $10^{-8}$ volts may be detected in a normal equipment. The final result is independent of the absolute calibration of the zero instrument.

A great number of compensation methods has been developed, and old methods are improved and modified for special applications, especially in the field of modern radio and radar techniques. In the following, we will restrict ourselves to the discussion of the principles underlying some basic methods, which are frequently used for the determination of d.c. resistances. For refinements, one should refer to handbooks covering the special fields under consideration.

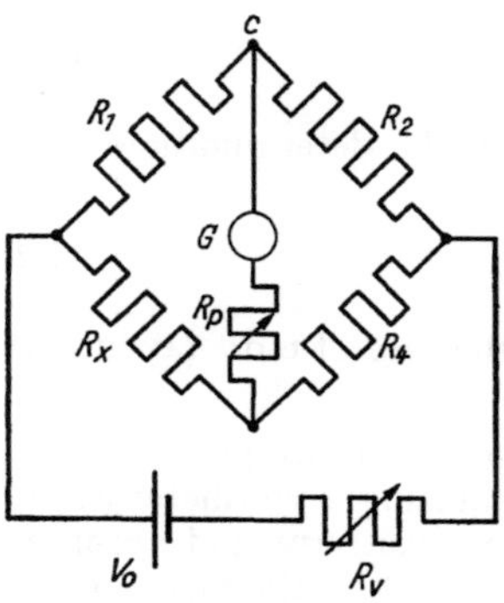

Fig. 1. The WHEATSTONE bridge.

*β 1) The* WHEATSTONE *bridge.* The simple bridge as suggested by CHRISTIE [*29*] and developed by WHEATSTONE, is still the most suitable method for easy determination of resistances from 10 to 10000 Ω with accuracy of the order of 0.1% [*39*]. A sketch of the bridge circuit has been given in Fig. 1. $R_x$ is the unknown resistance, $R_4$ a known, variable resistance or a standard resistance. $R_1$ and $R_2$ are two variable resistances, or a slide wire. $R_p$ is a resistance for protecting the galvanometer or micro-amperemeter $G$.

The bridge is fed from a battery with a resistor $R_v$ in series. $V_0$ and $R_v$ have to be adjusted in such a way that the energy developed in $R_x$ and the other resistances will not raise their temperature above a certain limit. Whilst this, of course holds for any measuring method, it is of special importance in compensation methods, where the material—and consequently the temperature coefficient of the unknown resistances may vastly differ from those of the known resistances. In the diagram of the circuit, we have for simplicity omitted the commutators, one in the battery branch, and another for interchanging the positions of $R_x$ and $R_4$. The insertion of these commutators ensures the elimination of thermo-e.m.f.'s and asymmetries in the $R_1R_2$-branch.

At zero deflection of a state of $G$ balance is reached for which:

$$R_x = \frac{R_1}{R_2} R_4. \tag{6.1}$$

In order to estimate the sensitivity of the bridge in terms of the galvanometer deflection $\alpha$ caused by a change $\delta R_x$ one has to calculate $I_g$, the current through $G$ when the bridge is not in balance. For these direct current networks THÉVENIN'S theorem is applicable and therefore

$$I_g = \frac{V_g}{R_{eq} + R_g}. \tag{6.2}$$

$V_g$ is the voltage across the galvanometer terminals and $R_{eq}$ is the equivalent resistance across these terminals when the battery (for which internal resistance is assumed negligible) is short circuited. $R_{eq}$ is given by:

$$R_{eq} = \frac{R_x R_4}{R_x + R_4} + \frac{R_1 R_2}{R_1 + R_2}. \tag{6.3}$$

Now the sensitivity $S_g$ of a moving coil galvanometer is defined as $S_g = \alpha_I / I_g$ where $I_g$ is the current in amperes through the coil and $\alpha_I$ is the deflection in mm. on a scale at a distance of 1 m., or, in the case of a pointer galvanometer, in scale divisions.

$S_g$ is mostly given for a current of 1 ampere, for a critically damped galvanometer, that is the total resistance in the galvanometer circuit $(R_g + R_{eq})$ equals the resistance value $R_c$, critical for electromagnetic damping[1]. Thus, for $I_g = 1$ amp, with $S_g = \alpha_1$, it follows that

$$\alpha_I = \alpha_1 I_g \tag{6.4}$$

The galvanometer deflections are then aperiodic in time.

For the case that the resistance $R$ in the galvanometer circuit differs from the critical value $R_c$, the deflection is given by[2]:

$$\alpha_I = \alpha_1 \sqrt{\frac{R}{R_c}}\, I_g. \tag{6.5}$$

Combining Eq. (6.2) with (6.5) we get for a change $\delta V_g$ across the galvanometer terminals effected by a $\delta R_x$ in $R_x$ a deflection $\delta\alpha$:

$$\delta\alpha = \alpha_1 \sqrt{\frac{R_{eq} + R_g}{R_c}}\, \frac{\delta V_g}{R_{eq} + R_g} = \alpha_1 \frac{\delta V_g}{\sqrt{R_c (R_{eq} + R_g)}} \tag{6.6}$$

in which the influence of $\delta R_x$ on the value of $R_{eq}$ has been neglected.

From the general expression (6.6), there follow two special cases:

(a) The galvanometer coil gets its maximum energy at a certain $V_g$ when $R_g = R_{eq}$, besides, so that $R_g + R_{eq} = 2R_{eq} = R_c$, in the case of a critically damped galvanometer. Then, Eq. (6.6) becomes:

$$\delta\alpha' = \alpha_1 \frac{\delta V_g}{R_c}. \tag{6.6a}$$

(b) $R_g = R_{eq}$, but $R_g + R_{eq} = 2R_{eq} \neq R_c$, then Eq. (6.6) becomes:

$$\delta\alpha'' = \alpha_1 \frac{\delta V_g}{\sqrt{2 R_c R_{eq}}}. \tag{6.6b}$$

Following JAEGER [*101a*] we put

$$R_4 = n R_x, \qquad R_1 = m R_x, \qquad R_2 = n m R_x \tag{6.7}$$

and add

$$R_g = f \cdot R_{eq}.$$

Calculating $\delta V_g$

$$\delta V_g = \frac{\delta R_x}{R_4 + R_x} V_0$$

and inserting (5.1), Eq. (6.6) can be written as:

$$\delta\alpha = \frac{\alpha_1}{S_x} I_x \sqrt{\frac{R_x}{R_c}}\, \frac{1}{\sqrt{f+1}} \sqrt{\frac{n}{(1+m)(1+n)}}. \tag{6.8}$$

---

[1] For many purposes the value $C_g = 1/S_g$, or galvanometer constant,—that is the number of amperes through the galvanometer circuit per unit deflection is preferable. The value of $C_g$ is mostly indicated on the instrument.

[2] For a derivation of (6.5) reference is made to the articles on electrical instruments, Vol. XXIII, this Encyclopedia.

1. For the case of maximum energy in $G$ and a critically damped circuit this becomes:

$R_{eq} = R_g (f = 1)$, and $R_c = 2R_{eq}$, thus

$$\delta\alpha' = \frac{\alpha_1}{S'_x} I_x \frac{1}{2(1+m)},$$

and for $\delta\alpha' = 1$, for example 1 mm. deflection,

$$S'_x = \frac{\alpha_1}{2} \frac{I_x}{1+m}, \tag{6.8a}$$

which is independent of $n$.

2. The same but not critically damped: $R_{eq} = R_g (f = 1)$, and $R_c \neq 2R_{eq}$ so that

$$\delta\alpha'' = \frac{\alpha_1}{S''_x} I_x \sqrt{\frac{R_x}{R_c}} \sqrt{\frac{n}{2(1+m)(1+n)}}$$

and for $\delta\alpha'' = 1$

$$S''_x = \alpha_1 I_x \sqrt{\frac{n R_x}{2R_c(1+m)(1+n)}}. \tag{6.8b}$$

For a bridge in symmetric balance $R_1 = R_2$, $(n = 1)$ and thus

$$\delta\alpha'' = \frac{\alpha_1}{S''_x} S_x \sqrt{\frac{R_x}{R_c}} \frac{1}{2\sqrt{1+m}}$$

and for $\delta\alpha'' = 1$

$$S''_x = \frac{1}{2} \alpha_1 I_x \sqrt{\frac{R_x}{R_c(1+m)}}. \tag{6.8c}$$

Assuming further that $m \ll 1$ (which means $R_x \gg R_1$ or $R_2$), it follows from (6.8a) and (6.8c) that:

$$\delta\alpha' = \frac{1}{2} \frac{\alpha_1}{S'_x} I_x \qquad \text{and for } \delta\alpha' = 1, \qquad S'_x = \frac{1}{2} \alpha_1 I_x, \tag{6.8a'}$$

$$\delta\alpha'' = \frac{1}{2} \frac{\alpha_1}{S''_x} S_x \sqrt{\frac{R_x}{R_c}}, \qquad \text{and for } \delta\alpha'' = 1, \qquad S''_x = \frac{1}{2} \alpha_1 I_x \sqrt{\frac{R_x}{R_c}}. \tag{6.8c'}$$

It is evident from Eqs. (6.8a') and (6.8c') that for large values of $R_x$, $S''_x$ for the non-critically damped galvanometer is larger than $S'_x$ for the critically damped (aperiodic) state. The gain in sensitivity in the non-aperiodic state is outweighed by the difficulties in the exact determination of $\delta\alpha'$. The maximum sensitivity in the critically damped state, with the maximum energy in $G$ can be found from (6.8a) by taking $m = 0$. Then

$$\delta\alpha' = \frac{\alpha_1}{S_{x,\max}} \frac{I_x}{2}, \qquad \text{or for } \delta\alpha' = 1, \qquad S_{x,\max} = \frac{1}{2} \alpha_1 I_x. \tag{6.9}$$

Table 1. *Sensitivities* $S_x = R_x/\delta R_x$ *for a* WHEATSTONE *bridge*. Galvanometer: $\alpha_1 = 10^9$ mm./amp.; $R_g = 30$ ohm; $R_1 = R_2 = 10$ ohm $(n = 1)$ and thus $R_x = R_4$; $I_x^2 \cdot R_x = 0.1$ watt.

| $R_c$ ohm | $R_x$ ohm | $I_x$ amp | $m$ | $R_{eq}$ ohm | $f$ | Case | $S_x$ | $\delta R_x$ ohm | galvanometer is: |
|---|---|---|---|---|---|---|---|---|---|
| 60 | $10^4$ | $3\times10^{-3}$ | $10^{-3}$ | 5005 | $6\times10^{-3}$ | (6.8) | $2.7\times10^7$ | $4\times10^{-4}$ | periodic |
| 60 | $10^2$ | $3\times10^{-2}$ | $10^{-1}$ | 55 | $6\times10^{-1}$ | (6.8) | $2.1\times10^7$ | $5\times10^{-6}$ | aperiodic, periodic side |
| 60 | 50 | $4.5\times10^{-2}$ | $2\times10^{-1}$ | 30 | 1 | (6.8a) | $1.8\times10^7$ | $2.5\times10^{6}$ | aperiodic |
| 60 | 1 | $3\times10^{-1}$ | 10 | 5.5 | 5.4 | (6.8) | $3.3\times10^6$ | $3\times10^{-7}$ | creeping |
| 60 | $10^{-2}$ | 3 | $10^3$ | 5 | 6 | (6.8) | $3.3\times10^5$ | $3\times10^{-8}$ | creeping |
| 100 | 50 | $4.5\times10^{-2}$ | $2\times10^{-1}$ | 30 | 1 | (6.8b) | $1.2\times10^7$ | $4\times10^{-6}$ | aperiodic, creeping side |

It follows from (6.8a) and (6.9) that for small resistances ($R_x \approx R_c$) the critically damped symmetric circuit has the maximum theoretically obtainable sensitivity. Since this is also the most convenient situation to work with, the critically damped state is mostly preferred.

The values of $(\delta R_x)_{\min}$, that is the smallest values of $\delta R_x$ which correspond to one mm. deflection of the galvanometer at a given energy developed in the resistance $R_x$, are, of course, of interest. For comparison these are given in the first table for some values of energy, $R_c$ and $R_x$.

In this case, and for the bridges described below, galvanometers have to be used for which a voltage across the terminals gives the largest possible current in the coil. These are voltage-sensitive galvanometers, having a low internal resistance. For more details we refer to vol. XXIII on electrical instruments. One of the inconveniences of the WHEATSTONE bridge is that the resistances of the connecting wires between the resistances are not eliminated. For precision measurements the general practice has been to construct standard resistances with current and potential terminals. They can be used in a WHEATSTONE bridge, and thus improve the determination of the value of an unknown resistance by the following method, as introduced by the British National Physics Laboratories[1] (see Fig. 2).

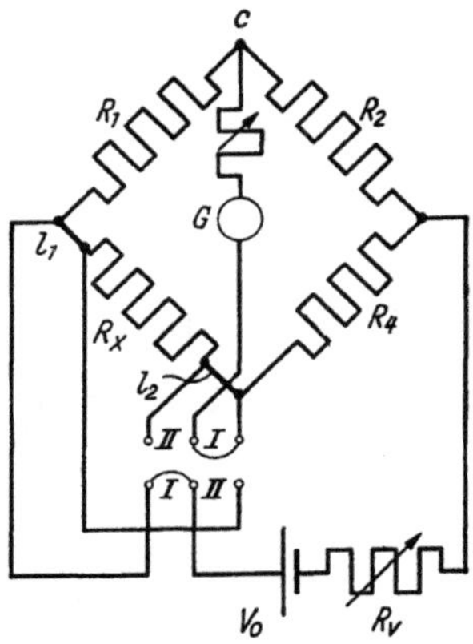


Fig. 2. Improved WHEATSTONE bridge.

In position I, $R_x + l_1 + l_2$ is in the bridge circuit, so that after balancing: $R_x + l_1 + l_2 = R_4 \frac{R_1}{R_2}$.

In position II, after balancing by change of $R_2 \to R_2'$, and of $R_4 \to R_4'$ we get $R_x = (R_4' + l_2)(R_1 + l_1)/R_2'$. Since the quantities are all known with the exception of $R_x$, $l_1$ and $l_2$, an estimate of the correction $l_1 + l_2$ can be made. Since $\frac{l_1 l_2}{R_2'} \ll l_2 \frac{R_1}{R_2'}$ there follows

$$(l_1 + l_2) = \left(1 + \frac{R_4' + R_1}{2 R_2'}\right)^{-1} R_1 \left(\frac{R_4}{R_2} - \frac{R_4'}{R_2'}\right). \tag{6.10}$$

For both bridges the accuracy depends mainly on the determination of the ratio $R_1/R_2$ when $R_4$ has a fixed value. For a slide wire arrangement, an accuracy of 1 in 2000 can be expected in the position of $C$. This sets also a limit to the lowest value of $R_x$ which can be measured with this accuracy using the unmodified bridge. The resistances of the wires leading to $R_x$ are of the order of $10^{-4}\,\Omega$, thus $R_x \approx 1\,\Omega$ can be measured with the estimated accuracy. Generally speaking, a galvanometer sensitivity of the order of $10^6$ mm./amp is sufficient for most practical purposes.

*β 2) The* THOMSON *bridge* is a modified WHEATSTONE bridge, especially developed for the determination of resistances smaller than 1 ohm. The standard resistance $R_4$ and the unknown $R_x$ have potential and current leads. The diagram of the bridge is given in Fig. 3.

One of the potential leads of $R_x$ and of $R_4$ is part of a branch in the circuit; the other one is connected in series with the resistances $a$ and $b$ which can be short circuited by a switch $d$.

By changing $R_1/R_2$ and $a/b$ until the opening or closing of $d$ has no influence on the zero deflection of $G$, the following condition for balance with $d$ closed is

[1] See ref. [*39*], p. 717.

reached:

$$R_x = \frac{R_1 R_4}{R_2} + \frac{b\,d}{a+b+d}\left(\frac{R_1}{R_2} - \frac{a}{b}\right). \tag{6.11}$$

Whilst $\frac{R_1}{R_2} = \frac{a}{b}$, this is the condition (6.1) for balance of a WHEATSTONE bridge.

To facilitate realization of this condition, the branches $R_1 + R_2$ and $a+b$ are made of slide wires with mechanically coupled contacts $A$ and $C$.

As regards an estimate of the sensitivity, the expressions derived for the WHEATSTONE bridge can be used, when $R_{eq}$, given by (6.3) is replaced by $R'_{eq} = R_{eq} + \frac{a\,b}{a+b}$ in the expressions for $\delta\alpha$. Again, as before:

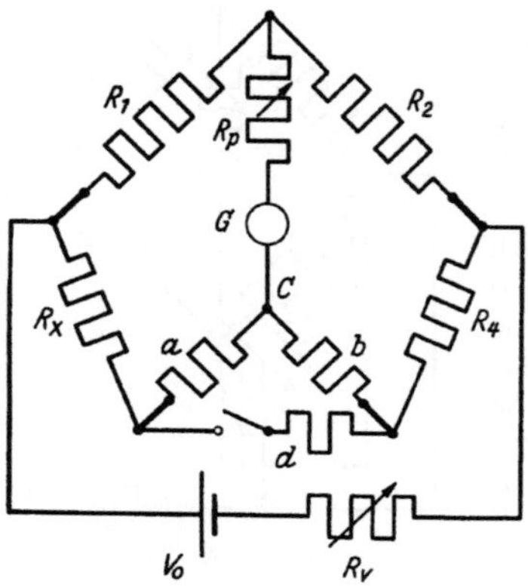


Fig. 3. The THOMSON bridge.

$$\delta\alpha = \alpha_1 \frac{\delta V_g}{\sqrt{R_c(R'_{eq} + R_g)}} \tag{6.12}$$

and for the case $R_g = R'_{eq} = \frac{1}{2} R_c$, it follows [cf. Eq. (6.6a)]:

$$\delta\alpha' = \alpha_1 \frac{\delta V_g}{R_c} = \alpha_1 \frac{\delta V_g}{2\left(R_{eq} + \frac{a\,b}{a+b}\right)}. \tag{6.12a}$$

It follows from Eqs. (6.6a) and (6.12a) that under identical conditions, that is for the same values of the resistances $R_1$, $R_2$, $R_x$, $R_4$ and the constants of the galvanometer,

$$(\delta\alpha)_{\text{Thom.}} < (\delta\alpha)_{\text{Wheat.}},$$

or,

$$(S_x)_{\text{Thom.}} < (S_x)_{\text{Wheat.}}.$$

This loss in sensitivity is, however, more than compensated by the larger accuracy in the determination of $R_x$, especially for small resistances ($R_x \approx 10^{-2}\,\Omega$) when the resistances of the potential leads to $R_x$ are not negligible.

Again it is the correct determination of the position of the contact $C$, which restricts the accuracy to $e_x \approx 2 \times 10^{-2}$ % under favourable conditions.

*β 3) Potentiometers.* As methods, developed for precision voltage measurements, potentiometers have been successfully used for comparison of a number of resistances connected in series in a circuit. The methods originate in the POGGENDORF method, which applies two separate circuits. Through a variable resistance in the secondary or compensating circuit (current $I_c$), a potential difference is adjusted which compensates the potential difference across an unknown resistance in the primary or measuring circuit (current $I_m$). Comparison with the potential difference across the known resistance in the primary circuit gives the value of the unknown resistance in terms of the known resistance. The principle of such a bridge is given in Fig. 4a. In position I of the switch $S_1$, the voltage $I_m R_s$ across a known or a standard resistance $R_s$ is compensated by sliding the contact $C$ so that $I_c R'_a = I_m R_s$, and in position II so that $I_c R''_a = I_m R_x$. Thus: $R_x = \frac{R''_a}{R'_a} R_s$. A repeated measurement of $I_m R_s$ affords a correction for an eventual change in $I_m$.

In each circuit is a commutator. By reversing both currents after each reading the stray e.m.f.'s are eliminated in the galvanometer circuit.

A displacement of the contact $C$ effects a change in the resistance of the galvanometer circuit, which affects the sensitivity of the galvanometer. In order to establish a constant sensitivity, the compensators are equipped with double slide wires or double decades, which realise a constant value of the resistance between $A$ and $S_1$, independent of the position of $C$ (Fig. 5).

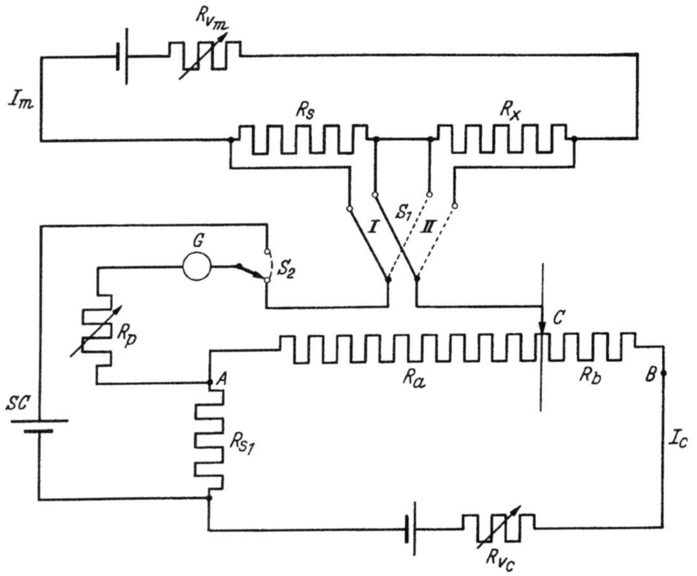


Fig. 4a. Principle of a compensator.

In general the compensators are constructed in such a way, that each position of $C$ corresponds to a decimal fraction of the potential difference across $AB$. This means that, for example for a five-decade instrument, one step in the last decade corresponds to $10^{-4}$ of the potential difference across one step in the first decade. For absolute measurements the compensator current $I_c$ has to be adjusted to $I_c = 10^{-n}$ amp ($n = 1, 2, 3$, or 4), which can be realised by varying $R_{vc}$ until $I_c R_{s1}$ equals the voltage of a standard cell $SC$ (Fig. 4a, $S_2$ in position not drawn). For a description of these compensators, such as the DIESSELHORST and the VERNIER bridges, the reader is referred to the volume on electrical instruments.

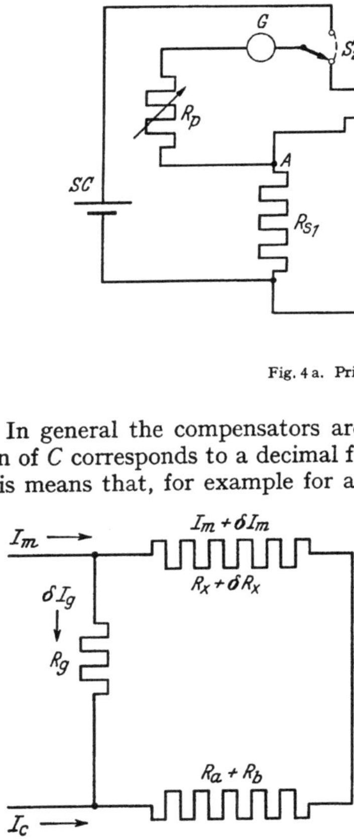


Fig. 4b. A substitution diagram for a compensator.

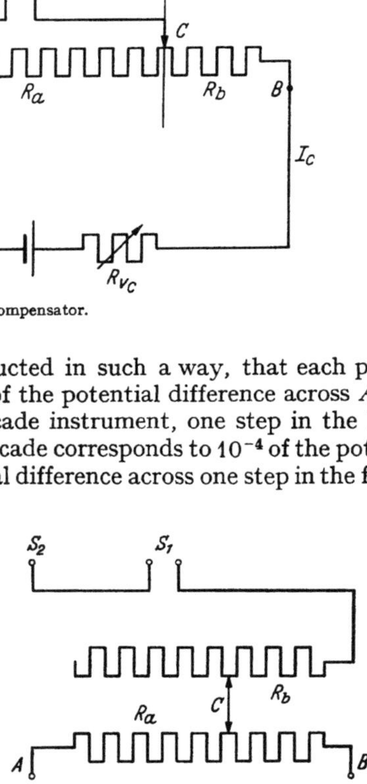


Fig. 5. Double slide wire arrangement.

The sensitivity of the method can be estimated easily. From the substitutional diagram Fig. 4b it follows that a $\delta R_x$ causes a $\delta I_g$ to flow through $G$, for which is: $\delta I_g(R_x+\delta R_x+R_d+R_g)=I_m\delta R_x$, when $R_d=R_a+R_b$. It has been supposed that $\delta R_x$ is too small compared with the total resistance in the $I_m$ circuit to affect the value of $I_m$.

Applying the general expressions (5.1) and (6.5), we find for the non-critically damped galvanometer:

$$\left.\begin{aligned} \delta\alpha &= \frac{\alpha_1}{S_x} I_m \frac{R_x}{\sqrt{R_c(R_x+R_d+R_g)}}, \\ \text{and for } \delta\alpha = 1 \qquad & \\ S_x &= \alpha_1 I_m \frac{R_x}{\sqrt{R_c(R_x+R_d+R_g)}}. \end{aligned}\right\} \tag{6.13}$$

For the critically damped galvanometer, that is $R_d+R_g=R_c$, and for the case $R_x \ll R_c$, it follows from Eqs. (5.1) and (6.4), and from the expression for $I_g$ in the present case:

$$\delta\alpha' = \frac{\alpha_1}{S'_x} I_m \frac{R_x}{R_c}, \qquad \text{and for } \delta\alpha' = 1, \qquad S'_x = \alpha_1 I_m \frac{R_x}{R_c}. \tag{6.13a}$$

For example, taking $R_c=60$ ohm, $\alpha_1=10^9$ mm./amp and a current $I_m$ corresponding to a power of 0.1 watt in $R_x$ the values for $S_x$, obtained with (6.13) and (6.13a), are compared with those from table 1 for the WHEATSTONE bridge.

Table 2. *Sensitivities* $S_x=\frac{R_x}{\delta R_x}$ *for a compensator.* Galvanometer: $\alpha_1=10^9$ mm./amp., $R_c=60$ ohm $I_m^2 R_x=0.1$ watt.

| $R_x$ ohm | $I_m$ amp. | $S_x$ | $\delta R_x$ ohm | equation | galvanometer is: | WHEATSTONE $S_x$ |
|---|---|---|---|---|---|---|
| $10^{-6}$ | 300 | $5\times10^3$ | $2\times10^{-10}$ | (6.13a) | aperiodic | — |
| $10^{-2}$ | 3 | $5\times10^5$ | $2\times10^{-8}$ | (6.13a) | aperiodic | $3.3\times10^5$ |
| 1 | 0.3 | $5\times10^6$ | $2\times10^{-7}$ | (6.13a) | aperiodic | $3.3\times10^6$ |
| $10^2$ | 0.03 | $3\times10^7$ | $3\times10^{-6}$ | (6.13) | periodic | $2.1\times10^7$ |
| $10^4$ | 0.003 | $1.4\times10^7$ | $7\times10^{-4}$ | (6.13) | periodic | $2.7\times10^7$ |

A comparison shows that, for a critically damped galvanometer, sensitivity is better obtained with a compensator for $R_x \lesssim 1$, whereas for $R_x \gg 1$, the WHEATSTONE bridge is more sensitive. A power of 0.1 watt in $R_x$ is seldom used. For example, resistances of the order of $10^{-2}$ ohm can be measured conveniently with $I_m=0.01$ amp., which results in an $S$-value of the order of $2\times10^3$ ($\delta R_x=0.05\%$ of $R_x$). Since, in the case of the potentiometer, $S_x \sim 1/R_c$, galvanometers with a small $R_c$ value are used. Generally potentiometer and galvanometer are so adjusted that for

$$R_x \ll R_c, \quad R_c = R_d + R_g \quad \text{and} \quad R_d \approx R_g.$$

The accuracy of the determination is defined by the determination of the ratio $R'_a/R''_a$ or the linearity and the calibration of the slide wire, or the decades. Most potentiometers have been checked up to an accuracy of 1 in 10000 or even higher for the more expensive instruments. Though the compensators are rather complicated instruments, the advantage of the method, besides its simplicity, is that no extra readings need to be taken for the determination of corrections for asymmetry and connecting leads.

*β 4) Differential bridge method.* As a last example, we mention a method which is now obsolete, but is easy to realise in any laboratory where a double coil galvanometer is available. The circuit is given in Fig. 6a. The deflections of the double coil galvanometer in positions I and II of the commutator are read.

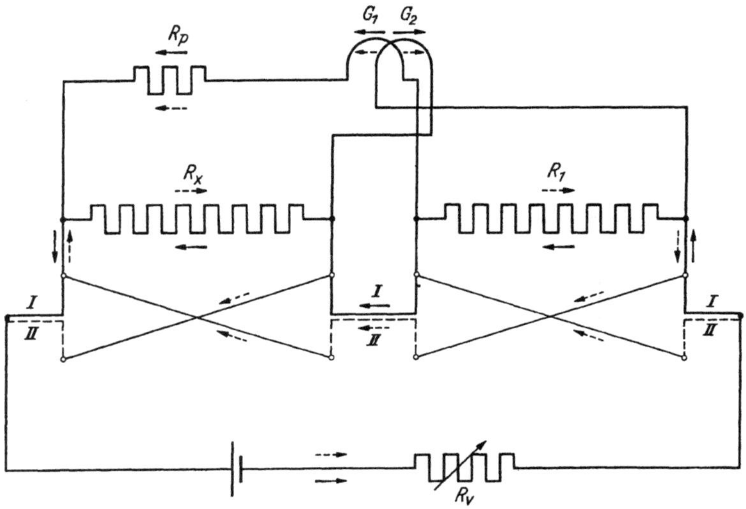


Fig. 6a. Differential bridge method.

When the deflection of the galvanometer is the same before and after commuting the currents in $R_x$ and $R_1$, then $R_x = R_1$. The directions of the current in the galvanometer coils remain the same before and after commuting. By changing $R_1$ and reading the change in deflection of $G$, the sensitivity of the galvanometer has to be determined after each change in the circuit.

For an estimate of $S_x$, the value of the equivalent resistance across each of the coils has to be determined. This is done by help of the substitution diagram, Fig. 6b. To the positions I and II of the commutator, there corresponds an interchange of $R_1$ and $R_x$. When $R_v \gg R_1$ or $R_x$, $R_{eq} \approx R_1$. A change $\delta R_x$ in $R_x$ results into a current $\delta I_{g1}$ in the coil $G_1$, given by

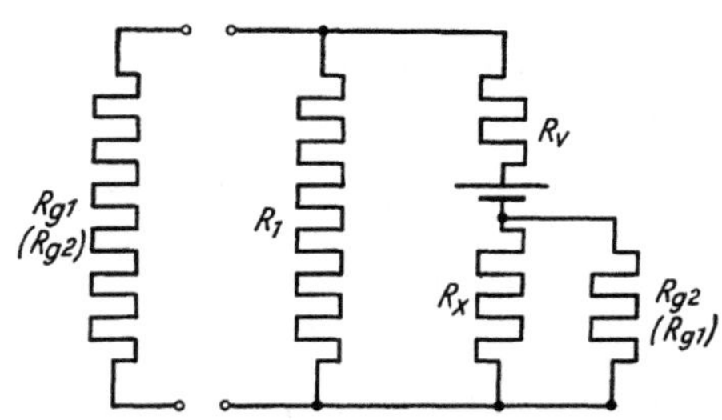


Fig. 6b. A substitution diagram for the differential bridge method.

$$\delta I_{g1} = I_m \frac{\delta R_x}{R_{c1}}, \quad \text{if} \quad R_{g1} + R_x = R_{c1}.$$

This leads to the same sensitivity as for the potentiometer (6.13).

The present method automatically eliminates the resistance of the leads to $R_x$ and $R_1$.

With equal currents through the coils, the galvanometer may show a deflection, but this can easily be corrected by an extra resistance in one of the branches ($R_p$, Fig. 6a). Though the method is simple, it has the disadvantages of a deflection method, viz. a smaller stability of a galvanometer in a deflected position, a possible departure from the linear relationship between current and deflection for large deflections, and possible abrupt or continuous changes in the sensitivity.

When a relative change of one of a pair of resistances as a function of a physical parameter, say temperature, has to be measured or has be followed continuously, the present method can be applied successfully, provided however, that the changes are too small to affect the critically adjusted galvanometer circuits[1]. For small values of $R_1$ and $R_x$ this condition is not easy to realise, and hence is the potentiometer method preferable in this case. The following differential method, using a single coil galvanometer and a potentiometer has been developed by BOWEN and RODENBACH [*17*] (Fig. 7).

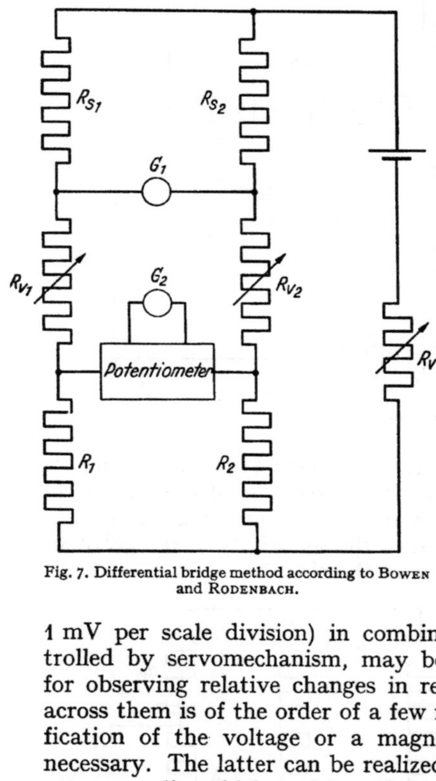

Fig. 7. Differential bridge method according to BOWEN and RODENBACH.

The galvanometer $G_1$ indicates the potential difference between corresponding points of the equal resistances $R_s$. By adjusting the variable $R_v$'s, the currents in both branches are equalized. When $R_1$ and $R_2$—being the resistances which are to be compared—are small relative to $R_{v1}$ and $R_{v2}$, changes $\delta R_1$ and $\delta R_2$ will not have a measurable influence on the equality of $I_1$ and $I_2$, but will give a measurable difference between the potentials over $R_1$ and $R_2$, which is measured by a potentiometer.

*γ) Miscellaneous methods.* All bridge methods mentioned above are methods for measuring small potential differences ($10^{-8}$ to $10^{-3}$ V). When used with a normal galvanometer, they have a lowest detectable limit of the order of $10^{-8}$ V. For recording purposes galvanometers (mostly with a sensitivity of 1 μV to 1 mV per scale division) in combination with automatic potentiometers, controlled by servomechanism, may be used. These instruments are very useful for observing relative changes in resistors, when the total potential difference across them is of the order of a few millivolts. For voltages below $10^{-8}$ V amplification of the voltage or a magnification of the galvanometer deflection is necessary. The latter can be realized optically by using a photoelectric cell with a narrow slit, which registers the movement of the image of the galvanometer mirror. By this method, a gain by a factor 100 can be obtained, before the natural noise of the indication of the galvanometer due to the brownian movement of the coil, is reached. Expressed in amperes, this is $\delta I_B = \sqrt{\frac{\pi \varkappa T}{R_c \tau_0}}$, which at $T = 290°$ K corresponds to $\delta I_B = 11.2/\sqrt{R_c \tau_0}$ amp. for electromagnetic damping only.

For a commercial galvanometer in a bridge with $R_c = 60\,\Omega$, $\tau_0 = 7$ sec., $\alpha_1 = 10^9$ mm./amp. this amounts to $\delta I_B \approx 0.5 \times 10^{-11}$ amp., a deflection $\delta\alpha = 5 \times 10^{-3}$ mm. or $3 \times 10^{-10}$ volts over the galvanometer circuit. For most resistance experiments other causes lead to much larger deflections.

Several other methods of increasing the sensitivity of galvanometers have been suggested recently. For example, a photoelectric amplifier with feed-back

[1] For resistance values of the order of 1 to 100 ohms, automatic and recording potentiometers are useful for this purpose.

circuits and a long period microamperemeter as a final indicator have been used. Then $\delta I_B$ can be reduced, and the overall stability of the instrument can be improved so that current readings can actually be taken with an average error of $\pm 1.0 \times 10^{-11}$ amp. or $6 \times 10^{-10}$ volt ([*93*], HILL and others).

Quite another solution to the problem of increasing the galvanometer sensitivity has been given by PIPPARD and PULLAN [*169*]. They constructed a low temperature magneto-meter with a superconducting coil and resistor of the order of $10^{-8}\,\Omega$ in the circuit at low temperatures. Thus low voltages give comparatively large currents in a coil for which $\delta I_B$ is small (low $T$). In this case, voltages as low as $10^{-12}$ V could be detected. The use is, however, naturally restricted to investigations at low temperatures.

Another way to increase the sensitivity is by amplifying the unknown voltage. There are at least two possibilities: an amplification of the difference between the unknown voltage and the compensating voltage (the combination of amplifier plus output indicator is then a "galvanometer"), or an amplification of the unknown voltage and a compensation of the result at the output. This problem has solutions to suit special requirements and a vast amount of recent work has been reported [*T 1*], [*188*], [*189*].

The most reproducible method seems to be a conversion of the small unknown d.c. voltage into an a.c. voltage, which is amplified by an a.c. amplifyer and finally converted back into a d.c. voltage. By a method based on this principle voltages of the order of $10^{-10}$ volts have been measured [*85*]. In these cases the sensitivity of the method is restricted by the brownian motion in the input resistor or in the first tube of the amplifier.

*δ) Some technical remarks, current stabilizing.* For accurate measurements with a compensator generally several precautions have to be taken. When the resistance $R_x$ is at a temperature different from room temperature, all connections in the galvanometer circuit have to be clean and of the same material in order to avoid large thermo-e.m.f.'s. In the compensator, these are of the order of $10^{-8}$ volts or less. Normal clean contacts may generate thermo-e.m.f.'s in the circuit which are of the order of some tenths of a microvolt. As long as the conditions do not change markedly, this is too low to influence the results. For best results, however, the equipment,—bridge and accessories—has to be placed in a temperature controlled room. Especially in low and high temperature resistance measurements, one has to look for the optimum conditions. Another possible nuisance might occur by an unstability of the measuring current $I_m$ and the compensator current $I_c$. As long as these currents are small, ($\leq 1$ mA), this is not a serious problem. A variable resistor of good quality and a good battery will give a constant current some hours after the circuits have been closed. The constancy of $I_c$ can be checked with the standard cell. If $I_c$ is constant, then the potential difference across $R_{s1}$ (Fig. 4a) will indicate the drift or fluctuation in $I_m$. Many devices and in particular thyratrons and radiotube circuits, have been suggested for stabilizing currents of the order of 1 amp. The experience is however, that a combination of an electronically regulated power supply and a potentiometer circuit may give rise to an erroneous pick-up of small currents through the galvanometer, which influences the desired highest sensitivity, unless rather complicated shielding and insulating measures are taken. A simple method, which has proved its worth may be recommended here[1].

[1] Some years ago Dr. J. SMIT drew my attention to this method, for which I have always been grateful. The simplicity of the method makes it probable, that many experimenters have used it without thinking it worthwhile to report.

The circuit consists (Fig. 8) of batteries (voltages $V_j$) parallel to the main battery (voltage $V_0$) and a number of voltage dividing resistors $R_j$ in between, such that the voltage of the last battery equals the output voltage ($V_n = IR$).

Let the resistances of the branch of the $j$-th cells be $R_{b_j}$. Then, it easily follows that a voltage change $\Delta V_0$ in the main battery causes a $\Delta V_n$ over $R_{b_n}$ given by

$$\Delta V_n = \frac{\Delta V_0 R_{b_1}}{R_1 + R_{b_0} + R_{b_1}} \prod_2^n \frac{R_{b_j}}{R_j + R_{b_j}},$$

or, since $R_{b_j}/R_j \ll 1$:

$$\Delta V_n = \Delta V_0 \prod_1^n \frac{R_{b_j}}{R_j}.$$

The current provided by the last battery is $\Delta V_n/R$. For example, taking $V_0 = 12$ volts, $V_1 = 8$ volts, $V_2 = 4$ volts. $I = 1$ A and $R_1 = R_2 = R_{\text{load}} = 4\,\Omega$, one

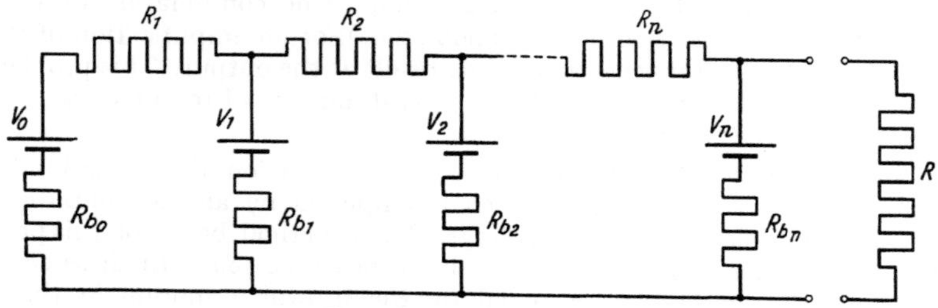


Fig. 8. Current stabilizing circuit.

finds with $R_{b_1} = R_{b_2} = 0.1\,\Omega$ that $\Delta V_2 = 6 \times 10^{-4}\,\Delta V_0$ or $\Delta I = 1.5 \times 10^{-4}$ A per volt change in $V_0$. The current in the second battery is then only 6 mA for one volt change in $V_0$.

The necessity of small values of $R_{b_j}$ that is the total internal resistance of each battery and its connections with the circuit is evident.

In practice, the circuit is adjusted by introducing a low resistance amperemeter in the branches 1 ... $n$ and varying $R_1, R_2, \ldots, R_n$ until no current flows in each branch.

**7. Alternating current methods.** Measurements have been carried out with a.c. methods when a desired small current through a resistance results in a very low voltage, or an extremely small resistance has to be measured, or when an automatic elimination of the influence of thermo e.m.f.'s is required.

In principle, the same methods are applicable to a.c. as to d.c. It has to be realised, however, that in the a.c. method the *impedance* of the resistance is measured and that in most cases a phase control unit has to be inserted in the circuit. Besides, shielding of the circuit against capacitive and inductive coupling is required.

*α) Modifications of the* WHEATSTONE *bridge.* Apart from the current-voltage ratio and interchange methods (cf. Sect. 6α) which, for a.c., is very inaccurate ($e \approx 10\%$) the earliest simple more reliable method is the application of a slightly modified WHEATSTONE bridge, in which, instead of a battery and a galvanometer, an a.c. supply and an a.c. zero indicator (head-phones, a.c. galvanometer, or electronic voltmeter) is used.

The errors are in general, larger than with the direct current bridges, unless special precautions are taken. With most bridges, the lowest resistance value which can be measured with an accuracy of ½ to 1% is of the order of 0.1 ohm.

*β) Compensators.* In principle, any a.c. method, developed in electronics for measuring potential differences, can be used after some modification for measurement and comparison of resistances, often with a sensitivity and accuracy comparable to or even surpassing the d.c. compensation (comparison) methods. The main difficulty exists in matching the input impedance of the bridge with the resistance which has to be investigated. Even a mere cataloguing of these methods, which have a large applicability in applied science, would go beyond the scope of this article. The reader is therefore referred to the original literature and handbooks in electronics and engineering.

As an example, we describe briefly a recently published method, developed especially by BROOM and CLOTHIER [*21*] for measuring small resistances of pure

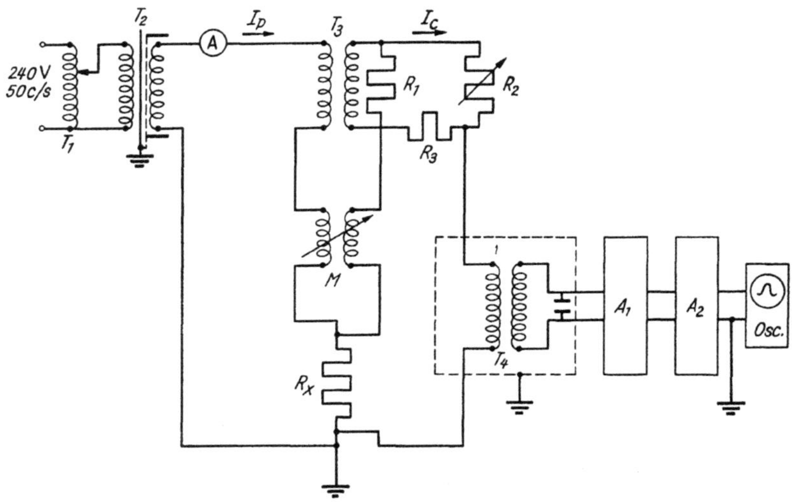


Fig. 9. An a.c. potentiometer according to BROOM and CLOTHIER.

metals, a diagram of which is given in Fig. 9. The current $I_p$, adjusted to a chosen value by means of two transformers $T_1$ and $T_2$, generates a voltage drop over $R_x$, which is compensated by a secondary circuit current $I_c$ coupled to $I_p$ by a normal current transformer $T_3$ (1:1). Besides, the secondary circuit is coupled to the primary circuit by a variable mutual inductor $M$. The resultant potential difference in the circuit $R_3 R_x M$ is across the primary of a transformer $T_4$. This difference is stepped up (transformer ratio 1:3500) and amplified by two amplifiers $A_1$ and $A_2$ and finally applied to an oscilloscope.

By varying the variable mutual inductance $M$, $V_1$, and $V_2$ are brought into phase. The ratio $I_p/I_c$ then is given by the transformer ratio $N$ of $T_3$. The value of $M$ does not enter the calculation of the result, which for zero deflection in the oscilloscope is given by: $R_x = \frac{R_1 R_3}{N(R_1 + R_2 + R_3)}$. The authors claim a possible detection of $10^{-10}$ V, an error $e = 0.02\%$ and a sensitivity $S_\lambda \approx 10^5$ for a power of 0.8 μ watt developed in $R_x$.

*γ) Induction methods.* The improvement in purification methods in metallurgy has resulted in the production of very pure metals. Any measurement of the properties of a specimen of such a very pure material forbids the introduction of impurities by welding or soldering even parts of the same material unto it.

Also, in cases where attaching of wires to the sample is not possible, resistance measurements should be made by using an inductive method. So far, inductive methods have been used only incidentally, one of the reasons being, that these methods allow only for relative measurements.

They consist, in principle, in measuring the self-inductance of one coil or the mutual inductance of two coaxial coils surrounding the specimen to be investigated. Their inductances are measured by a suitable a.c. bridge method. The eddy currents generated in the specimen affect the magnetic field of the coil or coils, which manifests itself as a change in inductance.

A cylinder of conducting material, diameter $d$, resistivity $\varrho$, placed in a coil with a field $H$ parallel to its axis, absorbs an energy $W_0 = \frac{\varrho}{4\pi} G(z) H^2$ ergs per cm.[3] [*27*] where $G$ is a combination of BESSEL functions with the variable $z^2 = \pi\omega d^2/\varrho$, $\omega$ and $d$ being respectively the a.c. frequency and diameter of the cylinder. If $z \ll 1$, $G(z) \sim z^6$, thus

$$W_0 \approx \frac{\pi^2}{4} \frac{\omega^3 d^6}{\varrho^2} H^2.$$

The value of $W_0$ can be determined by a wattmeter circuit, or by a bridge method, for which it is convenient to consider the loss as the dissipation of energy in a resistance $R_{\mathrm{eff}}$, in series with the self-inductance of the coil. A practical formula for $R_{\mathrm{eff}}$ is

$$R_{\mathrm{eff}} = 0.00658 \frac{d^2}{\varrho} \mu_{\max} f^2 \chi \text{ ohm}$$

in which $\mu_{\max}$ is the maximum permeability in the conductor, $f = \omega/2\pi$ the frequency of the a.c., and $\chi$ the inductive reactance of the coil[1].

A few applications of the above and analogous methods, are known (especially at low temperatures where $\varrho$ is small, and the loss relatively large at low frequencies).

VAN DEN BERG and VAN DER MAREL [*10*] checked the possible existance of a minimum in the $\varrho(T)$ curve at temperatures below 4° K, for a single crystal of gold, using a mutual inductive bridge (HARTSHORN bridge). In this bridge, the loss of flux in the inner one of a pair of coaxial coils is compensated by changing the number of turns in one coil of a mutual inductance. For the real part of the voltage compensation as well as for the imaginary part the authors prove that the number of turns in the mutual induction is proportional to $\sqrt{\varrho/f}$, (in their experiments $f = 400$ c./sec. and $f = 133\frac{1}{3}$ c./sec.). The methods are limited, however, by the penetration depth of the flux into the conductor, given by

$$\delta = 5033 \sqrt{\frac{\varrho}{\mu f}} \text{ cm}.$$

If the high frequency field confines itself only to a tiny surface ring, only a surface value of $\varrho$ is determined, which can be expected to be larger than the bulk value owing to physical and chemical surface contaminations. For the $133\frac{1}{3}$ c./sec. waves, the $\delta$-values at 300° K and 4° K respectively, are 0.4 cm. and 0.02 cm. for $\mu = 1$, $\varrho_{300^\circ \mathrm{K}} = 1 \times 10^{-6}$ ohm cm. and $\varrho_{4^\circ \mathrm{K}} = 2 \times 10^{-9}$ ohm cm.

Inductive methods have proved to be useful for determining the change in $\varrho$ at the melting point of a metal. Here the difficult problem is to introduce electrodes in such a way that the dimensions of the part whose resistivity is measured in the solid and liquid states, are unchanged.

[1] See ref. [*T 1*] and also [*133*].

REGEL [177] measured the ratio of $\varrho_{\text{liq}}/\varrho_{\text{solid}}$ of indium by another inductive method. A sphere was suspended by a fibre in a rotating homogeneous magnetic field. The magnetic field of the varying eddy currents tries to oppose the changes of the rotating external field, so that a resultant mechanical moment is created which gives the sphere a torque to be measured. The rotating field was produced by three coils 120° apart. The formulae for the moment are analogous to those for the energy loss due to eddy currents.

## III. The conductivity of pure metals at normal temperatures.

**8. The conductivity of elements at the ice-point.** In this section a summary will be given of the numerical values for the conductivity of pure metals at normal temperatures, particularly at 273.15° K. These values have only a practical significance, since from the physical standpoint, the icepoint is not a characteristic temperature for metals. As is usual (GRÜNEISEN, [G 1], [G 2]), we have separately tabulated the values of $\varrho_T$ (in ohm meters)[1], $\sigma_T$ (in ohm$^{-1}$ meters$^{-1}$)[1] and the temperature coefficient $\alpha_{\varrho,T}$ (in degree$^{-1}$), for isotropic metals in Table 3, mean values of these quantities for anisotropic metals in Table 4 and the principal values for anisotropic metals in Table 5. Unless stated otherwise, the values are taken from GRÜNEISEN's paper; for convenience we present them in the alphabetic order. The mean values of conductivity for quasi-isotropic wires of an anisotropic metal have been deduced from the principal conductivities by applying VOIGT's formula (3.3), for so far there are no experimental values available for polycrystalline material.

As a logical consequence of the interpretation of MATTHIESSEN's rule, written as

$$\varrho(T) = \varrho_m(T) + \varrho_i \tag{8.1}$$

— in which $\varrho_m(T)$ and $\varrho_i$ are the resistivities of the pure metal (ideal resistivity) and the temperature independent contributions from the impurities, respectively (see Sect. 27) — the values given in GRÜNEISEN's tables are $\varrho_m$-values and given as such in the present article. The number $\varrho_i$ can be found from the resistivity-temperature relationship[2].

The assumption that by this interpretation the ideal resistivity of the metal ($\varrho_m$) can be found is, however, rather questionable. For a great many types of impurities, low temperature measurements have shown that a $z$-value found at high temperatures differs from the $z$-value of the same wire as deduced from low temperature measurements. In the following we shall indicate by $\varrho_{i,0}$ a $\varrho_i$ value deduced from measurements below 10° K. This means that the quantity obtained by subtracting $\varrho_i$ from $\varrho_{\text{obs}}$ represents only the temperature dependent part of the resistivity of the given metal at the temperature of determination

---

[1] The index 273 for $T$ means the value of $\varrho$ and $\sigma$ at the normal ice point: $T = 273.15°$ K.

[2] If $\varrho_{\text{obs},T_1} = \varrho_{m,T_1} + \varrho_i$ and $\varrho_{\text{obs},T_2} = \varrho_{m,T_2} + \varrho_i$ are the resistivities at two different temperatures, and $\varrho_i \ll \varrho_{T_2}$ for $T_2 > T_1$, then introducing:

$$r_{\text{obs}} = \left(\frac{\varrho_{T_1}}{\varrho_{T_2}}\right)_{\text{obs}}, \quad r_m = \frac{\varrho_{m,T_1}}{\varrho_{m,T_2}} \quad \text{and} \quad z = \frac{\varrho_i}{\varrho_{T_1}}$$

it follows that

$$r_m = \frac{r_{\text{obs}} - z}{1 - z}.$$

In the simple case $\varrho_m \sim T$, $z$ can be found since $r_m = T_1/T_2$; in general, the expression for $r_m$ gives a temperature coefficient $\alpha = \frac{r_m}{r_{\text{obs}}}(1 - z)\,\alpha_m$.

and the order of magnitude of the ideal resistivity. Besides, this number strongly depends on the treatment of the material. Only where the real contribution of the temperature dependent part of the physical and chemical contaminations was known, the $\varrho_m$ for a metal could be derived. Thus the values in the Table 3, 4

Table 3a. *Resistivity, conductivity (at 273.15 °K) and temperature coefficient of resistivity for 273° K ≤ T ≤ 273° K for isotropic conducting metals.*

| Element | Crystal structure | $\varrho$ $10^{-8}$ ohm m. | $\sigma$ $10^{6}$ ohm$^{-1}$ m.$^{-1}$ | $\alpha_{\varrho, T}$ $10^{-3}$ degree$^{-1}$ | Ref. | Element | Crystal structure | $\varrho$ $10^{-8}$ ohm m. | $\sigma$ $10^{6}$ ohm$^{-1}$ m.$^{-1}$ | $\alpha_{\varrho, T}$ $10^{-3}$ degree$^{-1}$ | Ref. |
|---|---|---|---|---|---|---|---|---|---|---|---|
| Ag | f.c.c. | 1.50 | 66 | 4.10 | [*G 2*] | Mo | b.c.c. | 5.03 | 19.9 | 4.7 | [*67*] |
| Al | f.c.c. | 2.50 | 40 | 4.67 | [*G 2*] | Na | b.c.c. | 4.27 | 23.4 | 5.5 | [*G 2*] |
| Au | f.c.c. | 2.04 | 49 | 3.98 | [*G 2*] | Nb[4] | b.c.c. | 23.3 | 4.28 | 2.28 | [*18 e*] |
| Ba | b.c.c. | 36 | 2.8 | 6.1 | [*179*] | Ni | f.c.c. | 6.58 | 15.2 | 6.75 | [*G 2*] |
| Ca | f.c.c. | 3.6 | 28 | ≈4 | [*62*] | Pb | f.c.c. | 19.3 | 5.19 | 4.22 | [*G 2*] |
| Ce[4] | f.c.c. | 74.8 | 1.4 | 0.97 | [*18 c*] | Pd | f.c.c. | 9.77 | 10.2 | 3.8 | [*G 2*] |
| $Co_\alpha$ | f.c.c. | 5.2 | 19 | 6.58 | [*G 2*] | Po[2] | cubic | ≈45 | ≈2.3 | ≈4.6 | [*144*] |
| Cr | b.c.c. | 15.0 | 6.6 | | [*G 2*] | Pt | f.c.c. | 9.81 | 10.2 | 3.92 | [*G 2*] |
| Cs | b.c.c. | 19.0 | 5.2 | 5.0 | [*G 2*] | Rb | b.c.c. | 11.6 | 8.62 | 5.3 | [*G 2*] |
| Cu | f.c.c. | 1.55 | 64.5 | 4.33 | [*G 2*] | Rh | f.c.c. | 4.33 | 23 | 4.57 | [*G 2*] |
| $Fe_\alpha$ | b.c.c. | 8.71 | 11.5 | 6.57 | [*G 2*] | $Sr_\alpha$ | f.c.c. | 20 | 5.0 | ≈5 | [*179*] |
| Ir | f.c.c. | 4.74 | 21.1 | 4.33 | [*174*] | Ta | b.c.c. | 12.4 | 8.06 | 3.6 | [*G 2*] |
| K | b.c.c. | 6.3 | 16 | 5.4 | [*G 2*] | Th | f.c.c. | 19.1 | 5.24 | 3.3 | [*212*] |
| La | f.c.c. | 65 | 1.5 | | [*102*] | U | cubic | ≈25 | ≈4.0 | | [*209*] |
| Li | b.c.c. | 8.5 | 11.8 | 4.37 | [*G 2*] | V | b.c.c. | 18.2 | 0.549 | | [*172*] |
| $Mn_\alpha$ | cubic | 710 | 0.14 | 0.17[1] | [*24*] | W[3] | b.c.c. | 4.89 | 20.4 | 4.83 | [*67*] |
| $Mn_\beta$ | cubic | 91 | 1.1 | 1.4[1] | [*24*] | | | | | | |

and 5 are to be considered only as being the lowest for $\varrho$ (the highest for $\sigma$), and the highest for $\alpha$ for the given metal as taken from the literature.

The values for the semiconductors Ge, Si, Te and Se have been omitted. For, as far as the intrinsic conductivity range reaches into the room temperature region, the $\varrho$-values are much higher than those for the poorest metallic conductor. For example for intrinsic silicon $\varrho \approx 10^{-1}$ ohm m., which is about $10^5$-times the

*Table 3b.*

| | Bridgman | J., L., S. |
|---|---|---|
| Ce | 74.8 | 80—100 |
| La | 57.6 | 65 |
| Nd | | ≈ 75 |
| Pr | 69 | ≈ 80 |

*Table 3c.*

| | $\varrho_{obs}$ | $\varrho_m$ |
|---|---|---|
| Dy | 130±8 | 110 |
| Er | 175±5 | 160 |
| Gd | 145±5 | 140 |

resistivity of an element like Bi. We found no data for the metals astatine (At) and francium (Fr) and most of the rare earth group elements. From the tables it follows, that with the single exception of plutonium, all metals have positive temperature coefficients of resistivity in the region 273 to 373° K.

Some remarks may follow here about special cases and special methods of investigation.

*Silver and gold.* Without claiming al arge accuracy, the present author has found that for pure Johnson-Matthey material now available, $\varrho_{273}$ for pure annealed

[1] For 193° K < $T$ < 300° K.
[2] Obtained with films.
[3] Six-edged single crystal.
[4] At 303° K.

silver is $1.48 \times 10^{-8}$ ohm m. (with $\varrho_{i,0} = 9.5 \times 10^{-11}$ ohm m.) and for gold $\varrho_{273} = 2.02 \times 10^{-8}$ ohm m. (with $\varrho_{i,0} = 7.0 \times 10^{-11}$ ohm m.). The temperature coefficient of another wire of gold was $\alpha_{\varrho,T} = 4.09 \times 10^{-3}$ degree$^{-1}$ (with $\varrho_{i,0} = 4.0 \times 10^{-11}$ ohm m.). The values of $\varrho_{i,0}$ have been based on measurements down to 1.8° K [*70a*].

*Beryllium:* For a survey of the results obtained with Be, reference is made to POWELL's papers [*174b*]. Several authors have found $\varrho$-values which differ widely. A possible explanation can perhaps be found in POWELL's results. After giving his samples a heat treatment (heating to 975° K), he found for the sample with the lowest $\varrho$-value a decrease from 6.7 to $3.2 \times 10^{-8}$ ohm m. at 293° K. A supposed transition at 870° K, as suggested by others, was not found in POWELL's sample.

GRÜNEISEN and ADENSTED [*82*] found a temperature coefficient $\alpha_{\varrho,T} = 10 \times 10^{-3}$ degree$^{-1}$ for single crystals in the interval 273° K $\leq T \leq$ 293° K. They suggest that this large coefficient is a consequence of the high characteristic temperature (DEBYE temperature) of the metal, which is about 800° K at ordinary temperatures.

*Calcium.* The value $3.6 \times 10^{-8}$ ohm m. as given by FRANK and JEPPESEN [*62*] for $\varrho$ differs from the value $4.3 \times 10^{-8}$ ohm m. as given by RINCK [*179*].

*Cerium, Lanthanum, Neodymium and Prasedymium.* The 1927 experiments of BRIDGMAN [*18e*] have been repeated by JAMES, LEGVOLD and SPEDDING [*102*] in order to investigate possible crystallographic changes at low temperatures. The latter authors' results for $\varrho$ at 300° K are compared with the former ones of BRIDGMAN in Table 3b ($\varrho$ in $10^{-8}$ ohm m.).

An important feature, found by these authors is that, except in the case of Nd, where a rather good reproducible $\varrho(T)$ curve can be found, there is a large instability of the crystal structure of these rare earth elements. Lanthanum, at room-temperature, seems to consist of a mixture of f.c.c. (face-centered cubic) and c.p.h.

Table 4. *Mean resistivity, conductivity (at 273.15° K) and temperature coefficient of resistivity (for 273° K $\leq T \leq$ 373° K) for anisotropic conducting metals.*

| Element | Crystal structure | $\bar{\varrho}$ $10^{-8}$ ohm m. | $\bar{\sigma}$ $10^{7}$ ohm$^{-1}$ m.$^{-1}$ | $\bar{\alpha}_{\varrho,T}$ $10^{-3}$ degree$^{-1}$ | Ref. | Element | Crystal structure | $\bar{\varrho}$ $10^{-8}$ ohm m. | $\bar{\sigma}$ $10^{7}$ ohm$^{-1}$ m.$^{-1}$ | $\bar{\alpha}_{\varrho,T}$ $10^{-3}$ degree$^{-1}$ | Ref. |
|---|---|---|---|---|---|---|---|---|---|---|---|
| As[1] | hex. | 26 | 4.5 | | [*M 1*] | $Mn_{\gamma}$[1] | tetr. | 23 | 4.3 | 6.3[6] | [*24*] |
| Be[1] | hex. | 2.78 | 36 | 7.5 | [*174*] | Nd[1] | | ≈75 | 1.3 | | [*102*] |
| Bi | rhomb. | 107 | 0.93 | 4.45 | [*G 2*] | Os[1,7] | hex. | 95 | 1.05 | 4.2 | [*14*] |
| C | hex. | 39 | 2.6 | | [*183*] | Pr[1,3] | | 69 | 1.45 | 1.65 | [*18c*] |
| Cd | hex. | 6.73 | 14.8 | 4.26 | [*G 2*] | Pu[1] | rhomb. | 160 | 0.62 | −2.97 | [*197a*] |
| Dy[1] | c.p.h.[2] | 110 | 0.91 | 1.7 | [*134*] | Re[1] | hex. | 18.9 | 5.30 | 3.1 | [*149*] |
| Er[1] | c.p.h.[2] | 160 | 0.62 | 2.5 | [*134*] | Ru[1] | hex. | 6.67 | 15.0 | 4.5 | [*102*] |
| Ga[3] | rhomb. | 13.7 | 7.30 | 4.1 | [*174*] | Sb | rhomb. | 32.1 | 3.12 | 5.1 | [*G 2*] |
| Gd[1] | c.p.h.[2] | 140 | 0.71 | 3.45[4] | [*134*] | Sn | tetr. | 10.1 | 9.9 | 4.63 | [*G 2*] |
| Hf[1] | c.p.h. | 26.5 | 3.77 | 4.4 | [*137*] | Ti[1] | c.p.h. | 42 | 2.38 | 5.5 | [*78*] |
| Hg[5] | rhomb. | 21.2 | 4.71 | — | [*G 2*] | Tl | hex. | 15 | 6.7 | 5.2 | [*G 2*] |
| In[1] | tetr. | 8.2 | 12.2 | 5.1 | [*G 2*] | Zn | hex. | 5.45 | 18.3 | 4.20 | [*G 2*] |
| Mg | hex. | 3.94 | 25.4 | 4.2 | [*G 2*] | Zr[9] | hex. | 40.5 | 2.47 | 4.0 | [*2*] |

[1] Polycrystalline material as measured.
[2] See ref. [*5*]
[3] Polycrystalline, $\alpha$ for 90° K $< T <$ 273° K.
[4] For 253° K $< T <$ 303° K.
[5] At $T = 227.6$° K.
[6] For 193° K $< T <$ 300° K.
[7] At 293° K.
[8] At 303° K.
[9] Large crystals.

Table 5. *Principal resistivity, conductivity and ratio for axial conductivities (at 273.15° K) and temperature coefficient (for 273° K ≤ T ≤ 373° K) for anisotropic conducting metals.*

| Element | Crystal structure | $\varrho_{\parallel}$ $10^{-8}$ ohm m. | $\varrho_{\perp}$ $10^{-8}$ ohm m. | $\sigma_{\parallel}$ $10^{6}$ ohm$^{-1}$ m.$^{-1}$ | $\sigma_{\perp}$ $10^{6}$ ohm$^{-1}$ m.$^{-1}$ | $\frac{\sigma_{\parallel}}{\sigma_{\perp}}$ | $\alpha_{\parallel}$ $10^{-3}$ degree$^{-1}$ | $\alpha_{\perp}$ $10^{-3}$ degree$^{-1}$ | Ref. |
|---|---|---|---|---|---|---|---|---|---|
| As | hex. | >26 | 23.8 | <3.8 | 4.20 | 0.92 | | 4.79 | [*M 1*] |
| Be | hex. | 3.58 | 3.12 | 28.0 | 32.0 | 0.88 | 10[1] | | [*82*] |
| Bi | rhomb. | 127 | 99 | 0.785 | 1.01 | 0.78 | | | [*G 2*] |
| Cd | hex. | 7.73 | 6.35 | 12.9 | 15.7 | 0.87 | | | [*G 2*] |
| Ga[2] | rhomb. | 55.5 | 1: 17.3 | 1.80 | 5.77 | 0.31 | 4.05[3] | 1: 4.16 | [*174*] |
| | | | 2: 7.85 | | 12.7 | 0.14 | | 2: 4.12 | [*174*] |
| Hf | hex. | 32.7 | 32.0 | 3.05 | 3.13 | 0.98 | 3.51 | 4.50 | [*2*] |
| Hg[4] | rhomb. | 17.8 | 23.5 | 5.62 | 4.25 | 1.32 | | | [*G 2*] |
| Mg | hex. | 3.50 | 4.22 | 28.6 | 23.7 | 1.21 | | | [*76*] |
| Sb | rhomb. | 26.3 | 36.0 | 3.80 | 2.78 | 1.37 | | | [*176*] |
| Sn | tetr. | 9.09 | 13.08 | 11.1 | 7.60 | 1.46 | | | [*18e*] |
| Zn | hex. | 5.59 | 5.39 | 17.9 | 18.6 | 0.96 | | | [*18e*] |

(close packed hexagonal) modifications and is then rather stable. It was found also, that an f.c.c. modification of La changes within about 120 days into an c.p.h. modification with a corresponding change in $\varrho$ from 70 to $104 \times 10^{-8}$ ohm m. As the transition temperatures of Ce and Pr are below room temperature, it is obvious that a cooling and a successive heating will influence the $\varrho$-values of these crystallographically unstable elements.

*Dysprosium, Erbium and Gadolinium.* These c.p.h. (BANNISTER [*5*]) rare earth elements have been measured by LEGVOLD, SPEDDING, BARSON and ELLIOT [*134*] at roomtemperatures and at low temperatures. They give the following data (Table 3c)[5] for the observed values, the resistivity ($\varrho_{\text{obs}}$) and the deduced values ($\varrho_m$) in $10^{-8}$ ohm m. at 273° K, when the temperature independent contribution to the resistance is accounted for. The spectroscopic analysis of the materials showed weak lines of Mg, Co, Fe. The main impurities were Ho (0.1%), Ca, Sm (0.15%).

*Gallium.* Recent measurements of POWELL [*174a*] show a large anisotropy in the conductivity of Ga at room temperature, which was not found before. The crystal has a pseudo-tetragonal orthorhombic structure, with the principal conductivity ratio

$$\sigma_c : \sigma_a : \sigma_b = 1 : 3.2 : 7$$

($c$-axis is the pyramidal axis, $b$ and $a$ are the diagonal axes in the basic plane of the pyramid). POWELL in his experiments used glass tubes which were bent in such a way (see Fig. 10) that, after filling with liquid metal and crystallisation started by a seed crystal, a rod of solid metal was obtained in which the several sections had prescribed orientation of the crystallographic axis with respect to the rod axis. Fine platinum wires were fused into the glass to facilitate the resistivity measurement. This method for growing crystals of preferred orientation was used earlier in open moulds by SCHUBNIKOW, DE HAAS [*185*] and coworkers for bismuth and gallium.

*Hafnium.* For 99.0% Hf we listed the lowest value reported, namely that by LITTON [*137*]. DEEM [*35*] found for Hf with 2 weight % Zr that $\varrho_{293} = 341 \times 10^{-8}$ ohm m.

[1] For 273° K < $T$ < 293° K.
[2] At 293° K.
[3] For 90° K < $T_0$ < 273° K.
[4] At $T = 227.6°$ K.
[5] The author is indebted to Dr. LEGVOLD for supplying him with these data.

*Niobium*. Metals Reference Book [*M 2*] gives without any reference, $\varrho_{273} = 17 \times 10^{-8}$ ohm m., and $\alpha_{\varrho, T} = 3 \times 10^{-3}$ degree$^{-1}$.

*Lead*. POCHAPSKY [*171*] used a pulse-heating method for the measurement of the specific heat and resistivity of a lead wire. He finds $\alpha_{\varrho, T} = 4.08 \times 10^{-3}$ degree$^{-1}$, which is lower than the value given in Table 3.

*Polonium*. MAXWELL [*144*] measured the conductivity of films of Po, deposited in vacuo on a glass surface, on which had been mounted beforehand four tungsten electrodes. The deposit contained small amounts of Ca and Al as impurities. At room temperature the f.c.c. structure is not distinguishable from that of Pb. He found $\varrho_{273} = (42 \pm 10) \times 10^{-8}$ ohm m., $\alpha_{\varrho, T} = 4.6 \times 10^{-3}$ degree$^{-1}$.

A change into rhombohedric structure was found between 338 and 400° K, the transition temperature varying with the specimen. For this structure little change was found in $\varrho$, but a marked change in $\alpha$ to $7.0 \times 10^{-3}$ degree$^{-1}$.

*Plutonium*. The metal has five solid allotropic modifications at normal pressure. Below 390° K an orthorhombic structure is suggested (SMITH [*197a*], see also Sect. 13).

*Ruthenium*. JUSTI [*108*] used powder from HERAEUS. The powder was sintered at 1420° K in hydrogen atmosphere, and the process was continued in vacuo at $\approx 2500°$ K. The rod obtained has a density of 11.556 g./cm.$^3$ a little smaller than the density 12.304 g./cm.$^3$ found for bulk material. Determination of theresidual resistance at 20° K gave $\varrho_{273} = 6.670 \times 10^{-8}$ ohm m.; it appeared that $\alpha_{\varrho, T} = 4.49 \times 10^{-3}$ degree$^{-1}$. The $\varrho$-value is below that given by earlier results, namely, $7.6 \times 10^{-8}$ ohm m. The purity of the sample was claimed to be 99.99%.

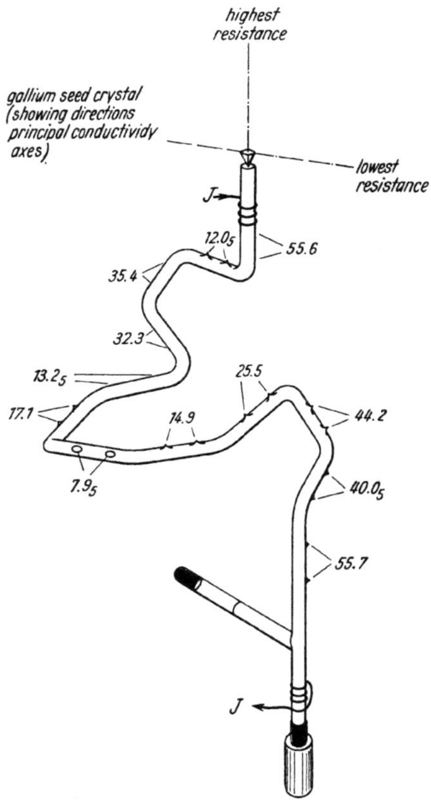


Fig. 10. Sketch of a gallium specimen in bent glass tubing and showing electrical resistivity ($10^{-8}\,\Omega$ m.) of each section at 293° K (POWELL [*174a*]).

*Thorium*. The values of $\varrho$ and $\alpha$ have been determined recently by BENDER [*8*] for 99.7% Th. He gives $\varrho_{273} = 24 \times 10^{-8}$ ohm m. and $\alpha_{\varrho, T} = 2.3 \times 10^{-3}$ degree$^{-1}$.

WEINER, CHIOTTI and WILHELM [*212*] found for pure Th, $\varrho_{273} = 19.1 \times 10^{-8}$ ohm m. and $\alpha_{\varrho, T} = 3.33 \times 10^{-3}$ degree$^{-1}$, whilst RENTSCHLER and MARDEN [*182*] found $\varrho_{290} = 18 \times 10^{-8}$ ohm m.

*Titanium*. Titanium has been intensively investigated during recent years for practical applications. The $\varrho$-value for the purest Ti is that of FAST [*59*].

GONSEN [*78*] found a large difference according as the titanium was reduced with iodine (in which case $\varrho_{273} = 42 \times 10^{-8}$ ohm m. and $\alpha_{\varrho, T} = 5.5 \times 10^{-3}$ degree$^{-1}$) or magnesium (in which case $\varrho_{273} = 55 \times 10^{-8}$ ohm m. and $\alpha_{\varrho, T} = 3.3 \times 10^{-3}$ degree$^{-1}$).

WEINER, CHIOTTI and WILHELM [212] looked for anomalies in the $\varrho(T)$ curve at high temperatures, and they used a wire with $\varrho_{273}=46\times10^{-8}$ ohm m. and $\alpha_{\varrho,T}=3.97\times10^{-3}$ degree$^{-1}$.

*Tungsten.* The data given are those taken from GEISS and VAN LIEMPT [67] who measured two types of single crystals. For a four-edged single crystal they found $\varrho_{273}=4.93\times10^{-8}$ ohm m.

*Uranium.* TYLER and WILSON [209] found that the resistivity between 20 and 300° K varied between $4.0\times10^{-8}$ and $29.2\times10^{-8}$ ohm m., in agreement with results of other authors, but in marked contradiction with the old MEISSNER and VOIGT [149] value $\varrho_{300}=430\times10^{-8}$ ohm m.

TYLER and WILSON conclude from measurement of the thermo-electric effect that the conductivity is mostly hole conductivity.

**9. The absolute value of the conductivity compared with theory. Monovalent metals.** A general expression for the conductivity can be found from the equation [cf. (3.1)] for a stationary current in an isotropic conductor

$$i = n\,e^2\,\tau\,E/m = \sigma\,E\,, \tag{9.1}$$

and it follows immediately that

$$\sigma = n\,e^2\,\tau/m. \tag{9.2}$$

In these expressions $n$ is the effective number of free electrons with charge $e$ and mass $m$ per unit volume, and $\tau$ the average time between two "collisions" with the metal ions.

For a current of a free electron gas, the time between two collisions with the lattice is

$$\tau = l/v\,, \tag{9.3}$$

where $l$ is the mean free path and $v$ is the mean velocity of the electrons in the direction of the field. In this case, $n$ can be calculated from the average number of free electrons per atom (valency) $f$, the AVOGADRO number, the density $d$ and the atomic weight $A$ of the element by the relation

$$n = (6f\,d/A)\times10^{29}\ \text{m.}^{-3}. \tag{9.4}$$

It has to be emphasized that Eq. (9.2) in its general form is valid for the metallic conductivity, whatever be the model chosen. Consequently, a mean free path and an average velocity can be defined as well. In the modern approach of the conductivity problem, however, the picture of a homely classical collision time has been abandoned. The time $\tau$ now is a relaxation time; that is, the time needed for a disturbed distribution of the electrons to approach equilibrium within $1/e$ times the original deviation after sudden removal of a system of external forces which determines the deviation from equilibrium[1]. Difficulties arise if a model for metallic conductivity is considered, which deviates from the simple model of a free electron gas. Then for a theoretical evaluation of $\tau$, apart from the consequences of the model, one has to consider the influence of the deviations from an ideal, that is, from a truly periodic and pure lattice at $T=0°$ K, with an infinite conductivity in a crude approximation. The more the state of the real metal is approached and factors such as the temperature motion of the lattice, the interaction of the electrons with this lattice, the deviations from lattice periodicity

[1] When $F$ and $F_0$ are the disturbed and undisturbed distribution functions of velocity-, space-, and time coordinates, then

$$(F-F_0)_t = (F-F_0)_{t=0}\,e^{-t/\tau}.$$

due to impurities, and deviations from the electron lattice interaction depending on the character of these impurities (COULOMB interaction) are considered, the more complicated will be the resulting expression for the conductivity.

For a survey of these theoretical problems, the reader is referred to the article by JONES [*J 1*] in this volume. The expressions (12.10) and (12.11) in that paper for the conductivity of a monovalent metal and for a model in which the electrons are considered to be perfectly free, and moving in the periodic lattice field with their possible energies at $T=0$ filling one half of the conduction band, allow for comparison with the experimental values for the conductivity. Inserting of the metal constants from other experimental evidences in these expressions leads to the result that for metals like silver and sodium the calculated values of $\sigma$ are about twice as large as the experimental ones. This illustrates the extremely complicated state of affairs in a metal: the best evaluation of the factors contributing to the electric resistivity and in particular the fact, that only an evaluation on the one-electron model picture (cf. FRIEDEL [*64*]) is technically possible in theoretical calculations, leads for the most simple, regular, monovalent metals to a $\sigma$ value which differs vastly from the experimental values. On the other hand, a better agreement between theory and experiment is reached when the qualitative dependence of the conductivity on factors such as temperature or magnetic field is described. Taking the general expression for the conductivity, and writing this expression[1] as:

$$\sigma = \text{const}\, \frac{e^2}{m^*} \left(\frac{\beta}{C_D}\right)^2 \frac{M_A\, \Theta_D^2}{V\,T}. \tag{9.5}$$

$M_A$, $V$, $\Theta_D$ being respectively the atomic weight, atomic volume and DEBYE characteristic temperature, $\beta/C_D$ and $m^*$ (effective mass) represent the magnitude of the binding of the electron to the lattice. For a free electron and monovalent metal $\beta$ has its maximum value with $\beta=1$ and $m^*=m$, whilst for stronger binding $\beta<1$ and $m^*>m$. $C_D$ is related to the kinetic energy $V_0(r)-E_0$ of the electron at the lowest state of energy in the atomic polyhedron, and has been derived by using the DEBYE assumptions for the atomic vibrations.

Thus, Eq. (9.5) contains the two important factors $\beta/C_D$ and $T/(M_A\,\Theta_D^2)$, which represent the binding of the electrons with the lattice and the squared amplitude of the atomic vibrations for high temperatures, respectively. It is obvious that the differences in electronic structure of the elements make themselves felt in this formula. One of the most important results of (9.5) is, that it is qualitatively in agreement with the experimentally found linear relationship between $\varrho$ and $T$ for temperatures from about $\Theta_D/2$ to the melting temperature (cf. Chap. IV) and with the law $\varrho \sim T^5$ which holds for some metals at temperatures below about 10° K.

**10. Divalent, multivalent and transition metals.** The discrepancy between the simplified theory and experiment for the values of $\sigma$ becomes stronger if divalent and transition metals are considered. The former, with their two *s*-electrons at $T=0$ occupy all possible energy states in the conduction band. At finite temperature these metals should have been insulators, unless an overlap between the filled band and the non-occupied band of next higher energy is assumed.

Perusal of the data for conductivity, Tables 3, 4, and 5, shows that for:

monovalent metals 5.3 (Cs) $\leq 10^{-6}\,\sigma \leq$ 67 (Ag),
divalent metals 4.2 (Hg) $\leq 10^{-6}\,\sigma \leq$ 28 (Ca) and for
transition metals 1.1 (Mn) $\leq 10^{-6}\,\sigma \leq$ 23 (Rh).

[1] Cf. Eq. (12.10), p. 250 in this volume.

There is a large difference in conductivity between the groups of the monovalent and the divalent metals, and a smaller difference between the latter and transition metals. For the large difference mainly the lower number of conduction electrons per atom in divalent and multivalent metals is responsible. For exceptional large differences in conductivity between the divalent and the transition metals MOTT's [*M 3*] suggestion has been accepted generally. It has been assumed that for the transition metals there is an overlap between the d- and the s-state with principal quantum numbers $n$ and $n+1$ respectively. (In the case of the divalent metals an overlap between the $n$, s- and the empty $n$, p-state is assumed.) Such an overlap means, that the s- and the d-band conductivities are reduced by mutual s-d-s scattering. Apart from the increase in the effective number of electrons when passing from a transition metal like Ni to the next element in the periodic system, Cu, the conductivity also increases by the removal of the s-d-s scattering. This latter effect has been confirmed by the study of copper-nickel and silver-palladium alloys[1]. That in the case of the transition metals the conductivity is present at all, is mainly due to the conductivity by s-electrons. The density of states in the d-band can be so large that the hole-conductivity is extremely small. Then, the short circuit made by the d-s transitions leads to the measured conductivity being small, since so few electrons are engaged in this transition. The s-d transitions reduce it further to values lower than could be expected from the effective number of electrons present in the s-band.

Considering the electrons in the two bands of divalent and multivalent metals, and designing their numbers, their relaxation times and their effective masses by $n_1$, $\tau_1$, $m_1^*$ and $n_2$, $\tau_2$, $m_2^*$, respectively, the total conductivity $\sigma$ may then be described by

$$\sigma = \left(\frac{n_1 \tau_1}{m_1^*} + \frac{n_2 \tau_2}{m_2^*}\right) e^2. \tag{10.1}$$

As long as the quantities involved can not be determined separately by experiment, this formula has a descriptive value only.

**11. The conductivity as related with the atomic numbers and electron configuration of the elements.** A comparison of (9.2) with (9.5) shows that for $T = \Theta_D$:

$$(n_{\text{eff}})_{\Theta_D} \sim \left(\frac{n_0}{C_D}\right)^2 \sim \frac{\sigma_\Theta}{M_A \Theta_D}. \tag{11.1}$$

Furthermore, the quantity

$$[\sigma]_v = \frac{\sigma_T}{M_A \Theta_D^2} \tag{11.2}$$

allows, for a given temperature, a comparison between the conductivity of the elements for equal atomic vibrations (MOTT, Sect. VII, 3 of [*M 3*]). Considering the arbitrary choice of temperature and volume for expressing the conductivity, SIMON [*194*] suggested the introduction of the atomic conductance $\Sigma_A$, which is the conductance of a parallelopiped of a single crystal similar to the elementary cell, and given by

$$\Sigma_{A,T} = \sigma_T V^{-\frac{1}{3}} \text{ ohm}^{-1}, \tag{11.3}$$

in which $V$ is the volume of one gram-atom.

Several authors have plotted the quantities (11.1), (11.2) and, for $T = \Theta_D$, (11.3) against the atomic number $Z$. From Eq. (11.1) GRÜNEISEN [*G 2*] calculates, by inserting the values of the constants for a given element, the values of $n_{\text{eff}}$ and he obtains for $C_D$ of monovalent metals, values of the order of the FERMI

[1] Cf. Chap. VIII.

energy, determined by

$$E_0 = \frac{h^2}{2m}\left(\frac{3}{8\pi}n\right)^{\frac{2}{3}}$$

(for free electrons).

A reproduction[1] of his graph is given in Fig. 11. The periodicity in $Z$ is obvious. This is also the case[2] for $[\sigma]_v$ and $\Sigma_{A,\Theta}$.

Another system of classification is possible. Recently SZABO and LAKATOS [*204*] have pointed out that in general plotting against the electronic configuration has advantages over a classification by the atomic number. The authors introduce a new periodic system based on the electronic configuration of the elements[3], and they demonstrate that the melting temperatures, the boiling temperatures, and the heats of sublimation are linear functions of the principle quantum number within the columns of their system. One of the important features of their

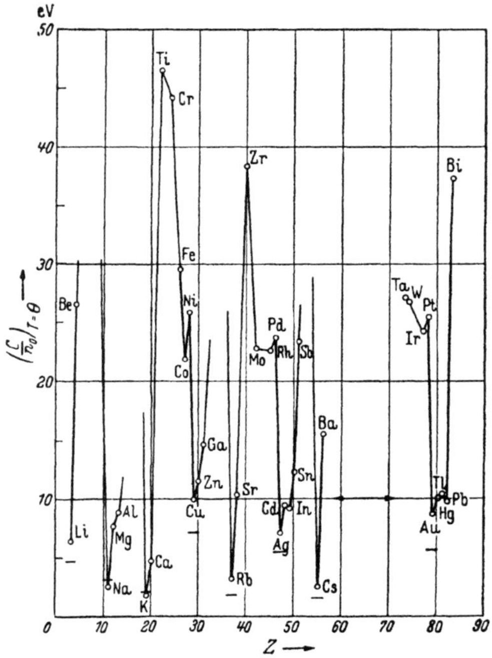


Fig. 11. Binding energy per electron $C/n_0$ against atom number $Z$. The horizontal marks at the $Z$ values for monovalent metals indicate the values for $E_0$ according to BETHE. After GRÜNEISEN [*92*], Fig. 7.

Table 6. *Values for $\sigma_{273}/M_A\,\Theta^2$; $\sigma_\Theta/M_A\,\Theta$ and $\Sigma_{A,\Theta}$ for monovalent metals.*

| Element | $M_A$ | $\Theta_D$ °K $\Theta_R$ | Ref. | $\sigma_{273}$ $10^6$ ohm$^{-1}$ m.$^{-1}$ | $\frac{\sigma_{273}}{M_A\,\Theta^2}$ | $\sigma_\Theta$ [6] $10^6$ ohm$^{-1}$ m.$^{-1}$ | $V$ $10^{-6}$ m.$^3$ | $\Sigma_{A,\Theta}$ $10^6$ ohm$^{-1}$ | $10^{-8}\frac{\sigma_\Theta}{M_A\,\Theta}$ |
|---|---|---|---|---|---|---|---|---|---|
| Li | 6.94 | 363 | [*G 2*] | 11.8 | 12.9 | 8.46 | 11.3 | 0.19 | 3.35 |
| | | 363 | [*139*] | | 12.9 | 84.6 | | 0.19 | 3.35 |
| Na | 23.0 | 160 | [*G 2*] | 23.4 | 39.8 | 42.7 | 23.7 | 1.25 | 11.6 |
| | | 202 | [*139*] | | 25.0 | 32.4 | | 0.93 | 7.0 |
| K | 39.1 | 100 | [*G 2*] | 16 | 40.6 | 47.8 | 45.4 | 1.70 | 12.2 |
| | | 163 | [*139*] | | 15.3 | 38.7 | | 1.37 | 6.12 |
| Cu | 63.6 | 335[4] | [*123*] | 64.5 | 9.0 | 37.9 | 7.14 | 0.74 | 1.78 |
| Rb | 85.4 | 68[5] | [*G 2*] | 8.62 | 21.8 | 39.9 | 62.2 | 1.50 | 6.86 |
| | | 175 | [*139*] | | 3.30 | 22.0 | | 0.88 | 1.47 |
| Ag | 108 | 210[4] | [*123*] | 66 | 14.0 | 89.2 | 10.3 | 1.94 | 3.93 |
| Cs | 133 | 54 | [*G 2*] | 5.2 | 13.6 | 30.1 | 71.0 | 1.25 | 4.19 |
| Au | 197 | 165 | [*K*] | 49 | 9.17 | 93.4 | 10.2 | 2.02 | 2.86 |

[1] As a rule, the diagrams given in the present article are contact reproductions of diagrams in the original papers and they will therefore contain unavoidable errors as to exact curvature and the position of experimental points. Some diagrams, which will be indicated as such, have been made by plotting data taken from tables.

[2] Cf. [*G 2*], Figs. 9 and 10.

[3] The system can be built up by putting the inert gases in a center column, putting to their right side the elements which add in each column first a s- and then a d-electron with the next principal quantum number and putting at the left side the elements building up their s- and p-levels before reaching the inert gas configuration of the same principle quantum number.

[4] As extrapolated to $T \to \infty$.

[5] From melting point data.

[6] For $\Theta > 173$ calculated with $\alpha_{\varrho,T}$, table 3. For $\Theta < 173$, $\sigma_\Theta$ taken from $G\,2$ or from recent experimental results for $\varrho_T$ or $R_T/R_{273}$.

system is, that some elements belong to another group as would follow from the commonly used system. For the metals they have evidence that magnesium

Table 7. *Values for $\sigma_{273}/M_A\Theta^2$, $\sigma_\Theta/M_A\Theta$ and $\Sigma_{A,\Theta}$ for multivalent metals.*

| Element | $M_A$ | $\Theta_D$ °K $\Theta_R$ | Ref. | $\sigma_{273}$ $10^6$ ohm$^{-1}$ m.$^{-1}$ | $\frac{\sigma_{273}}{M_A\Theta^2}$ | $\sigma_\Theta$ [2] $10^6$ ohm$^{-1}$ m.$^{-1}$ | $V$ $10^{-6}$ m.$^3$ | $\Sigma_{A,\Theta}$ $10^6$ ohm$^{-1}$ | $\frac{10^{-3}\sigma_\Theta}{M_A\Theta}$ |
|---|---|---|---|---|---|---|---|---|---|
| Be | 9.02 | 1160 | [*K*] | 36 | 2.96 | 4.70 | 4.97 | 0.08 | 0.45 |
| | | 652 | [*139*] | | 9.40 | 9.4 | | 0.16 | 1.6 |
| Mg | 24.3 | 330 | [*G 2*] | 25.4 | 9.57 | 20.4 | 13.9 | 0.49 | 2.54 |
| Al | 27.0 | 419[1] | [*123*] | 40 | 8.46 | 23.8 | 10.0 | 0.51 | 2.11 |
| Cs | 40.1 | 219 | [*G 2*] | 28 | 14.5 | 22.8 | 25.9 | 0.67 | 2.59 |
| | | 150 | [*139*] | | 31.2 | 54.4 | | 1.60 | 4.66 |
| Zn | 65.4 | 190[1] | [*122*] | 18.3 | 7.75 | 28.2 | 9.15 | 0.59 | 2.27 |
| Ga | 69.7 | 125 | [*G 2*] | 7.3 | 6.70 | 18.7 | 11.8 | 0.42 | 2.14 |
| Sr | 87.6 | 148 | [*G 2*] | 5.0 | 2.54 | 6.5 | 33.6 | 0.22 | 0.50 |
| | | 100 | [*139*] | | 6.55 | 7 [3] | | 0.24 | 0.8 |
| Cd | 112.4 | 300 | [*K*] | 14.8 | 1.47 | 13.5 | 13.0 | 0.32 | 0.40 |
| In | 114.8 | 109[1] | [*32*] | 12.2 | 8.95 | | 15.7 | | |
| Sn | 118.7 | 160[1] | [*122*] | 9.9 | 2.98 | 16.8 | 16.2 | 0.42 | 0.84 |
| Sb | 121.8 | 201 | [*G 2*] | 3.12 | 0.63 | 4.90 | 18.2 | 0.13 | 0.20 |
| Ba | 137.4 | 133 | [*G 2*] | 2.8 | 1.15 | ≈8 [3] | 36.6 | ≈0.32 | ≈0.4 |
| | | 100 | [*139*] | | 2.02 | ≈10 [3] | | ≈0.4 | ≈0.7 |
| Hg | 200.6 | 69 | [*G 2*] | 4.71 | | 17.9 | | 0.4 | 1.29 |
| Tl | 204.4 | 89[1] | [*123*] | 6.7 | 4.14 | 20.0 | 17.2 | 0.52 | 1.10 |
| Pb | 207.2 | 90 | [*K*] | 5.19 | 3.08 | 17.5 [4] | 18.2 | 0.46 | 1.05 |
| Bi | 209.0 | 120[1] | [*122*] | 0.93 | 0.31 | 2.17 [5] | 21.3 | 0.048 | 0.086 |

Table 8. *Values for $\sigma_{273}/M_A\Theta^2$, $\sigma_\Theta/M_A\Theta$ and $\Sigma_{A,\Theta}$ for the transition metals.*

| Element | $M_A$ | $\Theta_D$ °K | Ref. | $\sigma_{273}$ $10^6$ ohm$^{-1}$ m.$^{-1}$ | $\frac{\sigma_{273}}{M_A\Theta^2}$ | $\sigma_\Theta$ [2] $10^6$ ohm$^{-1}$ m.$^{-1}$ | V $10^{-6}$ m.$^3$ | $\Sigma_{A,\Theta}$ $10^6$ ohm$^{-1}$ | $\frac{10^{-3}\sigma_\Theta}{M_A\Theta}$ |
|---|---|---|---|---|---|---|---|---|---|
| Ti | 47.9 | 278 | [*K*] | 2.38 | 0.65 | 2.3 | 10.6 | 0.050 | 0.17 |
| V | 51.0 | 300 | [*M 1*] [6] | 0.59 | 0.13 | (0.59) | 8.5 | (0.012) | (0.39) |
| Cr | 52.0 | 403 | [*K*] | 6.6 | 0.79 | ≈3.1 | 7.33 | ≈0.060 | ≈0.15 |
| $Mn_\beta$ | 54.9 | 410 | [*K*] | 1.1 | 0.12 | 0.93 | 7.42 | 0.018 | 0.041 |
| $Fe_\alpha$ | 55.8 | 462[1] | [*124*] | 11.5 | 0.94 | 5.08 | 7.10 | 0.099 | 0.195 |
| $Co_\alpha$ | 58.9 | 445 | [*K*] | 19 | 1.65 | 9.0 | 6.76 | 0.170 | 0.343 |
| Ni | 58.7 | 413[1] | [*118*] | 15.2 | 1.61 | 7.8 | 6.58 | 0.145 | 0.322 |
| Zr | 91.2 | 270 | [*K*] | 2.47 | 0.37 | 2.5 | 14.0 | 0.060 | 0.101 |
| Nb | 92.9 | 250[1] | [*22*] | 4.28 | 0.86 | 5.6 | 11.0 | 0.124 | 0.239 |
| Mo | 96.0 | 425 | [*K*] | 19.9 | 1.15 | 12.3 | 9.25 | 0.258 | 0.300 |
| Ru | 102 | 426 | [*M 1*] [6] | 15.0 | 0.81 | 9.0 | 8.34 | 0.183 | 0.21 |
| Rh | 103 | 370 | [*G 2*] | 23 | 1.65 | 16.1 | 8.30 | 0.323 | 0.422 |
| Pd | 107 | 275 | [*K*] | 10.2 | 1.27 | 10.2 | 8.95 | 0.212 | 0.348 |
| La | 139 | 132 | [*K*] | 1.5 | 0.76 | — | — | — | — |
| Hf | 179 | 254 | [*M 1*] [6] | 3.77 | 0.33 | 4.12 | 15.2 | 0.102 | 0.091 |
| Ta | 181 | 247[1] | [*121*] | 8.06 | 0.73 | 8.91 | 10.9 | 0.198 | 0.199 |
| W | 184 | 380 | [*K*] | 20.4 | 0.77 | 13.7 | 9.5 | 0.290 | 0.196 |
| Re | 186 | 310 | [*M 1*] [6] | 5.30 | 0.30 | 4.8 | 9.05 | 0.100 | 0.083 |
| Ir | 193 | 316 | [*M 1*] [6] | 21.1 | 1.08 | 17.7 | 8.62 | 0.362 | 0.29 |
| Pt | 195 | 233[1] | [*123*] | 10.2 | 0.96 | 12.2 | 9.14 | 0.255 | 0.27 |

[1] As extrapolated to $T = \infty$.

[2] For $\Theta > 173$ calculated with $\alpha_{\varrho,T}$; for $\Theta < 173$ taken from [*G 2*] or recent results for $\varrho_T$ or $R_T/R_{273}$.

[3] Estimated from [*139*].

[4] [*88*].

[5] [*186*].

[6] $\Theta_R$-values.

should belong to the zinc group and that beryllium and aluminium are analogue elements. These arguments moreover seem to be justified by several chemical properties of the elements concerned and from the linearity between the several physical quantities mentioned, and the electron configuration.

Considering these arguments and the expectation that the conductivity may be more related to electron configuration than to mass numbers, a grouping of the metals according to the SZABO sceme seems to be worthwhile.

The values (9.8), (9.9) and (9.10) have been calculated for monovalent metals (Table 6), for multivalent metals (Table 7) and transition metals (Table 8). In

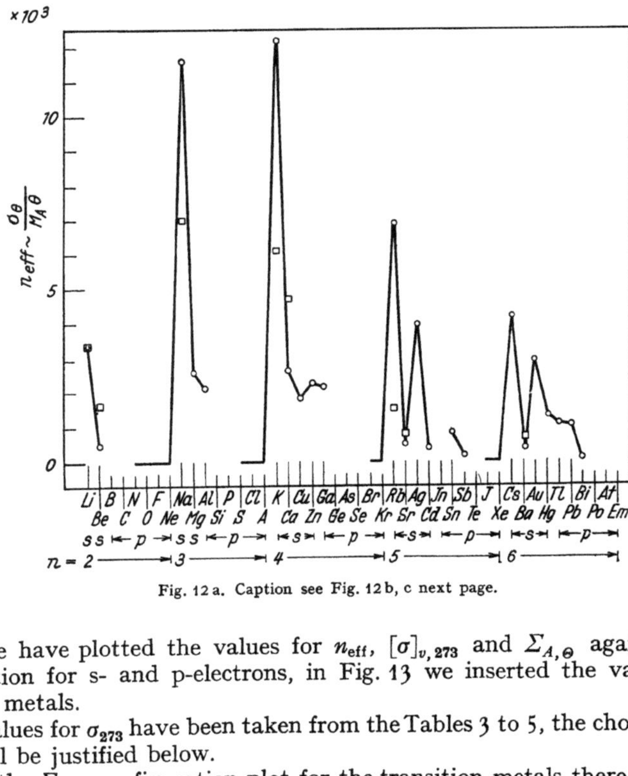


Fig. 12 a. Caption see Fig. 12 b, c next page.

Fig. 12 we have plotted the values for $n_{\text{eff}}$, $[\sigma]_{v,273}$ and $\Sigma_{A,\Theta}$ against electron configuration for s- and p-electrons, in Fig. 13 we inserted the values for the transition metals.

The values for $\sigma_{273}$ have been taken from the Tables 3 to 5, the choice of the $\Theta_D$ values will be justified below.

From the $\Sigma_{A,\Theta}$-configuration plot for the transition metals there follows that a linear increase of $\Sigma_{A,\Theta}$ with $n$ is not impossible for Ti-Zr-Hf, for V-Nb-Ta and for Ni-Pd-Pt. For the Ti- and Ni-groups a linearity might follow also for the quantities $n_{\text{eff}}$ (Fig. 13 a) and $[\sigma]_{V,273}$ (Fig. 13 b).

From Figs. 12 and 13 the following general conclusions on the conductivity can be drawn.

(1) The effective number of electrons (at $T=\Theta_D$), the conductivity for equal atomic vibrations (at $T=273°$ K), and the atomic conductance (at $T=\Theta_D$) show oscillations, periodic with the principal quantum number $n$. The first two values show a tendency to decrease with increasing $n$, whereas the last quantity has a more constant average value for $n \geq 3$. The quantities have a maximum for each $n$ for metals with one s-electron (Cu being an exception) and they decrease with increasing number of p-electrons.

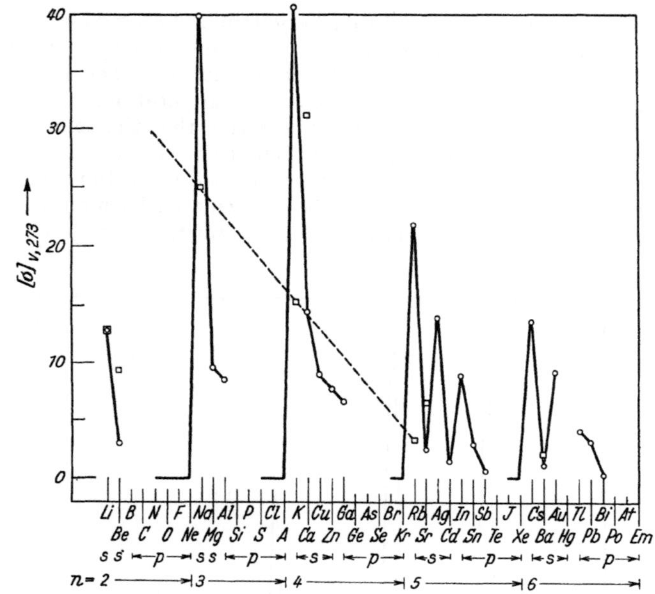


Fig. 12b.

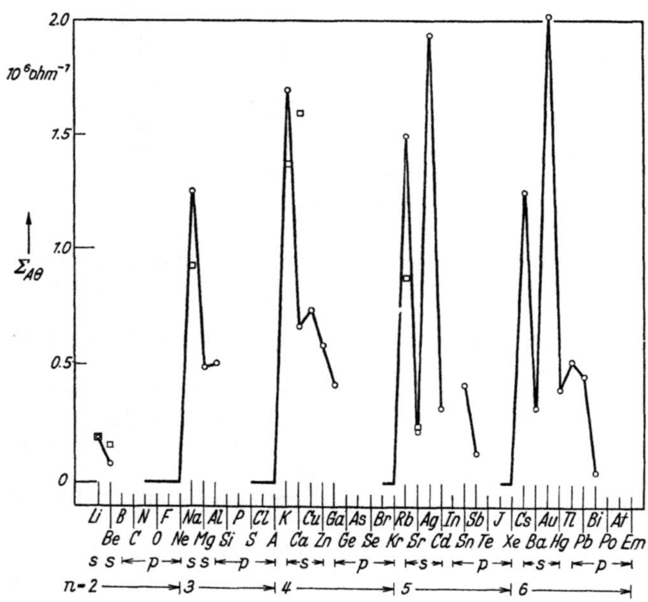


Fig. 12c.

Fig. 12a—c. The quantities $n_{\mathrm{eff}} \sim \frac{\sigma_\Theta}{M_A\,\Theta_D}$, $[\sigma]_{v,273}$ and $\Sigma_{A,\Theta}$ as taken from Tables 6, 7, 8, in graphs a, b and c, respectively, for s- and p-metals. For the alkali and earth alkaline metals the values $\Theta_D$ (⊙) and $\Theta_R$ (⊡) (MACDONALD and MENDELSSOHN [139]), have been used.

(2) Notwithstanding the lack of data for Sc, Y, Tc and Os, it follows from Fig. 13 that a periodicity with $n$ is evident for the quantities considered for the

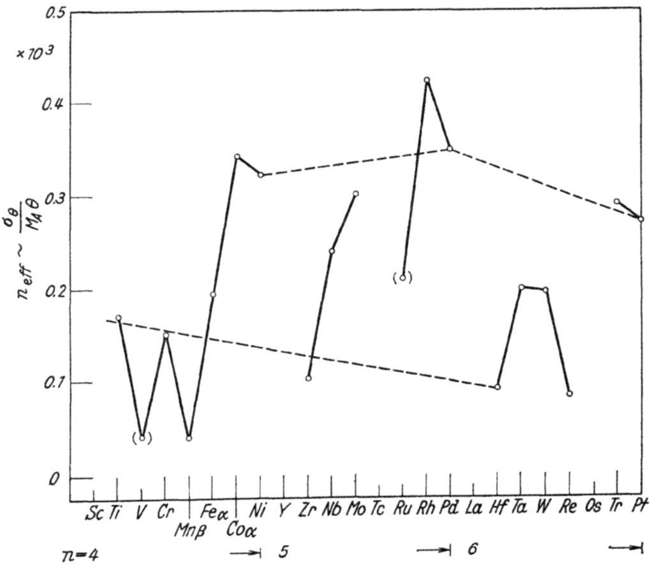


Fig. 13 a.

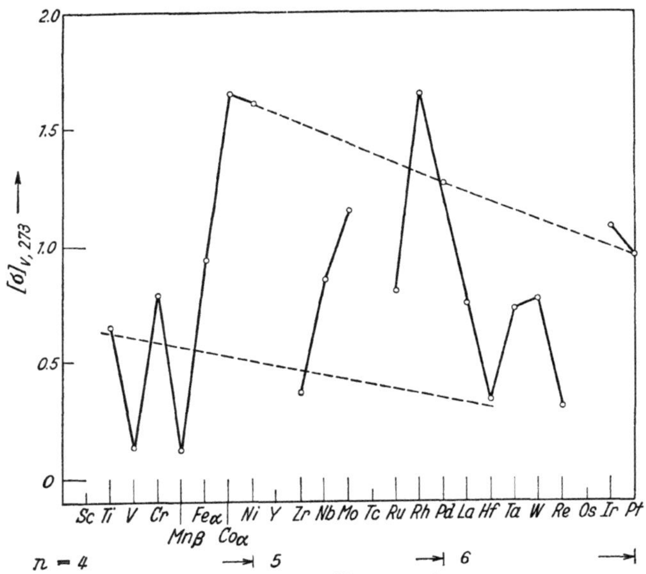


Fig. 13 b. Caption see Fig. 13 c next page.

transition metals too. Again, the number of effective electrons at $T = \Theta_D$ and the conductivity for equal atomic vibrations (at $T = 273°$ K) show a tendency to decrease with increasing $n$, whereas the atomic conductance clearly increases.

It has to be noted that the quantities involved show, for each value of $n$, maximum values for metals with 4 and 7 electrons in the d-band in the ionic state.

(3) All quantities calculated are smaller in the average for the transition- than for the s- and in some cases, the p-metals.

The above conclusions are related to the band model of the solid and they confirm the assumptions made in Sect. 10. In monovalent metals there is a large freedom of motion for one electron. The peaks in the atomic conductance diagrams are well defined. Then there is the large decrease in conductivity for the

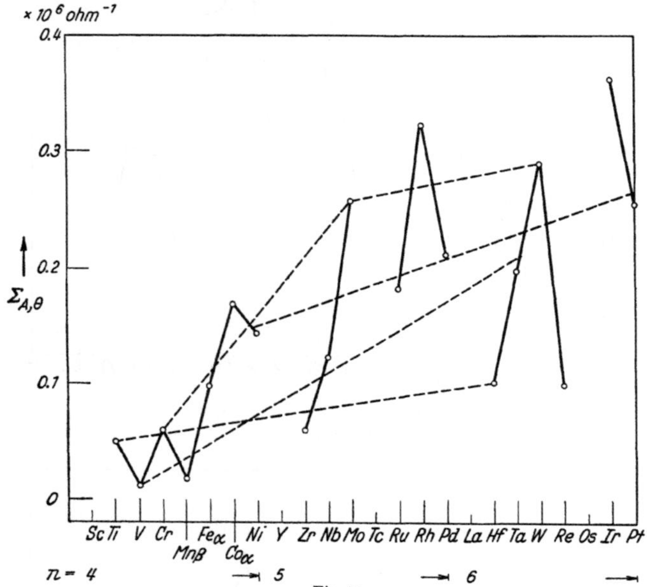


Fig. 13 c.

Fig. 13a—c. The quantities $n_{\text{eff}} \sim \frac{\sigma_\Theta}{M_A\,\Theta_D}$, $[\sigma]_{v,273}$ and $\Sigma_{A,\Theta}$ in graphs a, b and c respectively, for the transition metals (taken from Tables 6, 7 and 8).

elements with 2 electrons in the s-state. Here the number of effective electrons is smaller, and it is the overlap with the next empty p-band which controls the conductivity. In the case of the transition metals, the average atomic conductance increases in each group of elements during the filling of the d-band. The maximum $\Sigma_{A,\Theta}$-values for 4 and 7 d-electrons in a band are to be accepted as experimental fact. The atomic conductance increases with the principal quantum number in the transition metal group, whereas in the s-p group of metals it increases or decreases according as there is only one s-electron or two.

The periodicity should be accepted, without going into too much detailed explanation.

In calculating the values, a great insecurity is involved by the choice of the values for $\Theta_D$. According to GRÜNEISEN [G 2], the characteristic temperature in (9.5) and in the following expressions is the value of the DEBYE temperature as taken from the specific heat measurements[1]. While doing so, the temperature dependence of $\Theta_D$ has to be taken into account, and perhaps the best method is to extrapolate the $\Theta(T)$ relation to $T = \infty$ and to insert this value for $\Theta_D$.

[1] Cf. the discussion of the different ways to define the DEBYE temperature in M. BLACKMAN's article, vol. VII, part 1 of this Encyclopedia.

However, other suggestions have also been made. A $\Theta_R$ can be found as a parameter from GRÜNEISEN'S [G 2] expression for the resistance-temperature relation (see Sect. 12). These $\Theta_R$ values differ vastly in many cases from $\Theta_D$, which fact does not facilitate a physical understanding of the correlation between the characteristic temperature and the conductivity. For the case of alkali metals, the discrepancy between $\Theta_R$ and $\Theta_D$ may be attributed [193] to a more complicated picture of the lattice vibrations, which, apart from the DEBYE spectrum, also might have an anomalous effect.

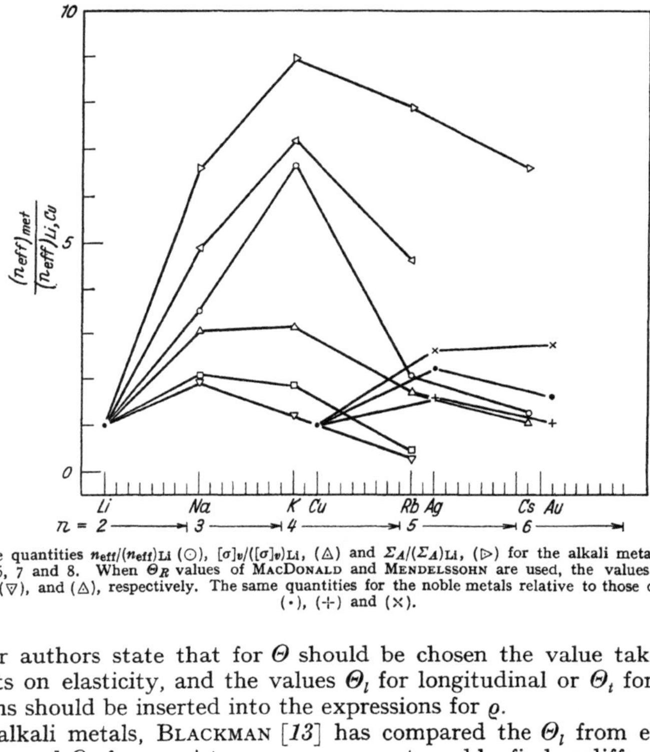


Fig. 14. The quantities $n_{eff}/(n_{eff})_{Li}$ (○), $[\sigma]_v/([\sigma]_v)_{Li}$, (△) and $\Sigma_A/(\Sigma_A)_{Li}$, (▷) for the alkali metals, as taken from the Tables 6, 7 and 8. When $\Theta_R$ values of MACDONALD and MENDELSSOHN are used, the values are designed by marks (⊡), (▽), and (△), respectively. The same quantities for the noble metals relative to those of Cu, are marked (·), (+) and (×).

Other authors state that for $\Theta$ should be chosen the value taken from experiments on elasticity, and the values $\Theta_l$ for longitudinal or $\Theta_t$ for transversal vibrations should be inserted into the expressions for $\varrho$.

For alkali metals, BLACKMAN [13] has compared the $\Theta_l$ from elasticity experiments and $\Theta_R$ from resistance measurements and he finds a difference between both values. Using the same experimental results, HOUSTON [97] claims that the choice of $\Theta$ is rather arbitrary and that by the introduction of a scattering parameter, indicating by its magnitude an isotropic or a forward scattering, very different $\Theta_l$- and $\Theta_t$-values can be inserted into the formula for $\varrho$ with equally good result.

Considering this discrepancy, we have used $\Theta_D$ values, taken from specific heat measurements and extrapolated to high temperatures as far as this could be done from available data[1]. Besides[2], in the case of alkali metals we used the

[1] I gratefully acknowledge help of Dr. P. H. KEESOM (Purdue University) who kindly wrote me some values, indicated by (K).

[2] The $\Theta_D$-value for indium is rather dubious, considering the results of CLUSIUS and SCHAGINGER [Z. angew. Physik **4**, 442 (1952)] who find from specific heat measurements taken at temperatures below 273° K that $\Theta'_D \rightarrow 0$ at those temperatures. When a considerable term, due to the electron specific heat is subtracted, a more or less constant value $\Theta_D \approx 140°$ K, is obtained.

$\Theta_R$ values from MACDONALD's [139] measurements since for these cases a deviation between $\Theta_D$ and $\Theta_R$ has been made understandable by SIMON's [193] suggestion.

In plotting the relative values (9.8), (9.9) and (9.10) with respect to those for Li and Cu, the advantage of the use of the electron configuration as abscissa is nicely demonstrated (Fig. 14) in the case of the monovalent metals. The correlation between the two groups of metals having one s-electron after a full p-band (Li-series) and after a full d-band (Cu-series) is obvious. Using MACDONALD's $\Theta_R$-values for the alkali metals instead of the $\Theta_D$-values, the qualitative agreement between the two groups is improved. Their use in the calculation of the quantities plotted in Fig. 12 does not change the qualitative character of the plots very much, with the exception, however, of the relative position of the quantities for rubidium.

## IV. Temperature dependence of the conductivity of pure metals.

**12. The general behaviour of resistance between 80° K and melting temperature.** In the preceding chapter we discussed the temperature dependence of the conductivity in the region of general practical interest, namely between

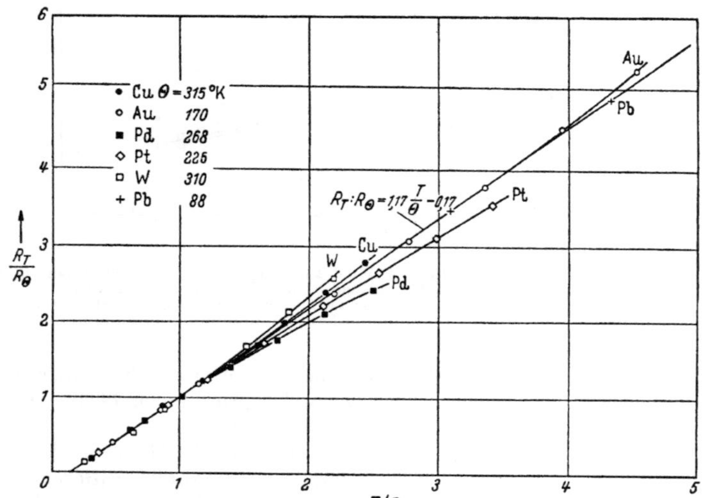


Fig. 15. The GRÜNEISEN-BORELIUS relation: $R_T/R_\Theta$ versus $T/T_\Theta$ for some cubic metals.

273° K and 373° K. For temperatures below and above these limits a linear $\varrho(T)$ relation was assumed for the calculation of $\sigma_\Theta$. Below about 150° K, the $\varrho(T)$ relation is no longer linear for many metals, the temperature coefficient for small intervals of $\Delta T$ becoming smaller than in the room temperature region.

The general picture of the influence of temperature for isotropic conducting metals is demonstrated by the GRÜNEISEN-BORELIUS relation [G 1], [G 2], [B 3], which suggests that the reduced resistance $R_T/R_\Theta$ is a linear function of the reduced temperature $T/\Theta$ (Fig. 15).

The straight line

$$\frac{R_T}{R_\Theta} = 1.17\frac{T}{\Theta} - 0.17 \qquad (12.1)$$

gives in a good approximation the resistance for an isotropic metal in the region $0.2 < T/\Theta < 1.2$. Deviations occur at the low and the high temperature extremes.

GRÜNEISEN [81] introduced an essentially experimental formula, afterwards confirmed by theoretical considerations

$$\varrho = \frac{B\,T}{\Theta} I\left(\frac{\Theta}{T}\right) \tag{12.2a}$$

where

$$I\left(\frac{\Theta}{T}\right) = 4\left(\frac{T}{\Theta}\right)^4 \int_0^{\Theta/T} \frac{\xi^5\,d\xi}{(e^{\xi}-1)(1-e^{-\xi})} \tag{12.2b}$$

with the special values

$$\left.\begin{aligned} I\left(\frac{\Theta}{T}\right) &\approx 1 && \text{and} \quad \varrho = \frac{B\,T}{\Theta} && \text{for} \quad T \gg \Theta, \\ I\left(\frac{\Theta}{T}\right) &\approx 498\,\frac{T^4}{\Theta^4} && \text{and} \quad \varrho = B'\,\frac{T^5}{\Theta^5} && \text{for} \quad T \ll \Theta. \end{aligned}\right\} \tag{12.2c}$$

A table for $\varrho$, calculated with (12.2) is given by MEISSNER[1].

An inspection of the $\Theta$-values given in our Tables 6 to 8 for the metals shows that, except for beryllium, a temperature of 100° K is still well above $0.2\,\Theta$, the approximate limit where the deviations from (12.2c) become too large. At still lower temperatures (12.2a) and (12.2b) yield $\varrho \sim T^5$, which has been found in several cases[2]. At high temperatures, the experimental values for $\varrho$ are larger than would follow from (12.2c). This effect has been ascribed to the decrease in $\Theta$ with the thermal expansion of the lattice[3]. Agreement with this theory has been found only for some cubic metals.

The resistivities of palladium and platinum at high temperatures are smaller than would follow from (12.2c) which, according to MOTT [156], might be explained considering that in the derivation of an expression for the conductivity such as (9.5) only the first constant term $\zeta_0$ has been inserted in the development of the FERMI energy in terms of temperature. He points out that for metals with a degeneracy temperature ($T_d \approx \zeta_0/\varkappa$) of the order of 10,000° K instead of the normal 50,000° K, the second order term decreases the $\varrho$-values at sufficiently high temperatures. A degeneracy temperature of 10,000° K for Pt would be considered very low, and, has not been confirmed experimentally. Another aspect of the temperature influence on $\zeta$ is that the temperature coefficient for the resistivity at the normal ice point is larger than would follow from a simple linearity between $\varrho$ and $T$ with a value of $\varrho = 0$ at 0° K for the ideal case. Under the latter assumptions, the temperature coefficient of the resistivity should be $\frac{1}{273} = 3.66 \times 10^{-3}$ degree$^{-1}$. From Tables 3 to 5 it follows that with the exceptions of Ce, Mn, Mg and Th of the group of isotropic conducting metals, and Dy, Er, Pr, Re and Pu of the group of anisotropic metals, all $\alpha$-values are larger than $\frac{1}{273}$. Iron with $\alpha_{\varrho,T} = 6.57 \times 10^{-3}$ degree$^{-1}$ and beryllium with $\alpha_{\varrho,T} = 7.5 \times 10^{-3}$ degree$^{-1}$ represent the maximum values. As a consequence, at low temperatures the $r_T = R_T/R_{273}$ values for Fe at 78° K and for Be at 80° K are 0.074 and 0.0128 respectively instead of being 0.30, according to $\alpha_{\varrho,T} = \frac{1}{273}$. For copper and

[1] Table 7 of ref. [M 1]. Tables for the integrals $I_n(\Theta/T)$, which are common in conductivity problems, are given in WILSON's book [W 1].

[2] For a discussion of the low temperature region, we refer to the contribution of D. K. C. MACDONALD, vol. XIV of this Encyclopedia.

[3] The reader is referred to Sect. 23 of JONES' article in this volume where this point is discussed.

Table 9. *Values of $R_T/R_{273}$ and $z = R_0/R_{273}$ for elements at temperatures between 80° K and $T_M$.*

| Metal | T °K | $R_T/R_{273}$ | $10^4 z$ | Ref. | T °K | $R_T/R_{273}$ | Ref. | T °K | $R_T/R_{273}$ | Ref. |
|---|---|---|---|---|---|---|---|---|---|---|
| | | | | *a) For cubic metals.* | | | | | | |
| Li[1] | 86.3 | 0.1514 | 58 | [149] | | | | 443 | 1.800 | [80] |
| Na[1] | 87.8 | 0.2304 | 32 | [149] | | | | | | |
| Al | 77.7 | 0.1008 | 65 | [149] | | | | 923 | 4.36 | [171] |
| K | 87.8 | 0.2635 | 69 | [214] | | | | | | |
| Ca[1] | 83.6 | 0.4582 | ≈2792 | [149] | | | | | | |
| Ti[1] | 78.5 | 0.2150[2] | | [30] | 373 | 1.469 | [30] | 1123 | 3.36 | [212] |
| V | 90.0 | 0.457 | (300) | [172] | | | | 773 | 2.18 | [172] |
| Cr[1] | 80 | 0.134 | 530 | [146] | | | | | | |
| | 90 | 0.122 | (160) | [172] | 773 | 2.77 | [172] | 1073 | 4.13 | [172] |
| Mn[1] | 88.9 | 0.9776 | 9581 | [149] | | | | | | |
| $Fe_{\alpha}$[1] | 78.2 | 0.0741 | 62 | [149] | | | | 1523 | 13.9 | [166] |
| | 80 | 0.0982 | | [95] | | | | 683 | 4.78 | [95] |
| Co[1] | 86.9 | 0.1829 | 426 | [149] | | | | | | |
| | | | | | | | | 673 | 4.56 | [95] |
| Ni | 87.4 | 0.1179 | 50 | [149] | | | | | | |
| | 81 | 0.2979 | | [95] | | | | 674 | 4.09 | [95] |
| Cu | 81.6 | 0.144[2] | 2.8[2] | [149] | | | | | | |
| | 81 | 0.152 | | [95] | | | | 673 | 2.76 | [95] |
| | | | | | | | | 673 | 2.77 | [159] |
| | | | | | | | | 1073 | 5.89 | |
| Rb[1] | 200 | 0.72 | ≈800 | [140] | | | | | | |
| | 87.8 | 0.3043 | 375 | [149] | | | | | | |
| Sr[1] | 86.3 | 0.3313 | 932 | [149] | | | | 773 | ≈3.5[3] | [179] |
| Nb | 90.0 | 0.403 | (2000) | [172] | 473 | 1.50 | [172] | 973 | 2.60 | [172] |
| Mo | 90.0 | 0.1705 | | | | | | 491 | 2.04 | [79] |
| | 86.9 | 0.1701 | 140 | [149] | | | | | | |
| | 80 | 0.1911 | | [95] | | | | 678 | 2.85 | [95] |
| Rh | 87.4 | 0.1466 | 30 | [149] | | | | | | |
| | | | (<100) | [172] | 773 | 3.56 | [172] | | | |
| Pd | 88.9 | 0.2220 | 56 | [149] | | | | | | |
| | 81 | 0.1963 | | [95] | | | | 673 | 2.355 | [95] |
| Ag | 90.2 | 0.2261 | | [113] | | | | 491 | 1.895 | [113] |
| | 87.4 | 0.2349 | 67.9 | [149] | | | | | | |
| | 82 | 0.2109 | | [95] | | | | 669 | 2.69 | [95] |
| | | | | | | | | 960 | 5.09 | [163] |
| Cs[1] | 82 | 0.265 | 170 | [146] | | | | | | |
| Ba[1] | 78 | 0.215 | 70 | [109] | 348 | 1.438 | [18e] | 653 | ≈5.7[3] | [179] |
| Pr[1] | 80 | ≈0.6[3] | | [102] | | | | | | |
| Ta | 88.3 | 0.2511 | 2 | [147] | | | | | | |
| | 81 | 0.2955 | | [95] | 473.2 | 1.661 | [95] | | | |
| W[1] | 90.2 | 0.1706 | | [79] | 373.2 | 1.476 | [79] | | | |
| | 87.4 | 0.1565 | 5.2 | [149] | | | | | | |
| | 80 | 0.1529 | | [95] | 672 | 3.02 | [95] | | | |
| | | | 600 | [170] | | | | 1273 | 7.02 | [170] |
| Ir | 77.7 | 0.1905 | 478 | [149] | | | | 773 | 3.26 | [172] |
| | 81 | 0.2257 | | [95] | 679 | 2.657 | [95] | | | |
| Pt[1] | 81 | 0.2060 | | [95] | 673 | 2.4984 | [95] | 717 | 2.65 | [147] |
| | 90.1 | 0.2466 | | [119] | | | | 900 | 3.25 | |
| | | | | | | | | 1233 | 4.25 | [154] |
| | | | | | | | | 1335 | 4.53 | |
| Au | 87.4 | 0.2553[4] | 2.9[4] | [149] | | | | | | |
| | 84.9 | 0.2480 | 10.9 | [149] | | | | | | |
| | 82 | 0.2411 | 10 | [95] | 670 | 2.669 | [95] | 1063 | 5.82 | [163] |
| Pb | 90.1 | 0.2975 | 5 | [88] | | | | 923 | 2.50 | [171] |
| | 84.8 | 0.2756 | <7 | [149] | | | | | | |
| | 81 | 0.2634 | 10 | [95] | 573 | 2.380 | [95] | | | |
| Th[1] | 86.1 | 0.2785 | 16 | [149] | 373 | 1.239 | [8] | 1373 | 3.108 | [8] |

[1] See Table 10. [2] Large crystals. [3] At $T > T_{\text{transf.}}$ [4] Single crystal.

*Table 9.* (Continued.)

| Metal | $T$ °K | $R_T/R_{273}$ | $10^4 z$ | Ref. | $T$ °K | $R_T/R_{273}$ | Ref. | $T$ °K | $R_T/R_{273}$ | Ref. |
|---|---|---|---|---|---|---|---|---|---|---|
| | | | | *b) For non cubic metals.* | | | | | | |
| Be[1] | 86.1 | 0.2419 | 1.2 | [149] | | | | 873 | 6.37 | [174b] |
| Mg | 88.2 | 0.2006 | 320 av[2] | [149] | 313 | 1.000 | [79] | 633 | 2.120 | [79] |
| Zn | 83.7 | 0.2351 | 18 ∥ | [149] | | | | | | |
| | 82.5 | 0.2141 | 17 ⊥ | [149] | | | | | | |
| | 80 | 0.2180 | av | [95] | 573 | 2.34 av | [95] | | | |
| Ga | 84.4 | 0.2340 | 100 $P_1$ | [89] | | | | | | |
| | 84.3 | 0.2204 | 70 $P_2$ | [89] | | | | | | |
| | 84.3 | 0.2255 | 50 $P_3$ | [89] | | | | | | |
| As | 85 | 0.195 | 127 | [146] | | | | | | |
| Zr | 88.2 | 0.2380 | 391 av | [149] | | | | 673 | 2.62 | [2] |
| | 77.4 | 0.1777[3] | av | [30] | 373 | 1.438[3] av | [30] | 1073 | 3.34 | [2] |
| Ru | 81.7 | 0.1734 | (680) | [108] | 373 | 1.449 | [108] | | | |
| Cd | 82.5 | 0.2617 | 1.5 ∥ | [149] | | | | | | |
| | 82.5 | 0.2542 | 4.7 ⊥ | [149] | | | | | | |
| | 81 | 0.2522 | av | [95] | 473 | 1.887 av | [95] | | | |
| In | 88.9 | 0.2567 | 25 av | [149] | | | | | | |
| Sn[1] | 88.2 | 0.2457 | 7.5 av | [149] | | | | | | |
| Sb | 88.4 | 0.2441 | 19 av | [149] | | | | | | |
| Nd | 80 | ≈0.52 | | [102] | | | | | | |
| Gd | 80 | ≈0.27 | | [134] | | | | | | |
| Dy | 80 | ≈0.48 | | [134] | | | | | | |
| Hf[1] | 90.1 | 0.3045 | 947 av | [87] | 673 | 2.58 ∥ | [2] | 1273 | 4.25 ∥ | [2] |
| Re | 77.7 | 0.161 | 170 av | [4] | 393 | 1.315 av | [3] | 2985 | 6.76 av | [3] |
| Hg | 90.1 | 0.258 | 18 av | [111] | 165.8 | 0.559 av | [111] | | | |
| Tl[1] | 88.6 | 0.2721 | 7 av | [112] | 195 | 0.6710 av | [112] | | | |
| | | | | | 313 | 1.00 av | [79] | 473 | 1.66 av | [79] |
| Bi | 83.7 | 0.2818 | 27 ∥ | [186] | 196 | 0.686 ∥ | [186] | | | |
| | 83.7 | 0.3308 | 42 ⊥ | [186] | 196 | 0.709 ⊥ | [186] | | | |
| | 81 | 0.3561 | | [95] | | | | 473 | 2.07 | [95] |

silver, which are the metals with lowest resistivity at 273° K, the corresponding values for $r_T$ are 0.144 (at 81.6° K) and 0.235 (at 87.4° K)[4]. This means that most metals, apart from contaminations, have a resistivity smaller than the $\varrho = T/273$ line in the $\varrho(T)$ diagram would indicate.

In Table 9 we have given the ratios $r_T$ for different metals at a temperature of about 80° K and a few higher temperatures.

It is interesting, that most metals follow, within 25%, the Grüneisen law (12.2) if for high temperatures a correction with higher order terms in $T$ is applied.

For a discussion of experimental results at low temperatures the knowledge of the residual resistance is of great importance. In the table the measured $r_T$-values are given. A $z$-value[5] has been inserted when $r$ has been measured down to 1.4° K (then $r_{1.4} \approx z$). By ($z$) is indicated a value which has been calculated from higher temperature measurements and thus only indicates the order of magnitude of $z$. Most data for 80° K have been taken from the results of Meissner and Voigt [149], obtained in the 1930's. For some cases resistances have

[1] See table 10.

[2] av: average or polycristalline; ∥: single crystal, $\varrho \parallel$ principal axis; ⊥: single crystals, $\varrho \perp$ principal axis.

[3] Crystals.

[4] The resistivity values of these metals at 80° K are roughly 0.35; 0.22; 0.64; 0.36 (in $10^{-8}$ ohm m.) respectively.

[5] Cf. note p. 155.

been measured with purer materials, but then mostly in the 20° K and lower temperature region. Since an interpolation for obtaining $r_{80}$ from 20° K yields a questionable result, we have not included these data.

In anisotropic metals, the ratio $\varrho_{\parallel}/\varrho_{\perp}$ of the resistivities parallel and perpendicular to the principal axis is temperature dependent too. This effect is illustrated in Fig. 16a for two pure single crystals of bismuth [*186*] by plotting $(R_T/R_{273})$ against temperature and in Fig. 16b for two single crystals of cadmium and zinc [*147*] by plotting $\varrho_{\parallel}/\varrho_{\perp}$ against $T$. It is obvious that this effect should be connected with the modes of thermal waves and elongation in the different crystallographic directions.

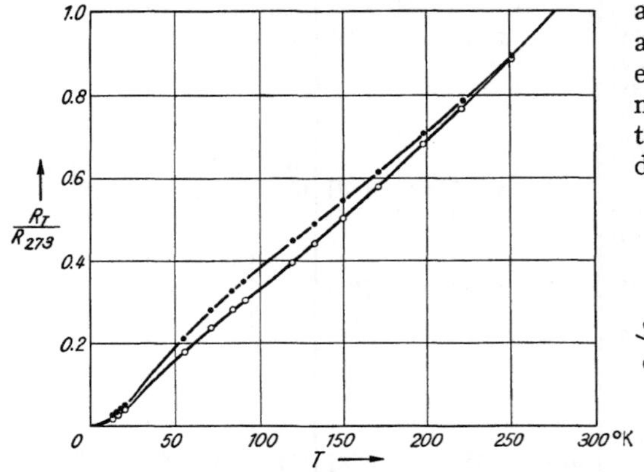


Fig. 16a. $(R_T/R_{273})_{\perp}$ and $(R_T/R_{273})_{\parallel}$ for bismuth (DE HAAS-SCHUBNIKOW), (○) crystal P 674, (•) crystal S 620.

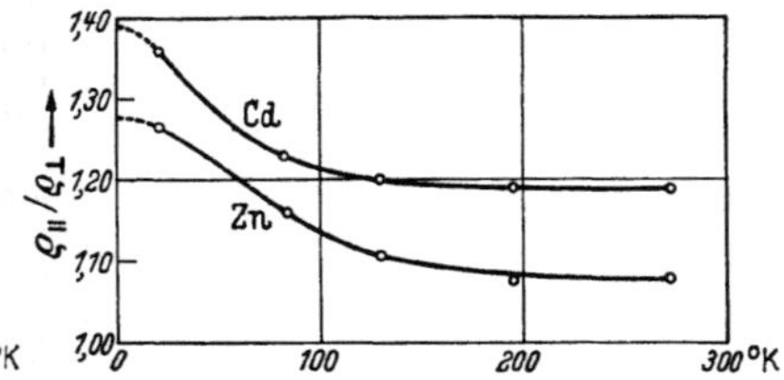


Fig. 16b. $\varrho_{\parallel}/\varrho_{\perp}$ for cadmium and zinc according to MEISSNER.

**13. Resistance anomalies due to crystallographic transitions.** For a great many metals more important deviations from Eqs. (12.2) occur in the temperature region under consideration. Changes in crystallographic or magnetic properties influence the electric resistivity. The experimentally proved correlation is a tool for detecting such changes by resistance measurements. Also in the case of alloys which will be discussed in Sect. 29; an anomalous behaviour of the resistance at a certain temperature is possible.

The importance of the lattice and magnetic transformations will be discussed elsewhere in this volume[1].

For the sake of completeness, a survey follows on the influence of the crystallographic transformation on the resistivity; experimental magnetic changes, for so far above 100° K, will be reported in Sect. 16 and 30. On the left side in Table 10 we have listed crystallographic changes in the metals, conformed by structural analysis, which partially have been taken from HUME-ROTHERY and RAYNOR's monograph [*H 2*] and from recent results of resistance measurements. On the right side are tabulated the results of the resistance measurements. Also listed are negative results in connection with supposed anomalies.

*Beryllium.* JAEGER and ZANSTRA [*100*], KOSOLAPOV and TRAPEZNIKOV [*128*] found changes in $R$ in their sample at 873° K which were confirmed later by SIDHU and HENRY [*191*]. A careful investigation of annealed material by POWELL [*174b*] did not show any change in the resistance at the said temperature.

*Chromium.* SÖCHTIG [*198*] and later ERFLING [*56*] investigated the occurance of a minimum in $\varrho$ at about 300° K, for which no plausible explanation has been given so far. Changes in the resistivity have been reported by FINE *et al.* [*60*] at 310° K, and by SULLY *et al.* [*201*] at 323° K which are connected with changes

[1] Cf. the preceding article by SLATER.

in other physical and mechanical properties as well. X-ray investigation, by FINE however, has not shown any change in crystallographic structure between 300 and 330° K.

WEINER, CHIOTTI and WILHELM [*212*], in order to check supposed anomalies in the $\varrho(T)$ curves, measured the resistances of *iron*, *titanium* and *thorium* between 273 and 1275° K. They connected the specimens in series with a Pt thermometer and they plotted $R_{\text{metal}}/R_{\text{Pt}}$ versus $R_{\text{Pt}}$. They could not find any anomalous behaviour. An analysis by the method of least squares on a separate $R(T)$ curve for each metal yielded a small anomaly for Ti at 608° K.

*Platinum.* HAM and SAMANS [*90*] report a change in $d\varrho/dT$ between 433 and 1075° K. For the use of a Pt wire as a resistance thermometer they suggest splitting the $R(T)$ curve into two straight lines instead of using one formula for the whole curve. Otherwise, a probable error of 2 at 600° K is introduced. This effect in platinum is connected with the experiments of WEINER *et al.*, described above.

*Plutonium.* The metal has five solid allotropic modifications at normal pressure. $\varrho(T)$, for $273°\text{ K} < T < 850°\text{ K}$ has been measured, a.o., by SMITH [*197a*]. His results are tabulated in Table 10 and given as a $\varrho(T)$-curve in Fig. 17.

From the figure follows, that a transformation cycle from α to ε-phase and back is not reversible as far as the absolute values of $\varrho$ are concerned.

Table 10. *Crystallographic transformations in pure metals, and related resistance effects.*

| Metal | Temperature of transformation °K | Transformation at cooling | Change in $R(T)$-curve | Temperatur °K | Ref. |
|---|---|---|---|---|---|
| | | *a) For elements cubic at room temperature.* | | | |
| Li | 65 | b.c.c.—c.p.h. | no anomaly | >20 | |
| | 65 | b.c.c.—f.c.c. after cold work | | | [*139*] |
| Na | below 78 | b.c.c.—f.c.c. | no anomaly | >20 | |
| Ca | 725 | c.p.h.—cubic? | $10^8\, dR/dT$ changes from | } 723 | [*179*] |
| | 575 | cubic?—f.c.c. | 1.28—5.78 | | |
| Cr | 310 | no change [1] | $dR/dT \approx$ zero | 315—273 | [*60*], [*201*] |
| Mn | 1407 | b.c.c.—f.c.c. | | | |
| | 1368 | f.c.c.—cub. 20 [2] | | | |
| | 1015 | cub. 20—cub. 58 [2] | | | |
| Fe | 1230 | f.c.c.—b.c.c. | no anomaly | 1275—273 | [*212*] |
| Rb | | | decrease in $dR/dT$ | 180—190 | [*139*] |
| Sr | 880 | b.c.c.—c.p.h. | discontinuity in $R(T)$ | 813—808 | [*179*] |
| | ≈500 | c.p.h.—f.c.c. | nick in $R(T)$ | 508 | [*179*] |
| Cs | | | hysteresis in $R(T)$ cycle | 273—100 | [*139*] |
| Ba | | | $10^8\, dR/dT$ changes from 12 to 80 | 648 | [*179*] |
| Ce | 70—180 | f.c.c.—c.p.h. | decrease in $R$ | ≈100 | [*102*] |
| Pr | <80 | f.c.c.—c.p.h. | increase in $dR/dT$ | ≈50 | [*102*] |
| W | | cubic—b.c.c. | no anomaly | 273—1100 | [*139*] |
| Pt | | | decrease in $dR/dT$ | 1075—433 | [*90*] |
| Th | | | no anomaly | 1275—273 | [*212*] |

[1] FINE [*60*] and SULLY [*201*] found changes in ductibility, internal friction, dilatation coefficient, electric resistance and thermo electric power, X-ray investigation showed no cristallographic transformation.

[2] 20 to 58 atoms in a unit cell.

*Table 10.* (Continued.)

| Metal | Temperature of transformation °K | Transformation at cooling | Change in $R(T)$-curve | Temperatur °K | Ref. |
|---|---|---|---|---|---|
| | | | *b) For elements non cubic at room temperature.* | | |
| Be | | | anomaly | 1146 | [*100*] |
| | | | | 1000—273 | [*174b*] |
| Ti | 1175 | b.c.c.—c.p.h. | 6% decrease in $R$ | 1155 | [*26*] |
| | | | flat maximum | | [*151*] |
| | | | anomaly | 608 | [*212*] |
| | | | pure Ti, sharp dip | 1012 | [*215*] |
| | | | impure Ti, flat max. | 1012 | [*215*] |
| Co | 1140 | b.c.c.—c.p.h. | max. and min. in $R$ | | |
| | | | hysteresis with temp. cycle | 1200—1000 | |
| Zr | ~1120 | b.c.c.—c.p.h. | dip in $R$ | 1123 | [*2*] |
| Sn | 291 | tetr.—diamond | changes into a semiconductor | | |
| La | 300 | f.c.c.—c.p.h. | decrease in $R$ | 300 | [*102*] |
| Hf | >130 | b.c.c. ?—c.p.h. | no anomaly found | 1273—273 | [*2*] |
| Tl | 540 | b.c.c.—c.p.h. | sharp discontinuity in $R$, hysteresis | 771 | [*187a*] |
| U | 1050 | b.c.c.—tetr. | | | |
| | 930 | tetr.—orthorh. | | | |
| Pu | 745 | b.c.c.—f.c.c. | $10^3 \alpha_{\varrho,T}$ increases from 0 to +0.15; $10^8 \varrho = 102$ ohm m. | 673 | [*197a*] |
| | 570 | f.c.c.—complex $\gamma$ | $10^3 \alpha$ decreases to 0 again; $10^8 \varrho = 110$ ohm m. | 573 | [*197a*] |
| | 470 | complex $\gamma$—complex $\beta$ | $\alpha$ remains 0; $10^8 \varrho = 110$ ohm m. | 473 | [*197a*] |
| | 390 | complex $\beta$—orthorh. | $10^3 \alpha$ decreases from 0 to −2.96; $10^8 \varrho = 150$ ohm m. | 298 | [*197a*] |

*Lanthanum.* The gradual phase changes of La, found by JAMES *et al.* [*102*] have been reported before (p. 157).

*Rare earth metals.* Cerium, praseodymium and neodymium are investigated from room temperature to 2.2° K, by JAMES, LEGVOLD and SPEDDING [*102*]. The samples were investigated by X-rays and successive phase transformations, coupled with hysteresis—like loops in the $\varrho(T)$-curves were found.

*Cerium* has a transition in the temperature region 70 to 180° K, connected with a sudden drop in resistance on cooling (to ≈60% of its value) between 100 and 120° K, a change attributed to a change from normal f.c.c. to c.p.h.[1]. The corresponding increase in resistivity was found, during heating at 170° K and was completed at about 220° K (Fig. 18a). Depending on the temperature-cycle made, and the annealing process, c.p.h. structures were found also at room temperature, these structures having a higher resistivity than the f.c.c. structure.

*Praseodymium.* On successive runs the resistivity increases, which indicates a transition of the cerium type.

*Neodymium.* No transformation of the c.p.h. sample down to about 24° K was found.

[1] Cooling to lower temperatures gave a transformation to a condensed f.c.c. form.

*Titanium.* BURGERS [26] among others found a decrease of 6% in the resistivity at the $\alpha$-$\beta$-transition temperature, $1155 \pm 20°$ K (c.p.h. → b.c.c.). MICHELS [151] for a wire of commercial material found a flat maximum in $R$ at this transition temperature, whereas WEINER [212] *et al.* (see above under platinum) found indications for an anomaly at 608° K only. The latter authors doubt whether anomalies at other temperatures (789, 907, 999 and 1075° K with a margin of $\pm 20°$ K in each case) exist as suggested in the literature. WYATT [215] reports an anomalous dip in $\varrho$ at 1012° K for pure Ti, a flat maximum occurs at this temperature for an impure sample.

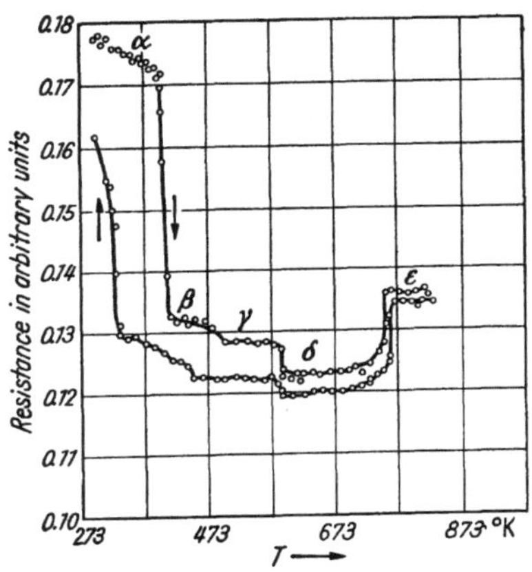


Fig. 17. The resistivity of plutonium between 273° K and its melting temperature (SMITH).

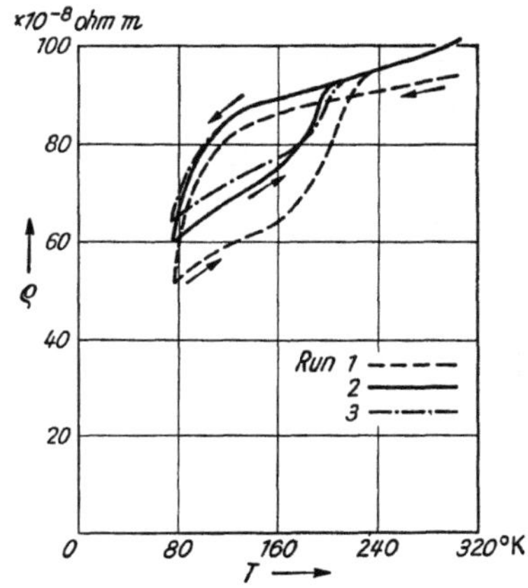


Fig. 18. Resistivity of cerium against temperature (JAMES, LEGVOLD and SPEDDING).

**14. Electrical conductivity of pure metals at and above the melting point.** A comparison between the small changes in several physical properties above and below melting point (for example, a small change in specific heat at atmospheric pressure at the melting point and the 2 to 6% change in atomic volume of the closepacked metals) and the relatively large changes at evaporation leads to the conclusion that melting is not such a drastic change as evaporation.

So far, several explanations have been given for the transformation at the melting point. For instance, the material seems to maintain its structure, but becomes more porous at melting under extinction of long range order (FÜRTH [65]); or the atoms at the lattice points start to move into interstitial places (LENNARD-JONES and DEVONSHIRE [135]); or the number of vacancies in the lattice increases by the heating process, so that finally the order collapses (FRENKEL [63]). These suggestions and further details have been described lately by FROST [F 1]. In the following, we shall give a short survey on the changes of conductivity at melting temperature and on the resistivity temperature relation above this temperature.

*α) Conductivity changes at the melting point.* It has been established that the value of the ratio between $\sigma_{\text{solid}} (\equiv \sigma_S)$ and $\sigma_{\text{liquid}} (\equiv \sigma_L)$ is practically restricted to $\frac{1}{2} \leq \sigma_S/\sigma_L \leq 2$ (cf. Table 11), and PERLITZ [168] classifies the metals, according to this ratio, as follows:

$\sigma_S/\sigma_L \approx \frac{1}{2}$ for open structure metals (Bi, Sb, Ga);

$\sigma_S/\sigma_L \approx \frac{3}{2}$ for b.c.c. metals (Na, Li, K, Rb, Cs and Al);

$\sigma_S/\sigma_L \approx 2$ for more close packed metals (Ag, Cd, Pb, Tl, Cu, Zn, Te, Sn and Au).

The volume of the open structure metals decreases at melting. A classification, based on the volume changes, electric conductivity, and specific heat has beer given by KUBASCHEWSKI [130] as follows: —

True metals $3\times 10^6 < \sigma < 7\times 10^7$ (Li, Na, K, Rb, Cs, Be, Cu, Ag, Au, Mg, Ca, Al, Co, Ni, Ti, Pb).

Meta-metals $10^6 < \sigma < 10^7$ (Zn, Cd, Hg, In).

Semi-metals $10^{-3} < \sigma < 10^6$ (C, Sn, Si, Ge, Sb, Bi, Te, Ga, Sc, P (black)).

Non-metals $\sigma < 10^{-4}$ (P, S, Se, Halogenes).

A somewhat modified classification into noble metals (Li, Na, Cu, K, Ag, Au, Mg, Al) meta-metals (Be, Zn, Cd, Hg, In, Tl Sn (white) Pb) semi-metals (B, graphite, Si, Ga, Ge, Sn (grey), Pb(black), As, Sb, Bi, Te, Po) and the non-metals has been suggested by KLEMM *et al.* [125]. The values of $\sigma_S/\sigma_L$ compiled in Table 12, comprise older results given in GRÜNEISEN's article [G 1] and mean values as given by POWELL [173]. A few suggestions have been made regarding the physical reason for the change in $\sigma$. FRÖHLICH [F2] and MOTT [155] first proposed that electron scattering with phonons of high energy would be responsible for the effect. In his paper MOTT suggested, that for $h\,T_M \gg h\nu_S$ ($T_M$ being the melting temperature, $\nu_S$ the harmonic lattice frequency in the solid state) the free energy at the fusion should be a minimum which leads, considering the atoms as harmonic oscillators, to:

Table 11: *Observed conductivity changes* $(\sigma_S/\sigma_L)_{\text{obs}}$ *(mean values) at the melting point, compared with the values calculated* $(\sigma_S/\sigma_L)_{\text{calc}}$ *with (14.1).*

| Metal | $T_M$ [1] °K | $L$ [1] kjoule/mol | $(\sigma_S/\sigma_L)_{\text{calc}}$ [2] | $(\sigma_S/\sigma_L)_{\text{obs}}$ | Ref. |
|---|---|---|---|---|---|
| Ag | 1234 | 11.4 | 2.09 | 1.9 | [G 1] |
| Al | 934 | 10.5 | 2.46 | 1.82 | [173] |
| Au | 1336 | 12.8 | 2.15 | 2.28 | [G 1] |
| Cs | 303 | 2.10 | 1.74 | 1.66 | [G 1] |
| Cu | 1357 | 13.0 | 2.16 | 2.07 | [G 1] |
| Fe | 1808 | 15.2 | 1.96 | 1.09 | [173] |
| K | 337 | 2.38 | 1.75 [155] | 1.55 | [G 1] |
| Li | 453 | 2.39 | 1.28 | 1.68 | [G 1] |
| Na | 370 | 2.65 | 1.77 [155] | 1.46 | [173] |
| Ni | 1728 | 17.7 | 2.27 | | |
| Pb | 601 | 4.98 | 1.99 | 1.98 | [173] |
| Rb | 312 | 2.20 | 1.76 [155] | 1.61 | [G 1] |
| Bi | 544 | 10.9 | 5.0 [155] | 0.47 | [126] |
| Cd | 594 | 6.41 | 2.3 [155] | 1.89 | [173] |
| Ga | 303 | 5.59 | 4.5 [155] | 1.67 | [173] |
| Hg | 234 | 2.33 | 2.23 [155] | 3.36 | [173] |
| | | | | 1.93 | [31] |
| In | 388 | 3.27 | 2.00 | 2.12 | [177] |
| Mg | 924 | 8.72 | 2.10 | 1.63 | [96] |
| Sb | 904 | 20.1 | 5.6 [155] | 0.71 | [173] |
| Sn | 505 | 7.08 | 3.06 | 2.11 | [173] |
| Tl | 722 | 4.31 | 1.56 | 2.0 | [173] |
| Zn | 693 | 7.20 | 2.3 [155] | 2.11 | [173] |

Table 12. *Correlation between* $\Delta\varrho/\varrho$ *and* $\Delta V/V$ *at* $T_M$ (SUTRA).

| | $\Delta\varrho/\varrho$ | $\Delta V/V$ | | $\Delta\varrho/\varrho$ | $\Delta V/V$ |
|---|---|---|---|---|---|
| Al | 0.79 | 0.060 | Bi | −0.75 | −0.034 |
| Au | 0.85 | 0.051 | Ga | −0.76 | −0.032 |
| Cu | 0.75 | 0.042 | Hg | 1.53 | 0.037 |
| K | 0.50 | 0.026 | Sb | −0.34 | −0.010 |
| Li | 0.49 | 0.017 | Sn | 0.78 | 0.028 |
| Pb | 0.66 | 0.035 | Zn | 0.80 | 0.042 |

$$\nu_L/\nu_S = e^{-\frac{1}{3}U/\varkappa T_M}$$

where $\nu_l$ is the oscillator frequency in the liquid and $U$ is the work which has to be done to put the atom in motion.

Since $\sigma \sim \nu^{-2}$ [9.5], it follows that:

$$\sigma_S/\sigma_L = e^{80\,L/T_M} \tag{14.1}$$

[1] $T_M$ and $L$ in kcal/mol are taken from Metals Referencebook 1948 [M 2].

[2] As taken from MOTT's paper [155] or calculated with (14.1) using tabulated data for $T_M$ and $L$ [M 2].

where $L$ is the latent heat of melting in kilo joules. MOTT concludes from a comparison between experimental and theoretical data, that, though the liquid is almost in an amorphous state (having no order what-so-ever), the conductivity is only determined by the increase of the phonon energy. The comparison between the data of the $\sigma$-ratios certainly justifies this conclusion (Table 11).

For the alkali metals the conditions $h\nu_S \ll \varkappa T_M$ is not satisfied, therefore a comparison with a little more complicated expression for $\sigma_S/\sigma_L$ has to be made which indeed results in somewhat decreasing the discrepancy:

| | Li | Na | K |
|---|---|---|---|
| $(\sigma_S/\sigma_L)_{\text{calc}}$ | 1.50 | 1.58 | 1.67 |
| $(\sigma_S/\sigma_L)_{\text{exp}}$ | 1.46 | 1.55 | 1.61 |

The theory is not able, however, to interpret the large differences between the experimental and theoretical ratios which are found in the cases of Hg, Bi, Ga and Sb. The only conclusion which from the viewpoint of the present theory could be made is, that for these metals, the number of effective electrons changes during the fusion process. Later development in the investigation of liquid metals by BRAGG interference methods as reported by GINGRICH [73] showed for alkali metals no differences in the BRAGG pattern for liquid and solid states, except that the liquid proved to be more microcrystalline than the solid. The grain sizes are of the order of 55, 57 and 130 atoms for Li, Na and K respectively. Since for a liquid in the neighbourhood of $T_M$ (as earlier experiments had already shown) $\varrho \sim T$, these results led GERSTENKORN and SAUTER [71] to the conclusion that the liquid with its microcrystalline structure might be a good example for large angle scattering of the conduction electrons by the lattice, followed by BRAGG reflections[1], together with a scattering determined by the grain size of the sample.

The latter influence could be expressed by a structure factor, which can be calculated for a solid, and which was estimated for liquids from the X-ray results. Taking one free electron per atom ($n/N=1$), or less than one ($n/N<1$) as optical data would suggest, the latter authors found for $\sigma_S/\sigma_L$:

| | $n=N$ | $n<N$ | Observed | Calculated by MOTT |
|---|---|---|---|---|
| Li | 1.62 | 2.04 | 1.68 to 1.96 | 1.50 |
| Na | 1.74 | 1.77 | 1.46 to 1.6 | 1.58 |
| K | 1.61 | 1.59 | 1.55 to 1.56 | 1.67 |

Since, as shown in the last but one column, experimental data on $\sigma_S/\sigma_L$ are not very exact, it is clear that a definite conclusion cannot easily be made. It is striking, however, that a model which assumes an amorphous state in the liquid (MOTT) and one which assumes a microcrystalline order (GERSTENKORN and SAUTER) (both taking into account the dependence of the scattering on the atomic vibrations) lead to the same result.

An approach from quite another viewpoint in order to explain the anomalous $\sigma_S/\sigma_L$ values of Hg and Bi, has been published by Miss SUTRA [202]. She doubts whether at the temperature of fusion the ion-electron dissociation or the thermal vibration of the atoms would change. The only quantities which seem to change are [cf. Eqs. (9.2) and (9.3)] $n$, the effective number of electrons per unit volume by a change of volume ($n \sim V^{-1}$), and $l$ the mean free path, by a change of the

[1] The so called $U$-processes, cf. Sect. 11 of JONES' article in this volume.

free volume $\varphi$ available for collisions. The latter might equal the volume of a gram-atom minus the volume of the $6.10^{23}$ ions. Since $l$ should vary as $\varphi^{\frac{1}{3}}$, $\sigma$ should vary at $T_M$ as $V^{-1}\varphi^{\frac{1}{3}}$. Analyzing the experimental results for $(\Delta V/V)_{T_M}$ and $\sigma_S/\sigma_L$ a proof of the relation between the volume change and conductivity change can be given (cf. Table 12) except in the case of Hg. Taking $\varphi_S = V_S - x$ and $\varphi_L = V_L - \gamma x$, she estimates the values of $\gamma$ (for Hg and Bi as 2.91 and $-21.4$, respectively) this being a measure of the change in the ion volume at liquefaction. This result implies that the radius of a Hg ion *increases* by a factor 1.4 and that of a Bi ion *decreases* by a factor 2.8 at fusion.

If this argument should hold, the ionic diameter of the Bi ion in the solid state should be extraordinarily large. These results are more or less supported by the evidence of the experimental values of the diamagnetic susceptibility which, at $T_M$, increases from $10^6\chi_S = -0.16$ to $10^6\chi_L = -0.19$ ($\chi_S/\chi_L = 0.78$) for Hg and decreases from $10^6\chi_S = -1.60$ to $10^6\chi_L = -0.08$ ($\chi_S/\chi_L = 20$) for Bi.

*β) Liquid metals.* The results of conductivity measurements for liquid metals are rather meagre. Since early measurements of BERNINI [*12*] (1904), BORNEMANN (1910), NORTHRUP [*163*] (1914), BRIDGMAN [*18*] (1922) and others, not many new results have been published (see Table 13).

Table 13. *Values of $R_T/R_{T_M}$ and $R_T/R_{273}$ (in italics) for some liquid metals.*

| Metal | $T_M$ | $T/T_M$ | | | | | | | | |
|---|---|---|---|---|---|---|---|---|---|---|
| | °K | 1.00 | 1.05 | 1.10 | 1.20 | 1.30 | 1.50 | 1.75 | 2.00 | Ref. |
| Li | 453 | 1.000 | 1.016 | 1.032 | | | | | | [*18*] |
| | | *3.549* | *3.610* | *3.674* | | | | | | |
| Na | 370 | 1.000 | 1.054 | 1.084 | | | | | | [*12*] |
| | | *1.894* | *1.997* | *2.095* | | | | | | |
| K | 337 | 1.000 | 1.081 | 1.164 | 1.325 | | | | | [*12*] |
| | | *1.894* | *2.190* | *2.31* | *2.51* | | | | | |
| Cu | 1357 | 1.000 | 1.025 | 1.051 | 1.103 | | | | | [*163*] |
| | | *12.28* | *12.57* | *12.90* | *13.52* | (*14.15*) | | | | |
| Ag | 1234 | 1.000 | 1.040 | 1.085 | 1.17 | 1.254 | | | | [*163*] |
| | | *10.0* | *10.4* | *10.9* | *11.7* | *12.5* | | | | |
| Au | 1336 | 1.000 | 1.030 | 1.058 | 1.121 | 1.184 | | | | [*163*] |
| | | *13.30* | *13.72* | *14.12* | *14.92* | *15.74* | | | | |
| Cd[1] | 594 | 1.000 | 0.995 | 0.992 | 0.995 | 1.000 | (1.022) | | | [*184*] |
| | | *5.03* | *4.99* | *4.98* | *4.99* | *5.03* | (*5.14*) | | | |
| In | 388 | 1.000 | 1.015 | 1.030 | 1.058 | 1.090 | 1.148 | 1.222 | 1.297 | [*184*] |
| | | *3.88* | *3.92* | *3.97* | *4.09* | *4.20* | *4.44* | *4.73* | *5.02* | |
| Mg[2] | 924 | 1.000 | 0.995 | 0.988 | 0.975 | (0.962) | | | | [*184*] |
| | | *7.36* | *7.31* | *7.27* | *7.17* | (*7.09*) | | | | |
| Sn | 505 | 1.000 | 1.012 | 1.026 | 1.054 | 1.083 | 1.138 | 1.205 | 1.275 | [*184*] |
| | | *4.49* | *4.55* | *4.61* | *4.74* | *4.86* | *5.12* | *5.42* | *5.73* | |
| Zn[3] | 693 | 1.000 | 0.985 | 0.972 | 0.960 | 0.955 | 0.965 | (0.99) | | [*184*] |
| | | *6.91* | *6.82* | *6.75* | *6.65* | *6.59* | *6.66* | (*6.84*) | | |

Since experiments with high temperature equipment are rather difficult and an investigation on liquid metals does not seem at first sight to be of great interest, the lack of new data is understandable. Perhaps new techniques for high temperature regulation will make this domain of investigation more accessible. The interest of the study of molten metals is illustrated by FRÖHLICH'S [*F 2*] remarks, namely, that a free electron moves in a solid under systematic deviations (forbidden zones of energy, due to a periodic lattice) whereas in a liquid, un-

[1] Minimum at $T \approx 650°$ K [*184*].
[2] Compare HORN'S results, which give for pure magnesium different results (Table 14).
[3] Minimum at $T \approx 940°$ K [*184*].

systematic deviations from the free electron model occur. This concept is based on the model of a liquid, where the structure is confined to a much smaller volume than the macroscopic volume of a solid metal. As X-ray analysis of liquid mercury has shown (MENKE [150]), the probability of finding an atom in an arbitrary direction from a given point is very different in a liquid from that in a solid metal (Fig. 19).

The electrical conductivity of liquid Hg has been investigated by JAEGER and VON STEINWEHR [101b] between 273.15° K and 373.15° K. For this temperature region they describe their results in the form:

$$\varrho_T = \varrho_{273}\left(1 + \alpha_1 (T - 273) + \alpha_2 (T - 273)^2\right),$$

with $\alpha_1 = 8.8915 \times 10^{-4}$, $\alpha_2 = 9.936 \times 10^{-7}$.

For temperatures between 273 and 370° K CLAY found [31] the coefficients $\alpha_1 = 7.886 \times 10^{-4}$ and $\alpha_2 = 3.36 \times 10^{-6}$. The temperature coefficient is low, as is found also for other liquid metals.

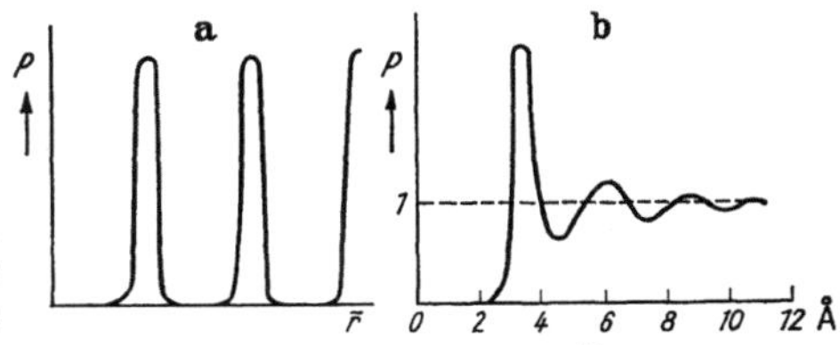


Fig. 19a and b. Probability $P$ to find at a given direction from an atom another atom in a solid (a) or a liquid (b) (MENKE).

Recently SCALA and ROBERTSON [184] investigated liquid Sn, Cd, Zn, In, Mg and Cu of high purity (99.996% for the impurest case, which was Sn). The metal is molten in mullite tubes, in which refractory insulating tubes are inserted. Molybdenum wires in these tubes serve as potential leads (Fig. 20.) The size factor of the tube with molten metal was determined by a determination of $\sigma$ of triple-distilled Hg at room temperature. Their results on pure metals and alloys demonstrate a close correspondence between the solid and the liquid state.

In Table 13 we have inserted some values of $r'_T = R_T/R_{T_M}$ and $r_T = R_T/R_{273}$ for the molten metals, including some of the old (1904) and the most recent data. The results have been represented in a plot (Fig. 21) of $r'_T$ against $T/T_M$, showing the existence of a high temperature minimum in the $R(T)$ relation. The data for magnesium are taken from SCALA and ROBERTSON's paper.

HORN [96] found that for 99.95% Mg the slope is positive, though very small, while for 99.52% and 98,67% Mg samples (main impurities Al, Fe, Mn, Cu, Si and Cr) the slopes were negative (Fig. 22).

Defining the temperature coefficient

$$\alpha_{\varrho_L \cdot T_M} = \frac{1}{\varrho_{L,T_M}} \frac{d\varrho_L}{dT} \qquad (14.2)$$

of the resistivity for a liquid metal, these values are given in Table 14 for those metals which show a linear $\varrho(T)$-relation. Especially for Cu, Ag and Au the temperature coefficients of resistivity of the liquid metals are an order of magnitude smaller in the liquid, than in the solid state.

Table 14. *Temperature coefficient of resistivity* $\left(\alpha_{\varrho_L} = \frac{1}{\varrho_{T_M}} \frac{d\varrho}{dT}\right)$ *for liquid metals at the melting temperature.*

| Metal | $T_M$ °K | $10^3 \alpha_L$ degree$^{-1}$ | Metal | $T_M$ °K | $10^3 \alpha_L$ degree$^{-1}$ |
|---|---|---|---|---|---|
| K | 337 | 4.82 | Cd | 594 | −0.18 [1] |
| Na | 370 | 3.04 | Au | 1336 | 0.461 |
| Cu | 1375 | 0.372 | Mg | 924 | ≈ 0 [2] |
| Zn | 693 | −0.43 [1] | In | 388 | 0.766 |
| Ag | 1234 | 0.692 | Sn | 505 | 0.547 |

[1] At $T_M$ only.

[2] Spectrographic pure material, $\alpha < 0$ for less pure material (HORN [96]).

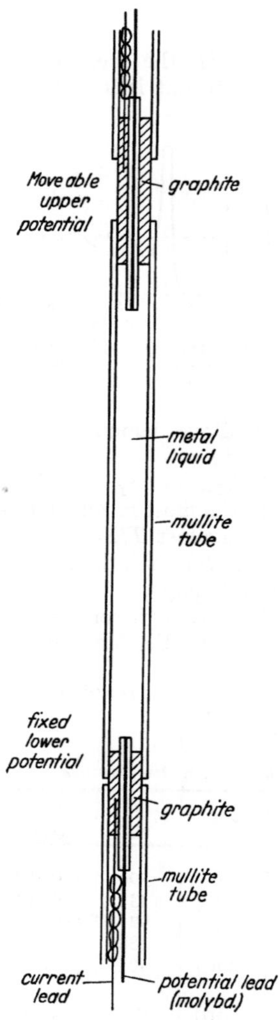


Fig. 20. Apparatus for measuring the resistivity of molten metals (SCALA and ROBERTSON).

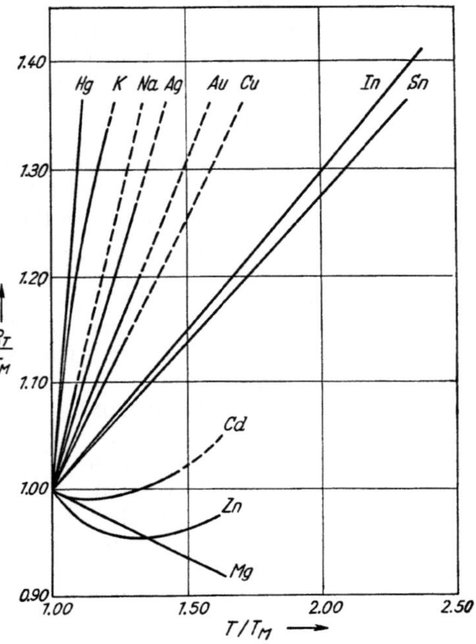


Fig. 21. $R_T/R_{T_M}$ against $T/T_M$ for some metals (data taken from original papers). Full line: measured region. K and Na: BERNINI; Cu, Ag and Au: NORTHRUP; In, Sn, Cd, Mg and Zn: SCALA and ROBERTSON; Hg: JAEGER and STEINBACH.

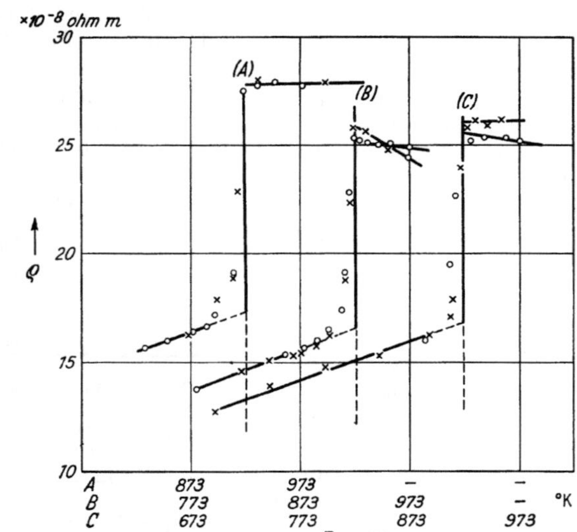


Fig. 22. A sketch of HORN's results for magnesium. (A): Spectrographically pure material 99.95% Mg. $\sigma_S/\sigma_L = 1.63$; (B): Commercially pure material 99.52% Mg. (C): Commercially pure material 98.67% Mg. Circles: cooling; crosses: warming.

## V. Magneto resistance effects.

**15. Non-ferromagnetic metals.** A moving electron in a metal will, if a transverse magnetic field is applied, describe a LORENTZ spiral around the field lines as axis. This results in a loss of momentum in the direction of the electric field and an increase of resistance. The deflected electrons will form charges of opposite sign at the boundary of the metal specimen and by COULOMB repulsion hinder the sideward motion of other electrons. Thus, for perfectly free electrons in a metal, the HALL potential is an automatic compensation for the effect of the magnetic field on the current and no increase in resistance would follow. The fact that a magneto-resistance effect has been found for all pure metals is another proof that the simple picture of free electrons fails to describe the phenomena. Even if the electrons are considered to move freely in a periodic COULOMB field, their scattering is such that the effect of the magnetic field on the velocity component under an angle with the current direction results in a resistance increase.

*α) Transverse effects* ($H \perp I$). The general evidence is that the increase $\Delta\varrho$ in resistivity is, for weak fields, proportional to $H^2$. Bismuth seems to be an exception since for very weak fields of the order of 100 Oe[1], there is an indication of a linearity between $\Delta\varrho$ and $H$ [*28*]. For strong fields KAPITZA's [*114*] experiments from room temperature down to 78° K have shown that for a great number of metals, $\Delta\varrho = aH^2 + bH$, the influence of the linear term being observable only for $H \gtrsim 100$ kOe and more.

Experiments at low temperature have shown that the coefficient of $H^2$ is inversely proportional to the resistance in zero field [*152*].

This, and other evidence implies that measurements of the magnetic resistance effects at low temperatures are of great interest for the following reasons:

(1) The large decrease of $\varrho$ for pure metals will yield relatively large $\Delta\varrho$-values in field strengths obtainable with normal magnets.

(2) Anisotropic resistance changes will be better exhibited at low than at high temperatures.

(3) The magnetic energy of the spin of an ion of a paramagnetic metal in fields of the order of 20 kOe at temperatures below 4° K is of the same order as the thermal energy.

In the present article a short summary of results obtained with metals at normal temperatures will be given, for a more detailed survey on the magneto-resistance effects the reader is referred to the low temperature volumes of this Encyclopedia[2].

We mentioned already KAPITZA's experiments. These experiments (carried out during the 1920's) are still of experimental interest because of the method used for obtaining strong fields (300 kOe) in iron-free coils. Large currents (7000 amp) were obtained by a discharge of a specially designed lead accumulator battery through a coil of 0.02 Ω resistance, dissipating a power of 1 megawatt in 3 millisec [*114a*]. A galvanometer, which projected a light beam on a moving photographic plate, could register the potential difference over the resistance and was also used for measuring the field by means of an induction coil. The results of measurements at room temperature and at 78° K may be summarized as follows. For the pure metals, KAPITZA found no saturation which, at that time, was a theoretically remarkable result [*114c*]. The field dependency of the

[1] 1 Oe = $10^3$ A/m. in MKSA units. Since only relative effects will be discussed in this chapter the author prefers the use of oersted as an unit for the magnetic field.

[2] Cf. vols. XIV and XV.

resistance is quadratic, but in strong fields the linear term plays an important role. He introduces a field $H_c$ which follows from the experimental $\Delta R(H)$-curve as the intersection of the asymptote of the curve with the field axis. and he describes his results with $H_c$ and a parameter $\beta$:

$$\Delta R_{H,T}/R_{0,T} = \beta\,(H - H_c) + \frac{\beta H^2}{3H_c} \quad \text{for} \quad H \geq H_c, \tag{15.1}$$

$$\Delta R_{H,T}/R_{0,T} = \beta\,\frac{H^2}{3H_c} \quad \text{for} \quad H \leq H_c. \tag{15.2}$$

To this formula, a correction has to be applied in $\beta$ so that it represents the relation for a pure metal. The parameter increases with decreasing temperature and is, for a given temperature, more or less related to the position of the element in the periodic system, as is the case with other physical quantities.

The elements Bi, Ce, Sb and Te have been measured at room temperature by TANABE [*206*] who, by short circuiting a generator, obtained fields of the order of 200 kOe in a large coil ($I \approx 10^4$ amp, for 10 milliseconds). A chopper amplifier was used to measure the voltage over the samples. TANABE finds a confirmation of the experimental expression (15.1) for these cases also.

Table 15a. *$\Delta R_{HT}/R_{0,T}$ for several metals according to* KAPITZA. Field strength 300 kOe, $H \perp I$.

| | 273° K | 78° K | | 273° K | 78° K |
|---|---|---|---|---|---|
| Li | | 0.15 | Sn | 0.02 | 0.23 [2] |
| Na | | 0.07 | Pb | < 0.01 | 0.05 [2] |
| Cu | | 0.47 | Zr | | 0.05 [2] |
| Ag | | 0.38 | Th | | 0.155 [2] |
| Au | | 0.28 | As [1] | 1.1 | 29 [2] |
| Be | 0.65 | 2.3 | Sb | 3.5 | 40 [2] |
| Mg | 0.17 | 2.84 | Bi | 3.7 | 1360 [2] |
| Zn | 0.06 | 0.915 | V | < 0.01 | 20.04 |
| Cd | | 1.00 | Ta | | ≈ 0.01 |
| Hg | | 0.05 [2] | Cr | 0.30 | 4.40 |
| Al | | 0.73 | Mo | | 0.915 |
| Ga | | 0.87 [2] | W | | 0.938 |
| In | | 0.14 [2] | Te | 0.36 | 0.39 [2] |
| Tl | | 0.16 [2] | Pd | | 0.10 |
| | | | Pt | | 0.75 |

Table 15b. *$\Delta R_H/R_{0,T}$ for bismuth, according to* KAPITZA.

| | | $H \perp$ bin. axis | $H \parallel$ bin. axis |
|---|---|---|---|
| $I \parallel$ principal axis<br>300 kOe $\perp I$ | 290° K<br>liquid air | 50<br>1650 | 38<br>1580 |
| | | $H \perp$ princ. axis | $H \parallel$ princ. axis |
| $I \perp$ principal axis<br>300 kOe $\perp I$<br>250 kOe $\perp I$ | <br>290° K<br>liquid air | <br>22<br>450 | <br>35<br>1500 |

In Table 15 we have summarized some of KAPITZA's results. A superposition of a quadratic and a linear term in $H$ is found to be true even at low temperatures, a description, however, of the results in terms of an $H_c$, which field has no understandable physical meaning, proved impossible (cf. [*M 1*]). KOHLER [*127a*] has suggested an expression in which $\Delta R(H)$ is described by a simple formula

$$(\Delta R_H/R_0)_T = (\Delta \varrho_H/\varrho_0)_T = f(H/\varrho_{0,T}). \tag{15.3}$$

The function $f$ depends on the angle between the field and the current and, in the case of single crystals, on the orientation of the crystallographic axes of the sample. In a later paper [*127b*], KOHLER by means of the two-band model developed a more detailed expression

$$(\Delta R_H/R_0)_T = \frac{a H^2}{1 + b H^2}. \tag{15.4}$$

[1] Single crystal.
[2] Temperature of liquid air.

in which the coefficients $a$ and $b$ are functions of the electron conductivity in the d- and s-bands, and of the number of electrons in these bands. Now a saturation effect is possible, unless the numbers of electrons in the two bands are equal (than $b=0$), which is the case for a divalent metal. See also ref. [*J 1*].

According to KOHLER, Eq. (15.3) holds in the case of very low temperatures, when the residual resistances $\varrho_i$ are small compared to the total resistance in zero field, or in the case when $\varrho_i$ is so large that no temperature dependent part of the resistance is present, or in the case of high temperatures. It is customary to plot the curves logarithmically as $(\Delta R_H/R_0)_T$ against $H/r_T$ or $H/r_\Theta$, where this latter ratio has to be considered with due regard to the uncertainties in the chosen $\Theta$-values. JUSTI expected [*107*] that by the introduction of $r_\Theta$ a universal diagram valid for all pure metals, could be made. The result is given in Fig. 23, which has been taken from one of KOHLER'S papers[1]. Though the result does not seem to answer the expectation of universality, the advantage of this diagram and of that with $H/r_T$ as abscissa is that the order of magnitude of the magneto-resistance increase can be read immediately if the $r_T$ or $r_\Theta$-value of a wire is known. Since in the value for $r_T$ (or $r_\Theta$) the impurity of the metal is accounted for, the diagram 23 is useful for practical purposes. It is interesting to note, that for large $H/r$ values the data comprise measurements taken in strong fields at room temperature (with, for example, $r_{273}=1$), but also taken in much weaker fields at low temperatures (for a pure metal, $r_T \approx 10^{-3}$).

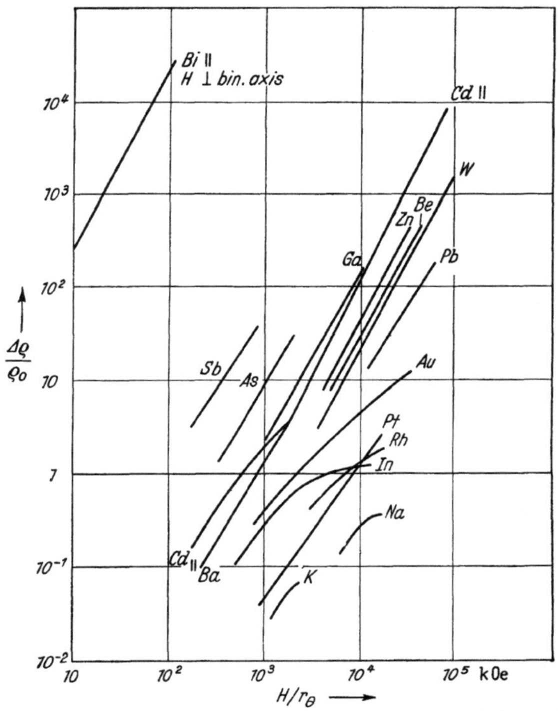

Fig. 23. $(\Delta\varrho_H/\varrho_0)_T$ against $H/r_\Theta$ (double logarithmic scales) for several metals, as taken from KOHLER.

*β) Longitudinal effects.* In general the longitudinal increase $\Delta R_{\parallel}$ is smaller than the transverse increase $\Delta R_{\perp}$, their ratio for pure noble metals being of the order $\frac{1}{2}$. The longitudinal increase follows a KOHLER diagram, different but similar to that for the transverse effect.

For single crystals of Bi, KAPITZA [*114b*] found a saturation in $\Delta R_{\parallel}$, when current and field were adjusted parallel to the principal axis. For the transverse effect only an indication in the slope of the $\Delta R(H)$ curve for saturation was found for $H \perp I$, $I$ parallel to the principal axis.

[1] Ref. [*127b*].

*γ) Anisotropy in the magneto-resistance.* Experiments at low temperatures have shown [*106*] that the magneto-resistance effect is anisotropic, the anisotropy for cubic metals being observable only below temperatures of about 70° K. For non-cubic crystals the anisotropy is a natural consequence of the anisotropic conduction of the crystal. With a crystal of cadmium JUSTI, KRAMER and SCHULZE found a large asymmetry in a field of 34 kOe at 273° K and 78° K [*110*]. The crystal was cut perpendicular to the principal axis, the current parallel to the direction of cutting. The resistance of the crystal was measured when the field was turned through 360° around the current direction as axis. Expressing the asymmetry in terms of the ratio $a = (\Delta \varrho_{\max}/\Delta \varrho_{\min})_{T \text{ and } H = \text{constant}}$, these authors found the values $a = 26$ and $a = 3.4$ for 273° K and 78° K, respectively. For a bismuth crystal with its length and direction of current parallel to the principal axis KAPITZA [*114b*] in 300 kOe found the values $a = 1.3$ and 1.05 at 270° K and 83° K, respectively. Extensive measurements on bismuth single-crystals have been made by STIERSTADT [*200*] at temperatures between room temperature and 70° K. A diagram for a cadmium crystal is reproduced in Fig. 24 [*110* (*Cd 47*)].

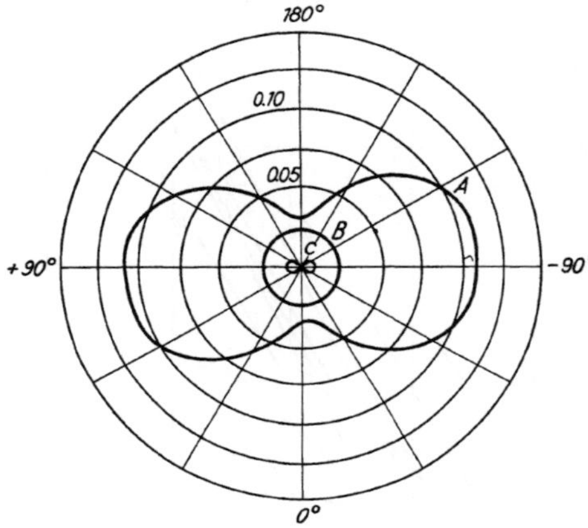


Fig. 24. $(\Delta \varrho_H/\varrho_0)_T$ as a function of field direction for Cd crystals (JUSTI *et al.*), $H = 34.4$ kOe. Curves A and C, the anisotropy effect at 78° K and 273° K for rod and current perpendicular to the principal axis; $H$ parallel to the principal axis at angles ±90°, $H$ perpendicular to the principal axis to angles 0 and 180° [Cd 47]. Curve B: isotropic effect for rod and current parallel to the principal axis [Cd 46].

The minimum in the anisotropy effect found in particular for cadmium perpendicular to its hexogonal axis at 20.4° K is, according to JUSTI, attributed to a co-operation of two anisotropy effects which seem to cancel each other in part at about 30° K. Normally, the anisotropy increases at low temperatures due to the fact that the anisotropy in electron binding, which remains a constant, has a relatively stronger effect on the resistance at low temperatures because of the decreasing influence of thermal motion. If the latter is anisotropic, as follows from an anisotropy in elastic properties, then at high temperatures both reasons for anisotropy are present. At low temperatures the influence of the latter will decrease.

For cubic crystals the anisotropy follows from the consideration, that a cube, mounted on a central axis parallel to a cubic plane (which axis may represent for example the direction of the current in a cubic crystal) will, after turning through 90°, be in the same position with respect to this axis, and any external field vector perpendicular to this axis as before, while for a cube mounted on a central axis parallel to a diagonal in one of the faces, rotation through 180° is needed to restore an identical position. For results of measurements reference is made to the chapter on low temperature resistance effects in volume XIV of this Encyclopedia.

**16. Ferromagnetic metals.** So far, we have excluded from this and the preceding chapter a discussion of the resistance effects in ferromagnetic pure metals. This is done since, apart from the conditions which control the conductivity in a non-magnetic metal, the properties of magnetic ions in a conductor may influence the resistivity, irrespective of the external field.

*α) No external magnetic field.* The difference between transition metals in the non-ferromagnetic and in the ferromagnetic state is that, in the latter, spontaneous magnetization occurs. Below the Curie temperature $\Theta_C$ characteristic for each metal, the magnetic moments of the magnetic ions are parallel in macroscopic domains, that is, their thermal oscillations no longer occur about

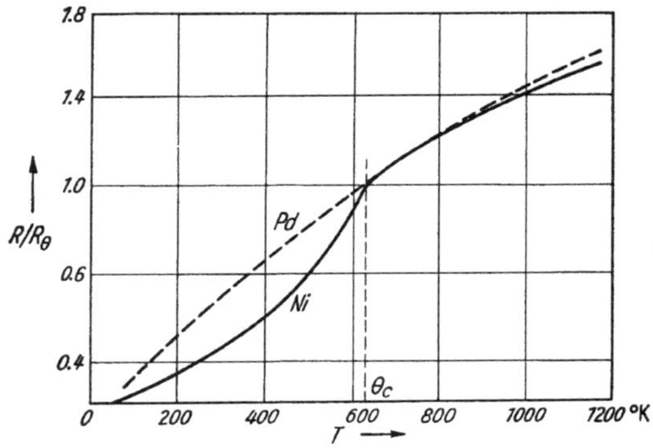


Fig. 25a. $R/R_{\Theta_c}$ against $T$ for nickel and palladium. $\Theta_c$ is the Curie temperature of Ni (631° K).

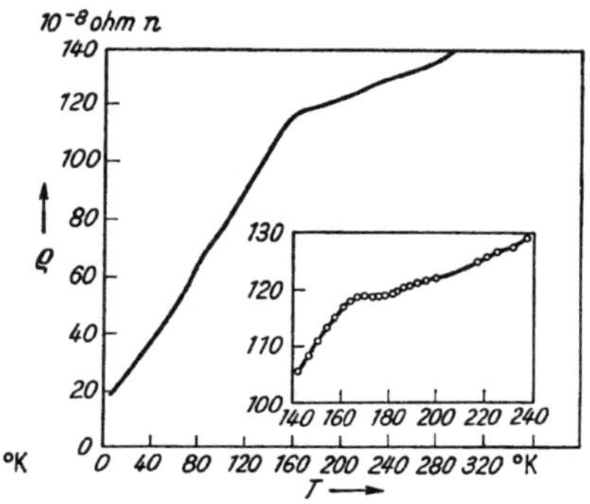


Fig. 25b. $\varrho$ against $T$ for dysprosium (Legvold *et al.*).

arbitrary positions but about preferred directions, determined for each domain by factors such as crystal orientation, domain size and strain. At temperatures low with respect to the Curie temperature, the thermal energy of the ions is smaller than magnetic interaction and each domain is saturated. The specimen as a whole may then have a resultant magnetization. If this occurs, the d-electrons are divided into two groups, with their spin parallel and antiparallel to the ion spin. The states of lowest energy will be completely occupied which means that the s-electrons have lost their ability of being scattered by the d-electrons with spin parallel to the spontaneous magnetization thus rendering a smaller resistivity than if such a scattering possibility were present. This is the accepted interpretation (Sect. 10) of the fact generally noticed in the $R(T)$-curves for a ferromagnetic metal (pure metal or alloy) that below $\Theta_C$, the resistivity is smaller than for a corresponding non-ferromagnetic transition metal. This is illustrated in Fig. 25a where the $R(T)$-curves for Ni and Pd are given (Becker and Döring [*B 1*], Bozorth [*B 4*]) and in Fig. 25b which is a $\varrho(T)$ curve for Dy (Legvold *et al.* [*134*]). The Curie temperatures for several metals following from resistance measurements (a) and magnetic measurements (b) (see Bozorth) are:

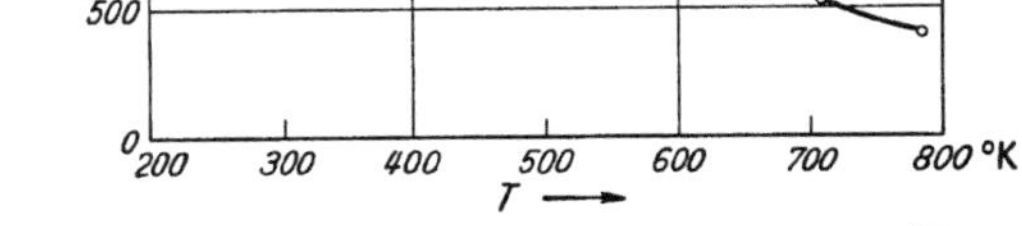


Fig. 26. The temperature coefficient of resistivity $\alpha_R = \frac{1}{R}\frac{dR}{dT}$ for electrolytical nickel ($\Theta_c = 628°$ K) (Gerlach).

| | Fe | Co | Ni | Gd | Dy | Er | |
|---|---|---|---|---|---|---|---|
| (a): $\Theta_c$ | 1033 [*166*], [*172*] | | 636 [*203*] | 289 [*134*] | 170 [*134*] | 80 [*134*] | °K |
| (b): $\Theta_c$ | 1043 | 1403 | 631 | 289 | 105 | — | °K |

The Curie temperature can be found more accurately by plotting the temperature coefficient of resistivity as a function of temperature. Gerlach in this way determined the $\Theta_C$-values for some nickel wires (Fig. 26) [*68*].

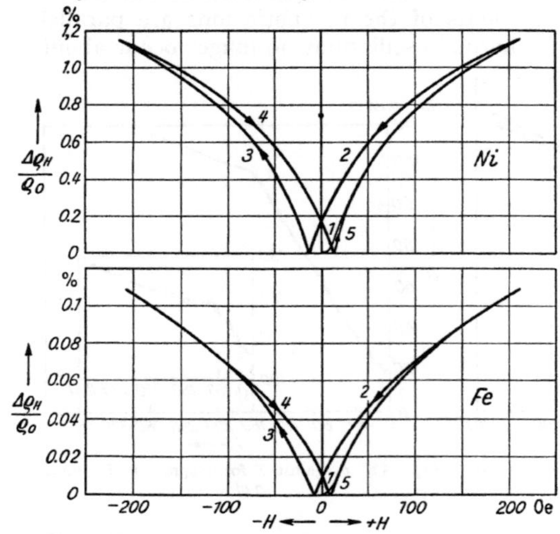


Fig. 27. Magneto resistance increase for wires of nickel and iron in a longitudinal field (Stierstadt).

The question whether for $T > \Theta_C$ the spontaneous magnetization has disappeared and that for $T = \Theta_C$ a discontinuous change occurs, has been solved by Svensson [*203*] by measuring the resistivity of several rather pure Ni wires after different annealing and de-gassing processes. Since no external fields have been applied in this method, the result, that $(dR/dT)$ is discontinuous for $T = \Theta_C$ is more conclusive than a measurement of the true magnetization. This discontinuous change at $\Theta_C$, has been found for other ferromagnetic metals by later authors.

*β) Resistivity in an external magnetic field.* The differences in resistivity between ferromagnetic and other metals are still more striking if an external field is applied. The existence of a parallelism between magnetization and electric resistivity is demonstrated by Fig. 27 (Stierstadt [*199*]) which is a graph of $\Delta R_H/R_0$ against $H$ for Ni and for Fe, and which is similar to the well-known hysteresis curves of magnetization.

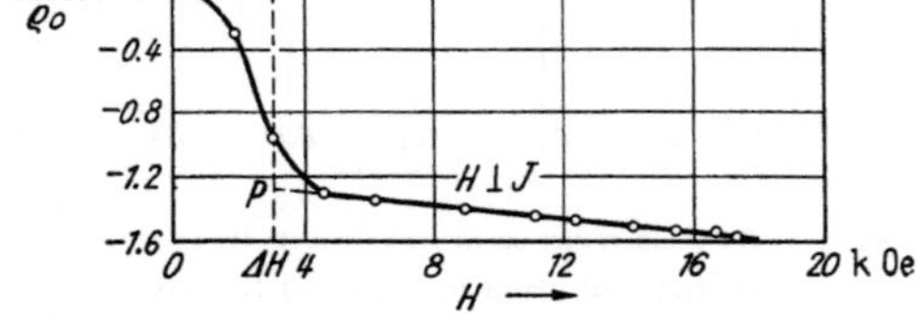


Fig. 28. Change of resistance of nickel in longitudinal and transverse fields (taken from Bozorth [B 4], Fig. 167).

For $T < \Theta_C$ the quantity $\Delta R(H)$ is positive for $H \parallel I$, and negative for $H \perp I$ (Fig. 28).

Again the domain theory allows an explanation of these curves. The resistivity increase consists besides the normal Lorentz increase, of a ferromagnetic part $(\Delta\varrho_H)_{\text{fer}}$, which is due to the total magnetization of the specimen. Now, the macroscopic specimen is considered to contain a great number of domains, each with its own resultant magnetization, in a direction making angles $\alpha_i$ with the current vector. For $(\Delta\varrho_H)_{\text{fer}}$, one may write

$$(\Delta\varrho_H)_{\text{fer}} = A + B\sum_i \cos^2\alpha_i. \qquad (16.1)$$

Since the resistance of a domain does not change by reversing the current, in (16.1) a linear term in $\alpha_i$ has to be omitted. For a uniform distribution of $\alpha_i$

at low or zero field $\cos^2 \alpha_i = \frac{1}{3}$, and the total magnetization of the specimen being zero $(\Delta \varrho_H)_{\text{fer}} = 0$.

When all domains are aligned parallel to the current, $\alpha = 0$ and $(\Delta \varrho_H)_{\text{fer}} = \Delta \varrho_l$, a maximum for the longitudinal effect. Thus (16.1) becomes:

$$(\Delta \varrho_H)_{\text{fer}} = \tfrac{3}{2} \Delta \varrho_l (\cos^2 \alpha - \tfrac{1}{3}) \tag{16.2}$$

and for the transverse effect ($\alpha = \pi/2$) the maximum

$$(\Delta \varrho_H)_{\text{fer}} = \Delta \varrho_t = -\tfrac{1}{2} \Delta \varrho_l. \tag{16.3}$$

At high temperatures, the LORENTZ effect is negligible as compared to the ferromagnetic effect, and the values $\Delta \varrho_l$ and $\Delta \varrho_t$ may be taken from Fig. 28, after accounting for the demagnetization of the specimen.

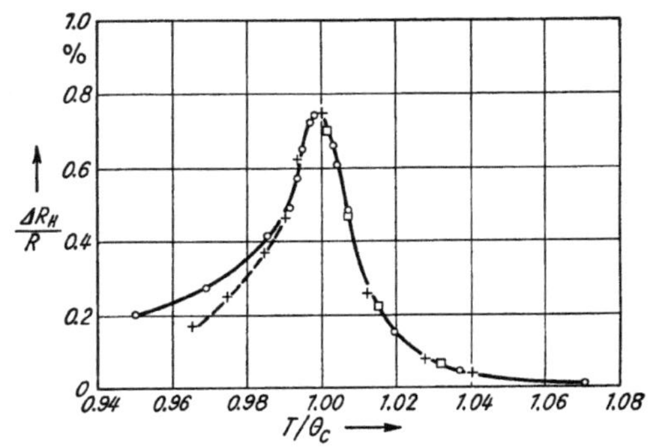


Fig. 29. Magneto resistance of iron in a longitudinal (○) and a transverse (□) field and magneto caloric effect (+, in arbitrary units) as a function of $T/\Theta_c$ (POTTER).

The true magnetic field $H$ in a magnetized body may be obtained from the applied field $H_a$ and the demagnetization factor $N/4\pi$, using the relation:

$$H = H_a - \Delta H = H_a - \frac{N}{4\pi}(B - H). \tag{16.4}$$

For Ni, $\Delta H = \frac{1}{2} B_s = 3050$ gauss saturation. Thus when $H_a = 3050$ gauss, the field $H = 0$ and there is no contribution to the increase of spontaneous magnetization and the value $\Delta \varrho_t$ (point P, Fig. 28) is found. For a wire for which a length to diameter ratio is 100, the demagnetization factor $N/4\pi \approx 10^{-4}$ when $H_a$ is parallel to the wire axis. Thus for $H_a = 0$, the $\Delta \varrho$ value is found (at the point Q, Fig. 28).

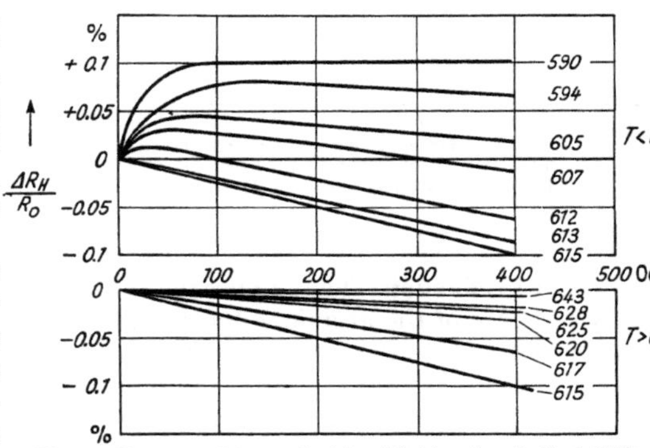


Fig. 30. Magneto resistance effect for nickel in a longitudinal field, below and above the CURIE-temperature (613° K). The numbers indicate the temperature in °K (GERLACH and SCHNEIDERHAN).

In the neighbourhood of the CURIE temperature, the resistivity increase is proportional to the energy which is connected with the formation of the intrinsic ferromagnetic state, as follows from experiments by POTTER [*172*], whose results for the measurements of the magneto-caloric energy near $\Theta_C$ are given in Fig. 29. In Fig. 30 the results of resistance measurements with nickel in the neighbourhood of $\Theta_C$ by GERLACH and SCHNEIDERHAN [*69*] are reproduced in order to illustrate the given discussions[1].

*γ) Resistivity of ferromagnetic crystals.* Magnetization experiments have shown that in ferromagnetic crystals there exists a direction of easiest magnetization, viz. the (100)-axis for iron and the (111)-axis for nickel. According to WEBSTER

[1] The magneto-resistance of cobalt at temperatures above 273° K is measured by R. DE MANDROT [Helv. phys. Acta **26**, 563 (1953)].

[*211*] for an iron crystal with magnetic field perpendicular to the current there exists a fourfold anisotropy in $\Delta R_H$ at 300° K for rods and current parallel to the (100)-axis, and a twofold anisotropy when grown parallel to the (110)-axis. In both cases, $\Delta R_H$ is negative for fields weaker than 5 kOe and also above 12 kOe. For fields between 5 kOe and 12 kOe the anisotropy increases, $\Delta R_H$ becoming positive for $I \parallel$ (100) and remaining negative for $I \parallel$ (110) or (111). For the case $H \parallel I$ the values $\Delta R_H$ are positive for the three directions, $\Delta R_H/R_0$ being 0.03, 0.21, and 0.33 for $I$ parallel to (100), (110) and (111), respectively. His results are reproduced in Fig. 31. In a strong field (154 kOe), Tanabe [*206*] for a crystal of iron at 300° K, having $I \parallel$ (100) and $H_{\parallel}$ (001) found $\Delta R_H/R_0 = 4 \times 10^{-3}$.

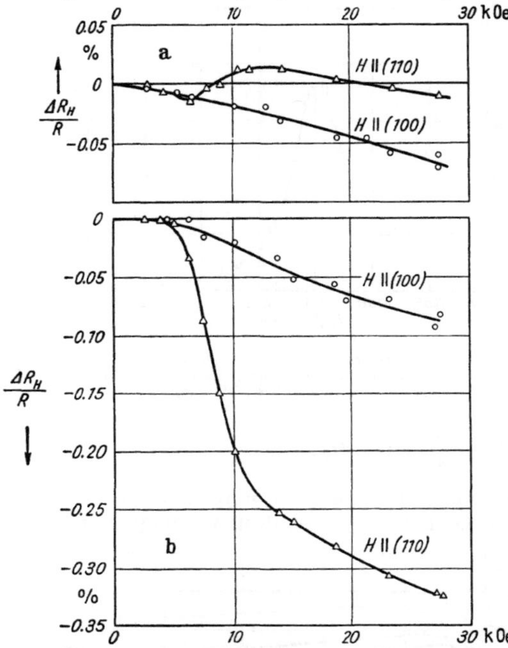


Fig. 31a and b. Magneto-resistance for iron crystals (Webster). (a) Crystal axis ∥ (100), current ∥ (100). (b) Crystal axis ∥ (110), current ∥ (110).

Kaya and Takahi [*116*] investigated single crystals of nickel. They found positive and negative values of $\Delta R_H$ for the transverse effect if the field was turned around an axis parallel to the current direction, the latter being parallel to the (100) and (110) axes, respectively, but a small constant negative value of the order of one per cent if the current was parallel to the (111)-axis. Calculations of the magneto-resistance effect, in particular the $\Delta \varrho_l$- and $\Delta \varrho_t$-values in terms of the angles between the current, field and crystallographic axes have been given by Döring [*42*] for nickel and Hironi and Hori [*94*] for iron.

## VI. The effect of pressure on the resistivity.

**17. Introduction.** The application of a hydrostatic pressure to a metal results in a decrease of the atomic volume, and to a decrease of the amplitudes of lattice vibrations. Since the resistivity is strongly dependent on the magnitude of these vibrations [cf. (9.5) and (9.6)], an analogous effect with respect to the resistivity can be expected as by lowering the temperature: that is, the resistivity will decrease. On the other hand, the electron energy bands and Brillouin zones may change their overlapping and their position with in the pattern of the crystallographic structure, thus influencing the resistivity in a manner not always predictable. After the first experiments at the beginning of this century [for example Chwolson (1881) Lissell (1902), Beckman (1911)] which yielded a resistivity decrease for the noble metals, several investigators have found that this decrease is not a general feature at all. Most results were obtained by Bridgman, whose high-pressure investigations on electric resistivity and other physical quantities have been an outstanding source of information for the last thirty years.

The experimental procedure consists in applying, by means of a suitable liquid, a hydrostatic pressure in a chamber where the resistance wire under investigation is mounted.

Apart from numerous difficulties arising from the high pressure technique as such, there is a natural limitation to the possibility of applying a hydrostatic pressure, since liquids under pressure will solidify above a certain pressure limit. Up to pressures of $3\times10^8$ kg./m.$^2$ at room temperature, a liquid like isopentane can be used. For higher pressures helium gas may be used, perhaps to about $10^9$ kg./m.$^2$, but BRIDGMAN already encountered enormous leakage difficulties when using this gas at $7\,.\,10^7$ kg./m.$^2$ at 90° K.

A solution has been found by applying mechanical pressure for the range between $3\times10^8$ and $10^9$ kg./m.$^2$, by using silver chloride as transmittant. In this case, however, one has to apply unknown corrections for shearing stress and deformation of the sample, a problem which BRIDGMAN solved experimentally by a determination of the resistivity in the pressure region between 2 and $5\times10^8$ kg./m.$^2$, by the hydrostatic and by the mechanical pressure method as well, and applying the correction factor thus determined to the results obtained at higher pressures. Though this method seems to be right in good approximation, the data for the highest pressures are to be considered as less accurate.

A scheme of the method [*18 l*] is given in Fig. 32. The two "carboloy" blocks $A$ and $B$ are pressed together in a hydraulic press. A thin spacing was left between them in which a carefully adjusted ring of pipestone prevented the AgCl from blowing out. The wire under investigation was embedded into the AgCl, and electric contact with the blocks was made mechanically. The specimen was connected in a single circuit, the potential over the wires being read with a normal high sensitivity ZERNIKE galvanometer. For the many technical and experimental problems which had to be solved, such as the determination of the actual pressure by a calibrated manganine gauge and the difficulties arising for the alkali metals in these experiments, we refer to the original paper. For calculation of the pressure coefficient of resistivity, defined as

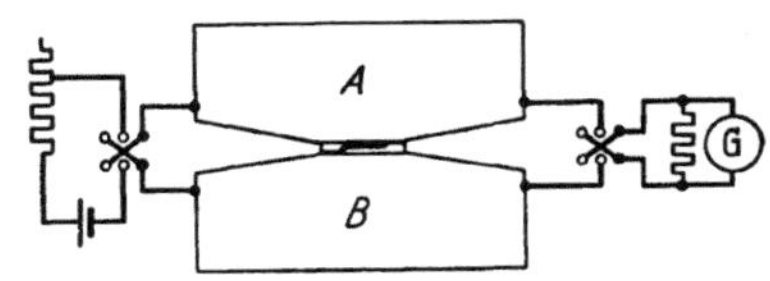


Fig. 32. Sketch of BRIDGMAN's measuring method (Fig. 2 of [*18 l*]).

$$\alpha_{\varrho,P} = -\frac{1}{\varrho}\frac{\partial\varrho}{\partial P}, \tag{17.1}$$

the volume change of the sample has to be taken into account. From the relation:

$$\frac{1}{R}\frac{\partial R}{\partial P} = \frac{1}{\varrho}\frac{\partial\varrho}{\partial P} + \frac{1}{l}\frac{\partial l}{\partial P} - \frac{1}{q}\frac{\partial q}{\partial P},$$

$l$ and $q$ being length and area of cross section, respectively, it follows for cubic metals for which $\frac{\partial \log q}{\partial P} = 2\frac{\partial \log l}{\partial P}$ that

$$\frac{1}{\varrho}\frac{\partial\varrho}{\partial P} = \frac{1}{R}\frac{\partial R}{\partial P} + \frac{1}{l}\frac{\partial l}{\partial P}. \tag{17.2}$$

For non-cubic metals, and for the case that the current is parallel to the principal axis, there follows the expression:

$$\left(\frac{1}{\varrho}\frac{\partial\varrho}{\partial P}\right)_{\parallel} = \left(\frac{1}{R}\frac{\partial R}{\partial P}\right)_{\parallel} - \left(\frac{1}{l}\frac{\partial l}{\partial P}\right)_{\parallel} + 2\left(\frac{1}{l}\frac{\partial l}{\partial P}\right)_{\perp}, \tag{17.3}$$

and if the current is perpendicular to the crystal axis one obtains:

$$\left(\frac{1}{\varrho}\frac{\partial\varrho}{\partial P}\right)_{\perp} = \left(\frac{1}{R}\frac{\partial R}{\partial P}\right)_{\perp} + \left(\frac{1}{l}\frac{\partial l}{\partial P}\right)_{\parallel}. \tag{17.4}$$

Thus, a knowledge of the compression coefficients $\left(\frac{1}{l}\frac{\partial l}{\partial P}\right)$ for the directions parallel and perpendicular to the principal axes is necessary. In the case of an quasi-isotropic non-cubic metal $-\frac{1}{l}\frac{\partial l}{\partial P} = -\frac{1}{3}\left(\frac{1}{V}\frac{\partial V}{\partial P}\right)$, where $\left(\frac{1}{V}\frac{\partial V}{\partial P}\right)$ is the volume compressibility.

**18. Experimental results.** BRIDGMAN determined the pressure effect at room-temperature for 47 metallic and 5 non-metallic elements up to $10^9$ kg./m.$^2$, employing the mechanical method, described before.

Table 16. *Pressure coefficient of resistance $\alpha_{R,P}$ (in $10^{-10}$ m.$^2$/kg.) according to* BRIDGMAN.

| Metal | Temp. °K | $p$ in $10^8$ kg./m.$^2$: 0 | 1 | 3 | |
|---|---|---|---|---|---|
| Li | 303 | −7.00 | −7.52 | −9.00 | I |
| Na | 303 | 58.8 | 23.6 | 4.04 | I |
| K | 303 | 134.4 | 30.0 | 0.88 | I |
| Rb | 303 | 154.0 | 14.4 | −28.8[1] | I |
| Be | 298 | 1.77 | 1.63 | 1.46 | II |
| Mg | 298 | 5.40 | 4.67 | 3.81 | II |
| Ca | 303 | −9.48 | −12.2 | −20.7 | I |
| Sr | 303 | −45.6 | −59.0 | −118.8 | I |
| Ba | 303 | 7.2 | 1.2 | −13.6[2] | I |
| Al | 301 | 4.29 | 4.06 | 3.60 | II |
| Ti | 296 | 1.19 | 1.12 | 1.02 | II |
| Cr | 298 | 22.2 | 17.3 | 8.96[3] | II |
| Fe | 303 | 2.42 | 2.26 | 1.90 | I |
| Co | 297 | 0.96 | 0.90 | 0.80 | II |
| Ni | 298 | 1.77 | 1.82 | 1.73[4] | II |
| Zr | 299 | 0.39 | 0.33 | 0.22 | II |
| Nb | 297 | 1.40 | 1.37 | 1.30 | II |
| Mo | 300 | 1.31 | 1.29 | 1.24 | II |
| Rh | 299 | 1.65 | 1.62 | 1.56 | II |
| Pd | 299 | 2.10 | 2.04 | 1.93 | II |

| Metal | Temp. °K | $p$ in $10^8$ kg./m.$^2$: 0 | 1 | 3 | |
|---|---|---|---|---|---|
| Ta | 302 | 1.62 | 1.62 | 1.55 | II |
| W | 302 | 1.33 | 1.31 | 1.25 | II |
| Ir | 296 | 1.39 | 1.37 | 1.33 | II |
| Pt | 296 | 1.92 | 1.88 | 1.78 | II |
| Cu | 303 | 1.92 | 1.80 | 2.42 | I |
| Ag | 303 | 3.48 | 3.28 | 2.60 | I |
| Au | 303 | 3.02 | 2.84 | 2.44 | I |
| Zn ∥[5] | 303 | 9.68 | 8.76 | 6.72 | I |
| Zn ⊥ | 303 | 5.28 | 4.40 | 2.84 | I |
| $Hg_{liq}$ | 303 | 23.1 | 17.0[6] | | I |
| $Hg_{sol}$ | 303 | — | | 12.0 | I |
| In | 296 | 1.25 | 1.09 | 0.85 | II |
| Sn ∥[7] | 303 | 10.0 | 9.0 | 6.1 | I |
| Sn ⊥ | 303 | 9.24 | 8.26 | 5.61 | I |
| Pb | 303 | 13.7 | 11.6 | 6.96 | I |
| Sb 1[8] | 303 | −9.84 | −14.8 | −2.80 | I |
| Sb ⊥[9] | 303 | −5.60 | −3.36 | +12.7 | I |
| Bi[10] | 303 | −14.8 | −18.5 | | I |
| U | 293 | 4.88 | 4.56 | 4.10 | III |
| Ce[11] | 297 | −4.1 | | 1.6 | II |
| Pr | 297 | 1.36 | 1.20 | 1.02 | II |
| Nd | 297 | 1.57 | 1.32 | 1.03 | II |

I BRIDGMAN 1938 [*18 h*].
II BRIDGMAN 1951 [*18 k*].
III BRIDGMAN 1952 [*18 l*].

[1] Minimum at $P = 1.5 \times 10^8$ kg./m.$^2$

[2] Break at $P = 1.7 \times 10^8$ kg./m.$^2$.

[3] Cusp at $P = 0.34 \times 10^8$ kg./m.$^2$ $\alpha_{R,P} = 28.8$.

[4] Cusp at $P = 1.0 \times 10^8$ kg./m.$^2$ $\alpha_{R,P} = 1.8^2$.

[5] Single crystal, angle between principal axis and length 17° for ∥, 87° for ⊥.

[6] Solidification at $P = 1.35 \times 10^8$ kg./m.$^2$.

[7] As [5] but angles 17° and 82°, respectively.

[8] Single crystal, angle between principal axis and length 41°; maximum at $P = 3.1 \times 10^8$ kg./m.$^2$.

[9] As [8] but angle is 87° and maximum at $P = 1.45 \times 10^8$ kg./m.$^2$.

[10] Single crystal, axis inclined 78° to length. $R_P/R_0$ decreases from 1.50 to 0.24 at $P = 2.47 \times 10^8$ kg./m.$^2$ This is followed by an increase of $R_P/R_0$ to 0.62 at $P \approx 2.62 \times 10^8$ kg./m.$^2$.

[11] Transition at $0.5 \times 10^8 < P < 1 \times 10^8$ kg./m.$^2$, jump in $R/R_0$ from 1.027 to 0.582.

The results are given in Fig. 33, the values have been taken from the tables in BRIDGMAN'S paper [*18l, m*]. Table 16 gives the values of the pressure coefficient of resistance, that is in analogy to (17.1):

$$\alpha_{R,P} = -\frac{1}{\Delta P}\frac{\Delta R}{R_0} \tag{17.5}$$

for several elements at room temperature for the pressures 0, $10^8$ and $3\times 10^8$ kg./m.² [*18h, k, l*], which have been found with a liquid as a transmittant, whereas Table 17 gives the average pressure coefficient (for the pressure range from zero to $0.7\times 10^8$ kg./m.²) at different temperatures [*18d, m*].

Table 17. *Average pressure coefficients of resistance $\alpha_{R,P}$ (in $10^{-10}$ m.²/kg.) for pressures up to $0.7\times 10^8$ kg./m.², at different temperatures, according to* BRIDGMAN [*18d*].

| Metal | 273.15° K | 194.75° K | 90.25° K |
|---|---|---|---|
| Mg | 4.39 | 4.49 | 5.89 |
| Al | 4.28 | 4.71 | 9.16 |
| Fe | 2.34 | 2.27 | 2.44 |
| Ni | 1.85 | 2.00 | 1.88 |
| Nb | 1.18 | 0.98 | 0.80 |
| Mo | 1.30 | 1.29 | 1.91 |
| Rh | 1.64 | 1.86 | 2.26 |
| Pd | 2.13 | 2.32 | 2.82 |
| Ta | 1.45 | 1.42 | 1.17 |
| W | 1.37 | 1.42 | 1.36 |
| Pt | 1.93 | 1.97 | 2.34 |
| Cu | 1.88 | 2.14 | 3.09 |
| Ag | 3.45 | 3.46 | 4.09 |
| Au | 2.94 | 2.97 | 3.27[1] |
| Pb | 12.99 | 12.88 | 12.76 |

Concerning the general conclusion on the pressure effect, let us quote BRIDGMAN's conclusion from his 1952 paper [*18l*] (see Table 16).

"Of the 47 metallic elements, 31[2] exhibit 'normal' behaviour of the resistance under pressure, that is, a resistance decreasing with pressure with upward curvature. Included in these 31 normal metals are four with changes, mercury with one freezing and one polymorphic transition, gallium with one freezing and one transition, thallium, with one transition and zirconium with one transition. The discontinuity of resistance at the phase changes of these four metals also behaves 'normally', that is, resistance follows the volume, decreasing with it, at the phase change. The high pressure phases of these four elements also show normal curvature (presumptive for the high pressure form of zirconium)."

BRIDGMAN then mentions the obtuse cusps at low pressures for chromium and nickel and continues with his conclusion about the abnormal metals:

"The 16[3] 'abnormal' metals show a wide variety of behaviour. In the first place, at the phase changes of these metals resistance does not universally follow the volume, but there are three examples where resistance increases with a decreasing volume: calcium (at 22000 kg./cm.²), barium and bismuth. The simplest 'abnormality' in a resistance increasing with pressure."

From earlier measurements (up to $3\times 10^8$ kg./m.²) BRIDGMAN had concluded that each metal would show an analogous behaviour as was found for some alkali metals (cf. Fig. 33), that is, a resistance decrease at low pressures, then a minimum, passing on to an increase of $R$ with pressure. The extension of the measurements up to $10^9$ kg./m.² has not proved this prediction. In the case of Cs there has been found a change of the first order at $4.5\times 10^8$ kg./m.² corresponding with a discontinuity in volume (BRIDGMAN [*18n*]). In the case of Sr, however, the maximum at $4.6\times 10^8$ kg./m.² does not seem to be connected with volume changes at all, such volume changes have, however, been found

[1] max. pressure $0.43\times 10^8$ kg./m.².

[2] Viz. the following metals: Cu, Ag, Au, Be, Mg, Zn, Cd, Hg, Al, In, Tl, Ti, Zr, Th, Sn, Pb, V, Nb, Ta, Cr, Mo, W U, Fe, Ni, Co, Rh, Pd, Ir, Pt, Ga ($p > 4\times 10^8$ kg./m.²).

[3] Viz. the metals Li, Na, K, Rb, Cs, Ca, Sr, Ba, La, Ce, Pr, Nd, As, Sb, Bi, Mn.

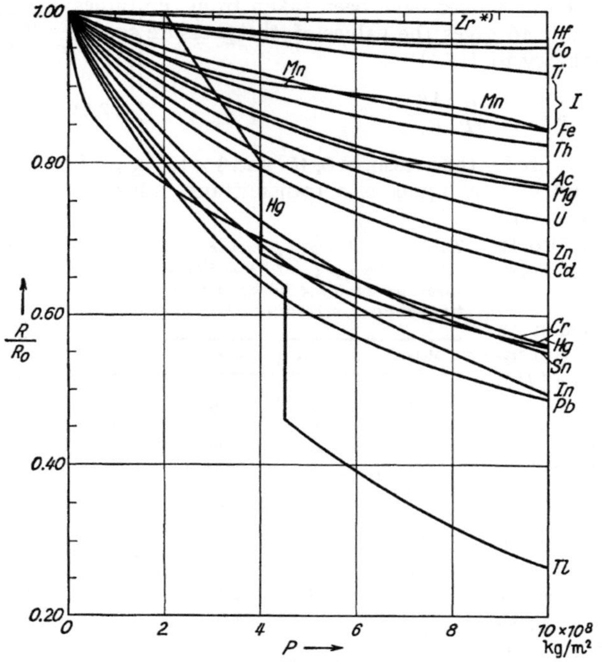


Fig. 33a.

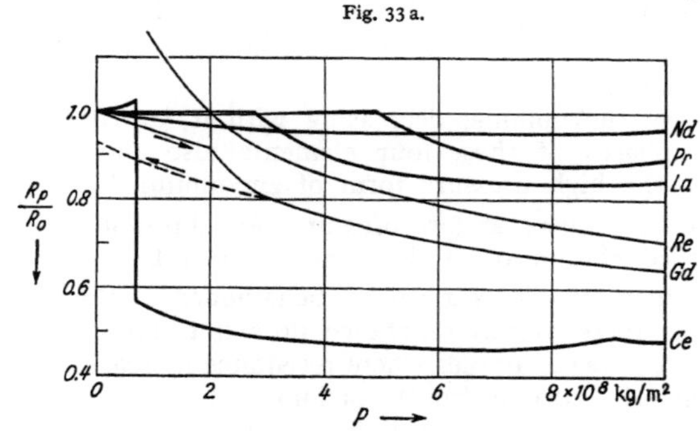


Fig. 33b.

at pressures of 2.5 and $6.5\times10^8$ kg./m.$^2$ [*18i*]. The abnormalities found for the rare earth metals are all different as Fig. 33b shows (cf. Table 9 of [*18l*]).

For arsenic, no good reproducible results were obtained. Bismuth has a number of polymorphic transitions, including two with large volume changes (3 to 5%) at 2.56 and $2.71\times10^8$ kg./m.$^2$ [*18i*]. The transitions with smaller volume changes at about 4.8, 6.5 and $9.0\times10^8$ kg./m.$^2$ did not show up in the resistance pressure curve. The abnormal behaviour of manganese needs a cautious consideration because of high impurity contents (Ni 1%, Cu 2%) of the specimen. For the non-cubic metals, effects related with crystal orientation occur. For some metals and some single crystals (Zn, Cd, Sn, Bi, Sb, As) BRIDGMAN [*18 e*, p. 95] investigated the influence of crystal orientation and temperature down to 90° K on the pressure effect. These experiments were carried out with pressures up to $0.7\times10^8$ kg./m.$^2$ and helium as a hydrostatic transmittant. We give the results in terms of the change of the anisotropy ratio $F=\varrho_\parallel/\varrho_\perp$, ($\varrho_\parallel$ and $\varrho_\perp$ being the resistivities for crystal rods with their axis parallel and perpendicular to the principal axis) per unit pressure, i.e. $\frac{1}{F}\frac{\partial F}{\partial P}$ in Table 18.

It follows that there is no large temperature influence, being largest for low melting metals [*18d, e*]. This is in line also with results to 475° K [*18m*]. Generally, except for niobium and tantalum, which were rather impure, an increase of the coefficient has been found with decreasing temperature. FISHER [*61*] found an increase for lead below 80° K, which implied a relation with a certain value of $T/\Theta_D$ as a critical value.

For single crystals, an interpretation of the results of resistance measurements in different directions with respect to the crystal axes is difficult, since the measured resistance of a single crystal rod has to be converted into the effect of pressure on the resistivity in the direction under investigation. Correction has to be applied for the effect of the pressure on the linear dimensions of the rod, and also for changes of the angle between the axis of the rod and the crystal axis, the latter being due to the anisotropy in the elastic coefficients.

BRIDGMAN [*18e*] has derived the fractional resistance changes for both directions. These corrections, depending on the differences between the elastic constants, are not always small. For cadmium, for example, they amount to 37% of the total effect[1]. From his data, it follows that pressure tends to equalize the resistance in different directions for Zn, Cd, and Sn, but to accentuate them for Bi and Sb. The equalizing effect becomes less marked at low temperatures for Zn and Cd (cf. Table 18).

For liquid metals (alkali metals and mercury) the ratio $R_{\text{liq}}/R_{\text{solid}}$ does not change under pressures up to $10^8$ kg./m.$^2$ [*18*]. From theoretical consideration the decrease of resistance with pressure can be explained and evaluated in order of magnitude. BRIGDMAN'S [*18d*] experimentally found relationship between

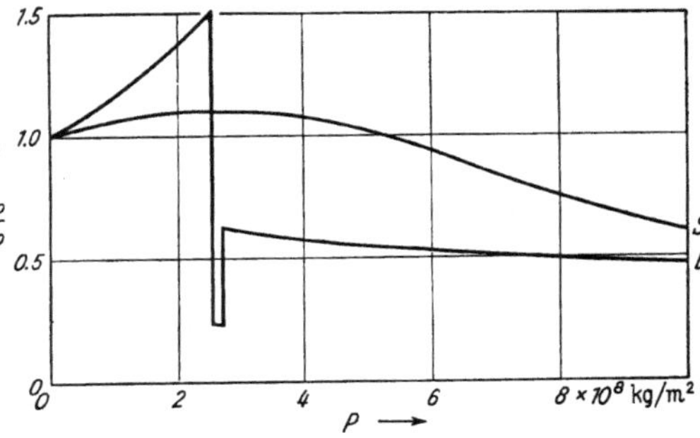


Fig. 33c.

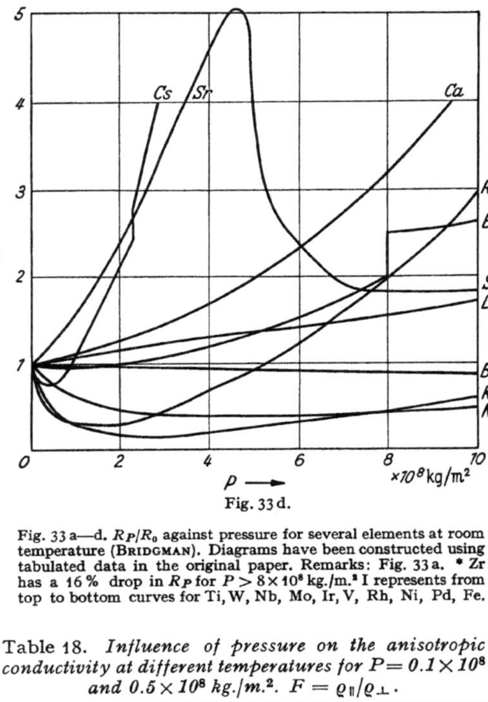


Fig. 33d.

Fig. 33a—d. $R_P/R_0$ against pressure for several elements at room temperature (BRIDGMAN). Diagrams have been constructed using tabulated data in the original paper. Remarks: Fig. 33a. * Zr has a 16% drop in $R_P$ for $P > 8 \times 10^8$ kg./m.$^2$ I represents from top to bottom curves for Ti, W, Nb, Mo, Ir, V, Rh, Ni, Pd, Fe.

Table 18. *Influence of pressure on the anisotropic conductivity at different temperatures for $P = 0.1 \times 10^8$ and $0.5 \times 10^8$ kg./m.$^2$. $F = \varrho_{\parallel}/\varrho_{\perp}$.*

| | $T = 195°$ K | | $T = 90°$ K | |
|---|---|---|---|---|
| | $\left(\frac{1}{F}\frac{\partial F}{\partial P}\right)_{0.1}$ | $\left(\frac{1}{F}\frac{\partial F}{\partial P}\right)_{0.5}$ | $\left(\frac{1}{F}\frac{\partial F}{\partial P}\right)_{0.1}$ | $\left(\frac{1}{F}\frac{\partial F}{\partial P}\right)_{0.5}$ |
| Zn | −3.91 | −3.35 | −3.37 | −2.92 |
| Cd | −3.18 | −2.50 | −1.99 | −1.79 |
| Sn | −0.19 | −0.07 | −0.88 | −1.08 |

$\Delta F/F$ values:

| | $P = 0.3$ | $P = 0.7$ | $P = 0.3$ | $P = 0.7$ |
|---|---|---|---|---|
| Bi | +0.032 | +0.097 | +0.053 | +0.067 |
| Sb | +0.058 | +0.157 | +0.040 | +0.107 |

[1] The fractional increase of resistance under unit increment of pressure arising from change of dimensions [*18e*] is:

$$2(S_{11} + S_{12})\cos^2\varphi + 2S_{13}\sin^2\varphi + S_{33}(1 - 2\cos^2\varphi);$$

that from the change of angle produced by pressure is:

$$2\,\frac{\varrho_{\perp} - \varrho_{\parallel}}{\varrho}\sin^2\varphi\cos^2\varphi\,(S_{33} + S_{13} - S_{11} - S_{12}),$$

$S_{ik}$ being elastic constants, $\varphi$ the angle between crystal axis and length.

pressure coefficient and temperature has been discussed by GRÜNEISEN [83] for normal metals under small pressure.

**19. Influence of tension.** So far, we have discussed the conductivity changes under the influence of an overall pressure. The effects due to a pressure applied in one direction can be investigated by applying a tension to a wire. In the direction of the tension an elongation takes place and perpendicular to it a contraction. The tensile force should be so low that the deformations remain elastic.

Suppose that on a parallelopiped of an isotropic conductor with edges $x_1$, $x_2$ and $x_3$, a tensile force $F_1$ is applied perpendicular to the plane $x_2 x_3$. This force causes an extension in the $x_1$-direction and contractions in the $x_2$- and $x_3$ direcotins. Since

$$\varrho_1 = R_1 \frac{x_2 x_3}{x_1},$$

we obtain

$$\frac{1}{\varrho_1}\frac{\partial \varrho_1}{\partial F_1} = \frac{1}{R_1}\frac{\partial R_1}{\partial F_1} - \frac{1}{x_1}\frac{\partial x_1}{\partial F_1} + \frac{1}{x_2}\frac{\partial x_2}{\partial F_1} + \frac{1}{x_3}\frac{\partial x_3}{\partial F_1}.$$

Since

$$\frac{1}{x_1}\frac{\partial x_1}{\partial F_1} = \frac{1}{E}, \qquad \frac{1}{x_2}\frac{\partial x_2}{\partial F_1} = \frac{1}{x_3}\frac{\partial x_3}{\partial F_1} = -\frac{\mu}{E}$$

($E$ and $\mu$ being respectively the elasticity modul and POISSON's ratio of extension and contraction) it follows for $(\alpha_{\varrho_1 F})_l$, the longitudinal tension coefficient of resistivity:

$$(\alpha_{\varrho_1 F})_l = \frac{1}{\varrho_1}\frac{\partial \varrho_1}{\partial F_1} = \frac{1}{R_1}\frac{\partial R_1}{\partial F_1} - \frac{1+2\mu}{E}.$$

For a tension $F_2$ in the $x_2$-direction $(\alpha_{\varrho_1 F})_t$, the transverse coefficient of resistivity can be found by an analogous procedure, to be given by

$$(\alpha_{\varrho_1 F})_t = \frac{1}{\varrho_1}\frac{\partial \varrho_1}{\partial F_2} = \frac{1}{R_1}\frac{\partial R_1}{\partial F_2} + \frac{1}{E}.$$

When $E$ and $\mu$ are known, the longitudinal and transversal coefficients can be found from the measured values of $\frac{1}{R_1}\frac{dR_1}{dF}$.

The physical analogy between a metal under hydrostatic pressure and under tension being evident, BRIDGMAN has suggested that the pressure coefficients $\alpha_{\varrho_1 P} = \frac{1}{\varrho}\frac{\partial \varrho}{\partial P}$ and the longitudinal and transverse resistivity coefficients for tension should be related. Experiments by TOMLINSON (1883) and BRIDGMAN [*18b*, p. 423] have shown, that $(\alpha_{\varrho_1 F})_l$ and $(\alpha_{\varrho_1 F})_t$ are positive for most metals investigated (compare Table 19, $F$ has been expressed in kg./m.$^2$). BRIDGMAN's results are given below.

Table 19. *Tension coefficients for resistivity* (BRIDGMAN, [*18b*]).

| Metal | $\frac{1}{E}\,10^{13}$ | $\mu$ | $10^{10}\,(\alpha_{\varrho_1 F})_l$ | $10^{13}\,(\alpha_{\varrho_1 F})_t$ |
|---|---|---|---|---|
| Cu | 0.81 | 0.34 | +1.75 | −2.4 |
| Ag | 1.26 | 0.38 | +2.86 | −0.04 |
| Au | 1.25 | 0.42 | +3.87 | −4.9 |
| Fe | 0.48 | 0.28 | +1.42 | +0.54 |
| Pd | 0.89 | 0.39 | +1.37 | +0.51 |
| Pt | 0.59 | 0.39 | +1.78 | +0.30 ($F_\parallel$ rolling)<br>+0.78 ($F_\perp$ rolling) |

For single crystals, ALLEN's results for bismuth are mentioned here. For a tension parallel to the axis of the crystal along which the current also flows, the coefficient depends on the orientation of principal and binary axes with respect to the direction of the current [*18b*]. BRIDGMAN suggests a correlating of the several coefficients by calculating a number of constants which connect the resistance with stress through equations similar to those which connect strain with stress.

## VII. Influence of lattice defects on the resistance.

**20. Earlier cold work and total recovery experiments.** Following BERGMANN (1889), COKE (1890) and CREDNER (1918) it has been common knowledge that, in general, the resistivity of a metal is higher when cold-worked than in an annealed state. Further mechanical evidence beside this general rule has led to casual investigation of cold work and recovery (HOLBORN [*95*] and, for example, by DAMMERS [*119*]) of platinum for selecting suitable resistance thermometer wires. DAMMERS found that annealing of platinum wires at several temperatures up to 1073° K resulted in a decrease of resistivity and an increase of the temperature coefficient at 273° K. Annealing above 1073° K, on the other hand, gave reversed results. A simple demonstration of this phenomenom has been described by MEISSNER [*M 1*].

More systematic investigations have been reported by TAMMANN and DREYER [*205*] and by WEYERER [*213*]. The increase in hardness and resistivity due to rolling at room temperature is reduced to zero by annealing at about 500° K for Cu, Ag, Au, Pt and Pd. Cu and Fe [*213*] have a minimum of room temperature resistivity when the annealing temperatures are respectively below 620° K and 1020° K, these latter being their recrystallization temperatures, as follows from X-ray evidence.

A table from MEISSNER [*M 1*] for the temperatures required for total mechanical and electrical recovery is given here (Table 20) as a useful outcome of the experiments described so far.

Table 20. *Annealing temperatures for total recovery for some metals (taken from* MEISSNER).

| Metal | Annealing temperature in °K | Metal | Annealing temperature in °K |
|---|---|---|---|
| Pb | 470 | Cu | 650 |
| Zn | 470 | Ag | 770 |
| Cd | 490 | Au | 670 |
| Bi | 470 | Pt | 1020 |
| Al | 520 to 620 | W | 1850 |
| Fe | 590 to 670 | | |

Results of experiments on the mechanical properties of solids led to an extensive study of crystal imperfections, such as dislocations, interstitial ions and vacancies[1]. According to TAYLOR [*207*], dislocations move in the crystal at high temperatures and may get stuck, for example, when two mutually perpendicular dislocations meet. Since their diffusion decreases at lower temperatures in any case, a permanent influence of dislocations on electrical resistivity at temperatures below the recrystallization temperature can be expected. A line or a plane of dislocations can be expected to influence an electron wave just as impurities occuring in a line of a plane. The same holds, in a first approximation, for some other imperfections such as interstitial ions and vacancies. In the following, experiments will be described which have been chosen more or less arbitrarily and which may give an impression of the phenomenons involved[2].

**21. Elastic and small plastic deformation.** A metal can be measurably deformed by extension, drawing (through dies) and by rolling. DRUYVESTEYN [*46*] and

[1] For a discussion of imperfections in solids and for further references see SLATER'S paper in the present volume and the detailed discussions in vol. VII of this Encyclopedia.

[2] Cf. [*B 5*], [*S 1*] for a more complete review.

coworkers developed a method for measuring the influence of deformation in the elastic range on the resistivity of foils. They designed an apparatus with which the total change in resistance after deformation could be separated into the two changes due to the deformations in and perpendicular to current direction. A foil $F$ (see Fig. 34) was glued on a copper plate ($PP'$) resting on two bars a and b, which were supported by knife edges. Pushing down the ends $P$ and $P'$ by means of a lever, a bending moment was applied to the plate and to the foil. Care was taken that the bars a and b had such dimensions, that they got the same curvature as the plate. Hence, a bending moment in the plate occurred only in one direction and not in a direction perpendicular to it. By varying the length direction—in which the foil was rolled—over 90° with the direction $PP'$ the

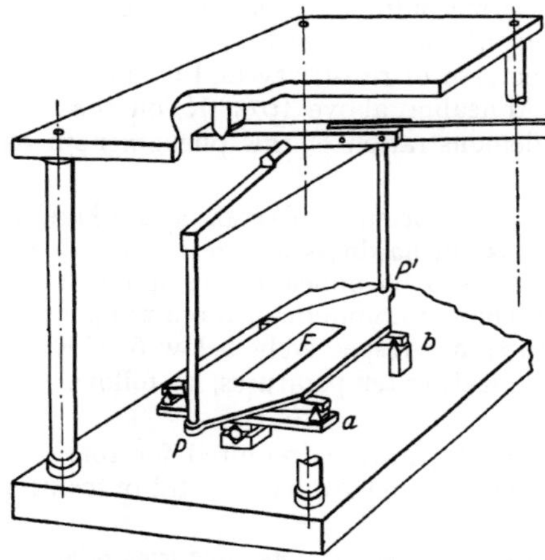


Fig. 34. DRUYVESTEYN's apparatus for deforming foils elastically.

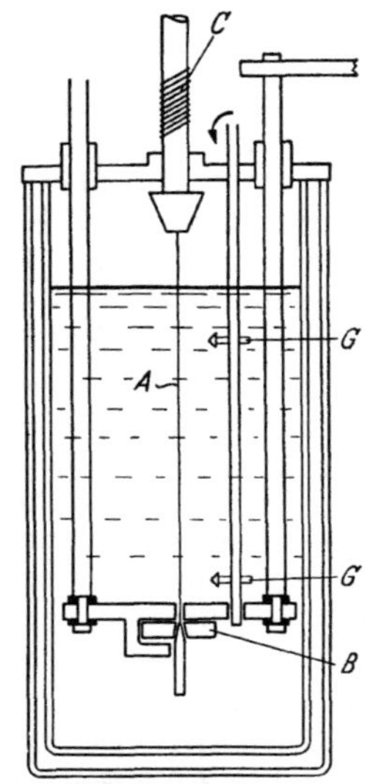


Fig. 35. Apparatus for drawing wires at controlled temperatures (BROOM).

anisotropic change of resistance due to elastic deformation could be studied for isotropic and anisotropic (single crystal) conductors as well. The resistance was measured in the length direction of the foil, giving a value for $\Delta R_{xx}$. The curvature of the surface was determined by measuring light beam deflections from mirrors glued on the plate. For some cases DRUYVESTEYN succeeded in determining the coefficients $k$ in the bi-tensor: $\Delta \varrho_{ij}/\varrho = k_{klij}\,\varepsilon_{kl}$, $\varepsilon_{kl}$ being the symmetric deformation tensor.

For extended wires of pure poly-crystalline metals, the following fractional increase of resistance has been found at room temperature for a 10%-extension [*B 5*].

| | Al | Fe | Cu | Ag | Au | Pt | |
|---|---|---|---|---|---|---|---|
| $10^2 \dfrac{\Delta\varrho}{\varrho}$ | 0.4 | 0.3 | 0.55 | 0.5 | 0.75 | | at 273° K |
| | | | 3.9 | 4.2 | 3.2 | 3.1 | at 90° K |

For lower deformation temperatures (AARTS-JARVIS [*1*] and MANINTVELD [*141*][1] the increase is larger as is demonstrated by some results of MANINTVELD in the table above.

[1] Fig. 4

**22. Deformation experiments in the plastic region.** Plastic deformation, affecting a less surveyable situation in the specimen, can be obtained by wire drawing and by rolling.

The effect of large deformations is best illustrated by BROOM's experiments [*20*] for two cubic metals, copper and nickel. His arrangement was as follows (Fig. 35): A wire $A$ was drawn upward by means of a screw mechanism ($c$) through a die $B$,

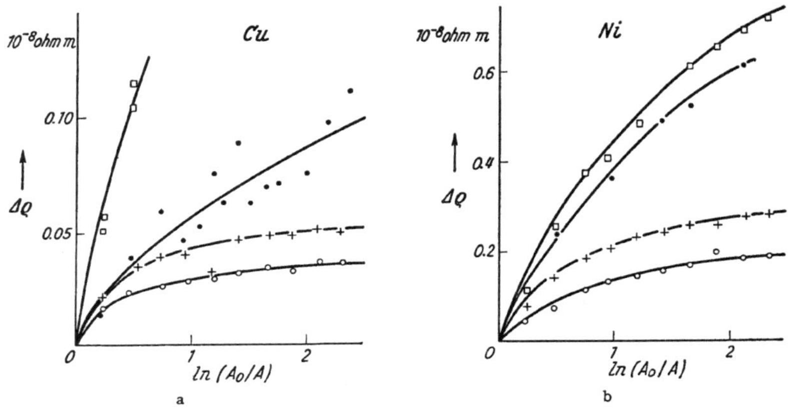


Fig. 36a and b. Effect of temperature and degree of deformation on the resistivity of copper and nickel (BROOM). Index: ○ 373° K, + 273° K, ● 195° K, □ 90° K.

that was placed in a liquid of desired temperature. Two knife edges $G$ connected to a KELVIN double bridge were mounted so that they could be moved on to the wire after a drawing. The resistance of the freshly drawn wire was measured. Then it was put into another dewar with liquid at the same temperature as in the drawing vessel and, keeping its one end at room temperature, moved into a smaller die, the latter then being placed into the drawing vessel ready for the next drawing experiment.

The part of the wire that had been at room temperature was cut off. Since the shifting from one bath to another lasted only two seconds, BROOM estimated that no recovery (see Sect. 25) could have taken place. The investigated specimens were wires of Al, Cu, Ni, Fe and of some alloys. Results for copper and nickel are given in Fig. 36. In Table 21 the values of maximal resistance increase at 90° K are tabulated. The relatively small increase of resistance at room temperature is due to recovery processes, which are faster at this temperature than at 90° K. A discussion of the recovery will be given in Sect. 25. PRY and HENNING [*175*] found that for very pure copper (99.999%), in which grains were rendered almost uniform by a suitably chosen heat treatment, the cold drawing experiments at 78 and 186° K showed that tensile stress for plastic flow, the resistance and the strain rate were uniquely related.

Table 21. *Maximum fractional increase $\Delta\varrho/\varrho$ for deformed (drawn) wire at 90° K* (BROOM).

| Metal | % deformation | $10^2\,\Delta\varrho/\varrho$ | $10^2\,\Delta\varrho/\varrho$ pro % deformation |
|---|---|---|---|
| Cu | 37 | 38 | 1.0 |
| Ni | 90 | 40 | 0.44 |
| Fe | 20 | 6 | 0.30 |
| 75/25 Brass | 90 | 69 | 0.76 |
| 50/50 Ag-Au | 37 | 4 | 0.11 |
| $Ag_3Mg$ ordered | 37 | 112 | 3.0 |
| $Ag_3Mg$ disordered | 50 | 11 | 0.22 |

In Fig. 37 we reproduce a diagram which gives the resistance increase as a function of the slip of cold worked copper wires, as measured by several authors. It is obvious, that either the experimental conditions have not been comparable, or the material used has been very different with respect to impurity content and imperfections besides those resulting from the deformation $\varepsilon$. In general the resistance increase $\Delta\varrho$ is given by:

$$\Delta\varrho = a\,\varepsilon^{p}$$

with $a \approx 0.1 \times 10^{-8}$ ohm m., $p \approx 1.3$ and $\varepsilon$ in per cent [1].

Torsion of the wire also leads to a resistance increase. VAN BUEREN and JONGENBURGER [25] found a 20% increase for polycrystalline silver when the wire was twisted at 20° K through an angle corresponding to an effective slip of 10%, this increase falling by an order of magnitude when copper and silver were twisted at 77° K.

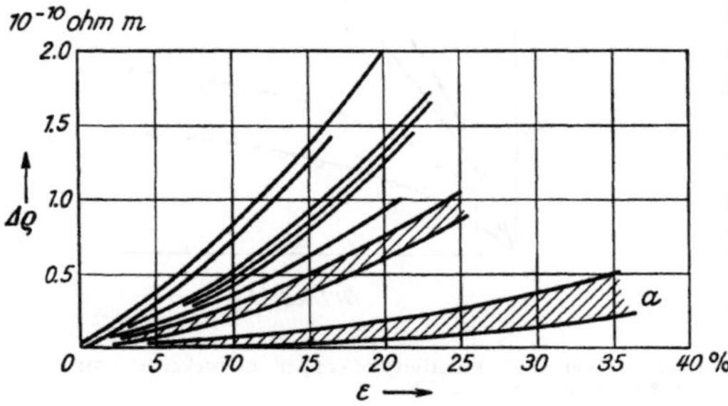


Fig. 37. A diagram, showing the resistivity increase as a function of the deformation for copper wires as measured by several authors. Polycrystalline wires, except for group *a* (diagram taken from VAN BUEREN [*B 6*]).

The extra resistance, due to coldwork, can be described formally as a resistance increase ascribed to an impurity. It is not clear, at the moment, whether MATTHIESSEN's rule, which is a consequence of this assumption, should be valid. According to LÜCKE [*138*] the rule should hold for copper, which material was investigated after different heat treatments before the cold drawing process. BERGHOUT [*11*] comes to the same conclusion for the metals Cu and Al, but he claims to have found deviations from the rule for Au and Ag. OGILVIE and CLOTHIER [*164*] claiming a sensitivity in $\varrho$ of $10^{-5}$, the same as claimed by LÜCKE, also find deviations from MATTHIESSEN's rule, for copper, aluminium, bronze and 80/20 brass. JONGENBURGER's experiments [*105*] are interesting in this connection, because for pure copper, drawn to 5 and 10% elongation, he finds that the magnetoresistance of cold worked wire does not follow the KOHLER relation (at low $H/r_T$ values) for pure copper. This might imply that the analogy between extra resistivity due to impurity and that due to cold work is not so good as one would like to believe.

For some metals a resistance decrease has been found after stretching or drawing the wire. For pure iron, COTTRELL and CHURCHMAN [*34*] found such an effect, which is probably a recovery effect (see Sect. 25). WEYERER [*213*] found a decrease for some commercial copper. For solid solutions of NiCr (ROHM [*181*]), AuCr and AuFe (LINDE [*136 b*]), FeAl (Al concentrations smaller than 6 atom %, THOMAS [*208 a*]), Fe 68—85, Cr 7—30, Al 5—7, Co 0—2 atom % (SCHULZE [*187 b*]) a decrease after coldworking has been found. SCHULZE suggests that, during the cold work, electrons trapped earlier in the d-band are moved into the s-band, thus yielding a better conductivity. During the recovery process these electrons should be trapped again in the d-band. Another explanation is (LINDE [*209*]) that coldwork increases the average number of electrons in the d-shell of displaced transition metal atoms. These atoms will partake less in the s-d-s scattering process, which is a dominant factor for the resistivity of the alloys containing paramagnetic components. As long as the decrease in scattering outweighs

[1] For single crystals: $a \approx 0.01 \times 10^{-8}$ ohm m., $p \approx 1.8$.

the increase of the scattering due to coldwork displacements, a resultant decrease in $\varrho$ will be found.

Coldwork introduces anisotropy in an isotrope conductor, as found by BROOM and CLOTHIER [21]. They measured the resistivity of wires drawn at room temperature, parallel to the direction of drawing $\varrho_{\parallel}$, and, also perpendicular to the wire axis, $\varrho_{\perp}$. Their results, given in Table 22 demonstrate the existence of an anisotropy.

**23. Deformation by quenching.** By quenching, large strain causes an increase in the concentration of vacancies and hence large increase in resistivity can be obtained. These experiments demand, because of the fast recovery at high temperatures, that the time for removing the wire from the high temperature to the equilibrium quenching temperature should be very short, whereas the time of stay at the quenching temperature should be reproducable for a series of measurements. KAUFMANN and KOEHLER [115] cooled wires of 99.999% Au by means of blasts of gaseous helium from different high temperatures to room temperature within 5 to 10 milliseconds. Then the wires were immersed in liquid nitrogen and the resistance was determined. They found $\Delta R/R = 0.82$ for a quenching from 1193° K. At temperatures between 251.5 and 270.5° K, about one third of the extra resistance disappeared, the other two thirds disappeared by annealing at temperatures between 400 and 770° K.

Table 22. *Comparison of $\varrho_{\parallel}$ and $\varrho_{\perp}$ for colddrawn wires* (BROOM and CLOTHIER).

| Metal | $\varrho_{\parallel}/\varrho_{\perp} - 1$ |
|---|---|
| Cu | $\leqq 0$ |
| Ni | $= 0$ |
| α Al bronze | $> 0$ |
| Cu 75 Ni 25 | $< 0$ |
| Cu 70 Ni 30 | $> 0$ |
| Cu 64 Ni 36 | $> 0$ |
| Cu 60 Ni 40 | $> 0$ |

**24. Irradiation effects**[1]. Besides mechanical coldwork, bombarding with fast charged or neutral particles results in resistance increase. For example, bombarding copper with 35 Mev α-particles (EGGLESTON [50]), Cu, Ag, Au, Ni, and Ta with 12 Mev deuterons (MARX *et al.* [142]) at temperatures between 123 and 133° K, Cu and Fe-Cu solid solutions with electrons (DENNEY [37], EGGER *et al.* [49]), and $Cu_3Au$ and Pt with neutrons (SIEGEL [192], DUGDALE [47]) led to increase the resistance, the magnitude of the effect being determined by the intensity and energy of the beam of bombarding particles[2]. During the passage of a fast particle, a local temperature effect occurs along the track of recoil atoms. The situation is frozen in, immediately, if the ambient temperature is low. Thus, part of the resistivity increase may be due to a quenching effect. By this reasoning SIEGEL [192] explained the greatest part of the resistivity increase by neutron bombardment ($3.3 \times 10^{23}$ neutrons/m.$^2$) of specimens of ordered and disordered $Cu_3$ Au. The resistivity of the ordered specimen increased by about 120%, whereas that of the disordered sample increased by less than 1%, which could be explained by the formation of $Hg^{198}$ by the reaction:

$$Au^{197} + n^1 \rightarrow Au^{198} \rightarrow Hg^{198} + e^- .$$

From X-ray results and other experiments on the influence of coldwork and quenching and from calculations by SEITZ (1949) and by SIEGEL it was concluded that only a small number of atoms was disordered by direct collisions with the incoming neutrons. For a more detailed account of experiments with this alloy, we refer to BROOM's monograph [*B 5*].

[1] These effects have been discussed in more detail by A. SEEGER in vol. VII, part 1, of this Encyclopedia.

[2] With an increase of resistivity an increase of volume is connected, which was found for copper at 77° K to be of the order 0.2% after 21 Mev-deuteron bombardment [145]. Increase disappeared after room temperature annealing (vacancies and interstitials).

**25. Recovery processes.** The total recovery by annealing to the recrystallization temperature has been discussed in Sect. 20. As to the physical microprocesses resulting in the resistance increase and recovery modern experiments have given a fund of evidence. The recovery experiments have enabled better understanding of the processes of coldwork and recovery than have experiments on the increase of resistance only. The interpretation of the results is usually not so simple, since the observed behaviour of the resistance depends on many other factors besides the coldwork. During the past few years many experiments have been done with pure copper, so that the results of different experimenters are better comparable.

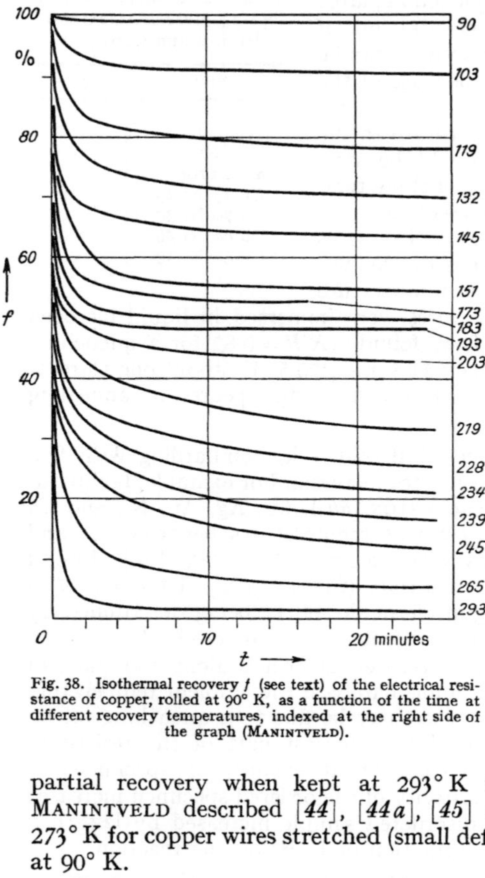


Fig. 38. Isothermal recovery $f$ (see text) of the electrical resistance of copper, rolled at 90° K, as a function of the time at different recovery temperatures, indexed at the right side of the graph (MANINTVELD).

The most important result is, that the recovery experiments have succeeded in distinguishing between the influences of the various imperfections on the resitivity. Further, as a quantitative result, recovery experimens yield information regarding the energies involved for annihilating imperfections. In the following, we give a general picture of the experiments, their results and the most important, but in many respects preliminary, conclusions.

It has been known for long—in particular, by the manufacturers of standard resistors—that at room temperature recovery occurs. That is, keeping a resistor at room temperature long enough, say, for a few years, will finally result in a stable value of its resistance lower than at the start. MOLENAAR and AARTS found [*153*], that a wire, coldworked at 90° K, showed a partial recovery when kept at 293° K for three hours. DRUYVESTEYN and MANINTVELD described [*44*], [*44a*], [*45*] a stepwise recovery between 90 and 273° K for copper wires stretched (small deformation) or rolled (large deformation) at 90° K.

Their experiment can be described as follows: a roller, with rolls of hardened steel, and of the dimensions of a match box, was placed in a vessel filled with liquid air. Rolling was applied between two points of the wire, at which potential leads had been soldered. After rolling, the wire was removed within 0.2 sec. to a vessel with a liquid at the chosen recovery temperature. It was kept there for a certain time, then returned to the liquid air and its resistance was then measured.

The results for the isothermal recovery are given in Fig. 38 (MANINTVELD [*141*]) as a function of $f = \frac{R - R_E}{R_0 - R_E}$ against time after coldwork at 90° K. $R_E$ is the resistance value after total recovery at 363° K, $R_0$ is the resistance value before the coldwork, $R$ is the resistance measured during the recovery experiment. $R$, $R_0$ and $R_E$ are always measured at 90° K. From Fig. 38 has been obtained Fig. 39 with curves for these large deformations similar to the results obtained from stretching experiments in which smaller deformations were applied. It is clear that two recovery steps are involved in the low temperature recovery. EGGLESTON [*51*] twisted Cu wires at 4° K, and found large scattering after annealing to 130° K and the results from Fig. 38 for annealing at higher temperatures. The recovery depends, however, on the initial (vacuum) annealing, prior to coldwork[1].

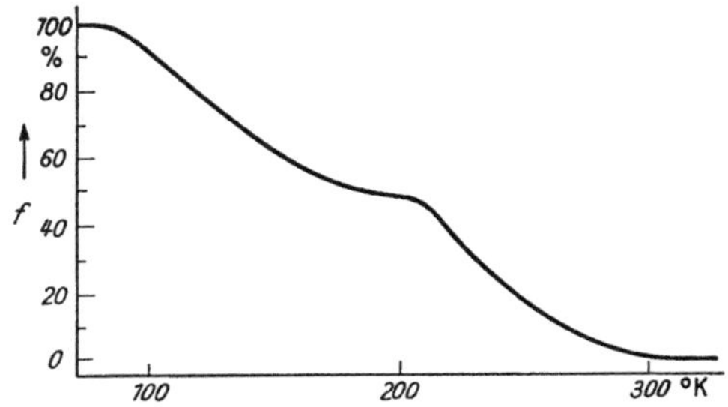


Fig. 39. Recovery $f$ of the resistance of copper, rolled at 90° K as a function of the annealing temperature for five minutes annealing.

Besides these two recovery phases ($P_{\text{II}}$ and $P_{\text{III}}$) another phase ($P_{\text{I}}$) has been found at lower temperatures (recovery after bombardment, COOPER *et al.* [*33*]) and two more phases ($P_{\text{IV}}$ and $P_{\text{V}}$) at higher temperatures, the phase at the highest temperature ($P_{\text{V}}$) corresponding to recrystallization.

Experiments have shown, that occurrence of these phases depends on the character of the coldwork, that is, whether the material has been mechanically deformed, bombarded, or quenched from high temperatures.

In his review article VAN BUEREN [*B 6*] gives a picture which clearly demonstrates the occurrence of these different phases (Fig. 40).

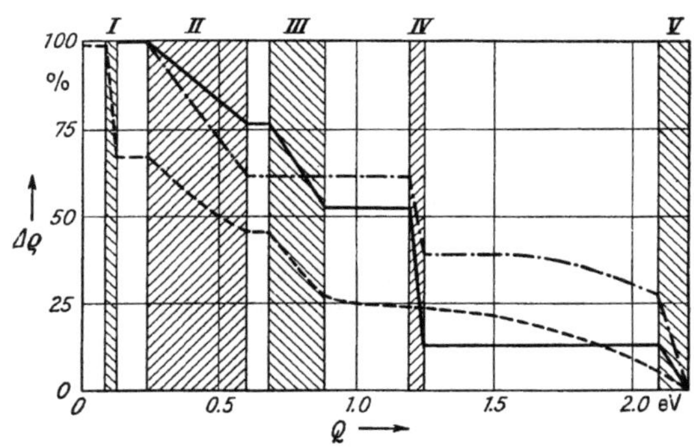


Fig. 40. The occurrence of recovery phases as connected with the character of the coldwork (taken from VAN BUEREN). Index: ——— recovery after deformation, —·—·— recovery after quenching, - - - - - - recovery after irradiation.

From the recovery curves the activation energies $Q$, involved in the recovery process, can be deduced. Under the assumption that the recovery process, due to diffusion of vacancies, interstitials and the like, allows a description analogous to that of self-diffusion in metals, a temperature dependent parameter

$$\tau = e^{Q/\varkappa T} \tag{25.1}$$

is introduced which follows from a logarithmic plot of $\Delta\varrho/\varrho$ against $1/T$. Assuming $\Delta\varrho/\varrho = f(t/\tau)$, then for a constant value of recovery, $\tau\, e^{-Q/\varkappa T}$ has to be a constant.

For any value of recovery it follows that $t_i = A\, e^{Q/\varkappa T_i}$ $(i = 1, 2 \ldots)$ and hence for the same value of recovery,

$$Q = \varkappa \frac{T_1 T_2}{T_1 - T_2} \log \frac{t_1}{t_2}. \tag{25.2}$$

[1] See Fig. 25 of [*141*].

As an average for stretching and rolling, there follow according to MANINTVELD the values $Q_{II}$ and $Q_{III}$ in eV for the phases $P_{II}$ and $P_{III}$:

| | $Q_{II}$ | $Q_{III}$ |
|---|---|---|
| Cu | 0.22 | 0.85 |
| Ag | 0.18 | 0.69 |
| Au | 0.29 | 0.69 |
| Pt | 0.22 | 0.99 |
| Temperature | 140 to 200° K | 240 to 300° K |

The same information can be obtained by recovery experiments after other types of coldwork. For example, for copper: MARX *et al.* [*142*] found from their radiation experiments $Q_{II}=0.20$ eV, $Q_{III}=0.9$ eV; EGGLESTON from coldwork [*51*] at 4° K: $Q_{II}=0.44$ eV, $Q_{III}=0.67$ eV; and from irradiation [*50*]: $Q_{III}=0.72$ eV; OVERHAUSER from irradiation [*165*]: $Q_{II}=0.2$ to 0.6 eV, $Q_{III}=0.68$ eV; KAUFMANN from quenching [*115*]: $Q_{II}=0.4$ eV.

In the case of gold, a large discrepancy is found as to the value of $Q_{III}$: 0.69 eV after rolling (MANINTVELD [*141*], 1.0 eV after irradiation (MARX *et al.* [*142*]); 0.40 eV after quenching (KAUFMANN, KOEHLER [*115*]). From EGGLESTON'S experiments with copper [*50*] bombarded with 35 MeV $\alpha$-particles below 120° K or coldworked at this temperature, it followed that the recovery below 200° K was the same for irradiation and coldwork, but differed above this temperature. He claims that at 297° K there is still an increase in $\varrho$ of 25% of the total increase caused by irradiation, whereas in the case of deformation by coldwork 50% of the total effect is left.

On the fourth phase ($P_{IV}$), temperature region 370 to 520° K, there has been reported by BOWEN, EGGLESTON, KROPSCHOTT [*16*] and by SMART, SMITH and PHILLIPS [*196*] for mechanically cold worked wires, with $Q_{IV}\approx 1.25$ eV. This recovery is not found as a distinct recovery phase after irradiation experiments [*165*]. The final restoration is obtained at temperatures of the order of 600° K, with a $Q_V\approx 2.12$ eV.

At temperatures above the fifth phase the growing crystals cause dislocations and large strains at their boundaries, these leading to an increase of resistance again.

The evidences mentioned clearly show the different mechanisms which are present. A detailed discussion on the connection between the resistance recovery phenomenon and the diffusion, X-ray and mechanical experiments will not be given here. The idea now generally accepted, is that for interpretation the recovery energies have to be compared with the energies of diffusion ($Q_d$) of the several imperfections in the metal. These latter have been calculated, among others, by HUNTER and NABARRO for dislocations [*98*], by KAUFFMANN and KOEHLER for vacancies [*115*], by BARTLETT and DIENES for pairs of vacancies [*6*], by HUNTINGTON and SEITZ for interstitials [*99*]. In Table 23 we give the values for the energies of formation and diffusion of the imperfections (see [*B 6*]).

A comparison of the data for $Q_d$ and for $Q_I \ldots Q_V$ may lead to the following interpretation of the recovery phases.

Phase I. According to COOPER [*33*] $Q_I\approx 0.1$ eV. Temperature region about 40° K. This might represent a diffusion of interstitials.

Phase II. $Q_{II}$ being of the order 0.2 to 0.6 eV, for this phase there would follow a diffusion of pairs of vacancies, or a diffusion of interstitials.

Table 23. *Energy of formation $E_f$ and of diffusion $Q_d$ in electronvolts for imperfections, and their contribution ($\delta\varrho$) to the electrical resistivity for one defect per unit of volume or crystal plane* (VAN BUEREN).

| Imperfection | $E_f$ eV | $Q_d$ eV | $\delta\varrho_{def}$ |
|---|---|---|---|
| Vacancy | 0.9 | 1.2 | $1.53\times10^{-23}$ ohm m./m.$^3$ |
| Vacancy pair | $\approx 1.6$ | 0.25 to 0.6 | $\approx 3\times10^{-23}$ ohm m./m.$^3$ |
| Interstitial | 5 to 10 | 0.1 to 0.25 | $5.7\times10^{-23}$ ohm m./m.$^3$ |
| Dislocation | $\approx 3$ (per layer of atoms) | $\approx 2$ | 0.4 to $1.5\times10^{-26}$ ohm m./m.$^2$ |

Phase III. The $Q$-value for this phase is $\approx 0.7$ eV and is thus easily related with one of the $Q_d$ values in Table 23. DRUYVESTEYN and coworkers [*141*] suggested a diffusion of vacancies, BRINKMAN *et al.* [*19*], however, would attribute this phase to a diffusion of interstitials.

Phases IV and V are the phases in which the dislocations will recover [*145*]. Since there is an indication that $P_{IV}$ is not found for bombarded wires, where the number of dislocations is expected to be small, $P_{IV}$ is perhaps a combinative effect, for example, a diffusion or absorption of vacancies by dislocations (VAN BUEREN [*B 6*]).

The different consequences of cold work and irradiation can be explained as follows. Cold work at low temperatures causes slip in the metal. As a consequence all imperfections, such as vacancies, interstitials and dislocations will occur, and three recovery steps may be observed in principle. By bombardment, however, only vacancies and interstitials could be expected to be formed. At a low temperature, the disorder caused by the recoil atoms after collision leaves a small number of quenched dislocations as a result of the bombardment (cf. EGGLESTON'S experiment [*50*]). Thus only two recovery steps will be found when the temperature is increased. If the irradiation experiments are carried out at sufficiently high temperatures, so that a subsequent recovery of dislocations and an annihilation of vacancies by interstitials take place during the experiments, bombardment will increase order. A great many experiments to study this phenomenon have been carried out with alloys, in particular, with such compounds as $Cu_3Au$ (cf. Sect. 29).

**26. Resistivity increase caused by the defects.** In analogy with the historically better known case of contamination of a pure metal by foreign atoms (Chap. VIII), a resistivity increase $\delta\varrho_{def}$ has been defined per unit imperfection. In a first approximation, this increase should be independent of temperature, that is, for the coldworked metal, the MATTHIESEN rule should hold. This is questionable according to measurements of OGILVIE and CLOTHIER [*164*]. For pure copper, commercial copper, and some copper alloys they found a deviation of about 0.1% in the slope of the coldworked samples with respect to that of the annealed sample between 90 and 290° K. It is possible, however, that with their method of measurement partial recovery at roomtemperature blurred their results.

We mentioned already JONGENBURGER'S results [*105*] for the magneto-resistance measurements, which may be interpreted as an essential difference between the influence of impurities and of defects on the electrical resistance. There is no harm, however, in giving $\delta\varrho_{def}$'s as approximate values, useful for practical purposes.

Vacancies: $\delta\varrho\approx 1.3\times10^{-8}$ ohm m. per atom per cent of vacancies in copper (JONGENBURGER, see [*B 2*]. Interstitial ions: $\delta\varrho$ has been estimated by DEXTER to be about $\delta\varrho \approx 0.6 \times 10^{-8}$ ohm m./at % [38] of interstitial ions in copper, silver or gold. JONGENBURGER in criticizing the calculations, derives a much larger value: $5 \times 10^{-8}$ ohm m./at % [104].

Dislocations: HUNTER and NABARRO gave a survey of the estimates for $\delta\varrho$ [98]. They draw attention to the anisotropy of the dislocation effect: there is no resistance increase in the direction of the dislocation axis, and the increase in the slip direction is one third of that perpendicular to the slip plane (cf. BROOM and CLOTHIER's experiments [21]). For a screw dislocation, there is no extra resistance parallel to the axis, but an isotropic change in directions perpendicular to the axis.

HUNTER and NABARRO's $\Delta\varrho$-values for copper are $\Delta\varrho = 0.59 \times 10^{-20} N\,\mu$ ohm m. for edge dislocations and $\Delta\varrho = 0.18 \times 10^{-20} N\,\mu$ ohm m. for screw dislocations ($N$ is the number of dislocation lines per m., $\Delta\varrho = N\delta\varrho$). For equal densities of edge and screw dislocations, they find that $\Delta\varrho = 0.39 \times 10^{-20} N\,\mu$ ohm m. or, for the common value of $5 \times 10^{16}$ dislocations/m.$^2$, $\Delta\varrho = 0.02 \times 10^{-2}\,\mu$ ohm m. In Table 23 the $\delta\varrho$-values are given for the different imperfections.

A great discrepancy between the various experimental results and calculations is noticeable and this part of the problem of the imperfections in metallic conductors is still under investigation by many experimenters. The experiments have shown, however, that unless stringent precautions are taken, in general the results of experiments on electrical properties of metals and alloys are apt to be influenced in a most disturbing manner by the presence of imperfections, which are not always under control of the experimenter.

## VIII. Resistivity of alloys.

### a) Properties of alloys with non magnetic components.

**27. General remarks.** The results of earlier systematic experiments by MATTHIESSEN and VOGT (1864) with alloys [143] led to the already discussed separation of the resistivity into a temperature dependent and a temperature independent part. The availability of purer metals as components and in particular the better possibility of making experiments at low temperatures made it possible to test the validity of this hypothesis more accurately. Generally speaking, the MATTHIESSEN rule (8.1) holds as long as the temperature independent part of the resistance is small compared to the temperature dependent part, and as long as the introduction of none of the components affects considerably the lattice of the solvent metal. The restrictions to the application of MATTHIESSEN's rule, based on theoretical considerations are given in more detail in JONES' article [J 1] in this volume.

Several authors have analysed (8.1) for practical applications. In particular for alloys, BORELIUS [B 3], [136a] distinguished two formal expressions for (8.1). For very dilute alloys his interpretation of the MATTHIESSEN rule is written as:

$$\left(\frac{\partial\varrho_1}{\partial T}\right) = \left(\frac{\partial\varrho_2}{\partial T}\right), \tag{27.1}$$

that is, derivatives of resistivity with respect to temperature of two specimens of the same metal, which differ in purity, are equal.

For binary alloys (27.1) reads

$$\left(\frac{\partial\varrho}{\partial T}\right)_{\text{alloy}} = a_1\left(\frac{\partial\varrho_1}{\partial T}\right) + a_2\left(\frac{\partial\varrho_2}{\partial T}\right) \tag{27.2}$$

in which $a_1$ and $a_2$ are the atom concentrations ($a_1 + a_2 = 1$) of the components 1 and 2. From (27.2) there follows

$$\varrho_{\text{alloy}(T)} = \varrho(0) + a_1 \varrho_1(T) + a_2 \varrho_2(T), \tag{27.3}$$

or, for a given ratio of the components in a binary alloy the difference between the observed resistivity and that calculated from the resistivities of the components, is temperature independent. This formulation is called by BORELIUS the MATTHIESSEN law. For both expressions, (27.1) and (27.3), it holds that they represent a first approximation only.

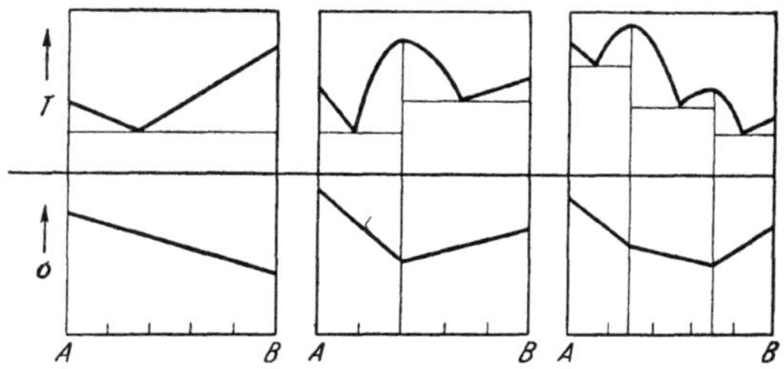


Fig. 41. Electrical conductivity of binary alloys *vs.* the concentration of one component in percent volume, for alloys immiscible in solid state (Figs. 41 and 42 taken from HANSEN *et al.*).

During the last twenty years resistivity measurements on alloys have become of ever growing interest, since experimental results have shown that crystallographic transformations, such as disorder-order transformations, can be detected rather sensitively by means of a determination of the resistivity. In particular, the reaction of alloys to coldwork and particle bombardment is reflected in their resistivity. Most of these experiments, however, have their main interest in the field of the structure of the solid state and will be discussed from this view point elsewhere in the present Encyclopedia. A concise discussion on the electric changes will be given in the present paper.

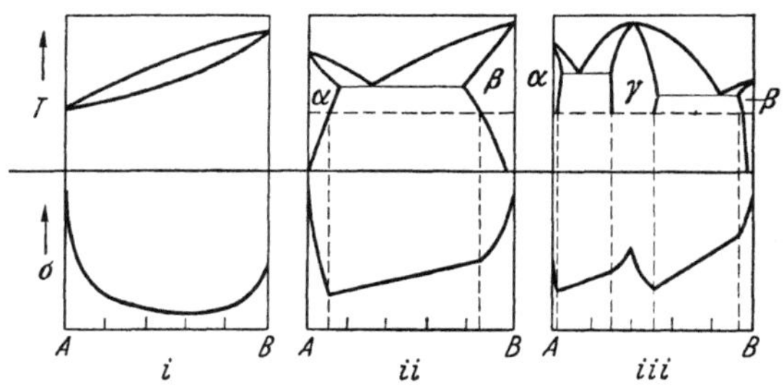


Fig. 42. Electrical conductivity of binary alloys *vs.* the concentration of one component in percent volume, for (i) continuous series of solid solutions; (ii) miscibility gap between primary solid solutions; (iii) primary solid solutions and intermediate phase of variable composition.

**28. Resistivity of binary solid solutions.** A general picture of the relation between the resistivity of binary metallic mixtures and the composition of the mixture is illustrated by Fig. 41 and 42 which are a representation of the LE CHATELIER-GUERTLER [*132*] rules. These rules are:

(1) the electrical conductivity and its temperature coefficient of heterogenous mixtures of two phases vary linear with the composition, measured in volume per cents (Fig. 41);

(2) the electrical conductivity and its temperature coefficient for solid solutions are always below those of the solvent metal and for continuous solid solutions the conductivity is a U-shaped curve in the conductivity-composition diagram (Fig. 42a).

As a function of temperature, $\varrho(T)$ relations may occur which differ markedly from the normal linear relations found for pure metals for temperatures above 80° K. In particular alloys with transition metals, and alloys having crystallographic or magnetic transformations, show such deviations. For small amounts of a solute in a solvent metal, forming dilute binary solid solutions, interesting results have been found.

*α) Dilute binary solid solutions. Influence of concentration and temperature.* Assuming additivity of the resistivity of the pure solvent and the solute metal, the observed resistivity $\varrho_{\text{obs}}$ can be written as

$$\varrho_{\text{obs}} = \varrho_m + \varrho_i . \tag{28.1}$$

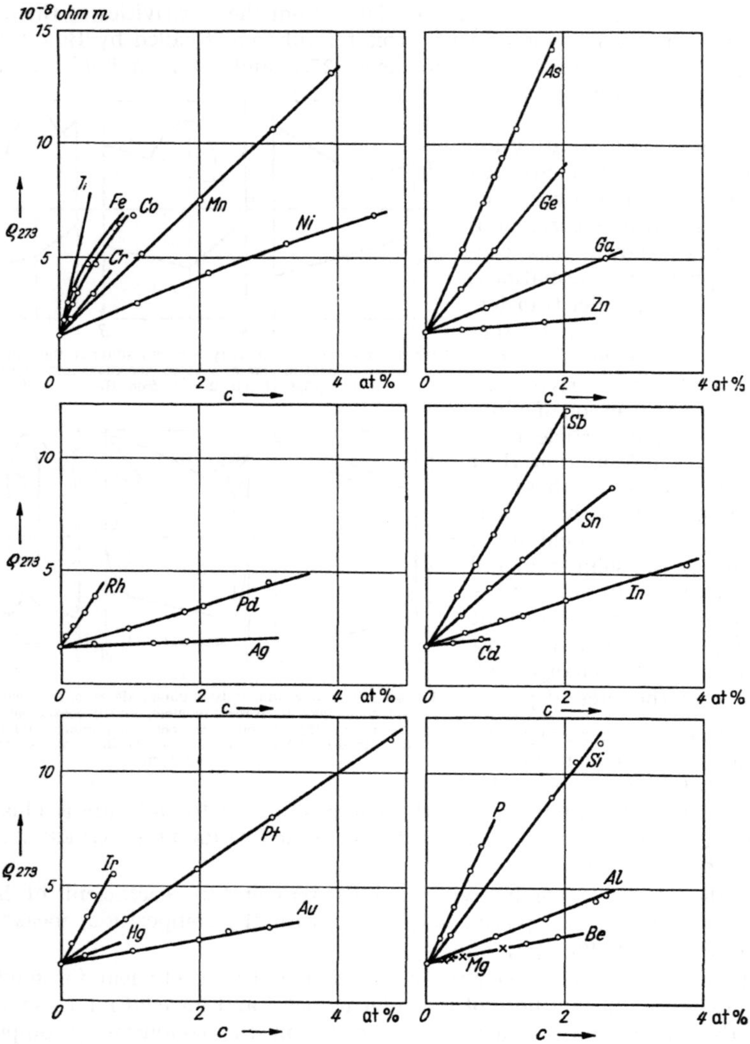


Fig. 43. Resistivity at 273° K for copper base binary alloys as a function of the concentration of the second component (LINDE).

In most cases, it has been found that $\varrho_i$ is related to the concentration of the solute by a simple relation

$$\varrho_i = c\,\delta\varrho_i + c^2\,\delta'\,\varrho_i + \cdots \tag{28.2}$$

in which $c$ is the amount of solute in atom per cent (at %), $\delta\varrho_i$ and $\delta'\varrho_i$ resistivity values. This relation, found by BENEDICKS [9] for some copper-silver and copper-gold alloys, has been investigated systematically by LINDE [136a] for a great number of alloys with a noble metal as a solvent. The departure from linearity for concentrations of the order of a few atom per cent is not negligible at room temperature. Thus, LINDE defines the atomic resistivity increase $\delta\varrho_i$ as

$$\delta\varrho_i = \lim_{c\to 0} \varrho_i/c. \tag{28.3}$$

For a number of copper base alloys LINDE'S results are reproduced in Fig. 43.

For the special case that solvent and solute metal are in the same column of the periodic table $\delta\varrho_i$ may be the same whether the metal $M_{\mathrm{I}}$ is solved in $M_{\mathrm{II}}$ or $M_{\mathrm{II}}$ is solved in $M_{\mathrm{I}}$. For example, $\delta\varrho_i = 0.36\times 10^{-8}$ ohm m. for silver in gold as well as for gold in silver. For such cases, the NORDHEIM relation [162]

$$\varrho_i \sim x(1-x) \tag{28.4}$$

holds for $0\leq x\leq 1$, where $x$ is the fractional concentration of the solute. This relation is theoretically based on a first order perturbation calculation and holds when the electron concentration and structural properties of the base metal are not modified by alloying.

Besides LINDE'S work we mention the earlier investigations by NORBURY and KUWADA [161], the systematic investigations by EUCKEN and SCHÜRENBERG [57] on lead base alloys, by ROBINSON and DORN [180], GULYAEV and TRUSOVA [84], on aluminium base alloys.

Furthermore the investigations by SMITH and PALMER (Cu—Al, Cu—Ni, Cu—Mn) [197b] and by NEMILOW (Pt—Ir, Fe—Pt) [158] give results on some special alloys. The results of LINDE, ROBINSON and DORN, EUCKEN and SCHÜRENBERG are listed in Table 24 where the values of $\delta\varrho_i$ and $\delta'\varrho_i$ are given in $10^{-8}$ ohm m./at. % and $10^{-8}$ ohm m./(at. %)$^2$.

NORBURY found [160] for the resistivity increase of alloys consisting of elements of different valency,—except for alloys with a transition metal as one of the components—, that the atomic resistivity increase changes proportionally with the squared difference of atomic number between solvent ($Z_m$) and solute ($Z_i$) metal. This rule has been checked by LINDE for his data on copper, silver and gold base alloys, by ROBINSON and DORN for the aluminium base alloys. The latter authors could represent LINDE'S and their own results by a relation

$$\delta\varrho_i = K_2 + K_1(Z_m - Z_i)^2 \tag{28.5}$$

where $K_2$ and $K_1$ are constant for a given solvent for each period from which the solutes are taken. They arrive at the conclusion that for aluminium as a solvent a valency of 2.5 has to be taken for the ion valency in the metallic state.

The deviations which occur when NORBURY'S rule is applied to transition metals is shown in Fig. 44 in which for some gold base alloys LINDE'S values for $\delta\varrho_i$ are inserted against $(Z_m - Z_i)^2$ as abcissae. The general procedure for measuring the $\delta\varrho_i$ values is to determine, at a given temperature, the resistivity of a specimen, and plot the results as a function of the concentration of the solute element. This value can be deduced from the weights of the materials during the manufacturing of the alloys. When small quantities are involved this is a rather inaccurate method, its success depending on the method of melting and homogenizing of the alloys in solid state. For not too small quantities of the

Table 24. *The atomic resistivity increase $\delta\varrho_i$ and $\delta'\varrho_i$ in $10^{-8}$ ohm m./at.% and $10^{-8}$ ohmm./(at. %)$^2$ for copper-, silver-, gold-, aluminium- and lead-base alloys.*

| Base Metal | Cu | | Ag | | Au | Al | Pb |
|---|---|---|---|---|---|---|---|
| | $\delta\varrho_i$ | $\delta'\varrho_i$ | $\delta\varrho_i$ | $\delta'\varrho_i$ | $\delta\varrho_i$ | $\delta\varrho_i$ | $\delta\varrho_i$ |
| Be | 0.62 | | | | | | |
| Mg | 0.65 | | 0.50 | | 1.30 | 0.46 | 4.0 $c<0.5$ |
| Al | 1.25 | 0.055 | 1.95 | | 1.87 | — | |
| Si | 3.95 | | | | | | |
| P | 6.7 | | | | | | |
| V | | | | | | 426 (?) | |
| Ti | | | | | 12.9 | | |
| Cr | (3.6) | | | | 4.25 | | |
| Mn | 2.90 | | 1.60 | | 2.41 | | |
| Fe | 9.3 | | | | 7.9 | | |
| Co | 6.35 | 1.3 | | | 6.1 | | |
| Ni | 1.25 | | | | 0.79 | | |
| Cu | — | — | 0.077 | | 0.45 | 0.73 | |
| Zn | 0.32 | | 0.64 | 0.023 | 0.95 | 0.023 | |
| Ga | 1.42 | 0.066 | 2.36 | 0.058 | 2.2 | | |
| Ge | 3.79 | 0.096 | 5.5 | 0.115 | 5.2 | 0.80 | |
| As | 6.8 | | 8.5 | | 8.0 | | |
| Rh | 4.40 | | | | 4.15 | | |
| Pd | 0.89 | | 0.44 | | 0.41 | | |
| Ag | 0.14 | | — | — | 0.36 | 1.14 | 0.11 $c<0.005$ |
| Cd | 0.30 | | 0.38 | 0.16 | 0.63 | 0.57 | 3.02 $c<0.65$ |
| In | 1.06 | 0.026 | 1.78 | 0.041 | 1.39 | | 1.13 |
| Sn | 2.88 | 0.094 | 4.36 | 0.10 | 3.36 | | 0.285 |
| Sb | 5.4 | 0.076 | 7.25 | | 6.8 | | 1.24 $c<0.13$ |
| Ir | 5.7 | | | | | | |
| Pt | 2.1 | | 1.6 | | 1.01 | | |
| Au | 0.55 | | 0.36 | | — | | |
| Hg | 1.0 | | 0.79 | | 0.44 | | 2.30 |
| Tl | | | 2.27 | 0.050 | 1.9 | | 0.635 |
| Pb | | | 4.65 | | (3.9) | | — |
| Bi | | | 7.3 | | (6.5) | | 0.96 |
| Temp. | 273° K | | | | | 297° K | 273° K |
| Ref.: | LINDE | | | | | ROBINSON | EUCKEN |

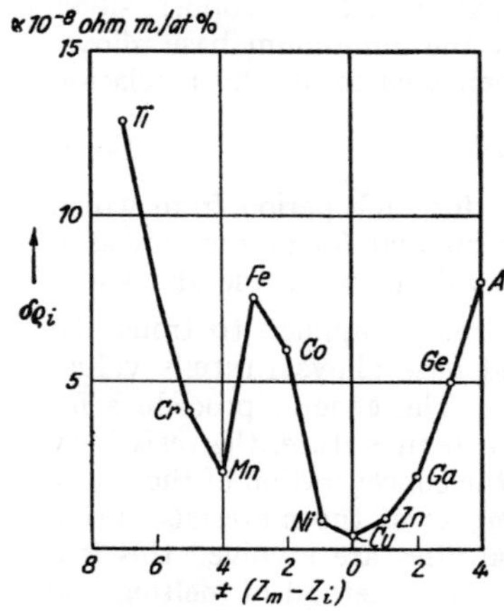


Fig. 44. $\delta\varrho_i$ for alloys with gold as a base metal against $Z_m - Z_i$ (taken from LINDE).

solute the composition of the alloy can be found from micro-spectrographic analysis. EUCKEN and SCHÜRENBERG determined $\delta\varrho_i$ from relative measurements at different temperatures of the resistance, under the assumption that Eq. (28.1) hold for the temperature region they considered. With Eq. (28.1), $r'_T$ given by

$$r'_T = \left(\frac{\varrho_T}{\varrho_{273}}\right)_{\text{obs}} = \frac{r_{m,T} + \dfrac{\varrho_{i,T}}{\varrho_{m,273}}}{1 + \dfrac{\varrho_{i,T}}{\varrho_{m,273}}}$$

is determined as a result of resistance measurements at $T°$ K and at $273°$ K. From this it follows that

$$\frac{\varrho_i}{\varrho_{m,273}} = \frac{r'_T - r_{m,T}}{1 - r'_T}. \tag{28.6}$$

From (28.6) $\varrho_{i,T}$ can be obtained after inserting the data for $\varrho_{m,273}$, taken from literature.

The error $\Delta_T$ which is given by

$$\Delta_T = \frac{\varrho_{i,273}}{\varrho_{m,273}} - \frac{\varrho_{i,T}}{\varrho_{m,273}}$$

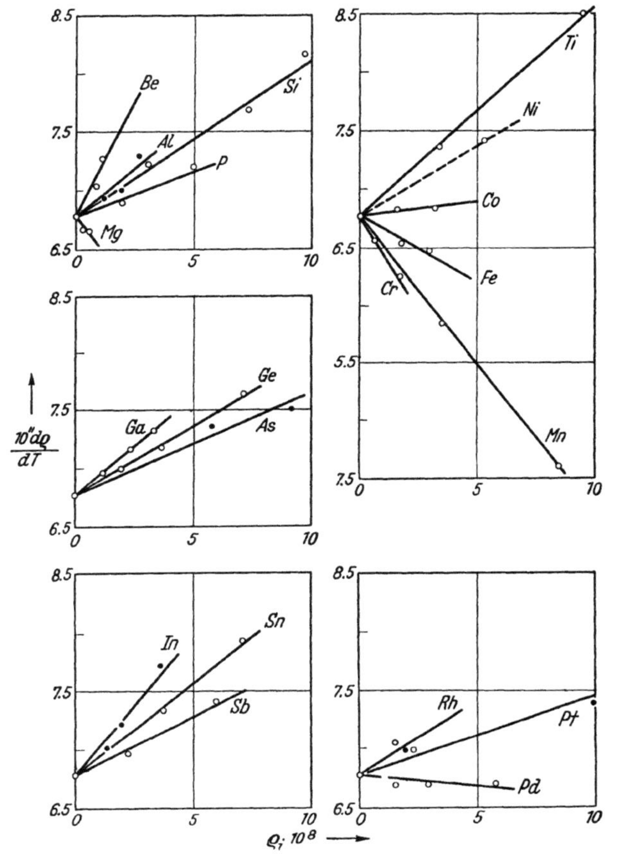


Fig. 45. $d\varrho/dT$ in ohm m./degree against $\varrho_i$ for several copper base alloys (LINDE).

proved to be measurable for temperatures below 20° K, which implies that the assumption of a temperature independent part $\varrho_i$ is no longer valid at these temperatures.

By differentiating (28.1) by $T$, it follows that

$$\left(\frac{\partial \varrho}{\partial T}\right)_{\text{obs}} = \left(\frac{\partial \varrho_m}{\partial T}\right) + c\left(\frac{\partial\, \delta \varrho_i}{\partial T}\right) = \left(\frac{\partial \varrho_m}{\partial T}\right) + \varrho_i\left(\frac{1}{\delta \varrho_{i,T}}\,\frac{\partial\, \delta \varrho_i}{\partial T}\right)$$

or

$$\left(\frac{\partial \varrho}{\partial T}\right)_{\text{obs}} = \left(\frac{\partial \varrho_m}{\partial T}\right) + \varrho_{i,T}\,\alpha_{\delta\varrho,T}, \tag{28.7}$$

in which the temperature coefficient for the atomic resistivity increase,

$$\alpha_{\delta\varrho,T} = \frac{1}{\delta \varrho_i}\,\frac{\partial\, \delta \varrho_i}{\partial T}, \tag{28.8}$$

is introduced analogous to the definition of the temperature coefficient of resistivity (cf. 2.10). From LINDE's measurements between 273° K and 373° K there follows linearity of $\left(\frac{\partial \varrho}{\partial T}\right)_{\text{obs}}$ with $\varrho_i$ (see Fig. 45 and the values of $\alpha_{\delta\varrho, T}$ given in Table 25).

Table 25. *Temperature coefficient $\alpha_{\delta\varrho, T}$ in degree$^{-1}$ of the atomic resistivity increase for copper, silver and gold-base dilute alloys* (LINDE).

| Solute metal | $10^4\,\alpha_{\delta\varrho, T}$ | | | Solute metal | $10^4\,\alpha_{\delta\varrho, T}$ | | |
|---|---|---|---|---|---|---|---|
| | Cu | Ag | Au | | Cu | Ag | Au |
| Be | 3.6 | — | — | P | 0.8 | — | — |
| Mg | (−2.3) | 1.4 | (−0.5) | As | 0.85 | 0.55 | — |
| Zn | — | 2.4 | 1.9 | Sb | 0.95 | 0.72 | — |
| Cd | — | 1.6 | 1.7 | Bi | — | 0.30 | — |
| Hg | — | 2.3 | — | Ti | 1.8 | — | — |
| Al | 1.6 | 1.7 | 1.5 | Cr | −3.2 | — | −2.5 |
| Ga | 1.6 | 1.5 | 2.4 | Mn | −2.65 | −1.9 | −0.92 |
| In | 2.3 | 1.5 | 2.6 | Fe | (−1.7) | — | — |
| Tl | — | 1.3 | — | Co | 0.3 | — | −2.6 |
| Si | 1.4 | — | — | Ni | 1.2 | — | 4.1 |
| Ge | 1.2 | 0.95 | — | Rh | (0.8) | — | — |
| Sn | 1.55 | 1.1 | 1.5 | Pd | −0.3 | −0.55 | — |
| Pb | — | 0.80 | — | Pt | 0.8 | 0.7 | 0.6 |

Apart from their magnitude, the negative values of $\alpha_{\delta\varrho}$ for some of the transition metals as solute demonstrate the speciality of these elements.

Concluding, it follows from experimental data that:

(1) the resistivity increase is for concentrations of the solute less than a few atom per cent, in general proportional to the concentration of dissolved component;

(2) with the exception of transition metals as a solute, NORBURY's rule is generally valid;

(3) the resistivity increase is temperature dependent, with a positive coefficient of the order $10^{-4}$ degree$^{-1}$;

(4) the resistivity increase of transition metals, dissolved in copper, silver or gold, deviates from NORBURY's rule, whereas negative instead of positive temperature coefficients may occur[1].

HANSEN, JOHNSON and PARKS [*91*] have drawn attention to a forgotten experimental relation, found by. DELLINGER [*36*] between the conductivity and the temperature coefficient of resistance (at 273° K) for a dilute alloy. According to DELLINGER conductivity at 273° K for commercial copper wires should be proportional to the temperature coefficient of resistivity of the alloy:

$$(\sigma_{273})_{\text{alloy}} = \beta\,\alpha_{\text{alloy, 273}}\,. \tag{28.9}$$

Considering LINDE's data and those found by other authors, HANSEN *et al.* find a confirmation of this rule by plotting $\sigma_{273}$ against $\alpha_{\text{alloy}}$ for a great number of alloys. They claim that, for a number of base metals, $\beta = 1/\varrho_m \alpha_\varrho$, the values for $\varrho_m$ and $\alpha_\varrho$ being taken at $T = 273°$ K. Using LINDE's data, a plot of $\sigma$ against $\alpha_\varrho$ has been reproduced here (Fig. 46) in order to illustrate the accuracy of

[1] At low temperatures complicated anomalies have been found in the resistance of dilute alloys of this type, which have been interpreted in terms of anomalous scattering by the magnetic solute ions. There is no obvious justification for using a similar interpretation for the high temperature effects (GERRITSEN and LINDE [*70a*]; KORRINGA and GERRITSEN [*70b*]).

Eq. (28.9). HANSEN *et al.* do not enter into the physical meaning of this effect, a simple calculation shows, that for the dilute alloys either the quantity $\varrho_i \alpha_{\delta\varrho}$ is much smaller than $\varrho_m \alpha_\varrho$, or that the values for $\alpha_{\delta\varrho}$ as given in Table 25 should hold only for $\varrho_i \to 0$, or $c \to 0$. It follows from the figure that certainly for low values of $\sigma_{\text{alloy}}$ and $\alpha_{\text{alloy}}$ the relationship (28.9) is not valid, whereas the linearity of $(d\varrho/dT)_{\text{alloy}}$ with $\varrho_i$ holds, as proved by LINDE.

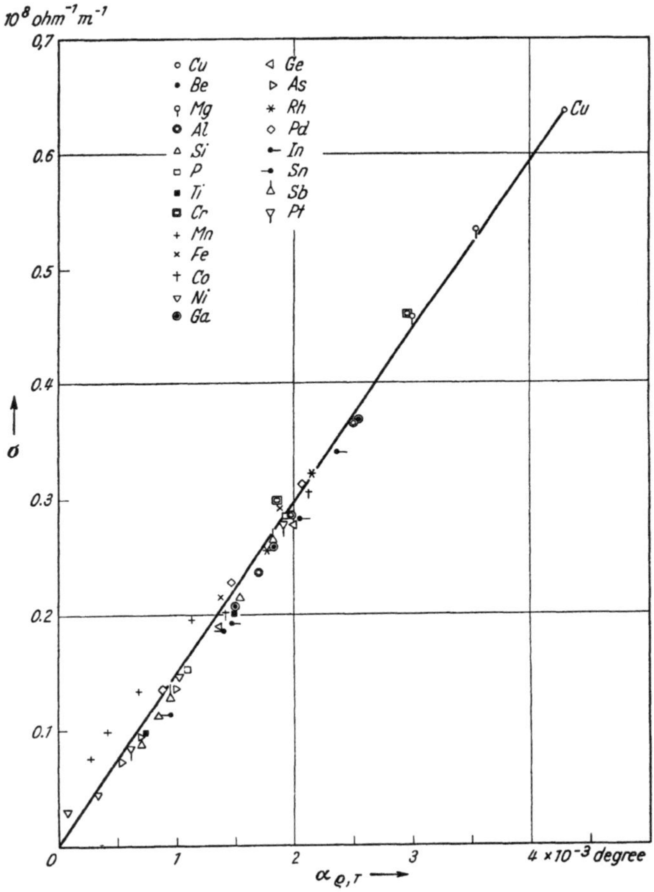


Fig. 46. Conductivity of copper base alloys against the temperature coefficient of resistivity (in $10^{-3}$ degree$^{-1}$). The values have been taken from LINDE ([*136a*], Table 7).

*β) Dilute binary solid solutions. Influence of pressure.* LINDE [*136a*] has investigated systematically at room temperature, up to pressures of $3 \times 10^7$ kg./m.$^2$, a number of dilute alloys with a noble metal as the solvent. One of his results has been to find an alloy suitable as material for a resistance manometer. LINDE's alloy, 15.9 at.% Mn in silver, shows a much smaller temperature dependence and a larger pressure coefficient than the manganine (which is a copper alloy) used so far, which had the disadvantage of an appreciable temperature coefficient. From his results LINDE derives the pressure coefficient of the atomic resistivity increase ($\alpha_{\delta\varrho,P}$) in an analogous way as the temperature coefficient (28.8).

Starting from (28.1) and (28.2) and differentiating ($T$=const), one obtains

$$\left(\frac{\partial \varrho}{\partial P}\right)_{\text{obs}} = \left(\frac{\partial \varrho_m}{\partial P}\right) + c\,\frac{\partial\,\delta \varrho_i}{\partial P},$$

whence it follows that the coefficient $\frac{\partial\,\delta\varrho_i}{\partial P}$ can be found from the $\left(\frac{\partial \varrho}{\partial P}\right)_{\text{obs}}$ *versus* $c$-curves. Then the pressure coefficient of resistivity $\alpha_{\delta\varrho, P}$ is

$$\alpha_{\delta\varrho, P} = \frac{1}{\delta\varrho_i}\,\frac{\partial\,\delta\varrho_i}{\partial P}, \qquad (28.10)$$

for $P \to 0$ for a given temperature.

In Fig. 47 we have given LINDE's results for $\left(\frac{\partial \varrho}{\partial P}\right)_{\text{obs}}$ for gold alloys, and in Table 26 the values for $\alpha_{\delta\varrho, P}$ have been inserted ($T \approx 273^\circ$ K). According to later investigations of LINDE for large Fe concentrations, the non-linear relations shown in Fig. 47 for AuFe alloys, are due to the continuous change from the paramagnetic to the ferromagnetic region of the alloy. In the case of the resistance at normal pressure, LINDE finds for this alloy a $\varrho$ *vs.* $c$-curve as it normally has been found for such transitions (Sect. 30). There is a discontinuity at 273° K for 23 at. % Fe. At 30 at. % Fe the value $\frac{\partial \varrho}{\partial P} = -150$ in the scale of Fig. 47.

Table 26. *The pressure coefficient $\alpha_{\delta\varrho, P}$ (in $10^{-10}$ m.²/kg.) for the atomic resistivity increase for some alloys at room temperature according to* LINDE.

| Solute metal | Solvent metal | | | Solute metal | Solvent metal | | |
|---|---|---|---|---|---|---|---|
| | Cu | Ag | Au | | Cu | Ag | Au |
| Mg | | (+1.7)[1] | | Ga | | −1.06 | −0.62 |
| Al | −0.39 | | | Ge | −0.55 | | |
| Ti | | | +0.04 | As | −0.42 | −0.36 | |
| Cr | | | +1.11 | Pd | +0.32 | +0.18 | |
| Mn | +2.55 | +2.78 | +1.38 | Sn | −0.70 | −1.03 | −0.74 |
| Fe | | | +0.2[2] | Sb | −0.65 | | |
| Co | | | −0.22 | Pt | +0.24 | −0.19 | −0.56 |
| Zn | | | +0.5 | Hg | | (−1.7)[3] | |

Other alloys, which show abnormalities in the volume pressure curve, as well as a great number of compounds have been investigated by BRIDGMAN [*18e, g, k*]. It falls beyond the scope of this paper to discuss these results here.

*γ) Concentrated binary solid solutions.* For concentrations of solute metals exceeding for instance 5 at. % in some cases, and for higher concentrations in all cases, the deviation from the assumed normal additivity of the impurity resistance becomes so large that the applicability of MATTHIESSEN's law is doubtful, and any theoretical interpretation, based on first or higher order perturbation, is doomed to fail. The concentrated alloys are, in general, of greater interest for solving technical rather than physical problems, as long as order-disorder and magnetic transformations are excluded from consideration.

We referred before to HANSEN *et al.* who found Eq. (28.9) holding for dilute solid solutions. It is remarkable that for continuous solid solutions, for example silver palladium alloys [*66*], a $\sigma_{\text{alloy}}$ *vs.* $\alpha_{\text{alloy}}$ plot gives a linear relation for all

[1] One alloy only, with 3.18 at % Mg.
[2] Extrapolated for $c \to o$, cf. Fig. 47.
[3] One alloy only, with 2.6 at % Hg.

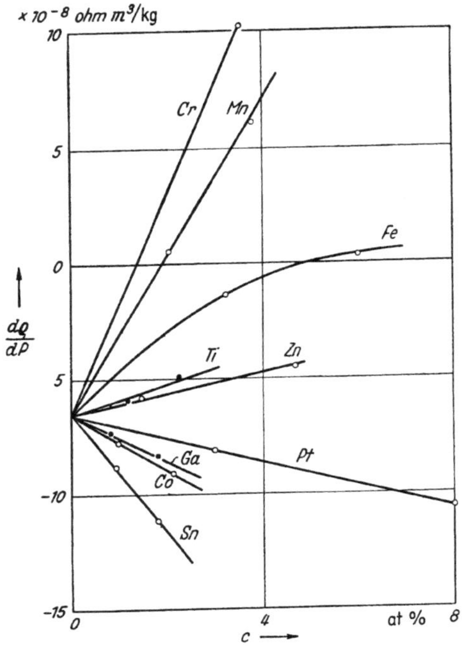


Fig. 47. $d\varrho/dP$ (in $10^{-18}$ ohm m.$^3$/kg.) against the concentration of several second components in gold base alloys (LINDE).

silver palladium solid solutions, split however into a silver-line and a palladium-line, defined by the coefficient $\beta=\sigma/\alpha$ for silver and palladium, respectively (see Fig. 48). For binary alloy systems in which the component metals are soluble in each

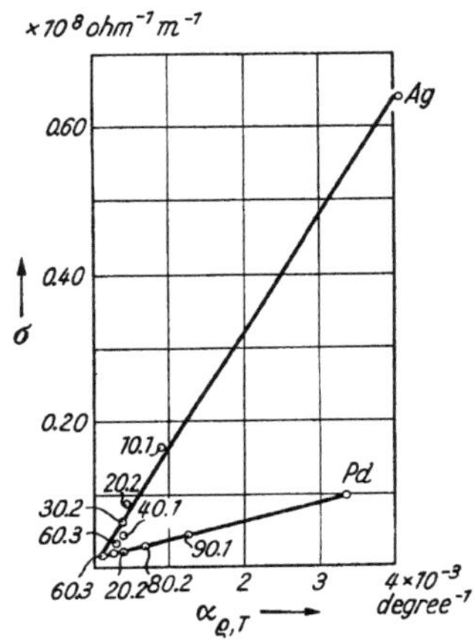


Fig. 48. Conductivity of silver-palladium alloys against the temperature coefficient of resistivity (taken from HANSEN *et al.*). The numbers indicate the palladium contents in at. %.

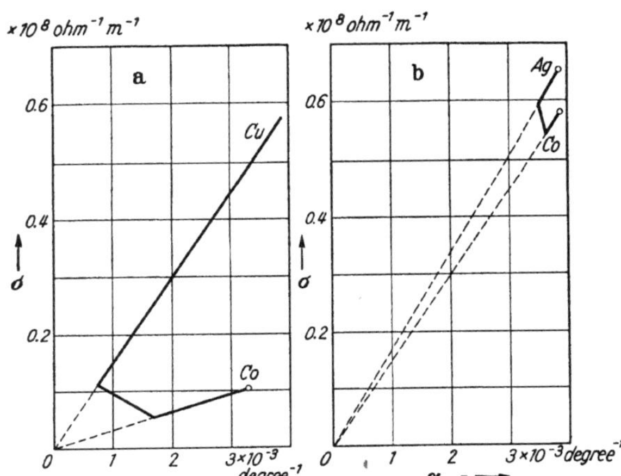


Fig. 49a and b. Conductivity against temperature coefficient of resistivity for: (a) copper-cobalt alloys; (b) copper-silver alloys (taken from HANSEN).

other to a certain extent and which do not form an intermediate phase, the diagram consists again of two straight lines representing the properties of the primary solid solutions (as in the case of silver-palladium) and a third straight line joining the points for the saturated solid solutions (see [*91*]). This is represented for the Cu—Co [*178*] and Cu—Ag [*131*] system in Fig. 49a and b.

Most concentrated unordered alloys with a non-conductor as a component, are semi-conductors. Exceptions are solid solutions of Zr—B ($\varrho_{273} = 30 \times 10^{-8}$ ohm m.) as found by GLASER [*74*] and the long known palladium-hydrogen alloys [*M 1*].

*δ) Influence of coldwork on the resistivity of binary alloys.* Mechanical deformation—or bombardment—experiments with binary alloys have become of great interest for the study of order and disorder phenomena. The change of resistivity is not a linear function of plastic deformation, and, as has been found for some cases is not always positive (Fig. 50 gives LINDE's results [*136b*] for Cu—Al and Au—Fe alloys).

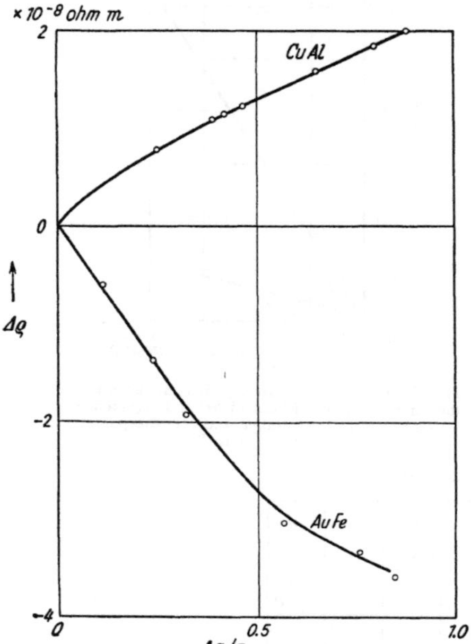


Fig. 50. Change of resistivity as a function of the degree of plastic deformation for a copper-aluminium alloy containing 10 at. % Al and a gold-iron alloy containing 8.7 at. % Fe, according to LINDE.

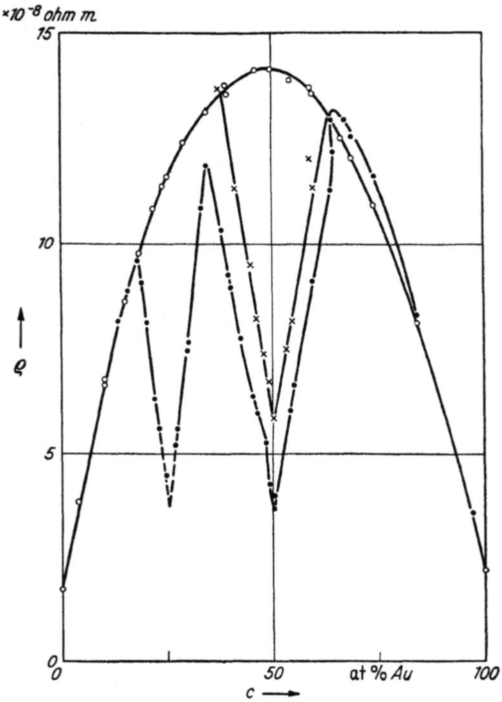


Fig. 51. Resistivity of gold-copper alloys: (○) quenched from 923° K; (●) annealed at 473° K (that is, cooling from 670 to 470° K in 380 hrs., at 473° K during 150 hours); (×) cooling from 670 to 470° K in 50 min. (LINDE).

Most results have been interpreted in terms of annihilating vacancies, interstitials and dislocations, whereas for the resistance decrease effects which so far seem to occur in alloys with a transition metal as one of the components only, assumptions have been made based on atomic rearrangement effects with resultant different distributions of s-electrons in the s- and d-bands of the transition metal components of the alloy.

LINDE found that cold worked Cu—Al and Ag—Sn alloys showed a smaller deviation from (28.1) and (28.2) than when in an annealed state. From this he concludes that in an annealed dilute alloy the solute atoms are not uniformly distributed, but that they might form clusters. Low temperature resistance properties of Ag—Mn and other alloys support this assumption [*70a*]. Bombarding experiments have often been carried out with alloys of a composition in the neighbourhood of an intermediate phase (or compound).

**29. Intermediate phases of binary solid solutions.** From the physical process which controls conductivity, it follows that the conductivity of a material will increase with increasing crystallographic order. This is demonstrated clearly by JOHANNSON and LINDE in their experiments [*103*] with a continuous solid solution of Au—Cu alloys. On quenching the alloys from the melt, a disordered solid solution is obtained, and the diagram of resistivity against composition shows a normal behaviour. An annealing of the samples at 473° K, which was followed by resistance measurements at normal temperature showed that resistivity for the compositions $Cu_3Au$ and CuAu (see Fig. 51) was decreased. Such effects have been found for many intermediate phases or intermetallic compounds[1] and gives, in combination with other physical informations (X-ray analysis, specific heat measurements), an insight into the structural properties of the compounds. That it is also possible to detect liquid compounds by resistance measurements has been proved recently by DRUGAS, REHM and WILKINSON for sodium-potassium solutions at 493° K [*43*].

The order-disorder phenomena are of general interest, and the resistance of an intermediate phase of an alloy is a suitable indicator for the degree of order present. The internal transformations, affected by particle bombardment may increase or decrease order.

For example, BLEWITT [*15*] bombarded the $Cu_3Au$ compound with 3 MeV neutrons, using fast particles ($2.5 \times 10^{15}$ neutrons/m.$^2$ sec.) and slow particles ($1 \times 10^{16}$ neutrons/m.$^2$ sec.) at 423° K. After bombarding through 500 hours, the resistivity of the sample carefully annealed and partially ordered prior to the experiment, decreased from $7.92 \times 10^{-8}$ ohm m. to $5.94 \times 10^{-8}$ ohm m.

GLICK *et al.* found that the decrease depends on the intensity of the flux [*75*]. A disordered compound of $Cu_3Au$ showed a decrease of resistance of the order of 7% when a flux intensity of $4 \times 10^{23}$ neutrons/m.$^2$ sec. was used. Increasing the flux to $6.5 \times 10^{23}$ neutrons/m.$^2$ sec., the resistance did not change. For an ordered $Cu_3Au$ alloy the flux of lower intensity decreased the resistance by 3%, whereas bombardment with the high intensive neutron beam increased the resistance by 23% above its original value. This probably means, that the number of dislocations,—the experiment being done at a temperature too low for a recovery of dislocation effects—, formed during the less intense irradiation, is too small to compensate for the ordering caused by the annihilation of already present vacancies and interstitials. With more intense flux, the influence of the dislocations outweighs that of the ordering, if present.

More recent experiments by DIXON, MEECHAN and BRINKMAN [*40*] with ordered $Cu_3Au$ under bombardment of 1 MeV electrons at 373 and 78° K gave a decrease in $\varrho$ of 3% for the high temperature bombardment and no decrease (less than 1%) for the low temperature bombardment. From this they conclude that the generated imperfections in their experiments are vacancies, which recombine at the lower temperature during the bombardment[2]. We have mentioned these experiments, only to give an impression of the type of work that can be donne with these intermediate phases.

## b) Properties of magnetic alloys.

**30. Dilute binary solid solutions.** The resistivity increase of dilute binary solid solutions, containing a transition metal as solute, is in general proportional

[1] The question, which of the expressions "intermediate phases" or "intermetallic compounds" is righteously used, is "to some extent a matter of definition" (see [*H 2*]).

[2] See also R. A. DUGDALE and A. GREEN [*48*].

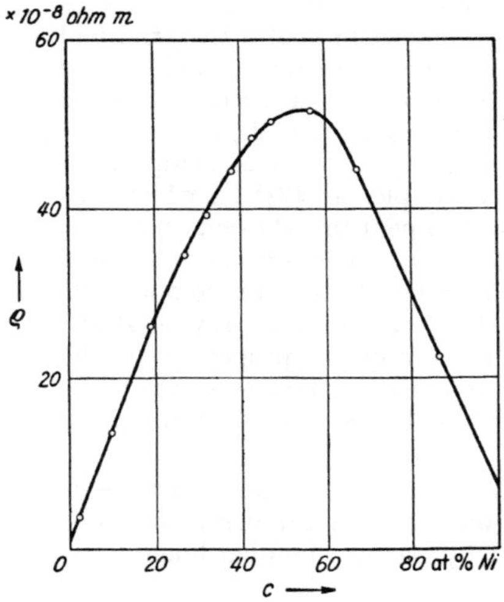


Fig. 52. Resistivity of copper-nickel alloys as a function of the nickel concentration in atom per cent. Data taken from KRUPKOWSKI and DE HAAS.

to the concentration, but fails NORBURY's rule. Apart from topics discussed above and from anomalous effects in $\varrho$ at low temperatures, these dilute solid solutions do not seem to show different behaviour compared with solid solutions with non-transition metals. SHIBUYA [*190*] concludes that for dilute Cu—Ni alloys, the number of effective electrons per unit volume decreases with increasing Ni concentration, which suggests the absorption of electrons into the d-band of the transition metal, as also may be seen from results of magnetic measurements[1] (see below).

**31. Concentrated binary solid solutions.** For concentrated solutions of a non-magnetic and a magnetic metal deviations from the normal behaviour in $\varrho$ can be expected to occur more easily.

For example, for a nickel ion a magnetic moment of 0.61 BOHR magnetons has been assumed (PAULING [*167*]), or, in its d-band, the ion has 0.61 non compensated spins per atom. It has been found that when alloying Ni with Cu the ferromagnetic properties disappear when the composition exceeds 60 at. % Cu-content. The first measurements which connected the magnetic and electrical properties with temperature, and below room temperature, were those of KRUPKOWSKI and DE HAAS [*129*]. The resistivity of the Cu—Ni wires showed kinks in the $\varrho(T)$ curves at temperatures which were lower for smaller Ni concentrations. These kinks represent the CURIE temperature, that is, the temperature below which the alloy under consideration becomes ferromagnetic.

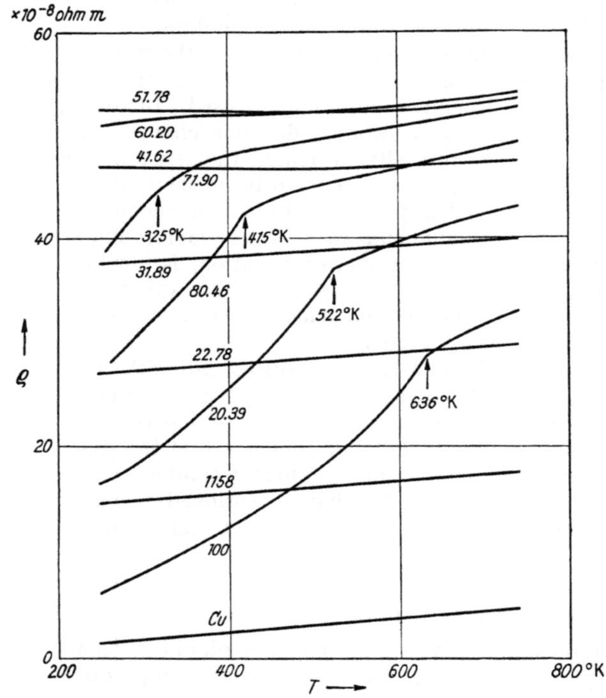


Fig. 53. Resistivity as a function of temperature for several copper-nickel alloys. The numbers indicate the nickel amount in atom per cent.

In the $\varrho(c)$-curve the effect is not so clearly visible. As an illustration of the resistance behaviour, which is typical for all binary alloys which are ferromagnetic

[1] See also [*H 2*], [*M 3*].

above a certain limiting value of the concentration of the magnetic component, we reproduce for copper nickel a resistivity-concentration curve obtained by KRUPKOWSKI and DE HAAS (Fig. 52) and a resistivity-temperature curve taken from SVENSSON [*203*] for the temperature region we are at present interested in (Fig. 53). Moreover, SVENSSON calculated from his measurements the value $\varrho_{\text{ferr}}$, that is the contribution of the ferromagnetism to the resistivity, given by a simple additivity rule:

$$\varrho_{\text{ferr}} = \varrho_{\text{observed}} - (\varrho_{\text{alloy}} + \varrho_i).$$

$\varrho_i$ is the residual resistance, $\varrho_{\text{alloy}}$ is the temperature dependant part of the resistance of the alloy, which SVENSSON calculates with the GRÜN EISEN formula. The values for $\varrho_{\text{ferr}}$ against $c$ are given in Fig. 54 in which also the CURIE temperatures are plotted which follow from the KRUPKOWSKI, DE HAAS, and SVENSSON results.

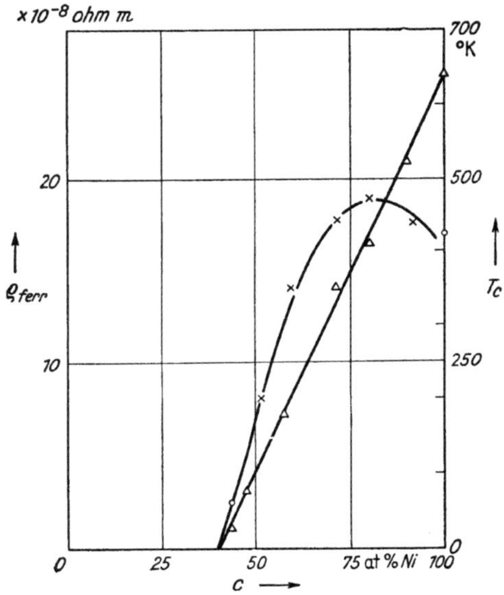


Fig. 54. $\varrho_{\text{ferr}}$ (×) and $\Theta_c$ (△) following from resistance measurements as a function of the nickel concentration in copper-nickel alloys.

These effects have been investigated for many alloys (such as Ag—Pd, Au—Pd) and are still often scope of investigation. We mention some of the more recent results: Ni—Si (DOMINICALI [*41*]), Ni—Al (GUSEVA [*86*]), Cr—Te[1] (FAKIDEW *et al.* [*58*]), Cu—Fe[2] (UCHIDA [*210*]). As new evidence we mention the detection of a NÉEL point, which indicates the temperature below which a material is anti-ferromagnetic, by resistance measurements. GIANSOLDATI and LINDE investigated Au—Mn alloys and $Au_3Mn$ compounds and found NÉEL temperatures of 500 and 150° K, respectively [*72*]. Their results are given in Fig. 55. Below these temperatures, the temperature coefficient of resistivity increases markedly to $4.1 \times 10^{-3}$ degree$^{-1}$ for AuMn and to $4.0 \times 10^{-3}$ degree$^{-1}$ for $Au_3Mn$ which is the order of magnitude for pure gold.

For concentrated alloys with a transition metal as a solute component, THOMAS suggests an universal $R(T)$-curve in the normal temperature region below the CURIE point of the alloy [*208b*]. As shown before, the temperature coefficient of the resistivity increase of this type of alloy is negative. With suitably chosen concentration of the transition metal under consideration, the temperature coefficient of the resistivity of the alloy can be made positive, zero, or negative for temperatures neither too far below nor above room temperature. At lower temperatures, the temperature coefficient of the alloy becomes positive again, so that the $R(T)$-relation is represented by an $S$-shaped curve.

THOMAS has collected data on Cu—Mn, Cu—Ni, Fe—Al, Fe—Si, Ni—Cr, and Ni—Cu—Zn alloys and he suggests a certain state of the electrons ($K$ state)

[1] Cr 51.5%, Te 48.5% is a conductor with $\varrho_{293^\circ\text{K}} = 5 \times 10^{-6}$ ohm m., CURIE point 331° K, whereas for example MnTe is an antiferromagnetic, semiconducting compound.

[2] Maximum in $\varrho$ at 0.26 at. % Fe. The effect is dubious since for smaller concentrations Fe precipitates (HETHORINGTON [*92*]).

to be responsible for this effect. Though there seems to be no theoretical explanation so far, it is a remarkable fact, as is demonstrated in Fig. 56, that the $R(T)$-curve of this type of binary alloys allows a generalization.

**32. Magneto-resistance of binary magnetic alloys.** For binary solid solutions with non-magnetic components, normal magneto-resistance effects as shown by

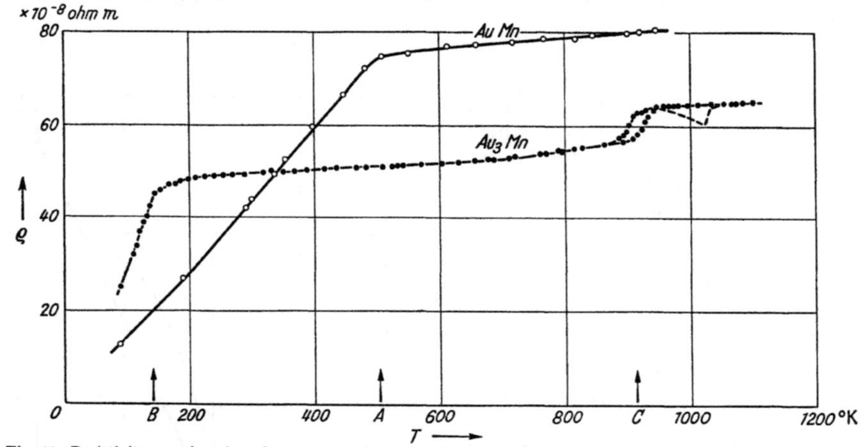


Fig. 55. Resistivity as a function of temperature for a gold-manganese alloy and a compound, according to GIANSOLDATI and LINDE. (A) NÉEL temperature for AuMn (500° K). (B) NÉEL temperature for $Au_3Mn$ (150° K). (C) Order-disorder transformation for $Au_3Mn$ (≈923° K).

KOHLER's diagram have been found (see Sect. 15). Because of the larger values of $R_0$, the resistance in zero field, the $\Delta R/R_0$ value is smaller for a diluted alloy than for the solvent metal. The $\Delta R/R_0$ values for the diluted alloy are generally on the KOHLER line for the solvent metal, or on a line parallel to it [*B 6*], [*70*]. Hence, the alloys with transition elements are of more interest, in this respect. The change of CURIE temperature with concentration, which for some cases is a small effect in the $R(T)$-curve, shows out better in magneto-resistance measurements. The normal magneto-resistance effects as described for pure ferromagnetic metals have also been found for alloys at temperatures higher than 80° K.

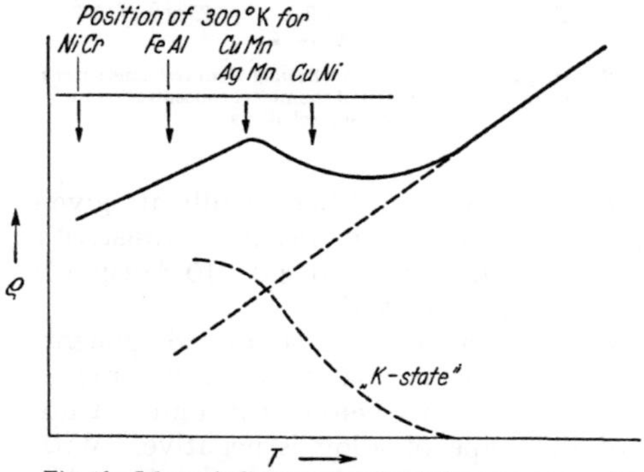


Fig. 56. Schematic diagram according to THOMAS, representing the resistance—temperature relation for some typical binary alloys.

The study of alloys is of interest for the influence of orientation of magnetic domains, connected with the structural properties of the alloys. The domain orientation depends, not only on the composition of the alloy, but also on the crystal- and magnetic anisotropies, and on such factors as strain. Metals with positive and negative magnetostriction increase or decrease their resistivity under longitudinal tension. This is illustrated in Fig. 57a and b for permalloy 89 (negative magnetostriction) and permalloy 78 (positive magnetostriction). A combination of a magnetic and a structural transformation, as follows from specific heat measurements, has been found in the case of Cr—Te. The resistance decreases almost in the same ratio for a transverse as for a longitudinal field at room temperature,—which is not far from the CURIE temperature (330° K), with a maximum effect at the CURIE temperature. Specific heat measurements have indicated that at 330° K a structural transformation takes place (see [*58*]).

**33. Alloys with more than two components.** Groups of ternary alloysystems have been investigated for practical purposes, in particular for the construction of standard resistors with a negligible temperature coefficient at room temperature. Examples are series of Cu—Mn alloys with addition of Al (MALMAN alloys) and of Cu—Ni alloys (manganin) with addition of Sn or Zn. Without claiming any completeness, we refer to earlier work published by the German Physikalisch

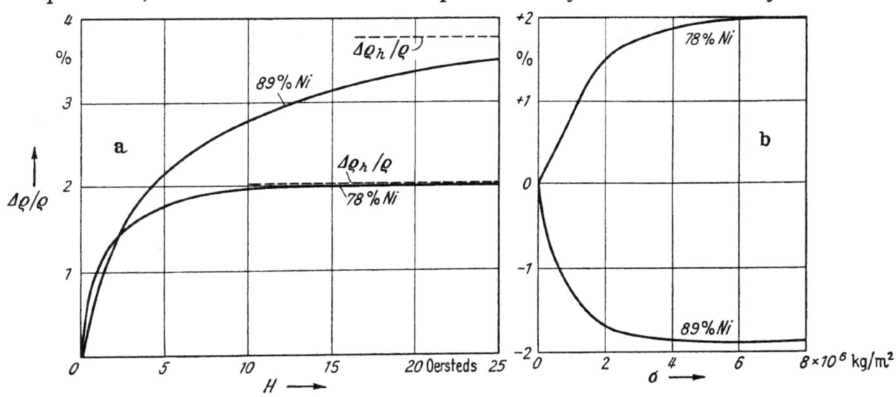


Fig. 57a and b. Comparison between (a) magneto-resistance increase and (b) resistance increase by tension for 78 permalloy (positive magneto striction) and 89 Permalloy (negative magnetostriction) (taken from BOZORTH [*B 4*]). For 78 permalloy, $\Delta\varrho/\varrho$ has the same limiting value for a magnetic field as for tension.

Technische Reichsanstalt and the more recent work of Japanese investigators (from the Engineer Research Institute of Kyoto University).

For magnetic alloys, the electrical resistance effects are of subsidiary interest, the main purpose generally being the construction of an alloy that meets certain demands as regards magnetic or mechanical properties.

For low temperature investigation, the adding of a third element to binary solid solutions is of interest in so far as it changes the electron/atom ratio of the material, which in turn causes changes in magnetic and electrical properties. With a Ti alloy containing three solute elements, Mn 4.7%, Al 3.99% C 0.14%, TYLER, NESBITT and WILSON [*209b*] succeeded in making metal with a constant resistivity between 20 and 300° K. Other examples will undoubtedly be found in the literature.

**34. Liquid alloys.** A current through a liquid binary solution changes the composition of the liquid at both electrodes. SKAUPY [*195*] gives an explanation for this effect (which was found for a solution of alkali metals in mercury) by suggesting that the alkali ions, which are surrounded by mercury ions in the case of a dilute solution, are pushed by the moving mercury ions towards the negative electrode, where consequently an increase in solute ion concentration will be found. The existence of an ion current is demonstrated by the evidence, that the internal friction of the liquid influences the conductivity. Since addition of alkali metals decreases this internal friction, the conductivity will increase.

CLAY [*31*] found analogous results for solutions of gold and silver in mercury.

For example, for 10.9 at. % gold the conductivity of the amalgams is 1.0069 times the conductivity of pure mercury at the same temperature.

*Acknowledgments.* The author is indebted to Dr. G. J. van den Berg and Dr. J. de Nobel for reading the manuscript and for their comments; and, for his help with language difficulties to Mr. M. R. Chari, M.Sc. Further he has appreciated Dr. H. Jones' willingness to send the manuscript of his theoretical article when starting the preparation of the present paper.

## General bibliography.

[*A 1*] Arkel, A. E. van: Reine Metalle. Berlin: Springer 1939.
[*B 1*] Becker, R., u. W. Döring: Ferromagnetismus. Berlin: Springer 1939.
[*B 2*] Boas, W., and J. K. Mackenzie: Progress in Metal-Physics, vol. 2 p. 90. London: Butterworth Scientific Publications 1950.
[*B 3*] Borelius, G.: Grundlagen des metallischen Zustandes. In Handbuch der Metallphysik, Bd. 1. Leipzig 1935.
[*B 4*] Bozorth, R. M.: Ferromagnetism. New York: Van Nostrand Co. Inc. 1951.
[*B 5*] Broom, T.: Adv. Physics **3**, 26 (1954).
[*B 6*] Bueren, H. G. van: Z. Metallkde. **46**, 272 (1955).
[*F 1*] Frost, B. R. T.: Progress in Metal Physics, vol. 5 p. 96. London: Pergamon Press Ltd. 1954.
[*F 2*] Fröhlich, H.: Elektronentheorie der Metalle. Berlin: Springer 1936.
[*G 1*] Grüneisen, E.: Handbuch der Physik, Bd. XIII. Berlin: Springer 1928.
[*G 2*] Grüneisen, E.: Ergebn. exakt. Naturw. **21**, 50 (1945).
[*H 1*] Haughton, J. L., and A. Prince: The Constitutional Diagrams of Alloys: A Bibliography. Inst. Met. Monograph 2, London, 1956.
[*H 2*] Hume-Rothery, W., and G. V. Raynor: The structure of metals and alloys. Inst. Met. Monograph 1, London, 1954.
[*J 1*] Jones, H.: Encycl. of Physics, vol. XIX. Berlin: Springer 1956.
[*J 2*] Justi, E.: Leitfähigkeit und Leitungsmechanismus fester Stoffe. Göttingen: Van den Hoeck & Ruprecht 1948.
[*L 1*] Lumsden, J.: Thermodynamics of Alloys. Inst. Met. Monograph 11, London, 1952.
[*M 1*] Meissner, W.: Handbuch der Experimentalphysik, Bd. XI/2. Leipzig: Akad. Verlagsgesellschaft 1935.
[*M 2*] Metals Reference Book (ed.: C. J. Smithells). London: Butterworth 1949.
[*M 3*] Mott, N. F., and H. Jones: Theory of the properties of metals and alloys . Oxford 1936.
[*S 1*] Seitz, F.: Adv. Physics **1**, 43 (1952).
[*T 1*] See, among others, Terman, F. E.: Radio Engineers Handbook. New York: McGraw-Hill Book Co. Inc. 1943.
[*V 1*] Voigt, W.: Lehrbuch der Kristallphysik. Leipzig: Teubner 1910.
[*W 1*] Wilson, A. H.: The Theory of Metals, 2nd ed. New York: Cambridge University Press 1954.

## Special references.

[*1*] Aarts, W. H., and R. H. Jarvis: Acta met. **2**, 87 (1954).
[*2*] Adenstedt, H. K.: Trans. Amer. Soc. Met. **44**, 949 (1952).
[*3*] Agte, C., H. Alterthum, K. Becker, G. Heyne u. K. Moers: Z. anorg. allg. Chem. **196**, 129 (1931).
[*4*] Aschermann, G., u. E. Justi: Phys. Z. **43**, 207 (1942).
[*5*] Bannister, J. R., S. Legvold, and F. H. Spedding: Phys. Rev. **94**, 1140 (1954).
[*6*] Bartlett, J. H., and G. J. Dienes: Phys. Rev. **89**, 848 (1953).
[*7*] Beck, P. A.: Adv. Physics **3**, 245 (1954).
[*8*] Bender, D.: Z. Metallkde. **40**, 257 (1949).
[*9*] Benedicks, C.: Jb. Rad. **13**, 351 (1916).
[*10*] Berg, G. J. van den and L. C. van der Marel: Publ. I.I.R. Proc. Grenoble meeting, Sept. 1954; p. 161.
[*11*] Berghout, C. W.: Physica, Haag **18**, 978 (1952).
[*12*] Bernini, A.: Phys. Z. **5**, 241, 406 (1904); **6**, 74 (1905).
[*13*] Blackman, M.: Proc. Phys. Soc. Lond. A **64**, 681 (1951).
[*14*] Blau, F.: Elektrotechn. Z. **25**, 198 (1905).
[*15*] Blewitt, T. H., and R. R. Coltman: Acta met. **2**, 549 (1954).
[*16*] Bowen, D., R. R. Eggleston and R. H. Kropschott: J. Appl. Phys. **23**, 630 (1952).
[*17*] Bowen, D., and G. W. Rodenbach: Acta met. **1**, 649 (1953).
[*18*] Bridgman, P. W.: Proc. Amer. Acad. **56**, 59 (1921).

[18a] Bridgman, P. W.: Proc. Amer. Acad. **57**, 131 (1922).
[18b] Bridgman, P. W.: Proc. Amer. Acad. **60**, 305, 423 (1925).
[18c] Bridgman, P. W.: Proc. Amer. Acad. **62**, 207 (1927).
[18d] Bridgman, P. W.: Proc. Amer. Acad. **67**, 305 (1932).
[18e] Bridgman, P. W.: Proc. Amer. Acad. **68**, 27, 95 (1933).
[18f] Bridgman, P. W.: Proc. Amer. Acad. **70**, 96 (1935).
[18g] Bridgman, P. W.: Proc. Amer. Acad. **70**, 285 (1935).
[18h] Bridgman, P. W.: Proc. Amer. Acad. **72**, 157 (1937).
[18i] Bridgman, P. W.: Proc. Amer. Acad. **74**, 425 (1942).
[18j] Bridgman, P. W.: Proc. Amer. Acad. **76**, 1 (1945).
[18k] Bridgman, P. W.: Proc. Amer. Acad. **79**, 149 (1951).
[18l] Bridgman, P. W.: Proc. Amer. Acad. **81**, 165 (1952).
[18m] Bridgman, P. W.: Proc. Amer. Acad. **82**, 83 (1953).
[18n] Bridgman, P. W.: Proc. Amer. Acad. **76**, 55 (1945).
[18o] Bridgman, P. W.: Phys. Rev. **42**, 848 (1932).
[19] Brinkman, J. A., C. E. Dixon u. C. J. Meechan: Acta met. **2**, 38 (1954).
[20] Broom, T.: Proc. Phys. Soc. Lond. B **65**, 871 (1952).
[21] Broom, T., and W. K. Clothier: Austral. J. Sci. Res. A **5**, 119 (1952).
[22] Brown, A., M.W. Zemansky and H. A. Boorse: U.S. Nat. Bur. Stand. Circ. **519**, 99 (1952).
[23] Bruggeman, D. A. M.: Ann. Phys. (5) **25**, 645 (1936).
[24] Brunke, F.: Ann. Phys. (5) **21**, 139 (1934).
[25] Bueren, H. G. van, and P. Jongenburger: Nature, Lond. **175**, 544 (1955).
[26] Burgers, W.: Proc. Akad. Amsterdam **37**, 515 (1936).
[27] Butterworth, S.: Proc. Phys. Soc. Lond. **3**, 312 (1921).
[28] Casimir, H. B. G., and A. N. Gerritsen: Physica, Haag **8**, 1107 (1941); Commun. Kamerlingh Onnes Lab. 265b[1]. — Donovan, B., and G. K. T. Conn: Nature, Lond. **162**, 336 (1948). — Phil. Mag. (7) **41**, 770 (1950).
[29] Christie: Roy. Soc. Phil. Trans. **1833**.
[30] Clausing, P., and G. Moubis: Physica (former ed.), Haag **7**, 245 (1927).
[31] Clay, J.: Physica, Haag **7**, 838 (1940).
[32] Clement, J. R., and E. H. Quinell: Phys. Rev. **92**, 258 (1953).
[32a] Clusius, K., u. L. Schaginger: Z. angew. Phys. **4**, 442 (1952).
[33] Cooper, H. G., J. S. Koehler and J. W. Marx: Phys. Rev. **94**, 496 (1954); **97**, 599 (1955).
[34] Cottrel, A. H., and A. T. Churchman: J. Iron Steel Inst. **162**, 271 (1949).
[35] Deem, H. W.: U.S. A. E. C. publ. 1953 (BMI-853).
[36] Dellinger, J. H.: J. Franklin. Inst. **170**, 213 (1910).
[37] Denney, J.: J. Bull. Amer. Phys. Soc. **27**, 9 (1952).
[38] Dexter, D. L.: Phys. Rev. **87**, 768 (1952).
[39] Dictionary of Applied Physics, vol. V, p. 683ff. London: MacMillan Co. Ltd. 1922.
[40] Dixon, C. E., C. J. Meechan and J. A. Brinkman: Phil. Mag. (7) **44**, 449 (1953).
[41] Dominicali, C. A., and F. A. Otter: J. Appl. Phys. **26**, 377 (1955).
[42] Döring, W.: Ann. Phys. **32**, 259 (1938).
[43] Drugas, P. G., I. R. Rehm, and W. D. Wilkinson: U.S.A. E. C. publ. 1953 (ANL-5115).
[44] Druyvesteyn, M. J., and J. A. Manintveld: Nature, Lond. **168**, 868 (1951).
[44a] Manintveld, J. A.: Nature, Lond. **169**, 623 (1952).
[45] Druyvesteyn, M. J.: Rev. gén. Sci. appl. **11**, 1 (1954).
[46] Druyvesteyn, M. J.: Physica, Haag **17**, 748 (1951).
[47] Dugdale, R. A.: Phil. Mag. (7) **43**, 912 (1953).
[48] Dugdale, R. A., and A. Green: Phil. Mag. (7) **45**, 163 (1954).
[49] Egger, D. T., and M. J. Lauberstein: Phys. Rev. **91**, 238 (1953).
[50] Eggleston, R. R.: Acta met. **1**, 679 (1953).
[51] Eggleston, R. R.: J. Appl. Phys. **23**, 1400 (1952).
[52]—[55] not in text.
[56] Erfling, H. D.: Ann. Phys. (5) **41**, 100 (1942).
[57] Eucken, A., u. H. Schürenberg: Ann. Physik (5) **33**, 1 (1938).
[58] Fakidov, J. G., N. P. Grazhdankina i A. K. Kikoin: Dokl. Akad. Nauk, SSSR. **68**, 491 (1949).
[59] Fast, D.: Z. anorg. allg. Chem. **241**, 42 (1939).
[60] Fine, M. E., E. S. Greiner and W. C. Ellis: J. Met. Trans. **191**, 56 (1951).
[61] Fisher, U.: Z. phys. Chem. Abt. B **8**, 207 (1930).
[62] Frank, V., and O. G. Jeppesen: Phys. Rev. **89**, 1153 (1950).
[63] Frenkel, J.: Kinetic theory of liquids. Oxford 1936.

[1] Commun. Kamerlingh Onnes Lab., Leiden, will be abbreviated in the following by: Commun.

[64] FRIEDEL, J.: Phil. Mag (7) **43**, 153 (1952).
[65] FÜRTH, R.: Proc. Cambridge Phil. Soc. **37**, 34, 252 (1941).
[66] GEIBEL, W.: Z. anorg. allg. Chem. **69**, 430 (1911); **70**, 240, 242 (1911).
[67] GEISS, W., u. J. A. M. VAN LIEMPT: Z. Metallkde. **17**, 194 (1925).
[68] GERLACH, W.: Phys. Z. **33**, 953 (1932).
[69] GERLACH, W., u. K. SCHNEIDERHAN: Ann. Phys. (5) **6**, 772 (1930).
[70a] GERRITSEN, A. N., and J. O. LINDE: Physica, Haag **17**, 573 (1951); Commun. 285c[1]. — GERRITSEN, A. N.: Physica **18**, 877 (1952); Commun 290d.
[70b] KORRINGA, J., and A. N. GERRITSEN: Physica, Haag **20**, 457 (1953); Commun Suppl. 106[1].
[71a] GERSTENKORN, H., u. F. SAUTER: Naturwiss. **38**, 158 (1951).
[71b] GERSTENKORN, H.: Ann. Physik (6) **10**, 49 (1952).
[72] GIANSOLDATI, A., et J. O. LINDE: J. Phys. Radium **16**, 341 (1955).
[73] GINGRICH, N. S.: Rev. Mod. Phys. **15**, 90 (1943).
[74] GLASER, F. W., and B. POST: J. Metals **5**, 1117 (1953).
[75] GLICK, H. L., F. C. BROOKS, W. F. WITRIG and W. E. JOHNSON: U.S. A.E.C. publ. 1952 (WAPD-T 7).
[76] GOENS, E., u. E. SCHMIDT: Naturwiss. **19**, 376 (1931).
[77] GOENS, E., u. E. GRÜNEISEN: Ann. Phys. (5) **14**, 164 (1932).
[78] GONSEN, B. W.: Titanium Rept. Symposium. O.N.R. Washington D.C. 1948.
[79] GRUBE, G., u. J. HILLE: Z. Elektrochem. **40**, 101 (1934).
[80] GRUBE, G., H. VOSSKÜHLER u. H. VOGT: Z. Elektrochem. **38**, 873 (1932).
[81] GRÜNEISEN, E.: Ann. Phys. (5) **16**, 530 (1933).
[82] GRÜNEISEN, E., u. H. K. ADENSTEDT: Ann. Phys. (5) **31**, 714 (1938).
[83] GRÜNEISEN, E.: Ann. Phys. (5) **40**, 543 (1941).
[84] GULYAEV, A. P., and E. F. TRUSOVA: Zhur. Tekh. Fiz. **20**, 66 (1950).
[85] GUNN, R.: Rev. Sci. Instrum. **9**, 267 (1938).
[86] GUSEVA, L. N.: Dokl. Akad. Nauk SSSR. **77**, 415 (1951).
[87] HAAS, W. J. DE, en J. VOOGD: Versl. Kon. Akad. Amsterdam **37**, 872 (1928); Commun. 194c[1].
[88] HAAS, W. J. DE, J. DE BOER and G. J. VAN DEN BERG: Physica, Haag **1**, 1115 (1933/34); Commun. 233b[1].
[89] HAAS, W. J. DE, and J. BLOM: Physica, Haag **4**, 767 (1937); Commun. 249c[1].
[90] HAM, W. R., and C. H. SAMANS: Science, Lancaster, Pa. **104**, 38 (1946).
[91] HANSEN, M., W. R. JOHNSON and J. M. PARKS: J. Met. Trans. A.S.M.E. (1951) 1184.
[92] HETHORINGTON, F. E., and J. REEKIE: J. Appl. Phys. **22**, 1293 (1951).
[93a] HILL, A. V.: J. Sci. Instrum. **25**, 225 (1948).
[93b] COPELAND, K., A. C. DOWNING and A. V. HILL: J. Sci. Instrum. **30**, 40, 44 (1953).
[94] HIRONI, T., and N. HORI: Sci. Rep. Tôhoku Imp. Univ. **30**, 125 (1942).
[95] HOLBORN, L.: Ann. Phys. **59**, 145 (1919). — Z. Physik **8**, 58 (1921).
[96] HORN, F. H.: Phys. Rev. **84**, 855 (1951).
[97] HOUSTON, W. V.: Phys. Rev. **88**, 1321 (1952).
[98] HUNTER, S. C., and F. R. N. NABARRO: Proc. Roy. Soc. Lond., Ser. A **220**, 542 1953).
[99] HUNTINGTON, H. B., and F. SEITZ: Phys. Rev. **61**, 311 (1941).
[100] JAEGER, F. M., and J. E. ZANSTRA: Proc. Acad. Amsterdam **36**, 636 (1933).
[101a] JAEGER, W.: Z. Instrumentenkde. **26**, 73 (1906).
[101b] JAEGER, W., u. H. v. STEINWEHR: Ann. Phys. (4) **43**, 1165 (1914).
[102] JAMES, N. R., S. LEGVOLD and E. H. SPEDDING: Phys. Rev. **88**, 1092 (1952).
[103] JOHANNSON, V., u. J. O. LINDE: Ann. Phys. (5) **25**, 1 (1936).
[104] JONGENBURGER, P.: Nature, Lond. **175**, 545 (1955).
[105] JONGENBURGER, P.: Supp. bull. I. I. F., Annexe. **3**, 411, 1955.
[106] JUSTI, E., u. H. SCHEFFERS: Phys. Z. **37**, 383, 475 (1936).
[107] JUSTI, E.: Forsch. u. Fortschr. **18**, 65 (1942). — Phys. Z. **41**, 486 (1940).
[108] JUSTI, E.: Z. Naturforsch. **4**a, 472 (1949).
[109] JUSTI, E., u. H. KRAMER: Phys. Z. **41**, 196 (1940).
[110] JUSTI, E., J. KRAMER u. R. SCHULZE: Phys. Z. **41**, 308 (1940).
[111] KAMERLINGH ONNES, H., and G. HOLST: Commun. 142a, 1914.
[112] KAMERLINGH ONNES, H., and W. TUYN: Commun. Suppl. 58, 1926.
[113] KANNULUIK, W. G.: Proc. Roy. Soc. Lond., Ser. A **141**, 159 (1933).
[114a] KAPITZA, P.: Proc. Roy. Soc. Lond., Ser. A **105**, 691 (1924).
[114b] KAPITZA, P.: Proc. Roy. Soc. Lond., Ser. A **119**, 358 (1928).
[114c] KAPITZA, P.: Proc. Roy. Soc. Lond., Ser. A **123**, 292, 342 (1929).
[115] KAUFMANN, J. W., and J. S. KOEHLER: Phys. Rev. **88**, 149 (1952).
[116] KAYA, S., and T. TAKAKI: Sci. Rep. Tokio Univ. **1936**, 314.
[117] KEESOM, P., and N. PEARLMAN: Vol. XIV, this encyclopedia.

[1] See note p. 223.

[118] KEESOM, W. H., and C. W. CLARK: Physica, Haag **2**, 557 (1935); Commun. 235e[1].
[119] KEESOM, W. H., and B. G. DAMMERS: Physica, Haag **2**, 1051 (1935); Commun. 239d[1].
[120a] KEESOM, W. H., and B. G. DAMMERS: Physica, Haag **2**, 1080 (1935); Commun. 239e (Pt 68)[1].
[120b] DAMMERS, B. G.: Thesis Leiden, Noord Hollandse Uitg. Mij. Amsterdam 1936.
[121] KEESOM, W. H., and M. DÉSIRANT: Physica, Haag **8**, 273 (1941); Commun. 257b[1].
[122] KEESOM, W. H., and J. N. VAN DEN ENDE: Proc. Acad. Amsterdam **34**, 210 (1931); **35**, 143 (1932); Commun. 213c, 219b.
[123] KEESOM, W. H., and J. A. KOK: Proc. Acad. Amsterdam **35**, 301 (1932). — Physica, Haag **1**, 175 (1933/34); **3**, 1035 (1936); **4**, 835 (1937); Commun. 219d; 230c; 245a; 248e[1].
[124] KEESOM, W. H., and B. KURRELMEYER: Physica, Haag **6**, 633 (1939); Commun. 257a[1].
[125] KLEMM, W., H. SPITZER, W. LINGENBERG u. H. J. JUNKER: Mh. **83**, 629 (1952).
[126] KNAPPWOST, A.: Z. Elektrochem. **57**, 618 (1953).
[127] KOHLER, M.: Ann. Phys. (5) **32**, 211 (1938); (6) **6**, 18 (1949).
[128] KOSOLAPOV, G. F., i A. K. TRAPEZNIKOV: J. exp. theor. Phys. USSR. **6**, 1163 (1936).
[129] KRUPKOWSKI, A., en W. J. DE HAAS: Versl. Kon. Akad. Amsterdam **37**, 714 (1928); Commun. 194a[1].
[130] KUBASCHEWSKI, O. R.: Trans. Faraday Soc. **45**, 931 (1949).
[131] KURNAKOW, N. S., N. PUSCHIN u. N. SENKOWSKI: Z. anorg. allg. Chem. **68**, 123 (1910).
[132] LE CHATELIER, H., and W. GUERTLER: See ref. [G 1], p. 39.
[133] LEGG, V. E.: Bell Syst. Techn. J. **15**, 39 (1936).
[134] LEGVOLD, S., F. H. SPEDDING, F. BARSON and J. F. ELLIOT: Rev. Mod. Phys. **25**, 129 (1953).
[135] LENNARD-JONES, J.E., and H. DEVONSHIRE: Proc. Roy. Soc. Lond., Ser. A **170**, 464 (1939).
[136a] LINDE, J. O.: Thesis, Lund 1939.
[136b] LINDE, J. O.: J. Appl. Sci. Res. B **4**, 73 (1954).
[137] LITTON, F. B.: J. Electrochem. Soc. **98**, 488 (1951).
[138] LÜCKE, K.: Z. Metallkde **42**, 1 (1951).
[139] MACDONALD, D. K. C., and K. MENDELSSOHN: Proc. Roy. Soc. Lond., Ser. A **202**, 104, 523 (1950).
[140] MACDONALD, D. K. C.: Phil. Mag. (7) **43**, 479 (1952).
[141] MANINTVELD, J. A.: Thesis, Excelsior, The Hague, 1954.
[142] MARX, J. W., H. C. COOPER and J. W. HENDERSON: Phys. Rev. **88**, 106 (1952).
[143] MATTHIESSEN, A., u. C. VOGT: Pogg. Ann. **122**, 19 (1864).
[144] MAXWELL, CH. R.: J. Chem. Phys. **17**, 1288 (1949).
[145] MCDONNEL, R., and H. A. KIERSTEAD: Phys. Rev. **93**, 247 (1954) (L).
[146] MCLENNAN, J. C., C. D. NIVEN and J. O. WILHELM: Phil. Mag. (7) **6**, 666, 672 (1928).
[147] MEISSNER, W.: Phys. Z. **27**, 725 (1926). — Z. Physik **38**, 647 (1926).
[148] MEISSNER, W., u. H. FRANZ: Z. Physik **63**, 558 (1930).
[149] MEISSNER, W., u. B. VOIGT: Ann. Phys. (5) **7**, 761, 892 (1930).
[150] MENKE, H.: Phys. Z. **33**, 593 (1932).
[151] MICHELS, W. C., and S. WILFORD: J. Appl. Phys. **20**, 1223 (1949).
[152] MILNER, C. J.: Proc. Roy. Soc. Lond., Ser. A **160**, 207 (1937).
[153] MOLENAAR, J., and W. H. AARTS: Nature, Lond. **166**, 690 (1950).
[154] MOSER, H.: Ann. Phys. (5) **6**, 852 (1930).
[155] MOTT, N. F.: Proc. Roy. Soc. Lond., Ser. A **146**, 465 (1934).
[156] MOTT, N. F.: Proc. Roy. Soc. Lond., Ser. A **153**, 699 (1936).
[157] NACHTIGALL, E.: Z. Metallkde. **43**, 23 (1952).
[158] NEMILOW, V. A.: Z. anorg. Chem. **204**, 46 (1932); **218**, 55 (1934).
[159] NICOLAI, G.: Phys. Z. **9**, 367 (1908).
[160] NORBURY, L.: Trans. Faraday Soc. **17**, 257 (1921).
[161] NORBURY, A. L., and K. KUWADA: Phil. Mag. (7) **4**, 1338 (1927).
[162] NORDHEIM, L.: Ann. Phys. (5) **9**, 607 (1931).
[163] NORTHRUP, E. F.: J. Franklin Inst. **177**, 1, 287 (1914); **178**, 85 (1914).
[164] OGILVIE, G. J., and W. K. CLOTHIER: Austral. J. Phys. **7**, 210 (1954).
[165] OVERHAUSER, A. W.: Phys. Rev. **90**, 393 (1953).
[166] PALLISTER, P. R.: J. Iron Steel Inst. **161**, 87 (1949).
[167] PAULING, L.: Phys. Rev. **54**, 899 (1938).
[168] PERLITZ, H.: Phil. Mag. (7) **2**, 1148 (1926).
[168a] PHILLIPS, V. A., and A. PHILLIPS: J. Inst. Met. **81**, 185 (1952).
[169] PIPPARD, A. P., and G. T. PULLAN: Proc. Cambridge Phil. Soc. **48**, 188 (1952).
[170] PIRANI, M. V.: Phys. Z. **13**, 753 (1912).
[171] POCHAPSKY, T. E.: Acta met. **1**, 747 (1953).

[1] See note p. 223.

[172a] POTTER, H. H.: Phil. Mag. (7) **13**, 238 (1932) (Fig. 4). — Proc. Phys. Soc. Lond. **49**, 671, (1937).
[172b] POTTER, H. H.: Proc. Phys. Soc. Lond. **53**, 695 (1941).
[173] POWELL, R. W.: J. Iron Steel Inst. **162**, 315 (1949).
[174a] POWELL, R. W.: Proc. Roy. Soc. Lond., Ser. A **209**, 525 (1951).
[174b] POWELL, R. W.: Phil. Mag. (7) **44**, 645, 772 (1953). — Nature, Lond. **164**, 153 (1949); **166**, 1110 (1950).
[175] PRY, R. H., u. R. W. HENNING: Acta met. **2**, 318 (1954).
[176] RAUSCH, K.: Ann. Phys. (6) **1**, 190 (1947).
[177] REGEL, A. R.: Zhur. Tekh. Fiz. **18**, 511 (1948).
[178] REICHARDT, G.: Ann. Phys. (4) **6**, 832 (1901).
[179a] RINCK, E.: C. R. Acad. Sci., Paris **192**, 421 (1931) (Ca).
[179b] RINCK, E.: C. R. Acad. Sci., Paris **193**, 1328 (1932) (Ba).
[179c] RINCK, E.: C. R. Acad. Sci., Paris **234**, 845 (1952) (Sr).
[180] ROBINSON, A. T., and J. E. DORN: J. Metals Trans. **3**, 457 (1951).
[181] ROHM, W.: Z. Metallkde. **19**, 196 (1927).
[182] RENTSCHLER, H. C., and J. W. MARDEN: Phys. Rev. **25**, 589 (1925).
[183] RYSCHKEWITZ, E.: Z. Elektrochem. **29**, 474 (1923).
[184] SCALA, E., and W. D. ROBERTSON: J. Metals **5**, 1141 (1953).
[185a] SCHUBNIKOW, L.: Proc. Acad. Amsterdam **33**, 130 (1930); Commun. 207a [1].
[185b] HAAS, W. J. DE, u. J. W. BLOM: Physica, Haag **1**, 134 (1933/34); Commun. 229b [1].
[185c] HAAS, W. J. DE, A. N. GERRITSEN u. W. H. CAPEL: Physica, Haag **3**, 1143 (1936); Commun. 247c [1].
[186] SCHUBNIKOW, L., u. W. J. DE HAAS: Proc. Acad. Amsterdam **33**, 350 (1930); Commun. 207c [1].
[187a] SCHULZE, A.: Z. angew. Chem. **42**, 1095 (1929).
[187b] SCHULZE, A.: Z. Metallkde. **42**, 120 (1951).
[188] Several papers in J. Sci. Instrum. e.g.: **22**, 48, 218 (1945); **23**, 173, 301 (1946).
[189] Several papers in Rev. Sci. Instrum. e.g.: **22**, 513 (1951); **25**, 902 (1954).
[190] SHIBUYA, Y.: Sci. Rep. Res. Inst. Tohoko Univ. (A) **6**, 199 (1954).
[191] SIDHU, S. S., and C. O. HENRY: J. Appl. Phys. **21**, 1036 (1950).
[192] SIEGEL, S.: Phys. Rev. **75**, 1823 (1949).
[193] SIMON, F.: Ergebn. exakt. Naturw. **9**, 223 (1930).
[194] SIMON, F.: Z. phys. Chem. **109**, 136 (1924).
[195] SKAUPY, F.: Z. Physik **3**, 178 (1920).
[196] SMART, J. S., A. A. SMITH and A. J. PHILLIPS: Trans. Amer. Inst. Met. Eng. **145**, 272 (1941).
[197a] SMITH, C. S.: Phys. Rev. **94**, 1068 (1954).
[197b] SMITH, C. S., and E. W. PALMER: Trans. Amer. Inst. Met. Eng. **117**, 225 (1935).
[198] SÖCHTIG, H.: Ann. Phys. (5) **38**, 97 (1940).
[199] STIERSTADT, O.: Phys. Z. **31**, 561 (1930).
[200] STIERSTADT, O.: Z. Physik **80**, 636 (1933); **85**, 310, 697 (1933).
[201] SULLY, A. H., E. A. BRANDES and K. W. MITCHELL: J. Inst. Met. **81**, 585 (1952/53).
[202] SUTRA, GENEVIÈVE: C. R. Acad. Sci., Paris **234**, 2589 (1952); **235**, 707 (1952); **236**, 1643, 2391 (1953).
[203] SVENSSON, B.: Ann. Phys. (5) **22**, 97 (1935); (5) **25**, 263 (1936).
[204] SZABO, Z. G., u. B. LAKATOS: Acta chim. Hung. **4**, 129 (1954). — Research **5**, 1952 (1952).
[205] TAMMAN, G., and K. L. DREYER: Ann. Phys. (5) **16**, 111 (1933).
[206] TANABE, Y.: Sci. Rep. Res. Inst. Tôhoku Univ. A **1**, 267, 275 (1949); **2**, 341, 531 (1950); **3**, 91 (1951).
[207] TAYLOR, G. I.: Proc. Roy. Soc. Lond., Ser. A **145**, 362, 388 (1934).
[208a] THOMAS, H.: Z. Metallkde. **41**, 185 (1950).
[208b] THOMAS, H.: Z. Physik **129**, 219 (1951).
[209a] TYLER, W. W., and A. C. WILSON: U.S. A. E. C. publ. 1952 (KAPL-802).
[209b] TYLER, W. W., A. C. WILSON and G. J. WOLGA: J. Metals **5**, 1238 (1953).
[209c] TYLER, W. W., L. B. NESBITT and A. C. WILSON: J. Metals **5**, 1104 (1953).
[210] UCHIDA, E., u. H. KONDO: Busseiron Kenkyu **59**, 88 (1953).
[211] WEBSTER, W.: Proc. Roy. Soc. Lond., Ser. A **113**, 196 (1927); **114**, 611 (1927). — Proc. Phys. Soc. Lond. **42**, 431 (1929/30).
[212] WEINER, L., P. CHIOTTI and H. A. WILHELM: U.S. A. E. C. publ. 1952 (ISC-305).
[213] WEYERER, H.: Z. Metallkde. **44**, 51 (1953).
[214] WOLTJER, H. R., and H. KAMERLINGH ONNES: Commun. 173a, 1924.
[215] WYATT, J. L.: J. Metals **5**, 903 (1953).

[1] See note p. 223.

# Theory of Electrical and Thermal Conductivity in Metals.

By

H. JONES.

With 19 Figures.

## I. The properties of conduction electrons in metals.

**1. The electron gas model.** The theory of electrical and thermal conductivity in metals, originally developed by DRUDE[1] in 1900 and by LORENTZ[2] in 1905, is based on the conception of the free electron gas. It is assumed that the valency electrons become detached from their atoms and can be regarded as free to move throughout the whole metal. In LORENTZ's theory, and also in the modern theory of conduction, the interaction between electrons is largely neglected. In recent years good reasons have been given for neglecting collisions between electrons in most conduction processes, and thus to a great extent LORENTZ's model, upon which existing conductivity theory is based, finds justification.

It has been shown by BOHM and PINES[3] that the free electrons in a metal undergo oscillations similar to the oscillations of the plasma in a discharge tube. The frequency of these oscillations is high and the associated quantum of energy is in the neighbourhood of 10 ev for most metals. The zero-point energy of these oscillations, half of which is potential energy, accounts for much of the COULOMB energy of interaction between the electrons, with the result that the remaining interaction, to be associated with the gas-like motion of the electrons, is of a short range nature given by

$$\frac{e^2}{r_{12}}\left\{1-\frac{2}{\pi}\,Si(k_c r_{12})\right\}, \tag{1.1}$$

where $r_{12}$ is the distance between the electrons, $Si(x)=\int\limits_0^x \frac{\sin u}{u}\,du$, and $k_c^{-1}$ is a length approximately equal to the average inter-electronic distance, e.g., about 2 Å.

In order to estimate the mean free path for collisions between particles with short range interactions similar to (1.1) we shall replace this rather complicated mathematical expression by the simpler form $\frac{e^2}{r_{12}}\exp\left(-\frac{r_{12}}{a}\right)$, where $a$ determines the range of the interaction and is of the order of an Ångström unit. The total effective cross-section for collision between such particles is given by[4]

$$\pi a^2\left\{\left(\frac{\pi a_H}{\lambda}\right)^2+\left(\frac{a_H}{4a}\right)^2\right\}^{-1}, \tag{1.2}$$

[1] P. DRUDE: Ann. Phys., Lpz. (4) **1**, 566 (1900).
[2] H. A. LORENTZ: Proc. Acad. Sci. Amst. **7**, 438, 585, 684 (1904/05).
[3] D. BOHM and D. PINES: Phys. Rev. **92**, 609 (1953).
[4] N. BOHR: The Penetration of Atomic Particles Through Matter (eqn. 1.4.6.). Copenhagen 1948.

where $\lambda$ is the DE BROGLIE wave-length of an electron with a velocity equal to the relative velocity of approach before collision, and $a_H$ is the radius of the first BOHR hydrogen orbit. Since $a$ is about 1 Å and $\lambda$ for a typical metal about 3 to 4 Å we see that the average cross-section is about 10 square Angstrom units, which implies an effective diameter for collision $d$ of about 3 Å.

At first sight this may seem rather a large diameter but the number of collisions per unit time is greatly reduced by the PAULI exclusion principle which results in a mean free path for electron-electron collisions $l_e$ which is very large compared with the mean free path for electron-lattice interaction. This can be seen as follows[1]: Consider an electron with momentum $\hbar \boldsymbol{k}_1$, where $\hbar$ is PLANCK's constant divided by $2\pi$ and $\boldsymbol{k}_1$ is the wave-vector, which is scattered by collision with a second electron through an angle $\vartheta$ without change of energy to a state $\boldsymbol{k}_1'$ as indicated in Fig. 1. Applying the laws of conservation of energy and momentum, we find that the wave-vector of the second electron $\boldsymbol{k}_2$ must lie

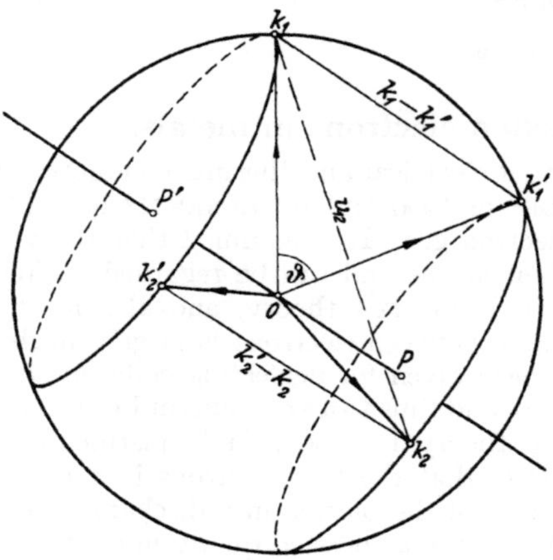


Fig. 1.

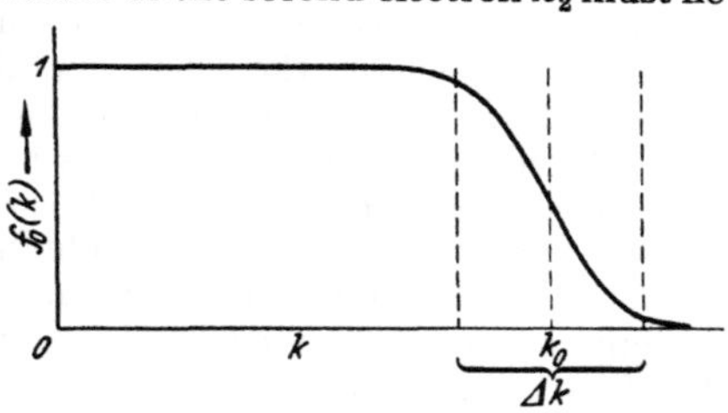


Fig. 2.

Fig. 1. Diagram to illustrate collisions between electrons. To conserve momentum $\boldsymbol{k}_1 - \boldsymbol{k}_1'$ and $\boldsymbol{k}_2' - \boldsymbol{k}_2$ must be generators of the cylinder with axis $P'OP$.

Fig. 2. FERMI distribution function. $\Delta k/k_0$ is of order $kT/\zeta$.

on the plane shown in Fig. 1 at right angles to $OP$. If the direction of $\boldsymbol{k}_1$ is taken as the direction of the $z$ component of momentum and the $x$ direction lies in the plane containing $\boldsymbol{k}_1$ and $\boldsymbol{k}_1'$, the equation of this plane, on which $\boldsymbol{k}_2$ must lie, is

$$k_{2z} = \left(\cot\frac{\vartheta}{2}\right) k_{2x} - k_1. \tag{1.3}$$

In reality the energies of the incident and scattered electrons will not be exactly equal, but they must be nearly so, on account of the nature of the distribution of electrons in momentum space. In equilibrium this distribution is expressed by the well-known FERMI function

$$f_0(\boldsymbol{k}) = \frac{1}{1 + e^{(E-\zeta)/kT}}, \tag{1.4}$$

where the energy $E$, for free electrons, is given by

$$E = \frac{\hbar^2}{2m} k^2, \tag{1.5}$$

and $\zeta$ is the partial potential per electron.

Fig. 2, which is a plot of $f_0$ against $k$, shows that all states having $k$ appreciably less than $k_0$ are fully occupied, and that $f_0$ falls from unity to almost

[1] Cf. W. G. BABER: Proc. Roy. Soc. Lond., Ser. A **158**, 383 (1937).

zero in a range $\Delta k$ where

$$\frac{\Delta k}{k_0} = \frac{\mathrm{k} T}{\zeta}. \tag{1.6}$$

Thus if $\boldsymbol{k}_1$ lies in a shell of radius $k_0$ and thickness $\Delta k$ the electron cannot lose more energy than an amount approximately equal to $\mathrm{k}T$, otherwise the state into which it might be scattered would be occupied. Similarly the second electron can, at most, lose an amount $\mathrm{k}T$ of energy. Hence we see that all the vectors $\boldsymbol{k}_1'$, $\boldsymbol{k}_2$ and $\boldsymbol{k}_2'$ must lie on a shell of thickness $\Delta k$ at the FERMI surface. If now we fix $\boldsymbol{k}_1$ and $\vartheta$, that is the direction of $\boldsymbol{k}_1'$, we see that $\boldsymbol{k}_2$ can originate only in a region having the shape of a torus of variable cross-section lying about the circle of centre $P$ as indicated in Fig. 1. The average cross-section must be of order $(\Delta k)^2$, one factor $\Delta k$ arising from the possible variation of $k_1'$ and the other factor from the possible variation, $\Delta k$, in $\boldsymbol{k}_2$ which lies in the plane of the circle. The volume of $k$-space therefore from which the second electron may come is of order $2\pi k_0 (\Delta k)^2$. The number of electrons available per unit volume of actual space which can result in a transition from a given $\boldsymbol{k}_1$ to a state making an angle $\vartheta$ with $\boldsymbol{k}_1$ is of order

$$\{1 - f_0(\boldsymbol{k}_1')\} f_0(\boldsymbol{k}_2) \{1 - f_0(\boldsymbol{k}_2')\} \frac{k_0 (\Delta k)^2}{2\pi^2}. \tag{1.7}$$

The factor $4\pi^3$ takes account of the density of states in $k$-space.

To find the number of collisions in time $dt$ we note that electron 2 must lie in an element of volume of real space equal to

$$\bar{v}_{12} d^2 \delta\omega\, dt, \tag{1.8}$$

if it is to suffer a collision with electron 1, where $\bar{v}_{12}$ is an average of the component of the relative velocity $v_{12}$ along the line of centres at the moment of impact, and $\delta\omega$ is the element of solid angle within which $\boldsymbol{k}_1'$ lies. Thus, multiplying (1.7) and (1.8) together, assuming an average value of $\frac{1}{2}$ for each of the FERMI functions and taking all angles into account, i.e. replacing $\delta\omega$ by $2\pi$ we find

$$\frac{\bar{v}_{12} d^2}{8\pi} k_0 (\Delta k)^2 \tag{1.9}$$

for the number of collisions which electron 1, in state $\boldsymbol{k}_1$, suffers per second. Now the total number of electrons per unit volume $n$ is given by $n = k_0^3/3\pi^2$ so that (1.9) can be written

$$\frac{3\pi}{8} \bar{v}_{12} d^2 n \left(\frac{\Delta k}{k_0}\right)^2. \tag{1.10}$$

If $l_e$ is the mean free path, the product of $l_e$ and the number of collisions a single electron suffers per second is equal to the average velocity of the electron. For the purpose of an order of magnitude calculation we can equate this average velocity to $\bar{v}_{12}$ and drop the numerical factor $\frac{3\pi}{8}$. Thus[1]

$$l_e \approx \left(\frac{k_0}{\Delta k}\right)^2 \frac{1}{n d^2} = \frac{1}{n d^2} \left(\frac{\zeta}{\mathrm{k} T}\right)^2. \tag{1.11}$$

For monovalent metals $n$ is equal to the reciprocal of the atomic volume, and, since $d$ is approximately equal to the interatomic distance, $(n d^2)^{-1}$ is also about 3 Å. As $\zeta$ is usually about 5 or 6 ev $\zeta/\mathrm{k}T \approx 6 \times 10^4/T$. Thus at room temperature $l_e \approx 10^{-3}$ cm., and at 6° K, for example, $l_e = 3$ cm.

[1] This result has been given by E. ABRAHAMS: Phys. Rev. 95, 839 (1954).

The interaction of electrons with the thermal motion of the lattice at room temperature leads to a mean free path of about 300 Å, which is more than a hundred times less than the electron-electron mean free path at the same temperature. On the other hand at very low temperatures the mean free path is determined by residual impurities and lattice defects which result in a mean free path far less than a centimetre. Thus we are able to conclude that collisions between electrons do not play a significant role in the conduction properties of metals, although it is just possible that, in the transition metals, collisions between $s$ and $d$ electrons have an effect at low temperatures[1]. Although the above conclusions have been reached for truly free electrons it seems unlikely that they would be significantly changed for the conduction electrons of actual metals whose wave functions have the BLOCH form. We see, therefore, that the short range interaction between electrons which results from the plasma-like oscillations, together with the FERMI statistics due to the exclusion principle justifies the model of the free particle gas first introduced by DRUDE and LORENTZ.

**2. Electronic states in a real metal.** Let $\boldsymbol{a}_1, \boldsymbol{a}_2, \boldsymbol{a}_3$ be the three vectors which determine the unit cell of a BRAVAIS lattice[2], so that any point of the lattice is specified by a vector

$$\boldsymbol{R}_n = n_1 \boldsymbol{a}_1 + n_2 \boldsymbol{a}_2 + n_3 \boldsymbol{a}_3, \tag{2.1}$$

where $n_1, n_2, n_3$ are integers, and let $\boldsymbol{b}_1, \boldsymbol{b}_2, \boldsymbol{b}_3$ be three vectors determined by the relations

$$\boldsymbol{a}_i \cdot \boldsymbol{b}_j = 2\pi\, \delta_{ij}, \tag{2.2}$$

where $\delta_{ij}$ is the KRONECKER delta. The vectors $\boldsymbol{b}_i$ are, therefore, the vectors of the unit cell of the reciprocal lattice multiplied by $2\pi$. The points in this lattice are specified by the vectors

$$\boldsymbol{K}_l = l_1 \boldsymbol{b}_1 + l_2 \boldsymbol{b}_2 + l_3 \boldsymbol{b}_3, \tag{2.3}$$

where $l_1, l_2, l_3$ are integers.

A region about the origin of the reciprocal lattice bounded by the planes

$$\boldsymbol{k} \cdot \boldsymbol{K}_l = \frac{1}{2} |\boldsymbol{K}_l|^2, \tag{2.4}$$

which are the planes normal to, and bisecting, the line from the origin to a neighbouring lattice point $\boldsymbol{K}_l$ is known as the domain of the BRILLOUIN zones of the lattice. Corresponding to any point $\boldsymbol{k}$ within this polyhedron of $k$-space there is a discrete spectrum of stationary states of the electron with energies which may be denoted by $E_n(\boldsymbol{k})$. For given $n$ the energy varies continuously with $\boldsymbol{k}$, and the set of energy states so defined constitutes one of the BRILLOUIN zones. In each BRILLOUIN zone the energy function $E_n(\boldsymbol{k})$ has the same symmetry as the reciprocal lattice.

An alternative specification of the stationary states can be obtained as follows: The second BRILLOUIN zone is divided by certain symmetry planes, and the pieces are placed around the first zone in such a way that states which had originally the same $\boldsymbol{k}$ in the two zones are now separated by a reciprocal lattice vector $\boldsymbol{K}_l$, and also so that the symmetry of the energy as a function of position in $k$-space maintains the symmetry of the lattice. The third zone is divided and placed around the second, again so that originally equal $\boldsymbol{k}$ values are now separated by one of the $\boldsymbol{K}_l$ vectors. This process can be continued,

[1] Cf. W. G. BABER: loc. cit.

[2] For definition of the BRAVAIS lattice, the reciprocal lattice, etc., cf. vol. VII, part 1, articles of H. JAGODZINSKI and G. LEIBFRIED.

and is unique, and leads to the conception of an "extended $k$-space" in which any state is specified simply by a given $\boldsymbol{k}$ vector[1]. The domain of $\boldsymbol{k}$ is now the whole of the space of the reciprocal lattice. This description is useful for metals with complicated structures, and for alloys, since it is often possible by qualitative or semi-quantitative methods to obtain an idea of the planes in extended $k$-space across which are the significant energy discontinuities.

The number of electronic states per unit volume of metal, for a given state of the spin, contained within an element of volume $d\boldsymbol{k}$ is $(2\pi)^{-3} d\boldsymbol{k}$. The total volume of the domain of the BRILLOUIN zone contains just one state per unit-cell of the BRAVAIS lattice. Hence for complex structures many BRILLOUIN zones are contained within the FERMI surface even when the atoms of the metal have low valency.

The wave function of a stationary state specified by a wave vector $\boldsymbol{k}$ has the BLOCH form, viz.

$$\psi_{\boldsymbol{k}}(\boldsymbol{r}) = e^{i\boldsymbol{k}\cdot\boldsymbol{r}}\, u_{\boldsymbol{k}}(\boldsymbol{r}) \tag{2.5}$$

where $u_{\boldsymbol{k}}(\boldsymbol{r})$ has the periodic symmetry of the space lattice, i.e. $u_{\boldsymbol{k}}(\boldsymbol{r}+\boldsymbol{R}_n) = u_{\boldsymbol{k}}(\boldsymbol{r})$ for all $\boldsymbol{k}$. We notice that

$$\psi_{\boldsymbol{k}}(\boldsymbol{r}+\boldsymbol{R}_n) = e^{i\boldsymbol{R}_n\cdot\boldsymbol{k}}\, \psi_{\boldsymbol{k}}(\boldsymbol{r}) \tag{2.6}$$

and if

$$\boldsymbol{k} = k_1\boldsymbol{b}_1 + k_2\boldsymbol{b}_2 + k_3\boldsymbol{b}_3, \tag{2.7}$$

$$\boldsymbol{R}_n\cdot\boldsymbol{k} = 2\pi(n_1k_1 + n_2k_2 + n_3k_3) \tag{2.8}$$

according to Eqs. (2.1) and (2.2).

**3. Properties of wave packets in metals.** α) *The average velocity associated with a* BLOCH *state.* We begin by proving that the average velocity $\bar{\boldsymbol{v}}_{\boldsymbol{k}}$ of an electron in a state $\boldsymbol{k}$ is given by[2]

$$\bar{\boldsymbol{v}}_{\boldsymbol{k}} = \frac{1}{\hbar}\nabla_{\boldsymbol{k}} E. \tag{3.1}$$

This result is suggested by the usual relation for the group velocity which for the $x$-component may be written, $v_x = \partial\omega_{\boldsymbol{k}}/\partial k_x$, where $\omega_{\boldsymbol{k}}$ is the angular frequency related to the energy by the equation $\hbar\omega_{\boldsymbol{k}} = E_{\boldsymbol{k}}$. The expression for the group velocity is derived, however, for wave packets constructed from plane waves, whereas (3.1) refers to a single BLOCH state. It is necessary therefore to examine this equation more closely.

For a wave function of the form (2.5) we find at once that $u_{\boldsymbol{k}}(\boldsymbol{r})$ must satisfy the equation

$$\nabla^2 u_{\boldsymbol{k}} + 2i\boldsymbol{k}\cdot\nabla u_{\boldsymbol{k}} + \frac{2m}{\hbar^2}\left\{E - \frac{\hbar^2k^2}{2m} - V(\boldsymbol{r})\right\} u_{\boldsymbol{k}} = 0, \tag{3.2}$$

where $V(\boldsymbol{r})$ is the potential energy of the electron in the metal, and also that

$$\int \psi_{\boldsymbol{k}}^* \nabla \psi_{\boldsymbol{k}}\, d\boldsymbol{r} = i\boldsymbol{k} + \int u_{\boldsymbol{k}}^* \nabla u_{\boldsymbol{k}}\, d\boldsymbol{r}. \tag{3.3}$$

In order to evaluate the integral on the right hand side of (3.3) we operate on (3.2) with $\nabla_{\boldsymbol{k}}$ and multiply from the left with $u_{\boldsymbol{k}}^*$, and thus obtain

$$\left.\begin{aligned} u_{\boldsymbol{k}}^*\left[\nabla^2 + 2i\boldsymbol{k}\cdot\nabla + \frac{2m}{\hbar^2}\left\{E - \frac{\hbar^2k^2}{2m} - V(\boldsymbol{r})\right\}\right]\nabla_{\boldsymbol{k}} u_{\boldsymbol{k}} + {}\\ + 2i\,u_{\boldsymbol{k}}^*\nabla u_{\boldsymbol{k}} + \left(\frac{2m}{\hbar^2}\nabla_{\boldsymbol{k}}E - 2\boldsymbol{k}\right)u_{\boldsymbol{k}}^* u_{\boldsymbol{k}} = 0.\end{aligned}\right\} \tag{3.4}$$

---

[1] The form of the resulting energy surfaces depends, of course, on which BRILLOUIN zone is selected as the "first".

[2] H. JONES and C. ZENER: Proc. Roy. Soc. Lond., Ser. A **144**, 101 (1934).

Since the operator contained in the square brackets is hermitian we find on integration that this term may be expressed as

$$\int \nabla_{\boldsymbol{k}} u_{\boldsymbol{k}} \left[\nabla^2 + 2i\,\boldsymbol{k}\cdot\nabla + \frac{2m}{\hbar^2}\left\{E - \frac{\hbar^2 k^2}{2m} - V(\boldsymbol{r})\right\}\right]^* u_{\boldsymbol{k}}^*\, d\boldsymbol{r},$$

which vanishes since the integrand contains the complex conjugate of (3.2). Hence we find

$$\int u_{\boldsymbol{k}}^* \nabla u_{\boldsymbol{k}}\, d\boldsymbol{r} = \frac{i m}{\hbar^2} \nabla_{\boldsymbol{k}} E - i\,\boldsymbol{k}, \tag{3.5}$$

and therefore from (3.3)

$$\int \psi_{\boldsymbol{k}}^* \frac{\hbar}{i} \nabla \psi_{\boldsymbol{k}}\, d\boldsymbol{r} = \frac{m}{\hbar} \nabla_{\boldsymbol{k}} E. \tag{3.6}$$

This equation is equivalent to (3.1) since on integrating the left hand side by parts we obtain

$$\frac{\hbar}{2mi}\int (\psi_{\boldsymbol{k}}^* \nabla \psi_{\boldsymbol{k}} - \psi_{\boldsymbol{k}} \nabla \psi_{\boldsymbol{k}}^*)\, d\boldsymbol{r} = \frac{1}{\hbar} \nabla_{\boldsymbol{k}} E. \tag{3.7}$$

*β) Motion of a wave packet in an external field.* We consider a wave packet, represented by a wave function $\psi(\boldsymbol{r}, t)$, constructed from BLOCH functions belonging to a single energy band. Thus

$$\psi(\boldsymbol{r}, t) = \int a(\boldsymbol{k}, t)\, \psi_{\boldsymbol{k}}(\boldsymbol{r})\, d\boldsymbol{k}, \tag{3.8}$$

where the $\psi_{\boldsymbol{k}}$ are determined by the SCHRÖDINGER equation which may be written as follows

$$H_0 \psi_{\boldsymbol{k}} = E(\boldsymbol{k})\, \psi_{\boldsymbol{k}}. \tag{3.9}$$

The energy $E(\boldsymbol{k})$ refers to a single BRILLOUIN zone.

We now show that, if we construct an operator $E(-i\nabla)$ by replacing $\boldsymbol{k}$ in $E(\boldsymbol{k})$ by $-i\nabla$, there will hold the equation

$$E(-i\nabla)\, \psi_{\boldsymbol{k}}(\boldsymbol{r}) = E(\boldsymbol{k})\, \psi_{\boldsymbol{k}}(\boldsymbol{r}). \tag{3.10}$$

To do this we expand $E(\boldsymbol{k})$ within the domain of the BRILLOUIN zone as a FOURIER series[1]:

$$E(\boldsymbol{k}) = \sum_{\boldsymbol{n}} c(\boldsymbol{n})\, e^{i\boldsymbol{R}_{\boldsymbol{n}}\cdot\boldsymbol{k}}, \tag{3.11}$$

where we have used the notation of (2.8).

Since $\exp(\boldsymbol{R}_{\boldsymbol{n}}\cdot\nabla)\,\psi_{\boldsymbol{k}}(\boldsymbol{r}) = \psi_{\boldsymbol{k}}(\boldsymbol{r}+\boldsymbol{R}_{\boldsymbol{n}})$ we have only to notice that $\psi_{\boldsymbol{k}}$ satisfies the BLOCH condition (2.6) to see that (3.10) is true. If the Hamiltonian in the presence of the external field is $H_0 + H_1$ we see that the motion of the wave packet (3.8) is determined by the following equation

$$\{E(-i\nabla) + H_1\}\, \psi(\boldsymbol{r}, t) = i\hbar \frac{\partial}{\partial t} \psi(\boldsymbol{r}, t), \tag{3.12}$$

since $E(-i\nabla)$ has the same effect as $H_0$ on all $\psi_{\boldsymbol{k}}$.

We now consider the case in which $H_1$ arises from a constant electric field $\boldsymbol{F}$ and a constant magnetic field $\boldsymbol{H}$, and we introduce the scalar potential $\varphi$, and the magnetic vector potential $\boldsymbol{A}$, by the equations:

$$\left.\begin{aligned} \boldsymbol{F} &= -\nabla\varphi, \\ \boldsymbol{H} &= \operatorname{curl}\boldsymbol{A}, \\ \operatorname{div}\boldsymbol{A} &= 0. \end{aligned}\right\} \tag{3.13}$$

[1] Cf. J. C. SLATER: Phys. Rev. 76, 1592 (1949).

Neglecting second powers of the magnetic field the perturbation $H_1$ is given by

$$H_1 = -\frac{e\hbar}{imc}\boldsymbol{A}\cdot\nabla - e\varphi. \tag{3.14}$$

According to (3.6) and (3.10) the matrix elements of $-\frac{e\hbar}{imc}\boldsymbol{A}\cdot\nabla$ with respect to the functions $\psi_{\boldsymbol{k}}$ are identical with the matrix elements of $-\frac{e}{c\hbar}\boldsymbol{A}\cdot\nabla_{\boldsymbol{k}}E(-i\nabla)$ with respect to the same functions. Since, neglecting second powers of $\boldsymbol{A}$, we may write

$$E(-i\nabla) - \frac{e}{c\hbar}\boldsymbol{A}\cdot\nabla_{\boldsymbol{k}}E(-i\nabla) = E\left(-i\nabla - \frac{e}{c\hbar}\boldsymbol{A}\right), \tag{3.15}$$

and writing $\boldsymbol{p} = -i\hbar\nabla$, Eq. (3.12) may be expressed

$$\left\{E\left(\frac{1}{\hbar}\boldsymbol{p} - \frac{e}{c\hbar}\boldsymbol{A}\right) - e\varphi\right\}\psi(\boldsymbol{r},t) = i\hbar\frac{\partial}{\partial t}\psi(\boldsymbol{r},t). \tag{3.16}$$

We now make use of the well-known result that a wave packet follows the motion of a particle determined by the classical Hamiltonian. In this case the Hamiltonian is

$$E\left(\frac{1}{\hbar}\boldsymbol{p} - \frac{e}{c\hbar}\boldsymbol{A}\right) - e\varphi(\boldsymbol{r}), \tag{3.17}$$

where $\boldsymbol{p}$ and $\boldsymbol{r}$ are conjugate dynamical variables. Applying the classical Hamiltonian equations we get the results

$$\hbar\boldsymbol{v} = \nabla_{\boldsymbol{k}}E\left(\frac{1}{\hbar}\boldsymbol{p} - \frac{e}{c\hbar}\boldsymbol{A}\right). \tag{3.18}$$

$$\dot{\boldsymbol{p}} - \frac{e}{c}\dot{\boldsymbol{A}} = \frac{e}{c}\boldsymbol{v}\times\boldsymbol{H} + e\boldsymbol{F}. \tag{3.19}$$

Differentiating (3.18) with respect to time, and using (3.19) we obtain

$$\left.\begin{aligned}\hbar^2\dot{\boldsymbol{v}} &= (\nabla_{\boldsymbol{k}}^2 E)\left(\dot{\boldsymbol{p}} - \frac{e}{c}\dot{\boldsymbol{A}}\right)\\ &= e\left\{\boldsymbol{F} + \frac{1}{c}\boldsymbol{v}\times\boldsymbol{H}\right\}\nabla_{\boldsymbol{k}}^2 E,\end{aligned}\right\} \tag{3.20}$$

where in $\nabla_{\boldsymbol{k}}^2 E$ we put $\boldsymbol{A} = 0$ in order to keep the expression for the acceleration to first powers only of the field strengths.

If now we wish to interpret the result (3.20) in terms of the motion of a wave packet in *k-space*, whose centre lies at $\boldsymbol{k}_0(t)$, we find on differentiating (3.1) with respect to time

$$\hbar\dot{\boldsymbol{v}}_{\boldsymbol{k}} = \dot{\boldsymbol{k}}_0\nabla_{\boldsymbol{k}}^2 E, \tag{3.21}$$

and, therefore, by comparison with (3.20)

$$\dot{\boldsymbol{k}}_0 = \frac{e}{\hbar}\left\{\boldsymbol{F} + \frac{1}{c}\boldsymbol{v}\times\boldsymbol{H}\right\}. \tag{3.22}$$

This shows that an electric field alone causes a wave packet in $k$-space to move forward with constant velocity. The effect of a magnetic field alone can be visualised as follows. Imagine a plane drawn through the point $\boldsymbol{k}_0$ in the Brillouin zone normal to the direction of $\boldsymbol{H}$. The point $\boldsymbol{k}_0$ then moves in this plane along a line of constant energy, as indicated in Fig. 3, in which the direction of $\boldsymbol{H}$ is supposed to be normal to the plane of the paper. Since the perpendicular distance in $k$-space between two neighbouring surfaces of constant energy, $E$ and

$E+\delta E$, is equal to $\delta E/|\nabla_k E|$, and the magnitude of $\dot{k}_0$ is proportional to the velocity $v$, it follows that points which originally lie between these two surfaces continue to do so as they circulate under the action of the magnetic field. Thus, for example, if all the states lying within the contour, indicated in Fig. 3, were fully occupied a magnetic field, acting alone, would cause no change in the distribution, but only a circulation of the points in $k$-space along stream-lines determined by the energy surfaces and the direction of the magnetic field.

We conclude this section by making an estimate of the effect of neglecting second powers of the magnetic field. In the expression for the square of the kinetic potential we have neglected the term in $|\boldsymbol{A}|^2$ in comparison with the scalar product of $\boldsymbol{A}$ and $\boldsymbol{p}$, i.e. we assume

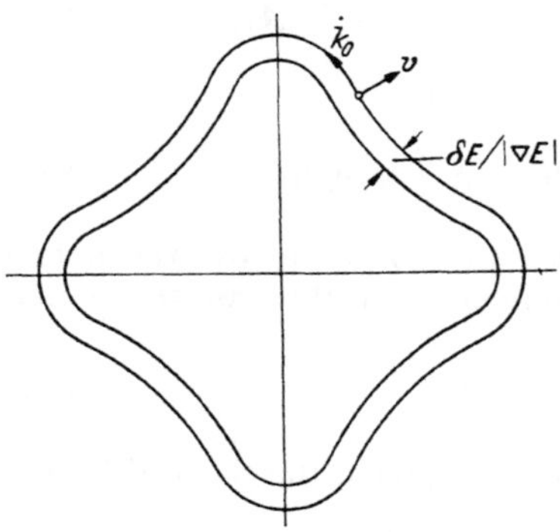


Fig. 3. Illustration of motion of wave packet in $k$-space. Direction of magnetic field is normal to the plane of paper. $\dot{k}_0$ moves along energy contour. Product of $\dot{k}_0$ and $\delta E/|\nabla E|$ is constant.

$$\frac{e}{2c}A^2 \ll \boldsymbol{A}\cdot\boldsymbol{p}. \tag{3.23}$$

Let $\bar{v}$ denote the magnitude of the velocity averaged with respect to all directions relative to $\boldsymbol{A}$ so that we can write

$$\frac{e}{2mc}A \ll \bar{v}. \tag{3.24}$$

For a uniform magnetic field the vector potential may be taken as $\boldsymbol{A}=\frac{1}{2}\boldsymbol{H}\times\boldsymbol{r}$, and if we assume, as we may, that at the initial instant at which the wave packet is defined, the average value of $\boldsymbol{r}$, over the wave packet, is zero, then at a short time $t$ later the magnitude of $\boldsymbol{r}$ will be approximately $\bar{v}t$, and, therefore, the magnitude of $\boldsymbol{A}$ of the order $H\bar{v}t$. Hence the condition (3.24) may be expressed

$$\frac{eHt}{2mc} \ll 1, \quad \text{or} \quad \omega t \ll 1, \tag{3.25}$$

where $\omega$ is the LARMOR angular frequency.

In the process of conduction we are concerned with the free motion of electrons for intervals of time no longer than the average time between collisions with the lattice. Thus, if $l$ denotes the mean free path, (3.25) is equivalent to $l \ll v/\omega$, where $v$ is the velocity of an electron at the FERMI surface. For a typical metal $v$ is about $2\times10^8$ cm./sec. and thus we conclude that the omission of second powers of the field will be justified provided $l$ is less than approximately $(20/H)$ cm., where $H$ is expressed in gauss. Except at very low temperatures and high field strengths this is generally the case.

## II. Statistical theory; the steady state equation.

**4. The BOLTZMANN equation.** In the previous sections we have shown that the quasi-free electrons in a metal can be regarded as an electron gas satisfying FERMI statistics, in which collisions between the electrons can be neglected to a high degree of approximation. We have seen also that an electron in a fixed space periodic field behaves with regard to an external field, like a free particle, provided the velocity is taken according to (3.1) and $\hbar\dot{\boldsymbol{k}}$ is regarded as the rate of change of momentum. We are thus entitled to use the methods of the kinetic theory of gases to study conduction processes. The principal difference, apart from the use of quantum statistics, lies in the fact that the scattering processes which lead to the steady state, arise from the motion, or irregularities, of the lattice, and not, as in a gas, from collisions between the particles.

We introduce a general distribution function defined so that

$$(4\pi^3)^{-1} f(\boldsymbol{k}, \boldsymbol{r}, t)\, d\boldsymbol{k}\, d\boldsymbol{r} \tag{4.1}$$

is the number of electrons which lie in an element $d\boldsymbol{r}$ of real space, and in $d\boldsymbol{k}$ of $k$-space at time $t$. In the equilibrium state this function becomes the FERMI distribution given by (1.4) and denoted by $f_0$, in which the partial potential $\zeta$ is defined in the absence of any external field, but may vary with $\boldsymbol{r}$ as a result of a temperature gradient. It is assumed that an element of volume $d\boldsymbol{r}$ can be chosen which is so small that the temperature and the external fields are sensibly constant within it, and yet is so large that it contains great numbers of electrons.

Those electrons which lie within $d\boldsymbol{k}\, d\boldsymbol{r}$ at time $t$ will, in accordance with LIOUVILLE's theorem, lie in an equal volume of six dimensional phase space about $\boldsymbol{k} + \dot{\boldsymbol{k}}\, dt$, $\boldsymbol{r} + \boldsymbol{v}\, dt$, at the time $t + dt$. The difference, therefore, between $f(\boldsymbol{k} + \dot{\boldsymbol{k}}\, dt,\ \boldsymbol{r} + \boldsymbol{v} dt,\ t + dt)$ and $f(\boldsymbol{k}, \boldsymbol{r}, t)$ must be due to the electrons scattered into the new positions in the time $dt$. We may write therefore

$$\frac{\partial f}{\partial t} + \dot{\boldsymbol{k}} \cdot \nabla_{\boldsymbol{k}} f + \boldsymbol{v} \cdot \nabla_{\boldsymbol{r}} f = \left(\frac{\partial f}{\partial t}\right)_{\text{coll}}, \tag{4.2}$$

where the symbol on the right hand side of (4.2) denotes the rate of change of $f$ due to scattering by the lattice.

To calculate this quantity we proceed as follows: Consider first a single electron and let it be in a state $\boldsymbol{k}$ at time $t$, then the probability of finding it in a state $\boldsymbol{k}'$ (assumed to be vacant) without change of spin direction, at a time $\delta t$ later, may be written $P(\boldsymbol{k}, \boldsymbol{k}')\, \delta t$. Thermal and impurity scattering in metals do not, in general, change the spin state of an electron so that only transitions of the above kind need to be considered. For the inverse transition the probability of the electron switching from $\boldsymbol{k}'$ to $\boldsymbol{k}$ in time $\delta t$ without change of spin, is $P(\boldsymbol{k}', \boldsymbol{k})\, \delta t$.

The number of electrons, per unit volume of metal, gained by $d\boldsymbol{k}$ from $d\boldsymbol{k}'$ in time $\delta t$ is[1]

$$\tfrac{1}{2}(4\pi^3)^{-2}\, \delta t\, P(\boldsymbol{k}', \boldsymbol{k})\, f(\boldsymbol{k}') \left(1 - f(\boldsymbol{k})\right) d\boldsymbol{k}'\, d\boldsymbol{k},$$

and the number of electrons lost from $d\boldsymbol{k}$ to $d\boldsymbol{k}'$ in $\delta t$ is

$$\tfrac{1}{2}(4\pi^3)^{-2}\, \delta t\, P(\boldsymbol{k}, \boldsymbol{k}')\, f(\boldsymbol{k}) \left(1 - f(\boldsymbol{k}')\right) d\boldsymbol{k}'\, d\boldsymbol{k}.$$

Thus we obtain for the right-hand side of (4.2)

$$\left(\frac{\partial f}{\partial t}\right)_{\text{coll}} = \frac{N\Omega}{8\pi^3} \int \left[P(\boldsymbol{k}', \boldsymbol{k})\, f(\boldsymbol{k}') \left(1 - f(\boldsymbol{k})\right) - P(\boldsymbol{k}, \boldsymbol{k}')\, f(\boldsymbol{k}) \left(1 - f(\boldsymbol{k}')\right)\right] d\boldsymbol{k}', \tag{4.3}$$

where $N$ is the number of atoms in the metal, and $\Omega$ the atomic volume. In equilibrium there exists detailed balancing and therefore

$$P(\boldsymbol{k}', \boldsymbol{k})\, f_0'(1 - f_0) = P(\boldsymbol{k}, \boldsymbol{k}')\, f_0\, (1 - f_0'), \tag{4.4}$$

in which we have simplified the notation by writing $f_0'$ to indicate $f_0(\boldsymbol{k}')$.

---

[1] The factor $\frac{1}{2}$ occurs because there are only $\frac{1}{8\pi^3}\, d\boldsymbol{k}$ states in $d\boldsymbol{k}$ available for each electron in $d\boldsymbol{k}'$.

Eq. (4.3) may therefore be expressed.

$$\left(\frac{\partial f}{\partial t}\right)_{\text{coll}} = \frac{N\Omega}{8\pi^3} \int P(\boldsymbol{k}, \boldsymbol{k}')\, f_0 (1 - f_0') \left[\frac{f'(1-f)}{f_0'(1-f_0)} - \frac{f(1-f')}{f_0(1-f_0')}\right] d\boldsymbol{k}'. \tag{4.5}$$

In problems of conductivity we are interested, primarily, in very small displacements from the equilibrium state. Thus we are justified in writing $f = f_0 + f_1$, and neglecting second powers of $f_1$. Hence

$$\frac{f'(1-f)}{f_0'(1-f_0)} - \frac{f(1-f')}{f_0(1-f_0')} = \frac{f_1'}{f_0'(1-f_0')} - \frac{f_1}{f_0(1-f_0)}, \tag{4.6}$$

and since

$$f_0(1-f_0) = -\,\mathrm{k}T\,\frac{d f_0}{d E}, \tag{4.7}$$

if we write

$$f_1 = -\,\chi(\boldsymbol{k})\,\frac{d f_0}{d E}, \tag{4.8}$$

we obtain finally

$$\left(\frac{\partial f}{\partial t}\right)_{\text{coll}} = \frac{N\Omega}{8\pi^3 \mathrm{k}T} \int P(\boldsymbol{k}, \boldsymbol{k}')\, f_0(1-f_0')\,\{\chi(\boldsymbol{k}') - \chi(\boldsymbol{k})\}\, d\boldsymbol{k}'. \tag{4.9}$$

Combining this equation with (4.2) and using (3.1) and (3.22), we have

$$\left.\begin{aligned} \frac{\partial f}{\partial t} + \frac{e}{\hbar}\left\{\boldsymbol{F} + \frac{1}{c\hbar}\nabla_{\boldsymbol{k}} E \times \boldsymbol{H}\right\}\cdot \nabla_{\boldsymbol{k}} f + \frac{1}{\hbar}\nabla_{\boldsymbol{k}} E \cdot \nabla_{\boldsymbol{r}} f \\ = \frac{N\Omega}{8\pi^3 \mathrm{k}T}\int P(\boldsymbol{k}, \boldsymbol{k}')\, f_0(1-f_0')\,\{\chi(\boldsymbol{k}') - \chi(\boldsymbol{k})\}\, d\boldsymbol{k}'. \end{aligned}\right\} \tag{4.10}$$

The solution of (4.10) would give the distribution function in the presence of an external field and temperature gradient, and hence is sufficient to determine the electrical and thermal conductivities and their dependence on a magnetic field, as well as all thermo-electric and thermo-magnetic effects. There is one basic assumption contained in (4.10) which probably excludes the phenomenon of super-conductivity from any theory based upon it. This assumption is that the thermal motion of the lattice may be regarded merely as an external influence on the electronic system giving rise to transitions between single-electron states. In fact, even in the ground state of the whole system, the zero-point motion of the lattice is able to affect the energy distribution of electronic states and hence the properties of the metal[1].

**5. Approximate solutions of the BOLTZMANN equation (4.10): The time of relaxation.** We shall be concerned entirely with metals which are in a steady state and thus we can put $\partial f/\partial t = 0$ in (4.10). This is usually the case even for metals in radio-frequency fields such as are used in experimental determinations of the skin effect. Also we are interested primarily only in the case of weak external fields and small temperature gradients, i.e., when OHM's law is applicable and the thermal current is proportional to the temperature gradient. Thus on the left hand side of (4.10) we can, in some cases, replace $f$ by $f_0$, the equilibrium value. This is not always possible however, for example, in the case of the magneto-resistance effect.

$P(\boldsymbol{k}, \boldsymbol{k}')$ is determined by the nature of the interaction between the electrons and the lattice irregularities, which may be thermal or due to foreign atoms, and can be regarded as a known function in (4.10). This equation is, therefore, an integral equation for the unknown $\chi(\boldsymbol{k})$. Let us imagine, now, that for given

[1] Cf. H. FRÖHLICH: Proc. Roy. Soc. Lond., Ser. A **223**, 296 (1954)

external conditions, $\boldsymbol{F}$, $\boldsymbol{H}$ and $\nabla T$ Eq. (4.10) has been solved. Using the solution $\chi(\boldsymbol{k})$ the integral on the right of (4.10) becomes a determined function of $\boldsymbol{k}$, which we can write $-f_1/\tau(\boldsymbol{k})$ thereby defining the function $\tau(\boldsymbol{k})$. It will be observed that $\tau(\boldsymbol{k})$ is not only a function of position in $k$-space, but also depends upon the external fields. For example, $\tau(\boldsymbol{k})$ for the case of an electric field acting alone may be quite different from $\tau(\boldsymbol{k})$ obtained when a temperature gradient exists but no electric current.

We shall now show that under certain conditions, to be specified, $\tau(\boldsymbol{k})$ is, in fact, to a good approximation, independent of the external fields and the temperature gradient. When this is the case we speak of $\tau(\boldsymbol{k})$ as the time of relaxation, since any arbitrary distribution $f_{\text{init}}$ will, under these conditions, relax exponentially to the equilibrium value when left to the action of the scattering processes alone:

$$f - f_0 = (f_{\text{init}} - f_0)\, e^{-t/\tau}. \tag{5.1}$$

If the energy of an electron in the final state after scattering, $E'$, differs by appreciably less than $\mathrm{k}T$ from the energy of the initial state $E$, then $\frac{1}{\mathrm{k}T} f_0(1-f_0')$ differs from $-\frac{df_0}{dE}$ by a factor

$$1 + \left(\frac{E'-E}{\mathrm{k}T}\right) f_0. \tag{5.2}$$

Thus, if $P(\boldsymbol{k},\boldsymbol{k}')$ is small except when $E'-E \ll \mathrm{k}T$, we can write Eq. (4.9) as follows

$$\left(\frac{\partial f}{\partial t}\right)_{\text{coll}} = -\chi(\boldsymbol{k})\,\frac{df_0}{dE}\,\frac{N\Omega}{8\pi^3}\int P(\boldsymbol{k},\boldsymbol{k}')\left(\frac{\chi(\boldsymbol{k}')}{\chi(\boldsymbol{k})} - 1\right) d\boldsymbol{k}', \tag{5.3}$$

and thus by (4.8)

$$\frac{1}{\tau(\boldsymbol{k})} = \frac{N\Omega}{8\pi^3}\int P(\boldsymbol{k},\boldsymbol{k}')\left(1 - \frac{\chi(\boldsymbol{k}')}{\chi(\boldsymbol{k})}\right) d\boldsymbol{k}'. \tag{5.4}$$

Hence, provided $\chi'/\chi$ is independent of the fields and temperature gradient, $\tau$ will be a true relaxation time.

Since thermal scattering takes place by the emission or absorption of sound wave quanta, and since the greatest energy of such a quantum is $\mathrm{k}\Theta_D$, where $\Theta_D$ is the DEBYE temperature, we see that the condition $(E'-E)/\mathrm{k}T < 1$ will always hold if $T > \Theta_D$. Moreover, if there is scattering by impurity atoms, $P(\boldsymbol{k},\boldsymbol{k}')$ will contain a contribution from elastic collisions for which $E' = E$, and this contribution satisfies the condition required for the existence of a relaxation time at any temperature. Thus we may summarise by saying that there exists a time of relaxation in pure metals at high temperatures, and in sufficiently impure metals at all temperatures.

**6. The electrical conductivity.** In this section we shall obtain a general expression for the electrical conductivity under the conditions, just stated, in which the concept of the time of relaxation may be used.

We will assume that the uniform electric field is given by $\boldsymbol{F} = (0, 0, F)$ and that no magnetic field or temperature gradient is present. On the left hand side of (4.10) we may replace $f$ by $f_0$ which is equivalent to neglecting powers of $F$ higher than the first, and we obtain

$$\frac{eF}{\hbar}\,\frac{\partial f_0}{\partial k_z} = -\frac{f - f_0}{\tau(\boldsymbol{k})}. \tag{6.1}$$

This equation may be expressed alternatively in the form:

$$f = f_0\left(k_x, k_y, k_z - \frac{eF\tau}{\hbar}\right), \tag{6.2}$$

which enables us to visualise the distribution function in the steady state while a current is flowing. If we assume $\tau$ to be a constant, independent of $\boldsymbol{k}$, and if we consider the electrons to be entirely free, so that the surfaces of constant energy are spheres with centres at the origin of $k$-space, then the distribution when a current is flowing is as indicated in Fig. 4. The dotted circle indicates the FERMI surface in the equilibrium condition and the full circle the limit of the occupied states under the action of the electric field. The distance between the centres $OO'$

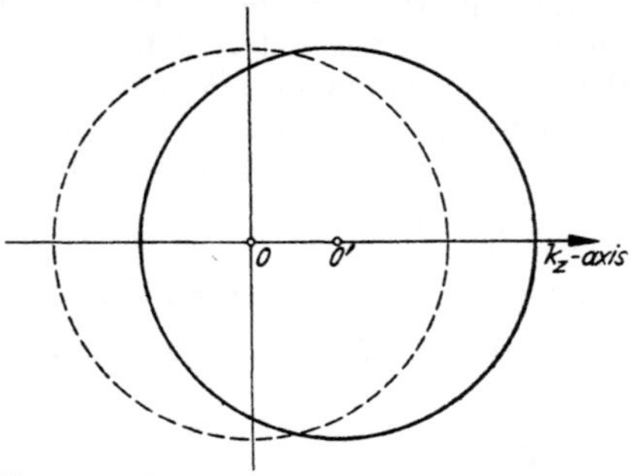


Fig. 4. Diagram to illustrate the displacement of the FERMI distribution in a metal carrying an electric current. $OO'$ is equal to $eF\tau$.

*Table 1.*

| Metal | $n \times 10^{-22}$ | $\sigma \times 10^{-17}$ (esu) At 0° C | Mean free path $l \times 10^{8}$ cm. |
|---|---|---|---|
| Li | 4.6 | 1.06 | 110 |
| Na | 2.5 | 2.09 | 350 |
| K | 1.3 | 1.47 | 370 |
| Rb | 1.1 | 0.78 | 220 |
| Cs | 0.85 | 0.49 | 160 |
| Cu | 8.5 | 5.76 | 420 |
| Ag | 5.8 | 6.12 | 570 |
| Au | 5.9 | 4.37 | 410 |

is equal to $eF\tau/\hbar$. The total momentum per unit volume of metal is therefore $eF\tau n$, where $n$ is the number of electrons per unit volume, and hence the current density $j_z$ is given by

$$j_z = \frac{e^2 F \tau n}{m}, \tag{6.3}$$

and the conductivity $\sigma$ by

$$\sigma = \frac{e^2 n \tau}{m}. \tag{6.4}$$

More generally if we do not make the assumption of free electrons, or of a constant $\tau$, we obtain from (6.1) and (3.1)

$$\sigma = \frac{j_z}{F} = \frac{e^2}{4\pi^3\hbar^2} \int \left(-\frac{df_0}{dE}\right) \tau(\boldsymbol{k}) \left(\frac{\partial E}{\partial k_z}\right)^2 \frac{dS\,dE}{|\nabla E|}, \tag{6.5}$$

where we have expressed the element of volume in $k$-space as an element of area on an energy surface, $dS$, times the normal distance between the surfaces $E$ and $E+dE$.

It is convenient to define a mean free path by the equation $l = v\tau$, where $v$ is the magnitude of the velocity at the FERMI surface. For free electrons $v$ is given by $v = (3\pi^2 n)^{\frac{1}{3}}\hbar/m$.

For monovalent metals the simple theory which results in Eq. (6.4) should be approximately correct and, therefore, by using the observed values of the conductivities, we are able to make estimates of the mean free paths in these metals. Table 1 gives the results for the alkali metals and for copper, silver and gold.

## III. Wave mechanics of the scattering of electrons.

**7. Determination of $\tau$ in terms of the matrix of the scattering perturbation.** To make the calculation of the transition probability feasible it is necessary to make the following assumptions. First, that the energy of the electrons depends only on the magnitude of the wave vector, thus

$$E = \frac{\hbar^2}{2m} |\boldsymbol{k}|^2,$$

where $m$ is a constant known as the effective mass; for completely free electrons it is ordinary electronic mass. Secondly, that the metal can be regarded as an isotropic elastic solid. These two conditions are never, in fact, satisfied together by any real metal. For instance, the first assumption may be valid for some of the alkali metals, notably sodium, but these metals are extremely elastically anisotropic; on the other hand, the nearly isotropic metals, e.g., aluminium, magnesium or tungsten, have complicated electronic structures.

The most important result of these two assumptions is that $P(\boldsymbol{k}, \boldsymbol{k}')$ becomes independent of the relative directions of $\boldsymbol{k}$ and the current, or of any fixed crystallographic direction, and depends only on the magnitudes of $\boldsymbol{k}$ and $\boldsymbol{k}'$ and the angle between these vectors[1]. Essentially the whole of the detailed theory of conductivity, so far developed, is based on this assumed property of $P(\boldsymbol{k}, \boldsymbol{k}')$. A second consequence of these assumptions is that $\tau$ can depend only on the magnitude of $\boldsymbol{k}$, and thus comparing (6.1) with (4.10) we see that $\chi$ must have the form

$$\chi(\boldsymbol{k}) = k_z C(E), \tag{7.1}$$

where $C(E)$ is some unspecified function of the energy. Thus Eq. (5.4) can be written

$$\frac{1}{\tau} = \frac{N\Omega}{8\pi^3} \int P(k, k'\,\vartheta) \left\{1 - \frac{k'_z C(E')}{k_z C(E)}\right\} d\boldsymbol{k}'. \tag{7.2}$$

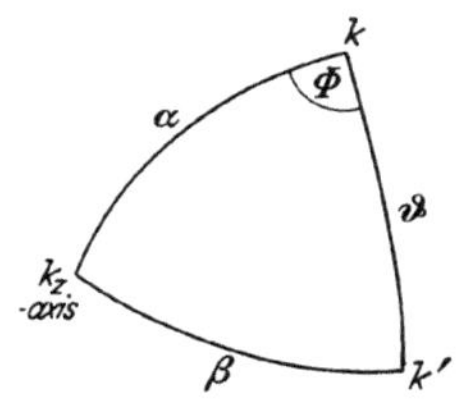


The accompanying diagram represents a spherical triangle determined by the vectors $\boldsymbol{k}$, $\boldsymbol{k}'$, and the $k_z$-axis. Let $\vartheta$ be the angle between $\boldsymbol{k}$ and $\boldsymbol{k}'$, $\alpha$ and $\beta$ the angles between $\boldsymbol{k}$ and $\boldsymbol{k}'$ and the $k_z$-axis respectively, and let $\varphi$ be the angle between the plane containing $\boldsymbol{k}$ and the $k_z$-axis, and the plane containing $\boldsymbol{k}$ and $\boldsymbol{k}'$. For this spherical triangle we have

$$\cos\beta = \cos\alpha\cos\vartheta + \sin\alpha\sin\vartheta\cos\varphi, \tag{7.3}$$

and thus

$$\frac{k'_z}{k_z} = \frac{k'\cos\beta}{k\cos\alpha} = \frac{k'}{k}(\cos\vartheta + \tan\alpha\sin\vartheta\cos\varphi). \tag{7.4}$$

For the element of volume at $\boldsymbol{k}'$ we can write

$$d\boldsymbol{k}' = k'^2\, dk' \sin\vartheta\, d\vartheta\, d\varphi. \tag{7.5}$$

Substituting (7.4) and (7.5) into (7.2) and noting that the integrand is independent of $\varphi$ we find

$$\frac{1}{\tau} = \frac{N\Omega}{4\pi^2} \int P(k, k', \vartheta) \left\{1 - \frac{k' C(E')}{k C(E)} \cos\vartheta\right\} k'^2\, dk' \sin\vartheta\, d\vartheta. \tag{7.6}$$

Now let $V_0(\boldsymbol{r})$ be the potential energy of an electron at the point $\boldsymbol{r}$ in the perfect crystal in which every atom is supposed to be at rest at its lattice point. Let $V(\boldsymbol{r}, t)$ be the potential energy of the electron at the same point, at an instant $t$, in the crystal which now contains irregularities produced by the thermal motion of the ions, or by impurities, defects, etc. The difference between these two energies

$$U(\boldsymbol{r}, t) = V(\boldsymbol{r}, t) - V_0(\boldsymbol{r}), \tag{7.7}$$

is the perturbation giving rise to transition between the electronic states.

---

[1] See, however, Sect. 11.

In order to keep the calculation as simple as possible at this stage, and, at the same time, to make clear some of its important features, we use the following device. We imagine that the displacements of all the atoms are "frozen in" at some instant, and we then calculate the scattering due to this fixed field. The value of the transition probability thus determined may then be averaged over all possible configurations of the lattice appropriate to a given temperature. This procedure, by making $U$ independent of time, automatically excludes all but elastic scattering.

A wave function in the disturbed lattice can be expressed as a linear sum of Bloch wave functions belonging to the perfect crystal:

$$\psi = \sum_{\boldsymbol{k}} b(\boldsymbol{k}, t)\, \psi_{\boldsymbol{k}}(\boldsymbol{r})\, e^{-\frac{i E_{\boldsymbol{k}} t}{\hbar}}, \tag{7.8}$$

where the $\psi_{\boldsymbol{k}}(\boldsymbol{r})$ are assumed to be normalised to the volume of the crystal. The assertion that at $t=0$ the electron is in the state specified by the wave vector $\boldsymbol{k}$ is expressed by putting $b(\boldsymbol{k}, 0) = 1$ and all $b(\boldsymbol{k}', 0) = 0$, $\boldsymbol{k}' \neq \boldsymbol{k}$. According to the well-known perturbation theory of wave mechanics we then find that after a short interval of time $t$,

$$|b(\boldsymbol{k}', t)|^2 = 4\,|U_{\boldsymbol{k}'\boldsymbol{k}}|^2 \frac{\sin^2\left(\frac{E_{\boldsymbol{k}} - E_{\boldsymbol{k}'}}{2\hbar}\right) t}{(E_{\boldsymbol{k}} - E_{\boldsymbol{k}'})^2}, \tag{7.9}$$

where

$$U_{\boldsymbol{k}',\boldsymbol{k}} = \int \psi^*_{\boldsymbol{k}'}\, U(\boldsymbol{r})\, \psi_{\boldsymbol{k}}\, d\boldsymbol{r}. \tag{7.10}$$

$|b(\boldsymbol{k}'\, t)|^2$ is interpreted to mean the probability of finding the electron in state $\boldsymbol{k}'$ at time $t$ after starting from state $\boldsymbol{k}$ at $t=0$. Hence the *rate* of increase of $|b(\boldsymbol{k}'\, t)|^2$ has the same meaning as our previously defined $P(\boldsymbol{k}, \boldsymbol{k}')$. It is given by

$$\frac{\partial}{\partial t}|b(\boldsymbol{k}', t)|^2 = \frac{2}{\hbar}\,|U_{\boldsymbol{k}',\boldsymbol{k}}|^2 \frac{\sin\left(\frac{E_{\boldsymbol{k}} - E_{\boldsymbol{k}'}}{\hbar}\right) t}{E_{\boldsymbol{k}} - E_{\boldsymbol{k}'}}. \tag{7.11}$$

Substituting this result in (7.6), replacing $k'\,dk'$ by $(m/\hbar^2)\,dE'$, and integrating over $dE'$ we find

$$\frac{1}{\tau} = \frac{m\,k\,N\Omega}{2\pi\hbar^3} \int_0^\pi |U_{\boldsymbol{k}',\boldsymbol{k}}|^2 (1 - \cos\vartheta) \sin\vartheta\, d\vartheta. \tag{7.12}$$

$\tau$ is, of course, independent of $N$ since, as we shall see, $|U_{\boldsymbol{k}',\boldsymbol{k}}|^2$ varies as $N^{-1}$ when evaluated with normalized wave functions.

In the above integration we have assumed that $\left\{\sin\left(\frac{E'-E}{\hbar}\right) t\right\} \big/ \{(E'-E)\pi\}$ behaves like a $\delta$-function with respect to $(E'-E)$, which ensures that the remainder of the integrand is evaluated at $E'=E$.

It is by no means obvious that this procedure is correct because the average life-time of a state near the Fermi surface is no longer than the time of relaxation $\tau$, and for some metals $\hbar/\tau$ is of order $\mathrm{k}T$. However this question has been discussed by Peierls[1] and more recently by van Wieringen[2], and the conclusion reached is that as far as this particular step is concerned the use of perturbation theory is justified. For further details the reader is referred to the original papers.

[1] R. Peierls: Helv. phys. Acta **7**, Suppl. 2, 24 (1934).
[2] J. S. van Wieringen: Proc. Phys. Soc. Lond. A **67**, 206 (1954).

**8. Calculation of the scattering matrix $\boldsymbol{U_{k'k}}$.** For monovalent metals a good approximation to the potential energy of a conduction electron in the perfect crystal, $V_0(\boldsymbol{r})$, is obtained by assuming that, within a symmetrically constructed atomic polyhedron about a particular ion, $V_0(\boldsymbol{r})$ is equal to the potential energy of the electron in the field of that ion alone. This approximation is particularly good for cubic metals which have very symmetrical atomic polyhedra since the field, due to the ion plus its electron, must be very small beyond the polyhedron itself. If the cell were spherical the field outside would be zero. Also on account of the correlation between the simultaneous positions of electrons it can be shown that the probability of two or more electrons being together within one cell is small.

Let $\boldsymbol{s}$ be a vector denoting the position of a lattice-point, so that its components are integral multiples of $\boldsymbol{a}_1$, $\boldsymbol{a}_2$, $\boldsymbol{a}_3$, then, within the polyhedron surrounding $\boldsymbol{s}$, $V_0(\boldsymbol{r}-\boldsymbol{s})$ is taken to be the potential energy of an electron at a distance $|\boldsymbol{r}-\boldsymbol{s}|$ from the centre of an isolated ion. Imagine now that all atoms except one, the $s$-th, are in their equilibrium positions, and that the $s$-th atom is displaced a small distance $\boldsymbol{R}(\boldsymbol{s})$ from the lattice point $\boldsymbol{s}$. We now assume that within the $\boldsymbol{s}$-th polyhedron the potential energy is still determined by the distance from the displaced ion and is therefore $V_0(\boldsymbol{r}-\boldsymbol{s}-\boldsymbol{R}(\boldsymbol{s}))$. The effect of the displacement in all other cells is neglected, thus

$$\left.\begin{aligned} U(\boldsymbol{r}) &= V_0(\boldsymbol{r}-\boldsymbol{s}-\boldsymbol{R}(\boldsymbol{s})) - V_0(\boldsymbol{r}-\boldsymbol{s}) \\ &= -\boldsymbol{R}(\boldsymbol{s}) \cdot \nabla V_0, \end{aligned}\right\} \tag{8.1}$$

within the $s$-th cell and is zero outside it if no other atom is displaced. When all atoms of the crystal are displaced the perturbation may still be taken as (8.1) where $V_0(\boldsymbol{r})$ is now the undisturbed periodic potential and $\boldsymbol{R}(\boldsymbol{s})$ depends upon $\boldsymbol{r}$ in such a way that, when $\boldsymbol{r}$ lies within the polyhedron about $\boldsymbol{s}$, $\boldsymbol{R}(\boldsymbol{s})$ is the displacement of that ion. The result of neglecting the effect of an ionic displacement beyond its own polyhedron is undoubtedly to underestimate the resistivity, but if we assumed that the displacement of ions did not effect the charge distribution of the conduction electrons in any way we should certainly overestimate it. We shall return to this subject in the discussion of BARDEEN'S calculation of the scattering matrix.

Before calculating the matrix elements $U_{\boldsymbol{k}'\boldsymbol{k}}$ we must consider more carefully the normalisation of the electronic wave functions. Imagine, in the first place, an infinite crystal, and suppose this to be divided up into large identical rhombohedra, whose edges lie in the directions of the lattice-vectors $\boldsymbol{a}_1$, $\boldsymbol{a}_2$, $\boldsymbol{a}_3$, and each of which contains N atoms. The assertion that in all states the crystal repeats identically in each rhombohedron can be regarded as the boundary conditions which determine the electronic wave functions and the lattice vibrations. If $\Omega$ denotes the atomic volume, the volume of the rhombohedral domain of the crystal is $N\Omega$ and we normalise all wave functions to this volume so that

$$\int_{N\Omega} |\psi_{\boldsymbol{k}}|^2 \, d\boldsymbol{r} = 1. \tag{8.2}$$

According to (8.1) and (2.5) we have therefore

$$\left.\begin{aligned} U_{\boldsymbol{k}',\boldsymbol{k}} &= -\int_{N\Omega} \left(\boldsymbol{R}(\boldsymbol{s}) \cdot \nabla V_0(\boldsymbol{r})\right) e^{i(\boldsymbol{k}-\boldsymbol{k}')\cdot\boldsymbol{r}} u_{\boldsymbol{k}'}^* u_{\boldsymbol{k}} \, d\boldsymbol{r} \\ &= -\sum_{\boldsymbol{s}} \boldsymbol{R}(\boldsymbol{s}) \, e^{i(\boldsymbol{k}-\boldsymbol{k}')\cdot\boldsymbol{s}} \cdot \int_{\Omega} \psi_{\boldsymbol{k}'}^* \psi_{\boldsymbol{k}} \nabla V_0(\boldsymbol{r}) \, d\boldsymbol{r}, \end{aligned}\right\} \tag{8.3}$$

where the integration in the second line is over a unit cell, and the summation is over the $N$ values of $\boldsymbol{s}$ in the basic rhombohedron of the lattice. We denote the integral, which is a vector, by $\boldsymbol{I}$; it is the same, of course, for any cell of the crystal. Hence we write

$$U_{\boldsymbol{k}',\boldsymbol{k}} = -\sum_{\boldsymbol{s}} e^{i(\boldsymbol{k}-\boldsymbol{k}')\cdot\boldsymbol{s}} \big(\boldsymbol{R}(\boldsymbol{s}) \cdot \boldsymbol{I}\big) \tag{8.4}$$

and therefore

$$|U_{\boldsymbol{k}',\boldsymbol{k}}|^2 = \sum_{\boldsymbol{s}} \sum_{\boldsymbol{t}} e^{i(\boldsymbol{k}-\boldsymbol{k}')\cdot(\boldsymbol{s}-\boldsymbol{t})} \big(\boldsymbol{R}(\boldsymbol{s}) \cdot \boldsymbol{I}\big)\big(\boldsymbol{R}(\boldsymbol{t}) \cdot \boldsymbol{I}\big)^*. \tag{8.5}$$

Now the direction of the vector $\boldsymbol{I}$ is fixed for given $\boldsymbol{k}$ and $\boldsymbol{k}'$, and, at high temperatures, we may assume that the relative directions of $\boldsymbol{R}(\boldsymbol{s})$ and $\boldsymbol{R}(\boldsymbol{t})$, i.e., the displacements in *different* cells, are randomly distributed, so that the average of $\big(\boldsymbol{R}(\boldsymbol{s}) \cdot \boldsymbol{I}\big)\big(\boldsymbol{R}(\boldsymbol{t}) \cdot \boldsymbol{I}\big)^*$, taken over all configurations which arise in the course of time, is zero. Within a single cell the average of the square of the cosine of the angle between $\boldsymbol{R}$ and $\boldsymbol{I}$ is $\frac{1}{3}$ and thus

$$|U_{\boldsymbol{k}',\boldsymbol{k}}|^2 = \sum_{\boldsymbol{s}} |\boldsymbol{R}(\boldsymbol{s}) \cdot \boldsymbol{I}|^2 = \frac{N}{3} \overline{R^2} I^2. \tag{8.6}$$

We can determine $\overline{R^2}$ on the basis of the EINSTEIN model. If $\nu_E$ denotes the natural frequency of an atom, the average potential energy, which is equal to $\frac{3}{2}\mathrm{k}T$ at high temperatures, may be written:

$$2\pi^2 \nu_E^2 M \overline{R^2} = \tfrac{3}{2}\mathrm{k}T, \tag{8.7}$$

where $M$ is the mass of the atom.

The EINSTEIN characteristic temperature is defined by

$$h\nu_E = \mathrm{k}\Theta_E, \tag{8.8}$$

and thus we have, according to (8.6), (8.7) and (8.8)

$$|U_{\boldsymbol{k}',\boldsymbol{k}}|^2 = \frac{N\hbar^2}{\mathrm{k}} \left(\frac{T}{M\Theta_E^2}\right) I^2. \tag{8.9}$$

**9. Scattering matrix for a DEBYE solid.** The theory proceeds as before to the point at which we obtain Eq. (8.4). Now, however, it is necessary to express the displacements $\boldsymbol{R}(\boldsymbol{s})$ in terms of the amplitudes of the elastic waves in the crystal. We write

$$\boldsymbol{R}(\boldsymbol{s}) = \frac{1}{\sqrt{N}} \sum_{\boldsymbol{q}} e^{i\boldsymbol{q}\cdot\boldsymbol{s}} \boldsymbol{a}(\boldsymbol{q}), \tag{9.1}$$

which is a linear transformation of the $N$ real vectors $\boldsymbol{R}(\boldsymbol{s})$ to the $N$ complex vectors $\boldsymbol{a}(\boldsymbol{q})$. $\boldsymbol{q}$ is a vector in reciprocal space, like $\boldsymbol{k}$, and its domain is the same fundamental BRILLOUIN zone which contains just $N$ distinct $\boldsymbol{q}$ vectors. The summation in (9.1) is over these $N$ values. In order to ensure that $\boldsymbol{R}(\boldsymbol{s})$ are real vectors we must have

$$\boldsymbol{a}(-\boldsymbol{q}) = \boldsymbol{a}^*(\boldsymbol{q}) \tag{9.2}$$

where the star denotes the complex conjugate.

From (8.4) we find

$$\left.\begin{aligned} U_{\boldsymbol{k}',\boldsymbol{k}} &= -\frac{1}{\sqrt{N}} \sum_{\boldsymbol{s}} \sum_{\boldsymbol{q}} e^{i(\boldsymbol{k}-\boldsymbol{k}'+\boldsymbol{q})\cdot\boldsymbol{s}} \big(\boldsymbol{a}(\boldsymbol{q}) \cdot \boldsymbol{I}\big) \\ &= -\sqrt{N}\big(\boldsymbol{a}(\boldsymbol{q}) \cdot \boldsymbol{I}\big), \end{aligned}\right\} \tag{9.3}$$

where, in the second line

$$\boldsymbol{q} = \boldsymbol{k}' - \boldsymbol{k} - \boldsymbol{K}_l \tag{9.4}$$

and $\boldsymbol{K}_l$ may be zero or is a vector in the reciprocal lattice defined by (2.3).

Averaging the square of $|U_{\boldsymbol{k}'\boldsymbol{k}}|$ over all directions and magnitudes $\boldsymbol{a}(\boldsymbol{q})$ relative to the fixed vector $\boldsymbol{I}$ we find

$$|U_{\boldsymbol{k}',\boldsymbol{k}}|^2 = \tfrac{1}{3} N |\overline{\boldsymbol{a}}(\boldsymbol{q})|^2 I^2, \tag{9.5}$$

which is analogous to (8.6).

In order to determine the value of $|\boldsymbol{a}(\boldsymbol{q})|^2$ we consider the actual oscillations of the lattice, and differentiate (9.1) with respect to time obtaining

$$\dot{\boldsymbol{R}}(\boldsymbol{s}) = \frac{1}{\sqrt{N}} \sum_{\boldsymbol{q}} i\, \omega_{\boldsymbol{q}}\, \boldsymbol{a}(\boldsymbol{q})\, e^{i\,\boldsymbol{q}\cdot\boldsymbol{s}}, \tag{9.6}$$

where $\omega_{\boldsymbol{q}}$ is the angular frequency associated with the normal mode of wave vector $\boldsymbol{q}$. If $M$ is the mass of the atom we find for the total kinetic energy

$$\left.\begin{aligned} \frac{1}{2} M \sum_{\boldsymbol{s}} \dot{\boldsymbol{R}}^2(\boldsymbol{s}) &= \frac{M}{2N} \sum_{\boldsymbol{q}} \sum_{\boldsymbol{q}'} \sum_{\boldsymbol{s}} \omega_{\boldsymbol{q}}\, \omega_{\boldsymbol{q}'}\, \boldsymbol{a}(\boldsymbol{q}) \cdot \boldsymbol{a}(\boldsymbol{q}')\, e^{i(\boldsymbol{q}+\boldsymbol{q}')\cdot\boldsymbol{s}} \\ &= \frac{1}{2} M \sum_{\boldsymbol{q}} \omega_{\boldsymbol{q}}^2 |\boldsymbol{a}(\boldsymbol{q})|^2, \end{aligned}\right\} \tag{9.7}$$

in which we have used (9.2).

At high temperatures we can make use of the fact that the energy is equally distributed amongst the normal modes and thus obtain

$$\tfrac{1}{2} M\, \omega_{\boldsymbol{q}}^2 |\overline{\boldsymbol{a}}(\boldsymbol{q})|^2 = \tfrac{3}{2} \mathrm{k} T. \tag{9.8}$$

The relation between angular frequency and wave number is

$$\omega_{\boldsymbol{q}} = c_s q, \tag{9.9}$$

where $c_s$ is the velocity of the wave, and $q$ is the magnitude of the wave vector $\boldsymbol{q}$. As in the DEBYE theory of specific heats the velocity $c_s$ is related to the maximum frequency $\nu_{\max}$ by the equation[1]

$$N\Omega \int_0^{\nu_{\max}} \frac{12\pi}{c_s^3} \nu^2\, d\nu = 3N, \tag{9.10}$$

which, expressed in terms of the DEBYE characteristic temperatures, $\Theta_D$, defined by

$$\hbar\, \omega_{\max} = \mathrm{k}\, \Theta_D, \tag{9.11}$$

gives

$$c_s = \left(\frac{2}{9\pi}\right)^{\frac{1}{3}} r_s \left(\frac{\mathrm{k}\,\Theta_D}{\hbar}\right), \tag{9.12}$$

where, as before, $r_s$ is the radius of the atomic sphere. Substituting (9.8) into (9.5), using (9.9) and (9.12), we obtain

$$|U_{\boldsymbol{k}',\boldsymbol{k}}|^2 = \frac{N\hbar^2}{\mathrm{k}} \left(\frac{T}{M\,\Theta_D^2}\right) \left(\frac{9\pi}{2}\right)^{\frac{2}{3}} \frac{I^2}{(r_s q)^2}. \tag{9.13}$$

This result differs from (8.9), based on the EINSTEIN model, in that $\Theta_D$ replaces $\Theta_E$, and scattering through small angles, i.e., when $q$ is small, is made relatively more probable by the occurence of the $q^{-2}$ factor.

---

[1] Cf. e.g. the article of M. BLACKMAN in vol. VII, part 1 of this Encyclopedia.

**10. Calculation of the vector integral *I*.** The integral in (8.3) can be simplified by noting that both wave functions $\psi_{\boldsymbol{k}}$ and $\psi_{\boldsymbol{k}'}$ satisfy the SCHRÖDINGER equation with the potential energy $V_0(\boldsymbol{r})$. The following identity may easily be verified:

$$\frac{\hbar^2}{2m}\left[\psi^*_{\boldsymbol{k}'}\nabla^3\psi_{\boldsymbol{k}} - \nabla\psi_{\boldsymbol{k}}\nabla^2\psi^*_{\boldsymbol{k}'}\right] = \psi^*_{\boldsymbol{k}'}\psi_{\boldsymbol{k}}\nabla V_0 + (E_{\boldsymbol{k}'} - E_{\boldsymbol{k}})\,\psi^*_{\boldsymbol{k}'}\nabla\psi_{\boldsymbol{k}}. \tag{10.1}$$

For those transitions for which $E_{\boldsymbol{k}'} = E_{\boldsymbol{k}}$ we obtain, therefore,

$$\boldsymbol{I} = \frac{\hbar^2}{2m}\int\left(\psi^*_{\boldsymbol{k}'}\frac{\partial}{\partial n}\nabla\psi_{\boldsymbol{k}} - \nabla\psi_{\boldsymbol{k}}\frac{\partial}{\partial n}\psi^*_{\boldsymbol{k}'}\right)dS, \tag{10.2}$$

where we have used GREEN's theorem, and $\partial/\partial n$ denotes differentiation along the outward normal at the surface element $dS$. We now introduce a number of approximations: First, we replace the actual atomic polyhedron by a sphere of equal volume, whose radius we denote by $r_s$; secondly, we assume that all wave functions are of the form

$$\psi_{\boldsymbol{k}} = e^{i\boldsymbol{k}\cdot\boldsymbol{r}}\,u(r) \tag{10.3}$$

where $u(r)$ is independent of $\boldsymbol{k}$, and depends only on the magnitude of $\boldsymbol{r}$. Moreover we assume that $u(r)$ satisfies the condition

$$\left(\frac{du}{dr}\right)_{r=r_s} = 0. \tag{10.4}$$

These assumptions are probably satisfactory for metallic sodium, and, to a lesser extent, for the other alkali metals, but are almost certainly very far from the truth in many cases: even for the noble metals copper, silver and gold.

Remembering that $u' = 0$, over the surface of integration; (10.2) becomes

$$\boldsymbol{I} = \frac{\hbar^2}{2m}\int i\boldsymbol{k}\left(\psi^*_{\boldsymbol{k}'}\frac{\partial\psi_{\boldsymbol{k}}}{\partial r} - \psi_{\boldsymbol{k}}\frac{\partial\psi^*_{\boldsymbol{k}'}}{\partial r}\right)dS + \frac{\hbar^2}{2m}\int\frac{\boldsymbol{r}}{r}e^{i(\boldsymbol{k}-\boldsymbol{k}')\cdot\boldsymbol{r}}\,u\,u''\,dS. \tag{10.5}$$

By applying GREEN's theorem to the first integral, and using the SCHRÖDINGER equation again, this integral may easily be shown to vanish when $E_{\boldsymbol{k}} = E_{\boldsymbol{k}'}$. To evaluate the second integral, take $\boldsymbol{k} - \boldsymbol{k}'$ as the direction of the polar axis and let $\boldsymbol{r}$ be inclined at an angle $\alpha$ to this axis. Then, since the component of $\boldsymbol{r}$ at right angles to $\boldsymbol{k} - \boldsymbol{k}'$ clearly cancels in the integration, $\boldsymbol{I}$ must lie in the direction of $\boldsymbol{k} - \boldsymbol{k}'$.

Hence

$$\boldsymbol{I} = \frac{\pi\hbar^2 r_s^2}{m}\,\frac{\boldsymbol{k}-\boldsymbol{k}'}{|\boldsymbol{k}-\boldsymbol{k}'|}\,(u\,u'')_{r_s}\int\limits_0^\pi e^{i|\boldsymbol{k}-\boldsymbol{k}'|r_s\cos\alpha}\cos\alpha\sin\alpha\,d\alpha \tag{10.6}$$

where $\alpha$ is the angle between $\boldsymbol{k} - \boldsymbol{k}'$ and $\boldsymbol{r}$. Therefore

$$\boldsymbol{I} = \frac{2\pi\hbar^2}{im}(\boldsymbol{k}-\boldsymbol{k}')\left\{\frac{\xi\cos\xi - \sin\xi}{\xi^3}\right\}r_s^3\,(u\,u'')_{r_s}, \tag{10.7}$$

where $\xi = |\boldsymbol{k} - \boldsymbol{k}'|\,r_s = 2k_0 r_s \sin\vartheta/2$, and $\vartheta$ is the angle between $\boldsymbol{k}$ and $\boldsymbol{k}'$.

Since $u(r)$ is the wave function for the state at the bottom of the conduction band, whose energy is $E_0$, we find

$$(u\,u'')_{r_s} = \frac{2m}{\hbar^2}\left(V_0(r_s) - E_0\right)u^2(r_s). \tag{10.8}$$

According to the calculations of WIGNER and SEITZ[1] for sodium $u^2$ is approximately constant over the atomic sphere, and therefore according to (8.2)

$$u^2(r_s) \approx (N\Omega)^{-1}. \tag{10.9}$$

Thus according to (10.7), (10.8) and (10.9), we find finally

$$\boldsymbol{I} = i\,(\boldsymbol{k} - \boldsymbol{k}')\,\frac{(V_0(r_s) - E_0)}{N}\left\{\frac{3\,(\sin\xi - \xi\cos\xi)}{\xi^3}\right\}. \tag{10.10}$$

Substituting this result into (8.9) and (9.13) we obtain the scattering matrix as a function of the scattering angle for the EINSTEIN and DEBYE models. It is convenient to introduce the variable $w$ defined by

$$w = \sin\frac{\vartheta}{2}, \tag{10.11}$$

so that $\xi = (18\pi)^{\frac{1}{3}} w$, for monovalent metals, and to define the functions:

$$f(w) = \sin(18\pi)^{\frac{1}{3}} w - (18\pi)^{\frac{1}{3}} w \cos(18\pi)^{\frac{1}{3}} w, \tag{10.12}$$

and

$$G(w) = \frac{1}{4\pi}\left\{\frac{V_0(r_s) - E_0}{\zeta}\right\}\frac{f(w)}{w^3}. \tag{10.13}$$

We then obtain for the EINSTEIN model

$$|U_{\boldsymbol{k}',\boldsymbol{k}}|^2 = \frac{2^{\frac{8}{3}}\pi^2}{3}\frac{\hbar^2}{k_0}\left(\frac{T}{M\,\mathrm{k}\,\Theta_E^2}\right)\zeta^2\{2^{\frac{4}{3}} w^2 G^2(w)\}, \tag{10.14}$$

and for the DEBYE model

$$|U_{\boldsymbol{k}',\boldsymbol{k}}|^2 = \frac{2^{\frac{8}{3}}\pi^2}{3}\frac{\hbar^2}{k_0}\left(\frac{T}{M\,\mathrm{k}\,\Theta_D^2}\right)\zeta^2\left\{\frac{|\boldsymbol{k} - \boldsymbol{k}'|^2}{q^2}\,G^2(w)\right\}. \tag{10.15}$$

**11. *N*-processes and *U*-processes.** The angular dependence of the scattering matrix, on the basis of the EINSTEIN model, is given by (10.14) for all angles between 0 and $\pi$, i.e. for all $w$ between 0 and 1, and this expression may be substituted into (7.12) and the relaxation time determined. The DEBYE model, on the other hand, leads to (10.15) which contains the factor $|\boldsymbol{k} - \boldsymbol{k}'|/q$, and this factor is not uniquely determined by $w$. The magnitude of the vector $\boldsymbol{k} - \boldsymbol{k}'$ depends only on the relative positions of $\boldsymbol{k}$ and $\boldsymbol{k}'$, and only on $w$ if both these vectors lie on the FERMI sphere. However $\boldsymbol{q}$, according to (9.4), depends also on the direction of $\boldsymbol{k}$ itself relative to the axes of the reciprocal lattice. The situation may be understood by reference to Fig. 5. This shows a section of the reciprocal lattice for a simple cubic crystal; $K_1$, $K_2$ etc. being the nearest reciprocal lattice points to the origin $O$. The domain of the BRILLOUIN zone is any one of the cubes centred on $O$, or any of the $K_n$. The electrons are assumed to have energies depending on the magnitude of the wave vector $\boldsymbol{k}$ only, and thus the FERMI surface is a sphere of radius $k_0$ and of volume equal to one half of that of the BRILLOUIN zone. We have to consider two similar wave-vector spaces: those of $\boldsymbol{q}$ and $\boldsymbol{k}$. At high temperatures we need to consider all points in the BRILLOUIN zone of $\boldsymbol{q}$, but for monovalent metals only the immediate neighbourhood of the FERMI sphere in $k$-space. $O$ denotes the origin of $q$-space, and $O'$ the origin of $k$-space. As the direction of the vector $\boldsymbol{k}$ varies $O'$ moves over a sphere of radius $k$ centred on $O$. The points $A$, $B$, $C$ show possible positions of the wave vector $\boldsymbol{k}'$ of the scattered wave. The main feature of the construction

[1] E. P. WIGNER and F. SEITZ: Phys. Rev. 43, 804 (1933).

is to show that although $\boldsymbol{q}$ must always lie within the BRILLOUIN zone, this zone, according to (9.4), can be centred on any one of the reciprocal lattice points. Thus if $\boldsymbol{k}'$ lies within the zone centred on $O$ we have $\boldsymbol{k}' - \boldsymbol{k} = \boldsymbol{q}$ and the factor $|\boldsymbol{k} - \boldsymbol{k}'|/q$ in (10.15) is unity. Such scattering processes have been called normal or $N$-processes. If, as at $A$, $\boldsymbol{k}'$ lies in the zone $K_1$, then

$$\left.\begin{aligned} \boldsymbol{q} &= \overrightarrow{K_1 O} + \overrightarrow{O A} \\ &= \boldsymbol{k}' - \boldsymbol{k} - \boldsymbol{K}_1, \end{aligned}\right\} \quad (11.1)$$

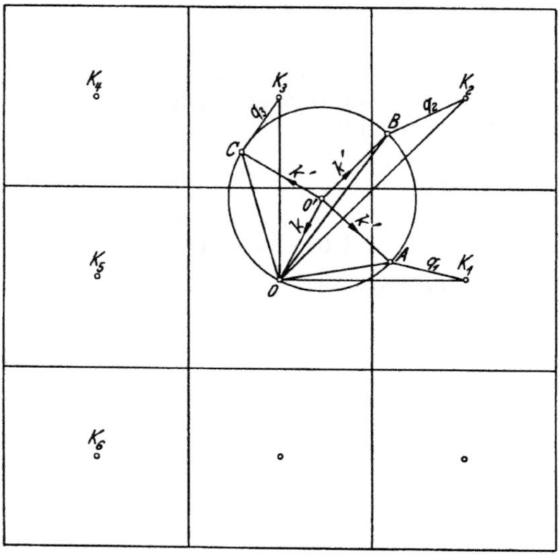


Fig. 5. Diagram to show the occurrence of $U$-processes. Points $O$, $K_n$ denote section of simple cubic reciprocal lattice. $O'$ is centre of the FERMI distribution, $O$, $A$, $B$, $C$ points on its surface. $q_1$, $q_2$, $q_3$ are possible lattice wave-vectors.

and the factor $|\boldsymbol{k} - \boldsymbol{k}'|/q$ is greater than unity. These scattering processes were called by PEIERLS "Umklapp-Prozesse" and are now frequently referred to by the abbreviated form $U$-processes[1].

The following facts may be noted: (i) For given $\boldsymbol{k}$ and $\boldsymbol{k}'$ the value of $\boldsymbol{q}$ is precisely determined, i.e., the solution of (9.4) is always unique. (ii) For given $k$ and $k'$, and the angle between them $\vartheta$, the scattering may be either $N$ or $U$, depending on the direction of $\boldsymbol{k}$. Thus, for example, if $OO'$ is rotated $A$ may easily be brought within the first zone. (iii) The factor $|\boldsymbol{k} - \boldsymbol{k}'|/q$ is always greater than unity for $U$-processes and may become quite large if the FERMI sphere is large enough almost to touch the BRILLOUIN zone planes in $k$-space. (iv) In $U$-processes $\boldsymbol{q}$ is not, in general, parallel to $\boldsymbol{k}' - \boldsymbol{k}$. The displacement of an atom at a lattice point $\boldsymbol{s}$ due to a single real wave, may be written $\delta \boldsymbol{R}(\boldsymbol{s})$, where

$$\delta \boldsymbol{R}(\boldsymbol{s}) = \frac{1}{\sqrt{N}} \{ e^{i \boldsymbol{q} \cdot \boldsymbol{s}} \boldsymbol{a}(\boldsymbol{q}) + e^{-i \boldsymbol{q} \cdot \boldsymbol{s}} \boldsymbol{a}^*(\boldsymbol{q}) \}. \quad (11.2)$$

In an isotropic elastic solid the lattice waves may be classified into longitudinal and transverse waves, according as $\delta \boldsymbol{R}(\boldsymbol{s})$ is parallel or perpendicular to $\boldsymbol{q}$. Thus in $N$-processes the transverse waves make no contribution to $U_{\boldsymbol{k}'\boldsymbol{k}}$ according to (9.3). On the other hand we see that in $U$-processes both longitudinal and transverse waves contribute to the scattering since, in this case, $\boldsymbol{I}$ is no longer parallel to $\boldsymbol{q}$. (v) The classification into $N$- and $U$-processes is very largely conventional and depends upon the validity of the assumption expressed by (10.3). If we are concerned with divalent metals, or even monovalent metals in which the energy surfaces deviate appreciably from the spherical form, then, whether we call a particular scattering process $N$ or $U$ is, to a large extent, arbitrary. This situation is illustrated in Figs. 6a and 6b. Fig. 6a shows the BRILLOUIN zone drawn in the conventional way with the state of lowest energy lying at $L$ the centre of the zone. The transitions $OB$ and $OC$ are $U$-processes

[1] R. PEIERLS: Ann. Phys., Lpz. (5) **12**, 154 (1932).

in this description. However, in a case of this sort where the zone is almost full, an equivalent, and probably more convenient description, is that shown in Fig. 6b, where the centre point $H$ is now the state of highest energy and the state of lowest energy lies at the equivalent corner points. The corresponding

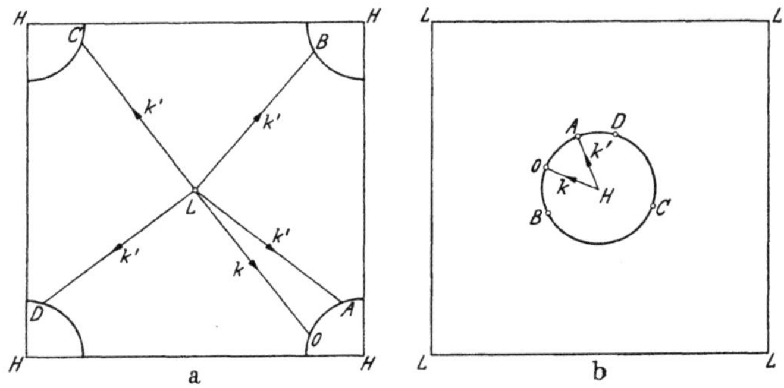


Fig. 6. Diagram to illustrate the equivalence of $N$- and $U$-processes in polyvalent metals. (a) and (b) are alternative descriptions of the same BRILLOUIN zone. In (a) transition $O$ to $D$ is a $U$-process, in (b) the same transition is an $N$-process.

transitions $OB$ and $OC$ are now $N$-processes. The most important consideration in determining whether to regard a given scattering process as $N$ or $U$ is the form of the vector $\boldsymbol{I}$. For example, if this integral takes a simpler form in terms of the wave vectors of the BRILLOUIN zone b than in terms of the wave vectors of a, then the description in terms of b will be most convenient. Metallic sodium is exceptional amongst metals in that the states at the FERMI surface are simply related to the ground state of the zone as indicated in (10.3). In this case the distinction between $N$- and $U$-processes is unambiguous.

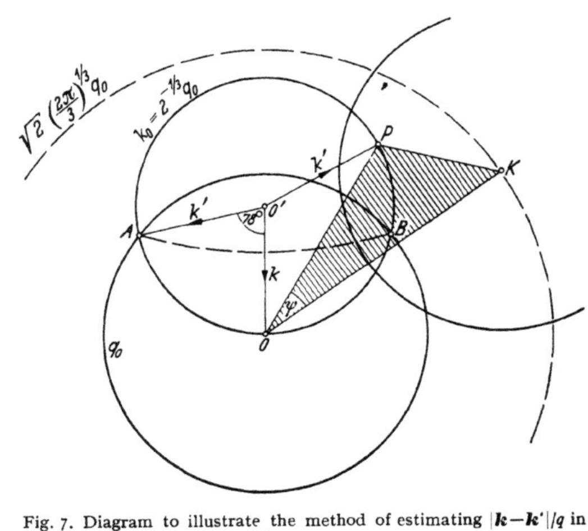


Fig. 7. Diagram to illustrate the method of estimating $|\boldsymbol{k}-\boldsymbol{k}'|/q$ in $U$-processes.

The appearance of the $U$-processes, which, as we shall see in the next section, are dominant at high temperatures, greatly complicates the theory of conductivity. Even with the assumptions of elastic isotropy and free electrons the transition probabilities $P(\boldsymbol{k}, \boldsymbol{k}')$ are no longer dependent only on the magnitudes of the vectors $\boldsymbol{k}$ and $\boldsymbol{k}'$ and the angle between them, but also on the direction of $\boldsymbol{k}$ relative to the axes of the reciprocal lattice. Without further approximations, therefore, the mathematics becomes intractable.

As a first approximation we replace the BRILLOUIN zone in $q$-space by a sphere of equal volume whose radius $q_0$ is given by

$$q_0 = \left(\frac{9\pi}{2}\right)^{\frac{1}{3}} \frac{1}{r_s}, \tag{11.3}$$

where $r_s$ is the radius of the atomic sphere. For a body-centred cubic space lattice the reciprocal lattice is face-centred cubic and there are twelve lattice points, such as $K$ of Fig. 7, lying at equal distances $OK$ from the origin given by

$$OK = \sqrt{2}\,(3\pi^2)^{\frac{1}{3}}/r_s. \tag{11.4}$$

Thus $OK/q_0 = 2^{\frac{5}{6}}\,3^{-\frac{1}{3}}\,\pi^{\frac{1}{3}} = 1.8094$, and the spheres about $O$ and $K$, therefore, overlap.

The radius of the FERMI sphere $k_0$ is less than $q_0$ because each state is doubly occupied on account of the electron's spin. Clearly

$$k_0 = 2^{-\frac{1}{3}}\,q_0. \tag{11.5}$$

In this model, therefore, $N$-processes occur for $\vartheta$ less than $\vartheta_0$ where

$$2k_0 \sin\left(\frac{\vartheta_0}{2}\right) = q_0 \tag{11.6}$$

i.e. when $w < w_0$, where $w_0 = 2^{-\frac{2}{3}} = 0.630$. The limiting angle, therefore, for $N$-scattering is 78° 6′. In Fig. 7 $\boldsymbol{k}$ indicates the initial state and we see that for $N$-scattering $\boldsymbol{k}'$ must lie on that part of the FERMI surface below the line $AB$. This is a fraction less than 0.4 of the whole surface.

In order to determine the probability of $U$-scattering we have to estimate the magnitude of the ratio $|\boldsymbol{k}' - \boldsymbol{k}|/q$ which occurs in Eq. (10.15). In the notation of Fig. 7 we have

$$\left.\begin{aligned}\frac{|\boldsymbol{k} - \boldsymbol{k}'|^2}{q^2} &= OP^2/KP^2\\ &= OP^2/\{OP^2 + OK^2 - 2OP\cdot OK\cos\psi\}.\end{aligned}\right\} \tag{11.7}$$

Using (11.6) we may write

$$OP = (1/a)\,(w/w_0)\,OK, \tag{11.8}$$

where $a = OK/q_0 = 1.8094$, and thus,

$$\frac{|\boldsymbol{k} - \boldsymbol{k}'|^2}{q^2} = \frac{(w/w_0)^2}{a^2 + (w/w_0)^2 - 2a\,(w/w_0)\cos\psi}. \tag{11.9}$$

That the right-hand side of this equation can be incorrectly greater than unity, when $w = w_0$, is due to the assumption of spherical "zones" and the consequent overlap between them. We can estimate an average value of $|\boldsymbol{k} - \boldsymbol{k}'|^2/q^2$, over all possible directions of $\boldsymbol{k}$ for given $w$, and with respect to a single reciprocal lattice point $K$, as follows: Keeping the relative directions of $\boldsymbol{k}$ and $\boldsymbol{k}'$ fixed, let the whole sphere $AOBP$ rotate about the fixed line $OK$, and then let $\psi$ vary from 0 to $\psi_0$, where $\psi_0$ is determined by the condition that $PK = q_0$. Thus

$$\left.\begin{aligned}\left\langle\frac{|\boldsymbol{k} - \boldsymbol{k}'|^2}{q^2}\right\rangle_{\text{av}} &= \frac{(w/w_0)^2}{(1-\cos\psi_0)}\int\limits_0^{\psi_0}\frac{\sin\psi\,d\psi}{a^2 + (w/w_0)^2 - 2a\,(w/w_0)\cos\psi}\\ &= \frac{2\left(\dfrac{w}{w_0}\right)^2\log\dfrac{1}{a-(w/w_0)}}{1-(a-w/w_0)^2}.\end{aligned}\right\} \tag{11.10}$$

Since, for a body-centred cubic space lattice, all reciprocal lattice points nearest the origin are equivalent, the above average should be satisfactory for all directions of $\boldsymbol{k}$. It must be too large for values of $w$ near $w_0$, on account of the overlapping spheres, but for values of $w$ nearer to unity it is probably satisfactory.

When $w=0.75$ the factor multiplying $(w/w_0)^2$ in (11.10) is 1.58, when $w=0.88$ it is 2.15, and when $w=1.0$ it is 3.16. The assumption made by BARDEEN[1], that in all $U$-processes $q$ may be replaced by $q_0$, is equivalent to assuming that the factor multiplying $(w/w_0)^2$ in (11.10) is equal to unity. Such an assumption considerably underestimates the scattering probability of the $U$-processes and leads to too low a value for the resistance.

**12. Formulae for the time of relaxation and the electrical conductivity.** If we introduce the expressions for $|U_{\boldsymbol{k}'\boldsymbol{k}}|^2$ from (10.14) and (10.15) into (7.12) we obtain the two results for the time of relaxation according to the EINSTEIN and the DEBYE models, respectively. We can write

$$\frac{1}{\tau}=\frac{\pi}{3}\frac{m}{\hbar}\left(\frac{T}{M\,\mathrm{k}\,\Theta_E^2}\right)C_E^2, \tag{12.1}$$

where

$$\left(\frac{C_E}{\zeta}\right)^2=64\int_0^1 w^5\,G^2(w)\,dw \tag{12.2}$$

for the EINSTEIN model, and

$$\frac{1}{\tau}=\frac{\pi}{3}\frac{m}{\hbar}\left(\frac{T}{M\,\mathrm{k}\,\Theta_D^2}\right)C_D^2, \tag{12.3}$$

where

$$\left(\frac{C_D}{\zeta}\right)^2=2^{\frac{14}{3}}\int_0^1 \frac{|\boldsymbol{k}-\boldsymbol{k}'|^2}{q^2}\,w^3\,G^2(w)\,dw \tag{12.4}$$

for the DEBYE model.

(12.3) is an equation given by SOMMERFELD and BETHE[2] and $C_D$ is the constant in terms of which it has become customary to express the coupling between the electrons and the thermal motion of the lattice. Comparing (12.2) and (12.4) it will be seen that the DEBYE model gives greater scattering at small angles than the EINSTEIN model, and also that (12.2) contains an extra numerical factor $2^{\frac{4}{3}}$.

We can reduce the integrals in (12.2) and (12.4) still further by introducing $f(w)$ defined by (10.12). We obtain

$$\left(\frac{2^{-\frac{1}{3}}\pi C}{V_0(r_s)-E_0}\right)^2=\begin{cases} 2^{\frac{4}{3}}\displaystyle\int_0^1 f^2(w)\,w^{-1}\,dw & \text{(EINSTEIN)}, \qquad (12.5)\\[2ex] \displaystyle\int_0^{2^{-\frac{2}{3}}} f^2(w)\,w^{-3}\,dw+\int_{2^{-\frac{2}{3}}}^1 \frac{|\boldsymbol{k}-\boldsymbol{k}'|^2}{q^2}\,f^2(w)\,w^{-3}\,dw & \text{(DEBYE)}. \qquad (12.6)\end{cases}$$

Figs. 8a and 8b show the plots of the integrands. Numerical integration gives the following results:

$$\int_0^1 f^2(w)\,w^{-1}\,dw=5.35,\qquad \int_{2^{-\frac{2}{3}}}^1 f^2(w)\,w^{-1}\,dw=3.87,\qquad \int_0^{-\frac{2}{3}} f^2(w)\,w^{-3}\,dw=6.34. \tag{12.7}$$

To estimate the second integral in (12.6) we take, first of all, $|\boldsymbol{k}-\boldsymbol{k}'|/q$ to be equal to $w/w_0$. This is BARDEEN'S assumption[1] that all $U$-processes may be considered to take place with $q=q_0$. This gives, according to (12.7), for the

[1] J. BARDEEN: Phys. Rev. **52**, 688 (1937).

[2] A. SOMMERFELD and H. BETHE: Handbuch der Physik von GEIGER und SCHEEL Bd. 24/2 S. 333—622. 1933.

value of the second integral $3.87 \times 2^{\frac{4}{3}} = 9.76$. From (11.10) we see, however, that $(w/w_0)^2$ should be multiplied by a factor which varies from about 1.58 to 3.16. If we take, as a rough approximation, an average value of 2.0 the second integral of (12.6) amounts to 19.5 and therefore the total integral is 25.8. Thus, finally, we may write

$$\frac{C}{\zeta} = \left(\frac{V_0(r_s) - E_0}{\zeta}\right) \begin{cases} 1.47 & \text{(EINSTEIN)} \\ 2.04 & \text{(DEBYE)}. \end{cases} \tag{12.8}$$

Although the constant $C/\zeta$ is larger for the calculation based on the DEBYE model, the relaxation times are more nearly comparable because of the different $\Theta$

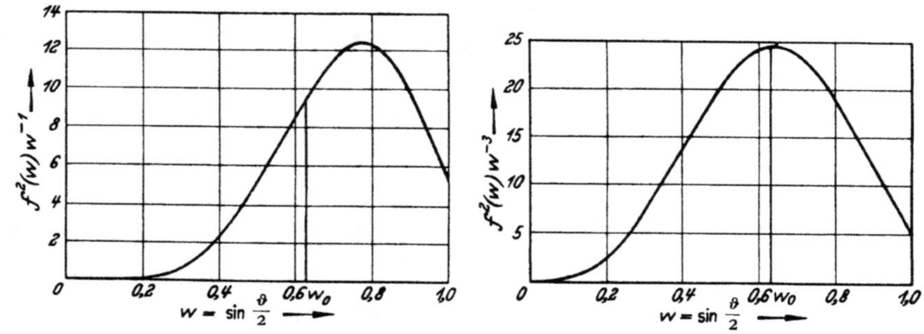


a b

Fig. 8a and b. a Graph of the integrand of (12.5). $w_0 = 2^{-\frac{2}{3}}$. b Graph of the function occurring in the integrand of (12.6).

values involved. For temperatures twice, or more, times as great as the characteristic temperature the ratio $\Theta_D/\Theta_E$ which ensures the same value of the specific heat, whether calculated according to the EINSTEIN or DEBYE expression, is 1.29. Thus we find

$$(C_D/\Theta_D)^2 = 1.15\,(C_E/\Theta_E)^2, \tag{12.9}$$

where the subscripts to the constant $C$ indicate the method of calculation to which it refers.

(12.9) shows that, at high temperatures, the EINSTEIN and DEBYE models lead to approximately the same value for the time of relaxation and, therefore, to approximately the same value of the conductivity. This might have been anticipated on general grounds, particularly as the perturbation arising from the displacement of any one atom has been assumed to be confined to its own atomic cell.

It will be observed that, in the DEBYE model, the $U$-processes are responsible for three quarters of the total scattering at high temperatures. This may be seen from (12.6) in which the second integral contributes a fraction 0.755 of the total according to the present estimate. Even with BARDEEN'S assumption, which certainly underestimates the effect of $U$-processes, these are responsible for over half (0.6) of the total scattering. The recognition of the large part played by $U$-processes at high temperatures is of importance when comparing the electrical and thermal conductivities at high and low temperatures, because, as we shall see, at low temperatures $U$-processes play no part in monovalent metals.

To obtain a formula for the resistivity we have only to substitute (12.1) or (12.3) into (6.4). For the DEBYE model we obtain

$$\frac{1}{\sigma} = \frac{\pi}{3}\,\frac{m^2\Omega}{e^2\hbar}\left(\frac{T}{M\,\mathrm{k}\,\Theta_D^2}\right) C_D^2, \tag{12.10}$$

where $\Omega$ denotes the atomic volume, and the equation is applicable to monovalent metals only. Since the mass of the electron $m$ only occurs through the derivative of the energy with respect to $k$ we can replace it by the effective mass if it is possible to retain the assumption of spherical energy surfaces in $k$-space. This modification of (12.10) might be useful for the alkali metals other than sodium.

Introducing the numerical values of the universal constants and using practical units we obtain

$$\frac{1}{\sigma} = 4.776 \cdot 10^{-2} \left(\frac{T}{M_A \Theta_D^2 \Omega^{\frac{1}{3}}}\right) \left(\frac{C_D}{\zeta}\right)^2 \text{ohms cm.}, \tag{12.11}$$

where $M_A$ is the atomic weight and $\Omega$ is the atomic volume in cubic Ångströms. For sodium $M_A = 23$, $\Theta_D = 150$, $\Omega^{\frac{1}{3}} = 3.40$ Å. Thus at room temperature the calculated resistivity is $7.41 \cdot 10^{-6}\,(C/\zeta)^2$ ohms cm. on the DEBYE model. The observed value is $4.27 \cdot 10^{-6}$ ohms cm. Thus to obtain agreement with experiment we must take $C_D = 0.76\,\zeta$, and hence from (12.8) we require that $V_0(r_s) - E_0 = 1.19$ ev, since $\zeta$ for sodium is 3.2 ev.

It is interesting to compare the value of $V_0(r_s) - E_0$, required to give agreement with the observed resistivity, with a direct calculation of this quantity along the lines of the WIGNER and SEITZ method. FRÖHLICH[1] gives the result

$$V_0(r_s) - E_0 = (r_s^2 - r_0^2)/r_s^3 \quad \text{RYDBERG units}, \tag{12.12}$$

where $r_s$ and $r_0$ are in BOHR units, and $r_0$ refers to the minimum of the curve showing $E_0$ as a function of the radius of the atomic sphere. Numerical values, according to FRÖHLICH's calculation, are: $r_s = 4.03$ BOHR units, and $r_0 = 3.24$ BOHR units, which, on substitution into (12.12) give $V_0(r_s) - E_0 = 1.2$ ev in almost exact agreement with the value required to give the observed resistivity.

**13. BARDEEN's calculation of the interaction constant.** Although the calculation of the resistivity given in the preceding section agrees well with the observed value, it should be remembered that it depends upon certain assumptions which are not easy to justify by *a priori* argument. In the first place the numerical value which should be taken for $\Theta_D$ is uncertain to the extent of several percent, and secondly the assumption that the effect of the displacement of an atom can be neglected outside its own cell can only be regarded as a rough hypothesis. This last assumption is likely to be invalid for displacements resulting from long wave-length lattice waves and is, therefore, probably inappropriate for the theory of scattering at very low temperatures.

We now discuss two other methods of computing the coupling between electrons and lattice vibrations.

BARDEEN[2] has given a self-consistent calculation of the scattering matrix based upon first order perturbation theory. He first determines a potential energy $V_i$ at a given point $\boldsymbol{r}$ arising from the displacements of *all* ions, assuming the electronic charge distribution to be unchanged. This leads to an expression similar to (8.1) but summed over all atoms $\boldsymbol{s}$. $V_i$, taken alone, would lead, of course, to far too large a value of the resistivity. BARDEEN next calculates the change in the electronic charge distribution resulting from $V_i$ in terms of the one-electron wave functions, $\psi_{\boldsymbol{k}}$, by first order perturbation theory. From this a second contribution to the potential, $V_\varrho$, is determined by the use of POISSON's equation; a contribution which is, of course, in the opposite sense to $V_i$. The

[1] H. FRÖHLICH: Elektronentheorie der Metalle, S. 276, eqn. (14). 1936.
[2] J. BARDEEN: Phys. Rev. **52**, 688 (1937).

self-consistency is introduced by determining the wave functions from the equation

$$\left\{-\frac{\hbar^2}{2m}\nabla^2+V_0+V_i+V_\varrho\right\}\psi_{\boldsymbol{k}}=E_{\boldsymbol{k}}\psi_{\boldsymbol{k}}, \tag{13.1}$$

in which $V_\varrho$ contains implicitly the wave functions $\psi_{\boldsymbol{k}}$.

The calculation is rather long and will not be reproduced in detail; the result, however, may be quoted by giving the function which replaces our previously defined $G(w)$; it is

$$G(w)=\frac{1}{6\pi}\left\{\frac{1+\left(\frac{V_0(r_s)-E_0}{\zeta}\right)\frac{3w^2}{\beta}}{\frac{1}{2}+\left(\frac{1-w^2}{4w}\right)\log\left(\frac{1+w}{1-w}\right)+\frac{2w^2}{\beta}}\right\}\frac{f(w)}{w^3}, \tag{13.2}$$

where $\beta=e^2k_0/\pi\zeta$. The constant $\beta$ indicates the part played by the interaction between electrons, since it is a ratio of a COULOMB energy to the FERMI energy. In fact, if $\beta$ is allowed to tend to zero, (13.2) becomes identical in form with our previously defined $G(w)$, Eq. (10.13). However, in BARDEEN'S expression $V_0(r_s)-E_0$ has a somewhat different meaning from that attached to this quantity in the previous section. Here it is determined from the equation

$$\nabla^2u+\frac{2m}{\hbar^2}\left(E_0-v(\boldsymbol{r})-\frac{3e^2}{2r_s}+\frac{e^2r^2}{2r_s^3}\right)u=0, \tag{13.3}$$

with the boundary condition $u'(r_s)=0$ as before, and where $v(\boldsymbol{r})$ denotes the potential energy due to the ion alone. In this case, therefore, the COULOMB term, arising from a uniform distribution of negative charge within the atomic sphere, is added to the potential energy due to the ion itself $v(\boldsymbol{r})$, and $V_0(r_s)$ refers to the sum of these two terms. $E_0$ is also differently defined since it now contains the COULOMB term, viz. the average of $\left(\frac{3e^2}{2r_s}-\frac{e^2r^2}{2r_s^3}\right)$. As $u$ is approximately constant over a large part of the atomic cell, this average must be nearly equal to $1.2\,e^2/r_s$ which is 0.59 Ryds. for sodium. $E_0$ determined with $v(\boldsymbol{r})$ alone, as in the WIGNER-SEITZ method, gives $-0.60$ Ryds. for this eigenvalue at $r_s=4.05$ BOHR units. Thus the eigenvalue of (13.3) must be quite small; only a fraction of an electron volt. At the boundary of the atomic cell $v(r)\approx -e^2/r$, therefore $V_0(r_s)\approx 0$, and hence $V_0(r_s)-E_0$ is just the negative of the eigenvalue of (13.3). For this quantity BARDEEN gives 0.2 ev. It is clear that $V_0(r_s)-E_0$ must always be positive since it is a measure of $u''(r_s)$ which must necessarily be positive.

The following table gives the values of $G(w)$ on the assumption $V_0(r_s)-E_0=0$.

*Table 2.*

| $w$ | 0 | 0.2 | 0.4 | 0.6 | 0.8 | 1.0 |
|---|---|---|---|---|---|---|
| $G(w)$ Eq. (13.2) | 1.0 | 0.900 | 0.661 | 0.397 | 0.241 | 0.096 |

It is of some interest to compare these values with those given by (10.13), which are shown in Table 3, in which $V_0(r_s)-E_0$ has the value 1.2 ev as derived in Sect. 12.

*Table 3.*

| $w$ | 0 | 0.2 | 0.4 | 0.6 | 0.8 | 1.0 |
|---|---|---|---|---|---|---|
| $G(w)$ Eq. (10.13) | 0.562 | 0.530 | 0.440 | 0.315 | 0.182 | 0.069 |

It will be observed that the values of $G(w)$ given by (10.13) are somewhat less than BARDEEN's values; the differences being greater at small values of $w$.

With $G(w)$ as given in Table 2 BARDEEN obtains excellent agreement with the observed resistivity. The reason why the ultimate result in BARDEEN's calculation agrees with that obtained in the previous section, using the smaller values of $G(w)$, is that BARDEEN underestimates the $U$-scattering. This occurs between $w = 0.63$ and $w = 1.0$ which, as Fig. 8a shows, is the most effective region of the whole range of $w$.

It is remarkable that the two calculations outlined in Sect. 10 and 13 should lead to such similar values for the scattering probabilities as shown by Tables 2 and 3. The two calculations emphasise complementary aspects of the coupling between the lattice motion and the electrons. In BARDEEN'S calculation the effect of exchange between the conduction electrons is neglected, which results in a very small contribution to the scattering from the region within the atomic cell of the displaced atom. In the method of Sect. 10, on the other hand, the effect of exchange is over-emphasised since it is assumed that the "FERMI hole" fills the whole atomic sphere.

In the next section we consider a third method of calculating the scattering of electrons by the thermal motion of the lattice which is most satisfactory for very low temperatures where only long wave-length lattice waves are excited.

**14. The deformation potential.** The method of calculating the electronic scattering to be described in this section was first given by BARDEEN and SHOCKLEY[1] and has been developed by HUNTER and NABARRO[2]. It is only applicable when the atomic displacements are the results of strains which vary slowly by comparison with the interatomic distance. When this is the case we can regard the crystal as a continuum, as far as its elastic properties are concerned, and we can define a dilatation $\Delta(\boldsymbol{r})$ at any point $\boldsymbol{r}$. Consider, now, an electron coming from an unstrained region of the crystal in a state specified by the wave vector $\boldsymbol{k}$. If we retain the assumption expressed by Eq. (10.3) the energy of the electron is given approximately by

$$E_{\boldsymbol{k}} = E_0(r_s) + \frac{\hbar^2}{2m} k^2. \tag{14.1}$$

As the electron moves from the unstrained region, where $r_s$ has its equilibrium value, to regions where $r_s$ takes slightly different values its energy changes through the change of $E_0(r_s)$. This change of energy may be regarded as the perturbing potential leading to the scattering. Thus we may write

$$U(\boldsymbol{r}) = \frac{r_s}{3} \frac{\partial E_0}{\partial r_s} \Delta(\boldsymbol{r}), \tag{14.2}$$

where we have made use of the relation

$$3 \frac{dr}{r} = \Delta. \tag{14.3}$$

Since we assume that the lattice can be treated as a continuum the expression (9.1), for the displacement, may be written

$$\boldsymbol{R}(\boldsymbol{r}) = \frac{1}{\sqrt{N}} \sum_{\boldsymbol{q}} e^{i\boldsymbol{q}\cdot\boldsymbol{r}} \boldsymbol{a}(\boldsymbol{q}), \tag{14.4}$$

[1] J. BARDEEN and W. SHOCKLEY: Phys. Rev. **80**, 72 (1950).
[2] S. C. HUNTER and F. R. N. NABARRO: Proc. Roy. Soc. Lond. **220**, 542 (1954).

and since

$$\varDelta(\boldsymbol{r}) = \operatorname{div} \boldsymbol{R}(\boldsymbol{r}), \tag{14.5}$$

we have

$$\varDelta(\boldsymbol{r}) = \frac{1}{\sqrt{N}} \sum_{\boldsymbol{q}} i\boldsymbol{q} \cdot \boldsymbol{a}(\boldsymbol{q})\, e^{i\boldsymbol{q}\cdot\boldsymbol{r}}. \tag{14.6}$$

Hence

$$\left.\begin{aligned} U_{\boldsymbol{k}'\boldsymbol{k}} &= \frac{r_s}{3\sqrt{N}} \frac{\partial E_0}{\partial r_s} \sum_{\boldsymbol{q}} i\,\boldsymbol{q} \cdot \boldsymbol{a}(\boldsymbol{q}) \int_{N\Omega} e^{i(\boldsymbol{k}-\boldsymbol{k}'+\boldsymbol{q})\cdot\boldsymbol{r}} (N\Omega)^{-1} d\boldsymbol{r} \\ &= -\frac{i r_s}{3\sqrt{N}} \frac{\partial E_0}{\partial r_s} \{\boldsymbol{a}(\boldsymbol{q}) \cdot (\boldsymbol{k} - \boldsymbol{k}')\}, \end{aligned}\right\} \tag{14.7}$$

where $\boldsymbol{q} = \boldsymbol{k}' - \boldsymbol{k}$ and, in accordance with (14.1), we regard $u(r)$ as constant and equal to $(N\Omega)^{-\frac{1}{2}}$. Comparing with (9.3) we see that in this theory $\boldsymbol{I}$ is given by

$$\boldsymbol{I} = \frac{i r_s}{3N} \frac{\partial E_0}{\partial r_s} (\boldsymbol{k} - \boldsymbol{k}'). \tag{14.8}$$

To determine $(\partial E_0/\partial r_s)$ we use the condition that the partial potential of the electrons throughout the inhomogeneously strained crystal is constant. Thus the height of the FERMI level, relative to a constant energy independent of position in the metal, remains unchanged as we pass through regions of differing strain. Hence $E_F$, given by

$$E_F = E_0(r_s) + \frac{\hbar^2}{2m} k_0^2, \tag{14.9}$$

must be independent of the dilatation $\varDelta$. Since $k_0^2 = (9\pi/4)^{\frac{2}{3}} r_s^{-2}$ we find

$$\frac{\partial E_0(r_s)}{\partial r_s} = \frac{2\zeta}{r_s}. \tag{14.10}$$

Substituting (14.10) into (14.8) and the resulting value of $\boldsymbol{I}$ into (9.13) we find again the expression (10.15) for $|U_{\boldsymbol{k}'\boldsymbol{k}}|^2$ in which now

$$G(w) = 1. \tag{14.11}$$

That $G$ is independent of $w$ is due to the initial assumption that the deformation potential may only be used for very long wave-length thermal waves. Thus only scattering through very small angles is contemplated in this theory, and hence the constant value of $G$ obtained should be compared with the previously determined functions in the limit as $w$ tends to zero. It will be observed that in this limit there is perfect agreement with BARDEEN's calculation. Although these two calculations differ in many respects they have one important feature in common: both take into consideration the change in the electronic charge distribution under the influence of the atomic displacements of the lattice. In the method of the deformation potential this arises from the assumption of the constant FERMI level. If we had assumed that the crystal remained electrically neutral in every atomic cell, independent of the strain, the FERMI level in regions where $\varDelta$ was large would be lower than in regions where $\varDelta$ was small. Hence there is, in fact, a flow of electrons from regions of negative to regions of positive dilatation.

If we had neglected the flow of electrons under the action of the strain we should have obtained $\frac{6}{5}\frac{\zeta}{r_s}$, for $\frac{\partial E_0}{\partial r_s}$, in place of $\frac{2\zeta}{r_s}$, which would have led to a value 0.6 for $G$. This is close to the value given in Table 2 for $w = 0$. The theory of Sect. 10 on which this table is based also implicitly contains the assumption of no flow of electrons.

## IV. The conductivity of impure metals and alloys.

**15. Resistance due to impurities; MATTHIESSEN's rule.** We consider a substitutional solid solution in which the concentration of solute atoms is so small that the direct influence of these atoms on each other can be neglected. We discuss, therefore, first the scattering of conduction electrons by a single impurity atom situated at one of the lattice points of the crystal.

Let $V_A(\boldsymbol{r})$ denote the potential energy of an electron in the neighbourhood of one of the ions of the solvent metal. The most satisfactory approximation for $V_A(\boldsymbol{r})$, in the case of monovalent metals, is obtained by taking the energy of an electron in the field of a single ion within each atomic cell. Now let $V_B(\boldsymbol{r})$ denote the potential energy in the neighbourhood of the solute atom. If we could neglect any accumulation of charge, arising from the conduction electrons, $V_B(\boldsymbol{r})$, within the atomic cell surrounding the impurity ion, would be the energy of an electron in the field of this singly charged ion alone. In general $V_B - V_A$ is not small but very large. For instance, when a gold atom is dissolved in copper, $V_B - V_A$, regarded as a potential well, is large enough to contain many bound stationary states. It is not, therefore, possible to construct a realistic theory by regarding $V_B - V_A$ as a perturbation, scattering free electrons represented by plane waves. To attempt to do so would not only be an unwarranted application of perturbation theory but would lead to conclusions at variance with experimental results. For example, one atomic percent of zinc in copper increases the resistance to a greater extent than one atomic percent of silver in copper, although the field of the singly charged zinc ion is much closer to that of the copper ion than the field of the silver ion; thus $V_B - V_A$ would be less for zinc than for silver.

It is clear that, with the true scattering potential $V_B - V_A$, the wave functions of the conduction electrons are modified in a drastic way inside the ionic core of the impurity atom. We shall show in the next section how this can be approximately taken into account, at least for monovalent solutes in monovalent solvents. Meanwhile, in order to construct a formal theory, we introduce the idea of an effective scattering potential $U^{(i)}(\boldsymbol{r})$ which can be assumed to be small enough to justify the use of first order perturbation theory. In addition, we assume that $U^{(i)}(\boldsymbol{r})$, is real and that, with respect to the centre of the impurity atom, it possesses all the symmetry elements of the lattice, including the inversion. The matrix of $U^{(i)}(\boldsymbol{r})$, with respect to unperturbed wave functions of the form (10.3), is as follows:

$$U^{(i)}_{\boldsymbol{k}'\boldsymbol{k}} = \int_{N\Omega} e^{i(\boldsymbol{k}-\boldsymbol{k}')\cdot\boldsymbol{r}}\, u_0^2(\boldsymbol{r})\, U^{(i)}(\boldsymbol{r})\, d\boldsymbol{r}. \tag{15.1}$$

Now for the most general BLOCH wave function we have

$$u_{\boldsymbol{k}}(-\boldsymbol{r}) = u_{-\boldsymbol{k}}(\boldsymbol{r}), \tag{15.2}$$

if the origin is chosen to lie at a lattice point. Thus for the ground state we have $u_0(-\boldsymbol{r}) = u_0(\boldsymbol{r})$. When the origin is at the impurity atom $U^{(i)}(-\boldsymbol{r}) = U^{(i)}(\boldsymbol{r})$. Hence, replacing $\boldsymbol{r}$ by $-\boldsymbol{r}$ in the integral of (15.1), which does not change its value, and using the above two results, we see that $U^{(i)}_{\boldsymbol{k}'\boldsymbol{k}}$ is equal to its complex conjugate and is therefore real.

The matrix element $U_{\boldsymbol{k}'\boldsymbol{k}}$, arising from the thermal motion, and calculated using the same assumption (10.3) regarding the wave functions, was found to be purely imaginary. Hence

$$|U_{\boldsymbol{k}'\boldsymbol{k}} + U^{(i)}_{\boldsymbol{k}'\boldsymbol{k}}|^2 = |U_{\boldsymbol{k}'\boldsymbol{k}}|^2 + |U^{(i)}_{\boldsymbol{k}'\boldsymbol{k}}|^2\,. \tag{15.3}$$

Thus the contributions to the transition probability $P(\boldsymbol{k}', \boldsymbol{k})$ from impurity and thermal scattering are additive, and we deduce immediately that the resistivity $\varrho$ can be written

$$\varrho = \varrho_t + \varrho_i, \tag{15.4}$$

where $\varrho_t$ is the resistance due to thermal motion and $\varrho_i$ the resistance due to impurity scattering which, according to the above theory, is independent of temperature since $U^{(i)}_{\boldsymbol{k}'\boldsymbol{k}}$ does not depend on the temperature. (15.4) is the formal expression of MATTHIESSEN's rule. The assumption underlying the theoretical derivation of this rule may be summarised as follows: (i) The temperature is high enough to justify treating all thermal scattering as elastic. (ii) The conduction electrons can be regarded as quasi-free with wave functions of the form (10.3). Without this assumption $U_{\boldsymbol{k}'\boldsymbol{k}}$ and $U^{(i)}_{\boldsymbol{k}'\boldsymbol{k}}$ would not, in general, satisfy (15.3) and interference between thermal and impurity scattering would take place with the result that (15.4) would fail[1]. (iii) The thermal motion of the lattice does not influence the effective scattering potential $U^{(i)}(\boldsymbol{r})$. (iv) The impurity atoms do not affect the thermal vibrations of the lattice and, therefore, the temperature dependent part of the scattering. (v) The addition of impurity atoms with valency different from the valency of the solvent does not change the effective number of conduction electrons.

None of these assumptions is likely to be strictly valid, and, in fact, accurate experimental work has shown that $d\varrho/dT$ varies systematically with the impurity concentration.

**16. Monovalent impurity in a monovalent metal.** In this section we make an approximate calculation of the magnitude of the increment in the resistance due to monovalent impurities in monovalent metals. The determination of the wave functions and self-consistent field for the valency electrons in metals containing impurities has proved to be a most intractable mathematical problem, and no satisfactory solution has yet been obtained. There are two difficulties: the accumulation of charge in the neighbourhood of the impurity atom affects the potential, not only within the cell of this atom, but also the potential in the cells of the neighbouring solvent atoms. Within a neighbouring cell the potential is asymmetrical with respect to the centre, and this greatly increases the difficulty of obtaining satisfactory wave functions. The second difficulty concerns the construction of the self-consistent field. Experience with the idealized problem of the square well shows that the wave functions, in the neighbourhood of the well, vary appreciably with the energy, as this passes through values corresponding to the width of the FERMI band. Thus many individual wave functions with different eigenvalues would be required to determine the charge density and hence the field.

It is clear that to make progress drastic approximations are necessary. For examples of approximate self-consistent solutions the reader is referred to the work of SEITZ and HUNTINGTON[2] who deal with the problem of the vacant lattice site, and to the work of HUANG[3] on the impurity problem treated by the FERMI-THOMAS method; also to several papers by FRIEDEL[4] on this subject.

[1] In the more detailed theory of conduction (cf. Sect. 35) in which energy exchanges with the lattice vibrations are taken into account, thermal scattering is inelastic and is, therefore, separated from the necessarily elastic scattering due to impurity atoms, whatever the nature of the matrix elements. However, the justification for regarding the impurity and thermal scattering as entirely distinct and, therefore, additive then depends on the validity of the assumption mentioned in the last paragraph of Sect. 7.

[2] F. SEITZ and H. B. HUNTINGDON: Phys. Rev. **61**, 315 (1942).

[3] K. HUANG: Proc. Phys. Soc. Lond. **60**, 161 (1948).

[4] e.g. J. FRIEDEL: Phil. Mag. **43**, 153 (1952).

The model which we adopt in this section was first suggested by MOTT[1] and has been developed by ARAFA[2] and has, at least, the merit of simplicity. We neglect altogether the effect of the accumulation of charge round the impurity atom. The potential in which the electron moves, therefore, is that of the $A$-ion in the $A$-cells, and of the $B$-ion in the $B$-cells. We define two functions $V(\boldsymbol{r})$ and $\Phi(\boldsymbol{r})$ as follows:

$$V(\boldsymbol{r}) = \begin{cases} V_A(\boldsymbol{r}) \text{ in } A \text{ cells} \\ V_B(\boldsymbol{r}) \text{ in } B \text{ cells,} \end{cases} \qquad \Phi(\boldsymbol{r}) = \begin{cases} 0 \text{ in } A \text{ cells} \\ U_0 \text{ in } B \text{ cells,} \end{cases} \tag{16.1}$$

where $U_0$ is a constant to be determined.

If we express the wave function as a product of two functions $\chi(\boldsymbol{r})\,u(\boldsymbol{r})$ the SCHRÖDINGER equation, with the above potential, may be written

$$u\left\{\nabla^2\chi + \frac{2m}{\hbar^2}(E-\Phi)\,\chi\right\} + \chi\left\{\nabla^2 u + \frac{2m}{\hbar^2}(E_0+\Phi-V)\,u\right\} = -2\nabla\chi\cdot\nabla u, \tag{16.2}$$

in which $(E_0+E)$ denotes the energy of the electron. For a pure metal the assumption that the wave functions have the form (10.3) is equivalent to neglecting the term on the right hand side of the equation. We make the same approximation here and, therefore, obtain the equations

$$\nabla^2 u + \frac{2m}{\hbar^2}(E_0+\Phi-V)\,u = 0, \tag{16.3}$$

and

$$\nabla^2\chi + \frac{2m}{\hbar^2}(E-\Phi)\,\chi = 0. \tag{16.4}$$

In the $A$-cells the solution of (16.3) is $u_0(\boldsymbol{r})$ which is the wave function of the lowest state, with energy $E_0$, in the conduction band of the pure solvent. In the $B$-cells $\Phi$ takes the constant value $U_0$, and the solution of (16.3) is denoted by $u_B$. We now find the value of $U_0$ which makes the normal derivative of $u_B$ zero over the surface of the polyhedron surrounding the $B$-atom. In practice, in accordance with the WIGNER-SEITZ approximation, it is sufficient to replace the polyhedron by a sphere of equal volume and make $(d u_B/d r)=0$ over the surface of this sphere. Clearly this construction makes $u_B$ join smoothly with $u_0(\boldsymbol{r})$, and we obtain a solution $u(\boldsymbol{r})$ of (16.3) which is continuous and has continuous first derivatives and, therefore, qualifies as a factor of a wave function. This function has, moreover, the important property that it is orthogonal to all the localised electronic states of the ions of both the $A$- and $B$-atoms. The physical meaning of this property is that the conduction electrons are kept away from the regions where $V_B - V_A$ is very large.

Without $\Phi$, Eq. (16.4) has solutions $\chi = e^{i\boldsymbol{k}\cdot\boldsymbol{r}}$ where, as formerly, $E=\hbar^2 k^2/2m$. We regard $\Phi$ in this equation as the perturbation which gives rise to the scattering. As unperturbed wave functions we take, therefore, $\psi_{\boldsymbol{k}}$ defined as follows:

$$\psi_{\boldsymbol{k}} = e^{i\boldsymbol{k}\cdot\boldsymbol{r}}\,u(\boldsymbol{r}), \qquad u(\boldsymbol{r}) = \begin{cases} u_0 \text{ in } A \text{ cells} \\ u_B \text{ in } B \text{ cells.} \end{cases} \tag{16.5}$$

The effective scattering potential $U^{(i)}(\boldsymbol{r})$, in this theory is therefore equal to the constant $U_0$ in the cell surrounding the impurity atom and zero elsewhere. Thus for the scattering matrix we have

$$U^{(i)}_{\boldsymbol{k}'\boldsymbol{k}} = \int_{\Omega} e^{i(\boldsymbol{k}-\boldsymbol{k}')\cdot\boldsymbol{r}}\,u_B^2(\boldsymbol{r})\,U_0\,d\boldsymbol{r}. \tag{16.6}$$

According to our method of normalizing wave functions (8.2) $\overline{u_0^2} = (N\Omega)^{-1}$, where the bar indicates the average over the atomic cell. It is clear from the method of

[1] N. F. MOTT: Proc. Camb. Phil. Soc. **32**, 281 (1936).
[2] M. K. I. ARAFA: Proc. Phys. Soc. Lond. B **62**, 238 (1949).

construction of $u(\boldsymbol{r})$, that $\overline{u_B^2}$ will not, in general be equal to $\overline{u_0^2}$. This quantity, $\overline{u_B^2}$, gives, in fact, an indication of the way the charge density varies in the neighbourhood of the impurity atom. Since $u_B(\boldsymbol{r})$ is generally almost constant over a large fraction of the atomic cell, it is a sufficient approximation to replace $u_B^2$ by $\overline{u_B^2}$ in (16.6). The integral can then be evaluated by elementary methods and we obtain

$$U_{\boldsymbol{k}'\boldsymbol{k}}^{(i)} = \frac{2}{q}\, U_0 \overline{u_B^2}\, r_s^3 \frac{f(w)}{w^3}, \tag{16.7}$$

where, as in the previous sections, $w$ denotes the sine of half the angle of scattering and $f(w)$ is the function defined by Eq. (10.12). It will be observed that impurity scattering has a somewhat similar angular dependence to thermal scattering at high temperatures.

Substituting (16.7) into (7.12), and supposing that the fraction of lattice sites occupied by impurity atoms is $x$, we find for the time of relaxation due to this type of scattering:

$$\frac{1}{\tau_i} = \frac{\pi \hbar x}{4 k_0 m \Omega} \left( \frac{U_0 \overline{u_B^2}}{\zeta \overline{u_0^2}} \right)^2 \int_0^1 f^2(w) \frac{dw}{w^3}. \tag{16.8}$$

The integrand of (16.8) is shown in Fig. 8b; the value of the integral is 12.66. Substituting (16.8) in (6.4), using the numerical values of the universal constants, and expressing the resistivity in micro-ohms cm. we have finally

$$\frac{1}{\sigma_i} = 132.2\, \Omega^{\frac{1}{3}} \left( \frac{U_0 \overline{u_B^2}}{\zeta \overline{u_0^2}} \right)^2 x \quad \text{micro-ohms cm.}, \tag{16.9}$$

where $\Omega^{\frac{1}{3}}$ is measured in Å. The following tables give numerical data relating to the metals copper, silver and gold.

We consider first the pair of metals silver and gold. It will be observed that 1% gold in silver produces the same increment of resistance as 1% silver in gold. This might have been anticipated from the preceding theory, since these two metals have almost exactly the same atomic volume and it is reasonable to conjecture therefore that $\overline{u_B^2} \approx \overline{u_0^2}$ in each case. No WIGNER-SEITZ solution for metallic gold is yet available, though a self-consistent field for the gold atom has recently been calculated[1]. It is not therefore possible at present to make a comparison between the directly calculated value of $U_0$, and the value, 1.8 ev required to explain the resistance increment. We can, however, make an estimate in the following way: For these two metals $U_0$ is just the difference in energy between the bottom of the conduction bands in the pure metals. Hence since the FERMI energy is the same in each case we may take $U_0$ as the difference between the total electronic energies. Fig. 9a shows how $U_0$ may be obtained. The energies at large $r_s$ are determined by the ionization potentials: 9.22 ev and 7.56 ev for

*Table 4.*

| | Cu | Ag | Au |
|---|---|---|---|
| $\Omega^{\frac{1}{3}}$ (Å) | 2.2729 | 2.5688 | 2.5639 |
| $\zeta$ (ev) | 7.056 | 5.525 | 5.546 |

*Table 5.*

| 1 At % solute | Solvent Cu | | Solvent Ag | | Solvent Au | | |
|---|---|---|---|---|---|---|---|
| | Ag | Au | Au | Cu | Cu | Ag | |
| $(1/\sigma_i)$ obs. | 0.14 | 0.55 | 0.36 | 0.077 | 0.45 | 0.36 | micro-ohms cm. |
| $U_0 \left(\overline{u_B^2}/\overline{u_0^2}\right)$ | 1.523 | 3.018 | 1.799 | 0.833 | 2.020 | 1.807 | ev |

[1] W. G. HENRY: Proc. Phys. Soc. Lond. A **67**, 789 (1954).

Au and Ag respectively, and the depths of the minima are given by the observed cohesive energies. On the basis of this estimate, therefore, $U_0 = 2.46$ ev. The reason why this value is somewhat too large may be that the observed cohesive energies in these metals, contain terms arising from attractions, due to polarization, between the ion cores. Hence the contributions from the conduction electrons alone might be less than those shown in Fig. 9a and, therefore, lead to a smaller value of $U_0$.

ARAFA has made detailed calculations for the pair of metals copper and silver and has obtained the following results: For copper in silver he finds $U_0 = -0.62$ ev and $\overline{u_B^2} \approx 1.2\,\overline{u_0^2}$, and for silver in copper $U_0 = +0.85$ ev and $\overline{u_B^2} \approx 0.8\,\overline{u_0^2}$. The reason for the increase of the wave function at a copper cell in the silver lattice is indicated in Fig. 9b. $r_1$ denotes the atomic radius in pure copper. When the

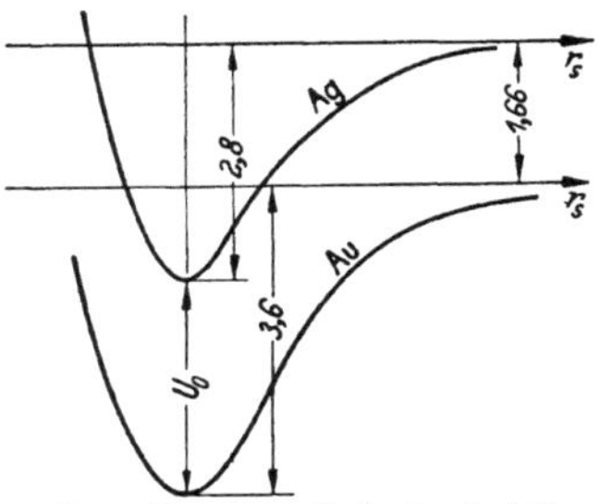


Fig. 9a. Illustration of method of calculating $U_0$. Difference between ionization potentials 1.66 ev. Cohesive energy of Au 3.6 ev, of Ag 2.8 ev per electron.

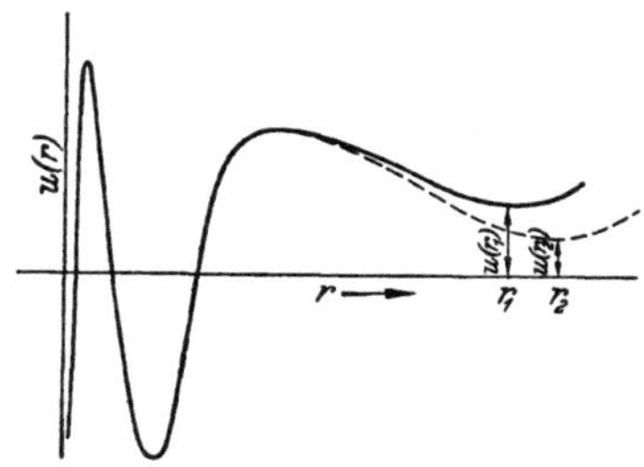


Fig. 9b. Diagrammatic representation of wave functions in the copper atomic sphere for two different eigenvalues. $r_1$ is the atomic radius for pure Cu, $r_2$ the atomic radius for pure Ag.

minimum occurs at a larger radius $r_2$ the value of $u(r_2)$ is less than $u(r_1)$ which refers to pure copper. However, in order to join smoothly to the wave function in the neighbouring cells of the silver lattice $u(r_2)$ must be made equal to $u(r_1)$. This requires that the whole function within the copper cell be multiplied by a factor $u(r_1)/u(r_2)$, and this leads to a value of $\overline{u_B^2}$ greater than $\overline{u_0^2}$.

It will be observed that for Cu in Ag there is very good agreement. The direct calculation gives 0.75 ev for $U_0(\overline{u_B^2}/\overline{u_0^2})$ whilst the value required to give the observed resistance is 0.83 ev. For Ag in Cu, however, the direct calculation gives too small a value, viz. 0.68 ev, to account for the resistance increment which would need 1.52 ev. This discrepancy might have the following explanation. It is known from the study of the elastic properties that, in the noble metals, the ionic cores are in close contact. Thus the replacement of a silver ion by a smaller copper ion is likely to produce less distortion than the substitution of a silver ion in the copper lattice. The outward displacement of the twelve copper ions surrounding the silver impurity must add a contribution to the resistance increment.

**17. Polyvalent impurities in monovalent metals.** The general effect of polyvalent impurities on the resistivity of the noble metals may be seen from the following data taken from LINDE's measurements[1]. $\delta\varrho_i$ is the increment to the resistivity of gold due to one atomic percent of solute expressed in micro-ohms cm.

| Solute | Ti | Cr | Mn | Fe | Co | Ni | Cu | Zn | Ga | Ge | As | |
|---|---|---|---|---|---|---|---|---|---|---|---|---|
| $\delta\varrho_i$ | 12.9 | 4.2 | 2.4 | 7.9 | 6.1 | 0.8 | 0.5 | 0.9 | 2.2 | 5.2 | 8.0 | micro-ohms cm. |

[1] A comprehensive collection of data referring to the resistivities of dilute solid solutions is given by J. O. LINDE. Elektrischer Widerstand verdünnter Cu-, Ag- und Au-Legierungen. Lund, Sweden: Gleerupska Univ.-Bokhandeln.

It will be observed that, for the elements to the right of Cu, $\delta\varrho_i$ increases approximately as $(v-1)^2$ where $v$ denotes the valency of the solute. This behaviour is to be expected on general grounds since, in the region beyond the ionic core of the impurity, the effective scattering potential must be proportional to $(v-1)$, and probably decreases exponentially with distance from the ion. The scattering is proportional to the square of the matrix element of this potential.

MOTT[1] has given a theory, based on the assumed potential $\{(v-1)\, e^2/r\}\, e^{-qr}$, using the BORN approximation in which the unperturbed wave functions are taken to be plane waves. To obtain agreement with experiment he has to assume that $q^{-1} \approx 0.3$ Å. Thus the screening length $q^{-1}$ must be assumed to be less than the radius of the ionic core of the solute. This rather awkward conclusion is probably due to the use of the BORN approximation.

The transition elements, titanium to nickel, give somewhat irregular resistivity increments. The large values for Fe and Co as compared with Cr, Mn, and Ni are interesting, but hitherto unexplained. Magnetic measurements show that in these dilute solid solutions the transition-metal atoms possess a resultant spin moment, and, in the case of Fe and Co, possibly also a resultant orbital moment. When scattering takes place at an ion having a resultant spin, it is no longer valid to assume that the conduction electrons are scattered without changing their spin states. In the most favourable case this kind of scattering could lead to a doubling of the additional resistance as compared with that due to ions with a similar scattering potential but no spin.

**18. Variation of the temperature coefficient of resistance with impurity content.** We have seen that, under certain assumptions stated in Sect. 15, the resistivity of a dilute solid solution may be written

$$\varrho = a\,T + b\,x, \tag{18.1}$$

where $a$ is the coefficient of $T$ in Eq. (12.11), and $b$ is the coefficient of $x$ (the atomic fraction of impurity) in Eq. (16.9). Hence, if $\varrho_0 = a_0 T$ denotes the resistivity of the pure solvent, we can write

$$\frac{T}{\varrho_0}\,\frac{\partial^2 \varrho}{\partial T\,\partial x} = \frac{1}{a_0}\,\frac{d a}{d x} + \frac{1}{a_0}\,\frac{d b}{d T}\,. \tag{18.2}$$

LINDE[2] has measured the value of $(\partial\varrho/\partial T)$, at various atomic fractions $x$, for a large number of alloys based on the noble metals. From these measurements we can readily obtain the pure number which forms the left hand side of Eq. (18.2). The following table shows the experimental results for copper-based alloys:

*Table* 6.

| | Be | Al | Si | Cr | Mn | Fe | Co | Ni | Ga | Ge | As | In | Sn | Sb |
|---|---|---|---|---|---|---|---|---|---|---|---|---|---|---|
| $\frac{T}{\varrho_0}\,\frac{\partial^2 \varrho}{\partial T\,\partial x}$ | 4.6 | 2.5 | 8.8 | −18.1 | −12.8 | −12.0 | 1.7 | 2.4 | 3.6 | 7.4 | 9.6 | 4.3 | 7.4 | 9.6 |

There are several factors in the coefficient $a$ which must depend upon $x$, and some in $b$ which must depend upon $T$, but rough estimates appear to lead to values for the right-hand side of (18.2) which are appreciably less than the numbers given in the above table. The most promising factor of $a$ apears to be $\Theta_D$. This would give a contribution $-(2/\Theta_D)\,(d\Theta_D/dx)$ to $(1/a_0)\,(d a/d x)$ according to (12.11). As we have seen $\Theta_D$ is proportional to the velocity of sound and this

[1] N. F. MOTT: Reference p. 257.
[2] J. O. LINDE: loc. cit.

in turn is proportional to the square root of the rigidity $\mu$. Now $\frac{1}{\mu}\frac{d\mu}{dx}$ has been measured for many alloys of the type under consideration and is generally found to be negative[1]. It thus leads to a positive contribution to the right-hand side of (18.2), but in magnitude it is too small to account at all quantitatively for the data of Table 6. The anomalous behaviour of Cr, Mn and Fe remains unexplained.

**19. Polyvalent metals: The two-zone theory.** It has not yet proved possible to construct a quantitative theory of the conductivity of polyvalent metals, since most of the approximations which render the theory manageable for monovalent metals become untenable in this case. Nevertheless it has proved possible to make interesting qualitative deductions regarding many metals which, at first sight, might have been thought to be prohibitively complicated. As the valency increases so does the intensity of the periodic crystalline field and thus the BRILLOUIN zones tend to separate from each other. In the divalent metals the separation between the first and higher zones is incomplete, and the FERMI surface has regions in at least two zones which play a significant part in the properties of the metal. The elements in the fourth column of the periodic table have their zones more widely separated, some of them, the semi-conductors Si and Ge are completely separated whilst others, e.g. Sn, Pb, Ti, Zr, Hf, show metallic properties and must, therefore, have FERMI surfaces, probably having very complicated forms, lying in several zones. The electronic structure of the transition metals is also likely to be extremely complicated. The magnetic properties of these metals, and the specific heats at very low temperatures, show that the $d$-shells are incomplete. The model based on the assumption that single-electron wave functions of the BLOCH form may be used also for these $d$-shell electrons has had considerable success, and we shall use this model to discuss the conductivity of these metals.

For the present we restrict our discussion of the conductivity to the case of normal or high temperatures in which the concept of the relaxation time is applicable. Without making any special assumptions the expression for the time of relaxation (5.4) can be simplified by introducing the scattering matrix from (7.11). Writing for $d\boldsymbol{k}'$ the equivalent volume element of $k$-space $dE'\,dS'/|\nabla E'|$, where $dS'$ denotes an element of area on the FERMI surface, we obtain

$$\frac{1}{\tau(\boldsymbol{k})} = \frac{N\Omega}{4\pi^3\hbar}\int |U_{\boldsymbol{k}'\boldsymbol{k}}|^2 \frac{\sin\left(\frac{E'-E}{\hbar}\right)t}{E'-E}\left(1-\frac{\chi(\boldsymbol{k}')}{\chi(\boldsymbol{k})}\right)\frac{dE'\,dS'}{|\nabla E(\boldsymbol{k}')|}. \tag{19.1}$$

Comparing (4.8) and (6.1) we find

$$\chi(\boldsymbol{k}) = e F v_z \tau(\boldsymbol{k}) \tag{19.2}$$

where $v_z$ denotes the $z$-component of the velocity. Thus on integrating (19.1) over $E'$ and using (19.2) we have

$$\frac{1}{\tau(\boldsymbol{k})} = \frac{N\Omega}{4\pi^2\hbar}\int |U_{\boldsymbol{k}'\boldsymbol{k}}|^2\left(1-\frac{\tau(\boldsymbol{k}')\,v_z'}{\tau(\boldsymbol{k})\,v_z}\right)\frac{dS'}{|\nabla E(\boldsymbol{k}')|}, \tag{19.3}$$

For our first model we consider a metal in which the FERMI surface lies only in two zones and forms in each zone a simply connected closed surface. We assume, further, that $U_{\boldsymbol{k}'\boldsymbol{k}}=0$ when $\boldsymbol{k}$ and $\boldsymbol{k}'$ lie in different zones, i.e. we neglect transitions between the two zones. Also we assume that in each zone $\tau(\boldsymbol{k})$ depends

[1] W. KOSTER: Z. Metallkde. **39**, 111 (1948). — A. D. N. SMITH: J. Inst. Met. **80**, 477 (1952). — C. ZENER: Acta crystallogr. **2**, 163 (1949).

only on the energy. Thus the integral Eq. (19.3) breaks up into two equations giving

$$\frac{1}{\tau_1} = \frac{N\Omega}{4\pi^2\hbar} \int |U_{\boldsymbol{k}'\boldsymbol{k}}|^2 \left(1 - \frac{v_z'}{v_z}\right) \frac{dS'}{|\nabla E(\boldsymbol{k}')|}, \tag{19.4}$$

where $\tau_1$ is the time of relaxation for the first zone, and an exactly similar equation for $\tau_2$ the time of relaxation in the second zone. $\tau_1$ and $\tau_2$ will not, of course, in general be equal. It is clear that, with the above assumptions, the metal will behave, as far as its conductivity is concerned, like two independent conductors connected in parallel. However, this model is of interest because the number of states enclosed within the FERMI surface, in either zone, need no longer equal one per atom but may be much less. We are thus led to enquire what the effect on the relaxation time will be of a greatly reduced FERMI surface.

We make the simplest possible assumptions, viz. that the energy of a state, within the zone under consideration, is given by $(\hbar^2/2m)\, k^2$, and that there are $n_0$ states per atom within this region of the FERMI surface.

The calculation of the matrix element $U_{\boldsymbol{k}'\boldsymbol{k}}$ is unaffected up to Eq. (9.13), but Eq. (10.10) for $\boldsymbol{I}$ is modified since now

$$k_0 r_s = \left(\frac{9\pi n_0}{4}\right)^{\frac{1}{3}}, \tag{19.5}$$

and therefore $\xi = (18\pi\, n_0)^{\frac{1}{3}}\, w$ in place of $(18\pi)^{\frac{1}{3}}\, w$. The most interesting case arises when $n_0$ is very small. We can then replace $3(\sin\xi - \xi\cos\xi)/\xi^3$ by unity, since $w \leq 1$, then (10.10) gives

$$|\boldsymbol{I}|^2 = \frac{1}{N^2}\, |\boldsymbol{k} - \boldsymbol{k}'|^2\, (V_0(r) - E_0)^2. \tag{19.6}$$

Thus with (9.13) we obtain

$$|U_{\boldsymbol{k}'\boldsymbol{k}}|^2 = \left(\frac{9\pi}{2}\right)^{\frac{2}{3}} \frac{\hbar^2}{N r_s^2} \left(\frac{T}{M\, \mathrm{k}\, \Theta_D^2}\right) (V_0(r_s) - E_0)^2, \tag{19.7}$$

where we have made use of the relation $\boldsymbol{k}' - \boldsymbol{k} = \boldsymbol{q}$. When $n_0$ is very small and, therefore, the radius of the FERMI sphere $k_0$ is also very small, it is clear from Fig. 5 that no $U$-processes are possible, while $N$-processes can occur for all $w$ from 0 to 1.

It will be observed that $|U_{\boldsymbol{k}'\boldsymbol{k}}|^2$ does not depend on $n_0$ in spite of the fact that $\boldsymbol{k}' - \boldsymbol{k}$ must always be small when $n_0$ is small. This is because, in the DEBYE model, the amplitude of the lattice waves contains the factor $1/q$ and is, therefore, large when $q$ is small. The EINSTEIN model is not applicable for the case of small $n_0$ for the same reason, essentially, that it leads to the wrong specific heat at low temperatures.

Substituting (19.7) into (7.12) and integrating over $\vartheta$ from 0 to $\pi$, we obtain for the relaxation time

$$\frac{1}{\tau} = (48\pi^2)^{\frac{1}{3}}\, \frac{k_0 m r_s}{\hbar} \left(\frac{T}{M\, \mathrm{k}\, \Theta_D^2}\right) (V_0(r_s) - E_0)^2. \tag{19.8}$$

It is interesting to compare the mean free path, defined by $l = v\,\tau$, of a monovalent metal, as obtained from (12.3), with the mean free path in the present case where there are only $n_0$ states per atom. From (12.3) we find for the monovalent metal

$$l = \left\{1.832\, \frac{\hbar^2}{m^2 r_s}\right\} \Big/ \left(\frac{T}{M\, \mathrm{k}\, \Theta_D^2}\right) C^2, \tag{19.9}$$

and from (19.8)

$$l = \left\{0.534\, \frac{\hbar^2}{m^2 r_s}\right\} \Big/ \left(\frac{T}{M\, \mathrm{k}\, \Theta_D^2}\right) C^2, \tag{19.10}$$

where we have made use of Eq. (12.8) to replace $V_0(r_s) - E_0$ by $C$ to facilitate comparison between the two expressions for $l$.

It is interesting to see that the mean free path remains approximately the same as in a monovalent metal, however small $n_0$ may be. Of course if $n_0$ were too small the gas would become non-degenerate and (19.10) would require modification; the main conclusion, however, still holds and the mean free path in an elementary semi-conductor such as germanium remains approximately of the same magnitude as that in a monovalent metal.

The preceding theory throws some light on the electrical properties of bismuth and allied metals. It has long been known that, to account for the diamagnetism, the change of resistance in a magnetic field, and the HALL effect in bismuth, it is necessary to assume, on the basis of a two-zone model, that the number of holes and electrons is extremely small: of the order $10^{-4}$ or $10^{-5}$ per atom[1]. In addition it has to be assumed that the effective mass of these electrons and holes is exceptionally small. A likely value of the effective mass is thought to be in the neighbourhood of $^1/_{10}$ of the electronic mass. Thus, according to (19.10), the mean free path of the conductions electrons in bismuth might be as much as 100 times as great as the mean free path in a typical monovalent metal, for example, gold. For gold at room temperature $l = 4.1 \times 10^{-6}$ cms.

From measurements of the anomalous skin effect in bismuth (cf. Sect. 47) PIPPARD and CHAMBERS[2] have estimated that the mean free path lies between $2 \times 10^{-4}$ and $4 \times 10^{-4}$ cms.; a result which is quite compatible with the preceding considerations. The neglect of inter-zone transitions, in the two-zone model, makes the theory unrealistic and leads to conclusions at variance with observed properties. The temperature variation of the resistivity of bismuth is a case in point. The primary cause of the deviation of the resistivity from linear dependence on the temperature is the failure of Eq. (9.8) as $T$ decreases in value. The quantization of the lattice vibrations necessitates the replacement of $\mathrm{k}T$ in (9.8) by $\hbar\omega_q/(e^{\hbar\omega_q/\mathrm{k}T} - 1)$ according to PLANCK's well-known theory of the oscillator. This modification alone is not, of course, sufficient to give the low temperature resistivity since we have neglected interchanges of energy between the lattice and the electrons, but it does indicate when the linear dependence of the resistivity on temperature fails. Clearly this occurs when $\hbar\omega_0/\mathrm{k}T$ is no longer small compared to unity, where $\omega_0$ is the largest angular frequency involved in the scattering processes. In a monovalent metal $\omega_0$ is equal to $c q_{\max}$ where $q_{\max}$ is the radius of the DEBYE sphere in $q$-space. In the model under consideration $\omega_0$ is $2ck_0$ where $k_0$ is the radius of the FERMI sphere in one zone. The ratio $2k_0/q_{\max}$ is equal to $(4n_0)^{\frac{1}{3}}$ and thus, if $n_0 \approx 10^{-4}$, the linearity of the resistivity with temperature extends to temperatures some ten times less than in a monovalent metal. Since for bismuth $\Theta_D \approx 110°$ K we should expect the linear relation down to about 10° K. This is not the case and the reason is evidently the existence of inter-zone transitions involving much larger changes in $k$ than $2k_0$.

**20. Deviations from MATTHIESSEN's rule for alloys with polyvalent solvents.** Deviations from MATTHIESSEN's rule can arise if the time of relaxation varies over the FERMI surface. The two zone model provides the simplest example of such a case[3]. If we neglect inter-zone transitions the total conductivity is the sum of the conductivities arising from each zone separately. When the metal contains impurities the relaxation time $\tau_1$ of a single zone is made up of two

[1] Cf. H. JONES: Proc. Roy. Soc. Lond., Ser. A **144**, 225 (1934). — E. H. SONDHEIMER: Proc. Roy. Soc. Lond., Ser. A **65**, 561 (1952).

[2] A. B. PIPPARD and R. G. CHAMBERS: Proc. Phys. Soc. Lond. A **65**, 955 (1952).

[3] E. H. SONDHEIMER and A. H. WILSON: Proc. Roy. Soc. Lond., Ser. A **190**, 435 (1947).

parts as follows:

$$\frac{1}{\tau_1} = \frac{1}{\tau_{t1}} + \frac{1}{\tau_{i1}}, \tag{20.1}$$

where $\tau_{t1}^{-1}$ arises from the scattering due to thermal motion, $\tau_{i1}^{-1}$ from impurity scattering, and it has been assumed that (15.3) holds for both zones separately. Thus the resistance from each zone is the sum of two resistances, $r_{t1}$ and $r_{i1}$, due respectively to thermal motion and impurities. Thus the total resistance of the metal is that of a network as shown in Fig. 10, where the subscripts 1 and 2 refer to zones 1 and 2 respectively. The resistance of the pure solvent is obtained by putting $r_{i1} = r_{i2} = 0$ and is, therefore, $r_{t1} r_{t2}/(r_{t1} + r_{t2})$, and the resistance due to impurities alone is given by putting $r_{t1} = r_{t2} = 0$ and is therefore $r_{i1} r_{i2}/(r_{i1} + r_{i2})$. We define the deviation from Matthiessen's rule $\Delta$ as the difference between the resistance between $A$ and $B$ and the sum of the above two resistances. $\Delta$ is, therefore, the difference between the resistance of the network when $C$ and $D$ are not connected, and when they are connected with zero resistance. It follows at once that $\Delta$ must be positive. At low temperatures $r_{t1}$ and $r_{t2}$ tend to zero and therefore $\Delta$ must tend to zero since the potentials at $C$ and $D$ are then equal and, therefore, independent of whether $C$ and $D$ are connected or not. At high temperatures when $r_{t1}$ and $r_{t2}$ are much greater than $r_{i1}$ and $r_{i2}$ we find

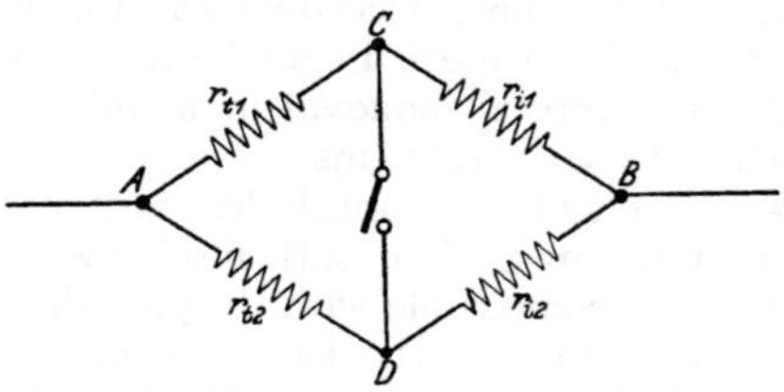


Fig. 10. Diagram to illustrate the deviation from Matthiessen's rule in the two-zone model.

$$\Delta = \frac{(r_{t1} r_{i2} - r_{t2} r_{i1})^2}{(r_{t1} + r_{t2})^2 (r_{i1} + r_{i2})}, \tag{20.2}$$

and, therefore, since both $r_{t1}$ and $r_{t2}$ are proportional to the temperature, $\Delta$ is constant. For all temperatures we see that the potentials at $C$ and $D$ are equal, and therefore $\Delta = 0$, if

$$\frac{r_{t1}}{r_{t2}} = \frac{r_{i1}}{r_{i2}}. \tag{20.3}$$

Thus the ratio of the thermal scattering to the impurity scattering in the first zone must differ from the same ratio in the second zone if deviations from Matthiessen's rule are to occur according to this model. It is to be emphasized that in this discussion we have considered only one possible cause of deviations from Matthiessen's rule; other causes which can lead to such deviations have already been listed in Sect. 15.

**21. The effective number of free electrons.** The way the resistivity of a metal varies with its position in the periodic table can be understood to some extent by the following considerations first given by Mott. Two factors dominate the conductivity of a metal: the magnitude of the time of relaxation, and the extent of the free Fermi surface. If in Eq. (6.5) we take $\tau(\boldsymbol{k})$ outside the integration and replace it by an average value $\bar{\tau}$ we can express the conductivity as follows

$$\sigma = \frac{e^2 \bar{\tau}\, n_{\text{eff}}}{m}, \tag{21.1}$$

where

$$n_{\text{eff}} = \frac{m}{4\pi^3 \hbar^2} \int \left(\frac{\partial E}{\partial k_2}\right)^2 \frac{dS}{|\nabla E|}. \tag{21.2}$$

$n_{\text{eff}}$ is spoken of as the effective number of free electrons per unit volume. When $E = \hbar^2 k^2/2m$ it is equal to $n$. In a polyvalent metal the extent of the free Fermi surface depends on how completely the Brillouin zones are filled, and clearly, according to (21.2), some indication of the extent to which the zones are filled is given by the value of $n_{\text{eff}}$. Relative values of $n_{\text{eff}}$ for different metals may be obtained from the observed conductivities of pure metals at room temperature, if we assume that the relative values of $\bar{\tau}^{-1}$ are determined mainly by the factor $T/M\Theta_D^2$, which is a measure of the square of the amplitude of the atomic vibrations.

The time of relaxation depends on many factors besides the amplitude of the atomic vibrations, for example, the density of electronic states, and the coupling between the electrons and the lattice motion. Nevertheless the relative values of $\sigma/M\Theta_D^2$ for different metals might be expected to bear roughly some relation to the electronic structure of the metal. Table 7 shows that this is the case.

It will be observed that there is some repetition of the values in the columns of the periodic table, and that, as might be anticipated, $n_{\text{eff}}$ has smaller values for polyvalent than monovalent metals.

*Table 7.*

| Metal | Conductivity at 0° C $\sigma\times 10^{-4}$ ohms cms. | $M$ | $\Theta_D$ | $(\sigma/M\Theta_D^2)\times 10^2$ |
|---|---|---|---|---|
| Li | 11.8 | 6.94 | 363 | 12.9 |
| Be | 18 | 9.02 | 1000 | 2.0 |
| Na | 23 | 23.0 | 202 | 24.0 |
| Mg | 25 | 24.3 | 357 | 8.1 |
| Al | 40 | 27.0 | 395 | 9.5 |
| K | 15.9 | 39.1 | 163 | 15.3 |
| Ca | 23.5 | 40.1 | 230 | 11.1 |
| Ti | 1.2 | 47.9 | 342 | 0.2 |
| Cr | 6.5 | 52.0 | 495 | 0.5 |
| Fe | 11.2 | 55.8 | 420 | 1.14 |
| Co | 16.0 | 58.9 | 401 | 1.7 |
| Ni | 16.0 | 58.7 | 375 | 1.9 |
| Cu | 64.5 | 63.6 | 333 | 9.1 |
| Zn | 18.1 | 65.4 | 213 | 6.1 |
| Ga | 2.45 | 69.7 | 125 | 2.25 |
| As | 2.85 | 74.9 | 291 | 0.45 |
| Rb | 8.6 | 85.4 | 85 | 14 |
| Sr | 3.3 | 87.6 | 171 | 1.3 |
| Zr | 2.4 | 91.2 | 288 | 0.32 |
| Mo | 23.0 | 96.0 | 380 | 1.7 |
| Ru | 8.5 | 101.7 | 426 | 0.46 |
| Rh | 22.0 | 102.9 | 370 | 1.6 |
| Pd | 10.0 | 106.7 | 270 | 1.3 |
| Ag | 66.7 | 107.9 | 223 | 12.4 |
| Cd | 15 | 112.4 | 172 | 4.5 |
| In | 12 | 114.8 | 198 | 2.7 |
| Sn | 10 | 118.7 | 260 | 1.2 |
| Cs | 5.6 | 132.9 | 54 | 14.0 |
| Ba | 1.7 | 137.4 | 113 | 1.0 |

**22. The transition metals.** It will be observed from the preceding table that the metals iron, cobalt and nickel have much higher resistivities than copper. Comparing the neighbouring metals, nickel and copper, we see that they have nearly equal atomic weights, the same crystal structure, not very dissimilar Debye temperatures and cohesive energies. The obvious difference is that nickel is ferromagnetic, due to the incomplete filling of the $3d$-shell in the metallic state, and copper is not. It is natural, therefore, to attribute the greater resistivity of nickel to the excess scattering of the conduction electrons resulting from the incomplete filling of the $d$-shell. Strong evidence in favour of this view will be adduced in Sect. 25 from the variation of the resistivity of silver-palladium alloys with composition.

Mott[1] has constructed a detailed theory of the properties of the transition metals on the assumption that the $3d$ electrons in the free atom are adequately described by single-electron Bloch wave functions in the metallic state. Independent detailed calculations by Howarth[2] and Fletcher[3] have confirmed

[1] N. F. Mott: Proc. Roy. Soc. Lond., Ser. A **153**, 699 (1936).
[2] D. J. Howarth: Proc. Roy. Soc. Lond., Ser. A **220**, 513 (1953).
[3] G. C. Fletcher: Proc. Roy. Soc. Lond., Ser. A **65**, 192 (1952).

MOTT's original assumption that the $d$-band is narrow, and that the density of states at the FERMI surface in the $d$-zone is much larger than the density of states on the other part of the FERMI surface, in what may be called the $s$-zone. HOWARTH gives 3.4 ev for the width of the $d$-band in copper and FLETCHER's value of $N_d(E)$, the density of states in nickel, is in remarkably good agreement with the observed electronic specific heat at low temperatures.

We now apply the two-zone model to estimate the resistivity. We make an assumption contrary to that of Sect. 20, viz. that the time of relaxation is dominated by inter-zone transitions: in this case, therefore, by $s$-$d$ transitions. This assumption is only partially justified by the known large value of $N_d(\zeta)$; it must depend also on $U_{\boldsymbol{k}'\boldsymbol{k}}$ being sufficiently large for inter-zone transitions. Unfortunately our knowledge of the electronic wave functions in the transition metals is not sufficient to enable reliable values of $U_{\boldsymbol{k}'\boldsymbol{k}}$ to be calculated.

The time of relaxation may be estimated by using Eq. (19.3) with the simplifying assumption that $\tau(\boldsymbol{k})$ is independent of $\boldsymbol{k}$ in each zone. If we denote the time of relaxation in the $s$-zone by $\tau_s$ and in the $d$-zone by $\tau_d$, we obtain from (19.3)

$$\frac{1}{\tau_s} = \frac{N\Omega}{4\pi^2\hbar}\left\{\int\limits_{s\text{-zone}} |U_{ss}|^2\left(1-\frac{v_z'}{v_z}\right)\frac{dS'}{|\nabla E'|} + \int\limits_{d\text{-zone}} |U_{sd}|^2\left(1-\frac{\tau_d v_z'}{\tau_s v_z}\right)\frac{dS'}{|\nabla E'|}\right\}, \tag{22.1}$$

$$\frac{1}{\tau_d} = \frac{N\Omega}{4\pi^2\hbar}\left\{\int\limits_{d\text{-zone}} |U_{dd}|^2\left(1-\frac{v_z'}{v_z}\right)\frac{dS'}{|\nabla E'|} + \int\limits_{s\text{-zone}} |U_{sd}|^2\left(1-\frac{\tau_s v_z'}{\tau_d v_z}\right)\frac{dS'}{|\nabla E'|}\right\}, \tag{22.2}$$

where $U_{ss}$ refers to transitions within the $s$-zone, $U_{dd}$ to transitions within the $d$-zone and $U_{sd}$ to inter-zone transitions. The first integral in (22.1) will contribute to $\tau_s^{-1}$ a term of approximately the size of the reciprocal of the relaxation time for copper. The second term in (22.1) should be considerably larger to account successfully for the high resistivity. Since it would be unrealistic to assume spherical energy surfaces in the $d$-zone, and still more unrealistic to assume that the $d$-zone wave functions could be expressed in the form (10.3), no real progress can be made in the calculation of $U_{sd}$ and $U_{dd}$ until more detailed knowledge of the electronic states is available. Here we shall assume simply that $U_{dd}$ and $U_{sd}$ are constants which can be taken outside the integration and that the first integral of (22.1) can be neglected. Hence, since the resultant of all $v_z'$ over the FERMI surface in either zone must be zero, we obtain

$$\frac{1}{\tau_s} = \frac{\pi\Omega N}{\hbar}|U_{sd}|^2 N_d(\zeta), \tag{22.3}$$

$$\frac{1}{\tau_d} = \frac{\pi\Omega N}{\hbar}\{|U_{dd}|^2 N_d(\zeta) + |U_{sd}|^2 N_s(\zeta)\}. \tag{22.4}$$

As we have already stated $N_d(\zeta) \gg N_s(\zeta)$, and thus if $|U_{dd}| \approx |U_{sd}|$ the two relaxation times $\tau_s$ and $\tau_d$ must be approximately equal. Because of the "tight-binding" of the $d$-electrons the average velocity associated with a state in the $d$-zone is much less than the average velocity of a state in the $s$-zone. Hence, if $\tau_s$ and $\tau_d$ are approximately equal, we conclude that the current is carried mainly by the $s$-electrons, and that to estimate the resistivity we need only consider these electrons and the single relaxation time $\tau_s$. The principal conclusion of this theory is that the effective relaxation time is inversely proportional to the density of states in the $d$-band.

**23. The resistivity at high temperatures.** For many metals the plot of the observed resistivity $\varrho$ against temperature is not exactly a straight line even in the high temperature region. For the noble metals $\varrho$ increases rather more rapidly than $T$; for some of the transition metals less rapidly than $T$. There are several reasons why the resistivity may be expected to deviate from strict proportionality with the temperature. We shall consider two: the variation of $\Theta_D$ through the effect of thermal expansion, and the finite energy interval, of order $\mathsf{k}T$, in which $f_0$ falls from unity effectively to zero.

Since, according to (12.10), $\varrho/T \propto \Theta_D^{-2}$ for pure metals we can write

$$\frac{d}{dT}\log\frac{\varrho}{T} = -2\frac{d\log\Theta_D}{dV}\frac{dV}{dT}, \tag{23.1}$$

where $V$ denotes the volume of one gram-atom of the metal. The coefficient of expansion $\alpha$, and the GRÜNEISEN constant $\gamma$ are defined by the equations

$$\alpha = \frac{1}{V}\left(\frac{\partial V}{\partial T}\right)_p, \qquad \gamma = -\frac{d\log\Theta_D}{d\log V}. \tag{23.2}$$

Hence

$$\frac{d}{dT}\log\left(\frac{\varrho}{T}\right) = 2\,\alpha\gamma, \tag{23.3}$$

and therefore

$$\frac{\varrho}{T} = \text{const}\,(1 + 2\alpha\gamma\,T), \tag{23.4}$$

where in all cases $2\alpha\gamma\,T$ is small compared to unity. $\gamma$ can easily be related to measured quantities as follows: In DEBYE's theory of the specific heat the entropy $S$ is a function of the single variable $\Theta_D/T$, and therefore

$$\left(\frac{\partial S}{\partial V}\right)_T = \frac{1}{T}\frac{d\Theta_D}{dV}S', \tag{23.5}$$

and

$$\left(\frac{\partial S}{\partial T}\right)_V = -\frac{\Theta_D}{T^2}S'. \tag{23.6}$$

Hence eliminating $S'$ we obtain

$$-\frac{1}{\Theta_D}\frac{d\Theta_D}{dV} = \frac{1}{C_v}\left(\frac{\partial S}{\partial V}\right)_T,$$

where $C_v$ is the specific heat per gram-atom at constant volume. The well-known MAXWELL relation gives

$$\left(\frac{\partial S}{\partial V}\right)_T = \left(\frac{\partial p}{\partial T}\right)_V = \frac{\alpha}{\varkappa_t}, \tag{23.7}$$

where $\varkappa_t = -\frac{1}{V}\left(\frac{\partial V}{\partial p}\right)_T$ is the isothermal compressibility. Thus

$$\gamma = \frac{\alpha V}{\varkappa_t C_v}. \tag{23.8}$$

For copper $2\alpha\gamma = 1.9\times10^{-4}$ according to (23.8), and this gives excellent agreement with the observed temperature variation of $\varrho$ according to (23.1). It appears, therefore that the variation of $\Theta_D$, due to thermal expansion, is sufficient to account for the $\varrho$, $T$ relation in the noble metals.

**24. Transition metals at high temperatures.** The expressions for the conductivity, given in the preceding sections, have been obtained by assuming that, in Eq. (6.5), $-(df_0/dE)$ could be regarded as a DIRAC $\delta$-function in $(E-\zeta_0)$. This is valid provided that no part of the integrand varies appreciably with $E$ in a range $\mathsf{k}T$. For the transition metals, however, we have seen that $\tau$ varies as $N_d^{-1}(E)$ and this may vary very rapidly in the neighbourhood of the FERMI surface. It is convenient to define a function $\sigma(E)$ by the equation

$$\sigma(E) = \frac{e^2}{4\pi^3\hbar^2}\int \tau(\boldsymbol{k})\left(\frac{\partial E}{\partial k_z}\right)^2 \frac{dS}{|\nabla E|}, \tag{24.1}$$

where the integration is over a surface on which the energy is constant and equal to $E$. The conductivity may thus be written

$$\sigma = \int_0^\infty \sigma(E)\left(-\frac{df_0}{dE}\right)dE. \tag{24.2}$$

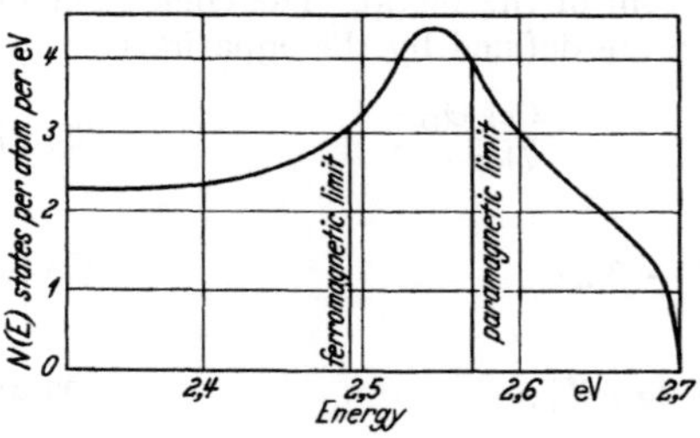


Fig. 11. Density of states curve for nickel. Energy in ev measured from the bottom of the $d$-band. The FERMI limits for paramagnetic and ferromagnetic nickel are shown.

This is a well-known type of integral and we may quote the following result

$$\sigma = \sigma(\zeta) + \frac{\pi^2}{6}(\mathrm{k}T)^2\left(\frac{d^2\sigma}{dE^2}\right)_\zeta, \tag{24.3}$$

in which fourth and higher powers of $\mathrm{k}T$ have been neglected. The dependence of $\zeta$ on $T$ may easily be obtained. Let $n_d$ be the total number of $d$ electrons, then

$$n_d = \int_0^\infty f_0(E)\, N_a(E)\, dE. \tag{24.4}$$

Carrying out an integration by parts (24.4), may be transformed into an integral of the type (24.2), from which, using the result analogous to (24.3) we obtain without difficulty

$$\zeta - \zeta_0 = -\frac{\pi^2}{6}(\mathrm{k}T)^2\left(\frac{1}{N_d}\frac{dN_d}{dE}\right)_{\zeta_0}. \tag{24.5}$$

From (24.3) and (24.5) therefore we have

$$\sigma = \sigma(\zeta_0) + \frac{\pi^2}{6}(\mathrm{k}T)^2\left\{\frac{d^2\sigma}{dE^2} - \frac{1}{N_d}\frac{dN_d}{dE}\frac{d\sigma}{dE}\right\}_{\zeta_0}. \tag{24.6}$$

According to the theory of Sect. 22 $\tau$ and therefore $\sigma(E)$ is proportional to $N_d^{-1}(E)$. Introducing this relation into (24.6) and writing the equation in reciprocal form, we find

$$\varrho(T) = \varrho(\zeta_0, T)\left[1 - \frac{\pi^2}{6}(\mathrm{k}T)^2\left\{3\left(\frac{1}{N_d}\frac{dN_d}{dE}\right)^2 - \frac{1}{N_d}\frac{d^2N_d}{dE^2}\right\}_{\zeta_0}\right]. \tag{24.7}$$

The density of states in the $d$-band appears to be a very complicated function of $E$. Fig. 11 shows the result of FLETCHER's calculation for the $d$-band of nickel[1]. It is clear that the factor in square brackets of (24.7) may either increase or decrease with $T$ depending upon the position of the FERMI limit in the $d$-band.

If we assume that in the neighbourhood of the FERMI surface

$$N_d = \text{const}\,(E_1 - E)^{\frac{1}{2}},$$

where $E_1$ denotes the energy at the head of the $d$-band, we find for the factor multiplying $\varrho(\zeta_0 T)$

$$1 - \frac{\pi^2}{6}\left(\frac{\mathrm{k}T}{E_1 - \zeta_0}\right)^2. \tag{24.8}$$

Since $E_1 - \zeta_0$ may be as small as a few tenths of an electron volt, this effect leads to the right order of magnitude for the deviation of $\varrho/T$ from constancy.

[1] G. C. FLETCHER: Proc. Phys. Soc. Lond. 65, 192 (1952).

If the FERMI limit occurs near a minimum in the density of states curve $(dN_d/dE)_{\zeta_0}$ will be small and $(d^2N_d/dE^2)_{\zeta_0}$ positive. Thus $\varrho/T$ would increase with $T$ in this case. It is possible that the temperature dependence of the resistivity of tungsten may be accounted for in this way.

**25. The resistivity of binary alloys.** In this section we shall be concerned primarily with binary systems which form continuous solid solutions, for example, silver-gold, potassium-rubidium, and silver-palladium. In Sect. 15, in which we considered dilute solid solutions, it was shown that the resistivity was composed of two parts: a temperature dependent part arising from the thermal motion of the lattice $\varrho_T$, and a temperature independent part arising from irregularities in the periodic field due to alloying $\varrho_0$. We discuss now how this second temperature independent part of the resistivity varies as we pass through the whole range of the continuous solid solution.

At each boundary of the phase diagram the resistivity $\varrho_0$ increases linearly with the atomic fraction of the second metal according to (16.9). Moreover we have seen that, if the two components have the same valency, we can expect the resistance increment to be similar at each end of the composition range. In the absence of any change in the electronic structure of the alloy, or of an ordering process, we can expect, therefore, that the resistivity $\varrho_0$ will vary approximately as $x(1-x)$, where $x$ denotes the atomic fraction of one component. This conjecture is supported by the following argument, first given by NORDHEIM[1]. Let $A$ and $B$ denote the two metals, and $x$ the atomic fraction of $A$ in the alloy. Let $V_A(\boldsymbol{r})$ denote the potential in an $A$-cell and $V_B(\boldsymbol{r})$ the potential in a $B$-cell. We now construct a purely artificial potential:

$$V(\boldsymbol{r}) = x\,V_A(\boldsymbol{r}) + (1-x)\,V_B(\boldsymbol{r}) \tag{25.1}$$

where $V(\boldsymbol{r})$ denotes the potential in each cell of the lattice. The difference between the actual potential in an $A$-cell and this average potential is

$$V_A(\boldsymbol{r}) - V(\boldsymbol{r}) = (1-x)\{V_A(\boldsymbol{r}) - V_B(\boldsymbol{r})\}, \tag{25.2}$$

and the difference between the actual potential in a $B$-cell and $V(\boldsymbol{r})$ is

$$V_B(\boldsymbol{r}) - V(\boldsymbol{r}) = -x\{V_A(\boldsymbol{r}) - V_B(\boldsymbol{r})\}. \tag{25.3}$$

Let

$$U(\boldsymbol{r}) = \sum_{\boldsymbol{s}} U(\boldsymbol{s}), \tag{25.4}$$

where $\boldsymbol{s}$ denotes the position of a lattice point, and $U(\boldsymbol{s})$ is equal to (25.2) when an $A$-atom is at $\boldsymbol{s}$, and is equal to (25.3) when a $B$-atom is at $\boldsymbol{s}$. We regard $U(\boldsymbol{r})$ as the potential which scatters the waves determined by the SCHRÖDINGER equation with the average potential (25.1). The square of the matrix element of $U(\boldsymbol{r})$, with respect to the states $\boldsymbol{k}$ and $\boldsymbol{k}'$, is therefore

$$|U_{\boldsymbol{k}'\boldsymbol{k}}|^2 = \sum_{\boldsymbol{s}} |U_{\boldsymbol{k}'\boldsymbol{k}}(\boldsymbol{s})|^2 + \sum_{\boldsymbol{s}\neq\boldsymbol{t}}\sum U_{\boldsymbol{k}'\boldsymbol{k}}(\boldsymbol{s})\,U^*_{\boldsymbol{k}'\boldsymbol{k}}(\boldsymbol{t}). \tag{25.5}$$

We neglect the second sum of (25.5) which is equivalent to assuming that $A$- and $B$-cells scatter independently. This can only be a rough approximation, though it is clear from (25.2) and (25.3) that the neglected sum consists of positive and negative terms. Since the number of $A$-atoms is proportional to $x$ and the

[1] L. NORDHEIM: Ann. Phys., Lpz. (5) 9, 607 (1931).

number of $B$-atoms to $(1-x)$ (25.5) may now be written

$$|U_{\boldsymbol{k}'\boldsymbol{k}}|^2 = \text{const}\{x(1-x)^2 + (1-x)\,x^2\} = \text{const}\, x(1-x). \tag{25.6}$$

Hence the residual resistance is proportional to $x(1-x)$. It is clear that the preceding theory can have, at most, a formal significance since, in the first place, in an alloy such as copper-gold, the unperturbed wave functions given by (25.1) can bear no relation to the true wave functions, and, secondly $V_A(\boldsymbol{r}) - V_B(\boldsymbol{r})$ is too large to be treated as a perturbation. On the other hand the method of Sect. 16 would be difficult to apply when there are approximately equal numbers of the two kinds of atoms.

The alloy systems silver-palladium and copper-nickel are of special interest since, for the palladium-rich and nickel-rich alloys, the $d$-band is not completely filled. We have seen that, when this is the case, the resistivity due to thermal scattering is enhanced by the inter-zone scattering between the $s$- and $d$-zones. As the concentration of silver or copper is increased the $d$-band fills, as is clearly shown by the magnetic properties of these alloys, and therefore the $s$-$d$ scattering diminishes. Fig. 12, taken from a paper by COLES[1] shows the experimentally determined resistivities for these two systems The deviation from the symmetrical form $x(1-x)$ is clearly compatible with the existence of excess resistivity, due to $s$-$d$ scattering, in alloys with greater than 40 atomic percent palladium. In the copper-nickel system the effect is not so clearly shown though the asymmetry is in the right sense. Copper-nickel differs from silver-palladium in that it shows paramagnetism for copper-rich alloys (more than 60 atomic percent. Cu), whilst silver-palladium is diamagnetic when the concentration of silver exceeds about 60 atomic percent. This indicates that in the copper-rich alloys some nickel atoms have complete $d$-shells, and therefore possess a resultant spin. This probably accounts for the lack of the sharp drop in the resistivity as the copper concentration exceeds fifty or sixty percent.

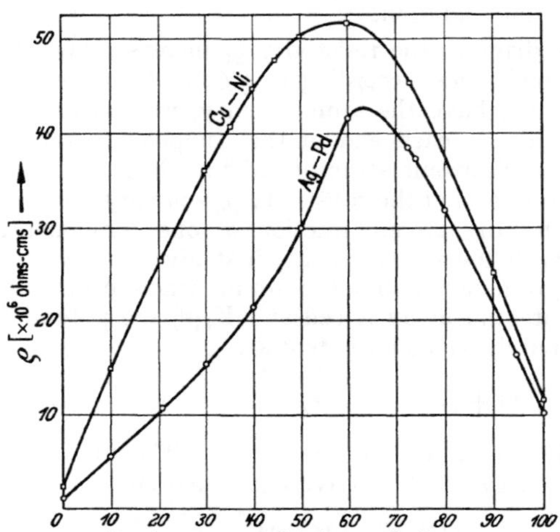


Fig. 12. Electrical resistivity $\varrho$ in ohms-cm. for copper-nickel and silver-palladium alloys. Results for silver-palladium refer to 0° C, and results for copper-nickel to 100° C.

That excess scattering apparently persists beyond the concentration of copper at which, on a simple theory, the $d$-band might be expected to be filled, is not conclusive evidence that an incomplete *d-band* exists in these alloys. Magnetic moments localized on the nickel atoms would undoubtedly lead to excess scattering, as already mentioned in Sect. 17. In this connection it might be worth emphasizing that the high resistivity of transition metals, and their alloys, is not due to the capture of $s$-electrons into $d$-band states. The number of $d \to s$ transitions is exactly equal to the number of $s \to d$ transitions. The effect might be best described as $s$-$s$ scattering through the intermediary action of the $d$-states.

[1] B. R. COLES: Proc. Phys. Soc. Lond. B **65**, 221 (1951).

## V. Thermal and electric conduction in the presence of a temperature gradient.

**26. Thermal conductivity and the thermo-electric effect.** We consider a steady state of the metal in which there exists simultaneously an electric and a heat current. The definition of the heat current requires care since it differs from the flow of energy. Our theory is based on the assumption that the currents are so small that the electronic system may be regarded as in local equilibrium at each point in the metal. We can, therefore, regard as given functions of position not only the temperature, but also the partial potential. Consider now a small region of the metal where the value of the partial potential is $\mu$, and suppose that the number of electrons in this region is increased by $\delta N$. The free energy of the region then increases by $\mu\,\delta N$. Thus we can define a free energy current as $\mu \boldsymbol{J}$ where $\boldsymbol{J}$ is the particle current. This is possible because, for given $\mu$, the free energy depends only on the number of particles in the system; if there were no particle current there could be no flow of free energy. This is not, of course, the case for the total energy $E$ which depends upon $T$ as well as $N$ for given $\mu$. $N$, it should be noted, is the number of particles in the system, not the number density. Thus a flow of energy is possible, without a particle current, as a result of a temperature gradient.

We assume that the thermal current and the particle current are in the same direction, and we take this direction as the $z$-axis; hence we can write

$$Q_z = E_z - \mu J_z, \tag{26.1}$$

where $E_z$ is the energy which passes through unit area, normal to the $z$-axis, per unit time, and $Q_z$ is the heat current.

If an electric field, is present and all energies are measured relative to a fixed level independent of position in the metal, the partial potential is given by $\mu = \zeta + e\varphi$, where $F = -d\varphi/dz$.

Boltzmann's Equation (4.2) applied to this particular case reads

$$\left(\frac{\partial f}{\partial t}\right)_{\text{coll}} = \frac{eF}{\hbar}\frac{\partial f_0}{\partial k_z} + \frac{1}{\hbar}\frac{\partial E}{\partial k_z}\frac{\partial f_0}{\partial z}, \tag{26.2}$$

where $f_0$ is given by (1.4), and we note that $\zeta$ not $\mu$ occurs in the expression for $f_0$. For the purpose of the present discussion we assume that the conditions necessary for the use of the concept of the relaxation time, as stated in Sect. 5, are fulfilled and, therefore, that we can equate the right hand side of (26.2) to $-f_1/\tau$. Hence we obtain

$$f_1 = -\frac{\tau}{\hbar}\frac{df_0}{dE}\frac{\partial E}{\partial k_z}\left\{eF + (E-\zeta)\,T\frac{d}{dz}\left(\frac{1}{T}\right) - \frac{d\zeta}{dz}\right\}. \tag{26.3}$$

The particle current and the heat current are given by

$$J_z = \frac{1}{4\pi^3\hbar}\int\left(\frac{\partial E}{\partial k_z}\right) f_1\, d\boldsymbol{k}, \tag{26.4}$$

$$Q_z = \frac{1}{4\pi^3\hbar}\int\left(\frac{\partial E}{\partial k_z}\right)(E(\boldsymbol{k})-\zeta)\, f_1\, d\boldsymbol{k}. \tag{26.5}$$

In the expression for $Q_z$, $E(\boldsymbol{k})$ is the energy of the electronic state, specified by $\boldsymbol{k}$, relative to the bottom of the conduction band. Thus the energy of the state $\boldsymbol{k}$, relative to the fixed energy level outside the metal, minus $\mu$ is just equal to

$E(\boldsymbol{k})-\zeta$. Substituting from (26.3) for $f_1$, we find

$$-J_z=\frac{L_{11}}{T}\frac{d\mu}{dz}+L_{12}\frac{d}{dz}\left(\frac{1}{T}\right), \tag{26.6}$$

$$Q_z=\frac{L_{21}}{T}\frac{d\mu}{dz}+L_{22}\frac{d}{dz}\left(\frac{1}{T}\right), \tag{26.7}$$

where

$$L_{11}=-\frac{T}{4\pi^3\hbar^2}\int\left(\frac{\partial E}{\partial k_z}\right)^2\tau\frac{df_0}{dE}\,d\boldsymbol{k}, \tag{26.8}$$

$$L_{12}=L_{21}=\frac{T}{4\pi^3\hbar^2}\int(E-\zeta)\left(\frac{\partial E}{\partial k_z}\right)^2\tau\frac{df_0}{dE}\,d\boldsymbol{k}, \tag{26.9}$$

$$L_{22}=\frac{T}{4\pi^3\hbar^2}\int(E-\zeta)^2\left(\frac{\partial E}{\partial k_z}\right)^2\tau\frac{df_0}{dE}\,d\boldsymbol{k}. \tag{26.10}$$

$L_{12}$ and $L_{21}$ are kinetic coefficients which are a measure of the interference between the heat current and the electric current. Their equality is an example of one of ONSAGER's relations[1]. Clearly the $L$ coefficients can be expressed in simpler terms. Let

$$K_n=\frac{1}{4\pi^3\hbar^2}\int E^n\left(\frac{\partial E}{\partial k_z}\right)^2\tau\frac{df_0}{dE}\,d\boldsymbol{k}, \tag{26.11}$$

then

$$L_{11}=TK_0, \tag{26.12}$$

$$L_{12}=L_{21}=-T(K_1-\zeta K_0), \tag{26.13}$$

$$L_{22}=T(K_2-2\zeta K_1+\zeta^2K_0). \tag{26.14}$$

We can now easily relate the experimentally determined quantities: the isothermal electrical conductivity, the thermal conductivity and the thermo-electric power, to the kinetic coefficients $L_{11}$, $L_{12}$, $L_{22}$ and, therefore, ultimately to the integrals $K_n$.

*α) The isothermal electrical conductivity.* $T$ and $\zeta$ are now independent of $z$, and the electrical current and the particle current are simply related as follows: $j_z=eJ_z$. Thus (26.6) gives

$$j_z=\frac{e^2L_{11}}{T}F, \tag{26.15}$$

and, therefore,

$$\sigma=\frac{e^2L_{11}}{T}=e^2K_0, \tag{26.16}$$

in agreement with (6.5).

*β) The thermal conductivity.* We eliminate $\frac{1}{T}\frac{d\mu}{dz}$ between (26.6) and (26.7) and obtain

$$Q_z=-\frac{L_{12}}{L_{11}}J_z+\left(L_{22}-\frac{L_{12}^2}{L_{11}}\right)\frac{d}{dz}\left(\frac{1}{T}\right). \tag{26.17}$$

The thermal conductivity $\varkappa$ is defined as follows

$$\varkappa=-\left(Q_z\Big/\frac{dT}{dz}\right)_{J_z=0}. \tag{26.18}$$

[1] For an excellent account of the application of ONSAGER's relations to thermoelectric effects the reader is referred to the paper of H. B. CALLEN: Phys. Rev. 73, 1349 (1948)

Thus

$$\left.\begin{aligned}\varkappa &= \frac{1}{T^2}\left(L_{22} - \frac{L_{12}^2}{L_{11}}\right)\\ &= \frac{1}{T}\left(K_2 - \frac{K_1^2}{K_0}\right).\end{aligned}\right\} \tag{26.19}$$

*γ) The thermo-electric power.* We begin by considering the generation of heat in a metal carrying both a heat current and an electric current. Although we are not dealing with a steady state we postulate that there shall be no accumulation of charge, and therefore the divergence of $\boldsymbol{J}$ is zero, i.e. $dJ_z/dz = 0$. The total energy is conserved and, therefore, the rate of increase of energy within an element of volume is related to the divergence of the energy flow by the equation of continuity:

$$\frac{\partial E_z}{\partial z} + \frac{\partial E}{\partial t} = 0. \tag{26.20}$$

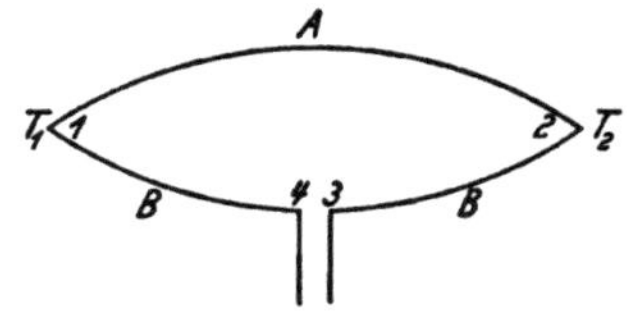


Fig. 13. Diagram to illustrate the calculation of the thermo-electric force.

Thus, from (26.1) and (26.20) we have

$$\frac{\partial E}{\partial t} = -\frac{dQ_z}{dz} - \frac{d\mu}{dz} J_z. \tag{26.21}$$

Differentiating (26.17) with respect to $z$, and substituting for $d\mu/dz$ from (26.6), we find

$$\frac{\partial E}{\partial t} = \frac{T}{L_{11}} J_z^2 + T\frac{d}{dT}\left(\frac{L_{12}}{TL_{11}}\right)\frac{dT}{dz} J_z - \frac{d}{dz}\left(L_{22} - \frac{L_{12}^2}{L_{11}}\right)\frac{d}{dz}\left(\frac{1}{T}\right). \tag{26.22}$$

When there is only a thermal current, so that $J_z = 0$, the metal is in a steady state in which there is no accumulation of energy. Thus the last term of (26.22) is zero when $J_z = 0$, and since the kinetic coefficients $L_{11}$ etc. do not contain the field this term must also vanish when $J_z$ is different from zero.

The first term of (26.22), which is equal to $(1/\sigma) j_z^2$, is the JOULE heat; the second term is the THOMSON heat, and the coefficient of $-\frac{dT}{dz} j_z$ is known as the THOMSON coefficient. The thermo-electric power $S$ is defined by the equation

$$\text{THOMSON coefficient} = T\frac{dS}{dT}, \tag{26.23}$$

and $S = 0$ when $T = 0$.

Thus comparing (26.22) and (26.23) we find

$$S = -\frac{L_{12}}{eTL_{11}}. \tag{26.24}$$

In terms of the $K_n$ integrals we have, according to (26.12) and (26.13)

$$S = \frac{1}{eT}\left(\frac{K_1}{K_0} - \zeta\right). \tag{26.25}$$

If the thermo-electric power has been determined for a standard metal by a direct measurement of the THOMSON coefficient, its value for any other metal may be obtained by making use of the SEEBECK effect. The accompanying sketch (Fig. 13) indicates a circuit composed of two metals $A$ and $B$. The junction 1 is maintained at temperature $T_1$ and the junction 2 at temperature $T_2$. At some point in the metal $B$, where the temperature is $T'$, the circuit is broken and the voltage $V$ between the points 3 and 4 is measured. Since in this circuit there is no current (26.6) becomes

$$\frac{d\mu}{dz} = \frac{L_{12}}{T L_{11}} \frac{dT}{dz}, \tag{26.26}$$

and, therefore, according to (26.24)

$$\frac{d\mu}{dz} = -e S \frac{dT}{dz}. \tag{26.27}$$

Since the partial potentials in the two metals must be equal at junction 1, and also at junction 2, we can integrate (26.27) and obtain the following three equations

$$\left.\begin{aligned} \mu_1 - \mu_4 &= -e \int_{T'}^{T_1} S_B \, dT, \\ \mu_2 - \mu_1 &= -e \int_{T_1}^{T_2} S_A \, dT, \\ \mu_3 - \mu_2 &= -e \int_{T_2}^{T'} S_B \, dT. \end{aligned}\right\} \tag{26.28}$$

Adding these equations, we obtain

$$\mu_3 - \mu_4 = e \int_{T_1}^{T_2} (S_B - S_A) \, dT. \tag{26.29}$$

Since the temperature is the same at 3 and 4, $\zeta_3 = \zeta_4$ and therefore $\mu_3 - \mu_4 = e(\varphi_3 - \varphi_4) = eV$.

Thus $V$, which is known as the thermo-electric force, is given by

$$V = \int_{T_1}^{T_2} (S_B - S_A) \, dT, \tag{26.30}$$

and therefore

$$\frac{dV}{dT_2} = S_B(T_2) - S_A(T_2). \tag{26.31}$$

Hence measurements of $V$ for different values of the temperature $T_2$ will determine $S_B$ if $S_A$ is known. The preceding derivation of (26.30) does not depend upon the interpretation given to the kinetic coefficients in terms of the electron theory. It does, however, depend upon the truth of the Onsager relation $L_{12} = L_{21}$, since (26.22) can only be written in that form provided this relation holds.

**27. Evaluation of the integrals; detailed formulae for the thermal conductivity and thermo-electric power.** Making use of the subsidiary function $\sigma(E)$, defined by (24.1) the integral $K_n$ may be written

$$e^2 K_n = \int_0^\infty \sigma(E) E^n \left(-\frac{df_0}{dE}\right) dE, \tag{27.1}$$

and, on comparing (27.1) with (24.2) and (24.3) we find

$$e^2 K_n = \sigma(\zeta)\zeta^n + \frac{\pi^2}{6} (\mathrm{k}T)^2 \left\{E^n \frac{d^2\sigma}{dE^2} + 2n E^{n-1} \frac{d\sigma}{dE} + n(n-1) E^{n-2} \sigma.\right\} \tag{27.2}$$

Substituting the appropriate values of $K_n$ into (26.19) and (26.25), it will be found that a good deal of cancellation occurs, and we obtain the following simple formulae for the thermal conductivity

$$\varkappa = \frac{\pi^2 \mathrm{k}^2 T}{3 e^2} \sigma, \tag{27.3}$$

and for the thermo-electric power

$$S = \frac{\pi^2 \mathrm{k}^2}{3 e} \left(\frac{1}{\sigma} \frac{d\sigma}{dE}\right)_\zeta T. \tag{27.4}$$

It is the term containing $(\mathrm{k}T)^2$ in (27.2) which determines $\varkappa$ and $S$. The first term $\sigma\zeta^n$ cancels in each expression. Thus the thermal conductivity and the thermo-electric power depend upon the variation of $\sigma(E)E^n$ in a range of energy of order $\mathrm{k}T$. The thermal conductivity, however, depends only on the variation of $E^n$; the derivatives of $\sigma(E)$ cancel exactly in (26.19). On the other hand $S$ involves $d\sigma/dE$ which makes it difficult to evaluate it with precision. This difficulty arises because the relaxation time $\tau$ is contained in the integrand of $\sigma(E)$. Consider a surface of constant energy $E$ in $k$-space which lies in the region of the Fermi 'tail', i.e. where $f_0$ is close to neither zero nor unity. Imagine now that the distribution of electronic states in this surface differs from the distribution appropriate to thermal equilibrium. The rate at which the equilibrium distribution is restored is determined by $\tau(E)$. The scattering which restores equilibrium involves, in reality, changes of energy of the electrons of order $\mathrm{k}T$. It seems reasonable to suppose that a calculation, based on elastic scattering only, would somewhat over-estimate the rate of variation of $\tau(E)$ with $E$ in the neighbourhood of the Fermi limit. Nevertheless such calculations are not without interest, and we shall now estimate the dependence of $\tau$ and $\sigma$ on $E$ for various models.

We begin with the relaxation time, for thermal scattering, of a monovalent metal in which the electronic energy $E$ is taken to be proportional to $k^2$. It has been stated that in this case $\tau \propto E^{\frac{3}{2}}$. This result, however, depends on the following assumptions: (i) that the scattering matrix element $U_{\boldsymbol{k}'\boldsymbol{k}}$ is equal to a constant times $(\boldsymbol{k}-\boldsymbol{k}')$ times the amplitude of the thermal wave of wave vector $\boldsymbol{q}$, where $\boldsymbol{q}=\boldsymbol{k}-\boldsymbol{k}'$; and (ii) that all $U$-processes in the scattering can be neglected. Since the amplitude of a lattice wave is proportional to $1/q$, this factor cancels with $|\boldsymbol{k}-\boldsymbol{k}'|$ and thus for all $N$-processes $U_{\boldsymbol{k}'\boldsymbol{k}}$ may be regarded as constant, according to (i), and taken outside the integration of (7.12). The upper limit of the angle of scattering $\vartheta$ now depends upon $E$ and is given by

$$2k \sin\frac{\vartheta}{2} = q_0, \tag{27.5}$$

where $q_0$ is the radius of the Debye sphere. Thus the integral of (7.12) is proportional to $k^{-4}$, and since there is a factor $k$ outside, we have altogether $\tau \propto k^3$, as already stated.

The assumptions (i) and (ii) are certainly untenable, but if we correct *both* by including the angular dependence of $U_{\boldsymbol{k}'\boldsymbol{k}}$, and by also including $U$-processes, the final result is not greatly different from that which has just been given. As was shown in Sect. 11 the inclusion of $U$-processes makes the calculation of $\tau$ complicated, but the final result is very similar to that obtained using an Einstein model. We shall, therefore, now calculate the dependence of $\tau$ on $k$ on the basis of this model. Comparing (7.12), (8.9) and (10.10), and remembering that on the Einstein model the factor $(\boldsymbol{k}-\boldsymbol{k}')$ does not cancel, we find

$$\frac{1}{\tau} \propto k \int\limits_{w=0}^{w=1} \frac{(\sin\xi - \xi\cos\xi)^2}{\xi^4} w^3\, dw, \tag{27.6}$$

where, as before, $\xi = 2 k r_s w$. If we write

$$F_1(x) = \int_0^x (\sin\xi - \xi\cos\xi)^2 \frac{d\xi}{\xi}, \tag{27.7}$$

then

$$\tau \propto k^3 / F_1(2 r_s k). \tag{27.8}$$

For the thermo-electric power the quantity of interest is $\left(\frac{k}{\tau}\frac{d\tau}{dk}\right)_{k_0}$ which gives the effective power of $k$ at the FERMI surface. From (27.8) we see that

$$\left(\frac{k}{\tau}\frac{d\tau}{dk}\right)_{k_0} = 3 - \left(\frac{x}{F_1}\frac{dF_1}{dx}\right)_{3.84}, \tag{27.9}$$

where 3.84 is the value of $2 k_0 r_s$ for a monovalent metal. $F_1(3.84) = 5.35$ and $(dF_1/dx)_{3.84} = 1.38$ and thus the second term of (27.9) is almost exactly equal to 1.0 giving $\left(\frac{k}{\tau}\frac{d\tau}{dk}\right)_{k_0} = 2.0$. It may be seen immediately from (24.1) that for the ideal monovalent metal

$$\sigma(E) \propto |\boldsymbol{k}|^3 \tau, \tag{27.10}$$

and thus, according to the above result, $\left(\frac{1}{\sigma}\frac{d\sigma}{dE}\right)_\zeta = \frac{5}{2\zeta}$. Using this result, we obtain for the thermo-electric power according to (27.4)

$$S = -6.110 \times 10^{-2} \frac{T}{\zeta\,(\text{e.v.})} \text{ micro-volts per degree.} \tag{27.11}$$

In the following table the first line gives the values of $dS/dT$ for the alkali metals in the liquid state, as determined experimentally[1]. Below are the corresponding values according to Eq. (27.11). It will be observed that the theoretical values are considerably too small, and, in the case of lithium, of the wrong sign.

| | Li | Na | K | Rb | Cs |
|---|---|---|---|---|---|
| $dS/dT$ obs. | 0.030 | −0.048 | −0.043 | −0.085 | −0.076 micro-volts |
| $dS/dT$ calc. | −0.013 | −0.020 | −0.029 | −0.034 | −0.040 per degree² |

Direct calculations by the cellular method have shown that the electronic states in lithium differ in important respects from those in sodium[2], and are not, therefore, at all well represented by the free electron approximation. The anomalous sign of $S$ may arise from this cause.

The thermo-electric powers for sodium and potassium have been measured by MACDONALD and PEARSON[3] in the temperature range 4° to 80° K. They find a linear relation between $S$ and $T$ for sodium with $S/T = -0.04$ micro-volts per degree[3]. This is similar to the observed value of $S/T$ for this metal at higher temperatures.

**28. Thermo-electric power on the two-zone model.** We shall assume that the FERMI surface in each zone is spherical, and that the electronic energies in the neighbourhood of these surfaces are given by

$$E = \frac{\hbar^2}{2 m_1} k^2, \tag{28.1}$$

[1] A. SOMMERFELD: Naturwiss. **22**, 49 (1934).

[2] Cf. B. SCHIFF: Proc. Phys. Soc. Lond. A **67**, 2 (1954). — D. J. HOWARTH and H. JONES: Proc. Phys. Soc. Lond. A **65**, 355 (1952).

[3] D. K. C. MACDONALD and W. B. PEARSON: Proc. Roy. Soc. Lond., Ser. A **219**, 373 1953).

and

$$E = E_0 - \frac{\hbar^2}{2m_2} k^2, \tag{28.2}$$

where $m_1$ and $m_2$ are effective masses.

In certain metals, for example bismuth, the number of "electrons", with energies given by (28.1) and the number of "holes" with energies given by (28.2), are very small. If we neglect the effect of holes altogether we obtain a formula for $S$ similar to (27.11) except that $\zeta$ now represents the energy interval between the centre of the spherical region in the first zone and the FERMI surface. To account for the magnetic properties of bismuth, we have seen that it is necessary to assume that the number of electrons per atom $n_0$, in the above sense, is of order $10^{-4}$, and that the effective mass is of order $10^{-1}$ electronic masses, Since the radius of the FERMI sphere varies as $n_0^{\frac{1}{3}}$ we see that the magnitude of $\zeta$ may be 10 to 50 times smaller than in a monovalent metal. Hence $S/T$ may be expected to be of order unity in micro-volts per degree squared. In fact, the thermo-electric power of bismuth at 0° C is $-110$ parallel to the principal axis, and $-54$ perpendicular to this axis in these units. It thus appears likely that the exceptionally large value of the thermo-electric power in bismuth, and allied metals, is correctly interpreted in terms of a small effective number of conduction electrons having small effective masses.

In metals in which hole conduction is predominant, we can neglect altogether the effect of the electrons in the zone for which (28.1) is applicable, and we obtain a thermo-electric power of the opposite sign. This follows at once from the fact that Eq. (27.4) contains $e$ only to the first power. Antimony is an example of a metal with a large positive thermo-electric power. The exceptional magnitude of $S$ can be accounted for in the same way as in the case of bismuth. The sign, however, cannot be predicted without detailed knowledge of the electronic structure.

**29. Thermo-electric powers of the noble metals.** The thermo-electric powers of the metals copper, silver and gold are positive and rather small in magnitude[1]. This fact has led to doubts concerning the validity of the preceding theory since these metals are monovalent and their other properties are compatible with a normal, approximately spherical, FERMI surface. The thermo-electric power is, however, very sensitive to the form of the FERMI surface and it is not necessary to have "hole-conduction" to obtain a positive value as the following argument shows[2].

The noble metals have face-centred cubic structures, and the FERMI surface lies close to the octahedral planes of the BRILLOUIN zone; appreciably closer than the FERMI surface lies to the {110} planes of the BRILLOUIN zone of the alkali metals. Fig. 14a shows diagramatically the FERMI surface bulging towards the bounding planes at $ABCD$. At these points $\nabla E(\boldsymbol{k})$ is very small, and this leads to a peak in the density of states curve as shown in Fig. 14b. Now the time of relaxation is determined by the following integral

$$\frac{1}{\tau} = \frac{N\Omega}{4\pi^2\hbar} \int |U_{\boldsymbol{k}'\boldsymbol{k}}|^2 \left(1 - \frac{k_z'}{k_z}\right) \frac{dS'}{|\nabla E'|}, \tag{29.1}$$

in which the scattering matrix $U_{\boldsymbol{k}'\boldsymbol{k}}$ may be assumed to have the same dependence on $(\boldsymbol{k}-\boldsymbol{k}')$ as in a typical monovalent metal. A reasonable assumption,

[1] Observed values at 0° C are: Cu 1.73, Ag 1.3, Au 1.7, micro-volts per degree.

[2] H. JONES: Proc. Phys. Soc. Lond. A **68**, 1191 (1955).

therefore, for the dependence of $\tau$ on $E$ would be to write

$$\tau \propto E^{\frac{3}{2}}/N(E)\,, \tag{29.2}$$

which would reduce to the previously obtained result (27.9) when $N(E) \propto \sqrt{E}$, as for the standard free electron model. In the integral (24.1) for $\sigma(E)$, $\tau(\boldsymbol{k})$ is multiplied by $dS/|\nabla E|$ and it might appear that the effect of the small $|\nabla E|$ would just cancel the effect of the increasing $N(E)$, since both arise from the same cause: the close approach of the energy surface to the zone boundaries. However, the integral for $\sigma(E)$ contains the factor $(\partial E/\partial k_z)^2$ and this is also small

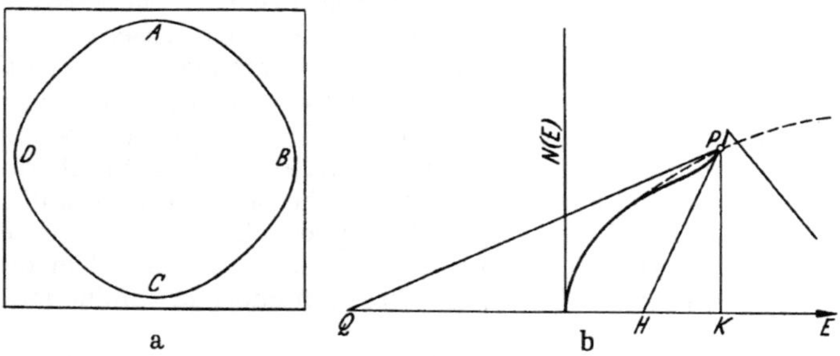


Fig. 14. Diagram relating to the thermo-electric effect in the noble metals. $QP$ is the tangent to the parabolic $N(E)$ curve for free electrons, $HP$ the tangent to the true $N(E)$ curve.

in the neighbourhoods of $ABCD$, so that the dominant parts of the integrand, in this case, come from the regions between $A$ and $B$, $B$ and $C$, etc. Thus we may still write

$$\sigma(E) \propto \tau(E)\, E^{\frac{3}{2}} \propto E^3/N(E)\,, \tag{29.3}$$

and therefore

$$\left(\frac{1}{\sigma}\,\frac{d\sigma}{dE}\right)_\zeta = \frac{1}{\zeta}\left\{3 - \left(\frac{E}{N(E)}\,\frac{dN(E)}{dE}\right)_\zeta\right\}. \tag{29.4}$$

$\left(\frac{E}{N(E)}\,\frac{dN(E)}{dE}\right)_\zeta = \frac{1}{2}$ for a parabolic density of states curve, and is equal to $\frac{1}{2}\,\frac{QK}{HK}$ at the point $P$ of Fig. 14b, where $HK$ is the sub-tangent to the full curve at $P$ and $QK$ the sub-tangent to the parabola. It is clear, therefore, that if the density of states curve is rising sufficiently rapidly near the Fermi limit, the thermo-electric power may become positive, even though the distortion of the Fermi surface from the spherical form is not exceptionally large.

**30. Effect of impurities on the thermo-electric power.** We begin by determining the dependence of the relaxation time, for impurity scattering, on $k$ the radius of the sphere in $k$-space on which the energy is equal to $E$. Eq. (16.6) for the scattering matrix element may be written

$$\left.\begin{aligned} U^{(i)}_{\boldsymbol{k}'\boldsymbol{k}} &= U_0\,\overline{u_B^2}\int_\Omega e^{i(\boldsymbol{k}-\boldsymbol{k}')\cdot\boldsymbol{r}}\,d\boldsymbol{r}\\ &= 4\pi r_s^3\,U_0\,\overline{u_B^2}\left(\frac{\sin\xi - \xi\cos\xi}{\xi^3}\right),\end{aligned}\right\} \tag{30.1}$$

where $\xi = 2r_s k w$, and $w$ is the sine of half the angle of scattering. Hence, from Eq. (7.12), we have

$$\left.\begin{aligned} \frac{1}{\tau_i} &\propto k\int_0^1 \frac{(\sin\xi - \xi\cos\xi)^2}{\xi^6}\,w^3\,dw\\ &\propto k^{-3}\,F_3(2k r_s)\,,\end{aligned}\right\} \tag{30.2}$$

where

$$F_3(x) = \int\limits_0^x (\sin\xi - \xi\cos\xi)^2 \frac{d\xi}{\xi^3}. \tag{30.3}$$

Thus, if $k_0$ denotes the radius of the FERMI sphere for an idealized monovalent metal, we have

$$\left(\frac{k}{\tau_i}\frac{d\tau_i}{dk}\right)_{k_0} = 3 - \left(\frac{F_3}{x}\frac{dF_3}{dx}\right)_{3.84} = 2.6. \tag{30.4}$$

$\tau_i$, therefore, varies somewhat more rapidly with $k$ than the relaxation time for thermal scattering. In fact, however, $\tau_i$ may vary even more rapidly with $k$ than is indicated by (30.4). This is due to the very approximate nature of the theory of Sect. 16, and in particular to the neglect of the term $2\nabla\chi\cdot\nabla u$ in Eq. (16.2), which is not justifiable even in pure metals except, perhaps, in the case of metallic sodium. It might be anticipated that a more precise theory would lead to a scattering matrix similar to (30.1) but in which the constant factor $U_0\overline{u_B^2}$ is replaced by one depending on $k$.

A sharply increasing density of states at the FERMI limit would, of course, add a negative term to (30.4) just as in the case of thermal scattering. In the absence of more detailed knowledge of the electronic wave functions in pure metals and alloys it is impossible to reach precise conclusions regarding the thermo-electric power. However, within the framework of the existing theory it is possible to account for many effects with quite reasonable assumptions. Thus, if the hypothesis of the preceding section with regard to the cause of the positive thermo-electric power of pure copper is accepted, the fact that small amounts of impurity change the thermo-electric power to negative values at low temperatures can readily be understood on the assumption that $\tau_i$ increases with $k$ at the FERMI limit.

If $\tau_t$ denotes the relaxation time for thermal scattering and $\tau_i$ that for impurity scattering, the combined scattering leads to a relaxation time $\tau$ given by

$$\frac{1}{\tau} = \frac{1}{\tau_t} + \frac{1}{\tau_i}, \tag{30.5}$$

and we have seen that

$$\frac{1}{\tau_t} = \alpha T, \qquad \frac{1}{\tau_i} = \beta x, \tag{30.6}$$

where $\alpha$ and $\beta$ are positive constants, and $x$ denotes the atomic fraction of impurity. Thus taking logarithms and differentiating we get

$$\frac{k}{\tau}\frac{d\tau}{dk} = \frac{1}{(\alpha T + \beta x)}\left[\alpha T\left(\frac{k}{\tau_t}\frac{d\tau_t}{dk}\right) + \beta x\left(\frac{k}{\tau_i}\frac{d\tau_i}{dk}\right)\right]. \tag{30.7}$$

This suggests the following general expression for the thermo-electric power of monovalent metals containing small amounts of impurity

$$S = \frac{\pi^2 \mathrm{k}^2 T}{3e}\left(\frac{AT + Bx}{\alpha T + \beta x}\right), \tag{30.8}$$

where $A$ and $B$ are constants depending on $\left(\frac{k}{\tau_t}\frac{d\tau_t}{dk}\right)$ and $\left(\frac{k}{\tau_i}\frac{d\tau_i}{dk}\right)$ respectively. $B$ might be expected to be positive in most cases, while $A$ should be negative for the noble metals. (30.8) would not, of course, be applicable at very low temperatures.

**31. Thermo-electric power of the transition metals and their alloys**[1]. It is not at present possible to give an adequate detailed theory of the thermo-electric effect in metals having such complicated electronic structures as the transition metals and their alloys. However, the dominant effect of *s-d* transitions on the resistivity in such cases might be expected to affect the thermo-electric power in a recognisable way. We have seen that the relaxation time for electrons in the *s*-zone is largely determined by the density of states in the *d*-band according to the relation

$$\tau_s \propto \frac{1}{N_d(E)} . \tag{31.1}$$

If, neglecting other factors, we write $\sigma(E) = \text{const}\, N_d^{-1}(E)$ and if we suppose that near the limit $E_1$ of the *d*-band $N_d(E) \propto \sqrt{(E_1 - E)}$, then clearly

$$\left(\frac{1}{\sigma}\frac{d\sigma}{dE}\right)_\zeta = \frac{1}{2(E_1 - \zeta)} . \tag{31.2}$$

Thus we may expect $S$ to be large and negative, and if the *d*-band is gradually filled, as, for example, by adding silver to palladium, the value of $-S$ will increase steadily to the limit at which the *d*-band is complete. Thereafter we should expect values similar to those of a typical monovalent metal with impurity scattering.

Fig. 15, as given by COLES[2], shows the observed thermo-electric power for the binary systems, silver-palladium, and copper-nickel. It will be observed that the silver-palladium alloys behave roughly in the way expected, but that the copper-nickel alloys do not.

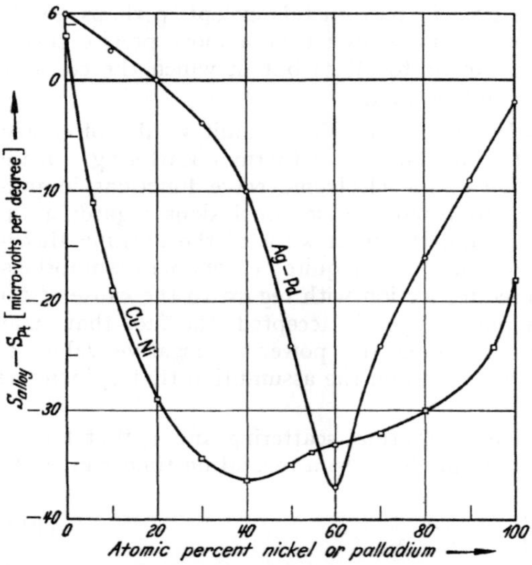


Fig. 15. Thermo-electric powers of the alloys copper-nickel and silver-palladium relative to platinum as functions of the composition.

**32. The distribution function for thermal conduction.** It is of interest to consider the nature of the distribution function in a metal carrying a thermal current but no electric current. If we put $J_z = 0$ in Eq. (26.6) we find

$$eF = \left(\frac{d\zeta}{dT} - \frac{L_{12}}{TL_{11}}\right)\frac{dT}{dz} , \tag{32.1}$$

and on substituting this into Eq. (26.3), and expressing $L_{12}/L_{11}$ in terms of the $K_n$ integrals, we find

$$f_1 = -\frac{\tau}{T}\frac{dT}{dz}\frac{1}{\hbar}\frac{\partial E}{\partial k_z}\frac{df_0}{dE}\left(\frac{K_1}{K_0} - E\right). \tag{32.2}$$

Neglecting *second* powers of $(\mathrm{k}T/\zeta)$, as compared to unity, $K_1/K_0$ may be replaced by $\zeta$. Also if $k_0$ denotes the radius of the sphere in $k$-space on which the energy

[1] For interesting work on the thermo-electric power of alloys the reader is referred to a paper by J. FRIEDEL: J. Phys. Radium **14**, 561 (1953).

[2] B. R. COLES, cf. reference p. 270.

is $\zeta$, and $k$ that on which the energy is $E$ we may write

$$\zeta - E = \frac{\hbar^2 k_0}{m}(k_0 - k), \tag{32.3}$$

so long as $(k_0 - k)/k_0 \ll 1$. Thus we obtain

$$f_1 = \left(\frac{\tau \hbar k_0}{m}\right) \frac{1}{T} \frac{dT}{dz} (k - k_0) \frac{\partial f_0}{\partial k_z}, \tag{32.4}$$

and therefore

$$f(k_x, k_y, k_z) = f_0(k_x, k_y, k_z + \delta k_z), \tag{32.5}$$

where

$$\delta k_z = \left(\frac{l}{T} \frac{dT}{dz}\right)(k - k_0), \tag{32.6}$$

since the mean free path $l$ is, by definition, equal to $(\tau \hbar k_0/m)$. Thus the distribution of electronic states on the sphere of radius $k_0$ is unaffected. In the presence

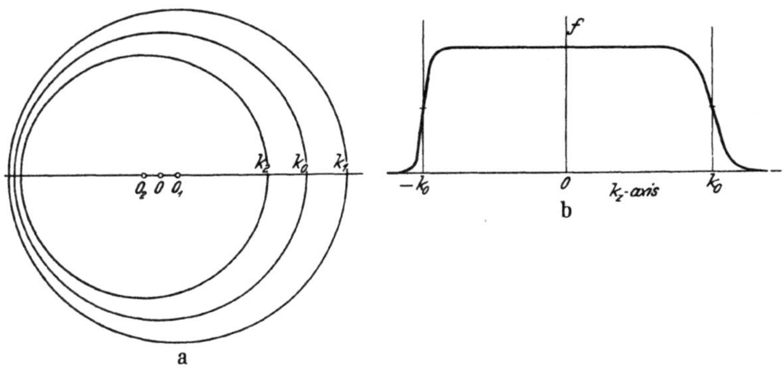


Fig. 16. Distribution function in a metal carrying a thermal current but no electric current. (a) Spheres on which the density of representative points is constant. (b) Section of the distribution function along the $k_z$-axis.

of a temperature gradient there is a uniform distribution on a sphere of radius $k_1 > k_0$ whose centre is now displaced to the point $O_1$, as shown in Fig. 16a. Similarly there is a uniform distribution of states on the sphere of radius $k_2 < k_0$ centred on $O_2$. Since $\frac{l}{T}\frac{dT}{dz}$ is equal to the ratio of the temperature drop, in a distance equal to the mean free path, to the ambient temperature, $\delta k_z$ is always very much less than the difference between the radii. The nature of the distribution function is shown in Fig. 16b which is a plot of $f$ along the $k_z$-axis.

There exists an interesting difference between the distribution functions for thermal and electrical conductivity which can be seen by comparing Figs. 4 and 16. Collisions between electrons could not destroy the total electrical current, on the other hand collisions between electrons could restore the distribution indicated in Fig. 16 to the thermal equilibrium distribution. It thus appears that the agreement between the observed value of the WIEDEMANN-FRANZ ratio at normal temperatures, and the theoretical value given by (27.3), is strong evidence that collisions between electrons play no important part in conductivity at these temperatures.

## VI. Conduction at low temperatures.

**33. BLOCH's equation: The variational method.** The theory of conduction at low temperatures is inevitably more complicated mathematically than the theory for high temperatures. This is due to the fact that interchanges of energy between the electrons and the lattice play an essential part in the phenomenon. Physically, however, the theory is simpler in important respects, since, at least for the idealized monovalent metal, no $U$-processes can occur at low temperatures.

We begin by formulating BLOCH's integral equation for electrical conduction in the presence of a uniform temperature gradient. For the present we are not concerned with the effect of a magnetic field, and consequently we put $H=0$ in Eq. (4.10). Introducing the partial potential $\mu$, where $d\mu/dz=-eF+d\zeta/dz$, the electric field and the temperature gradient being chosen to have the direction of the $z$-axis, (4.10) may be written

$$\frac{\hbar}{m} k_z \frac{df_0}{dE}\left\{\frac{d\mu}{dz} - T(E-\zeta)\frac{d}{dz}\left(\frac{1}{T}\right)\right\} = \Lambda\,\chi(\boldsymbol{k}), \tag{33.1}$$

where $\Lambda$ is an integral operator defined by the equation

$$\Lambda\,\chi(\boldsymbol{k}) = \frac{N\Omega}{8\pi^3 \mathrm{k}T}\int P(\boldsymbol{k},\boldsymbol{k}')\, f_0(1-f_0')\,\{\chi(\boldsymbol{k})-\chi(\boldsymbol{k}')\}\,d\boldsymbol{k}'. \tag{33.2}$$

The application of BOLTZMANN's Eq. (4.2) to yield (33.1), (33.2) was first made by BLOCH, and we refer to this equation as BLOCH's integral equation[1]. Since $\mu$ and $T$ are independent thermodynamic variables we can break up Eq. (33.1) into two independent equations by the substitution

$$\chi = \frac{\hbar}{m}\left\{-\frac{d\mu}{dz}\varphi_1 + T\frac{d}{dz}\left(\frac{1}{T}\right)\varphi_2\right\}, \tag{33.3}$$

and we obtain

$$-k_z\frac{df_0}{dE} = \Lambda\,\varphi_1, \tag{33.4}$$

and

$$-k_z(E-\zeta)\frac{df_0}{dE} = \Lambda\,\varphi_2. \tag{33.5}$$

Comparing Eqs. (26.3) and (33.3) we see that at high temperatures $\varphi_1$ goes over into $k_z\tau$ and $\varphi_2$ into $k_z(E-\zeta)\,\tau$. At low temperatures it will be shown that the dominant terms in $\varphi_1$ and $\varphi_2$ still retain this form except that the factor $\tau$ is no longer the same constant in the two cases. The reason for this difference can readily be understood. $\varphi_1$ corresponds to a displacement of the FERMI distribution like that shown in Fig. 4 for electrical conduction; $\varphi_2$ to that shown in Fig. 16 for thermal conduction. At high temperatures both displacements relax at the same rate, but at low temperatures the $\varphi_1$ displacement relaxes more slowly than the $\varphi_2$ displacement. The reason for this is that, at low temperatures, an electron can only be scattered through a very small angle in a single process. Thus to restore the equilibrium distribution from that indicated by Fig. 4 each electron must suffer many more individual collisions than is necessary to restore equilibrium from the displacement indicated by Fig. 16.

We can readily obtain expressions for the kinetic coefficients $L_{11}$, $L_{12}$ and $L_{22}$ in terms of the functions $\varphi_1$ and $\varphi_2$. Substituting (33.3) into Eqs. (26.4) and (26.5) we obtain for the particle and heat currents

$$-J_z = -\frac{\hbar^2}{4\pi^3 m^2}\int k_z\,\varphi_1\frac{df_0}{dE}\,d\boldsymbol{k}\left(\frac{d\mu}{dz}\right) + \frac{\hbar^2}{4\pi^3 m^2}\int k_z\,\varphi_2\frac{df_0}{dE}\,d\boldsymbol{k}\left(T\frac{d}{dz}\left(\frac{1}{T}\right)\right), \tag{33.6}$$

$$Q_z = \frac{\hbar^2}{4\pi^3 m^2}\int k_z(E-\zeta)\,\varphi_1\frac{df_0}{dE}\,d\boldsymbol{k}\left(\frac{d\mu}{dz}\right) - \frac{\hbar^2}{4\pi^3 m^2}\int k_z(E-\zeta)\,\varphi_2\frac{df_0}{dE}\,d\boldsymbol{k}\left(T\frac{d}{dz}\left(\frac{1}{T}\right)\right). \tag{33.7}$$

[1] One of the earliest successful applications of Wave Mechanics to the problem of electronic conduction in metals was made by W. V. HOUSTON: Z. Physik **48**, 449 (1928); Phys. Rev. **34**, 279 (1929).

Hence, by comparing with Eqs. (26.6) and (26.7), we find

$$L_{11} = -\frac{T\hbar^2}{4\pi^3 m^2}\int k_z\,\varphi_1\frac{df_0}{dE}\,d\boldsymbol{k}, \tag{33.8}$$

$$L_{12} = \frac{T\hbar^2}{4\pi^3 m^2}\int k_z\,\varphi_2\frac{df_0}{dE}\,d\boldsymbol{k}, \qquad L_{21} = \frac{T\hbar^2}{4\pi^3 m^2}\int k_z(E-\zeta)\,\varphi_1\frac{df_0}{dE}\,d\boldsymbol{k}, \tag{33.9}$$

$$L_{22} = -\frac{T\hbar^2}{4\pi^3 m^2}\int k_z(E-\zeta)\,\varphi_2\frac{df_0}{dE}\,d\boldsymbol{k}. \tag{33.10}$$

Eqs. (26.16), (26.19) and (26.24) for the electrical conductivity, the thermal conductivity and the thermo-electric power, in terms of the kinetic coefficients, remain, of course, valid for all temperatures.

We now verify that $L_{12}=L_{21}$ in accordance with the requirement of ONSAGER's relation.

Let

$$V(\boldsymbol{k},\boldsymbol{k}') = \frac{N\Omega}{8\pi^3 \mathrm{k}T}\,P(\boldsymbol{k},\boldsymbol{k}')\,f_0(1-f_0'). \tag{33.11}$$

Eq. (4.4) may then be written

$$V(\boldsymbol{k},\boldsymbol{k}') = V(\boldsymbol{k}',\boldsymbol{k}). \tag{33.12}$$

Thus

$$(\varphi_1,\varphi_2) \equiv \int \varphi_1\Lambda\,\varphi_2\,d\boldsymbol{k} = \iint V(\varphi_2-\varphi_2')\,\varphi_1\,d\boldsymbol{k}\,d\boldsymbol{k}', \tag{33.13}$$

and

$$(\varphi_2,\varphi_1) \equiv \int \varphi_2\Lambda\,\varphi_1\,d\boldsymbol{k} = \iint V(\varphi_1-\varphi_1')\,\varphi_2\,d\boldsymbol{k}\,d\boldsymbol{k}', \tag{33.14}$$

where the bracket symbol (..,..) is to be regarded as defined by either of these identities and may be applied to any pair of functions. Separating (33.13) into two integrals, and interchanging $\boldsymbol{k}$ with $\boldsymbol{k}'$ in the second, we see immediately that

$$(\varphi_1,\varphi_2) = (\varphi_2,\varphi_1). \tag{33.15}$$

Multiplying (33.4) by $\varphi_2$, (33.5) by $\varphi_1$ and integrating each with respect to $d\boldsymbol{k}$ we obtain the desired result: $L_{12}=L_{21}$.

To solve the integral equations (33.4) and (33.5) we make use of the simple and elegant procedure first introduced by M. KOHLER[1] and known as the variational method. We deal in detail with (33.4); the analogous solution of (33.5) may then be written down immediately.

Let $\psi(\boldsymbol{k})$ be one of a set of functions all of which satisfy the equation

$$\int\psi\left(\Lambda\,\psi + k_z\frac{df_0}{dE}\right)d\boldsymbol{k} = 0. \tag{33.16}$$

Clearly $\varphi_1$ is one function of this set which has the special property of making the integrand vanish at every value of $\boldsymbol{k}$ according to (33.4). The other functions of the set have not, in general, this property.

Multiplying (33.4) by $\psi(\boldsymbol{k})$ and integrating we have

$$\int\psi\left(\Lambda\,\varphi_1 + k_z\frac{df_0}{dE}\right)d\boldsymbol{k} = 0. \tag{33.17}$$

Hence comparing (33.16) and (33.17) we find

$$(\psi,\psi) = (\psi,\varphi_1). \tag{33.18}$$

[1] M. KOHLER: Z. Physik **125**, 679 (1949).

Since $\Lambda$ is a linear operator, it follows that

$$\left.\begin{aligned}(\varphi_1-\psi,\ \varphi_1-\psi) &= (\varphi_1,\varphi_1)+(\psi,\psi)-(\varphi_1,\psi)-(\psi,\varphi_1)\\ &= (\varphi_1,\varphi_1)-(\psi,\psi),\end{aligned}\right\} \tag{33.19}$$

where we have made use of (33.15) and (33.18). It is clear from (33.13) and (33.14) that the commutability expressed by (33.15) applies to any pair of functions, not only to $\varphi_1$ and $\varphi_2$. We now show that the left hand side of (33.19) is essentially positive. For any function $\chi(\boldsymbol{k})$ we have

$$\left.\begin{aligned}(\chi,\chi) &= \iint V(\chi-\chi')\,\chi\,d\boldsymbol{k}\,d\boldsymbol{k}'\\ &= \iint V(\chi'-\chi)\,\chi'\,d\boldsymbol{k}\,d\boldsymbol{k}'\\ &= \tfrac{1}{2}\iint V\,(\chi-\chi')^2\,d\boldsymbol{k}\,d\boldsymbol{k}',\end{aligned}\right\} \tag{33.20}$$

where in the second line we have interchanged $\boldsymbol{k}$ and $\boldsymbol{k}'$, and made use of (33.12). Since $V$ is necessarily positive we have established the result. Hence, by (33.19)

$$(\varphi_1,\varphi_1)\geq(\psi,\psi) \tag{33.21}$$

for all $\psi$ satisfying (33.16). Thus KOHLER's principle states that of all functions satisfying (33.16) that which makes $(\psi,\psi)$ a maximum is the solution of (33.4).

To apply this principle we use as variation functions polynomials in $(E-\zeta)$. To economise symbols it is advantageous to use $\eta=(E-\zeta)/\mathrm{k}T$ in place of $(E-\zeta)$ itself.

Thus we write

$$\psi = k_z\sum_{r=0}^{n} a_r\,\eta^r, \tag{33.22}$$

where the $(n+1)$ coefficients $a_r$ are to be taken as the parameters of the variation problem. The final value taken by each one of these coefficients $a_r$ will depend upon the number of them used in the variation function, i.e. on $n$, though to avoid encumbering the symbols this is not shown explicitly.

Introducing (33.22) into (33.16) we find

$$\sum_{r=0}^{n}\sum_{s=0}^{n} a_r\,a_s\,d_{rs} = \sum_{r=0}^{n}\alpha_r\,a_r, \tag{33.23}$$

where

$$\alpha_r = -\int k_z^2\,\eta^r\,\frac{df_0}{dE}\,d\boldsymbol{k}, \tag{33.24}$$

and

$$d_{rs} = \int k_z\,\eta^r\,\Lambda\,k_z\,\eta^s\,d\boldsymbol{k}. \tag{33.25}$$

On account of the commutability property of the bracket expression $d_{rs}=d_{sr}$.

We have now to find the values of the parameters $a_r$ which maximize

$$\sum_{r=0}^{n}\sum_{s=0}^{n} a_r\,a_s\,d_{rs} \tag{33.26}$$

subject to the subsidiary condition (33.23). To do this we write

$$y(a_0,a_1,\ldots,a_n) = \sum_{r=0}^{n}\sum_{s=0}^{n} a_r\,a_s\,d_{rs} + \lambda\sum_{r=0}^{n}\alpha_r\,a_r, \tag{33.27}$$

and put

$$\frac{\partial y}{\partial a_r}=0, \tag{33.28}$$

for all $r$. Hence

$$2\sum_{s=0}^{n} a_s d_{rs} + \lambda \alpha_r = 0, \tag{33.29}$$

and on multiplying by $a_r$ and summing we get

$$2\sum_{r=0}^{n}\sum_{s=0}^{n} a_r a_s d_{rs} + \lambda \sum_{r=0}^{n} \alpha_r a_r = 0. \tag{33.30}$$

Comparing with (33.23), to determine the multiplier, we find $\lambda = -2$. Thus from (33.29) we obtain a set of $(n+1)$ linear equations for the $(n+1)$ values of $a_r$ which satisfy the stationary condition for $(\psi, \psi)$, and the subsidiary Eq. (33.16):

$$\alpha_r = \sum_{s=0}^{n} a_s d_{rs}. \tag{33.31}$$

To obtain the solution of (33.5) we write

$$\beta_r = -\int k_z^2 \eta^r (E-\zeta) \frac{d f_0}{dE} d\boldsymbol{k}, \tag{33.32}$$

and, for the variation function,

$$\psi = k_z \sum_{r=0}^{n} b_r \eta^r, \tag{33.33}$$

from which, exactly as above, we have

$$\beta_r = \sum_{s=0}^{n} b_s d_{rs}. \tag{33.34}$$

**34. Expressions for the kinetic coefficients.** Before evaluating the integrals $\alpha_r, \beta_r$ and $d_{rs}$ we express the kinetic coefficients in terms of them, restricting ourselves, in the interests of simplicity, to the case $n=2$.

Thus, with the approximations

$$\varphi_1 = (a_0 + a_1 \eta) k_z, \tag{34.1}$$

$$\varphi_2 = (b_0 + b_1 \eta) k_z, \tag{34.2}$$

we find, on substitution into (33.8), (33.9), and (33.10), and using (33.24) and (33.32)

$$L_{11} = \frac{T\hbar^2}{4\pi^3 m^2} (a_0 \alpha_0 + a_1 \alpha_1), \tag{34.3}$$

$$L_{12} = -\frac{T\hbar^2}{4\pi^3 m^2} (b_0 \alpha_0 + b_1 \alpha_1), \tag{34.4}$$

$$L_{22} = \frac{T\hbar^2}{4\pi^3 m^2} (b_0 \beta_0 + b_1 \beta_1). \tag{34.5}$$

We now obtain the values of $a_0, a_1, b_0, b_1$ as determined by the variational method i.e. from Eqs. (33.31) and (33.34). They are

$$a_0 = \frac{\alpha_1 d_{01} - \alpha_0 d_{11}}{d_{10}^2 - d_{11} d_{00}}, \qquad a_1 = \frac{\alpha_0 d_{10} - \alpha_1 d_{00}}{d_{10}^2 - d_{11} d_{00}}, \tag{34.6}$$

$$b_0 = \frac{\beta_1 d_{01} - \beta_0 d_{11}}{d_{10}^2 - d_{11} d_{00}}, \qquad b_1 = \frac{\beta_0 d_{10} - \beta_1 d_{00}}{d_{10}^2 - d_{11} d_{00}}. \tag{34.7}$$

This completes the solution of the BLOCH equation, since the kinetic coefficients, and, therefore, the electrical and thermal conductivities and the thermo-electric power, are now expressed in terms of definite integrals. When these integrals are evaluated, and only the leading terms for low temperatures retained, it will be seen that the expressions for the kinetic coefficients simplify considerably.

**35. Calculation of the transition probability $P(\boldsymbol{k}, \boldsymbol{k}')$.** To evaluate the integrals $d_{rs}$ it is necessary to calculate the transition probability $P(\boldsymbol{k}, \boldsymbol{k}')$. The interaction between electrons and long wave-length lattice waves is most accurately described by the method of the deformation potential, which, as we have seen, is equivalent to BARDEEN's method in this limit, and takes account of the redistribution of electric charge under the deformation produced by the lattice wave.

According to Eqs. (14.2) and (14.10) the scattering potential is given by

$$U(\boldsymbol{r}) = \frac{2\zeta}{3} \Delta(\boldsymbol{r}), \tag{35.1}$$

where the dilatation $\Delta(\boldsymbol{r})$ is given in terms of the amplitudes of the lattice waves by (14.6)

We have now to calculate the matrix elements of $U(\boldsymbol{r})$, not only with respect to the stationary states of the electrons but also with respect to the stationary states of the lattice waves. If $\psi_{\boldsymbol{k}}$ and $\psi_{\boldsymbol{k}'}$ denote the wave functions of the initial and final electronic states, we have, according to (14.7),

$$\int \psi_{\boldsymbol{k}'}^* U(\boldsymbol{r}) \psi_{\boldsymbol{k}} d\boldsymbol{r} = \frac{i}{\sqrt{N}} \left(\frac{2\zeta}{3}\right) \boldsymbol{a}(\boldsymbol{q}) \cdot (\boldsymbol{k}' - \boldsymbol{k}) \tag{35.2}$$

where $\boldsymbol{q} = \boldsymbol{k}' - \boldsymbol{k}$. Thus only the component of $\boldsymbol{a}(\boldsymbol{q})$ in the direction of the wave vector $\boldsymbol{q}$ is of interest to us in the calculation of $P(\boldsymbol{k}, \boldsymbol{k}')$. At this stage we are compelled, in the present state of the theory, to make the invalid assumption that the crystal may be regarded as elastically isotropic. Then, neglecting anharmonic terms, the longitudinal and transverse waves form independent normal modes of vibration. Thus, as far as the present theory of electronic conduction at low temperatures is concerned, we can neglect entirely the transverse waves and quantize only the longitudinal motion.

The strain energy function $W(\boldsymbol{r})$ for dilatation waves is given by the simple expression

$$W(\boldsymbol{r}) = \tfrac{1}{2} \varrho c_s^2 \Delta^2(\boldsymbol{r}), \tag{35.3}$$

where $\varrho$ denotes the density of the solid, and $c_s$ the velocity of the longitudinal waves of long wave-length. The potential energy of a volume $N\Omega$ of the solid is, therefore, according to (14.6), given by

$$\int\limits_{N\Omega} W(\boldsymbol{r})\, d\boldsymbol{r} = -\frac{\varrho c_s^2}{2N} \sum_{\boldsymbol{q}} \sum_{\boldsymbol{q}'} (\boldsymbol{q} \cdot \boldsymbol{a}(\boldsymbol{q}))\, (\boldsymbol{q}' \cdot \boldsymbol{a}(\boldsymbol{q}')) \int\limits_{N\Omega} e^{i\boldsymbol{r}\cdot(\boldsymbol{q}+\boldsymbol{q}')}\, d\boldsymbol{r}. \tag{35.4}$$

The only non-vanishing terms of the double sum are those for which $\boldsymbol{q} + \boldsymbol{q}' = 0$, and since $\boldsymbol{a}(-\boldsymbol{q}) = \boldsymbol{a}^*(\boldsymbol{q})$ we have

$$\left.\begin{aligned} (\boldsymbol{q} \cdot \boldsymbol{a}(\boldsymbol{q}))\, (\boldsymbol{q}' \cdot \boldsymbol{a}(\boldsymbol{q}')) &= -(\boldsymbol{q} \cdot \boldsymbol{a}(\boldsymbol{q}))\, (\boldsymbol{q} \cdot \boldsymbol{a}^*(\boldsymbol{q})) \\ &= -q^2\, a(\boldsymbol{q})\, a^*(\boldsymbol{q}), \end{aligned}\right\} \tag{35.5}$$

where $a(\boldsymbol{q})$ denotes the component of $\boldsymbol{a}(\boldsymbol{q})$ along the direction of $\boldsymbol{q}$. Hence, since $c_s q = \omega_q$ and $\varrho\Omega = M$ is the mass of a single atom, the potential energy may be written

$$\tfrac{1}{2} M \sum_{\boldsymbol{q}} \omega_{\boldsymbol{q}}^2 \, a(\boldsymbol{q}) \, a^*(\boldsymbol{q}). \tag{35.6}$$

The kinetic energy of the longitudinal waves is, according to (14.4) given by

$$\left.\begin{aligned} \frac{1}{2}\varrho \int_{N\Omega} \dot{R}^2 \, d\boldsymbol{r} &= \frac{\varrho}{2N} \sum_{\boldsymbol{q}} \sum_{\boldsymbol{q}'} \dot{a}(\boldsymbol{q}) \, \dot{a}(\boldsymbol{q}') \int_{N\Omega} e^{i\boldsymbol{r}\cdot(\boldsymbol{q}+\boldsymbol{q}')} \, d\boldsymbol{r} \\ &= \frac{1}{2} M \sum_{\boldsymbol{q}} \dot{a}(\boldsymbol{q}) \, \dot{a}^*(\boldsymbol{q}). \end{aligned}\right\} \tag{35.7}$$

Since each term of the sums (35.6) and (35.7) is unchanged by replacing $\boldsymbol{q}$ by $-\boldsymbol{q}$, we may write the potential and kinetic energies respectively as

$$M \sum_{(\frac{1}{2})} \omega_{\boldsymbol{q}}^2 \, a(\boldsymbol{q}) \, a^*(\boldsymbol{q}), \quad \text{and} \quad M \sum_{(\frac{1}{2})} \dot{a}(\boldsymbol{q}) \, \dot{a}^*(\boldsymbol{q}), \tag{35.8}$$

where the summations are over one half of the BRILLOUIN zone in $q$-space. The component of the amplitude $a(\boldsymbol{q})$ is complex; thus introducing real variables by the substitution

$$a(\boldsymbol{q}) = Q_1 + i\, Q_2, \tag{35.9}$$

where the dependence on $\boldsymbol{q}$ need not be explicitly shown, we find for ex- expressions (35.8)

$$M \sum_{(\frac{1}{2})} \omega_{\boldsymbol{q}}^2 (Q_1^2 + Q_2^2), \quad \text{and} \quad M \sum_{(\frac{1}{2})} (\dot{Q}_1^2 + \dot{Q}_2^2). \tag{35.10}$$

The conjugate momenta $P_1$ and $P_2$ are obtained by differentiating the kinetic energy with respect to $\dot{Q}_1$ and $\dot{Q}_2$, and thus the Hamiltonian $H$ for the dilatational waves may be written

$$H = \sum_{(\frac{1}{2})} \left[\left(\frac{1}{4M} P_1^2 + M \omega_q^2 Q_1^2\right) + \left(\frac{1}{4M} P_2^2 + M \omega_q^2 Q_2^2\right)\right]. \tag{35.11}$$

The $N$ individual terms of (35.11) are Hamiltonians of simple harmonic oscillators of angular frequency $\omega_q$ and mass $2M$. Using well-known results we have, therefore, for the total energy

$$E_{\text{tot}} = \sum_{\boldsymbol{q}} \hbar \, \omega_{\boldsymbol{q}} (n_{\boldsymbol{q}} + \tfrac{1}{2}),$$

and for the matrix elements of any $Q$

$$\left.\begin{aligned} Q(n_{\boldsymbol{q}}, n_{\boldsymbol{q}} - 1) \\ Q(n_{\boldsymbol{q}}, n_{\boldsymbol{q}} + 1) \end{aligned}\right\} = \left(\frac{\hbar}{4\,\omega_{\boldsymbol{q}} M}\right)^{\frac{1}{2}} \begin{cases} \sqrt{n_{\boldsymbol{q}}} \\ \sqrt{n_{\boldsymbol{q}} + 1}. \end{cases} \tag{35.12}$$

Hence, for the matrix elements of the complex amplitude $a(\boldsymbol{q})$, we find

$$\left.\begin{aligned} |a(\boldsymbol{q})_{n_{\boldsymbol{q}}, n_{\boldsymbol{q}}-1}|^2 \\ |a(\boldsymbol{q})_{n_{\boldsymbol{q}}, n_{\boldsymbol{q}}+1}|^2 \end{aligned}\right\} = \frac{\hbar}{2\,\omega_{\boldsymbol{q}} M} \begin{cases} n_{\boldsymbol{q}} \\ n_{\boldsymbol{q}} + 1. \end{cases} \tag{35.13}$$

Combining this result with Eq. (35.2) we obtain

$$|(n_{\boldsymbol{q}}', \boldsymbol{k}' \,|\, U \,|\, n_{\boldsymbol{q}}, \boldsymbol{k})|^2 = \frac{\hbar}{2\varrho \, c_s N\Omega} \left(\frac{2\zeta}{3}\right)^2 q \begin{cases} n_{\boldsymbol{q}}, & n_{\boldsymbol{q}}' = n_{\boldsymbol{q}} - 1 \\ n_{\boldsymbol{q}} + 1, & n_{\boldsymbol{q}}' = n_{\boldsymbol{q}} + 1. \end{cases} \tag{35.14}$$

The change in the total energy of the system, electrons plus lattice, in the transition $n_{\boldsymbol{q}}, \boldsymbol{k} \to n'_{\boldsymbol{q}}, \boldsymbol{k}'$ may be expressed as follows:

$$\text{Final energy} - \text{Initial energy} = E_{\boldsymbol{k}'} - E_{\boldsymbol{k}} + \hbar\,\omega_{\boldsymbol{q}}(n'_{\boldsymbol{q}} - n_{\boldsymbol{q}}). \tag{35.15}$$

The upper line of (35.14) denotes a transition in which the lattice emits a quantum of vibrational energy, and the lower line to a transition in which the lattice absorbs a quantum of vibrational energy.

Eq. (7.11) for the transition probability may be applied to the present case if we replace $|U_{\boldsymbol{k}'\boldsymbol{k}}|^2$ by (35.14) and $(E_{\boldsymbol{k}'} - E_{\boldsymbol{k}})$ by (35.15). For the total system of electrons and lattice the perturbation giving rise to transitions, viz. (35.1), still does not contain the time explicitly. Hence

$$\frac{\partial}{\partial t}|b(n'_{\boldsymbol{q}}, \boldsymbol{k}, t)|^2 = \frac{2}{\hbar}|(n'_{\boldsymbol{q}}, \boldsymbol{k}'|U|n_{\boldsymbol{q}}, \boldsymbol{k})|^2 \frac{\sin\frac{t}{\hbar}\{E_{\boldsymbol{k}'} - E_{\boldsymbol{k}} + \hbar\,\omega_{\boldsymbol{q}}(n'_{\boldsymbol{q}} - n_{\boldsymbol{q}})\}}{E_{\boldsymbol{k}'} - E_{\boldsymbol{k}} + \hbar\,\omega_{\boldsymbol{q}}(n'_{\boldsymbol{q}} - n_{\boldsymbol{q}})}. \tag{35.16}$$

As in Sect. 7 we assume that $t$ may be imagined large enough to justify the use of the following rule:

$$\int_{-\infty}^{\infty} f(z)\frac{\sin\frac{t}{\hbar}z}{z}\,dz = \pi f(0), \tag{35.17}$$

and thus, remembering that the left hand side of (35.16) may be identified with $P(\boldsymbol{k}, \boldsymbol{k}')$, we obtain

$$\left.\begin{aligned} P(\boldsymbol{k}, \boldsymbol{k}')\,dE_{\boldsymbol{k}'} &= \frac{\pi}{\varrho\,c_s N\Omega}\left(\frac{2\zeta}{3}\right)^2 q\,\times \\ &\times\{n_{\boldsymbol{q}}\,\delta(E_{\boldsymbol{k}'} - E_{\boldsymbol{k}} - \hbar\,\omega_{\boldsymbol{q}}) + (n_{\boldsymbol{q}}+1)\,\delta(E_{\boldsymbol{k}'} - E_{\boldsymbol{k}} + \hbar\,\omega_{\boldsymbol{q}})\}\,dE_{\boldsymbol{k}'}, \end{aligned}\right\} \tag{35.18}$$

where $\delta(x)$ denotes the DIRAC $\delta$-function.

We shall now introduce an assumption, the validity of which has been questioned[1]: that $n_{\boldsymbol{q}}$ may be allowed to take the value appropriate to conditions of thermal equilibrium between the lattice waves. Thus we write

$$n_{\boldsymbol{q}} = \frac{1}{e^{\hbar\,\omega_{\boldsymbol{q}}/\mathrm{k}T} - 1}, \tag{35.19}$$

and, if we use the variable $x = (\hbar\,\omega_{\boldsymbol{q}}/\mathrm{k}T)$, and the previously introduced $\eta = (E - \zeta)/\mathrm{k}T$, we have finally

$$P(\boldsymbol{k}, \boldsymbol{k}')\,dE_{\boldsymbol{k}'} = \frac{\pi}{\hbar\,\varrho\,c_s N\Omega}\left(\frac{2\zeta}{3}\right)^2 \mathrm{k}T\,x\left\{\frac{\delta(\eta' - \eta - x)}{e^x - 1} + \frac{\delta(\eta' - \eta + x)}{1 - e^{-x}}\right\}d\eta'. \tag{35.20}$$

**36. Evaluation of the integrals $d_{rs}$.** From Eqs. (33.2) and (33.25) we find

$$\frac{8\pi^3\,\mathrm{k}T}{N\Omega}\,d_{rs} = \int k_z\,\eta^r\,d\boldsymbol{k}\int P(\boldsymbol{k}, \boldsymbol{k}')\,f_0(1 - f'_0)\,(k_z\,\eta^s - k'_z\,\eta'^s)\,d\boldsymbol{k}'. \tag{36.1}$$

Since $P(\boldsymbol{k}, \boldsymbol{k}')$ depends only on $k$, $k'$ and $q$ we can use the expression given in (7.5) for $d\boldsymbol{k}'$, and Eq. (7.4) for $k'_z/k_z$; the integration over $\cos\varphi$ then vanishes and we obtain

$$\left.\begin{aligned} \frac{8\pi^3\,\mathrm{k}T}{N\Omega}\,d_{rs} &= 2\pi\int k_z^2\,\eta^r\,d\boldsymbol{k}\iint P(\eta, \eta', x)\,f_0(1 - f'_0)\,\times \\ &\times\left(\eta^s - \frac{k'}{k}\,\eta'^s\cos\vartheta\right)k'^2\,dk'\sin\vartheta\,d\vartheta. \end{aligned}\right\} \tag{36.2}$$

---

[1] Cf. R. PEIERLS: Ann. Phys., Lpz. **4**, 121; **5**, 244 (1930); **12**, 154 (1932). Also P. G. KLEMENS: Proc. Phys. Soc. Lond. A **64**, 1030 (1951).

From the relation $\boldsymbol{q} = \boldsymbol{k}' - \boldsymbol{k}$ we get

$$\left.\begin{aligned} \frac{k'}{k}\cos\vartheta &= \frac{1}{2}\left(1 + \frac{k'^2}{k^2} - \frac{q^2}{k^2}\right) \\ &= \frac{1}{E}\left\{E + \frac{\mathsf{k}T}{2}(\eta' - \eta) - \frac{(\mathsf{k}T)^2 x^2}{4mc_s^2}\right\}, \end{aligned}\right\} \tag{36.3}$$

and differentiating this with $\eta'$ constant we obtain, on multiplying by $kk'$,

$$k'^2 \sin\vartheta\, d\vartheta = \frac{kk'}{E}\frac{(\mathsf{k}T)^2}{2mc_s^2}\, x\, dx. \tag{36.4}$$

Now on evaluation the double integral in (36.2) becomes a function of $\eta$ only, i.e. of $k$ only, and thus in the first integral we can replace $k_z^2$ by $\frac{1}{3}k^2$ and $d\boldsymbol{k}$ by $4\pi k^2\, dk$. Hence, with (36.3) and (36.4), (36.2) becomes

$$\left.\begin{aligned} d_{rs} = \frac{\mathsf{k}T N\Omega}{6\pi\hbar^2 c_s^2}\int \frac{k^5}{E^2}\eta^r\, dk \iint P f_0 (1 - f_0') \times \\ \times \left\{E\eta^s - \eta'^s\left(E + \frac{\mathsf{k}T}{2}(\eta' - \eta) - \frac{(\mathsf{k}T)^2 x^2}{4mc_s^2}\right)\right\} dE'\, x\, dx. \end{aligned}\right\} \tag{36.5}$$

Substituting for $P\,dE'$ from (35.20) we find

$$\left.\begin{aligned} d_{rs} = \frac{(\mathsf{k}T)^2}{6\hbar^3 c_s^4 \varrho}\left(\frac{2\zeta}{3}\right)^2 \int \frac{k^5}{E^2}\eta^r\, dk \iint \frac{\left\{E\eta^s - \eta'^s\left(E + \frac{\mathsf{k}T}{2}(\eta' - \eta) - \frac{(\mathsf{k}T)^2 x^2}{4mc_s^2}\right)\right\}}{(1 + e^{\eta})(1 + e^{-\eta'})} \times \\ \times \left\{\frac{\delta(\eta' - \eta - x)}{e^x - 1} + \frac{\delta(\eta' - \eta + x)}{1 - e^{-x}}\right\} d\eta'\, x^2\, dx. \end{aligned}\right\} \tag{36.6}$$

The double integral alone may be written, after integration with respect to $\eta'$,

$$\left.\begin{aligned} \int\limits_0^{\Theta/T} \frac{\left\{E\eta^s - (\eta + x)^s\left(E + \frac{\mathsf{k}T}{2}x - \frac{(\mathsf{k}T)^2 x^2}{4mc_s^2}\right)\right\}}{(1 + e^{\eta})(1 + e^{-\eta - x})(e^x - 1)} x^2\, dx + \\ + \int\limits_0^{\Theta/T} \frac{\left\{E\eta^s - (\eta - x)^s\left(E - \frac{\mathsf{k}T}{2}x - \frac{(\mathsf{k}T)\, x^2}{4mc_s^2}\right)\right\}}{(1 + e^{\eta})(1 + e^{-\eta + x})(1 - e^{-x})} x^2\, dx, \end{aligned}\right\} \tag{36.7}$$

where $\Theta = \hbar\,\omega_{\max}/\mathsf{k}$, and $\omega_{\max}$ is the highest angular frequency of the longitudinal waves under consideration. If $x$ is changed to $-x$ in the second integral, the integrand becomes identical with that of the first integral. Changing the order of the limits $-\Theta/T$ and 0, and at the same time replacing $e^x - 1$ by $|1 - e^x|$, the two integrals can be combined into a single integral between the limits $-\Theta/T$ and $\Theta/T$.

It is convenient to introduce constants $\alpha$ and $\gamma$ defined by

$$\alpha = \frac{\mathsf{k}T}{\zeta}, \quad \text{and} \quad \gamma^2 = \frac{(\mathsf{k}T)^2}{2mc_s^2\zeta}. \tag{36.8}$$

After some straightforward simplification we obtain the following expression for $d_{rs}$:

$$d_{rs} = \frac{2m^3\zeta(\mathsf{k}T)^3}{3\hbar^9 c_s^4\varrho}\left(\frac{2\zeta}{3}\right)^2 \int\limits_{-\infty}^{\infty}\int\limits_{-\Theta/T}^{\Theta/T} \frac{\left\{(1 + \alpha\eta)\eta^s - (\eta + x)^s\left(1 + \alpha\eta + \frac{\alpha}{2}x - \frac{\gamma^2}{2}x^2\right)\right\}}{(1 + e^{-\eta})(1 + e^{\eta + x})|1 - e^{-x}|}\eta^r x^2\, d\eta\, dx. \tag{36.9}$$

The lower limit of the integration over $d\eta$ is strictly $-\zeta/\mathrm{k}T$, but for monovalent metals at low temperatures no appreciable approximation is involved in replacing this by $-\infty$. The $\Theta$ which occurs in the limits of the integration over $dx$ is *not* the DEBYE characteristic temperature $\Theta_D$ used in the high temperature theory of conductivity. It is given by

$$\Theta = \frac{\hbar c_s}{\mathrm{k}} \left(\frac{6\pi^2}{\Omega}\right)^{\frac{1}{3}}, \tag{36.10}$$

where $c_s$ is the velocity of the longitudinal waves.

To complete the integration of (36.9) we evaluate the following integrals

$$I_{nm} = \int_{-\infty}^{\infty} d\eta \int_{-\Theta/T}^{\Theta/T} \frac{\eta^n x^m \, dx}{(1+e^{-\eta})(1+e^{\eta+x})\,|1-e^{-x}|}. \tag{36.11}$$

We make use of the following identity

$$\frac{1}{(1+e^{-\eta})(1+e^{\eta+x})} = \frac{1}{e^x - 1}\left\{\frac{1}{1+e^{\eta}} - \frac{1}{1+e^{\eta+x}}\right\}. \tag{36.12}$$

Let

$$g_n(x) = \int_{-\infty}^{\infty} \left(\frac{1}{1+e^{\eta}} - \frac{1}{1+e^{\eta+x}}\right) \eta^n \, d\eta, \tag{36.13}$$

and therefore

$$I_{nm} = \int_{-\Theta/T}^{\Theta/T} \frac{x^m g_n(x)\, dx}{(e^x-1)\,|1-e^{-x}|}. \tag{36.14}$$

Integrating (36.13) by parts, and thereafter writing $\eta - x$ for $\eta$ in the second (now convergent) integral we find

$$g_n(x) = \frac{1}{(n+1)} \int_{-\infty}^{\infty} \frac{\eta^{n+1} - (\eta - x)^{n+1}}{(1-e^{\eta})(1+e^{-\eta})}\, d\eta. \tag{36.15}$$

If

$$c_\nu = \int_{-\infty}^{\infty} \frac{\eta^\nu \, d\eta}{(1+e^{\eta})(1+e^{-\eta})}$$

we have the following well-known results:

$$c_0 = 1, \qquad c_2 = \frac{\pi^2}{3}, \qquad c_4 = \frac{7\pi^4}{15}, \qquad c_6 = \frac{31\pi^6}{21},$$

and all $c_\nu$ with odd $\nu$ are zero.

Thus without difficulty we find

$$\left.\begin{aligned} g_0(x) &= x, \\ g_1(x) &= -\frac{1}{2} x^2, \\ g_2(x) &= \frac{1}{3}(\pi^2 x + x^3), \\ g_3(x) &= -\frac{1}{4}(2\pi^2 x^2 + x^4), \\ g_4(x) &= \frac{1}{5}\left(\frac{7\pi^4}{3} x + \frac{10}{3}\pi^2 x^3 + x^5\right). \end{aligned}\right\} \tag{36.16}$$

If

$$J_p\left(\frac{\Theta}{T}\right)=\int_0^{\Theta/T}\frac{x^p\,dx}{(e^x-1)\,(1-e^{-x})}\,, \tag{36.17}$$

it follows at once that

$$\int_{-\Theta/T}^{\Theta/T}\frac{x^p\,dx}{(e^x-1)\,|1-e^{-x}|}=\left(1+(-1)^{p+1}\right)J_p. \tag{36.18}$$

Hence from (36.14) and (36.16) we find

$$\left.\begin{aligned}
I_{0m}&=\left(1+(-1)^m\right)J_{m+1},\\
I_{1m}&=-\frac{1}{2}\left(1-(-1)^m\right)J_{m+2},\\
I_{2m}&=\left(1+(-1)^m\right)\left\{\frac{\pi^2}{3}J_{m+1}+\frac{1}{3}J_{m+3}\right\},\\
I_{3m}&=-\left(1-(-1)^m\right)\left\{\frac{\pi^2}{2}J_{m+2}+\frac{1}{4}J_{m+4}\right\},\\
I_{4m}&=\left(1+(-1)^m\right)\left\{\frac{7\pi^4}{15}J_{m+1}+\frac{2\pi^2}{3}J_{m+3}+\frac{1}{5}J_{m+5}\right\}.
\end{aligned}\right\} \tag{36.19}$$

The values of the integrals $d_{rs}$ can now be written down without difficulty. For brevity we denote the constant before the integral of (36.9) by $B$, thus

$$B=\frac{2m^3\zeta(\mathrm{k}T)^3}{3\hbar^q c_s^4\varrho}\left(\frac{2\zeta}{3}\right)^2, \tag{36.20}$$

and therefore,

$$\left.\begin{aligned}
d_{00}&=\gamma^2 B J_5,\\
d_{01}&=\frac{1}{2}\alpha B J_5,\\
d_{11}&=\left(1+\frac{\pi^2}{3}\gamma^2\right)B J_5-\frac{1}{6}\gamma^2 B J_7.
\end{aligned}\right\} \tag{36.21}$$

The integrals $\alpha_r$ and $\beta_r$ may readily be evaluated. They are as follows:

$$\alpha_r=g\left\{c_r+\tfrac{3}{2}\alpha\,c_{r+1}+\tfrac{3}{8}\alpha^2 c_{r+2}-\tfrac{1}{16}\alpha^3 c_{r+3}\ldots\right\}, \tag{36.22}$$

where

$$g=\frac{2\pi}{3}\left(\frac{2m}{\hbar^2}\right)^{\frac{3}{2}}\zeta^{\frac{3}{2}}=\frac{4\pi^3 m\,n}{\hbar^2}, \tag{36.23}$$

and $n$ is the number of electrons per unit volume; and

$$\beta_r=\mathrm{k}T\,\alpha_{r+1}. \tag{36.24}$$

Neglecting terms containing second powers of $\alpha$ we find

$$\left.\begin{aligned}
&\alpha_0=g,\quad \alpha_1=\frac{\alpha}{2}\pi^2 g,\quad \alpha_2=\frac{\pi^2}{3}g,\\
&\beta_0=\mathrm{k}T\alpha\frac{\pi^2}{2}g,\quad \beta_1=\mathrm{k}T\frac{\pi^2}{3}g.
\end{aligned}\right\} \tag{36.25}$$

The integrals $d_{rs}$ contain the parameter $\gamma$ which like $\alpha$ vanishes as $T$ tends to zero. For monovalent metals it may readily be seen, using (36.10), that

$$\frac{\alpha}{\gamma}=2^{-\frac{1}{3}}\left(\frac{\mathrm{k}\Theta}{\zeta}\right), \tag{36.26}$$

from which we see that $\gamma = 2^{\frac{1}{3}}\, T/\Theta$. Although $\Theta$ is not the DEBYE temperature it is certainly of that magnitude, and therefore for temperatures in the liquid helium range $\gamma$ is generally in the neighbourhood of $10^{-2}$. Hence, for temperatures for which the theory is valid, we are justified in neglecting terms like $\frac{\pi^2}{3}\gamma^2$ in comparison with unity. Thus we may write

$$d_{00} = \gamma^2 B\, J_5, \quad d_{01} = \tfrac{1}{2}\alpha\, B\, J_5, \quad d_{11} = B\, J_5. \tag{36.27}$$

Using (36.24) and (36.26), we obtain from (34.6) and (34.7) the coefficients which determine the functions $\varphi_1$ and $\varphi_2$ according to (34.1) and (34.2):

$$\left.\begin{aligned} a_0 &= \frac{g}{\gamma^2 B\, J_5}, & a_1 &= -\frac{\alpha}{2}\, a_0, \\ b_0 &= \frac{\pi^2}{3}\, \mathrm{k} T\, \alpha\, a_0, & b_1 &= \frac{\pi^2}{3}\, \mathrm{k} T\, \gamma^2\, a_0. \end{aligned}\right\} \tag{36.28}$$

In addition to neglecting $\gamma^2$ compared with terms of order unity we also neglect all terms containing $(\alpha/\gamma)^2$ as a factor. This factor is independent of temperature but, for monovalent metals, is of order $10^{-4}$. With these approximations we obtain the following expressions for the kinetic coefficients:

$$\left.\begin{aligned} L_{11} &= \frac{T\hbar^2}{4\pi^3 m^2}\, g\, a_0, \\ L_{12} &= -\frac{T\hbar^2}{4\pi^3 m^2}\left(\frac{\pi^2}{3}\,\frac{(\mathrm{k}T)^2}{\zeta}\right) g\, a_0, \\ L_{22} &= \frac{T\hbar^2}{4\pi^3 m^2}\left(\frac{\pi^2}{3}\,\mathrm{k}T\,\gamma\right)^2 g\, a_0. \end{aligned}\right\} \tag{36.29}$$

The conductivities and the thermo-electric power may be obtained immediately by substituting (36.29) into (26.16), (26.19) and (26.24). First, however, it is of interest to consider the nature of the distribution function in two cases: (i) of an isothermal electrical current, and (ii) of a thermal current with no particle current. For case (i) we find, on substituting (34.1) into (33.3) and putting $dT/dz = 0$

$$f_1 = -\frac{eF}{\hbar}\, a_0\, \frac{\partial f_0}{\partial k_z}. \tag{36.30}$$

Thus we may write

$$f = f_0(k_x, k_y, k_z + \delta k_z), \tag{36.31}$$

where $\delta k_z = -\frac{eF}{\hbar}\, a_0$.

Since $eF/\hbar = \dot{k}_z$ we may regard $a_0$ as the relaxation time for this displacement. The distribution function for electrical conduction at low temperatures is similar, therefore, to that for high temperatures as indicated in Fig. 4; the time of relaxation, however, has quite a different temperature dependence. For case (ii) we put $J_z = 0$ in (26.6) and obtain therefore

$$\frac{d\mu}{dz} = \frac{L_{12}}{T L_{11}}\,\frac{dT}{dz}, \tag{36.32}$$

which on substitution into (33.3) gives

$$f_1 = \frac{1}{\hbar}\,\frac{\partial f_0}{\partial k_z}\left\{\frac{L_{12}}{L_{11}}\,(a_0 + a_1\eta) + b_0 + b_1\eta\right\}\frac{1}{T}\,\frac{dT}{dz}. \tag{36.33}$$

Using (36.28) and (36.29) the two terms $b_0$ and $L_{12}\, a_0/L_{11}$ are seen to cancel exactly and we find readily that the distribution function may be written in the form

(36.31) where

$$\delta k_z = (k - k_0) \frac{l}{T} \frac{dT}{dz}, \tag{36.34}$$

and

$$l = \frac{\hbar k_0}{m} \left( \frac{\pi^2}{3} \gamma^2 a_0 \right). \tag{36.35}$$

We have, therefore, again the same kind of displacement of the FERMI distribution as was described in Sect. 32. $\hbar k_0/m$ is the velocity of the electrons at the FERMI limit, and therefore $\frac{\pi^2}{3} \gamma^2 a_0$ may be regarded as the time of relaxation for this displacement.

**37. Electrical conductivity at low temperatures.** We now write down the complete expression for the electrical conductivity at low temperatures. According to (26.16) and (36.29)

$$\sigma = \frac{e^2 \hbar^2}{4 \pi^3 m^2} g\, a_0, \tag{37.1}$$

and on substituting the value of $g$ from (36.22) we get

$$\sigma = \frac{e^2}{m} \left( \frac{k_0^3}{3\pi^2} \right) a_0 = \frac{e^2 n\, a_0}{m}, \tag{37.2}$$

where $n$ is the number of electrons per unit volume, in agreement with the interpretation of $a_0$ as the time of relaxation for electrical conduction. $a_0$ itself is given by (36.28) and (36.20) and may be written

$$a_0 = \frac{8\pi \sqrt{2}\, \hbar^4 \zeta^{\frac{3}{2}} c_s^6 \varrho\, m^{\frac{1}{2}}}{(2\zeta/3)^2 (\mathrm{k} T)^5 J_5}, \tag{37.3}$$

If we use (36.10) and the relation $(\Omega k_0^3/3\pi^2) = 1$ and remember that $\Omega \varrho = M$ the mass of the atom, we find

$$\frac{1}{a_0} = \frac{\pi}{3} \frac{m}{\hbar} \left( \frac{T}{M \mathrm{k} \Theta^2} \right) \zeta^2 \left\{ \left( \frac{T}{\Theta} \right)^4 4 J_5 \right\}. \tag{37.4}$$

When $T \gg \Theta$ it may easily be seen from (36.17) that the factor in curly brackets is equal to unity. Thus comparing with (12.3) we see that $a_0$ may be formally identified with $\tau$ at high temperatures if $\Theta$ is assumed equal to $\Theta_D$ and $C$ assumed equal to $\zeta$. There is no justification for either of these assumptions. If the angular dependence of the scattering is neglected, i.e. $G(w)$ put equal to unity in (12.4), and the integration taken only to $w_0$, i.e. neglecting $U$-processes, then $C_D = \zeta$, as may easily be verified. In fact, as we saw in Sect. 12, for sodium $C_D = 0.76\,\zeta$ is the result required to give agreement with the observed resistivity; a result which is also compatible with direct calculations by the WIGNER-SEITZ method. For real crystals the ratio of $\Theta$ to $\Theta_D$ depends in a complicated way on the form of the frequency spectrum, in many cases $\Theta_D$ is less than $\Theta$; even for an isotropic elastic solid they are not equal.

By using the results (12.3) and (37.4) it is possible to compare the electrical conductivity at a high temperature $T_1 > \Theta_D$ with the conductivity at a very low temperature $T_2 \ll \Theta_D$. We find easily

$$\frac{\sigma(T_1)}{\sigma(T_2)} = 497.6 \left\{ \left( \frac{\zeta}{C_D} \right)^2 \left( \frac{\Theta_D}{\Theta} \right)^6 \right\} \frac{T_2}{T_1} \left( \frac{T_2}{\Theta_D} \right)^4, \tag{37.5}$$

since $J_5(\infty) = 124.4$. The factor in curly brackets is frequently omitted, its value being assumed to be unity. As we have seen, this is an unwarranted assumption, though, in fact, in the special case of gold it appears to be true. It should be remembered that the factor $\zeta/C_D$ appearing in (37.5) arises from our assumption that the method of the deformation potential, or the equivalent BARDEEN method, is satisfactory at low temperatures. More generally this factor should be replaced by the ratio of two coupling constants, one appropriate to low and one to high temperature scattering.

**38. Thermal conductivity at low temperatures**[1]. Substituting Eqs. (36.29) into (26.9) and neglecting $\alpha^2$ by comparison with $\gamma^2$, we find for the thermal conductivity

$$\varkappa = \frac{\hbar^2}{4\pi^3 m^2 T}\left(\frac{\pi^2}{3}\,\mathrm{k}T\gamma\right)^2 g\,a_0 \tag{38.1}$$

and thus for the WIEDEMANN-FRANZ ratio, using (37.1),

$$\frac{\varkappa}{\sigma T} = \frac{\pi^2 \mathrm{k}^2}{3e^2}\left(\frac{\pi^2}{3}\gamma^2\right) \tag{38.2}$$

This ratio, which varies as $T^2$, differs from the constant value for high temperatures by the factor $\frac{\pi^2}{3}\gamma^2$ which, as might have been anticipated, is the ratio of the two relaxation times, appropriate to low temperatures, one referring to the thermal current and the other to the electrical current.

We can obtain the ratio of the thermal conductivity at a high temperature $T_1$, when it is constant, to the thermal conductivity at a low (variable) temperature $T$ by using (37.5) and (38.2). We find

$$\left.\begin{aligned}\frac{\varkappa(T_1)}{\varkappa(T)} &= \frac{\sigma(T_1)}{\sigma(T)}\,\frac{T_1}{T}\left(\frac{3}{\pi^2\gamma^2}\right)\\ &= \left\{\left(\frac{\zeta}{C_D}\right)^2\left(\frac{\Theta_D}{\Theta}\right)^4\right\}\frac{95\cdot 3T^2}{\Theta_D^2},\end{aligned}\right\} \tag{38.3}$$

where we have made use of the relation $\gamma = 2^{\frac{1}{3}}T/\Theta$. Comparisons between experimentally determined thermal conductivities and a formula similar to (38.3) have frequently been made[2]. In the formula in question the factor of (38.3) in curly brackets is assumed to be equal to unity and a factor $(n/n_a)^{\frac{2}{3}}$ is added. $(n/n_a)$ is intended to denote the number of conduction electrons per atom, and it arises from the $\zeta$ in the denominator of $\gamma^2$. To get agreement with experimental values by such a method $n/n_a$ has to be assumed less than 1/10 which is quite unrealistic for monovalent metals, particularly for sodium and copper where recent calculations have shown that the value of $\zeta$ in the metal is remarkably close to the value obtained assuming completely free electrons.

The $\Theta$ which occurs in (38.3) is determined by the velocity of the longitudinal waves alone, according to (36.10). To estimate its value in relation to $\Theta_D$ is not easy for real metals, since the monovalent metals such as sodium or copper are very anisotropic elastically and, therefore, an averaging process, over different directions relative to the crystal axes, is necessary. BLACKMAN[3] has carried out such a calculation and his values, denoted by $\Theta_L$, for sodium and copper are shown in the first column of the accompanying table.

The third column should be multiplied by $(\zeta/C_D)^2$, according to (38.3), to correspond to the observed figures of the fourth column. For sodium $(\zeta/C_D)^2$ is 1.7 according to the estimate made in Sect. 12; this would give a calculated

[1] For an alternative method of calculating the thermal conductivity at low temperatures the reader is referred to the paper by R. E. B. MAKINSON: Proc. Cambridge Phil. Soc. **34**, 474 (1938).
[2] See, for example J. K. HULM: Proc. Roy. Soc. Lond., Ser. A **204**, 98 (1950).
[3] M. BLACKMAN: Proc. Phys. Soc. Lond. A **64**, 681 (1951).

value of $6.6 \times 10^{-4}$ for the coefficient of $T^2$ in (38.3) as compared with the observed value $3.8 \times 10^{-4}$. Using BARDEEN's estimate of $C_D/\zeta$, obtained from the observed resistivity of copper, we find $(\zeta/C_D)^2 = 0.57$, and thus the corresponding comparison for copper is $0.61 \times 10^{-4}$ (calculated) and $0.23 \times 10^{-4}$ (observed).

The difference between the two $\Theta$-values would also, of course, affect the ratio of the electrical conductivities at high and low temperatures according to (37.5). Unfortunately, in many cases, it is difficult to determine unambiguously the purely thermal part of the resistivity at very low temperatures[1]. Apart from the resistance minimum, which is hitherto unexplained, it frequently happens that the temperature dependence of the resistivity is not precisely as $T^5$, and, naturally, this at once destroys the possibility of quantitative comparison with the foregoing theory.

*Table 8.*

| Metal | $\Theta_L$ | $\Theta_D$ | $\left(\frac{\Theta_D}{\Theta_L}\right)^4 \frac{95 \cdot 3}{\Theta_D^5}$ | $\left(\frac{1}{T^2} \frac{\varkappa(\infty)}{\varkappa(T)}\right)_{\text{Obs.}}$ |
|---|---|---|---|---|
| Na | 282 | 160 | $3.86 \times 10^{-4}$ | $3.8 \times 10^{-4}$ |
| Cu | 540 | 310 | $1.08 \times 10^{-4}$ | $0.23 \times 10^{-4}$ |

**39. The thermo-electric power at low temperatures.** Introducing the expressions (36.29) for the kinetic coefficients into Eq. (26.24) we find for the thermo-electric power

$$S = \frac{\pi^2 \mathrm{k}^2 T}{3 e \zeta}. \tag{39.1}$$

It is difficult to assess the degree of applicability of this formula to real metals. At normal temperatures the thermo-electric power may be strongly affected by variations of $\nabla_{\boldsymbol{k}} E$ over the FERMI surface, as was shown in Sect. 29, even though deviations from the spherical form are not excessive. In the theory for low temperatures the constancy of $\nabla_{\boldsymbol{k}} E$ over the FERMI surface is assumed from the start, which makes it difficult to estimate the effect of relaxing this assumption.

**40. Effect of impurities at low temperatures.** We have to consider two perturbations acting simultaneously and giving rise to transitions between the stationary states of the total system: electrons plus lattice. One of these is given by (35.2) and contains the amplitude of the lattice waves. This leads to the transition probability (35.18) which is finite only when the lattice changes its state of motion, absorbing or emitting a phonon. The other perturbation does not contain the amplitude of the lattice waves and leads to a transition probability which is finite only when the lattice motion remains in the same stationary state. This perturbation adds to (35.18) the following term

$$\frac{2\pi}{\hbar} |U^{(i)}_{\boldsymbol{k}'\boldsymbol{k}}|^2 \, \delta(E' - E) \, dE', \tag{40.1}$$

where $U^{(i)}_{\boldsymbol{k}'\boldsymbol{k}}$ is given by (16.7). On substituting the complete expression for $P(\boldsymbol{k}', \boldsymbol{k})$ into (4.9) the integral breaks up into two parts. One part is the contribution to $(\partial f/\partial t)_{\text{coll}}$ from the thermal scattering and is unaffected by the additional perturbation; it may be denoted by $-\Lambda\chi$ in accordance with (33.2). The second part arises from (40.1); here the scattering is purely elastic, and may occur through large angles according to (16.7). For any displacement of the electronic distribution in $k$-space given by $(f - f_0)$ equals $k_z$ times some function of $E$ this second integral reduces to the form (7.12). Thus we can write

$$\left(\frac{\partial f}{\partial t}\right)_{\text{coll}} = \left(\frac{1}{\tau_i} \frac{d f_0}{dE} - \Lambda\right) \chi(\boldsymbol{k}), \tag{40.2}$$

[1] Cf. D. K. C. MACDONALD and K. MENDELSSOHN: Proc. Roy. Soc. Lond., Ser. A **202**, 103 (1950).

where $\tau_i$ denotes a relaxation time resulting from impurity scattering alone. Hence all results, previously obtained for the conductivity of pure metals at low temperatures, can be taken over into the theory which includes the effects of impurity scattering by replacing the operator $\Lambda$ by the operator $\Lambda - \frac{1}{\tau_i}\frac{df_0}{dE}$. We have, therefore, to replace the integrals $d_{rs}$ by $d_{rs} + e_{rs}$, where

$$e_{rs} = -\int k_z^2 \eta^{r+s} \frac{1}{\tau_i}\frac{df_0}{dE}\, d\boldsymbol{k}. \tag{40.3}$$

Expanding $1/\tau_i(E)$ as a power series in $(E-\zeta)$ we find

$$\frac{1}{\tau_i(E)} = \tau_i^{-1} + \mathrm{k}T\left(\frac{d\tau_i^{-1}}{dE}\right)_\zeta \eta + \frac{(\mathrm{k}T)^2}{2}\left(\frac{d^2\tau_i^{-1}}{dE^2}\right)_\zeta \eta^2 \ldots, \tag{40.4}$$

and, therefore, by (33.24) we have

$$e_{rs} = \tau_i^{-1}\alpha_{r+s} + \mathrm{k}T\left(\frac{d\tau_i^{-1}}{dE}\right)_\zeta \alpha_{r+s+1} + \frac{(\mathrm{k}T)^2}{2}\left(\frac{d^2\tau_i^{-1}}{dE^2}\right)_\zeta \alpha_{r+s+2} \cdots. \tag{40.5}$$

Thus, replacing $d_{rs}$ by $d_{rs} + e_{rs}$ in (34.6) and (34.7), and substituting in (34.3), (34.4) and (34.5) we find, after some calculation, the following values for the kinetic coefficients

$$\frac{4\pi^3 m^2}{\hbar^2 T} L_{11} = \frac{\alpha_0^2}{d_{00} + e_{00}} = \frac{g}{\left(\frac{1}{a_0} + \frac{1}{\tau_i}\right)}, \tag{40.6}$$

$$\left.\begin{aligned} \frac{4\pi^3 m^2}{\hbar^2 T} L_{12} &= -\frac{\alpha_0\{\beta_1(d_{10}+e_{10}) - \beta_0(d_{11}+e_{11})\} - \alpha_1\beta_1(d_{00}+e_{00})}{(d_{11}+e_{11})(d_{00}+e_{00})} \\ &= -\frac{\pi^2 \mathrm{k}^2 T^2}{3\zeta} g \frac{\left\{\frac{3}{\pi^2\gamma^2 a_0} + \frac{1}{\tau_i}\left(\frac{3}{2} + \left(\frac{E}{\tau_i}\frac{d\tau_i}{dE}\right)_\zeta\right)\right\}}{\left(\frac{1}{a_0} + \frac{1}{\tau_i}\right)\left(\frac{3}{\pi^2\gamma^2 a_0} + \frac{1}{\tau_i}\right)}, \end{aligned}\right\} \tag{40.7}$$

$$\frac{4\pi^3 m^2}{\hbar^2 T} L_{22} = \frac{\beta_1^2}{d_{11} + e_{11}} = \frac{\frac{\pi^2}{3}(\mathrm{k}T)^2 g}{\left(\frac{1}{\tau_i} + \frac{3}{\pi^2\gamma^2 a_0}\right)}. \tag{40.8}$$

These values of the kinetic coefficients have been obtained by making the same approximations as were used to obtain the formulae (36.29) which are applicable to pure metals. It will be recalled that $a_0$ is the relaxation time for a displacement arising from an electrical current alone, and $\frac{\pi^2}{3}\gamma^2 a_0$ is the relaxation time for a displacement arising from a thermal current alone. As might have been anticipated $L_{11}$ contains the sum of the reciprocals of $a_0$ and $\tau_i$, $L_{22}$ the sum of the reciprocals of $\frac{\pi^2\gamma^2}{3} a_0$ and $\tau_i$, and $L_{12}$, which arises from the interaction between the electrical and thermal currents, the product of these two sums.

Expressions for the conductivities and the thermo-electric power follow immediately.

*α) Electrical conductivity.* From (26.16) and (40.6) we find

$$\frac{1}{\sigma} = \frac{m}{e^2 n}\left(\frac{1}{a_0} + \frac{1}{\tau_i}\right), \tag{40.9}$$

where the number of electrons per unit volume $n$ has been introduced in the same way as in (37.2). Hence, to the approximation to which we have evaluated $L_{11}$, MATTHIESSEN's rule is applicable also at low temperatures. Had we retained the terms of order $\alpha^2$, a term representing a deviation from MATTHIESSEN'S rule would have appeared in (40.9). Although the complete expression for $1/\sigma$ can easily by obtained from the formulae already given, we shall not discuss this question further since it seems doubtful whether the effect is at all closely related to the observed deviations from MATTHIESSEN'S rule. The reader is referred to the papers of KOHLER[1] and of SONDHEIMER[2] for a full discussion of this matter.

*β) Thermal conductivity.* We obtain a formula for the thermal conductivity at low temperatures by substituting (40.8) into (26.19):

$$\frac{1}{\varkappa} = \frac{3m}{\pi^2 \mathrm{k}^2 n} \frac{1}{T} \left( \frac{1}{\tau_i} + \frac{3}{\pi^2 \gamma^2 a_0} \right). \tag{40.10}$$

Hence we see that the thermal resistivity is the sum of two terms, one due to the thermal scattering of the electrons; the other due to impurity scattering. Since $\gamma^2 a_0 \propto T^{-3}$, according to (37.3) and (36.8), and since $\tau_i$ is independent of temperature, the thermal resistivity may be expressed as follows:

$$\frac{1}{\varkappa} = A T^2 + \frac{B}{T}, \tag{40.11}$$

where $A$ and $B$ may be obtained by comparison with (40.10). This formula is only valid in the limit of low temperatures.

The theory of the preceding sections can be extended to higher temperatures if $U$-processes are neglected, and no account is taken of the angular dependence of the scattering matrix, i.e., the function $G(w)$ of Sect. 10 is regarded as constant. It is then found that the thermal conductivity at first falls from its constant value at high temperatures as the temperature is lowered. It passes through a minimum as the temperature is further decreased; rises to a maximum and finally falls linearly to zero as the temperature tends to zero. The linear dependence on temperature in the neighbourhood of the absolute zero is due to the effect of impurities as indicated by the term $T/B$ in (40.11).

Experimental determinations of the thermal conductivity do not show the presence of the minimum just referred to, and it seems likely that this result of the theory is spurious, and arises from the two invalid assumptions already stated. For a fuller discussion of this question the reader is referred to a recent paper by ZIMAN[3] which deals with the electrical and thermal conductivity of monovalent metals over the full temperature range.

*γ) Thermo-electric power.* Substituting (40.7) and (40.6) into (26.24) we find the following formula for the thermo-electric power

$$S = \frac{\pi^2 \mathrm{k}^2 T}{3 e \zeta} \frac{\left\{ 1 + \frac{\pi^2 \gamma^2 a_0}{3 \tau_i} \left( \frac{3}{2} + \left( \frac{E}{\tau_i} \frac{d\tau_i}{dE} \right)_\zeta \right) \right\}}{1 + \frac{\pi^2}{3} \frac{\gamma^2 a_0}{\tau_i}}. \tag{40.12}$$

At very low temperatures where $\frac{\pi^2}{3} \gamma^2 a_0 \gg \tau_i$ (40.12) becomes

$$S = \frac{\pi^2 \mathrm{k}^2 T}{3 e \zeta} \left( \frac{3}{2} + \left( \frac{E}{\tau_i} \frac{d\tau_i}{dE} \right)_\zeta \right). \tag{40.13}$$

When the concentration of impurity is very small it is possible for $\tau_i \gg \frac{\pi^2}{3} \gamma^2 a_0$, although the temperature is still low enough to ensure the applicability of the theory of Sect. 35. In this case (40.12) reduces to (39.1).

[1] M. KOHLER: cf. reference p. 283.
[2] E. H. SONDHEIMER: Proc. Roy. Soc. Lond., Ser. A **203**, 75 (1950).
[3] J. M. ZIMAN: Proc. Roy. Soc. Lond., Ser. A **226**, 436 (1954).

## VII. Conduction in the presence of a magnetic field.

**41. Galvomagnetic effects in metals.** First we consider the nature of the displacement of the FERMI distribution under conditions such as arise in an experimental determination of the change of resistance in a magnetic field, or of the HALL coefficient. Eq. (3.22) shows how the wave vector of an electronic state changes under the combined action of an electric field $\boldsymbol{F}$ and a magnetic field $\boldsymbol{H}$. If $\tau$ denotes a time of relaxation, the displacement of the FERMI distribution will be such as would be produced by each occupied point of $k$-space being displaced through a distance equal to

$$\frac{e\tau}{\hbar}\boldsymbol{F} + \frac{e\tau}{c\hbar}\boldsymbol{v_k}\times\boldsymbol{H}. \tag{41.1}$$

The first term of this displacement is the same for all occupied points of $k$-space, and alone would lead to the distribution like that shown in Fig. 4. The second term is a displacement along a surface of constant energy. In most experimental conditions this second term is enormously greater than the first. Thus the force $e\boldsymbol{F}$ is equal to $1.602\times10^{-12}F$ dynes if $F$ is in volts per cm., and $evH/c$ is equal to $1.602\times10^{-20}vH$ dynes, if $H$ is in Gauss and $v$ in cms. per second. For an electron at the FERMI surface $v$ is in the neighbourhood of $2\times10^{8}$ cm./sec., and if $H$ is, say, 1000 gauss we find for the force due to the magnetic field $3.2\times10^{-9}$ dynes. On the other hand, due to the high conductivity of many metals $F$ is, in general, quite small. Thus in copper at room temperature the electrical resistivity is $1.68\times10^{-6}$ ohm-cms. and, therefore, even at a current density of 100 amperes per cm.$^2$ $F$ is only $1.68\times10^{-4}$ volts/cm., and hence the force on the electron is $2.7\times10^{-16}$ dynes, or about $10^{-7}$ times less than the force due to the magnetic field.

The displacement in $k$-space, in the direction of the electric field, is always very small. For instance, the change in energy of a state $\boldsymbol{k}$ due to this displacement is generally much less than $\mathsf{k}T$, even at low temperatures. This may be seen as follows: Let $\boldsymbol{k}$ be in the direction of $\boldsymbol{F}$, which ensures the greatest possible change of energy, and multiply numerator and denominator of the first term of (41.1) by $v_{\boldsymbol{k}}$. Then, since the mean free path $l$ is equal to $v_{\boldsymbol{k}}\tau$, we may write

$$(\delta k)_F = \frac{eFl}{\hbar^2 k/m}, \tag{41.2}$$

where we have assumed that the velocity is given by $\hbar k/m$, as for free electrons.

Thus the change of energy due to the displacement $(\delta k)_F$ is $eFl$, and since $eF\approx10^{-16}$ dynes and $l\approx10^{-4}$ cms., we see that even at very low temperatures, for example in the liquid helium range, $eFl$ is very much less than $\mathsf{k}T$.

Although the displacement due to the electric field is small, the magnetic field may cause the $\boldsymbol{k}$ vectors to move considerable distances, between collisions, over the FERMI surface. Let $t_1$ denote the average time between successive scattering processes. The angle through which a vector $\boldsymbol{k}$ is turned in time $t_1$, assuming $v_{\boldsymbol{k}}$ to be perpendicular to the direction of $\boldsymbol{H}$ is given by

$$\frac{\delta k}{k} = \frac{eHt_1}{mc}, \tag{41.3}$$

where again we have taken the velocity to be equal to $\hbar k/m$. If $\delta k$ given by (41.3) is much less than $|\boldsymbol{k}'-\boldsymbol{k}|$ due to scattering, it is reasonable to assume that the time of relaxation will be unaffected by the magnetic field. Thus a theory of the galvomagnetic effects based on the use of a relaxation time should be valid, not only at high temperatures, but also at low temperatures provided we use $a_0$ given by (37.3) for the isothermal effects. The same conclusion has been drawn by SONDHEIMER and WILSON[1] as a result of a detailed analysis of a specialized model.

At low temperatures the relaxation time $a_0$ is much larger than the average time between collisions $t_1$ because it requires many collisions with small $q$ to transfer the representative point of an electron from one side of the FERMI distribution to the other. In a magnetic field, however, the representative point ($\boldsymbol{k}$-vector) can be transferred across the FERMI distribution by the action of the field. Thus the time required to destroy the electric current after the removal of the electric field would be much less than $a_0$. It appears, therefore, that unless $\delta k$ is smaller than the average value of $|\boldsymbol{k}' - \boldsymbol{k}|$ a theory based on the use of the relaxation time $a_0$ is likely to be invalid. It may easily be seen that the above condition is equivalent to the requirement

$$\frac{e H t_1}{2 m c} < \frac{T}{\Theta_D}. \tag{41.4}$$

**42. Isothermal conductivity in a magnetic field.** The theory which we shall develop in this and succeeding sections is based on the use of the concept of the time of relaxation. For isothermal conditions, using Eqs. (4.8) and (4.10) we find

$$\frac{e}{\hbar}\left\{\boldsymbol{F} + \frac{1}{c\hbar}\nabla E \times \boldsymbol{H}\right\} \cdot \nabla f = -\frac{f - f_0}{\tau} = \frac{\chi}{\tau}\frac{d f_0}{dE}, \tag{42.1}$$

in which no assumption has yet been made concerning $\tau$ as a function of $\boldsymbol{k}$.

Changing the order in the vector product we may write

$$\nabla E \times \boldsymbol{H} \cdot \nabla f = \boldsymbol{H} \cdot \nabla f \times \nabla E = -\boldsymbol{H} \cdot \boldsymbol{\Omega} f, \tag{42.2}$$

where $\boldsymbol{\Omega}$ is an operator defined by the equation

$$\boldsymbol{\Omega} = \nabla E \times \nabla. \tag{42.3}$$

$\boldsymbol{\Omega}$ has the property of commuting with any function of the energy. Thus if $S$ depends only on $E$

$$\boldsymbol{\Omega} S f = S \boldsymbol{\Omega} f. \tag{42.4}$$

Introducing

$$f = f_0 - \chi \frac{d f_0}{dE}, \tag{42.5}$$

into the left hand side of (42.1) we find, since $f_0$ and $d f_0/dE$ depend only on $E$,

$$\chi = \frac{e\tau}{\hbar} \boldsymbol{F} \cdot \nabla E + \frac{e\tau}{c\hbar^2} \boldsymbol{H} \cdot \boldsymbol{\Omega} \chi, \tag{42.6}$$

where we have neglected the product of $\boldsymbol{F}$ and $\chi$, and have cancelled $d f_0/dE$ through the equation. The product $\chi \boldsymbol{F}$ is neglected because it would depend upon second and higher powers of $F$.

We denote the Cartesian components of the vectors by the subscripts 1, 2, 3, and thus (42.6) may be written

$$\chi = \frac{e}{\hbar} \sum_n F_n \left(\tau \frac{\partial E}{\partial k_n}\right) + \frac{e}{c\hbar^2} \sum_n H_n \tau \Omega_n \chi. \tag{42.7}$$

[1] E. H. SONDHEIMER and A. H. WILSON: Proc. Roy. Soc. Lond., Ser. A **190**, 435 (1947).

This equation may be solved by iteration[1]. To second powers of the components of $\boldsymbol{H}$ the solution is

$$\left.\begin{aligned}\chi = \frac{e}{\hbar}\sum_n F_n\left(\tau\frac{\partial E}{\partial k_n}\right) + \frac{e^2}{c\hbar^2}\sum_{n,m} H_n F_m\left(\tau\Omega_n\,\tau\frac{\partial E}{\partial k_m}\right)\\ + \frac{e^3}{c^2\hbar^5}\sum_{n,m,l} H_n H_m F_l\left(\tau\Omega_n\tau\Omega_m\tau\frac{\partial E}{\partial k_l}\right).\end{aligned}\right\}\tag{42.8}$$

It is unnecessary to add the solution of the homogeneous part of (42.6), since the only physically significant solution of Eq. (42.6) with $\boldsymbol{F}=0$ is $\chi=0$.

We now calculate the components of the current. These are given by

$$j_r = \frac{e}{4\pi^3\hbar}\int\left(-\frac{df_0}{dE}\right)\frac{\partial E}{\partial k_r}\chi\,d\boldsymbol{k}.\tag{42.9}$$

Substituting for $\chi$ from (42.8) we find readily that

$$j_r = \sum_n \sigma_{rn}F_n + \sum_{n,m} A_{rnm}H_n F_m + \sum_{n,m,l} B_{rnml}H_r H_m F_l,\tag{42.10}$$

where $\sigma_{rn}$, $A_{rnm}$ and $B_{rnml}$ are given by

$$\sigma_{rn} = \frac{e^2}{4\pi^3\hbar^2}\int\left(-\frac{df_0}{dE}\right)\frac{\partial E}{\partial k_r}\tau\frac{\partial E}{\partial k_n}d\boldsymbol{k},\tag{42.11}$$

$$A_{rnm} = \frac{e^2}{4\pi^3 c\hbar^4}\int\left(-\frac{df_0}{dE}\right)\tau\frac{\partial E}{\partial k_r}\Omega_n\tau\frac{\partial E}{\partial k_m}d\boldsymbol{k},\tag{42.12}$$

$$B_{rnml} = \frac{e^4}{4\pi^3 c^2\hbar^6}\int\left(-\frac{df_0}{dE}\right)\tau\frac{\partial E}{\partial k_r}\Omega_n\tau\Omega_m\tau\frac{\partial E}{\partial k_l}d\boldsymbol{k}.\tag{42.13}$$

Certain properties of these integrals may be deduced as follows: Let $Z(\boldsymbol{k})$ be any function of $\boldsymbol{k}$ which is everywhere finite, then

$$\int\frac{df_0}{dE}\Omega_n Z(\boldsymbol{k})\,d\boldsymbol{k} = 0,\tag{42.14}$$

for each component of $\Omega$. To obtain (42.14) we observe first that the integrand may be written $\Omega_n\left(\frac{df_0}{dE}Z\right)$ since $\frac{df_0}{dE}$ depends only on $E$ and therefore commutes with $\boldsymbol{\Omega}$; secondly that

$$\Omega_n\left(\frac{df_0}{dE}Z\right) = \frac{\partial}{\partial\xi}\left(\frac{df_0}{dE}Z\frac{\partial E}{\partial\eta}\right) - \frac{\partial}{\partial\eta}\left(\frac{df_0}{dE}Z\frac{\partial E}{\partial\xi}\right),\tag{42.15}$$

where $\xi$ and $\eta$ are the components of $\boldsymbol{k}$ appropriate to $\Omega_n$. Thus on integration both terms vanish separately since $(df_0/dE)=0$ at both limits. Hence, if $X(\boldsymbol{k})$ and $Y(\boldsymbol{k})$ are two finite, but otherwise arbitrary functions of $\boldsymbol{k}$, we have

$$\int\left(-\frac{df_0}{dE}\right)X(\boldsymbol{k})\,\Omega_n Y(\boldsymbol{k})\,d\boldsymbol{k} = -\int\left(-\frac{df_0}{dE}\right)Y(\boldsymbol{k})\,\Omega_n X(\boldsymbol{k})\,d\boldsymbol{k}.\tag{42.16}$$

Using this result, (42.13) may be written

$$B_{rnml} = -\frac{1}{4\pi^3}\left(\frac{e^2}{c\hbar^3}\right)^2\int\left(-\frac{df_0}{dE}\right)\tau\left(\Omega_m\tau\frac{\partial E}{\partial k_l}\right)\left(\Omega_n\tau\frac{\partial E}{\partial k_r}\right)d\boldsymbol{k},\tag{42.17}$$

[1] Cf. H. JONES and C. ZENER: Proc. Roy. Soc. Lond., Ser. A **145**, 268 (1934).

where $\Omega_m$ and $\Omega_n$ operate only within the brackets by which they are enclosed. We thus deduce the following general properties

$$\sigma_{rs} = \sigma_{sr}, \tag{42.18}$$

$$A_{rnm} = -A_{mnr}, \quad A_{rnr} = 0, \tag{42.19}$$

$$B_{rnml} = B_{lmnr}, \tag{42.20}$$

and that $B_{rnnr}$ is always negative.

We now consider certain special cases of interest.

*α) Cubic crystals.* If we choose the coordinate axes to coincide with the axes of the cubic crystal, $E(\boldsymbol{k})$ and $\tau$ are invariant with respect to all operations of the cubic group. With respect to these operations $\partial E/\partial k_n$ has the same symmetry as $k_n$, and $\Omega_1$ the same symmetry as $k_2 k_3$, $\Omega_2$ as $k_1 k_3$ and $\Omega_3$ as $k_1 k_2$. Thus from (42.11) we find

$$\sigma_{rs} = 0 \quad \text{if} \quad r \neq s, \quad \text{and} \quad \sigma_{11} = \sigma_{22} = \sigma_{33}. \tag{42.21}$$

From (42.12) we find that the only $A_{rnm}$ different from zero are

$$A_{123} = A_{231} = A_{312} = -A_{321} = -A_{132} = -A_{213}, \tag{42.22}$$

and, from (42.17), that the only non-vanishing $B_{rnml}$ are

$$\left.\begin{aligned} &B_{1221} = B_{2112} = B_{3113} = B_{1331} = B_{2332} = B_{3223}, \\ &B_{1212} = B_{2121} = B_{1313} = B_{3131} = B_{2323} = B_{3232}, \\ &B_{1122} = B_{1133} = B_{2233} = B_{2211} = B_{3311} = B_{3322}, \\ &B_{1111} = B_{2222} = B_{3333}. \end{aligned}\right\} \tag{42.23}$$

With (42.21), (42.22) and (42.23) we find for the component $j_1$ of the current

$$\left.\begin{aligned} j_1 = {} & \sigma F_1 + A_{123}(H_2 F_3 - H_3 F_2) + B_{1221} H^2 F_1 + \\ & + H_1 (B_{1212} + B_{1122}) \boldsymbol{H} \cdot \boldsymbol{F} + H_1^2 F_1 (B_{1111} - B_{1212} - B_{1221} - B_{1122}), \end{aligned}\right\} \tag{42.24}$$

where we have written $\sigma$ for any of the three non-vanishing components of $\sigma_{rs}$. Expressions for the components $j_2$ and $j_3$ may be obtained in a similar manner, and the current may be written in vector form[1]

$$\boldsymbol{j} = \sigma \boldsymbol{F} + a \boldsymbol{H} \times \boldsymbol{F} + b H^2 \boldsymbol{F} + p (\boldsymbol{H} \cdot \boldsymbol{F}) \boldsymbol{H} + q M \boldsymbol{F}, \tag{42.25}$$

where $M$ is a tensor whose components relative to the crystal axes are given by

$$M = \begin{pmatrix} H_1^2 & 0 & 0 \\ 0 & H_2^2 & 0 \\ 0 & 0 & H_3^2 \end{pmatrix}. \tag{42.26}$$

The constants $a$, $b$, $p$ and $q$ are given at once by comparison with (42.24). It is possible to express the current in the form (42.25) because, as may be seen from (42.22) and (42.23), $A_{123}$ etc. have the same values for each component of $j$. In addition it will be observed that

$$b + p + q = B_{1111}. \tag{42.27}$$

If the fields $\boldsymbol{F}$ and $\boldsymbol{H}$ are held fixed and the crystal rotated, (42.25) shows that $\boldsymbol{j}$ will vary unless $q = 0$, since the components of $\boldsymbol{H}$ along the crystal axes change with the rotation. Only if $q$ happened to be zero would a single cubic crystal appear isotropic in this effect.

[1] This formula was first obtained by F. SEITZ: Phys. Rev. **79**, 372 (1950).

We shall now obtain explicit expressions for the change of resistance in a magnetic field in the transverse and longitudinal positions, and for the HALL coefficient.

For the transverse effect the idealized arrangement is indicated in the accompanying sketch (Fig. 17). We put $H_1 = H_2 = 0$, $F_3 = 0$ and $j_2 = j_3 = 0$. It should be noticed that this represents a special choice of the field directions with respect to the crystal axes.

We then have

$$\left.\begin{aligned} j_1 &= \sigma F_1 + A_{132} H F_2 + B_{1331} H^2 F_1, \\ 0 &= \sigma F_2 + A_{231} H F_1 + B_{2332} H^2 F_2. \end{aligned}\right\} \tag{42.28}$$

Since $A_{132} = -A_{231} = a$, and $B_{1331} = B_{2332} = b$ we find

$$\frac{F_2}{F_1} = \frac{a}{\sigma} H, \tag{42.29}$$

to first powers of $H$, and

$$\frac{j_1}{F_1} - \sigma = \left(\frac{b\sigma + a^2}{\sigma}\right) H^2. \tag{42.30}$$

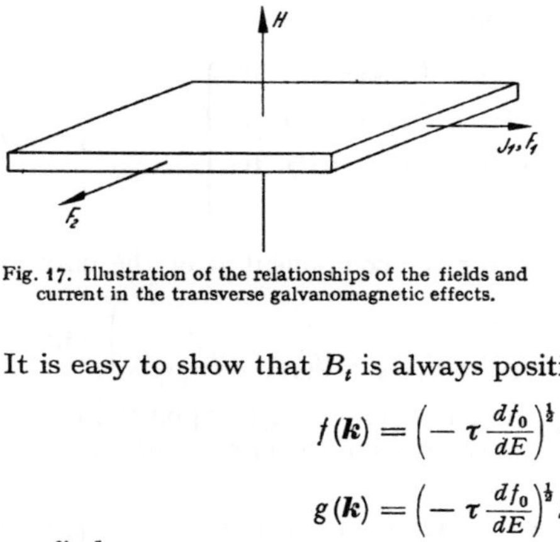


Fig. 17. Illustration of the relationships of the fields and current in the transverse galvanomagnetic effects.

Thus the isothermal HALL coefficient $A_H = F_2/H j_1$, and the change of resistance $\Delta\varrho$ are given by

$$A_H = \frac{a}{\sigma^2}, \tag{42.31}$$

and

$$\frac{\Delta\varrho}{\varrho} = B_t H^2, \tag{42.32}$$

where

$$B_t = -\frac{\sigma b + a^2}{\sigma^2}. \tag{42.33}$$

It is easy to show that $B_t$ is always positive. If we write

$$f(\boldsymbol{k}) = \left(-\tau \frac{d f_0}{dE}\right)^{\frac{1}{2}} \frac{\partial E}{\partial k_2}, \tag{42.34}$$

$$g(\boldsymbol{k}) = \left(-\tau \frac{d f_0}{dE}\right)^{\frac{1}{2}} \Omega_3 \left(\tau \frac{\partial E}{\partial k_1}\right), \tag{42.35}$$

we find

$$B_t = \left(\frac{e}{c\hbar^2}\right)^2 \left\{\frac{\int f^2 d\boldsymbol{k} \int g^2 d\boldsymbol{k} - [\int f g\, d\boldsymbol{k}]^2}{[\int f^2 d\boldsymbol{k}]^2}\right\}, \tag{42.36}$$

and thus, by the well-known SCHWARZ inequality, the right hand side of (42.36) is essentially positive. A magnetic field, therefore, according to this theory, always increases the resistance, however $E$ and $\tau$ depend on $\boldsymbol{k}$.

The integral $a$ and, therefore, the HALL coefficient may be either positive or negative according to the nature of the functions $E$ and $\tau$ at the FERMI surface. Making the special assumptions: (i) that $\tau$ depends only on $E$, and (ii) that terms of order $(\mathrm{k}T/\zeta)^2$ can be neglected, we find

$$\sigma = \frac{e^2\tau}{4\pi^3\hbar^2} \int \left(\frac{\partial E}{\partial k_1}\right)^2 \frac{dS}{|\nabla E|}, \tag{42.37}$$

$$a = \frac{e^3\tau^2}{4\pi^3 c\hbar^4} \int \frac{\partial E}{\partial k_1} \Omega_3 \frac{\partial E}{\partial k_2} \frac{dS}{|\nabla E|}, \tag{42.38}$$

$$b = -\frac{e^4\tau^3}{4\pi^3 c^2\hbar^6} \int \left(\Omega_3 \frac{\partial E}{\partial k_2}\right)^2 \frac{dS}{|\nabla E|}. \tag{42.39}$$

To obtain the values of $A_H$ and $B_t$ it is necessary to know how $E$ depends on $\boldsymbol{k}$. We consider, first, the consequences of the simplest possible assumption, namely that

$$E = E_0 + \frac{\hbar^2}{2m} k^2, \tag{42.40}$$

where $m$ is an effective mass which may be either positive or negative. As a result of this assumption

$$\Omega_3 \frac{\partial E}{\partial k_2} = \frac{\hbar^2}{m} \frac{\partial E}{\partial k_1}, \tag{42.41}$$

and, therefore,

$$a = \frac{e\tau\sigma}{mc}, \quad \text{and} \quad b = -\left(\frac{e\tau}{mc}\right)^2 \sigma. \tag{42.42}$$

Thus $B_t$ vanishes according to (42.33), and we find for $A_H$, since $\sigma = e^2 \tau n / |m|$,

$$A_H = \frac{1}{ecn}\left(\frac{|m|}{m}\right). \tag{42.43}$$

As $e$ is the charge on the electron, i.e. a negative number, we see that $A_H$ is negative if $m$ is positive, and positive if $m$ is negative. When a BRILLOUIN zone is almost completely filled the FERMI surface may be represented, in an idealized model, by (42.40) with $E = \zeta$, and $m$ negative. In such a model conduction may be thought of as due to positive holes, and in this case $A_H$ is positive.

In Eq. (42.1), and therefore throughout this section, $\boldsymbol{H}$ denotes the magnetic field in gauss, $e$ and $\boldsymbol{F}$ the charge and electric field in electrostatic units. Thus $A_H$, as given by (42.43), must be multiplied by $c$ to express it entirely in electrostatic units. Frequently the reciprocal of $A_H$ is expressed in coulombs per cubic centimetre. In these units (42.43) may be written

$$A_H^{-1} = \pm 1.602 \times 10^{-19}\, n \text{ coulombs per cm.}^3, \tag{42.44}$$

where the positive sign is associated with negative $m$ and vice versa.

In order to obtain a finite result, in the first approximation, for the change of resistance coefficient $B_t$ we must make a less restrictive assumption than that implied by (42.40). From (42.37), (42.38) and (42.39) we find

$$B_t = \left(\frac{e\tau}{c\hbar^2}\right)^2 \frac{\int \left(\frac{\partial E}{\partial k_1}\right)^2 \frac{dS}{|\nabla E|} \int \left(\Omega_3 \frac{\partial E}{\partial k_2}\right)^2 \frac{dS}{|\nabla E|} - \left\{\int \left(\frac{\partial E}{\partial k_1} \Omega_3 \frac{\partial E}{\partial k_2}\right) \frac{dS}{|\nabla E|}\right\}^2}{\left\{\int \left(\frac{\partial E}{\partial k_1}\right)^2 \frac{dS}{|\nabla E|}\right\}^2}. \tag{42.45}$$

The integrals of (42.45) depend only on the form of the energy surfaces in $k$-space and should not, therefore, be sensitive to small amounts of impurity in the metal. $\tau$ on the other hand is inversely proportional to the resistivity $\varrho$ and would, therefore, vary rapidly with impurity content. Moreover, as we have already seen, (42.45) should be valid, under the assumptions stated, for low as well as high temperatures provided the correct relaxation time is used, i.e., $a_0$ for thermal scattering at low temperatures. This suggests, therefore, that in some cases $\Delta\varrho/\varrho$ should be a unique function of $H/\varrho$ for all temperatures satisfying (41.4), and for various impurity contents. Many metals appear to follow this rule as JUSTI and SCHEFFERS[1] have shown. The theoretical result (42.45) does, however, depend on the assumption that $\tau$ has the same value at each point of the FERMI surface and it seems unlikely that this can be strictly correct.

[1] E. JUSTI and H. SCHEFFERS: Phys. Z. 39, 105 (1938).

**43. The longitudinal magneto-resistance effect.** If both the magnetic and electric fields are in the same direction, along one of the cubic axes, we may put $H_1 = H_2 = 0$ and $F_1 = F_2 = 0$ in (42.10) and so obtain

$$\frac{j_3}{F_3} - \sigma = B_{3333} H^2, \tag{43.1}$$

and hence, according to (42.17)

$$\frac{\Delta\varrho}{\varrho} = \frac{H^2}{4\pi^3}\left(\frac{e}{c\hbar^3}\right)^2 \frac{1}{\sigma}\int\left(-\frac{df_0}{dE}\right)\tau\left(\Omega_3\,\tau\,\frac{\partial E}{\partial k_3}\right)^2 d\boldsymbol{k}. \tag{43.2}$$

If, again, $\tau$ is assumed to be a function of $E$ only and we neglect terms of order $(\mathrm{k}T/\zeta)^2$ we find for the longitudinal magneto-resistance coefficient $B_l$

$$B_l = \left(\frac{e\tau}{c\hbar^2}\right)^2 \frac{\int\left(\Omega_3\frac{\partial E}{\partial k_3}\right)^2 \frac{dS}{|\nabla E|}}{\int\left(\frac{\partial E}{\partial k_1}\right)^2\frac{dS}{|\nabla E|}}. \tag{43.3}$$

It is clear that if $E$ depends on $\boldsymbol{k}$ in the way given by (42.40) $B_l$ is zero whether the electron gas is degenerate or not. For real metals, however, there is no reason to expect $B_l$ to be appreciably smaller than the transverse coefficient $B_t$.

It may be noticed that $B_l$ and $B_t$ would not vanish, even when the energy depends only on the magnitude of $\boldsymbol{k}$, if $\tau$ varied over the FERMI surface. This could easily arise as a result of the elastic anisotropy of the metal, and it may be that in a metal such as sodium the variation of $\tau$ with direction relative to the crystal axes is the dominant effect.

**44. The two-zone model.** Many polyvalent metals, for example bismuth, show a tendency for the magneto-resistance to reach saturation in high magnetic fields, i.e. for $\Delta\varrho/\varrho$ to tend towards a constant value. Attempts have been made to account for this effect, and the variation of the HALL coefficient with magnetic field in terms of the two-zone model[1]. We now give an outline of this theory.

The FERMI surface is supposed to consist of two distinct spherical regions on which the electronic energies are given by

$$E = E_i + \frac{\hbar^2}{2m_i}k^2, \qquad (i = 1, 2), \tag{44.1}$$

where, in general, $m_1$ is taken to be positive and $m_2$ negative; $E_1$ and $E_2$ are two constants.

Using (44.1) the basic Eq. (42.1) breaks up into two equations, one for $\chi_1$ referring to points on the FERMI surface in the first zone, and one for $\chi_2$ referring to points on the FERMI surface in the second zone. When $\boldsymbol{H} = (0, 0, H)$ and $\boldsymbol{F} = (F_x, F_y, 0)$ these equations may be written

$$\chi_i = \frac{e\hbar\tau_i}{m_i}(k_x F_x + k_y F_y) + \frac{eH\tau_i}{cm_i}\left(k_x\frac{\partial\chi_i}{\partial k_y} - k_y\frac{\partial\chi_i}{\partial k_x}\right), \qquad (i = 1, 2), \tag{44.2}$$

where $\tau_1$ and $\tau_2$ are assumed to be constants referring to zones 1 and 2 respectively. With the definition

$$\omega_i = \frac{eH\tau_i}{cm_i}, \qquad (i = 1, 2),$$

[1] H. JONES: Proc. Roy. Soc. Lond., Ser. A **155**, 653 (1936). — E. H. SONDHEIMER and A. H. WILSON: Cf. reference p. 299.

the solution of (44.2) may be written down without difficulty as follows:

$$\chi_i = \frac{e\hbar\tau_i}{m_i}\{k_x(F_x+\omega_i F_y)+k_y(F_y-\omega_i F_x)\}\Big/(1+\omega_i^2). \tag{44.3}$$

On substituting $\chi_i$ into the integral for the current components (42.9), and stipulating that $j_y=0$ we find

$$j_x = \frac{\sigma_1(F_x+\omega_1 F_y)}{1+\omega_1^2} + \frac{\sigma_2(F_x+\omega_2 F_y)}{1+\omega_2^2}, \tag{44.4}$$

$$0 = \frac{\sigma_1(F_y-\omega_1 F_x)}{1+\omega_1^2} + \frac{\sigma_2(F_y-\omega_2 F_x)}{1+\omega_2^2}, \tag{44.5}$$

where

$$\sigma_i = \frac{e^2 n_i\tau_i}{|m_i|}, \qquad (i=1,2). \tag{44.6}$$

The above equations follow without detailed calculation since it is clear that, when $H=0$, the coefficient of $F_x$ in (44.4), for example, must be $\sigma_1+\sigma_2$. $n_i$ is the number of states enclosed by the FERMI sphere in zone $i$. If $m_1$ is positive $n_1$ is the number of electrons in zone 1, and if $m_2$ is negative $n_2$ is the number of holes in zone 2.

If $\Delta j_x = j_x - F_x(\sigma_1+\sigma_2)$ denotes the change in the current due to the magnetic field, we obtain from (44.4) and (44.5) after some algebra

$$-\frac{\Delta j_x}{j_x} = \frac{\Delta\varrho}{\varrho} = \frac{\sigma_1\sigma_2(\omega_2-\omega_1)^2}{(\sigma_1+\sigma_2)^2+(\sigma_1\omega_2+\sigma_2\omega_1)^2}, \tag{44.7}$$

$$\frac{F_y}{j_x} = \frac{(\sigma_1\omega_1+\sigma_2\omega_2)+\omega_1\omega_2(\sigma_1\omega_2+\sigma_2\omega_1)}{(\sigma_1+\sigma_2)^2+(\sigma_1\omega_2+\sigma_2\omega_1)^2}. \tag{44.8}$$

These equations hold whatever the signs of $m_1$ and $m_2$ may be, but the most significant results are obtained by taking $m_1$ positive and $m_2$ negative. Let therefore

$$\frac{\omega_1}{\sigma_1} = \left(\frac{H}{ec}\right)\frac{1}{n_1}, \quad \text{and} \quad \frac{\omega_2}{\sigma_2} = -\left(\frac{H}{ec}\right)\frac{1}{n_2}, \tag{44.9}$$

and we find from (44.7) and (44.8) the following expressions for the HALL coefficient and the magneto-resistance effect

$$A_H = \frac{1}{ec}\left\{\frac{\left(\frac{\sigma_1^2}{n_1}-\frac{\sigma_2^2}{n_2}\right)-\frac{\sigma_1^2\sigma_2^2}{n_1 n_2}\left(\frac{1}{n_1}-\frac{1}{n_2}\right)\left(\frac{H}{ec}\right)^2}{(\sigma_1+\sigma_2)^2+\sigma_1^2\sigma_2^2\left(\frac{1}{n_1}-\frac{1}{n_2}\right)^2\left(\frac{H}{ec}\right)^2}\right\}, \tag{44.10}$$

$$\frac{\Delta\varrho}{\varrho} = \frac{\sigma_1\sigma_2\left(\frac{\sigma_1}{n_1}+\frac{\sigma_2}{n_2}\right)^2\left(\frac{H}{ec}\right)^2}{(\sigma_1+\sigma_2)^2+\sigma_1^2\sigma_2^2\left(\frac{1}{n_1}-\frac{1}{n_2}\right)^2\left(\frac{H}{ec}\right)^2}. \tag{44.11}$$

$\sigma_1$ and $\sigma_2$ are the conductivities in the absence of a magnetic field due to electrons and holes respectively, and, if the proper expressions are used for these conductivities, (44.10) and (44.11) should be applicable at low as well as high temperatures. Implicit in the derivation of these formulae, however, is the assumption that the times of relaxation are unaffected by the magnetic field. The increase of resistance arises solely from the rotation of the electronic distribution in $k$-space out of the exact direction of the applied electric field. At low temperatures it seems unlikely, for reasons already given, that the above-mentioned assumption can be true, and thus, although (44.10) and (44.11) have been formally deduced without restriction on the magnitude of $H$ they are probably invalid at low temperatures when the field becomes too large.

As the early experimental work of KAPITZA[1] showed, in the case of some metals, for example cadmium, $\Delta\varrho/\varrho$ appears to increase linearly with $H$ over a considerable range. Such a relationship cannot be accounted for by a formula of the type (44.11) whatever values of the parameters may be used. A more complicated form of the FERMI surface would, of course, lead to a formula different from (44.11), but the very striking linearity of the effect observed by KAPITZA remains essentially unexplained.

The galvomagnetic properties of bismuth, containing small amounts of impurity, provide an interesting application of formulae (44.10) and (44.11). For ideally pure bismuth it appears to be a reasonable conjecture that the number of holes and electrons should be taken to be equal in a two-zone model. Since the valency of bismuth is five, an atomic fraction $x$ of a tetravalent metal would leave $(5-x)$ valency electrons per atom on the average. If the dominant BRILLOUIN zone is exactly filled by 5 electrons per atom we are led to expect that

$$n_2 - n_1 = \frac{x}{\Omega}, \tag{44.12}$$

where $\Omega$ is the atomic volume, $n_1$ the number of electrons and $n_2$ the number of holes per unit volume. If the value of $x$ is very small, much less than either $n_1$ or $n_2$ in pure bismuth, then (44.10) and (44.11) may be written

$$A_H = \frac{A - \Omega C x H^2}{1 + x^2 C H^2}, \tag{44.13}$$

$$\frac{\Delta\varrho}{\varrho} = \frac{B H^2}{1 + x^2 C H^2}, \tag{44.14}$$

where $A$, $B$ and $C$ are constants with respect to $H$. These formulae do appear to give an approximate representation of the actual behaviour of bismuth containing small traces of impurity[2].

**45. The electronic thermal conductivity in a magnetic field.** If it may be assumed that the concept of the relaxation time is applicable, the formal derivation of the electronic thermal conductivity presents no difficulties. It may be shown that the WIEDEMANN-FRANZ law holds for all values of the magnetic field when there is a single relaxation time for both electrical and thermal conduction. Hence, the formulae already given for the dependence of $\sigma$ on $H$ give, at once, the dependence of the thermal conductivity, $\varkappa$, upon $H$.

At low temperatures, as we have seen, there are two different times of relaxation: one associated with the displacement arising in electrical conductivity, and another associated with thermal conductivity. Any phenomenon, such as the thermo-electric effect, which involves simultaneously both types of displacement must inevitably lead to intricate theory for low temperature effects. In the present case, however, we are concerned only with one type of displacement of the FERMI distribution, and the modificiations produced in it by the presence of the magnetic field. Hence, provided we use the correct relaxation time, viz. $\frac{\pi^2}{3}\gamma^2 a_0$, the theory for thermal conduction, analogous to that developed in Sect. 42 and 44 for electrical conduction in a magnetic field, should be applicable.

---

[1] P. KAPITZA: Proc. Roy. Soc. Lond., Ser. A **123**, 292 (1929).

[2] H. JONES: Cf. reference p. 304.

Thus in the two-zone theory, for example, it is only necessary to replace $\sigma_i$ $(i=1,2)$ by $\frac{3\varkappa_i}{T}\left(\frac{e}{\pi \mathrm{k}}\right)^2$ to obtain the corresponding formulae for thermal conduction in a magnetic field. Here $\varkappa_1$ and $\varkappa_2$ are the thermal conductivities due to electrons and holes respectively, in the absence of a magnetic field.

Making the above substitution in Eq. (44.11) we obtain therefore

$$\frac{\varkappa_0-\varkappa}{\varkappa}=\frac{\varkappa_1\varkappa_2\left(\frac{\varkappa_1}{n_1}+\frac{\varkappa_2}{n_2}\right)^2\left(\frac{3eH}{\pi^2\mathrm{k}^2cT}\right)^2}{(\varkappa_1+\varkappa_2)^2+\varkappa_1^2\varkappa_2^2\left(\frac{1}{n_1}-\frac{1}{n_2}\right)^2\left(\frac{3eH}{\pi^2\mathrm{k}^2cT}\right)^2}. \tag{45.1}$$

It will be observed that, at low temperatures, the WIEDEMANN-FRANZ law will no longer hold, according to (44.11) and (45.1). This is because in (44.11) the time of relaxation is $a_0$, and in (45.1) it is $\frac{\pi^2}{3}\gamma^2 a_0$. It is of some interest to consider the change in the WIEDEMANN-FRANZ ratio with magnetic field strength in this specialized two-zone model[1]. For this purpose we make the simplest possible assumptions, viz., $n_1=n_2=\frac{1}{2}n$, $\sigma_1=\sigma_2=\frac{1}{2}\sigma_0$ and $\varkappa_1=\varkappa_2=\frac{1}{2}\varkappa_0$. We then readily obtain

$$\left(\frac{\varkappa}{\sigma T}\right)_H=\left(\frac{\varkappa_0}{\sigma_0 T}\right)\left\{\frac{1+\left(\frac{H}{enc}\right)^2\sigma_0^2}{1+\left(\frac{3eH}{\pi^2\mathrm{k}^2ncT}\right)^2\varkappa_0^2}\right\}. \tag{45.2}$$

As we have seen $\frac{\varkappa_0}{\sigma_0 T}$ is constant and equal to $\frac{1}{3}\left(\frac{\pi \mathrm{k}}{e}\right)^2$ at high temperatures, and varies as $T^2$ at low temperatures for pure metals; being then less in magnitude than the high temperature ratio. In the limit of high fields, according to (45.2), we get

$$\left(\frac{\varkappa}{\sigma T}\right)_{H\to\infty}=\left(\frac{\pi^2\mathrm{k}^2}{3e^2}\right)^2\frac{\sigma_0 T}{\varkappa_0}, \tag{45.3}$$

and, as we have just seen, this ratio is greater than $(\pi^2\mathrm{k}^2/3e^2)$ at low temperatures. Thus, in general, we can expect the WIEDEMANN-FRANZ ratio to increase with $H$ at low temperatures. According to the measurements of GRÜNEISEN and his co-workers, this appears to be the case in fact.

The precise behaviour of the WIEDEMANN-FRANZ ratio, as a function of the magnetic field strength, depends, of course, in the two-zone theory, on the particular values given to the parameters $n_1, n_2, \sigma_1, \sigma_2, \varkappa_1$, and $\varkappa_2$. Since the two-zone model itself is rarely a good representation of actual metals, any agreement with experimental results obtained by assigning special values to the above parameters must be viewed with great caution.

**46. The effect of quantization of the electronic orbits.** We conclude the discussion of galvomagnetic effects in metals by mentioning briefly one of the basic assumptions implicit in the whole of the preceding theory, and which is likely to fail at low temperatures and high field strengths. In Sect. 3$\beta$ we showed that the wave vector of the centre of a wave packet, constructed from BLOCH wave functions, would change with time according to (3.22), but only for times much less than the reciprocal of the LARMOR frequency. If the $\boldsymbol{k}$-vector rotates through an appreciable angle between successive collisions of the electron it would be

[1] Many experimental investigations on the influence of a magnetic field on the electrical and thermal conductivities of metals at low temperatures have been carried out by GRÜNEISEN and his co-workers. Cf. Ann. Phys., Lpz. (5) **31**, 714 (1938); **32**, 219 (1938).

necessary to consider the quantization of the motion in the magnetic field. It is this quantization which leads to the DE HAAS-VAN ALPHEN effect, and which, even at high temperatures, is the origin of the LANDAU-PEIERLS diamagnetism.

The measurements of the magneto-resistance of bismuth at low temperatures by ALERS and WEBBER[1], following earlier work by DE HAAS, BLOM and SCHUBNIKOW[2], appear to show evidence of this quantum effect. At 4.2° K, and with the magnetic field perpendicular both to the principal axis and to one of the binary axes, it was found that the resistivity did not rise smoothly with $H$ but passed through a number of maxima and minima. No detailed theory of this effect has yet been given, but its connection with the DE HAAS-VAN ALPHEN effect seems evident. The limitation just mentioned, and expressed by (3.25), is one more reason, in addition to those already given, for regarding formulae such as (44.11) and (45.1) as giving no more than an indication of what might happen at high field strengths in polyvalent metals.

## VIII. The anomalous skin effect.

**47. Classical theory of the normal effect.** Interesting advances in our knowledge of the anomalous skin effect have been made in recent years, largely through the experimental work of PIPPARD[3] and the theoretical studies of REUTER and SONDHEIMER[4]. It now appears possible that investigations of this kind may lead to useful information regarding the nature of the FERMI surface in metals. We shall, therefore, give an account of the theory of the anomalous skin effect in this section, beginning with an outline of the classical theory of the normal effect.

Imagine a semi-infinite block of metal bounded by the $xy$ plane, as indicated in the accompanying sketch (Fig. 18) and suppose that radiation, propagated in the direction of the positive $z$-axis, falls on this plane; the polarization being such that the electric vector lies in the direction of the $x$-axis. If the angular frequency of the radiation is $\omega$, the electric and magnetic field strengths within the metal may be written

$$F_x = F(z)\, e^{i\omega t}, \quad \text{and} \quad H_y = H(z)\, e^{i\omega t}. \tag{47.1}$$

For this case MAXWELL's equations read

$$\frac{\partial F_x}{\partial z} = -\frac{1}{c}\dot{H}_y, \tag{47.2}$$

$$-\frac{\partial H_y}{\partial z} = \frac{1}{c}\dot{F}_x + \frac{4\pi}{c} j_x, \tag{47.3}$$

and if we assume that the current density, $j_x$, is determined at each point in the metal by the electric field at that point, in accordance with OHM's law, i.e.

$$j_x = \sigma_0 F_x, \tag{47.4}$$

where $\sigma_0$ is the bulk conductivity, we obtain from (47.1), (47.2), and (47.3)

$$\frac{d^2 F}{dz^2} + \frac{\omega^2}{c^2} F = \frac{4\pi\sigma_0\omega i F}{c^2}. \tag{47.5}$$

[1] P. B. ALERS and R. T. WEBBER: Phys. Rev. **84**, 863 (1951).
[2] W. J. DE HAAS, J. W. BLOM and L. W. SCHUBNIKOW: Physica, Haag **2**, 907 (1935).
[3] A. B. PIPPARD: Proc. Roy. Soc. Lond., Ser. A **191**, 385 (1947); **203**, 98, 195 (1950).
[4] G. E. H. REUTER and E. H. SONDHEIMER: Proc. Roy. Soc. Lond., Ser. A **195**, 336 (1948).

The ratio of the real to the imaginary part of the coefficient of $F$ in (47.5) is $\omega/4\pi\sigma_0$, and this number is negligibly small for the micro-wave range and pure metals at low temperatures. We shall therefore, henceforth, neglect the term $\omega^2 F/c^2$, and on writing

$$F(z) = F_0 e^{-(z/\delta)(1+i)}, \tag{47.6}$$

we obtain

$$\delta = \frac{c}{\sqrt{2\pi\sigma_0\omega}}. \tag{47.7}$$

The length $\delta$ is known as the "classical skin depth", and this may become comparable with the mean free path of the conduction electrons in pure metals, at very low temperatures, and with micro-waves of a few centimetres in wave length. When this condition arises (47.4) is no longer valid; the current density at any point then depends upon the field at other points. This is the origin of the anomalous skin effect.

Before developing the electron theory of this effect, we consider the nature of the measurements actually made in the experimental work. The POYNTING vector of the radiation falling on the metal bounded by the $xy$ plane is

$$\frac{c}{4\pi} F_x H_y, \tag{47.8}$$

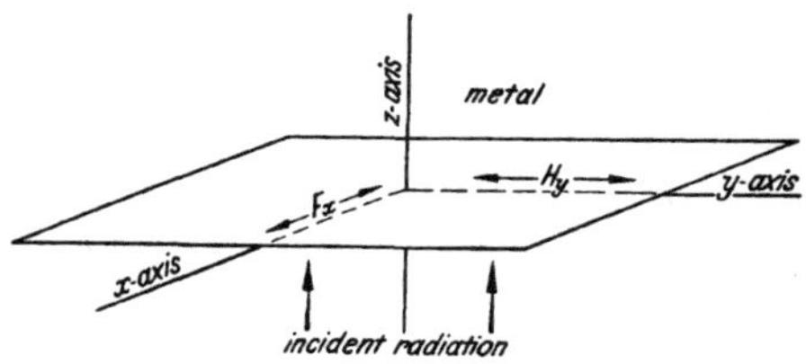


Fig. 18. Diagram to show the field directions of the incident radiation in relation to the chosen coordinate system.

and the real part of this is, therefore, the energy dissipated in the metal as JOULE heat per unit time per unit area. We may define the surface resistance $R$ by equating the real part of (47.8) to the real part of

$$R\left\{\int_0^\infty j_x(z)\,dz\right\}^2 \tag{47.9}$$

since the factor in brackets denotes the total current per unit area.

Integrating (47.3) over positive values of $z$, i.e., within the metal, and neglecting the term $\dot{F}_x/c$, which is small at the frequencies under consideration, we have

$$H_y(0) = \frac{4\pi}{c}\int_0^\infty j_x(z)\,dz, \tag{47.10}$$

where $H_y(0)$ denotes the value of the magnetic field at the surface of the metal. Thus $R$ is equal to the real part of

$$F_x(0)\Big/\int_0^\infty j_x(z)\,dz, \tag{47.11}$$

and since, according to (47.2), $H_y(0) = -\frac{c}{i\omega}F_x'(0)$, we may express $R$ as the real part of

$$-\frac{4\pi\omega i}{c^2}\frac{F_x(0)}{F_x'(0)}. \tag{47.12}$$

In practice the surface resistance is obtained by measuring the selectivity, $Q$, of a resonator constructed of the metal under investigation. If $\omega_0$ is the natural angular frequency of the resonator, and $\omega_0 + \frac{1}{2}\Delta\omega$ is the angular frequency of the forced oscillation such that

$$F_0^2(\omega_0 + \tfrac{1}{2}\Delta\omega) = \tfrac{1}{2}F_0^2(\omega_0), \tag{47.13}$$

where $F_0$ is the amplitude of the electric field within the resonator, then

$$Q = \omega_0/\Delta\omega. \tag{47.14}$$

Although the measurement of $Q$ is standard practice under normal conditions, at temperatures in the liquid helium range special devices are necessary, and the reader is referred to the original papers of Pippard for details of the experimental arrangement. As may be readily understood $\Delta\omega$ is proportional to the rate of dissipation of energy, i.e. to $R$, and thus $Q$ is a measure of the surface conductivity $R^{-1}$.

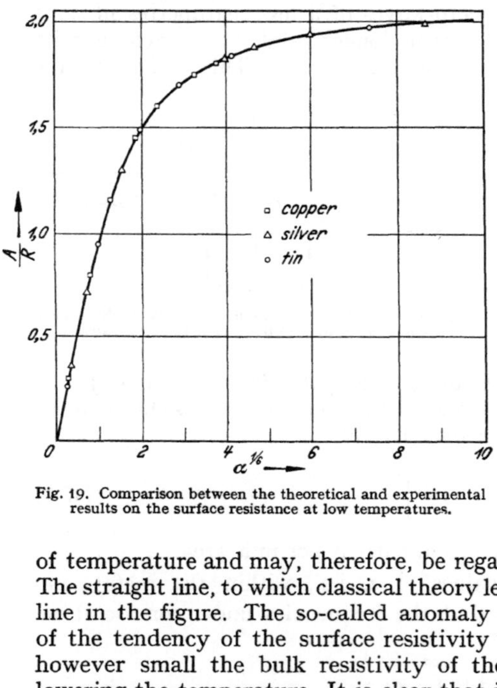


Fig. 19. Comparison between the theoretical and experimental results on the surface resistance at low temperatures.

From (47.12) and (47.6) we find at once

$$R = \frac{2\pi\omega}{c^2}\,\delta = \sqrt{\frac{2\pi\omega}{c^2\sigma_0}}, \tag{47.15}$$

where, to obtain the second equality, we have used (47.7). Thus if the selectivity is measured at various temperatures, and also the direct current conductivity $\sigma_0$, it is possible to construct a plot of $Q$ against $\sqrt{\sigma_0}$. According to Eq. (47.15) this should be a straight line. Fig. 19 shows actual results obtained by Chambers[1]. The abscissa indicates the numerical values of $\alpha^{\frac{1}{6}}$, a quantity fully defined later but which, we may note now, is proportional to $\sqrt{\sigma_0}$. $A$ is independent of temperature and may, therefore, be regarded as a constant in this connection. The straight line, to which classical theory leads, agrees with the initial slope of the line in the figure. The so-called anomaly of the skin effect consists, therefore, of the tendency of the surface resistivity to maintain a finite, constant, value, however small the bulk resistivity of the metal may become as a result of lowering the temperature. It is clear that impurities, or distortions of the metal, concentrated in a thin surface layer would simulate exactly such an effect, but it is now believed that these sources of uncertainty have been eliminated.

We now give a short account of the theory developed by Reuter and Sondheimer[2].

**48. Electron theory of the anomalous effect.** This theory is based on the following assumptions: First, that it is permissible to use the concept of the relaxation time, even though the theory is to be applied to metals at very low temperatures, and to an effect which depends upon the possibility of changing the bulk conductivity with temperature. Thus, when we introduce a mean free path through the relation $l = v\tau$, where $\tau$ is the time of relaxation and $v$ the mean velocity of the electrons at the Fermi surface, this $l$ should be regarded as having only a formal significance; here it is not the average distance travelled by an electron

[1] R. G. Chambers: Nature, Lond. **165**, 239 (1950).
[2] Cf. reference p. 308.

between successive scattering processes. Secondly, the scattering at the surface of the metal is taken account of formally by introducing a reflection coefficient. Two extreme cases may be considered: one in which electrons are assumed to be reflected elastically at the surface, imagined to be a true plane, or alternatively that all electrons are scattered diffusely at the surface with complete loss of any drift velocity acquired in an external field. In the following account we shall adopt the first alternative as this leads to somewhat simpler analysis, and does not involve the omission of any essential feature of the calculation.

BOLTZMANN'S Equation (4.2) may be written

$$\frac{\partial f_1}{\partial t} + \frac{f_1}{\tau} + v_z \frac{\partial f_1}{\partial z} = \left(-\frac{d f_0}{dE}\right) e\, v_x F(z), \tag{48.1}$$

where $F(z)$ denotes the $x$-component of the electric field. Since $f_1$ oscillates with angular frequency $\omega$ we can replace $\partial f_1/dt$ by $i\omega f_1$ and so obtain from (48.1)

$$\frac{\partial f_1}{\partial z} + a f_1 = b F(z), \tag{48.2}$$

where

$$a = \frac{1}{\tau v_z}(1 + i\omega\tau), \tag{48.3}$$

and

$$b = \left(-\frac{d f_0}{dE}\right) e \frac{v_x}{v_z}. \tag{48.4}$$

Although $\omega\tau$ is not always negligible compared with unity, it frequently is so under experimental conditions, and we shall neglect this term in the subsequent theory.

The solution of (48.2) may be written

$$f_1 = b \int^{z} F(\xi)\, e^{a(\xi - z)}\, d\xi. \tag{48.5}$$

We have now to express the condition that at the boundary $z=0$, the number of electrons with velocity components $v_x, v_y, v_z$ is exactly equal to the number with velocity components $v_x, v_y, -v_z$. Consider, first, the electrons which approach the boundary $z=0$ from above in the metal. For these electrons $v_z$ is negative and therefore $a$ is negative. Thus if we write

$$f_1 = b \int_{\infty}^{z} F(\xi)\, e^{a(\xi - z)}\, d\xi, \qquad (v_z < 0), \tag{48.6}$$

this ensures that $f_1$ is a convergent solution of (48.2) and vanishes as $z \to \infty$, i.e. at large distances from the boundary the electrons are in thermal equilibrium and are unaffected by the field.

To construct the solution for positive values of $v_z$ we consider (48.6) at $z=0$. If we replace $v_z$ by $-v_z$ we obtain

$$\left.\begin{aligned} f_1(v_x, v_y, -v_z, 0) &= -b \int_{\infty}^{0} F(\xi)\, e^{-a\xi}\, d\xi \\ &= b \int_{-\infty}^{0} F(-\xi)\, e^{a\xi}\, d\xi. \end{aligned}\right\} \tag{48.7}$$

Hence, if we define $F(-\xi)$ to be equal to $F(\xi)$, we may write

$$f_1 = b \int_{-\infty}^{z} F(\xi)\, e^{a(\xi - z)}\, d\xi, \qquad (v_z > 0), \tag{48.8}$$

and this function which satisfies (48.2) also now satisfies the requirement at $z=0$ called for by the assumption of specular reflection at the surface of the metal. (48.6) and (48.8) together now determine the distribution function for all velocities, and it will be necessary to find solutions for the field, $F(z)$, which ensure that $f_1$, given by (48.8), vanishes as $z$ tends to $+\infty$.

The current density may be obtained by substituting the expressions for $f_1$ into the following equation

$$j_x = \frac{e}{4\pi^3}\int v_x f_1 \, d\boldsymbol{k}. \tag{48.9}$$

If we assume that the conduction electrons may be treated as free electrons, and if we introduce polar coordinates in velocity space, i.e.

$$v_x = v\sin\vartheta\cos\varphi, \qquad v_z = v\cos\vartheta, \tag{48.10}$$

$j_x$ may be written

$$j_x = \frac{e\,m}{2\pi^3\hbar^3}\iiint E\, f_1 \sin^2\vartheta\cos\varphi\, d\vartheta\, d\varphi\, dE. \tag{48.11}$$

Substituting from (48.6) and (48.8) for $f_1$, and carrying out the integrations with respect to $d\varphi$ and $dE$ we obtain

$$\left.\begin{aligned} j_x = \frac{e^2 m\zeta}{2\pi^2\hbar^3}\Bigg[\int_0^{\pi/2}\frac{\sin^3\vartheta}{\cos\vartheta}\int_{-\infty}^{z} F(\xi)\, e^{\left(\frac{\xi-z}{l}\right)\sec\vartheta}\, d\xi\, d\vartheta\, + \\ + \int_{\pi/2}^{\pi}\frac{\sin^3\vartheta}{\cos\vartheta}\int_{\infty}^{z} F(\xi)\, e^{\left(\frac{\xi-z}{l}\right)\sec\vartheta}\, d\xi\, d\vartheta\Bigg].\end{aligned}\right\} \tag{48.12}$$

Changing the variable from $\vartheta$ to $\sec\vartheta$, and making use of the function $g(u)$ defined by

$$g(u) = \int_1^{\infty}\left(\frac{1}{s} - \frac{1}{s^3}\right) e^{-|u|s}\, ds, \tag{48.13}$$

(48.12) may easily be reduced to the following equation

$$j_x = \frac{e^2 m\zeta}{2\pi^2\hbar^3}\int_{-\infty}^{\infty} F(\xi)\, g\left(\frac{\xi - z}{l}\right) d\xi. \tag{48.14}$$

The factor before the integral of (48.14) is equal $(3\sigma_0/4l)$, as may easily be seen by using (6.4) and remembering that $n = (k_0^3/3\pi^2)$. The Eq. (48.14) replaces OHM's law, as used in the classical theory of the skin effect. The function $g(u)$ diverges as $-\log|u|$ for $u\to 0$, and falls away rapidly as $u$ increases. An elementary integration shows that

$$\int_{-\infty}^{\infty} g(u)\, du = \tfrac{4}{3}. \tag{48.15}$$

Thus, if $l$ is very small compared with the range in which $F(\xi)$ varies appreciably, we may take $F(\xi)$ outside the integral, as $F(z)$, and thus obtain $j_x(z) = \sigma_0 F(z)$.

From MAXWELL's equations we find

$$\frac{d^2 F}{dz^2} = \frac{4\pi\omega i}{c^2}\, j_x, \tag{48.16}$$

which, together with (48.14) gives

$$\frac{d^2 F}{dz^2} = \frac{3i}{2\delta^2 l}\int_{-\infty}^{\infty} F(\xi)\, g\left(\frac{\xi - z}{l}\right) d\xi. \tag{48.17}$$

It is advantageous to introduce non-dimensional variables defined as follows:

$$\xi = l\eta, \quad z = l y, \tag{48.18}$$

and thus lengths are measured in units of the mean free path. Also writing

$$\alpha = \frac{3}{2}\frac{l^2}{\delta^2}, \tag{48.19}$$

we obtain from (48.17)

$$\frac{d^2 F}{d y^2} = i\alpha \int_{-\infty}^{\infty} F(\eta)\, g(\eta - y)\, d\eta. \tag{48.20}$$

To obtain the complete solution of this equation is a difficult matter, and we refer the reader to the original paper of Reuter and Sondheimer[1] for details. An expression for $F(0)/F'(0)$ which is sufficient to determine the surface resistance can, however, be deduced without too much difficulty as follows:

Let

$$\Phi(t) = \int_{-\infty}^{\infty} F(\xi)\, e^{-i\xi t} d\xi = 2\int_{0}^{\infty} F(\xi) \cos \xi t\, d\xi, \tag{48.21}$$

and

$$\varkappa(t) = \int_{-\infty}^{\infty} g(\xi)\, e^{-i\xi t} d\xi. \tag{48.22}$$

Integrating the second integral of (48.21) twice by parts we obtain

$$\int_{-\infty}^{\infty} F''(\xi)\, e^{-i\xi t} d\xi = -t^2 \Phi(t) - 2F'(0). \tag{48.23}$$

Multiplying (48.20) by $e^{-iyt}$ and integrating with respect to $y$ we find

$$\left.\begin{aligned} \int_{-\infty}^{\infty} F''(y)\, e^{-iyt} dy &= i\alpha \int_{-\infty}^{\infty} F(\eta)\, e^{-i\eta t} d\eta \int_{-\infty}^{\infty} g(y-\eta)\, e^{-i(y-\eta)t} dy \\ &= i\alpha\, \Phi(t)\, \varkappa(t). \end{aligned}\right\} \tag{48.24}$$

Hence (48.23) and (48.24) give

$$\Phi(t) = -\frac{2F'(0)}{t^2 + i\alpha\varkappa(t)}, \tag{48.25}$$

and therefore since the Fourier transform of (48.21) is

$$F(y) = \frac{1}{\pi}\int_0^{\infty} \Phi(t) \cos y t\, dt \tag{48.26}$$

we have, for $y = 0$,

$$\frac{F(0)}{F'(0)} = -\frac{2}{\pi}\int_0^{\infty} \frac{dt}{t^2 + i\alpha\varkappa(t)}. \tag{48.27}$$

According to (47.12) and (48.27) the surface resistivity, $R$, is equal to the real part of

$$\frac{8\omega l i}{c^2}\int_0^{\infty} \frac{dt}{t^2 + i\alpha\varkappa}, \tag{48.28}$$

[1] Cf. reference p. 308.

where the factor $l$ arises when we re-express $F/F'$ in centimetres. Thus

$$R = \frac{8\omega l}{c^2} \int_0^\infty \frac{\alpha \varkappa}{t^4 + \alpha^2 \varkappa^2} dt. \tag{48.29}$$

$\varkappa(t)$ may be expressed explicitly in terms of $t$ by substituting the integral (48.13) for $g$ into (48.22). An elementary integration then gives

$$\frac{1}{2} \varkappa(t) = \left(\frac{1}{t} + \frac{1}{t^3}\right) \operatorname{arc\,tan} t - \frac{1}{t^2}. \tag{48.30}$$

For $t=0$, $\varkappa = \frac{4}{3}$ and decreases steadily as $t$ increases, varying asymptotically as $\pi/t$. When $\alpha$ is very large the significant parts of the integral (48.29) arise from large values of $t$ and, therefore, we may replace $\varkappa$ by its asymptotic value. A simple reduction then gives

$$\left.\begin{aligned} R_\infty &= \frac{4\omega l}{c^2} (\alpha\pi)^{-\frac{1}{3}} \int_0^\infty \frac{dx}{1+x^3} \\ &= \frac{8\pi\omega l}{3\sqrt{3}\, c^2} (\alpha\pi)^{-\frac{1}{3}} = \frac{8}{9} \left(\frac{\sqrt{3}\,\pi\omega^2 l}{c^4 \sigma_0}\right)^{\frac{1}{3}}. \end{aligned}\right\} \tag{48.31}$$

Since $\sigma_0 = (e^2 n l/m v)$, in which the only factor varying with temperature, or small amounts of impurity, is $l$, it follows that the surface resistivity in the extreme anomalous limit, $R_\infty$, is constant. The theory, therefore, accounts very successfully for the outstanding feature of the experimental observations.

Writing $\alpha$ in the form

$$\alpha^{\frac{1}{3}} = \left(\frac{3\pi\omega l^2}{c^2 \sigma_0^2}\right)^{\frac{1}{3}} \sqrt{\sigma_0}, \tag{48.32}$$

and defining the constant $A$ by the equation

$$A = \left(\frac{2\sqrt{6}\,\pi^2 \omega^2 l}{c^4 \sigma_0}\right)^{\frac{1}{3}}, \tag{48.33}$$

(48.29) may be written in the form

$$\frac{A}{R} = \frac{\pi\sqrt{3}}{4\sqrt{2}} f(\alpha), \tag{48.34}$$

where

$$\frac{1}{f(\alpha)} = \alpha^{\frac{1}{3}} \int_0^\infty \frac{\alpha \varkappa\, dt}{t^4 + \alpha^2 \varkappa^2}. \tag{48.35}$$

The limiting cases of very small and very large values of $\alpha$ have already been given. If we put $\varkappa = \frac{4}{3}$ in (48.35) we get the classical result which in the present notation reads

$$\frac{A}{R} = \alpha^{\frac{1}{3}} \qquad \text{(if } \alpha \text{ tends to zero)}, \tag{48.36}$$

and with $\varkappa = \pi/t$ we get

$$\frac{A}{R_\infty} = \frac{9\pi^{\frac{1}{3}}}{4\sqrt{2}} = 2.330 \qquad \text{(if } \alpha \text{ tends to infinity)}. \tag{48.37}$$

The complete relation (48.34) can be obtained by a number of numerical integrations for specified values of $\alpha$.

The experimentally obtained curves are plots of $1/R$ against $\sqrt{\sigma_0}$. Thus, by fitting the theoretical to the experimental curve, the value of $A$ may be determined and hence the ratio $\sigma_0/l$.

The theory which we have outlined is based on the assumption of regular reflection at the surface. It is found that somewhat better agreement with experiment is obtained if the alternative assumption of completely diffuse scattering is adopted. The curve shown in Fig. 19 is the theoretical result with this assumption, and the excellent agreement with the experimentally determined points will be noted.

With the assumption of completely free electrons we may write

$$\frac{\sigma_0}{l} = \left(\frac{8\pi}{3}\right)^{\frac{1}{3}} \frac{e^2}{h} n^{\frac{2}{3}}, \qquad (48.38)$$

| | Metal | | | | | |
|---|---|---|---|---|---|---|
| | Cu | Ag | Au | Sn | Hg | Al |
| $n/n_a$ | 1.0 | 0,68 | 0.60 | 1.10 | 0.23 | 0.37 |

and thus the anomalous skin effect provides a means of estimating $n$, the number of free electrons per unit volume. The above figures have been given by SONDHEIMER[1].

$n_a$ is the number of atoms per unit volume and, therefore, $n/n_a$ is the number of electrons per atom. The figures for the polyvalent metals can hardly have as much significance as those for the monovalent metals, since the assumption that the electronic energy depends only on the magnitude of the wave vector must inevitably fail in polyvalent metals and, therefore, much of the detailed theory becomes inapplicable.

## General References.


1933. SOMMERFELD, A., u. H. BETHE: In GEIGER-SCHEELS Handbuch der Physik, Bd. 24/2, S. 333—622. Full account of the BLOCH theory of conductivity as developed up to 1932.

1934. MOTT, N. F., and H. JONES: Theory of the Properties of Metals and Alloys. Oxford: Clarendon Press. Contains simplified account of the theory of conductivity at normal temperatures, including thermal conductivity, thermo-electric effects and galvo-magnetic phenomena. No detailed work on low temperature conductivity. Full account of conduction properties of alloys. Theory of transition metals, and of metals in the bismuth group.

1936. FRÖHLICH, H.: Elektronentheorie der Metalle. Berlin: Springer. Pages 156—224 deal with the theory of metallic conductivity. Contains short and clear accounts of all the important effects. A simplified treatment is given of low temperature conductivity.

1940. SEITZ, F.: The Modern Theory of Solids. McGraw-Hill Book Company. Pages 516—547 devoted specifically to metallic conductivity. Covers the main features of the general theory of conduction.

1952. MACDONALD, D. K. C.: Properties of Metals at Low Temperatures. Progr. Met. Phys. **3**. Mainly an account of experimental work but relates results to existing theory.

1953. WILSON, A. H.: The Theory of Metals. Cambridge University Press. The most detailed general account of the mathematical theory of conductivity now available. Full treatment of low temperature effects. Thermal conductivity, and galvo-magnetic effects at low temperatures. Detailed account of the variational method.

(In press) BARDEEN, J.: Conduction: Metals and Semi-conductors. Handbook of Physics. Edited by E. U. CONDON. Good general account without detailed mathematics. Gives the non-specialist ready access to the more important results.


[1] E. H. SONDHEIMER: Phil. Mag., Suppl. **1**, 1 (1952).

# Photoconductivity.

By

G. F. J. Garlick.

With 72 Figures.

## I. Introduction.

**1. Brief historical review.** It is useful to set the perspective of this article, which will be mainly concerned with developments in photoconductivity over the last decade or so, by giving a short review of the earlier beginnings of interest in the phenomenon. The first observation of photoconductivity in solids was probably that of Smith[1] on the fall in electrical resistance of selenium when illuminated by visible light. Following this a large interest developed in this property of selenium and by 1920 a considerable volume of literature on its properties was in existence. However, most of the experimental results were not amenable to fundamental interpretation. In 1927 Doty[2] gave a list of 1536 articles and well over a hundred patents dealing with selenium.

The discovery of the photovoltaic effect, i.e. the occurrence of a voltage across a solid or liquid boundary with an electrode on illumination, was made much earlier by Becquerel[3]. He used platinum or silver halide coated silver electrodes in a sulphuric acid solution. Commercial development of photovoltaic cells using selenium on iron or cuprous oxide on copper has occurred over many years and modern cells of these types can give potentials in excess of 0.1 volts on illumination. To these cells may now be added those using germanium, silicon and other semiconductors of recent development, giving even higher photovoltages.

Qualitative searches for photoconductivity in materials other than selenium were made very early in the present century. Case[4] observed photoconduction in many different sulphides, notably $Sb_2S_3$, $Bi_2S_3$, $MoS_2$, $Ag_2S$, $PbS_2$ $AsS_6$, $AgSbS_3$, $Pb_2Sb_2S_5$ etc. Many sulphides were also investigated by Coblenz and Kahler[5]. Other photoconductors known before 1920 include $Cu_2O$, $HgCl_2$, $I_2$, $S_2$ and silver salts. These earlier studies have been reviewed by Hughes and DuBridge[6]. The first quantitative measurements of photoconductivity, capable of theoretical treatment, were made by Gudden and Pohl from 1920 onwards. These have been reviewed by several authors, in particular, Hughes and DuBridge[6], Hughes[7], Nix[8] and by Gudden[9]. A more modern theoretical treatment of their results has been given by Mott and Gurney[10]. The measure-

[1] W. Smith: Nature, Lond. **7**, 303 (1878).
[2] M. F. Doty: Selenium, A list of references. New York: Library (1927).
[3] E. Beqcuerel: C. R. Acad. Sci. Paris **9**, 144, 561 (1839).
[4] T. W. Case: Phys. Rev. **9**, 305 (1917).
[5] W. W. Coblenz and H. Kahler: U.S. Bur. Stand. Bull. **15**, 121 (1919).
[6] A. L. Hughes and L. A. DuBridge: Photoelectric Phenomena New York 1932.
[7] A. L. Hughes: Rev. Mod. Phys. **8**, 294 (1936).
[8] F. C. Nix: Rev. Mod. Phys. **4**, 723 (1932).
[9] B. Gudden: Lichtelektrische Erscheinungen. Leipzig 1928.
[10] N. F. Mott and R. W. Gurney: Electronic Processes in Ionic Crystals. Oxford 1940

ments were made on natural specimens of diamond, zinc blende (ZnS), greenokite (CdS) and cinnabar (HgS) and in 1925 the first systematic study of selenium was published by GUDDEN and POHL[1]. Because of difficulty in obtaining natural crystals of the above materials free from contaminating impurities, and because of the difficulty in making synthetic crystals of more than microcrystalline dimensions, GUDDEN and POHL and some of their co-workers turned their attention to alkali halides containing colour centres and to the silver halides. It was these studies which opened the way to a more precise understanding of the photoconduction processes in solids.

The late war with its demands for infra-red detectors of high sensitivity led to a revival of interest in photoconductors. In Germany effort was concentrated on sulphides and resulted in production of cells of lead sulphide and selenide with spectral response extending into the 3 μ region of wavelength[2]. Thalofide (thallium sulphide) cells were developed in the U.S.A.[3] but have a less impressive spectral response. Since the war research on lead sulphide, selenide and telluride has increased enormously in many countries and together with a practical development of high sensitivity cells there has been an intensive study of their fundamental mechanism. The latter has been made possible by the production of large single crystal specimens. In addition to the interest in the above sulphides three other major studies may be mentioned. They are the investigation of photoconducting and photovoltaic processes in a) germanium, silicon and some other monatomic solids, b) cadmium and zinc sulphides, c) intermetallic compounds more specifically the III to V compounds (so called because of their position in the periodic table).

**2. Some early theories of photoconductivity.** The first attempts to obtain a consistent theory of photoconduction in crystals were published by GUDDEN and POHL in 1923[4]. The theory is best explained by reference to the simple diagram of Fig. 1 which shows schematically the photoconducting crystal mounted in between suitable electrodes and connected in series with a battery and galvanometer (or in many cases an electrometer), the latter recording the motion of free charges produced by light in the crystal. Such charges may be negative or positive or both. Drift of charges in the field applied to the cystal is opposed by the random thermal motion of the lattice which scatters them. In general the charge carriers will not move all the way to an electrode but will suffer recombination or trapping at specific places in the crystal before reaching the electrode. Trapping without recombination and return to the normal unexcited state will cause a volume space charge to build up in the crystal. Thus after prolonged illumination saturation effects may be observed due to the effect of the space charge field on that applied to the crystal. As is now known, surface barriers will also play a large part in determining the photocurrent because of

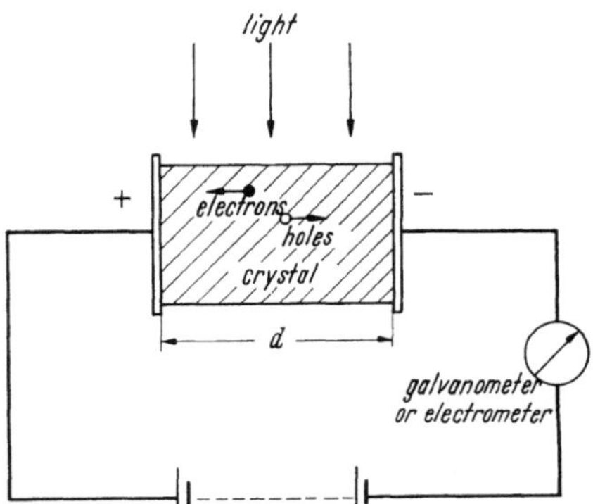


Fig. 1. Schematic experimental arrangement for measurement of photoconductivity.

[1] B. GUDDEN and R. W. POHL: Z. Physik **35**, 243 (1925).
[2] G. JOOS (Senior author): Physics of Solids II. Fiat Rev. German. Sci. Wiesbaden 1948.
[3] A. HIPPEL et coll.: J. Chem. Phys. **14**, 355, 370 (1946).
[4] B. GUDDEN and R. W. POHL: Z. Physik **16**, 170 (1923); **17**, 331 (1923).

their effect on carrier recombination and the formation of a surface space charge (see later sections).

We consider the simple case when trapping is negligible and a field $E$ is applied to a crystal of thickness $d$. Suppose that each charge carrier liberated by illumination of the crystal travels a distance $x$ along the field direction before it recombines with a vacant atom or ion level or with a carrier of opposite sign. If $N$ carriers of each sign are produced by a given illumination the charge passage recorded will be given by:

$$Q = N e (x_- + x_+)/d \tag{2.1}$$

where $e$ is the carrier charge and the suffix to $x$ denotes the sign of carrier. If all the liberated charges reach their respective electrodes then:

$$Q = N e \tag{2.2}$$

or for the case where only carriers of one sign are mobile:

$$Q_- \text{ or } Q_+ = N e/2. \tag{2.3}$$

The above theory neglects the fact that the applied field $E$ determines the distance $x$ travelled before recombination.

In general it should be assumed that each carrier has a mean life time $\tau$ and travels a mean distance $\delta$ along the field before recombination where

$$\delta = E \mu \tau \tag{2.4}$$

$\mu$ being the carrier mobility. Eq. (2.4) may be rewritten thus:

$$\delta = E e l/6 k T p \sigma \tag{2.5}$$

where $l$ is the mean free path of the carrier between lattice collisions, $k$ is BOLTZMANN'S constant, $T$ the absolute temperature, $p$ the number of recombination sites per unit volume and $\sigma$ the capture cross section of each site for a free carrier. The mean distance travelled along the field was termed the „*Schubweg*“ by GUDDEN and POHL. For large fields $\delta$ will be equal to or greater than the interelectrode distance i.e. the current saturates. If $E_s$ is the saturation field then we may write

$$\delta/d = E/E_s = V/V_s \tag{2.6}$$

where $V$ and $V_s$ are applied potentials.

For a uniformly excited crystal all carriers produced within a distance $\delta$ of their respective electrode will reach that electrode. In the simple case when only one type of carrier is mobile:

$$Q = N \frac{(d-\delta)}{d} \cdot \frac{\delta}{d} + \int_0^\delta \frac{Ne}{d} \cdot \frac{x}{d} \cdot dx, \tag{2.7}$$

$$Q = \frac{Ne\delta}{d}\left(1 - \frac{\delta}{2d}\right) = \frac{NeV}{V_s}\left(1 - \frac{V}{2V_s}\right). \tag{2.8}$$

Since $\delta = E\mu\tau$ then $\delta = \mu\tau V/d$ and so for saturation when $Q = Ne/2$

$$V_s = d^2/\mu\tau, \tag{2.9}$$

i.e. the saturation potential varies as the square of the crystal thickness between electrodes. The graphical form of Eq. (2.8) is given in Fig. 2.

At low fields the recorded charge passing for a given illumination increases linearly with field subsequently showing saturation. At any field value the charge passing is proportional to the intensity of illumination for the above simple model. Curves of the type shown in Fig. 2 were obtained experimentally for the silver halides by LEHFELDT[1] and in some cases more detailed studies were made possible by light probe techniques in which only a thin slice of crystal was illuminated at a selected distance from the electrodes.

In their experiments GUDDEN and POHL found a rapid fall in photocurrent under steady illumination due to trapping and space charge formation. They found that the space charge could be dissipated by heating or by illumination with long wavelength visible or near infra-red radiation. Thus in order to measure what they called the true primary photocurrent they used "flash" illumination with heating or infra-red removal of space charge between each measurement. For zinc sulphide and other materials they were able to show that quantum equivalence occurred i.e. each light quantum absorbed produced one free carrier (or two if of opposite sign and both mobile).

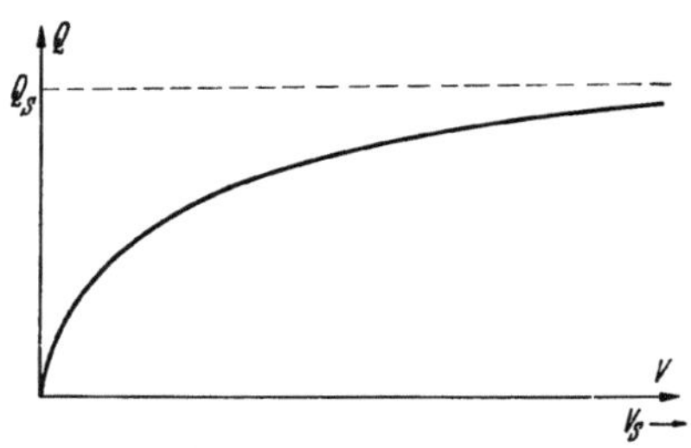

Fig. 2. Dependence of charge $Q$ recorded in external circuit of Fig. 1 on voltage $V$ applied to cryst 1.

Further results of the above simple model were elaborated by REIMANN[2]. In the case of steady illumination the creation of new free carriers will be balanced by recombination. If for such equilibrium excitation there are $n$ carriers of one sign or if the latter are not mobile then there must be $n$ recombination sites. If the rate of production of carriers of one sign by light is $J$ per sec per unit volume and the recombination coefficient is $\beta$ then in equilibrium we have:

$$\frac{dn}{dt} = 0 = J - \beta n^2. \tag{2.10}$$

$\beta$ is given by $u \cdot \sigma$ where $u$ is the thermal velocity of the carrier and $\sigma$ the capture cross section for a recombination centre. In equilibrium we thus have $J = \beta n^2$ and so the photocurrent density $(n e \mu E)$ is proportional to the square root of the light intensity. The decay of the photocurrent is obtained by putting $J = 0$ in Eq. (2.10) and $\frac{dn}{dt} \neq 0$. Then by solution:

$$n = n_0/(1 + \beta n_0 t). \tag{2.11}$$

$n_0$ being the initial number of carriers at the beginning of the decay. The latter is hyperbolic in form and the life time of carriers can only be specified by some arbitrary condition. The simplest condition is to specify the time for the number of charge carriers to be halved i.e.

$$t_{\frac{1}{2}} = 1/\beta n_0. \tag{2.12}$$

Many photoconducting phosphors were found to exhibit this bimolecular form for the decay of photocurrent. However, it is known now that such a form is largely due to the slow escape of carriers from trapping states with subsequent recombination at other sites. This is easily seen in the case of sulphide phosphors whəre $t_{\frac{1}{2}}$ can be of the order of seconds, which could not be explained by known values of $u$ and $\sigma$ and so of $\beta$.

[1] W. LEHFELDT: Göttinger Nachr., Fachgr. II **1**, 171 (1935). See also N. F. MOTT and R. W. GURNEY: Electronic processes in ionic crystals. Oxford 1940.

[2] A. L. REIMANN: Nature, Lond. **140**, 501 (1937).

GUDDEN and POHL[1] divided up the photocurrent into primary and secondary processes since the first stages of illumination gave linear behaviour and subsequent stages gave more complex relations between current and light intensity. Such distinctions are not necesary in more modern treatment as will be shown in later sections.

An interesting case using the simple model above may be considered as a conclusion to this section, namely, the space charge limited photocurrent in a crystal with only one sign of carrier mobile. If the carriers are produced near one electrode and travel across to the opposite electrode (i.e. electrons would originate at cathode and move to anode) then MOTT and GURNEY[2] give the maximum possible current density passing through the crystal as:

$$i = 10^{-13}\, V^2 \mu\, \varkappa / d^3 \text{ amp./cm.}^2 \tag{2.13}$$

with usual notation and with $\varkappa$ being the dielectric constant.

**3. Coverage of text.** In the text which follows an extensive coverage of photoconducting solids is provided but no attempt has been made to give a completely exhaustive account of all known photoconductors. The main sections are concerned with materials which have received the major attention during the last several years.

## II. Modern theories of photoconduction in crystals.

### a) Electronic energy states in crystals.

**4. Energy states in perfect crystals.** The recent developments of quantitative theories of photoconduction and related phenomena in solids have been largely due to the establishment of satisfactory models for the energy states of electrons in crystalline solids. Such models are dealt with elsewhere in this volume. That of most interest in dealing with photoconduction is known often as the "collective" electron model and will be briefly described below.

A perfect crystal lattice possesses a perfectly periodic field in the absence of thermal and zero point lattice vibrations. The wave functions of constituent electrons were first determined by BLOCH who obtained solutions of SCHRÖDINGER'S wave equation first of all for a one dimensional lattice. The developments of this approach, particularly those relevant to the present subject, have been described by several workers[3–6].

The permitted energy states for electrons in the perfect lattice fall into quasi-continuous levels forming bands separated by regions of forbidden energy states as shown in Fig. 3a.

In photoconductivity we are concerned only with electrons in the highest occupied bands i.e. those originally forming the valence electrons of the constituent atoms or ions. For the case of an insulator the highest occupied band is completely occupied as shown in Fig. 3b. This model for the crystal electrons, while specifying their energies leaves their positions undefined, the latter extending throughout the crystal. Application of an electric field to such a system produces no net current since no empty electron states exist of near energy to those already filled. If the field is strong enough then the electrons may be

[1] B. GUDDEN and R. W. POHL: Z. Physik **6**, 248 (1921).
[2] N. F. MOTT and R. W. GURNEY: Electronic Processes in Ionic Crystals. Oxford 1940.
[3] N. F. MOTT and R. W. GURNEY: Electronic Processes in Ionic Crystals. Oxford 1940.
[4] F. SEITZ: Modern theory of Solids. New York 1940.
[5] A. H. WILSON: Theory of Metals, 2nd edit. Cambridge 1953.
[6] W. SHOCKLEY: Electrons and Holes in Semi Conductors. New York 1950.

raised to the unoccupied band above known as the conduction band (ZENER effect[1]) and will then take part in conduction. Vacant levels in the filled band will also cause conduction the effect being that of carriers of positive charge with a mass approximating to that of an electron and known as *positive holes*. An important point in conduction in solids is that the effective mass of electrons or positive holes may be very different from that of a free electron. As a general rule the mass of an electron in the conduction band when only a small number are present and degeneracy is absent (i.e. classical gas statistics apply) is about that of a free electron (however, see Sect. 51) while the positive hole mass is generally greater. Electrons may also be raised from filled band to conduction band by thermal activation. In this case the solid is known as a semi-conductor of *intrinsic* type. If the raising of electrons is accomplished by photon absorption then *photoconduction* occurs of *intrinsic* type (to distinguish it from that in imperfect crystal lattices). The photon energy must be equal to, or greater than, the width of the forbidden band. In addition, optical transitions due to photon absorption (or emission i.e. luminescence) in solids are subject to certain selection rules. The selection rule applicable in the above model is that the wave numbers in the initial and final states are the same. For this reason and because of the low density of states, transitions from the top of the filled band to the bottom of the conduction band may be forbidden. It is thus seen that even in a perfect crystal the optical absorption limit and the threshold wavelength for photoconduction need not correspond to the width of the forbidden zone. The absorption spectrum and spectral response for photoconduction will be of complex form dependent on transition probabilities and on density distributions of states in the filled and conduction bands. The above model is difficult to relate to experiment for yet another reason. The absorption coefficients for lattice photoconduction are very high ($\sim 10^5$ mm.$^{-1}$) and so only surface excitation is possible with light quanta. The surface conditions are never those of an ideal lattice. In any case such ideal lattices are not met with in practice. We, therefore, consider the effect of lattice defects and impurities on the theoretical model.

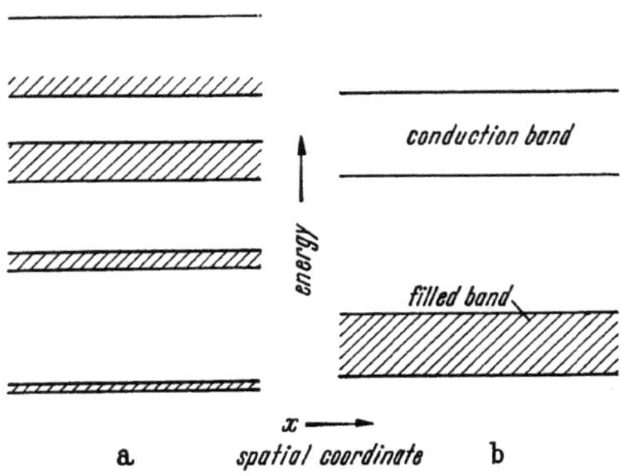

Fig. 3a and b. a Electronic energy band scheme for a perfect crystal; b Energy band scheme for an intrinsic photoconductor.

**5. Energy states in imperfect crystals.** We may apply the above model to crystals containing atomic defects in the lattice (vacant sites or interstitial atoms and/or containing impurity atoms if the numbers of such entities are small compare with the number of lattice atoms (i.e. less than one defect or impurity per $10^4$ lattice atoms). The crystal lattice still gives rise to the energy band system but each defect or impurity causes additional and localised energy states which are of most importance when they occur in the forbidden zone between filled and conduction bands, as shown in Fig. 4. These additional levels may or may not be filled in the normal unexcited state of the crystal, depending on the relative densities of levels and available electrons. E.g., Reduction of

[1] Cf. the article of W. FRANZ on dielectric breakdown in vol. XVII of this Encyclopedia.

stoichiometric zinc oxide will give effectively an excess of zinc with an electron available for each localised level due to an excess zinc atom. For $Cu_2O$ layers on copper excess oxygen provides, by creation of cuprous ion vacancies, unoccupied levels near to the filled band[1]. In general we may describe the imperfect crystal schematically as in Fig. 4 with an arbitrary distribution of defect or impurity levels. Occupancy of levels is determined by application of FERMI statistics (see e.g. pp. 325—326 and 327—328) and the limit of occupied levels at absolute zero is known as the FERMI *limit* (as in metals) and is shown by the broken line in Fig. 4. Semi-conduction and photoconduction in such systems can result from transitions of electrons from occupied defect or impurity levels into the conduction band or by transitions of holes from empty localised levels into the filled band. The two types of process giving rise to conduction form a convenient designation for the semi- or photo-conductor. When excitation provides conduction electrons the material is known as $N$-type. When free holes are produced, it is known as $P$-type[2]. In some cases both $P$ and $N$ type conduction may exist. In this case the carriers of lowest concentration are known as the *minority* carriers. It will be already obvious that the presence of localised electron states will affect the photoconductivity markedly. The spectral response will be extended to longer wavelengths and the carrier life time and mobility will be affected by electron and/or hole capture in localised states leading in some cases to recombination. Any unoccupied levels act as electron traps and any filled states as positive hole traps. We now consider the modification of the simple theory of photoconductivity necessary when trapping states are to be considered.

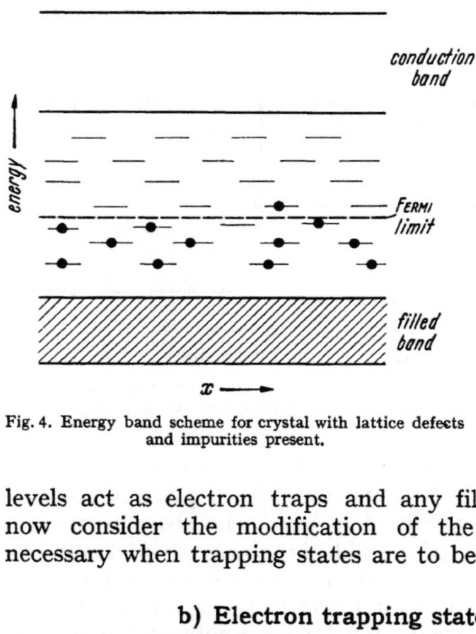


Fig. 4. Energy band scheme for crystal with lattice defects and impurities present.

### b) Electron trapping states and photoconductivity.

**6. Effect on intrinsic photoconduction.** If vacant, localised levels are present in the forbidden energy band and are all of the same "depth" below the conduction band then a simple treatment of their effect on photoconduction may be given. If $J$ electrons are excited per second per cm.$^3$, and there are at any instant $n$ conduction electrons and $m$ trapped electrons in the $M$ traps per cm.$^3$ then there are $(n+m)$ positive holes. We may write the kinetic equations as follows:

$$\frac{dn}{dt} = J - \beta n (n+m) - \beta n (M-m) + \gamma m \ldots, \tag{6.1}$$

$$\frac{dm}{dt} = \beta n (M-m) - \gamma m - x m (m+n) \ldots . \tag{6.2}$$

The recombination coefficients for holes and electrons and for electrons and empty traps are assumed equal for simplicity. $xm(m+n)$ represents the trapped

[1] N. F. MOTT and R. W. GURNEY: Electronic Processes in Ionic Crystals. Oxford 1940.

[2] A more detailed discussion is given by J. C. SLATER, in this volume, and by O. MADELUNG in vol. XX of this Encyclopedia.

electron-positive hole recombination which for the present will be assumed negligible, while $\gamma m$ represents the thermal activation of trapped electrons back into the conduction band ($\gamma$ is given by $s \exp(-\mathcal{E}/kT)$, $\mathcal{E}$ being the trap "depth" and $s$ a frequency constant $\sim 10^8$ to $10^{12}$ per sec.).

When the trap depth is large (or at low temperatures) and so $\gamma$ is small the traps will fill to saturation during excitation. When $\mathcal{E}$ is very small (or at high temperatures) the time spent by electrons in traps is negligible. For equilibrium excitation and neglecting the term $x m (m+n)$ we have:

$$\frac{dn}{dt} = 0 = J - \beta n (n + m) \ldots, \qquad (6.3)$$

$$\frac{dm}{dt} = 0 = \beta n (M - m) - \gamma m \ldots. \qquad (6.4)$$

For deep traps $\gamma$ is negligible and we have trap saturation ($M = m$), so:

$$J = \beta n (n + M) \ldots. \qquad (6.5)$$

At low intensities ($J$ small) $n \ll M$, and so:

$$n = J/\beta M \ldots. \qquad (6.6)$$

At high intensities $n \gg M$, and so:

$$n = \sqrt{J/\beta}. \qquad (6.7)$$

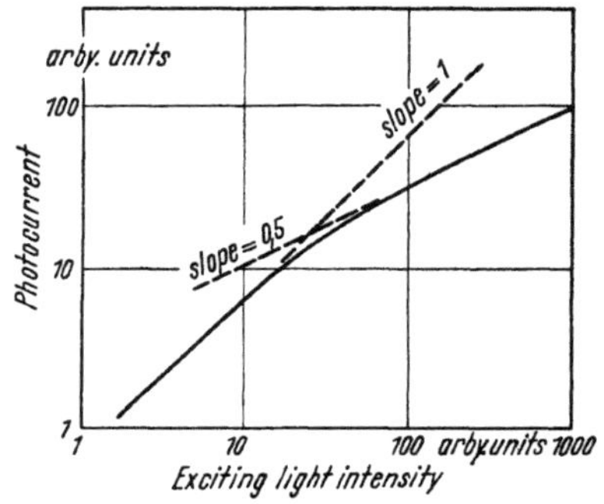


Fig. 5. Variation of steady state photocurrent with exciting light intensity for a CdS crystal (after Broser and Warminsky).

Since the photocurrent is proportional to $n$ if hole mobility is small we have for the deep trap case a linear variation of photocurrent with light intensity at low intensities and a square root relation at high intensities of excitation. These relations are shown for a photoconducting crystal of cadmium sulphide in Fig. 5. We see from Eq. (6.6) that when traps are present and at low intensities of excitation the number of charge carries is lower than in the absence of traps by the factor $M$. Since $M$ may be $10^{16}$ per cm.$^3$ or more the reduction in response is very large indeed.

The response time of the photoconductor will also be affected by the trapping states. If the excitation is removed, then, for deep traps Eqs. (6.1) and (6.2) give for low intensities

$$dn/dt = -\beta n M \qquad (6.8)$$

or

$$n = n_0 e^{-\beta M t}. \qquad (6.9)$$

That is, the decay of current is exponential with a time constant $\tau = 1/\beta M$. Thus trapping states produce a low time constant (i.e. fast response) but any advantage from this is offset by the large reduction in response efficiency. At high intensities the time constant is as in Eq. (2.12) and so

$$t_{\frac{1}{2}} = 1/\beta n_0 = 1/\sqrt{J\beta}, \qquad (6.10)$$

i.e. the time constant decreases with intensity of exciting light. For intermediate trap depths the above simple treatment indicates that the initial decay of current depends on recombination processes but later stages are dependent on the slow escape of trapped electrons into the conduction band before recombination (cf. experimental results for germanium in Sect. 21). Moss[1] has applied the above theory to the case where the levels in the forbidden zone are occupied and some thermal activation occurs at ordinary temperatures, i.e. the crystal

[1] T. S. Moss: Photoconductivity in the Elements. London 1952.

shows semiconduction. If it is assumed that electrons are excited per second by thermal activation and in the dark $n_0$ electrons are in the conduction band, the same dependence of photocurrent on light intensity is shown as before but the photosensitivity and response time are affected at lower intensities. The time constant $t_{\frac{1}{2}}$ is now given by:

$$t_{\frac{1}{2}} = 1/\beta(M + 2n_0). \tag{6.11}$$

The net photocurrent represented by $\Delta n$ is given by $J \cdot t_0$ and so the shorter time constant also involves a lower sensitivity. The value of $n_0$ will rise exponentially with temperature and so $t_{\frac{1}{2}}$ will decrease rapidly with temperature with consequent reduction in photoconduction sensitivity. This is characteristic of many experimental cases (see later sections).

**7. General treatment of trapping states and photoconductivity.** In most photoconductors we have to deal with a distribution of trapping states with some normally occupied. General approaches have been made to the theoretical problems involved notably by Rose[1] and by Broser and Broser-Warminsky[2] the latter also considering the comprehensive treatment of associated luminescence at the same time[3]. We consider first the approach due to Rose and begin with the effect of traps of one depth.

*α) Traps of one depth.* If a free carrier moves under an applied field then its mobility is determined by lattice collisions and will be denoted by $\mu_0$. If, however, it is trapped in transit for a mean time $t$ then this must be added to the transit time and so, the effective mobility $\mu$ is lower than that in the absence of trapping. With trapping the mobility is given by Rose as:

$$\mu = \frac{\text{average distance travelled along field between lattice collisions}}{\text{mean time in trap/number of lattice collisions between trapping acts}}, \tag{7.1}$$

$$= \frac{\text{average distance etc.} \times \text{time between trapping acts}}{\text{mean time in traps} \times \text{time between lattice collisions}}, \tag{7.2}$$

$$= \frac{s\, e^{-\mathcal{E}/kT}}{u\, \sigma M} \cdot \mu_0, \tag{7.3}$$

with previous notation.

From Mott and Gurney[4] the value of $s/\sigma$ is estimated to be $\sim 10^{26}$ and for room temperature $u \sim 10^7$ cm./sec. and so:

$$\mu = \frac{10^{19}}{M} e^{-\mathcal{E}/kT} \cdot \mu_0. \tag{7.4}$$

Now for a trap density of $10^{16}$/cm.$^3$ (a low value for most crystals) and for $\mathcal{E} \sim 0.5$ eV we have

$$\mu \sim 10^{-5} \mu_0 \tag{7.5}$$

showing the profound effect trapping can have on the effective mobility and therefore on the magnitude of the photoconductivity.

*β) Distribution of trapping states.* As shown by Rose, it is easier to consider the effect of trapping states as reducing the effective number of carriers rather than their normal mobility. It is also convenient to express the photocurrent density in terms of the life time of a free carrier $\tau$ and its transit time $t_r$

[1] A. Rose: RCA-Review **12**, 362 (1951).
[2] I. Broser and R. Broser-Warminsky: Brit. J. Appl. Phys. Suppl. **4**, S. 90 (1955).
[3] See the author's further contribution to vol. XXVI of this Encyclopedia.
[4] N. F. Mott and R. W. Gurney: Electronic Processes in Ionic Crystals. Oxford 1940.

between electrodes. If $i$ is the photocurrent density then the number of electronic charges passing per second

$$= i/e = \frac{\tau}{t_r} \cdot J \cdot \Omega, \tag{7.6}$$

where $\Omega$ is the volume of crystal and $J$ is assumed uniform over the whole volume. Among the quantities in Eq. (7.6) the transit time, the current, the intensity and volume are experimentally observable and so values of $\tau$ can be calculated and compared with observed time constants for photocurrent decay. If the values are equal, then maximum performance is occurring but in general the observed time constant is greater than that due to trapping. As ROSE points out, the reciprocity between sensitivity of a photoconductor and its response time, expected from the theory above in Sect. 2 and 6, need not apply if the effects of traps are properly considered.

The quantity $\tau/t_r$ can be considered as a gain factor for the photoconduction. In the ideal case, the life time will be equal to the transit time, the carrier then being lost at an electrode. However, if carrier replenishment occurs at the other electrode the effective life time can be much greater than the transit time and a current gain above that for quantum equivalence is obtained.

In most cases to which ROSE has applied his theory the positive holes are considered to have low mobility and so are regarded as fixed. Such fixation may also occur by hole trapping, e.g. in phosphors where holes are trapped in emission centres at which sites recombination takes place by capture of conduction electrons.

Following previous arguments the case of equilibrium excitation is considered. The recombination depends on the number of free electrons $n$ and the number of holes $p$ and the equilibrium conditions give:

$$J = \beta n p = u \sigma n p \tag{7.7}$$

and

$$p = n + m \tag{7.8}$$

where $m$ is the number of trapped electrons. Thus

$$n = J/u \sigma p. \tag{7.9}$$

In equilibrium the constant exchange between conduction electrons and traps does not affect the above equation but will affect the experimentally observed time constant. The mean life time of the free carrier $\tau$ is given by $1/\beta p$. The transit time is given by $t_r = d^2/V\mu_0$ and so the factor $\tau/t_r$ in Eq. (7.6) is given by

$$\tau/t_r = V\mu_0/d^2 u \sigma p \tag{7.10}$$

and the photocurrent by

$$i = \frac{e J \Omega \mu_0}{d^2 u \sigma p} = \frac{e J \Omega \mu_0}{d^2 u \sigma (n + m)}. \tag{7.11}$$

In the case for no trapping

$$i = \frac{e J \Omega \mu_0}{d^2 u \sigma n} \tag{7.12}$$

indicating that with traps present giving $n < p$ the effect is to give a lower sensitivity.

We now have to introduce a distribution of trapping states since the particular distribution will determine the relation between $n$ and $p$ and the observed time response of the photocurrents. In all normal photoconductivity the part of the trap distribution which is effective is that near to the FERMI limit. During excitation this limit will move up to a new steady state value set by the equilibrium between thermal activation of the trapped electrons and their replanishment by electrons from the conducting band. The number of holes $p$ is given by $(n+m)$ and $m$ is given by the integration of levels between the steady state FERMI limit and the normal FERMI limit. Thus as excitation intensity is increased the increase in photocurrent will depend on the trap distribution. We may quote from ROSE's results for specific trap distribution:

Uniform trap distribution: In this case the photocurrent is insensitive to temperature and increases linearly with light intensity. The time constant varies inversely as the light intensity and will always be greater than the life time $\tau$ of the conduction electrons. When the distribution is uniform below the conduction band but the FERMI limit lies below the distribution then the photocurrent increases as the square root of the light intensity. However, this case is very different from the simple bimolecular case as the number of conduction electrons is very different from the number of positive holes. However, $n$ and $p$, though not equal, increase proportionally with each other. The time constant turns out to be the same as in the absence of traps although the latter have lowered the overeall photosensitivity.

For an *exponential distribution of traps* below the conduction band i.e. $M_{\mathcal{E}}\, d\mathcal{E} \propto e^{-\alpha \mathcal{E}}\, d\mathcal{E}$ where $\alpha$ is a constant, the current varies as a power of the excitation intensity given by $T_1/(T+T_1)$ where $T_1$ is a fixed temperature determined by the constant $\alpha$ for the exponential distributions. Thus for all temperatures below $T_1$ the power is between 0.5 and unity. If $T > T_1$ then the 0.5 value is retained. The time constant for this case is, as in the case of a uniform trap distribution, very little affected by the trap distribution although photosensitivity has similarly been reduced.

**8. Positive hole motion and photoconductivity.** In the above theories it has been convenient to consider positive hole motion as a negligible quantity. In general for such photoconductors as cadmium sulphide this seems justified under normal conditions. However, when the photoconductor is also illuminated with infra-red light the photocurrent is often reduced rather than enhanced although there is plenty of evidence that electrons are removed from trapping states. In this case it seem likely that the infra-red radiation frees positive holes trapped at recombination sites and these migrate to and recombine with trapped electrons thus providing a drain or sink for conduction electrons. This quenching effect on the photoconductor is, however, dependent on wavelength of infra-red radiation as shown by the experiments of KLICK and LAMBE[1] on CdS crystals. 1 μ radiation caused a quenching, but 3 μ radiation an enhancing of photocurrent. In the latter case electron ejection from traps into the conduction band by the radiation appears to be the chief process.

**9. Conductivity decay and "glow curves".** We have already seen that the time constant of the photoconduction will in some cases depend on the slow escape of trapped electrons into the conduction band. The detailed form of this decay is of importance particularly in the case of photoconducting phosphors where correlation with phosphorescence decay is being investigated. The thermoluminescence curve for phosphors[2] also has its equivalent in some photoconductors. After excitation giving rise to photoconductivity at low temperatures, the warming of a photoconductor in the dark will cause a release of trapped electrons and so a transient conduction current whose variation with temper-

[1] C. C. KLICK and J. LAMBE: Phys. Rev. **98**, 909 (1955).
[2] Discussed in the writer's contribution to vol. XXVI of this Encyclopedia.

ature will depend on the trap distribution and also, of course, on the variation of mobility with temperature. The phenomenon is important in photoconducting phosphors where it can be used as an independent indication of the trap distribution. Theoretically it has been treated by Broser and Broser-Warminsky[1] in a comprehensive way although earlier, Herman and Meyer[2] gave equations for a single activation energy of the trapped electrons using the thermoluminescence "glow curve" theory of Randall and Wilkins[3] which assumes monomolecular kinetics for the electron recombination process at emission centres and an absence of retrapping of electrons. The theory given by Broser and Broser-Warminsky avoids some of the difficulties raised by the restricted monomolecular theory and gives relations between photoconductivity and "current glow curves" and light emission characteristics in luminescent photoconductors such as cadmium and zinc sulphides.

To make their theory as comprehensive as possible Broser and Broser-Warminsky assume a general distribution of defect or impurity levels in the forbidden zone and assume that emission or recombination centres are normally occupied levels i.e. those below the Fermi limit at absolute zero. Thus the number of these centres $A$ is given by:

$$A = \int_{\mathcal{E}_2}^{\zeta_0} M_{\mathcal{E}}\, d\mathcal{E} \tag{9.1}$$

where $M_{\mathcal{E}}$ is the number of levels per unit energy range. $\mathcal{E}_2$ is the energy at the top of the filled band and $\zeta_0$ is the energy at the Fermi limit ($\mathcal{E}$ is energy as before). The number of trapping states $M$ i.e. number of levels above the Fermi limit:

$$= \int_{\zeta_0}^{\mathcal{E}_1} M_{\mathcal{E}}\, d\mathcal{E} \tag{9.2}$$

where $\mathcal{E}_1$ is the energy at the bottom of the conduction band. If $m_{\mathcal{E}}$ is the number of occupied traps per unit energy range then the total number of trapped electrons is given by:

$$m = \int_{\zeta_0}^{\mathcal{E}_1} m_{\mathcal{E}}\, d\mathcal{E}. \tag{9.3}$$

For neutrality we must have:

$$n - p = \int_{\mathcal{E}_2}^{\zeta_0} (M_{\mathcal{E}} - m_{\mathcal{E}})\, d\mathcal{E} - \int_{\zeta_0}^{\mathcal{E}_1} m_{\mathcal{E}}\, d\mathcal{E} \tag{9.4}$$

where $n$ is the number of conduction electrons and $p$ is the number of positive holes in the filled band. If $J$ electrons and $J$ holes are created per cm.$^3$ per sec. by light excitation then the kinetic equations for the electrons and holes are:

$$\frac{dn}{dt} = J - n \int_{\mathcal{E}_2}^{\mathcal{E}_1} \beta (M_{\mathcal{E}} - m_{\mathcal{E}})\, d\mathcal{E} + \int_{\mathcal{E}_2}^{\mathcal{E}_1} \gamma\, m_{\mathcal{E}}\, d\mathcal{E}, \tag{9.5}$$

$$\frac{dp}{dt} = J - p \int_{\mathcal{E}_2}^{\mathcal{E}_1} \alpha\, m_{\mathcal{E}}\, d\mathcal{E} + \int_{\mathcal{E}_2}^{\mathcal{E}_1} \delta (M_{\mathcal{E}} - m_{\mathcal{E}})\, d\mathcal{E} \tag{9.6}$$

---

[1] I. Broser and R. Broser-Warminsky: Brit. J. Appl. Phys. Suppl. **4**, 90 (1955).
[2] R. C. Herman and C. F. Meyer: J. Appl. Phys. **17**, 743 (1946).
[3] J. T. Randall and M. H. F. Wilkins: Proc. Roy. Soc. Lond., Ser. A **184**, 390 (1954)

where $\beta$ is the recombination coefficient for electrons, $\gamma$ is the escape coefficient for trapped electrons, $\alpha$ the recombination coefficient for holes and $\delta$ the escape coefficient for trapped holes. These constants may have to be modified in specific cases. The above equations neglect band to band and interlevel transitions. Using Eqs. (9.4), (9.5) and (9.6) we obtain the time dependence of the trapped electron density:

$$\frac{dm_{\mathcal{E}}}{dt} = \beta\, n\, M_{\mathcal{E}}\left(1 + \frac{\delta}{\beta n}\right) - \beta\, n\, m_{\mathcal{E}}\left(1 + \frac{\alpha p}{\beta n} + \frac{\gamma}{\beta n} + \frac{\delta}{\beta n}\right). \tag{9.7}$$

For steady state conditions we have:

$$\frac{dm_{\mathcal{E}}}{dt} = 0$$

and so

$$\frac{m_{\mathcal{E}}}{M_{\mathcal{E}}} = \left(1 + \frac{\alpha p + \gamma}{\beta n + \delta}\right). \tag{9.8}$$

In the dark, Eq. (9.8) must give the FERMI distribution:

$$\frac{m_{\mathcal{E}}}{M_{\mathcal{E}}} = \left[1 + (F/n)\exp\{-(\mathcal{E}_1 - \mathcal{E})/kT\}\right]^{-1} \tag{9.9}$$

where

$$F = 2\,(2\pi\, m\, kT/h^2)^{\frac{3}{2}}.$$

We thus find an interrelation between $\beta$ and $\gamma$ and between $\alpha$ and $\delta$

$$\gamma = \beta \cdot F \exp\{-(\mathcal{E}_1 - \mathcal{E})/kT\}, \tag{9.10}$$

$$\delta = \alpha F \exp\{-(\mathcal{E} - \mathcal{E}_2)/kT\}. \tag{9.11}$$

Assuming the usual form for $\delta$ and $\gamma$ from theory we may write:

$$\gamma = \gamma_0 \exp\{-(\mathcal{E}_1 - \mathcal{E})/kT\} \quad \text{and} \quad \delta = \delta_0 \exp\{-(\mathcal{E} - \mathcal{E}_2)/kT\} \tag{9.12}$$

and assume that in Eqs. (9.10) and (9.11) $\alpha$ and $\beta$ vary only slowly with $\mathcal{E}$ and may here be taken as constant. With Eqs. (9.4), (9.10) and (9.11) we can modify Eqs. (9.5) and (9.6) to give:

$$\frac{dn}{dt} = J - \beta\, n\,(n + M - p) + \beta\, n \int_{\mathcal{E}_2}^{\mathcal{E}_1} (F/n)\, m_{\mathcal{E}} \exp\{-(\mathcal{E}_1 - \mathcal{E})/kT\}\, d\mathcal{E}, \tag{9.13}$$

$$\frac{dp}{dt} = J - \alpha\, p\,(p + A - n) + \alpha\, p \int_{\mathcal{E}_2}^{\mathcal{E}_1} (F/p)\,(M_{\mathcal{E}} - m_{\mathcal{E}}) \exp\{-(\mathcal{E} - \mathcal{E}_2)/kT\}\, d\mathcal{E} \tag{9.14}$$

which equations together with Eq. (9.7) give the general mathematical expression for finding the free electron and hole densities and hence the photoconductivity or dark conductivity decay and heating curves under any specified conditions and for any distribution of levels in the forbidden zone.

BROSER and BROSER-WARMINSKY then derive the steady state behaviour for more specific conditions, in particular for the case where the free holes are unimportant in the conduction process and where semiconduction is negligible (i.e. $\mathcal{E}_1 - \zeta_0$ large). We may select their derivations for non-steady states as in conductivity decay and heating experiments. The important point which they make is that the steady state variation of photoconductivity with exciting light intensity and temperature (the same observation holds for the related luminescence in photoconductors showing emission as well) is closely related to the form of the decay and "glow curves" since, as seen also from the theory of ROSE above, both depend on the trap distribution. Such relations can be established without a detailed knowledge of the specific trap distribution. For the conditions $n \gg p$ (also $\beta n \gg \alpha p$) and no dark current, Eqs. (9.13) and (9.14) yield:

$$\left.\begin{aligned}\frac{dn}{dt} + \int_{\zeta_0}^{\mathcal{E}_1} \frac{dm_{\mathcal{E}}}{dt} \frac{d\mathcal{E}}{[1 + (n/F)\exp\{(\mathcal{E}_1 - \mathcal{E})/kT\}]} \\ = J - \beta n (n + m) + \beta n \int_{\zeta_0}^{\mathcal{E}_1} \frac{M_{\mathcal{E}}\, d\mathcal{E}}{[1 + (n/F)\exp\{(\mathcal{E}_1 - \mathcal{E})/kT\}]}\end{aligned}\right\} \tag{9.15}$$

and assuming a slow variation of $M_{\mathcal{E}}$ with $\mathcal{E}$ and the validity of Eq. (9.9) for steady and non-steady states then

$$\frac{dm_{\mathcal{E}}}{dt} \approx \frac{M_{\mathcal{E}}(F/n)\exp\{-(\mathcal{E}_1 - \mathcal{E})/kT\}}{[1 + (F/n)\exp\{-(\mathcal{E}_1 - \mathcal{E})/kT\}]^2} \left\{\frac{1}{n} \cdot \frac{dn}{dt} - \frac{(\mathcal{E}_1 - \mathcal{E})}{kT^2} \cdot \frac{dT}{dt}\right\} \tag{9.16}$$

and by integration of Eq. (9.15):

$$\frac{dn}{dt}\left\{1 + \frac{M_{\mathcal{E}} kT}{2n}\right\} - \frac{dT}{dt} \cdot \frac{M_{\mathcal{E}} k}{2} \log(F/n) = J - \beta n \{n + m - M_{\mathcal{E}} kT \log(F/n)\}. \tag{9.17}$$

This gives the relation between $n$, $J$, $t$ and $T$ for the case of $N$-type conduction. For the decay of current after excitation to equilibrium with light intensity $J_0$ we have from Eq. (9.17):

$$\frac{dn}{dt} \cdot \frac{M_{\mathcal{E}} kT}{2n} - n\beta\{n + m - M_{\mathcal{E}} kT \log(F/n)\} = -n f(n) \tag{9.18}$$

giving the approximate solution:

$$n f(n) \approx \frac{J_0}{1 + \frac{2 J_0 t}{M_{\mathcal{E}} kT}}. \tag{9.19}$$

Reverting to the simple theories at the beginning of this section the quantity $f(n)$ represents the departure from the simple bimolecular recombination process. It is found from the variation of $n$ with $J$ in the steady state

$$f(n) = J/n_{(\text{steady state})}. \tag{9.20}$$

For the heating curve of conductivity, Eq. (9.17) is written:

$$\frac{dT}{dt} \cdot \frac{M_{\mathcal{E}} k}{2} \log(F/n) = n\beta\{n + m - M_{\mathcal{E}} kT \log(F/n)\} = n f(n) \tag{9.21}$$

assuming a slow variation of $M_{\mathcal{E}}$ with $\mathcal{E}$. In this case we obtain for the conductivity

$$\frac{\sigma_{t(T)}}{\sigma_T} = \frac{k}{2\tau_0} \cdot \frac{dT}{dt} \cdot M_{\mathcal{E}} \log(F/n) \tag{9.22}$$

where $\sigma_{t(T)}$ is the "glow" conductivity measured at time $t$ after excitation ceases and at temperature $T$ while $\sigma_T$ is the conductivity obtaining at temperature $T$ for steady state excitation of intensity $J_0$.

We thus see that the theory gives relations between steady state conditions although solution for absolute behaviour would be formidable except for simple trap distributions. It should also be noted that retrapping is implicitly taken care of in this treatment. Using the above theory, experimental decay curves can be used to derive the "glow curve" and this can be compared with the experimental heating curves[1]. Examples of glow curves are given in Sect. 39.

### c) The barrier layer theory of photoconductivity.

**10.** In later sections we shall be considering photoconductors in the form of thin microcrystalline layers rather than large crystals. In addition such materials as lead sulphide, selenide and telluride, germanium and silicon have appreciable semiconduction at their operating temperature. For such photoconductors the theories given above, which may be termed "recombination" theories, are inadequate. A different explanation has been put forward to explain their behaviour. It is suggested that the crystalline boundaries give rise to potential barriers or that there will be local fluctuations in defect or impurity levels giving transitions from $P$ to $N$ conduction. In such cases light absorption is effective when it occurs near the potential barriers at which rapid separation of the electron hole pairs occurs to offset recombination. This separation will cause a lowering of the potential barrier and so allow more conduction electrons to flow across the barrier. The process is thus one of current modulation by the action of light absorbed near the barriers. In the volume of each crystallite recombination is sufficiently rapid to give a negligible photocurrent. Qualitative discussion of the barrier modulation process has been given by several workers notably Sosnowski and coll.[2], Rittner[3] and by Schwarz[4]. Gibson[5] has given a quantitative treatment which is outlined below. The whole process demands the use of the metal-semiconductor contact theory based on the energy band model[6].

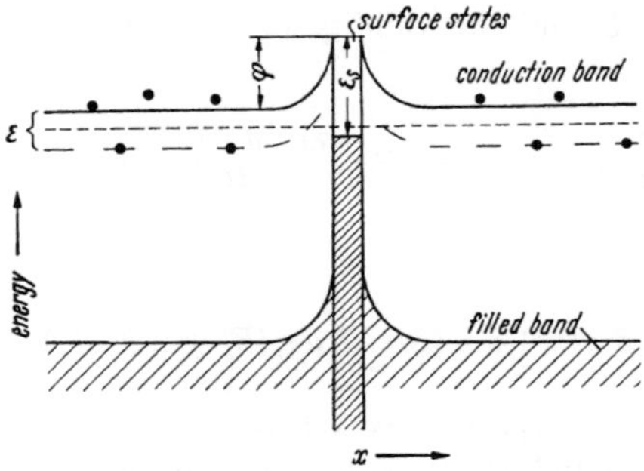


Fig. 6. Potential barrier between two $N$-type crystallites showing surface states for trapping electrons.

Gibson chooses a barrier model as shown in Fig. 6 which is a probable form for an intercrystalline boundary and might be termed a double Schottky barrier which has a maximum height of $\varphi$ above the bottom of the conduction band. Surface states are assumed to exist at the boundary, the nature of which will be discussed in later sections for specific photoconductors. The photoconductor is assumed to be $N$-type with a set of electron donor levels of a single activation energy for simplicity. The surface states are assumed to have an energy limit $\mathcal{E}_s$ below the top of the barrier. The Fermi limit will lie between the donor levels and the conduction band as shown in metal-semiconductor contact theory[6]. Let the density of donor levels be $M$ per cm.$^3$ at a depth $\varepsilon$ below the conduction band and the area density of the surface states be $S_0$ per cm.$^2$. In the dark

[1] I. Broser and R. Broser-Warminsky: Brit. J. Appl. Phys. Suppl. **4**, S. 90 (1955).
[2] L. Sosnowski, J. Starkiewiscz and O. Simpson: Nature, Lond. **159**, 818 (1947).
[3] E. S. Rittner: Science, Lancaster, Pa. **111**, 685 (1950).
[4] E. Schwarz: Nature, Lond. **162**, 614 (1949).
[5] A. F. Gibson: Proc. Phys. Soc. Lond. B **64**, 603 (1951).
[6] N. F. Mott and R. W. Gurney: Electronic Processes in Ionic Crystals. Oxford 1940.

let the number of surface states emptied be $\Delta S_0$ which is given by:

$$\Delta S_0 = S_0 \exp\left[(\varphi + \varepsilon - \mathcal{E}_s)/kT\right] \tag{10.1}$$

where $(\varphi + \varepsilon - \mathcal{E}_s)$ is the depth of the surface states below the FERMI limit and is assumed to be much greater than $kT$. If the donor levels are very near to the conduction band, $\varepsilon$ is very small and so the number of conduction electrons will not vary much with temperature. BARDEEN[1] has shown that $\varphi$ is given by:

$$\varphi = (S_0 + \Delta S_0)^2 e^2/2\varkappa M \tag{10.2}$$

where $\varkappa$ is the dielectric constant. The surface charge $Q$ is given by:

$$Q = (S_0 - \Delta S_0)\, e = (2\varkappa M \varphi)^{\frac{1}{2}}. \tag{10.3}$$

If illumination lowers the barrier height by $\Delta\varphi$ to $\varphi - \Delta\varphi$ and as a result a total of $\Delta S_0 + \Delta S$ surface states are empty ($\Delta S$ of them due to the light) then:

$$\Delta Q = e\,\Delta S \quad \text{and} \quad \varphi = \frac{Q\,\Delta Q}{\varkappa M} \tag{10.4}$$

and

$$\Delta S = \frac{\varkappa M \Delta\varphi}{e\,Q} = \frac{\varkappa M \Delta\varphi}{e^2 (S_0 - \Delta S_0)} \approx \frac{\varkappa M \Delta\varphi}{e^2 S_0} = H \cdot \Delta\varphi \tag{10.5}$$

where

$$H = \varkappa M/e^2 S_0.$$

Now if $n$ electrons per cm.$^2$ flow across the barrier per second and $n$ varies with barrier height, thus:

$$n = n_0 \exp(-\varphi/kT) \tag{10.6}$$

then for a small change $\Delta\varphi$ in $\varphi$ and using Eq. (10.5) we have:

$$\frac{\Delta n}{n} = -\frac{\Delta S}{HkT}. \tag{10.7}$$

In the dark thermal equilibrium gives

$$S_0 \cdot \nu \cdot \exp(-\mathcal{E}_s/kT) = n \cdot \Delta S_0\, a \tag{10.8}$$

where $a$ is the capture cross section of an empty surface state for an electron and $\nu$ is a frequency constant. If illumination empties $J$ surface states per second per cm.$^2$, then:

$$J + S_0 \nu \exp(-\mathcal{E}_s/kT) = a\,(n + \Delta n)\,(\Delta S_0 + \Delta S) \tag{10.9}$$

and so:

$$J = a\,\Delta n\,(\Delta S_0 + \Delta S + HkT) \approx \Delta n \left\{\frac{\nu S_0 \exp(-\mathcal{E}_s/kT)}{n} + HkT\right\} \tag{10.10}$$

if $\Delta S \ll \Delta S_0$, i.e. for relatively low illumination intensities.

The net photocurrent $\Delta i$ is proportional to $\Delta n$, i.e., it is the difference in currents flowing in the barrier whereas $n$ gives the sum of these currents. From rectifier theory[2] we have:

$$\frac{e\,n}{\Delta i} = \frac{\exp(eV/kT) + 1}{\exp(eV/kT) - 1}. \tag{10.11}$$

$V$ being the potential difference across the barrier. We thus see that $\Delta i$ is proportional to $J$ at low intensities. We can obtain the time constant $\tau$ which is given by:

$$\tau = \frac{HkT}{aHkT + S_0 \nu \exp(-\mathcal{E}_s/kT)} \tag{10.12}$$

[1] J. BARDEEN: Phys. Rev. **71**, 717 (1947).
[2] N. F. MOTT and R. W. GURNEY: Electronic Processes in Jonic Crystals. Oxford 1940

and for steady illumination we have

$$\Delta i = \tau \cdot J. \tag{10.13}$$

Thus if $\Delta n \propto J$ then $\tau$ does not change with light intensity.

The above theory, as was stressed by GIBSON[1], gives important information regarding the relation between photoconduction sensitivity and the response time. This may be compared with the relations resulting from the previous "recombination" theories.

We may define the photoconduction sensitivity ($\eta$) as the fractional change in conductivity per unit change in light intensity:

$$\eta = d\,(\Delta i/i)/d J \tag{10.14}$$

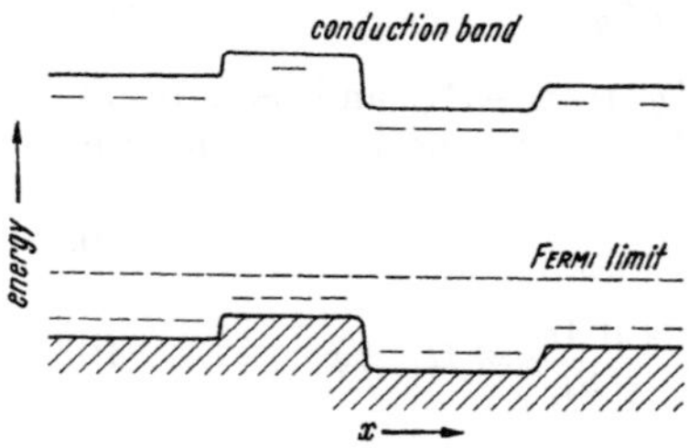

Fig. 7. Effect of varying acceptor level density on the band system of a $P$-type crystal.

and from the barrier modulation theory above we have

$$\eta = \tau/n\,kT, \tag{10.15}$$

From the recombination theories above, at least in their simple form:

$$\frac{\Delta i}{e} = C \cdot \tau \cdot J, \tag{10.16}$$

and so

$$\eta = e\,\tau/i \tag{10.17}$$

where $C$ usually defines the ratio of the number of positive holes to the number of conduction electrons. In the barrier theory, $C$ is proportional to $i$ and the latter does not therefore appear in the expression for the sensitivity. Thus the two theories give different expressions for the sensitivity and offer as shown, in Sect. 34 below, a means of distinguishing the mechanism appropriate to any specific photoconductor by the experimental behaviour.

The type of potential barrier shown in Fig. 6 is not the only one which can be envisaged. In experiments on lead sulphide layers the maximum sensitivity is obtained by oxidation of the original $N$-type material until it is just $P$-type. SCHWARZ[2] suggested that the barrier is formed by oxide layers at the crystalline interfaces. JAMES[3] and RITTNER[4] have assumed that $P$-type levels diffuse through the layers and give local fluctuations in level density so that the energy bands are displaced as shown in Fig. 7 to keep the FERMI level constant throughout the crystal. Fluctuations in $P$ and $N$ states will give $P$-$N$ junctions either sharp or diffuse.

Further discussion of such barriers is reserved for specific sections. Meanwhile we consider another photoconduction phenomenon which is dependent on potential barriers.

### d) Theory of the photovoltaic effect in solids.

**11.** The earliest discoveries of the photovoltaic effect showed that the photovoltage produced by illumination was associated with a potential barrier. This barrier can be either within the material, e.g. a $P$-$N$ junction or at a semiconductor-metal contact. Usually it is found that the photovoltage $V_L$ (measured

[1] A. F. GIBSON: Proc. Phys. Soc. Lond. B **64**, 603 (1951).
[2] E. SCHWARZ: Nature, Lond. **162**, 614 (1949).
[3] H. M. JAMES: Science, Lancaster, Pa. **110**, 254 (1949).
[4] E. S. RITTNER: Science, Lancaster, Pa. **111**, 685 (1950).

for open circuit conditions) is a logarithmic function of the exciting light intensity $J$ of the form:

$$V_L \propto \log(1 + \alpha\sigma) \tag{11.1}$$

where $\alpha$ is a constant. In general the photovoltaic effect is simpler to treat theoretically since the voltage can be measured at zero current and is not critically dependent on barrier shape or on carrier mobility. The logarithmic relation was obtained theoretically by GORTER, BROER and SNOEK[1] but we give below a treatment due to MOSS[2] for the case of a sharp $P$-$N$ junction. However, it can be applied to other potential barriers with relatively small modification.

The energy diagram for the $P$-$N$ junction is shown in Fig. 8. The FERMI limit is unaltered throughout. In the $N$-type section its position relative to the energy $\mathcal{E}_1$ of the bottom of the conduction band is given by:

$$\psi_N = \mathcal{E}_1 - kT \log(F/n) \tag{11.2}$$

and in the $P$-type relative to the energy $\mathcal{E}_2'$ of the top of the filled band by:

$$\psi_P = \mathcal{E}_2' + kT \log(F/P) \tag{11.3}$$

where $n$ and $p$ are the free electron and free hole densities and

$$F = 2(2\pi m kT/h^2)^{\frac{3}{2}} \tag{11.4}$$

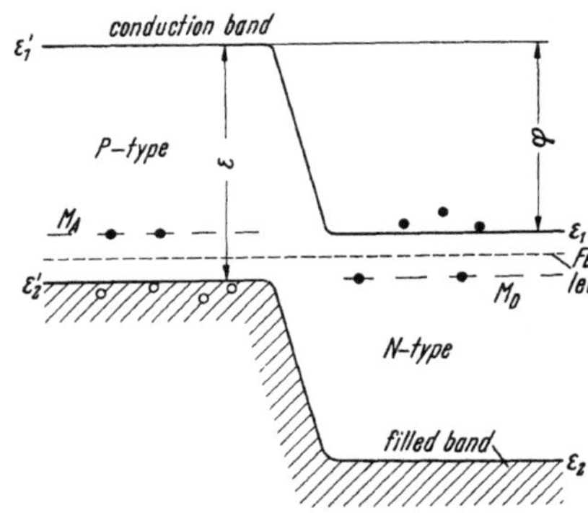

Fig. 8. Energy diagram for $P$-$N$ junction in a photoconductor.

as above in Eq. (9.9).

If there are $M_D$ donor levels per cm.$^3$ in the $N$-type region and $M_A$ acceptor levels per cm.$^3$ in the $P$-type region, these levels being very "shallow", then $n = M_D$ and $p = M_A$. Equating $\psi_N$ and $\psi_P$ together we have:

$$\mathcal{E}_2' = \mathcal{E}_1' - \mathcal{E} = \mathcal{E}_1 - kT \log(F^2/M_D \cdot M_A) \tag{11.5}$$

where $\mathcal{E}$ is the energetic width of the forbidden zone, and so:

$$\varphi = \mathcal{E}_1' - \mathcal{E}_1 = \mathcal{E} - kT \log(F^2/M_A M_D) \tag{11.6}$$

$$= \mathcal{E} - 2kT \log(F/M) \tag{11.7}$$

if it is assumed that $M_A \approx M_D$

$\varphi$ differs from $\mathcal{E}$ by about 0.1 eV for normal impurity concentrations. Using rectifier theory for the potential barrier[3] the current across the $P$-$N$ junction will be given by:

$$\left.\begin{aligned} i = i_e \exp(-\varphi/kT)\,[1 - \exp(-eV/kT)] + \\ + i_p \exp(-\varphi/kT)\,[1 - \exp(-eV/kT)] \end{aligned}\right\} \tag{11.8}$$

where $i_e$ and $i_p$ are the electron and hole current densities respectively and $V$ is the applied voltage across the junction.

If illumination frees $J$ electrons and $J$ holes per cm.$^3$ per second, then we can set up recombination equations. We assume a number $A$ of thermal excitations

[1] C. J. GORTER, L. J. F. BROER and A. P. SNOEK: Physica, Haag **11**, 401, 426 (1946).
[2] T. S. MOSS: Photoconductivity in the Elements. London 1952.
[3] N. F. MOTT and R. W. GURNEY: Electronic Processes in Ionic Crystals. Oxford 1940.

per cm.[3] per second giving $n_0$ carriers so that for $M$ impurity centres in the dark:

$$A = \beta n_0 (n_0 + m) \tag{11.9}$$

and on illumination to steady state conditions:

$$A + J = \beta n (n + m) \tag{11.10}$$

Thus:

$$\frac{\Delta n}{n} = \frac{J}{\beta n (n + M)}. \tag{11.11}$$

For electrons the sensitivity $\eta_e$ is given by:

$$\eta_e = \frac{\Delta n}{J n} = \frac{1}{\beta n (M_D + 2n)} \tag{11.12}$$

and for holes

$$\eta_p = \frac{\Delta p}{J p} = \frac{1}{\beta p (M_A + 2p)}. \tag{11.13}$$

For fully ionised impurity levels as assumed above and in the $P$-type conductor we have:

$$\eta(P)_p = \frac{1}{2\beta p^2} = \frac{1}{2\beta M_A^2} \tag{11.14}$$

and

$$\eta(P)_e = \frac{1}{\beta n M_A} \quad \text{since} \quad n \ll M_A. \tag{11.15}$$

If $n_i$ is the density of intrinsic carriers in either region it can be shown[1] that $n \cdot p = n_i^2$ and so:

$$n = n_i^2/p \quad \text{and} \quad \eta(P)_e = \frac{1}{\beta n_i^2}. \tag{11.16}$$

Now $M_A^2 \gg n_i^2$ and so $\eta(P)_e \gg \eta(P)_p$. Similarly in the $N$-type conductor $\eta(N)_p \gg \eta(N)_e$ showing that the photoconduction sensitivity is determined by the minority carriers. The current at the barrier on illumination will be given by:

$$\left.\begin{aligned} i_L = i_e \exp(-\varphi/kT)\,[1 + J\,\eta(P)_e - \exp(-eV/kT)] + \\ + i_p \exp(-\varphi/kT)\,[1 + J\,\eta(N)_p - \exp(-eV/kT)] \end{aligned}\right\} \tag{11.17}$$

and when the current is zero the photovoltage $V_L$ is given by:

$$V_L = -\frac{kT}{e} \log(1 + \eta J) \tag{11.18}$$

where

$$\eta = \frac{i_e \eta(P)_e + i_p \eta(N)_p}{i_e + i_p}. \tag{11.19}$$

If recombination coefficients for holes and electrons are the same then

$$\eta(P)_e = \eta(N)_p = \frac{1}{\beta n_i^2} \tag{11.20}$$

and also $\beta n_i^2 = A$ (see Sect. 6) since $m = 0$ when $n = n_i$. Thus $\eta = 1/A$ and now:

$$V_L = -\frac{kT}{e} \log(1 + J/A) \tag{11.21}$$

---

W. SHOCKLEY: Electrons and Holes in Semi Conductors. New York 1950.

which at low light intensities becomes

$$V_L = -\frac{kT}{eA} \cdot J. \tag{11.22}$$

Agreement of Eq. (11.21) with experiment has been verified by several workers over wide ranges of light intensities of the order of $10^6:1$.

For a semiconductor-metal barrier a similar equation may be established.

## e) The spectral dependence of photoconductivity.

**12. Spectral response and the energy band model.** The energy band model described above has obvious implications with respect to the spectral response of photoconductivity. For an intrinsic conductor the long wavelength limit will be set approximately by the width of the forbidden energy zone $(\mathcal{E}_1 - \mathcal{E}_2)$. However, transitions from the top of the filled band to the bottom of the conduction band are normally forbidden unless, as is so in practice, the crystal contains defects or impurities or the lattice is perturbed by thermal vibrations. The possible transitions giving rise to photoconductivity in the general case can be listed as follows:

*i)* Lattice band to band transitions producing both free electrons and free positive holes.

*ii)* Transitions of electrons to the conduction band from normally occupied defect or impurity states or from trapping states previously filled by higher energy excitation (*N*-type conduction).

*iii)* Transitions of electrons from the filled band to unoccupied defect or impurity levels (*P*-type conduction).

*iv)* Excitation in the matrix lattice producing excitons (bound electron hole-pairs) which are subsequently dissociated to give free carriers. It is quite difficult to distinguish this process in experiment.

The most interesting region of spectral response is that around the absorption edge, although in many cases photoconductivity is observed at very much longer wavelengths. (For example, silicon with absorption edge at 1 $\mu$ can show response at 38 $\mu$ at very low temperatures[1]. In the case of lattice transitions, absorption coefficients are very high and excitation is confined to a very thin surface layer of the material. This gives a relatively high carrier concentration at quite low excitation densities and so increases the probability of recombination with consequent reduction in carrier life time and resulting photoconduction sensitivity. This may be offset in some cases by high diffusion coefficients for the carriers, in which case the response will not fall off very rapidly with decrease in wavelength through the absorption edge. For example, we may compare the cases of germanium and diamond. For germanium the diffusion coefficient $D$ is $\sim 80$ cm.$^2$/sec. and the recombination coefficient for electrons $\beta$ is $\sim 10^{-11}$ cm.$^3$/sec., while for diamond the values are $D \sim 20$ cm.$^2$/sec. and $\beta \sim 10^{-6}$ cm.$^3$/sec. Thus for germanium the photosensitivity is maintained at shorter wavelengths. The photosensitivity for surface excitation will also be deleteriously affected by recombination of carriers in surface states. In the case of germanium the effect of surface states on carrier life time is well known and has been well explored experimentally as discussed later in Sect. 21.

[1] E. BURSTEIN, J. J. OBERLY and J. W. DAVISSON: Phys. Rev. **89**, 331 (1953).

For the case of intrinsic photoconduction, Moss[1] has constructed a semi-empirical model for the spectral response. He assumes that the response or sensitivity $\eta_{\mathcal{E}}$ is given by the formula:

$$\eta_{\mathcal{E}} = \frac{1}{1 + \exp\{c(\mathcal{E}_0 - \mathcal{E})\}} \tag{12.1}$$

where $c$ is a constant, $\mathcal{E}$ is the photon energy at wavelength $\lambda$ and $\mathcal{E}_0$ that at the *threshold* wavelength for response. The form of this relation is shown in

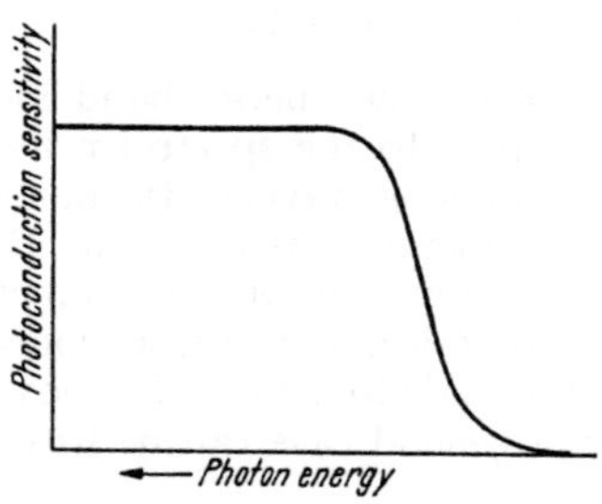


Fig. 9. Representation of dependence of photocurrent on photon energy.

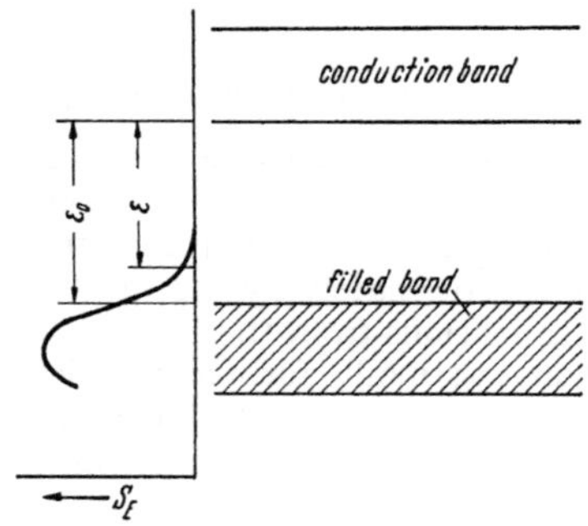


Fig. 10. Distribution of energy levels in the filled band of a photoconductor giving photoconduction response of Fig. 9 (after Moss ref. 3).

Fig. 9. Moss then shows that this type of response spectrum can arise from a distribution of energy levels in the filled band of the form:

$$S_{\mathcal{E}}\, d\mathcal{E} = \frac{G \cdot c \exp\{c(\mathcal{E}_0 - \mathcal{E})\}\, d\mathcal{E}}{[1 + \exp\{c(\mathcal{E}_0 - \mathcal{E})\}]^2} \tag{12.2}$$

as shown graphically in Fig. 10. For such a distribution the number of electrons in equilibrium in the conduction band due to thermal excitation is:

$$n \propto \exp(-\mathcal{E}_0/2kT). \tag{12.3}$$

The quantity $\mathcal{E}_0$ is the same as that which would be obtained from variation of semiconduction or HALL coefficient with temperature. Putting $\mathcal{E} = \mathcal{E}_0$ in Eq. (12.1) shows that for this value the photoconduction response is one half the maximum value. This treatment is only strictly valid for a non-polar crystal in which FRANCK-CONDON shifts are absent[2] but it applies approximately in other cases, e.g. PbS, as shown by Moss[3]. The $\lambda_{\frac{1}{2}}$ value in any case provides a useful specification of the photoconduction threshold. (For behaviour of polar crystals see treatment of oxides, Sects. 42 to 46).

Defects and impurities provide absorption centres for lower energy photons that those absorbed by the lattice. Since the concentration of the centres is only of the order of one per $10^4$ atoms, absorption coefficients are correspondingly smaller ($\sim 1$ mm.$^{-1}$) and so volume excitation of the photoconductor is possible with relatively low carrier concentrations and high life times. Thus in imperfect or impurity activated crystals the photoconduction response often shows a maximum outside the absorption edge (e.g. in cadmium sulphide). Only one type of carrier is mobile and the space charge due to the fixed carriers will in some cases facilitate entry of carriers of opposite sign from the appropriate electrode to replace carriers absorbed at the opposite electrode. In this case

[1] T. S. Moss: Photoconductivity in the Elements. London 1952.
[2] N. F. MOTT and R. W. GURNEY: Electronic Processes in Ionic Crystals. Oxford 1940.
[3] T. S. Moss: Photoconductivity in the Elements. London 1952.

we have the effective carrier life time greater than the transit time and so current gain occurs. This seems a more reasonable interpretation of gain effects e.g. in CdS, than a collisional multiplication of carriers since relatively low applied fields produce such gain compared with other cases such as light amplification and electroluminescence in similar crystals[1].

It is possible to obtain long wavelength response in crystals by previous excitation at shorter wavelengths, the latter raising carriers into relatively shallow traps from which they can then be ejected by absorption of low energy photons. This effect is well known in such photoconductors as CdS where short wavelength excitation ($\sim 0.55\,\mu$) can produce a photosensitivity response up to $6\,\mu$[2].

**13. Temperature dependence of spectral sensitivity.** For most photoconductors, lowering of temperature results in a small shift of the response limit to shorter wavelengths, the amount being of the order of $10^{-4}$ eV/°C. However, in certain cases, notably those of PbS, PbSe, PbTe, $Cd_3As_2$ and Te there is a shift to longer wavelengths as the photoconductor is cooled. The temperature dependence of spectral response will be determined by several factors. For an intrinsic conductor increase in temperature may broaden the energy bands with resultant narrowing of the forbidden zone and/or the whole band system may show relative movement in energy. Moss has discussed both these effects in some detail[3].

Energy band broadening with increase of temperature will arise from increase in lattice vibration energy and will give $d\mathcal{E}/dT$ negative. Band shifts may arise from increase in lattice constant again giving a negative value for $d\mathcal{E}/dT$. In the first case of broadening due to lattice vibrations theoretical treatment applicable to polar crystals has been given by RADKOWSKY[4], while FAN[5] has obtained theoretical results which correlate with those obtained experimentally for silicon and germanium. BARDEEN and SHOCKLEY[6] have calculated the effect due to lattice expansion. In both treatments the spectral shifts with temperature depend on carrier mobility. More quantitatively, the higher the mobility the smaller the shift as shown e.g. by germanium.

**14. Photoconductivity and refractive index.** It was postulated by KURRELMEYER[7] and by GUDDEN[8] that for a solid to be an intrinsic photoconductor its refractive index should be greater than 2. MOSS[8,9] has recently developed this idea in more detail by a consideration of the energy states likely for defects or impurities in solids. An electron trapped in such centres may have a hydrogen like system of energy levels except that system will be reduced in energy by a factor $1/\varkappa^2$ where $\varkappa$ is the effective *dielectric constant* at the site and will be assumed proportional to the square of the refractive index. If optical absorption occurs in these energy states and subsequent thermal activation frees the excited carrier then the necessary energy is proportional to $1/\varkappa^2$ or the wavelength threshold for photoconduction will vary as the 4th power of the refractive index $r$. Moss finds from comparison of experimental results that there is an approximate connection between the two quantities for a wide variety of photoconductors.

1 See the present writer's contribution to vol. XXVI of this Encyclopedia.
2 C. C. KLICK and J. LAMBE: Phys. Rev. **98**, 909 (1955).
3 T. S. MOSS: Photoconductivity in the Elements. London 1952.
4 A. RADKOWSKY: Phys. Rev. **73**, 749 (1948).
5 H. Y. FAN: Semi Conducting Materials. London 1951.
6 J. BARDEEN and W. SHOCKLEY: Phys. Rev. **80**, 72 (1950).
7 B. KURRELMEYER: Phys. Rev. **30**, 893 (1927).
8 B. GUDDEN: Lichtelektrische Erscheinungen. Leipzig 1928.
9 T. S. MOSS: Proc. Phys. Soc. Lond. A **63**, 167 (1950).

The relation is

$$\frac{r^4}{\lambda_{\frac{1}{2}}} \approx 77. \tag{14.1}$$

Variation of refractive index with temperature may explain the anomalous behaviour of the spectral response of the PbS type crystal as shown by MOSS. Beginning with the LORENTZ-LORENZ formula:

$$\frac{\varkappa - 1}{\varkappa + 2} = B \cdot N\alpha \tag{14.2}$$

where $B$ is a constant, $N$ is the number of molecules per cm.$^3$, and $\alpha$ is the polarizability of the ions involved. By differentiation:

$$\frac{d\varkappa}{dT} = \frac{1}{3}(\varkappa - 1)(\varkappa + 2)\frac{1}{N} \cdot \frac{dN}{d\tau} \tag{14.3}$$

and if $V$ is the volume we can put

$$\frac{1}{3N} \cdot \frac{dN}{dT} = -\frac{1}{3V}\frac{dV}{dT} = a \tag{14.4}$$

where $a$ is the linear coefficient of expansion. From which

$$d\varkappa/dT \approx -\varkappa^2 a.$$

For PbS,

$$\varkappa = (3.9)^2 \quad \text{and} \quad a = 20 \times 10^{-6}/{}^\circ\text{K}, \tag{14.5}$$

giving

$$\frac{d\varkappa}{dT} \approx 5 \times 10^{-3}/{}^\circ\,\text{K}$$

and

$$\lambda_{90^\circ\,\text{K}}/\lambda_{920\ \text{K}} = \left(1 + \frac{\Delta\varkappa}{\varkappa}\right)^2 = 1.13 \tag{14.6}$$

which is of the same order as that obtained experimentally. High refractive index and high expansion coefficient are needed for the effect to be appreciable compared with the other effect discussed above.

**15. Limiting sensitivity of photoconductive cells.** The ultimate sensitivity of a photoconductor is of considerable practical importance when it is used as a detector, particularly of thermal radiation. This limit, in the absence of electrical noise in the photoconductor, is set by the fluctuations in radiation falling on the material. The problem has been considered by several workers particularly MOSS[1], FELLGETT[2] and CLARK JONES[3].

FELLGETT's caculations, when compared with experimental results for a PbS layer, give a minimum detectable wattage only 2.3 times the theoretical limit set by such fluctuations and for a PbTe cell an even lower value of 1.9. This means that at the spectral response curve maximum the quantum efficiency is near unity. In practice the limiting sensitivity is determined by cell current noise which has a $1/\nu$ frequency spectrum. Cooling the cell will reduce this noise and is essential in order to approach the theoretical limit. The operational usefulness of a photoconductive detector is discussed in detail by the authors cited and will not be considered here.

---

[1] T. S. MOSS: J. Opt. Soc. Amer. **40**, 603 (1950).
[2] P. FELLGETT: J. Opt. Soc. Amer. **39**, 970 (1949).
[3] R. CLARK JONES: Adv. Electronics **5**, 1 (1953).

### f) The photo-electro-magnetic effect.

**16.** If a photoconducting crystal is placed in a magnetic field and illuminated at right angles to the field as shown in Fig. 11 then a voltage appears across the specimen in a direction perpendicular to the light beam and the field. The theory of this effect has been given by Moss, Pincherle and Woodward[1]. In brief, if the specimen of length $l$ between electrodes absorbs $J$ quanta per second in its surface then carriers produced diffuse into the specimen and are deflected through the Hall angle $\Theta$ given by $\arctan(\mu B)$ where $B$ is the magnetic induction. The electron carriers in a life time $\tau$ move a distance $(D_e \tau)^{\frac{1}{2}}$ where $D_e$ is the diffusion coefficient given by $\frac{\mu_e k T}{e}$, $\mu_e$ being the electron mobility. For small fields the distance moved between electrodes is $\mu_e B (D_e \tau)^{\frac{1}{2}}$ giving a charge distribution $\mu_e B (D_e \tau)^{\frac{1}{2}} e/l$ to the external circuit. Inclusive of the free hole contribution, the short circuit current is given by:

$$i_s = \eta \frac{B e J}{l} \left[\mu_e (D_e \tau)^{\frac{1}{2}} + \mu_p (D_p \tau)^{\frac{1}{2}}\right] \tag{16.1}$$

where $\eta$ is the quantum efficiency for carrier production. Using a combined $\mu$ and $D$

$$i_s = \eta \frac{B e J}{l} \mu^{\frac{3}{2}} \left(\frac{k T \tau}{e}\right)^{\frac{1}{2}}. \tag{16.2}$$

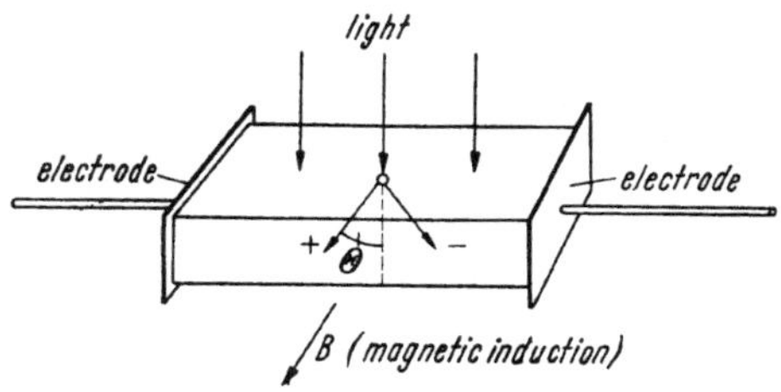

Fig. 11. Schematic diagram for the photo-electro-magnetic effect.

For high magnetic fields $i_s$ is reduced by a factor $\{1 + (\mu B)^2\}$ which is negligible for PbS, but considerable for germanium.

At high light intensities the photoelectro-magnetic effect saturates due to surface recombination of carriers at the back surface. The saturation field $E_s$ there $= B \cdot s$ where $s$ is the surface recombination velocity and so provides a simple means of finding $s$.

If the photocurrent $i$ is measured simultaneously then

$$i = \eta J e \mu \tau E/l \tag{16.3}$$

and so

$$\eta = \frac{l \cdot E}{J B^2 e \mu D} \left(\frac{i_s^2}{i}\right) \tag{16.4}$$

and

$$\tau = \frac{B^2 D}{E^2} \left(\frac{i^2}{i_s^2}\right). \tag{16.5}$$

The photo-electro-magnetic effect provides a means of measuring very short life times $< 10^{-9}$ secs.

## III. Photoconductivity in monatomic solids.

**17. Position in periodic table.** The occurrence of photoconductivity in monatomic solids has been the subject of a very exhaustive study and review by Moss[1] and his book gives a complete and detailed account of the field up to 1951 inclusive. In the present section a much briefer review is attempted but more

[1] T. S. Moss, L. Pincherle and A. M. Woodward: Proc. Phys. Soc. Lond. B **66**, 743 (1953).

recent data included, particularly in the cases of germanium and silicon which have become of great importance in the development of the crystal triode or transistor. Moss has shown that the elements of interest, forming monatomic solids, lie in a particular part of the periodic table from Group III to Group VI. They comprise boron; carbon, silicon, germanium, tin, phosphorus, arsenic, antimony; sulphur, selenium, tellurium and finally iodine. These solids satisfy the requisite rule of Gudden and Pohl (see Moss[1]) that the refractive index of a possible intrinsic photoconductor shall be greater than two. In the above groups antimony is a doubtful photoconductor and the effect has not been observed in tin (grey form). Among the above list of solids Moss has made the first photoconduction measurements in boron, arsenic, phosphorus, tellurium and iodine.

### a) Photoconductivity in boron.

**18.** Considerable difficulty has been experienced by Moss[1] in making photosensitive boron layers by direct evaporation, due to various causes such as high melting point and low vapour pressure. Layers having any sensitivity show an intrinsic photoconductivity limit at about 1 μ with subsidiary peaks at longer wavelengths. Similar results have been obtained more recently by Lagrenaudie[2] who gives the absorption edge as at 1 μ wavelength. Semi-conduction is $P$-type and indicates impurity states with 0.06, 0.3, 0.4 and 0.6 eV activation energies. The last two values may be compared with those for the subsidiary photoconduction response found by Moss in the 2.2 to 2.6 μ region. The most sensitive layers made by the latter were deposited by decomposition of diborane ($B_2H_6$) and having a dark resistance of $\sim 80\,\text{M}\Omega$ at room temperature. Both semi- and photoconduction were proportional to the applied field and the photocurrent was linear with light intensity, the time constant being less than one millisecond. Lagrenaudie[2] gives additional information to that of Moss such as the refractive index which is 3.5 and the dielectric constant (permittivity) for radiofrequencies which is about 12.

### b) Photoconductivity in diamond, silicon and germanium.

**19. Diamond.** Diamond, silicon, germanium and grey tin crystallise in the same zinc blende cubic structure, the basic unit of which is a tetrahedral group of atoms. Systematic studies of diamond were made by Gudden and Pohl[3] and later by Robertson, Fox and Martin[4]. Following more recent studies by the Indian school, notably those of Pant[5], interest in diamond was revived by its performance as a bombardment induced conductivity counter for nuclear particles and radiation[6].

Diamonds which show photoconductivity are those designated by Robertson, Fox and Martin as type II having transparency as far into the ultra violet as 2300 Å. The intrinsic photoconduction rises to a peak at this wavelength falling off on either side, as discussed in Sect. 12 above, and as shown in Fig. 12. The photocurrent in diamond is subject to space charge effects as first observed by Gudden and Pohl[3]. These workers found that infra red radiation could

[1] T. S. Moss: Photoconductivity in the Elements. London 1952.

[2] J. Lagrenaudie: J. Phys. Radium **13**, 554 (1952); **14**, 14 (1953). — J. Chim. phys. **50**, 629 (1953).

[3] B. Gudden and R. W. Pohl: Z. Physik **16**, 170 (1923); **17**, 331 (1923).

[4] R. Robertson, J. J. Fox and A. E. Martin: Phil. Trans. Roy. Soc. Lond., Ser. A **232**, 463 (1934).

[5] D. D. Pant: Proc. Ind. Acad. Sci. **19**, 315 (1944).

[6] R. Hofstadter: Nucleonics **4**, 2, 29 (1949) (review).

dissipate the space charge and increase the steady state photocurrent by a factor of two. In the past this has been assumed to be due to the release of trapped positive holes but in more recent experiments KLICK and MAURER[1] find negligible hole mobility in diamond. It is more likely that the effect arises from a release of trapped electrons from the states about 0.3 eV below the conduction band reported by MCKAY[2]. As seen in the theoretical Sect. 8 and 12 above, release of positive holes in an $N$-type conductor usually causes quenching of the photocurrent by making them available for recombination with electrons. This effect is observed in $N$-type germanium doped with various impurities and discussed below in Sect. 21. It is known that there are absorption bands in type II

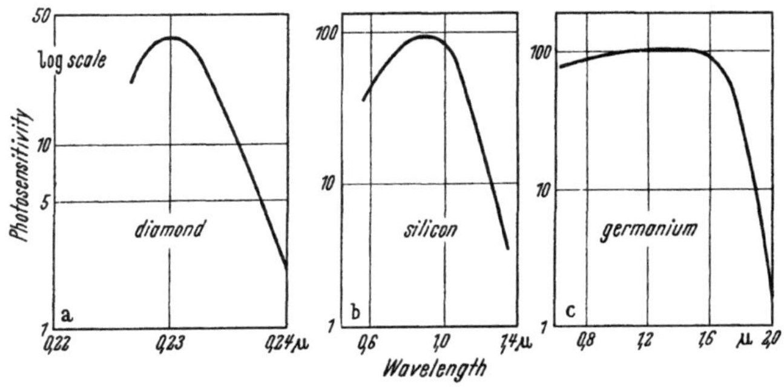


Fig. 12a—c. Spectral response of photoconduction in a) diamond, b) silicon, c) germanium at room temperature.

diamonds at 3 μ and 4.1 to 4.8 μ (~0.4 and 0.3 eV respectively) but no clear correlation between these bands and the „photocurrent stimulation" wavelengths has yet been reported.

In photoconducting diamonds electron mobilities of ~900 cm.$^2$/volt:sec are found and various workers agree on a carrier life time of about $10^{-8}$ sec. CHYNOWETH[3] estimates that in his specimens there were about $10^{16}$ traps per cm. and obtains a recombination coefficient of $10^{-8}$ cm.$^3$/sec. TROTT[4] using $\beta$ particle excitation found during warming of his specimens response peaks at 300 and 400° K and estimated trap depths of 0.5 and 0.8 eV. However, his calculations, based on phosphor glow curve theory[5], assume escape frequency constants for thermal activation of trapped electrons which are not necessarily characteristic of type II diamonds.

The photocurrent in diamonds is a linear function of light intensity and increases exponentially with temperature. The linear behaviour with intensity can be linked with the presence of strong trapping effects as discussed in Sect. 6 and the same is true of the dependence on temperature over the range where trapped electrons become thermally unstable (250 to 400° K). $\beta$ particle bombardment induced conductivity pulses throw light on the texture of diamonds according to CHAMPION[6]. Good counting diamonds appear to have regions of perfect crystallinity separated by thinner barriers of imperfect structure the perfect crystals being about 10 μ in size.

[1] C. C. KLICK and R. J. MAURER: Phys. Rev. **76**, 179 (1949); **81**, 124 (1951).
[2] K. G. MCKAY: Phys. Rev. **77**, 816 (1950).
[3] A. G. CHYNOWETH: Phys. Rev. **83**, 254, 264 (1951).
[4] N. G. TROTT: Proc. Roy. Soc. Lond., Ser. A **220**, 498 (1953).
[5] J. T. RANDALL and M. H. F. WILKINS: Proc. Roy. Soc. Lond., Ser. A **184**, 366 (1945).
[6] F. C. CHAMPION: Proc. Roy. Soc. Lond., Ser. A **220**, 485 (1953).

The photovoltaic effect in diamond was investigated by ROBERTSON, FOX and MARTIN[1] and later by MENDELSSOHN and DEMBER (see MOSS[2]). The latter found photovoltages as high as 0.75 volts in some specimens. The nature of the electrode material is not stated by MOSS.

**20. Silicon.** *α) Photoconductivity*. Silicon is rapidly becoming a very important semiconductor because of its potentialities for power rectification, crystal triode operation and its application as a solar energy generator. It is more difficult to purify and to grow in single crystal form than germanium but such problems are already being overcome.

Investigations of deposited layers of silicon by TEAL, FISHER and TREPTOW[3] showed that the intrinsic photoconduction response peaked at 0.9 μ with a $\lambda_{\frac{1}{2}}$ value of 1.1 μ as shown in Fig. 12b above. The absorption edge of the silicon lattice is also about 1.1 μ. Response times were of the order of 200 μ sec. and linearity between photocurrent and light intensity was observed.

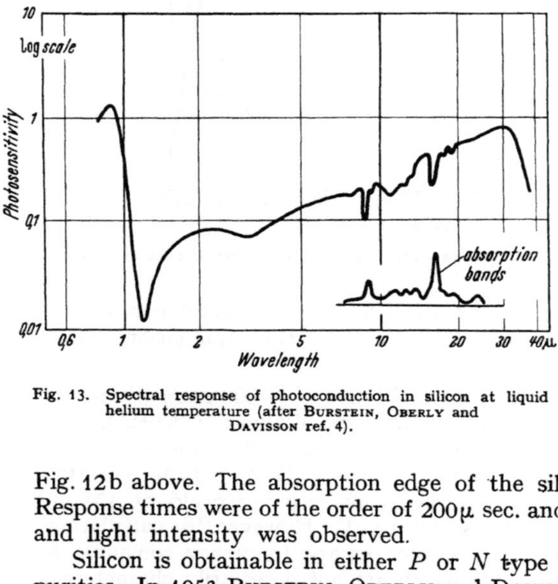


Fig. 13. Spectral response of photoconduction in silicon at liquid helium temperature (after BURSTEIN, OBERLY and DAVISSON ref. 4).

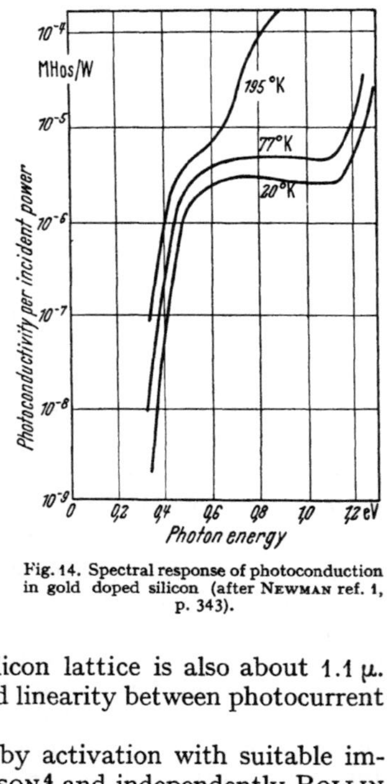


Fig. 14. Spectral response of photoconduction in gold doped silicon (after NEWMAN ref. 1, p. 343).

Silicon is obtainable in either $P$ or $N$ type by activation with suitable impurities. In 1953 BURSTEIN, OBERLY and DAVISSON[4] and independently ROLLIN and SIMONS[5] found that at liquid helium $N$-type silicon gave a photoconduction response out to 38 μ. Fig. 13 shows the results obtained by BURSTEIN *et coll.* for a specimen with $4 \times 10^{15}$ carriers per cm.$^3$ at room temperature. The peak response at long wavelengths gives an optical activation energy of ~0.04 eV compared with that of 0.06 eV obtained for the thermal ionisation energy from data due to PEARSON and BARDEEN[6]. Between 8 and 24 μ the photo response spectrum shows a reduction with fine structure in it which can be accounted for by the lattice vibrations which are excited by some of the incident radiation thus competing with photo ionisation processes.

---

[1] R. ROBERTSON, J. J. FOX and A. E. MARTIN: Phil. Trans. Roy. Soc. Lond., Ser. A **232**, 463 (1934).
[2] T. S. MOSS: Photoconductivity in the Elements. London 1952.
[3] G. K. TEAL, J. R. FISHER and A. W. TREPTOW: J. Appl. Phys. **17**, 879 (1946).
[4] E. BURSTEIN, J. J. OBERLY and J. W. DAVISSON: Phys. Rev. **89**, 331 (1953).
[5] B. V. ROLLIN and E. L. SIMONS: Proc. Phys. Soc. Lond. B **66**, 161 (1953).
[6] G. L. PEARSON and J. BARDEEN: Phys. Rev. **75**, 865 (1949).

NEWMAN[1] has investigated photoconductivity in gold activated silicon crystals and his spectral response curves are given in Fig. 14. TAFT and HORN[2] deduced from resistivity measurements that gold provided donor levels 0.33 eV above the filled band in $P$-type silicon. This value agrees well with the threshold for photoconduction and a simple type of process is indicated. Response is linear with light intensity and the time constant is about 10 μ-sec. Down to temperatures of 20° K no evidence is found for any trapping effects.

*β) Photovoltaic and other effects in silicon.* The photovoltaic effect at $P$-$N$ junctions in silicon is large enough and the spectral response suitable for the possible use of silicon photovoltaic cells as solar energy converters. CUMMEROW[3] has estimated a possible power conversion efficiency of 17% but obtained 4% experimentally. CHAPIN, FULLER and PEARSON[4] give corresponding values of 22% und 6% respectively. They attribute the discrepancy between theoretical and experimental values to reflexion of incident light at the surface, high recombination rate for carriers at the surface and resistive loss if the $P$-$N$ junction is too near the surface. RITTNER[5] suggests an optimum theoretical efficiency of 25%.

Some attempts have been made recently to investigate the photo-electromagnetic effect in silicon by BULLIARD[6] (see also section below on photo-electromagnetic effect in germanium) but no quantitative treatment of experimental data is given. LOFERSKI and RAPPAPORT[7] have measured the electron voltaic effect in silicon as a function of radiation damage by high energy electrons (for more details see section on germanium below). It is also interesting to note that cyclotron resonance experiments have been made on silicon at 4° K using photoconduction excitation by infra red radiation to produce the required carriers[8].

**21. Germanium.** Germanium is perhaps the most explored semiconductor so far developed and reference should be made to O. MADELUNG's article in this volume for other aspects of its properties and development than photoconductivity and related photoeffects. Investigations of the latter previous to 1951 have been reviewed by MOSS[9]. The spectral response curve for intrinsic photoconduction in germanium as obtained by BECKER[10] and confirmed by various other workers is as shown in Fig. 12c above. It shows close agreement with the optical absorption spectrum. It was shown by GOUCHER[11] that the quantum efficiency for the photoconduction was about unity at the optimum sensitivity. This also been confirmed in many later measurements. Carrier life times, as seen in later discussion are dependent on the particular specimen (on $N$ or $P$ type impurity inclusions, surface states, geometry etc.). From more recent and very extensive work since 1951 we may select the following topics for more detailed treatment.

α) The study of photoconduction in germanium doped with various impurities.

β) The development and physical characteristics of the various types of phototransistor.

[1] R. NEWMAN: Phys. Rev. **94**, 15, 30 (1954).
[2] E. A. TAFT and F. H. HORN: Phys. Rev. **93**, 64 (1954).
[3] R. L. CUMMEROW: Phys. Rev. **95**, 561 (1954).
[4] D. M. CHAPIN, C. S. FULLER and G. L. PEARSON: J. Appl. Phys. **25**, 676 (1954).
[5] E. S. RITTNER: Phys. Rev. **96**, 1708 (1954).
[6] H. BULLIARD: Phys. Rev. **94**, 1564 (1954).
[7] J. J. LOFERSKI and P. RAPPAPORT: Phys. Rev. **98**, 1861 (1955).
[8] R. N. DEXTER and B. LAX: Phys. Rev. **96**, 223 (1954).
[9] T. S. MOSS: Photoconductivity in the Elements. London 1952.
[10] J. A. BECKER: Electr. Engng. **68**, 937 (1949).
[11] F. S. GOUCHER: Phys. Rev. **78**, 646, 816 (1950).

γ) The problem of carrier life time measurements and the importance of surface and trapping states.

δ) Photo- and electron-voltaic and photo-electro-magnetic effects in germanium.

α) *Photoconductivity due to impurities in germanium.* Following similar studies on silicon, BURSTEIN and his associates[1] have investigated the photoconduction of $N$ and $P$ type germanium containing neutral impurities at liquid helium temperatures. Some of their results are given in Fig. 15. For copper impurity response occurs as far as 29 μ (0.043 eV) while for zinc the response limit is beyond 38 μ. A similar limit occurs for elements of Groups III and V of the periodic table as impurities. In the latter case low applied fields are necessary since electrical breakdown occurs for appreciable field strengths. The activation energies for the groups III and V elements in germanium have been obtained from HALL coefficients by GEBALLE and MORIN[2]. As in the case of silicon the long wavelength response of photoconduction is affected by the lattice vibration absorptions which occur in the 15 to 35 μ region for germanium and are shown in Fig. 15d.

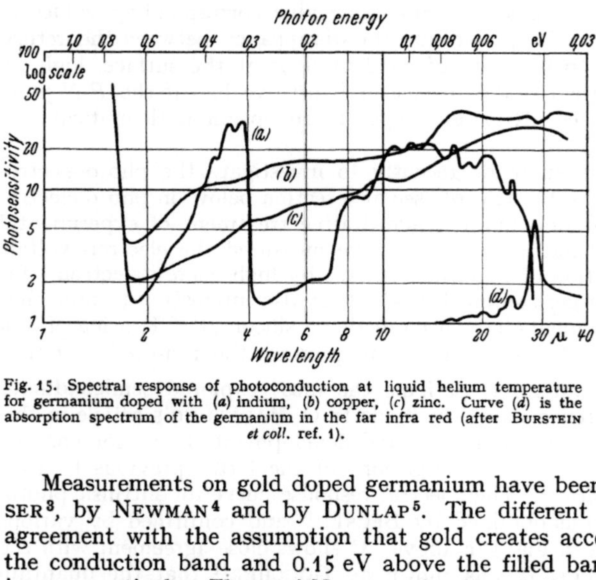


Fig. 15. Spectral response of photoconduction at liquid helium temperature for germanium doped with (a) indium, (b) copper, (c) zinc. Curve (d) is the absorption spectrum of the germanium in the far infra red (after BURSTEIN *et coll.* ref. 1).

Measurements on gold doped germanium have been made by FAN and KAISER[3], by NEWMAN[4] and by DUNLAP[5]. The different results are in reasonable agreement with the assumption that gold creates acceptor levels 0.2 eV below the conduction band and 0.15 eV above the filled band due to $Au^{2-}$ and $Au^{-}$ ions respectively. Those of NEWMAN for $N$ and $P$ types at 22° K are given in Fig. 16. indicating the different thresholds expected for the $N$ and $P$ types. NEWMAN observes optical quenching effects in $N$ type specimens which he considers are connected with optical ejection of holes from hole traps and consequent recombination with electrons to reduce the number of carriers and hence the photocurrent (see theortical Sect. 6 and 12). The photocurrent is linear with light intensity.

The transition elements iron, cobalt and nickel provide an interesting group of impurities in germanium. Measurements of photoconduction at low temperatures in $P$ and $N$ type material doped with these elements have been made by NEWMAN and TYLER[6] (iron), TYLER, NEWMAN and WOODBURY[7] (cobalt and

[1] E. BURSTEIN et coll.: Phys. Rev. **93**, 65 (1954).
[2] T. H. GEBALLE and F. J. MORIN: Phys. Rev. **95**, 1085 (1954).
[3] H. Y. FAN and W. KAISER: Phys. Rev. **93**, 977 (1954).
[4] R. NEWMAN: Phys. Rev. **94**, 278 (1954).
[5] W. C. DUNLAP: Phys. Rev. **97**, 614 (1955).
[6] R. NEWMAN and W. W. TYLER: Phys. Rev. **96**, 882 (1954).
[7] W. W. TYLER, R. NEWMAN and H. H. WOODBURY: Phys. Rev. **97**, 669 (1955); **98**, 461 (1955).

nickel). We give in Fig. 17 the spectral response curves for iron doped specimens of $P$ and $N$ type which are quite similar for the nickel and cobalt doped specimens. In all cases the impurity appears to provide two acceptor levels one below the conduction band and the other above the filled band. The respective activation energies $\mathcal{E}_c$ and $\mathcal{E}_f$ are given in the following table:

Table 1.

| Impurity | $\mathcal{E}_c$ (eV) | $\mathcal{E}_f$ (eV) |
|---|---|---|
| Fe . . . . | 0.27 | 0.34 |
| Co . . . . | 0.31 | 0.25 |
| Ni . . . . | 0.30 | 0.22 |

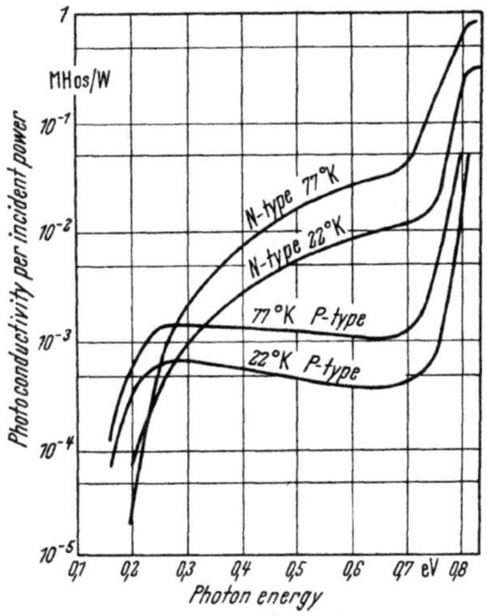


Fig. 16. Spectral response of photoconduction in gold doped germanium crystals of $P$ and $N$ type (after NEWMAN ref. 4, p. 344).

In $N$ type specimens much higher photosensitivity is obtained than in $P$ type, linear with light intensity over a wide range, and there is a linear relation between photosensitivity and response time. Time constants of up to 700 milliseconds are obtained. Quenching effects are observed in $N$ type specimens. Fig. 18 shows the type of quenching spectra obtained for iron impurity for different levels of steady photocurrent but with a constant quenching radiation intensity

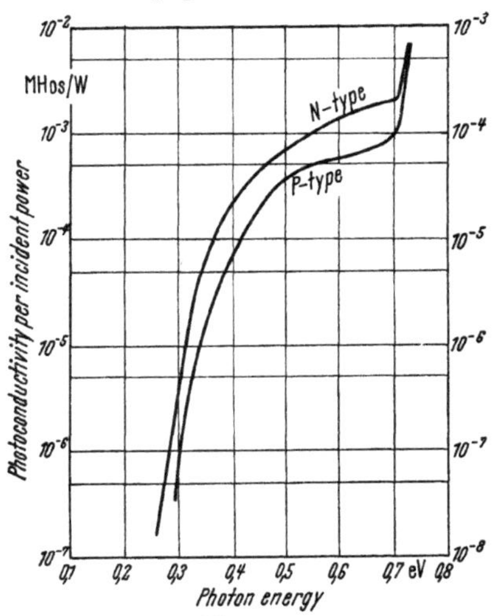


Fig. 17.

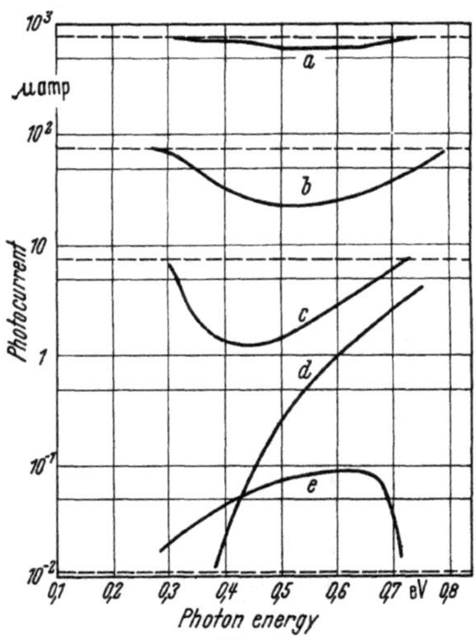


Fig. 18.

Fig. 17. Spectral response of photoconduction in iron doped germanium crystals of $P$ and $N$ type (after TYLER, NEWMAN and WOODBURY ref. 7, p. 344).

Fig. 18. Quenching spectra for photoconductivity in $N$-type germanium doped with iron. *a*, *b* and *c* Spectra for different steady state levels of photocurrent produced by radiation at shorter wavelengths (shown by broken lines). *d* Normal spectral response for specimen. *e* Power spectrum for quenching radiation (after TYLER, NEWMAN and WOODBURY ref. 7, p. 344).

(incident energy shown in curve e). As in the case of gold doped specimens quenching is attributed to release of trapped positive holes by the quenching radiation. However, as shown in Fig. 18 the quenching spectrum is not very well defined and appears to peak at about 0.4 to 0.5 eV rather than at 0.2 eV which would correspond to the 0.2 eV acceptor level. Further studies of the quenching appear necessary.

*β) The phototransistor.* The characteristics of the three types of phototransistor now available have been described by SHIVE[1]. They are shown schematically in Fig. 19.

The point contact phototransistor. This consists of a wafer of germanium with a point contact in one surface. Radiation is incident on the opposite

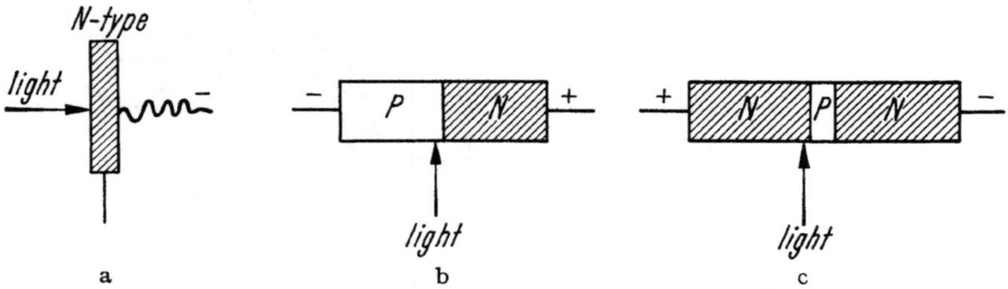


Fig. 19a–c. Schematic diagram of main types of phototransistor. a) Point contact phototransistor. b) *P-N* junction phototransistor. c) *N-P-N* photomultiplier transistor.

surface. The system is given a potential bias so that as a diode rectifier it would be operating in the reverse or high resistance direction. The photosensitivity is greatest when the light is incident directly opposite the contact, falling off as the incidence region is removed from this position[2]. Quantum efficiencies of three are possible for this phototransistor.

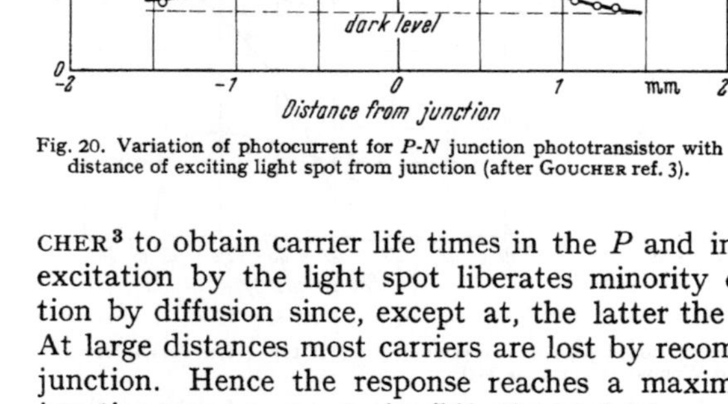


Fig. 20. Variation of photocurrent for *P-N* junction phototransistor with distance of exciting light spot from junction (after GOUCHER ref. 3).

The *P-N* junction phototransistor. Radiation falls on a junction of *P* and *N* type material. Positive bias is applied to the *N* type material so that the junction operates as a diode in the reverse direction. Fig. 20 shows the effect of moving a small light spot (0.005 inches diameter) across the junction on the photoresponse. This experiment was used by GOUCHER[3] to obtain carrier life times in the *P* and in the *N* type material. Photoexcitation by the light spot liberates minority carriers which reach the junction by diffusion since, except at, the latter the electric field strength is small. At large distances most carriers are lost by recombination and do not reach the junction. Hence the response reaches a maximum for light incident on the junction or very near to it. Life times of 17 μ sec. and 80 μ sec. in *P* and *N* material respectively were obtained. From the junction theory developed by

[1] J. N. SHIVE: J. Opt. Soc. Amer. **43**, 239 (1953).
[2] J. N. SHIVE: Bell. Labor. Rec. **25**, 289 (1950).
[3] F. S. GOUCHER: Phys. Rev. **81**, 475 (1951).

SHOCKLEY[1] it follows that the width of the peak in Fig. 20 is proportional to the square root of minority carrier life time on the appropriate side of the peak. *P-N* junction specimens show a quantum efficiency of about unity.

The *N-P-N* photomultiplier transistor. This phototransistor consists of a sandwich layer of *P* type material between *N* type germanium and is biased as shown in Fig. 19c so that the normal dark current of electrons flows right to left. The *P-N* junction on the right hand side offers a high impedance to the electrons. Light falls on the other junction liberating electrons and holes. The action of the phototransistor is explained by reference to the energy band diagram for the biased system as shown in Fig. 21.

This is the diagram for the now well known *P-N* hook[1]. Liberated holes will flow into the *P* type layer and become trapped in the "hook". Their resulting space charge lowers the right hand barrier and permits greater electron flow. We thus have a barrier modulated photocurrent which is also multiplied one positive hole giving rise to a flow of many electrons. The effect is determined by the diffusion length of the liberated holes and the width of the *P* type layer. SHOCKLEY, SPARKS and TEAL[2] have given the following formula for the effective quantum efficiency ($\eta$):

$$\eta = 1 + \frac{\sigma_e L_{pe}}{\sigma_p w} \tag{21.1}$$

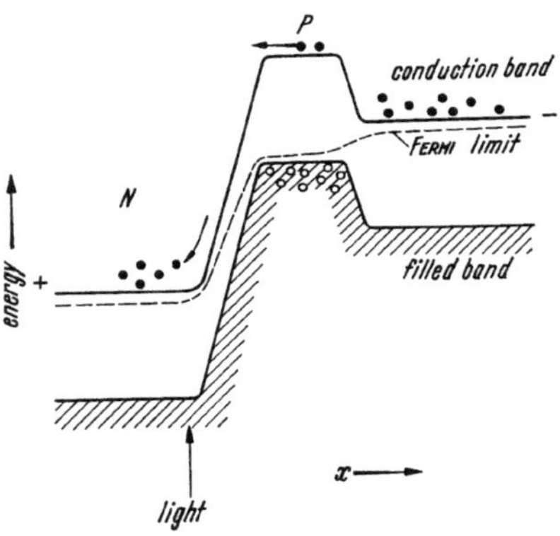


Fig. 21. Energy band diagram for *N-P-N* photomultiplier transistor.

where $\sigma_e$ is the conductivity of the *N* type material on the left hand side, $\sigma_p$ is that of the *P* type layer, $L_{pe}$ is the diffusion length for holes in the *N*-type material on the left and $w$ is the width of the *P*-type layer. Quantum yields of up to one thousand are possible in the multiplying phototransistor. The response time of the device is not a simple function of carrier life time but depends on the life time of the trapped holes in the *P*-type layer which is dependent on their electron capture, on loss of trapped holes and on their thermal activation into the *N*-type material.

It is interesting to note that for all three types of phototransistor the photoeffects are not very much affected by temperature although the dark currents flowing are, as expected, very much dependent on temperature. Spectral response is as in Fig. 12c for the intrinsic photoconduction in germanium.

*γ) Carrier life time, trapping and surface states* It is well known in transistor physics that carrier life times are strongly influenced by the action of surface states[3]. Photoconduction and the photovoltaic effect in germanium are also markedly affected by surface states. The general effect is that surface recombination of carriers is more rapid than that in the bulk material and so carrier life times are reduced. At surfaces energy states exist which can trap either kind of carrier. They may be divided into those which provide recombination sites and those which are "safe traps". The latter will modulate surface barrier heights as discussed in Sect. 10 above. In recent studies attempts have been

[1] W. SHOCKLEY: Electrons and Holes in Semi Conductors. New York 1950.
[2] W. SHOCKLEY, M. SPARKS and G. K. TEAL: Phys. Rev. **83**, 151 (1951).
[3] W. SHOCKLEY: Electrons and Holes in Semi Conductors. New York 1950.

made to use the effects of controlled surface treatments on photoconduction and photovoltaic potentials to distinguish between bulk and surface properties For example FAN, NAVON and GEBBIE[1] find that the initial rapid decay of photocurrent is strongly influenced by surface treatment but the longer duration tail of the decay is unaffected and can be ascribed to trapping of carriers in the crystal volume. HENISCH, REYNOLDS and TIPPLE[2] use the decay of photocurrent in germanium with a high bulk life time for carriers to estimate the surface recombination velocity under different conditions of humidity for the atmosphere surrounding the specimen. In other experiments, MOORE and WEBSTER[3] scan a small light spot over a "floating *P-N* junction" on a germanium filament. As the spot moves over the surface opposite the junction, the photocurrent response time is increased due to the relative absence of surface recombination.

Further information on surface states from study of photoeffects has been obtained by BARDEEN and MORRISON[4]. They measure the contact potential of germanium samples relative to platinum (using the KELVIN reed method) and the change in the potential due to light and its dependence on the degree of oxidation of the surface. It is concluded that adsorbed oxygen ions forming a surface potential barrier are neutralised by light liberated positive holes which decrease the barrier height. The photoconduction response is relatively slow and is determined by the rather slow electron transfer from volume to surface. The contact potential difference and its change with light are strongly affected by oxidation but the latter has a relatively small effect on recombination rates as shown by BRATTAIN and BARDEEN[5]. The contact potential difference is increased by light in *P*-type specimens and decreased in *N*-type material which supports the hypothesis that the holes neutralise the oxygen ions.

*δ) Photo- and electron voltaic, and photo-electro-magnetic effects in germanium.* Photovoltaic measurements on *P-N* junctions made by BECKER and FAN[6] have shown that the theoretical formula for the photovoltaic effect given in Eq. (11.20) above is accurately obeyed over a light intensity range of nearly $10^6:1$. The response time and magnitude of the photovoltage increase rapidly with fall in temperature which is also predicted by the above theory.

The photo-electro-magnetic effect in germanium has been investigated by MOSS, PINCHERLE and WOODWARD[7], by AIGRAIN and BULLIARD[8] and by BULLIARD[9,10]. MOSS *et coll.* obtained photovoltages for the effect as high as 0.5 eV. At low illumination levels the effect is independent of surface recombination but at higher levels saturation effects occur due to this process. BULLIARD[9] has shown that the initially rapid rise of the voltage when illumination begins is followed by a fall to a lower steady state value and that the ratio of peak to steady state values increases with intensity of illumination as shown in the oscillograph curves of Fig. 22. The peak effect is explained as follows by BULLIARD who also gives a quantitative expression for the peak voltage:—The initial photovoltage is due to the rapid diffusion of liberated carriers. The resulting voltage gives rise to photoconduction which builds up more slowly and

[1] H. Y. FAN, D. NAVON and H. GEBBIE: Physica, Haag **20**, 855 (1954).
[2] H. K. HENISCH, W. N. REYNOLDS and P. M. TIPPLE: Physics **20**, 1033 (1954).
[3] A. R. MOORE and W. M. WEBSTER: Physica, Haag **20**, 1046 (1954).
[4] J. BARDEEN and S. R. MORRISON: Physica, Haag **20**, 873 (1954).
[5] W. H. BRATTAIN and J. BARDEEN: Bell. Syst. Techn. J. **32**, 1 (1953).
[6] M. BECKER and H. Y. FAN: Phys. Rev. **78**, 301 (1950).
[7] T. S. MOSS, L. PINCHERLE and A. M. WOODWARD: Proc. Phys. Soc. Lond. B **66**, 743 (1953).
[8] P. AIGRAIN and H. BULLIARD: C. R. Acad. Sci. Paris **236**, 595 (1953).
[9] H. BULLIARD: Phys. Rev. **94**, 1564 (1954).
[10] H. BULLIARD: Ann. Phys. Paris **9**, 52 (1954).

due to its direction must offset the original photoelectromagnetic current. At high intensities the photoconductance is appreciable compared with the existing dark conductance and so will influence the photoelectromagnetic effect more strongly and thus increase the peak: steady state ratio. Grinding the unilluminated surface of the specimen increases the photoelectromagnetic effect in the steady state reducing the carrier life time by the consequent increase in surface states and rate of recombination.

We have already noted the use of the electron-voltaic effect in silicon to measure the effects of high energy electron bombardment[1]. The same workers made measurements on germanium and from them estimated the minimum energy for displacement of lattice atoms, which is 23 eV. It is interesting to note that the quantity measured was the short circuit current $i_s$ for the electron voltaic effect. This is proportional to the square root of the minority carrier life time $\tau$. This life time is given by:

$$\frac{1}{i_s} \propto \frac{1}{\sqrt{\tau}} \propto N_r \sigma f(\mathcal{E} - \zeta_0) \qquad (21.2)$$

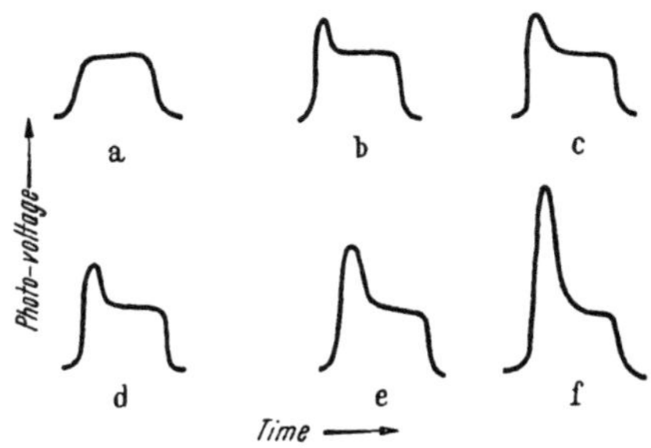

Fig. 22a–f. Oscillograph records of photovoltage for the photoelectro-magnetic effect in germanium. Curves (a) to (f) are for progressively increas ingintensity of incident light (after BULLIARD ref. 9, p. 348).

where $N_r$ is the number of recombination centres and is proportional to bombarding electron dosage, $\sigma$ is their capture cross section and $f(\mathcal{E} - \zeta_0)$ a function of the separation between the energy level $\mathcal{E}$ of the centres produced and the FERMI level $\zeta_0$ of the specimen. These experiments have another implication, namely, that electron bombardment can provide a means of introducing controlled numbers of defects into photoconductors. The effect of the latter on photoconduction processes can thus be quantitatively studied. Such techniques warrant further exploitation and for other photoconductors.

### c) Photoconduction in phosphorus, arsenic and antimony.

**22.** All the information on photoconduction in these solids is due to Moss[2]. Red phosphorus in deposited layers is found to be photosensitive with spectral response peaks at 0.4 μ (2.6 eV) and 0.79 μ (1.56 eV). From a study of several cells with phosphorus layers Moss concludes that the 0.4 μ peak is associated with a yellow form of phosphorus. Variation of photocurrent with light intensity in one case followed a 0.68 power law but in others the low intensity linear and the high intensity square root relation obtained as implied in the simple recombination theory of Sect. 2 and 6 above. Absorption bands for the layers agree with the spectral response.

Grey arsenic evaporated in vacuo on to cool surfaces formed photoconductive layers with linear response to light intensity. Spectral response showed peaks at 1 and 1.5 μ which agree with optical absorption data.

In the case of antimony Moss found it difficult to distinguish whether a true photoresponse or a resistance bolometer effect was being measured. The layers had to be evaporated on to cooled substrates and stored in dry ice as they reverted to metallic form above 0° C.

---

[1] J. J. LOFERSKI and P. RAPPAPORT: Phys. Rev. **98**, 1861 (1955).
[2] T. S. MOSS: Photoconductivity in the Elements. London 1952.

### d) Photoconduction in sulphur, selenium and tellurium.

**23. Sulphur.** The photoconductivity of sulphur was investigated by KURRELMEYER[1]. The dark resistivity of sulphur is $\sim 10^{18}\,\Omega$ cm. and measurements were difficult. The photoresponse was found to be linear with applied field and light intensity and showed spectral response peaks at 0.47 and 0.55 μ. There is, according to MOSS (ref. 6, sect. 22), good agreement between the optical and thermal activation energies ($\sim$2.5 eV).

In more recent studies NADZHAKOV and ANDREICHIN[2] and NADZHAKOV and KASHUKEEV[3] have studied the photovoltaic effect for sulphur with various electrode metals the photovoltages forming a series correlating with the contact potentials series of the metals. The photovoltaic effect showed a slow increase with time and slow decay in the dark. High purity samples retained their photovoltages indefinitely in the dark thus forming photoelectrets. Exposure to suitable radiation can remove the effect.

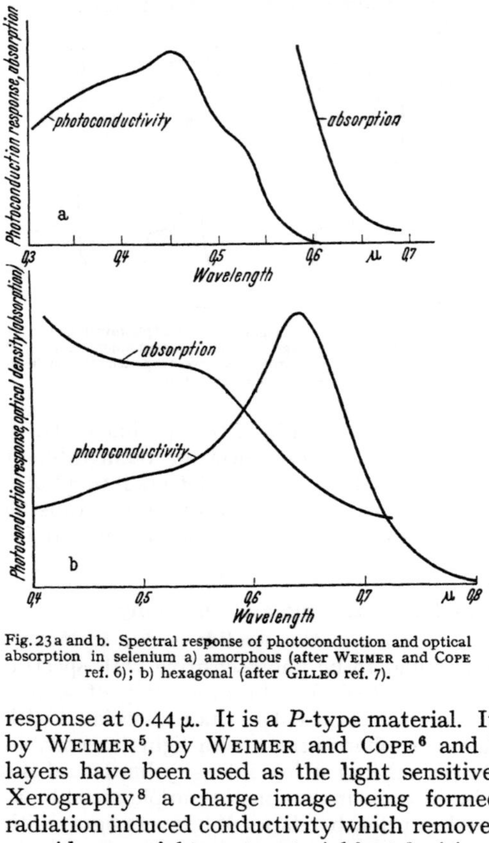


Fig. 23a and b. Spectral response of photoconduction and optical absorption in selenium a) amorphous (after WEIMER and COPE ref. 6); b) hexagonal (after GILLEO ref. 7).

**24. Selenium.** Selenium exists in four different forms: The common form as used in the ordinary selenium barrier layer photovoltaic cell is hexagonal and metallic in appearance. Photoconductivity response peaks in the red region. Monoclinic α and β form crystals are red and were studied by GUDDEN and POHL[4]. They show a response peak at 0.7 μ. The fourth form of selenium is amorphous, red and glassy. It is a reasonably good insulator in the dark but only quite recently has it been shown to exhibit photoconductivity with a peak response at 0.44 μ. It is a *P*-type material. It has also been investigated recently by WEIMER[5], by WEIMER and COPE[6] and by GILLEO[7]. Amorphous selenium layers have been used as the light sensitive material in the new technique of Xerography[8] a charge image being formed on the layer by X-ray or other radiation induced conductivity which removes previously induced charge. It also provides a useful target material for television pick up tubes of the "videcon" type.

The spectral response of photoconductivity and the optical absorption in amorphous selenium do not correlate well as shown by Fig. 23 from the results of WEIMER and COPE[6]. Photoconduction is also sustained for thick layers even

[1] B. KURRELMEYER: Phys. Rev. **30**, 893 (1927).
[2] G. NADZHAKOV and R. ANDREICHIN: Izv. Bulg. Akad. Nauk. **2**, 293 (1951).
[3] G. NADZHAKOV and N. T. KASHUKEEV: Izv. Bulg. Akad. Nauk **2**, 321 (1951).
[4] B. GUDDEN and R. W. POHL: Z. Physik **35**, 243 (1925).
[5] P. K. WEIMER: Phys. Rev. **79**, 171 (1950).
[6] P. K. WEIMER and A. D. COPE: RCA-Review **12**, 314 (1951).
[7] M. A. GILLEO: Report XXXV Lab for Insulation Res., M.I.T. 1951.
[8] R. M. SCHAFFER and C. D. OUGHTON: J. Opt. Soc. Amer. **38** 991 (1948).

when the excitation is confined to one surface and the quantum efficiency approaches unity. The photocurrent-voltage curves suggest that the currents are space charge limited. At low intensities the photocurrent is almost linear with light intensity but at high intensities saturation effects occur, probably because of space charge limitation. Time constants of the order of 50 μ sec. are obtained at room temperature. The photosensitivity improves rapidly with increasing temperature below 30° C and suggests that trapping effects are limiting the current. WEIMER and COPE[1] have observed that addition of a percent or so impurities, such as antimony and cadmium produces large increase in photosensitivity and dark current and the time constant increases to many seconds. An enhanced response to red light can also be obtained.

In the case of hexagonal selenium greater correlation is obtained between photoconduction spectral response and the optical absorption as shown by GILLEO's results in Fig. 23. The photosensitivity peaks as the absorption coefficient falls thus following the theoretical behaviour suggested in Sect. 12 and usually found in materials such as cadmium sulphide. Addition of tellurium as impurity increases the spectral response limit into the 0.9 μ region.

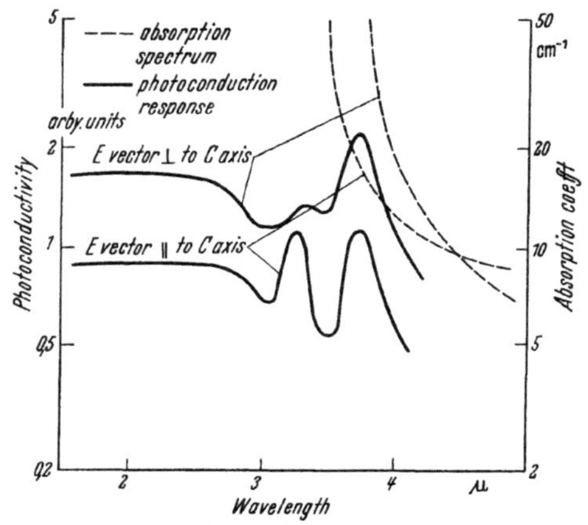


Fig. 24. Response of photoconduction and optical absorption of tellurium crystals and their dependence on the polarisation of the incident radiation (after LOFERSKI ref. 3).

**25. Tellurium.** Photoconductivity in tellurium was first observed by MOSS[2] using layers obtained by distillation of very pure tellurium. The photoresponse shows peaks at 1 and 1.6 μ and the conduction appears to be *P*-type. The threshold at 3.3 μ agrees with the absorption edge. The time constant is of the order of 400 μ sec. and the photocurrent shows a square root relation at high intensities.

LOFERSKI[3] has recently made measurements of the optical absorption and photoconductivity in single crystal needles of tellurium 0.2 mm. × 1.5 cm. grown by BRIDGMAN's technique[4]. By using polarised light he has shown that the absorption spectrum and spectral response of photoconductivity depend on the direction of the electric vector of the incident radiation with respect to the *c*-axis of the crystal. His results are collected in Fig. 24. The photocurrent is linear with applied voltage and also with light intensity for a wavelength of 3.6 μ (low absorption coefficient). For shorter wavelengths (i.e. high excitation densities) a square root relation is followed. The peaks at 3.24 and 3.72 μ in the response curves also show anisotropic properties. However, they move in the same way as the absorption edges with temperature although the photoconduction threshold $\lambda_{\frac{1}{2}}$ moves to longer wavelengths as the temperature is lowered. The time constant is less than 1 millisec. according to LOFERSKI's measurements.

LOFERSKI has also made Te—Se alloys. Addition of Se reduces the value of the threshold response wavelength.

[1] P. K. WEIMER and A. D. COPE: RCA-Review **12**, 314 (1951).
[2] T. S. MOSS: Photoconductivity in the Elements. London 1952.
[3] J. J. LOFERSKI: Phys. Rev. **93**, 707 (1954).
[4] P. A. BRIDGMAN: Proc. Amer. Acad. Sci. **60**, 305 (1925).

### e) Photoconductivity in iodine.

**26.** Iodine can be formed in thin plates by melting and pressing between glass plates or by simply melting a layer on to a surface. The spectral response of photoconduction shows peaks at 0.45 and 0.85 μ. Fig. 25 gives the spectral response together with the absorption spectrum for a particular layer thickness. The first peak is associated with molecular absorption while the second peak is due to the energy levels produced in the solid state. Moss suggests an energy band model for this molecular crystal but the absorption edge shown in Fig. 25 is not fixed but moves back towards that for vapour absorption as the layer thickness is decreased. It is of interest to note that in solid form iodine can be excited in the 0.7 μ region to give luminescence emission in a broad band at 1.5 μ[1]. Thus neither the longer wavelength peak of photoconduction response or the excitation and emission spectra of luminescence are characteristic of the iodine molecule but of the crystalline state. The photocurrent has a time response of about millisec., is linear with light intensity at low levels but shows the usual saturation effect at high intensities.

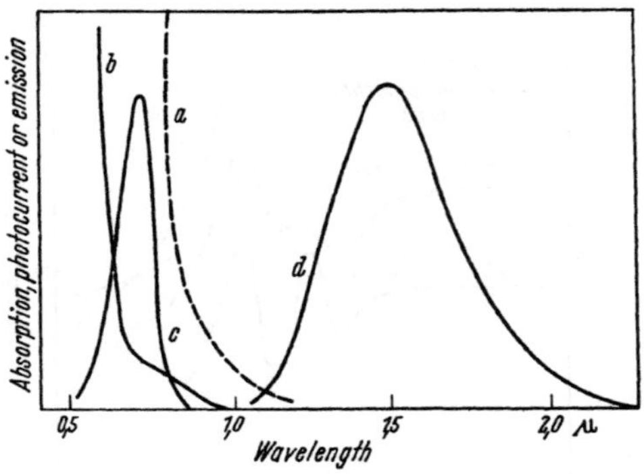


Fig. 25. Characteristics of iodine crystals. *a* Optical absorption spectrum. *b* Spectral response of photoconduction. *c* Excitation spectrum for luminescence. *d* Luminescence emission spectrum.

### f) General comments on photoconduction in monatomic solids.

**27.** The relation between threshold response wavelength and refractive index discussed above in Sect. 14 has been shown by Moss[2] to hold quite well for the monatomic solids above, with the exception of diamond. Bearing in mind the other abnormal characteristics of diamond among its neighbours in the periodic system this is not surprising. It is still not clear what the essential physical difference is between type I and type II diamonds and why only type II are photoconducting. The influence of included impurities has not been entirely eliminated from the possible causes.

## IV. Photoconductivity in binary compounds.

### a) The alkali halides.

**28. Alkali halide colour centres.** The alkali halides provide an attractive group of materials for photoconductivity studies, as realised by POHL and his school. Crystals additively coloured by heating in their metal vapour exhibit photoconduction properties of almost "ideal" kind when illuminated by light lying in the colour centre absorption band produced by the excess metal. The earlier studies of POHL and his associates[3] have been reviewed by MOTT and GURNEY[4] and later by SEITZ[5]. A more recent review has also been given by SEITZ[6]. To

[1] M. J. DUMBLETON: Proc. Phys. Soc. Lond. B **68**, 63 (1955).
[2] T. S. MOSS: Photoconductivity in the Elements. London 1952.
[3] R. W. POHL: Proc. Phys. Soc. Lond. **49**, (extra part), 3 (1937).
[4] F. SEITZ: Rev. Mod. Phys. **18**, 384 (1946).
[5] N. F. MOTT and R. W. GURNEY: Electronic Processes in Ionic Crystals. Oxford 1940.
[6] F. SEITZ: Rev. Mod. Phys. **26**, 7 (1954).

appreciate the researches involving photoconduction we first provide a brief resume of the nature of colour centres in the alkali halides.

The interesting electronic properties of the alkali halides depend on the presence of lattice ion vacancies which when occupied by electrons or holes give rise to absorption bands in the ultraviolet, visible and near infrared regions of the spectrum. The earliest centre investigated, that designated as the F centre, is produced by an electron in a negative ion vacancy. When the vacancy contains two electrons an F′ centre is formed. Aggregation of negative ion vacancies leads to other bands (M, R, N) at longer wavelengths. Positive ion vacancies or associates of such vacancies produce the V bands when occupied by positive holes. X-ray or high energy particle bombardment will produce vacancies of both kinds but additive coloration, i.e. heating in the metal or halogen vapour produces negative ion or positive ion vacancies respectively. We give further details of the various colour centres as required below.

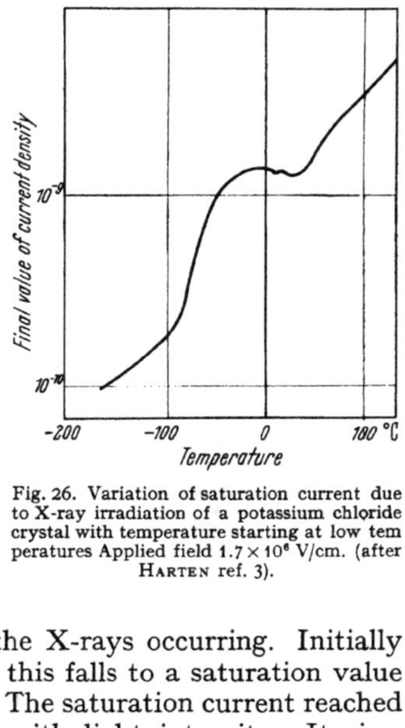

Fig. 26. Variation of saturation current due to X-ray irradiation of a potassium chloride crystal with temperature starting at low tem peratures Applied field $1.7 \times 10^6$ V/cm. (after HARTEN ref. 3).

**29. Photoconduction in uncoloured crystals.** The lattice absorption edge of alkali halides such as KCl and NaCl lies in the far ultraviolet region below 1800 Å and experimental studies become rather difficult. Early work by SMAKULA[1] indicated that absorption in the tail of the fundamental band produced no photoconductivity. This has been confirmed more recently by APKER and TAFT[2]. It is now concluded that excitons are formed by such absorption. However, free charge carriers have been produced in the alkali halides by high energy radiation or particles. HARTEN[3] used crystals of KCl 0.3 mm. thick with applied fields of 15000 V/cm. and irradiated them with X-rays of wavelength 0 2 to 1 Å and intensity $\sim 10^{-4}$ watts/cm.$^2$, 25% absorption of the X-rays occurring. Initially on irradiation a relatively large current flows but this falls to a saturation value at a rate which increases with rise in temperature. The saturation current reached is approximately linear with field strength and with light intensity. It rises rapidly with temperature rise, as shown in Fig. 26 due to HARTEN, and there is a corresponding rise in the F centre concentration due to the X radiation. At $-80°$ C the increasing production of F centres causes a decrease in the carrier range because of their trapping effect. Conductivity pulses due to α-particle bombardment of uncoloured NaCl were observed by WITT[4] when the particles were incident through the negative electrode. This corroborates other workers' conclusions that the mobility of positive holes is much smaller than that of electrons in the alkali halides. In addition it was shown by VAN HEERDEN[5] that α-particles are less efficient than electrons in producing conductivity pulses. This is presumably due to the very dense ionisation channel along the α-particle track which upsets the applied field strength in that region. Production of colour centres reduces the α-particle pulses as found above for X-ray induced conductivity.

[1] A. SMAKULA: Z. Physik **45**, 1 (1927).
[2] L. APKER and E. TAFT: Phys. Rev. **79**, 964 (1950); **81**, 698 (1951); **82**, 814 (1951).
[3] H. U. HARTEN: Z. Physik **126**, 619 (1949).
[4] H. WITT: Z. Physik **128**, 442 (1950).
[5] P. J. VAN HEERDEN: Diss. Utrecht 1945.

**30. Photoconduction in additively coloured alkali halides.** If crystals such as KCl are heated in the alkali metal vapour to produce negative ion vacancies and then quickly cooled the colouration is characteristic of the simple F centre. Slow cooling allows association and aggregation of vacancies to produce the more complicated R and M centres and even colloidal metal in the crystal.

*α) F-band photoconduction.* Irradiation in the F band liberates electrons and causes photoconduction. There are no free holes produced and so in the simplest case recapture in the negative ion vacancy is the final fate of the liberated electrons. The process can be examined in more detail by varying the temperature. It is convenient to consider the variation of the quantity $\eta\delta$ where $\eta$ is the quantum efficiency for production of free electrons and $\delta$ is their mean range in unit field. As shown by Eq. (2.4) above, $\delta$ is the product of carrier life time and mobility. Fig. 27 shows schematically the usual behaviour of $\eta\delta$ with respect to temperature and indicates four distinct regions of interest.

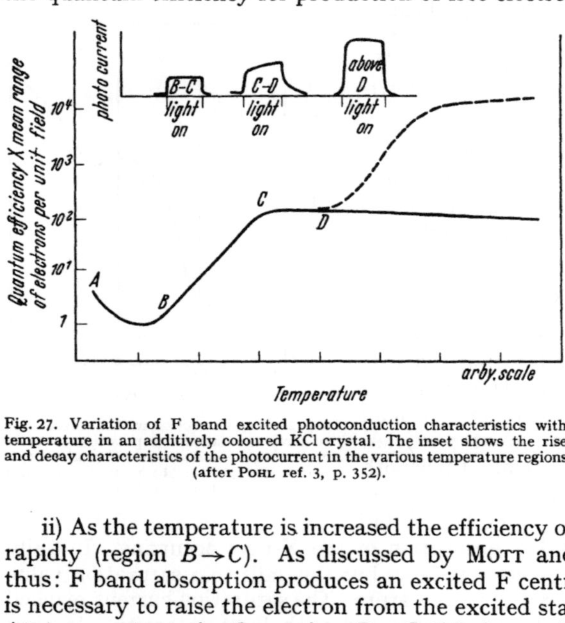


Fig. 27. Variation of F band excited photoconduction characteristics with temperature in an additively coloured KCl crystal. The inset shows the rise and decay characteristics of the photocurrent in the various temperature regions (after POHL ref. 3, p. 352).

i) At the lowest temperature (region $A \rightarrow B$) the photocurrent depends on liberation of electrons from impurities or residual F centre aggregates in the crystal. Very few studies have been made of the wavelength dependence of this photoconduction.

ii) As the temperature is increased the efficiency of carrier production increases rapidly (region $B \rightarrow C$). As discussed by MOTT and GURNEY this is explained thus: F band absorption produces an excited F centre but a further small energy is necessary to raise the electron from the excited state into the conduction band. At temperatures in the region $B \rightarrow C$ this is supplied as a thermal activation with the usual probability $s \cdot \exp(-\mathcal{E}/kT)$ (see Sect. 6 above).

iii) A range ($C \rightarrow D$) is then encountered in which a slow decrease of the quantity $\eta\delta$ occurs the main process being liberation of electrons from F centres and either recombination with vacant F centres or capture in F centres to form F′ centres.

iv) At higher temperatures the photosensitivity rises due to the thermally increased range of the electrons when the F′ centres can no longer retain them. The rapid initial rise of photocurrent with illumination occurs in less than $2 \times 10^{-5}$ secs. in KCl. The thermal activation energy for the F′ centre is found to be 0.42 eV in KCl with a frequency constant $s \approx 10^{11}$ sec.$^{-1}$. Mobility measurements have been made on NaCl by REDFIELD[1] giving values of ~260 cm.$^2$ per volt · sec. at 82° K and ~40 cm.$^2$/volt · sec. at 200° C. Measurements at room temperature by MACDONALD and ROBINSON[2] using an a.c. HALL measurement technique yield a value of the order of 12.5 cm.$^2$/volt · sec.

[1] A. REDFIELD: Phys. Rev. **91**, 244, 753 (1953); **94**, 526, 537 (1954).
[2] J. R. MACDONALD and J. E. ROBINSON: Phys. Rev. **95**, 44 (1954).

*β) Photoconduction in the M band etc.* When additively coloured crystals of KCl are irradiated in the F band at room temperature the M band at 8200 Å is formed and is assumed to be an F centre combined with a pair of vacancies. Photoconduction is produced by irradiation in the M band[1]. The photocurrent is linear with absorbed energy but the quantum efficiency is only about one tenth that for F band photoconduction. Irradiation in the M band can create other centres (R and N). F centres act as traps for the electrons excited from M centres and so become F′ centres. For this reason the M band photocurrent shows the thermally increased range effect when the temperature is such that the F′ centres become unstable.

Large aggregates or colloidal specks give rise to absorption bands and irradiation in them produces photocurrents as shown in early experiments by GYULAI[2]. The quantum efficiencies encountered are low as found by SCOTT *et coll.*[3] and the spectral response for the photoconduction is broader than the optical absorption band.

*γ) Crystals with excess halogen.* In crystals heated in the halogen vapour positive ion vacancies are created and hole conduction would be expected. DUTTON and MAURER[4] have studied the electrical current flowing with a constant applied field when V bands are thermally bleached by warming the crystals of KCl and KBr. The $V_1$ band (positive ion vacancy plus an electron) bleaches thermally at $-188°$ C in KBr and in KCl at $-145°$ C and the disappearance of the band is accompanied by a current pulse. A similar but larger pulse occurs when the $V_4$ band (antimorph of M centre) bleaches at $-27°$ C in KCl. At $-48°$ C in KBr there is a large current pulse which does not correspond to bleaching of a known band. The current pulses are also accompanied by thermoluminescence emission but in some cases the latter occurs without a current pulse.

Photoconductivity by irradiation in the $V_1$ bands and H band (hole trapped at a pair of neutral vacancies) has been investigated by TEEGARDEN and MAURER[5] but no liberation of carriers was observed, in contrast to the release of charge in the above thermal bleaching experiments. DUTTON and MAURER[6] found a burst of charge for H band irradiation when heating KCl from very low temperatures up to $-183°$ C.

*δ) Photo-dielectric effects in alkali halides.* MACDONALD[6] has found that at low alternating applied field frequencies coloured crystals of KBr irradiated in the F band show an increase in dielectric constant (real part) from 5 to 500. The effect is due to the space charge build up by excited electrons which cannot leave the crystal. MACDONALD[7] has given a detailed theoretical treatment of the effect.

## b) Lead sulphide, selenide and telluride.

**31. General characteristics.** The sulphide, selenide and telluride of lead form a semiconductor group with optical absorption edges in the infra-red region. They can be prepared as *P* or *N* type and when suitably formed as layers show marked photoconductivity. Photoconducting cells of lead sulphide were developed

---

[1] C. BARTH: See SEITZ, Rev. Mod. Phys. **18**, 384 (1954).

[2] Z. GYULAI: Z. Physik **35**, 411 (1925); **37**, 889 (1926).

[3] A. B. SCOTT et coll.: Phys. Rev. **79**, 341, 346 (1950); **83**, 982 (1951). — J. Chem. Phys. **20**, 529 (1952).

[4] D. DUTTON and R. J. MAURER: Phys. Rev. **90**, 126 (1953).

[5] K. TEEGARDEN and R. J. MAURER: Z. Physik **138**, 284 (1954).

[6] J. R. MACDONALD: Phys. Rev. **85**, 381 (1952). — J. Chem. Phys. **23**, 275 (1955).

[7] J. R. MACDONALD: Phys. Rev. **92**, 4 (1953).

in Germany during the late war[1,2]. The potentialities of lead selenide and telluride as infrared detectors with a longer wavelength threshold than lead sulphide were also realised. After the war considerable effort was given both in Great Britain and in the USA. to the development of such cells[3] and to a study of the fundamental processes involved. Detailed discussion of researches of more recent origin, including those on single crystals have been given by SMITH[4, 5, 6].

It is convenient to consider the work on microcrystalline layers separately from that on single crystals.

**32. Photoconducting lead sulphide layers.** Photosensitive lead sulphide layers may be obtained by chemical deposition in the presence of an oxidising agent[7] or by vacuum evaporation of pure lead sulphide[8]. Evaporation yields layers of low resistivity, $N$-types and showing negligible photoconductivity. By subsequent heating in a low pressure of oxygen they change to $P$-type and the photoconduction is at an optimum when the $P$-type condition is just reached and the layer has a high resistivity. The layer is mounted in vacuo with suitable electrodes and, if required, provision for cooling. Cells of high efficiency (see Sect. 16 above) can be made which operate at room temperature. The sulphide layer is about 1 μ thick and is microcrystalline with a grain size of 0.1 to 1 μ. Some recent studies by FELTYNOWSKI, GLASS, PIWKOWSKI and JORUN[9] indicate crystallites of 1000 Å size with deformed crystal lattices and the presence of compounds other than lead sulphide. SMITH[5] mentions observations in which traces of PbO and $PbSO_4$ occur in the photosensitive layers. REUTER and STEIN[10] suggest that for $N$-type PbS having excess lead rapid oxidation to PbO

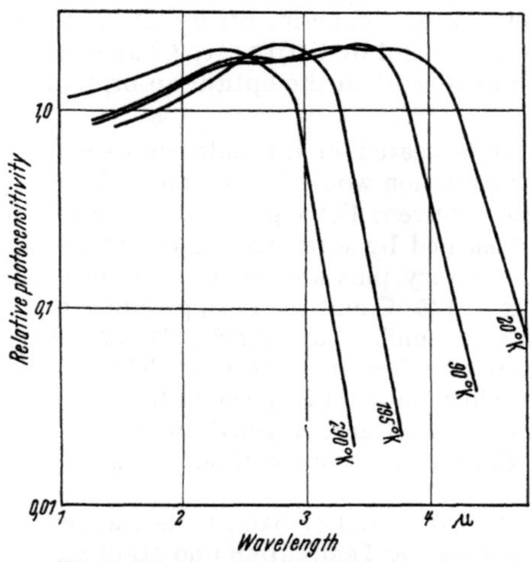


Fig. 28. Spectral response of photoconduction in a lead sulphide layer (after MOSS ref. 1, p. 357).

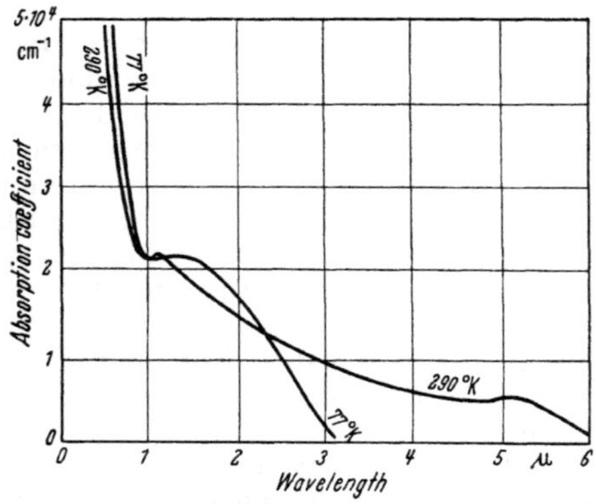


Fig. 29. Optical absorption spectra for a lead sulphide layer (after GIBSON ref. 3, p. 357).

---

[1] G. JOOS: Physics of Solids II. Fiat Rev. German. Sci. Wiesbaden 1948.
[2] A. ELLIOTT: Electrons and their application in Science and Industry, p. 97. London 1947.
[3] G. G. B. M. SUTHERLAND and E. LEE: Rep. Progr. Phys. **11**, 144 (1948).
[4] R. A. SMITH: Semi-cond. Materials, edit. H. K. HENISCH, p. 198. London 1951.
[5] R. A. SMITH: Adv. Physics **2**, 321 (1953).
[6] R. A. SMITH: Physica, Haag **20**, 910 (1954).
[7] L. SOSNOWSKI, J. STARKIEWISCZ and O. SIMPSON: Nature Lond. **159**, 818 (1947).
[8] E. SCHWARZ: Nature, Lond. **162**, 614 (1948).
[9] A. FELTYNOWSKI, I. GLASS, T. PINKOWSKI and A. JORUN: Bull. Acad. polon. Sci., Cl. 3. **2**, 389 (1954).
[10] B. REUTER and R. STEIN: Physica, Haag **20**, 1092 (1954).

occurs and that maximum photosensitivity is obtained at a stage when reaction between PbO and sulphur occurs.

Photosensitive layers of PbS have a spectral response as shown in Fig. 28 due to Moss[1]. The response threshold moves to longer wavelengths as the temperature decreases. The straight line feature of the response "edge" fits the Moss theory[2] given above in Sect. 12 [Eq. (12.1)] while the temperature dependence seems likely to be due to the change in refractive index with temperature[2] already discussed in Sect. 14. The threshold value $\lambda_g$ is little affected by impurities present and seems to be characteristic of the lead sulphide lattice. At shorter wavelengths the photosensitivity per incident quantum is sensibly constant and the quantum efficiency near unity. GIBSON measured the optical absorption spectra of PbS, PbSe and PbTe layers[3] and found little correlation with the spectral response of photoconductivity as shown by the results for PbS layers in Fig. 29. Below a wavelength of 1 μ absorption is intense ($\sim 10^5$ mm.$^{-1}$) but at longer wavelengths a tail stretches to 6 μ with a rather high absorption coefficient ($\sim 10^3$ mm.$^{-1}$). At low temperatures the tail retracts to about 3 μ. These measurements are difficult to interpret because of light scattering by the microcrystals. Measurements of HALL coefficients (see ref. 3 on p. 358) give a value of the intrinsic energy gap in lead sulphide of 1.2 eV which would appear to agree with the rise in absorption at 1 μ measured by GIBSON. We discuss this problem below in connection with results for single crystals.

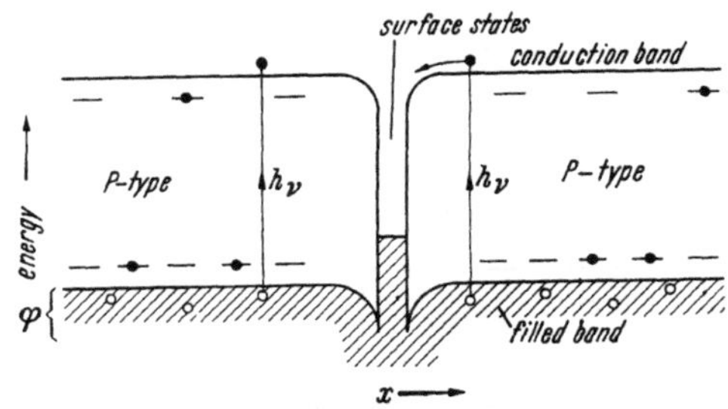


Fig. 30. Energy band diagram for barrier between two P-type crystallites with surface states present.

From various experimental studies of lead sulphide layers it has been concluded that the process of photoconduction is one of potential barrier modulation, the theory of which is given above in Sect. 10. It has been assumed by some workers that oxidation treatment produces *P-N* junctions between microcrystals[4]. SMITH[5] suggests an alternative model since the *P-N* barrier height would be large. He adopts a model due to MITCHELL and SILLARS[6] which assumes each crystallite to be *P*-type but with intercrystalline barriers produced by oxidation which have surface states capable of trapping positive holes. The energy diagram for the barrier is shown in Fig. 30. The detailed theoretical treatment of GIBSON[7] suggests a barrier height of 0.2 eV which is reasonable for the *P-P* barrier but too low for a *P-N* junction (cf. Figs. 8 and 30).

The primary source of electrons and holes due to illumination is not certain but the electrons produced, which are the minority carriers, will reach the surface states and neutralise some of the positive holes thus reducing the barrier height and modifying the standing dark current.

**33. Single crystals of lead sulphide.** The above model for PbS layers is further resolved by inclusion of results for single crystals. For such specimens there is

[1] T. S. MOSS: Proc. Phys. Soc. Lond. B **62**, 741 (1949).
[2] T. S. MOSS: Photoconductivity in the Elements. London 1952.
[3] A. F. GIBSON: Proc. Phys. Soc. Lond. B **63**, 756 (1950).
[4] L. SOSNOWSKI, J. STARKIEWISCZ and O. SIMPSON: Nature, Lond. **159**, 818 (1947).
[5] R. A. SMITH: Semi-cond. Materials, edit. H. K. HENISCH, p. 198. London 1951.
[6] E. W. J. MITCHELL and R. W. SILLARS: Proc. Phys. Soc. Lond. B **62**, 509 (1949).
[7] A. F. GIBSON: Proc. Phys. Soc. Lond. B **64**, 603 (1951).

correlation between optical absorption spectrum and the spectral response in the above layers. Theoretical studies of the energy band model for single crystals[1] have prompted some further discussion of the layer problem by SMITH[2]. He calls into question the values of the energy gap for single crystals (~1.12 eV) obtained from HALL coefficient measurements by PUTLEY[3] and suggests that at the temperatures necessary for measurement the activation energy is not independent of temperature and neither is the ratio of electron and positive hole mobilities. In addition there is evidence of irreversible changes in the crystals over the measuring range. Greater importance is therefore attached to the optical measurements and the intrinsic energy gap for lead sulphide is assumed to be 0.34 eV.

Photoconduction in single crystals is of relatively low efficiency but some recent studies by SOULE and CASHMAN[4] on cleavage pieces of lead sulphide crystals have thrown light on the barrier mechanism of photoconduction. Using a moving point probe on the crystal the potential distribution as a function of distance from a fixed electrode is measured. This reveals places where potential barriers occur. Scanning over the same terrain with a light probe shows peaks in the photoconduction response as a function of distance from the fixed electrode which coincide with the potential barriers. Additional measurements of thermoelectric power for a point probe show that the crystal is $P$-type and so the barriers are most likely to be due to gradations in positive hole density. These results support the $P$-$P$ barrier theory described above.

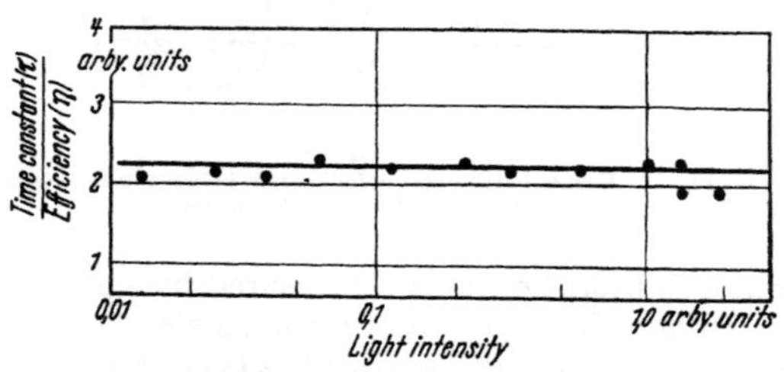


Fig 31. Variation of time constant/efficiency with light intensity for a lead sulphide layer (after GIBSON ref. 5).

**34. Life time measurements on lead sulphide.** The response time of photoconduction in PbS layers was extensively examined by GIBSON[5] in order to determine the validity of the barrier layer theory. As described in Sect. 10 above, the relation between photosensitivity ($\eta$) and response time $\tau$ provides a means of distinguishing between the barrier layer theory and "recombination" theories (e.g. as in Sect. 6 to 9). Eq. (10.15) above shows that the photosensitivity is proportional to the time constant but independent of light intensity in the barrier theory. Fig. 31 from GIBSON's results shows a constant value of the ratio of $\tau/\eta$ for a wide range of light intensities as expected from the barrier layer theory. A further test is provided by the dependence of time constant and sensitivity on temperature ($T$). From the barrier theory, [Eq. (10.15)] $\tau T/\eta$ should be independent of temperature. In experiment it alters by a factor of about 1.4 between 90° and 290° K. From recombination theories $\tau n/\eta$ should be a constant where $n$ is the number of carriers in equilibrium governed by Eq. (10.6) or not too high a ligth intensity. This quantity $\tau n/\eta$ changes by a factor of 22 over the range 90 to 290° K and so correlation is found for the barrier theory rather than for "recombination" theories. Time constants at 90° K are longest

[1] D. G. BELL, D. M. HUM, L. PINCHERLE, D. W. SCIAMA and P. M. WOODWARD: Proc. Roy. Soc. Lond. Ser. A **217**, 71 (1953).
[2] R. A. SMITH: Physica, Haag **20**, 910 (1954).
[3] E. H. PUTLEY: Proc. Phys. Soc. Lond. B **65**, 388, 736, 992 (1952).
[4] D. E. SOULE and R. J. CASHMAN: Phys. Rev. **93**, 635 (1954).
[5] A. F. GIBSON: Proc. Phys. Soc. Lond. B **64**, 603 (1951).

in chemically deposited layers being as much as milliseconds. In vacuum evaporated layers somewhat smaller values obtain. In single crystals much smaller values are found. Using the photo-electromagnetic effect Moss[1] has measured the lifetime of carriers in low resistivity samples of lead sulphide ($\sim 0.01\ \Omega$cm.) and obtains values of $\sim 10^{-9}$ sec. The lifetime $\tau$ is proportional to the square of the resistivity i.e. inversely proportional to the square of carrier concentration over a wide range. Life time measurements from other methods such as light spot scanning or study of the photocurrent rise time, give results in approximate agreement with the relation.

Some other studies of photocurrent decay in PbS layers at 90° K CHASMAR and GIBSON[2] have revealed a long period decay (hours) in these layers which they associate with surface trapping states of 0.3 eV for electrons. In general the much longer time constants for layers compared with single crystals suggest that the inertia effect is associated with the barrier. A similar effect was noted above in the case of surface states due to oxygen on germanium.

**35. Photovoltaic effect in PbS.** Lead sulphide shows a photovoltaic effect at barriers as would be expected. Recent work by BERLAGA and STRAKHOV[3] show that large photo e.m.f.'s can be obtained ($\sim$2 V) from layers evaporated in a low pressure of $CO_2$ ($\sim$0.1 mm. Hg) at 250° C and subsequently heat treated at 500° C before cooling.

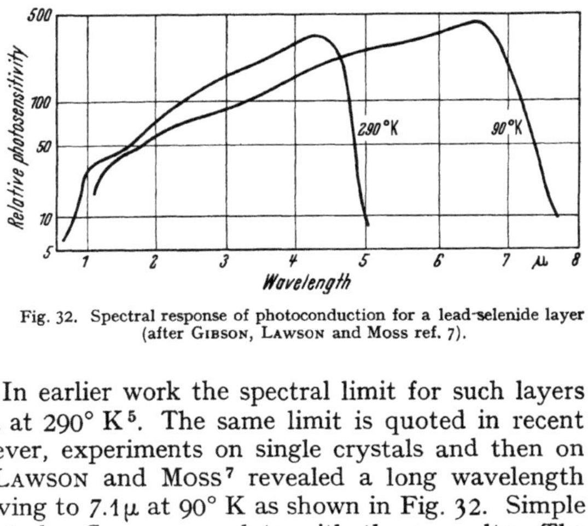


Fig. 32. Spectral response of photoconduction for a lead-selenide layer (after GIBSON, LAWSON and MOSS ref. 7).

**36. Photoconductivity in lead selenide.** Photoconducting layers of lead selenide are prepared in similar ways to those for lead sulphide. The layer must be brought to the $P$ form by oxidation[4]. In earlier work the spectral limit for such layers was assumed to be about 3.6μ at 290° K[5]. The same limit is quoted in recent studies by CHECINSKA[6]. However, experiments on single crystals and then on sensitised layers by GIBSON, LAWSON and MOSS[7] revealed a long wavelength threshold of 4.7μ at 290° K moving to 7.1μ at 90° K as shown in Fig. 32. Simple crystal absorption measurements by GIBSON correlate with these results. The details of the differences between the longer wavelength response layers and those usually obtained have not been published. Time constants for the photoconduction have not been extensively studied but those for the layers with $\lambda_{\frac{1}{2}} \sim 3.5\mu$ have recently been measured by SOSNOWSKI and CHMIELEWSKI[8]. LEE[9] has found a slow decay of photocurrent in PbSe layers at 90° K which is similar to that observed in PbS layers by CHASMAR and GIBSON[2].

[1] T. S. MOSS: Physica, Haag **20**, 989 (1954).
[2] R. P. CHASMAR and A. F. GIBSON: Proc. Phys. Soc. Lond. B **64**, 595 (1951).
[3] R. Y. BERLAGA and L. P. STRAKHOV: Ž. tekh. Fyz. **24**, 943 (1954).
[4] R. A. SMITH: Adv. Physics **2**, 321 (1953).
[5] G. JOOS: Physics of Solids II. Fiat Rev. German Sci. Wiesbaden 1948.
[6] H. C. CHECINSKA: Acta pays. polon **12**, 194 (1953).
[7] A. F. GIBSON, W. D. LAWSON and T. S. MOSS: Proc. Phys. Soc. Lond. A **64**, 1054 (1951).
[8] L. SOSNOWSKI and M. CHMIELEWSKI: Bull. Acad. polon. Sci., Cl. 3 **1**, 119 (1953).
[9] P. A. LEE: Canad. J. Phys. **31**, 1023 (1953).

**37. Photoconductivity in lead telluride.** Lead telluride layers are rendered photoconducting by the same process as for lead sulphide and selenide a change from $N$ to $P$ type occurring[1, 2, 3].

The spectral response of lead telluride is shown in Fig. 33 due to Moss[4]. The layers are not photosensitive at room temperature. Studies of photocurrent decay have been made by SCANLON *et coll.*[5] who find a two stage process the initial decay being exponential (monomolecular) and later stages being hyperbolic (bimolecular). Experiments on HALL coefficients during photoconduction by LEVY[6] show that illumination increases carrier concentration but has only a secondary effect on carrier mobility.

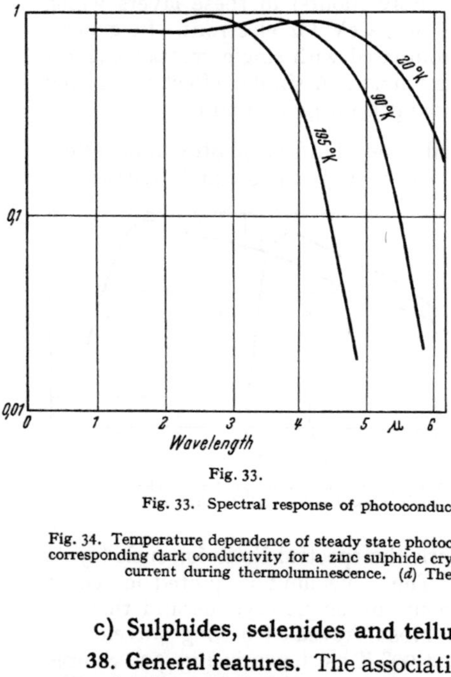


Fig. 33.

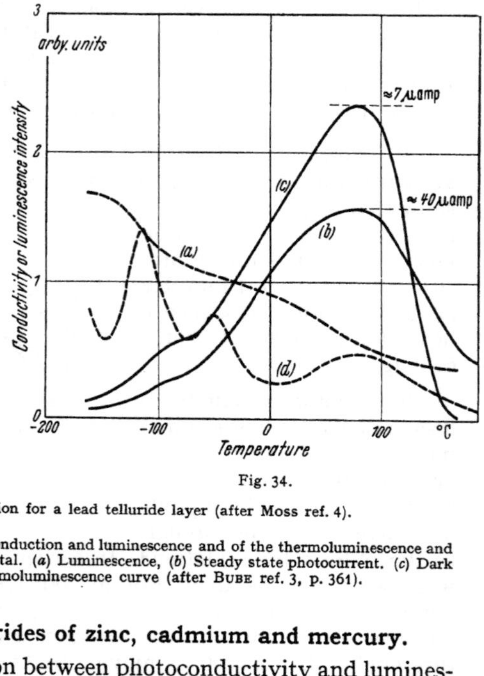


Fig. 34.

Fig. 33. Spectral response of photoconduction for a lead telluride layer (after Moss ref. 4).

Fig. 34. Temperature dependence of steady state photoconduction and luminescence and of the thermoluminescence and corresponding dark conductivity for a zinc sulphide crystal. (*a*) Luminescence, (*b*) Steady state photocurrent. (*c*) Dark current during thermoluminescence. (*d*) Thermoluminescence curve (after BUBE ref. 3, p. 361).

## c) Sulphides, selenides and tellurides of zinc, cadmium and mercury.

**38. General features.** The association between photoconductivity and luminescence in zinc and cadmium sulphide phosphor has been of great interest for many years. GUDDEN and POHL[7] investigated natural crystals of ZnS (blende-cubic form), CdS (greenokite) and HgS (cinnabar)[8] and also polycrystalline zinc sulphide phosphors. A strong revival of interest in these solids has taken place in recent years due to the development of techniques for obtaining single crystals. Interest in this section will be confined to single crystal specimens. The selenides and tellurides form a link with other semiconductors and recently some have been found to show luminescence in the infra red region.

[1] See footnote 4, p. 359.
[2] D. E. BODE and H. LEVINSTEIN: Phys. Rev. **96**, 259 (1954).
[3] H. CHECINSKA and L. SOSNOWSKI: Bull. Acad. polon. Sci., Cl. 3 **2**, 383 (1954).
[4] T. S. MOSS: Proc. Phys. Soc. Lond. B **62**, 741 (1949).
[5] W. W. SCANLON, R. L. PETRITZ and F. L. LUMMINS: Phys. Rev. **86**, 659 (1952).
[6] J. L. LEVY: Phys. Rev. **92**, 215 (1953).
[7] B. GUDDEN and R. W. POHL: Z. Physik **2**, 364 (1920); **5**, 176 (1921); **23**, 417 (1922).
[8] H. ROSE: Z. Physik **6**, 174 (1921).

**39. Zinc sulphide.** Zinc sulphide crystallises in cubic or hexagonal form the transition temperature being at about 1020° C. As the solid sublimes and dissociates at crystallising temperatures (1000 to 1200° C) production of large crystals of controlled composition is difficult. In recent years large advances have been made in the knowledge of the physical chemistry of the zinc and cadmium sulphides by KROGER and his collegues[1]. These have shown that many of the previous ideas on these materials have to be revised and that many previous results have little fundamental value because of the unknown crystal constitution. The characteristics of luminescence and photoconductivity are now known to be markedly dependent on the existence of lattice ion vacancies the generation of which is controlled by introduction of specific impurities (e.g. $Cu^+$, $Ag^+$, $Cl^-$, $Br^-$, $Al^{3+}$, $Sc^{3+}$, $Ga^{3+}$ ions etc.) or by oxidation or reduction of the crystals. The most complete study of the relation of electrical conductivity to crystal composition has been made for cadmium sulphide[2] but most of the results are relevant to the whole group of sulphides, selenides and tellurides. The zinc and cadmium sulphides are always $N$-type.

Using ZnS crystals made by evaporation of zinc metal in a stream of $H_2S$ and $H_2$ BUBE[3] investigated their photoconductivity characteristics. The crystals, which were cubic, showed a low intensity green emission under 3650 Å excitation which suggests that it was slightly reduced[1] Ohmic contact was made to the crystal by applying silver electrodes and removing the usual contact potential barriers by heating to 650° C. Fig. 34 shows the dependence of photoconduction (curve b) and green luminescence (curve a) on temperature and also variation of current with temperature obtained by warming the previously excited crystal in the dark. The broken curve (d) gives the thermoluminescence not for the same crystal but for another taken from the same batch. The steady photocurrent at any temperature is obviously controlled by the trapping processes and increases as the trapped electrons become thermally activated. The relatively sharp fall at higher temperatures is probably due to the effect of a marked fall in conduction electron mobility with temperature due to lattice scattering. This is also suggested by a similar rapid fall in the "current glow curve" (c). At low temperatures the photocurrent varied as a power of the exciting intensity between 0.5 and 1.0 due to the predominance of trapping effects. At higher temperatures a 0.5 power law held, the crystal thus showing the theoretically expected behaviour (Sect. 6). BUBE found that the photosensitivity was greatest when the crystal was in vacuo but decreased markedly when moist air was present. A corresponding decrease in time constant was also observed. This points to a surface recombination process as for germanium above due to trapping sites provided by adsorbed oxygen.

PIPER[4] measured the spectraresponse of photoconductivity, the absorption and other characteristics of single crystals of zinc sulphide grown in vacuo[5] or in hydrogen[6]. These specimens showed no visible luminescence and were colourless. The spectral response of photoconductivity shown in Fig. 35 peaks at the absorption edge. Fall off at shorter wavelengths is ascribed to increased rate of recombination in surface layers. The intrinsic semiconduction characteristics yield an energy gap for the forbidden zone of 3.77 eV which agrees well with the values from absorption edge and photoconductivity measurements.

[1] F. A. KROGER: Brit. J. Appl. Phys. Suppl. **4**, S 58 (1955).
[2] F. A. KROGER, H. J. VINK and J. BOOMGARD: Z. phys. Chem. **203**, 1 (1954). — Physica, Haag **20**, 1095 (1954).
[3] R. H. BUBE: Phys. Rev. **83**, 393 (1951).
[4] W. W. PIPER: Phys. Rev. **92**, 23 (1953).
[5] W. W. PIPER: J. Chem. Phys. **20**, 1343 (1952).
[6] D. C. REYNOLDS and S. J. CZYZAK: Phys. Rev. **79**, 543 (1950).

Recent absorption measurements by VAN DOORN[1] give a value extrapolated to zero temperature of $\sim 3.75$ eV. This agreement with the value from thermal measurement is important as it suggests that the ZnS lattice binding may be mostly homopolar in character. The photocurrent time constant at 290° K was $\sim 1000$ sec. i.e. trap controlled. In other studies PIPER and ROTH[2] obtained single crystal whiskers of zinc sulphide free of dislocations and stacking faults which showed a much faster decay, presumably due to absence of trapping states. CZYZAK *et coll.*[3] obtained either cubic or hexagonal crystals of zinc sulphide with

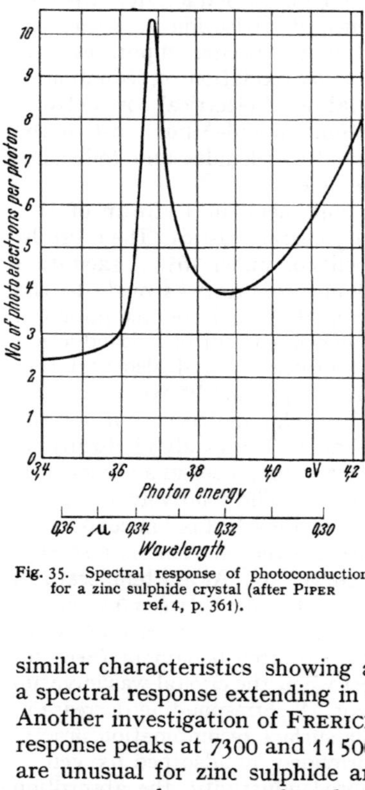


Fig. 35. Spectral response of photoconduction for a zinc sulphide crystal (after PIPER ref. 4, p. 361).

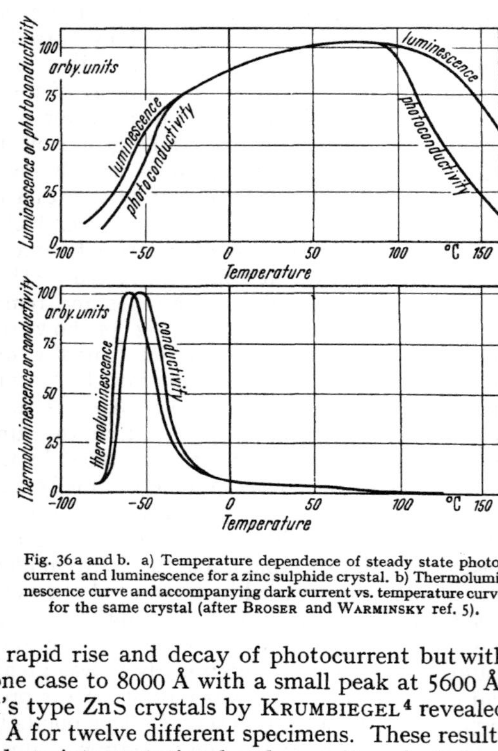


Fig. 36a and b. a) Temperature dependence of steady state photocurrent and luminescence for a zinc sulphide crystal. b) Thermoluminescence curve and accompanying dark current vs. temperature curve for the same crystal (after BROSER and WARMINSKY ref. 5).

similar characteristics showing a rapid rise and decay of photocurrent but with a spectral response extending in one case to 8000 Å with a small peak at 5600 Å. Another investigation of FRERICH's type ZnS crystals by KRUMBIEGEL[4] revealed response peaks at 7300 and 11 500 Å for twelve different specimens. These results are unusual for zinc sulphide and no interpretation has been offerred of these response peaks at wavelengths remote from the absorption edge. The strong green fluorescence of the specimens suggests a reduced crystal and therefore donor levels exist in the forbidden zone which may explain the response curves.

Following the development of general theoretical formulae for photoconduction in sulphides by BROSER and BROSER-WARMINSKY[5] the authors provided experimental results for zinc sulphide crystals. As seen in Sect. 9, it should be possible to find correlation between photoconduction and luminescence for particular conditions. Fig. 36a shows the steady state photoconduction and luminescence

[1] C. Z. VAN DOORN: Physica, Haag **20**, 1155 (1954).
[2] W. W. PIPER and W. L. ROTH: Phys. Rev. **92**, 503 (1953).
[3] S. J. CZYZAK, D. C. REYNOLDS, R. C. ALLEEN and C. C. REYNOLDS: J. Opt. Soc. Amer. **44**, 864 (1954).
[4] J. KRUMBIEGEL: Naturwiss. **39**, 447 (1952).
[5] I. BROSER and R. BROSER-WARMINSKY: Brit. J. Appl. Phys. Suppl. **4**, S 90 (1955).

of a ZnS crystal in their dependence on temperature. Below in Fig. 36b are the thermoluminescence and dark "current glow curves" for the same crystal after excitation at 90° K. The correlation between current and emission is marked in both (a) and (b). The steady state photocurrent rises with temperature as the traps indicated by the curves of Fig. 36b become effectively "shallow". At high temperatures the conductivity is reduced more than luminescence probably because of a fall in carrier mobility, with temperature.

With respect to decay of photocurrent and of luminescence, Fig. 37 constructed from the results of BROSER and BROSER-WARMINSKY[1] shows the relatively rapid decay of luminescence. Qualitatively this can be explained by the fact that the photocurrent is a linear function of the free carrier concentration while the luminescence, which involves recombining carriers, is proportional to a power of the carrier concentration greater than unity and reaching two for a simple bimolecular process. However, under certain conditions, as shown by the theoretical treatment of BROSER and BROSER-WARMINSKY (see Sect. 9, above), the photocurrent and luminescence decay rates may be identical. This will happen if both steady photoconduction and luminescence increase linearly with exciting intensity i.e. if recombination and the free carrier population is determined by the number of carriers which can escape from traps, there being a relatively large number trapped and therefore a correspondingly large number of empty recombination centres. In comparison with cadmium sulphide carrier mobilities in zinc sulphide are about a factor of ten lower.

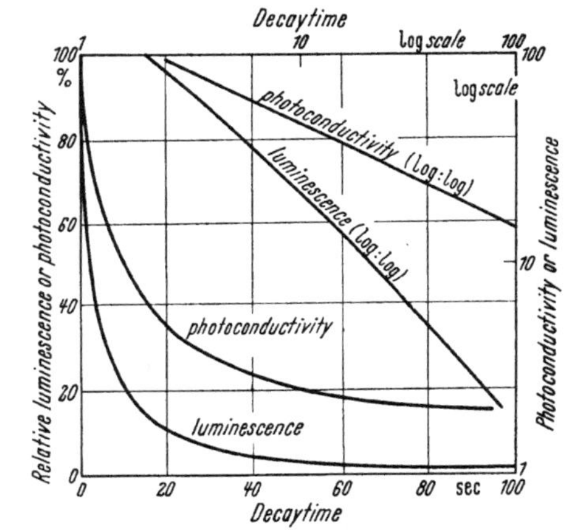


Fig. 37. Decay of conductivity and luminescence with time after excitation of a zinc sulphide crystal (after BROSER and WARMINSKY ref. 5, p. 363).

**40. Cadmium sulphide.** Extensive studies of CdS crystals have been made by many different workers and it is only possible here to review some of the investigations of fundamental interest. In 1947 FRERICHS[2,3] examined the photoconduction in the crystals of CdS made by evaporating cadmium in a stream of $H_2$ and $H_2S$. He found a spectral response peaking just outside the crystal absorption edge and also for high applied fields a multiplication process giving effective quantum efficiencies of $\sim 10^4$. His measurements were extended by KALLMANN and WARMINSKY[4] and others[5,6] and over the last few years a considerable number of papers have been published dealing with the optical and electrical properties of cadmium sulphide. It is necessary to select topics of interest and to confine attention to some of the more important investigations.

*α) Spectral dependence of photoconduction.* The spectral response of photoconduction in CdS shows a marked dependence on crystal constitution. As an

[1] See reference 5, p. 362.
[2] R. FRERICHS: Phys. Rev. **72**, 594 (1947).
[3] R. FRERICHS: Phys. Rev. **76**, 1871 (1949).
[4] H. KALLMANN and R. WARMINSKY: Ann. Phys., Lpz. **4**, 69 (1948).
[5] J. FASSBENDER and O. HACHENBERG: Ann. Phys., Lpz. **6**, 229 (1949).
[6] J. FASSBENDER and H. LEHMANN: Ann. Phys., Lpz. **6**, 214 (1949).

example of qualitative results Fig. 38 shows the response curves obtained by GOBRECHT and BARTSCHAT[1]. Curves (a) and (b) are for an "impure" crystal using two different thicknesses of specimen while curve (c) is for a pure crystal and shows a small tail in the response outside the absorption edge. The difference between curves (a) and (b) in the tail region is mainly due to the larger total light absorption in the thick crystal at longer wavelengths (low absorption). As seen in curve (c) the response falls rapidly at the shorter wavelengths than the absorption edge. As in the case of zinc sulphide above, this is probably due to increased recombination in surface states. Investigation of surface effects

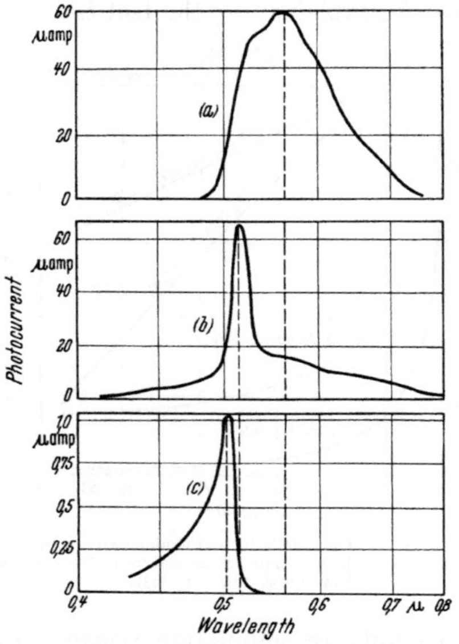


Fig. 38. Spectral response of photoconduction for different cadmium sulphide crystals. (*a*) Impure crystal—thick specimen. (*b*) Impure crystal—thin specimen. (*c*) Pure crystal (after GOBRECHT and BARTSCHAT ref. 1).

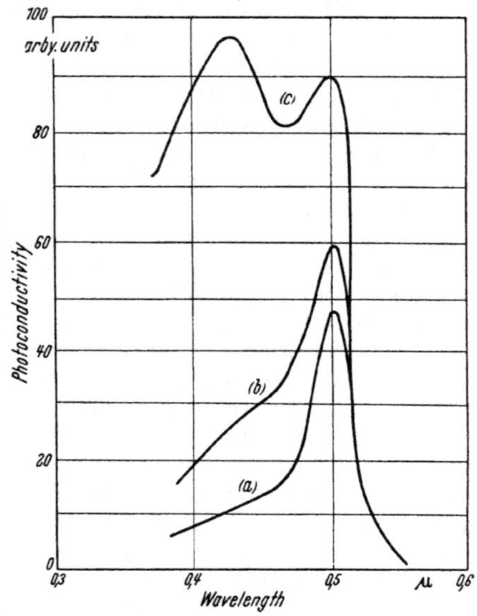


Fig. 39. Variation of photoconduction response in a layer of cadmium sulphide for different degrees of humidity in the surrounding atmosphere. (*a*) Rough vacuum pressure of water vapour. (*b*) Improved vacuum. (*c*) High vacuum (after BUBE ref. 2).

in photoconduction response have been made by BUBE[2]. Fig. 39 shows the spectral response of photoconductivity for a CdS layer subjected to different degrees of humidity in the surrounding atmosphere. For a high vacuum a relatively constant response is shown at the shorter wavelengths, but for rough vacua it falls off rapidly at short wavelengths. GOBRECHT and BARTSCHAT have reported an anisotropic spectral response for photoconduction in CdS crystals. Fig. 40 shows the photoconduction response curves for the electric vector of the incident radiation parallel to and perpendicular to the *c* axis of the crystal respectively. The small shift agrees with the shift of the absorption edge with the direction of the electric vector.

KLICK[3] has measured the change in spectral response of CdS with temperature down to 4° K. His measurements for response and optical absorption are shown in Fig. 41. At 4° K there is a long wavelength peak at 2.48 eV in the luminescence excitation spectrum for which there is no corresponding photoconduction. KLICK

[1] H. GOBRECHT and A. BARTSCHAT: Z. Physik **136**, 224 (1953).
[2] R. H. BUBE: J. Chem. Phys. **21**, 1409 (1953).
[3] C. C. KLICK: Phys. Rev. **89**, 274 (1953).

discusses the possibility of this peak being due to an exciton process but this is a doubtful possibility as the band is broad and shows no structure. A notable feature of the change in photoconduction response with temperature is the improvement in response for wavelengths shorter than the absorption edge as the

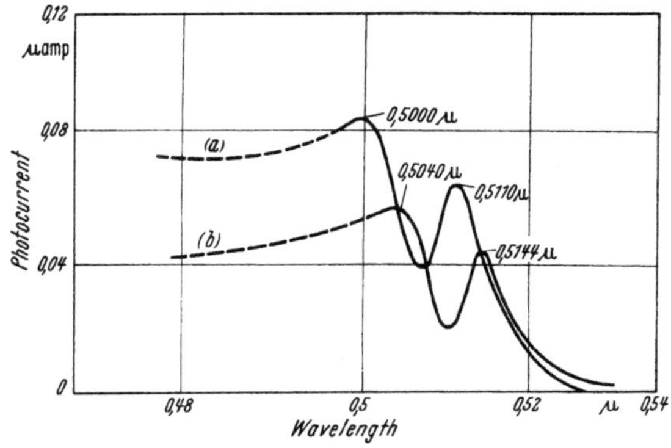


Fig. 40. Anisotropic behaviour in the optical response of photoconduction for a cadmium sulphide crystal. (a) Electric vector of incident light parallel to *c* axis of crystal. (b) Electric vector of incident light perpendicular to *c* axis of crystal (after GOBRECHT and BARTSCHAT ref. 1, p. 364).

temperature is lowered. This might be associated with an increase in mobility of carriers allowing the latter to move into the crystal away from surface recombination centres. In a recent paper LAMBE[1] has measured the variation in time constant of the photocurrent with wavelength of exciting light. He concludes that the carrier mobility is independent of wavelength and that the

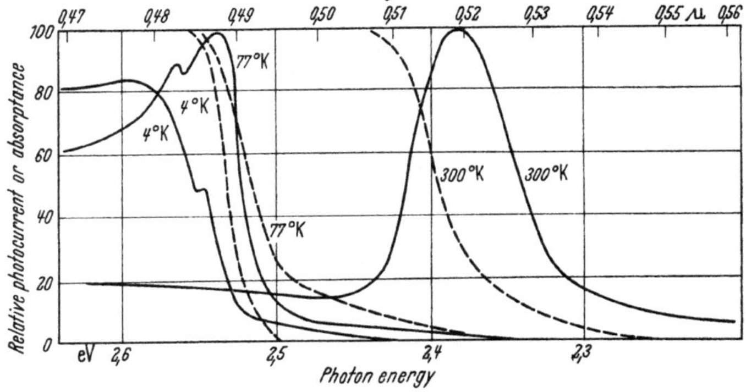


Fig. 41. Temperature dependence of spectral response of photoconduction and of absorption for a cadmium sulphide crystal. ——— photoconduction, ------ optical absorption (after KLICK ref. 3, p. 364).

spectral response curve for photoconduction is mainly determined by the dependence of carrier life time on absorption coefficient, i.e. on the distance from the surface at which the carriers are freed.

REYNOLDS, CZYZAK, ALLEN and REYNOLDS[2] have studied the effect of oxygen impurity (up to 2%) in CdS on the optical absorption and photoconduction

[1] J. LAMBE: Phys. Rev. **98**, 985 (1955).

[2] D. C. REYNOLDS, S. J. CZYZAK, R. C. ALLEN and C. C. REYNOLDS: J. Opt. Soc. Amer. **45**, 136 (1955).

response. The evidence for an entry of oxygen into the bulk crystal was based on crystallographic data, the lattice contracting with increasing oxygen. The alternative explanation is that the oxidation generates vacancies in the lattice[1,2] causing the observed contraction. An absorption tail develops for oxygenated crystals and the photoresponse increases in the tail region. These authors also show that the response curve obtained depends on whether the monochromator wavelength traverse is made from short to long or long to short wavelengths and also on the applied electric field strength. Response extends to longer wavelengths when the traverse is from short to long wavelength. This suggests a trapping of carriers released by the higher energy photons which are then reliberated by the lower energy photons. Insufficient details are given to substantiate this conclusion.

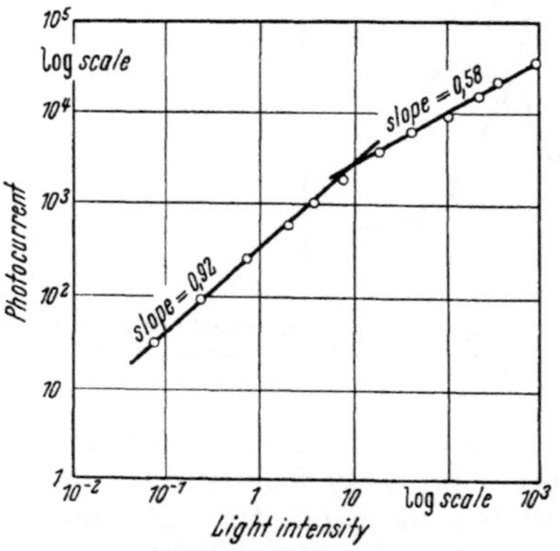


Fig. 42. Dependence of steady state photocurrent on incident light intensity for a cadmium sulphide crystal (after SMITH ref. 4).

*β) Variation of photocurrent with light intensity.* In the theoretical treatment of Sect. 7 we have seen that the relation between photocurrent and light intensity is dependent on the trap distribution in the crystal. Measurements on cadmium sulphide by various workers provide interesting verification of this dependence. From the simple bimolecular recombination theory inclusive of one depth of trapping states we have a linear relation for low intensities and a square root relation for high intensities The experimental correlation with these relations has been given already in Fig. 5 above.

In cases of complex trap distributions the power of the relation is not necessarily one or one half, as shown by ROSE[3] (Sect. 7 above). In Fig. 42 we have an example for a CdS crystal which shows a 0.92 power relation at low intensities and a 0.58 power relation at high intensities[4].

*γ) Time constants and trapping effects in CdS.* Cadmium sulphide crystals usually contain large numbers of normally unoccupied trapping states for carriers. Since these lie at an electron volt or so and below the conduction band they have a marked effect on photoconduction current decay and the variation of photocurrent with temperature. Detailed studies of the energy distributions may be made by current decay measurements, by variation of photocurrent with temperature and excitation intensity and also by study of the dark current warming curve as described above in Sect. 9 and for ZnS crystals in Sect. 39. In general it may be stated that to investigate trapping states in the bulk crystal the current warming curves give the most useful information. For trapping states at the crystal surface the decay constant, spectral response measurements and, if possible, studies of the photovoltaic effects are most useful.

Measurements of "current glow curves" for CdS have been made by BROSER and WARMINSKY[5,6], by MUSCHEID[7] and by BUBE[8]. MUSCHEID has demonstrated

[1] F. A. KROGER: Brit. J. Appl. Phys. Suppl. **4**, S 58 (1955).
[2] F. A. KROGER, H. J. VINK and J. BOOMGARD: Z. phys. Chem. **203**, 1 (1954). — Physica, Haag **20**, 1095 (1954).
[3] A. ROSE: RCA-Review **12**, 362 (1951).
[4] R. W. SMITH: RCA-Review **12**, 350 (1951).
[5] I. BROSER and R. BROSER-WARMINSKY: Brit. J. Appl. Phys. Suppl. **4**, S 90 (1955).
[6] I. BROSER and R. WARMINSKY: Z. Physik **133**, 340 (1952).
[7] W. MUSZHEID: Ann. Physik **13**, 322 (1953).
[8] R. H. BUBE: J. Chem. Phys. **23**, 18 (1955).

the effect of oxygen in the CdS crystal on the current glow curve above room temperature. In other studies[1] he obtained increases in semiconductivity on removal of oxygen by heating in vacuo but it is difficult to alow that these changes are due to diffusion of oxygen out of the bulk material. For the same reason the increase in magnitude of the current glow curve with a short treatment of CdS crystals in oxygen cannot be associated with production of trapping states in the bulk crystals.

As for zinc sulphide there is often correlation between the temperature variation of the steady state photoconduction and the "current glow curve" (cf. Fig. 34) but in other cases correlation appears to be absent. We select examples from Bube's results. His crystals are of two main types, those with relatively low halogen impurity content which are relatively insulating, and those with large halogen content which are good semiconductors at ordinary temperatures. Fig. 43 shows his results for an insulating crystal with $\sim 10^{16}$ trapping states. The glow current peaks at 3 μamp. the steady state photocurrent being of the order of hundreds of microamperes. In this case the normal photocurrent falls rapidly with temperature and shows a change of slope in the region of the "glow current" peak. At 100° C there is a rise of photocurrent. In this specimen there is a marked lack of correlation. The calculated trap depth for the current peak was 0.36 eV and the frequency constant $s \sim 3 \times 10^5$/sec. Bube gives the capture cross section for recombination centres as $5 \times 10^{-20}$ cm.$^2$ and for traps as $5 \times 10^{-21}$ cm.$^2$.

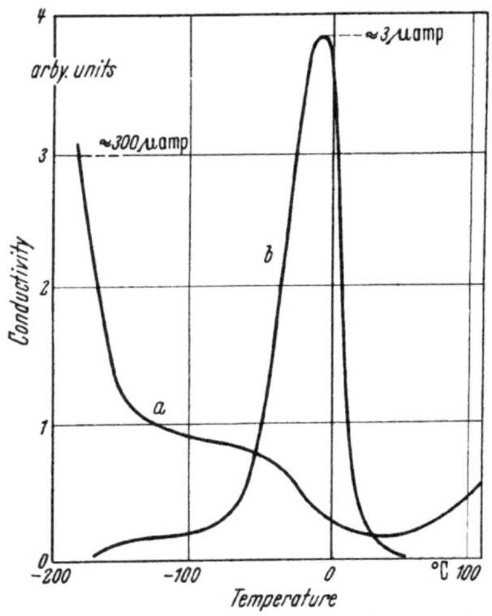


Fig. 43. *a* Variation of steady state photocurrent with temperature for a semiconducting cadmium sulphide crystal. *b* Variation of dark conductivity with temperature after steady state excitation at 90° K for the samecrystal (warming rate 0.77°/sec.) (after Bube ref. 8, p. 366).

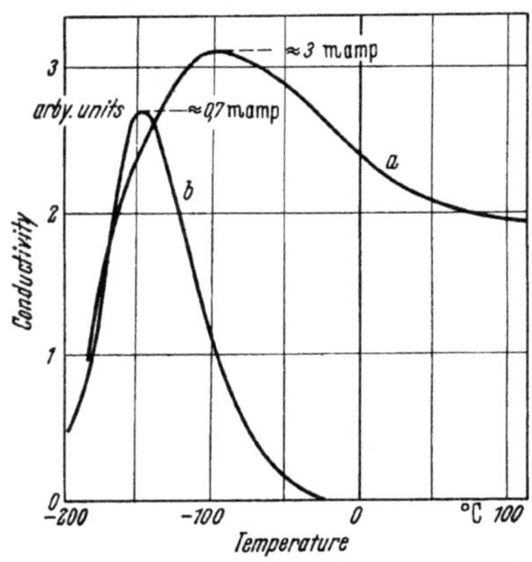


Fig. 44. *a* Variation of steady state photocurrent with temperature for an "insulating" cadmium sulphide crystal. *b* Variation of dark conductivity with temperature after steady state excitation at 90° K for the same crystal (warming rate 0,77°/sec.) (after Bube ref. 8, p. 366).

For a semiconducting crystal the corresponding curves are given in Fig. 44. In this case the trap depth is very small $\sim 0.06$ eV and the current glow peak correlates with a rise in steady photocurrent as the traps become shallow. In this specimen recombination centre and trap capture cross sections are $10^{-24}$ cm.$^2$ and $10^{-23}$ cm.$^2$ respectively i.e. several orders lower than in the more insulating specimen. The number of traps is much higher $\sim 3 \times 10^{19}$/cm.$^3$ in the semiconducting crystal. A surprisingly low value of the frequency constant $s \sim 10$/sec. is found for this crystal.

[1] W. Muscheid: Ann. Phys., Lpz. **13**, 305 (1953).

Using high light intensities to saturate the trapping states BUBE found carrier mobilities ~20 cm.²/volt. · sec. for the insulating type of crystal. From similar measurements of decay constant SMITH[1] obtained mobilities from 40 to 200 cm.² per volt · sec. The time constants were of the order of millisecs. and in one case varied approximately as the inverse of the photocurrent magnitude. This would be expected since at low currents (and light intensities) the time constant is governed by trapping processes. These values agree with those obtained by other workers notably GILDART and EWALD[2] and by FASSBENDER and LEHMANN[3]. The latter developed a method of computing the mobility from measurements of the relative d.c. and a.c. components of photocurrent with the frequency of a sine wave modulated light source used to excite the crystal.

In recent papers NIEKISCH[4] has shown that the modulated light method does not yield the true mobility but the effective mobility controlled by trapping processes. Elaboration of theory and experiment by NIEKISCH has provided a means of using the variation of the a.c. component of the modulated light excited photocurrent with temperature to determine the distribution of trapping states, the conduction electron concentration and the mobility (~200 cm.²/volt.sec. at 180° K). The technique is a powerful one and might be applied to other types of photoconductors.

Trapping states in CdS crystals can also be investigated by observing the effect of infra-red irradiation on the conductivity of a previously excited specimen. This process has already been referred to in Sect. 8 and 12 above. KLICK and LAMBE[5] have observed a photoresponse in a silver activated CdS crystal reaching out to 6 μ after previous and shorter wavelength excitation. The photocurrent is quenched by 1 μ radiation and enhanced by 3 μ radiation. They assume that the 1 μ radiation frees trapped positive holes which remove carrier electrons while the 3 μ radiation only liberates trapped electrons.

*δ) Barrier and photovoltaic effects in CdS.* Many photoconductivity measurements in the past on single crystals have been of little fundamental value because of the failure to ensure ohmic contact between the crystal and its electrodes. The problem has been considered for CdS in a recent paper by SMITH[6]. Indium and gallium electrodes are found to provide ohmic contact to the crystals and are essential for studies of bulk conductivity. However, provision of one ohmic and one non-ohmic contact (e.g. Ag or Au) enables the photo-effects to be studied at a cadmium sulphide-metal barrier which shows rectification effects. A study of space charge limited currents in CdS crystals with such barriers has already been reported[7, 8]. From the photoconductivity view point interest lies in the photovoltaic effect at barriers. This has been observed for CdS crystals by REYNOLDS *et coll.*[9] using an indium low resistance contact and a barrier contact metal of silver, gold or platinum. Voltages of 0.5 V and short circuit current densities of up to 300 ma/cm.² were observed. The spectral response is shown in Fig. 45. The peak at 7000 Å, outside the absorption edge, is unexpected as the absorption coefficients are low and carriers are produced at large distances

---

[1] R. W. SMITH: RCA-Review **12**, 350 (1951).
[2] L. GILDART and A. W. EWALD: Phys. Rev. **83**, 359 (1951).
[3] J. FASSBENDER and H. LEHMANN: Ann. Phys., Lpz. **6**, 214 (1949).
[4] E. A. NIEKISCH: Ann. Phys., Lpz. **15**, 279, 288 (1955).
[5] C. C. KLICK and J. LAMBE: Phys. Rev. **98**, 909 (1955).
[6] R. W. SMITH: Phys. Rev. **97**, 1525 (1955).
[7] R. W. SMITH and A. ROSE: Phys. Rev. **97**, 1531 (1955).
[8] A. ROSE: Phys. Rev. **97**, 1538 (1955).
[9] D. C. REYNOLDS, G. LEIES, L. L. ANTES and R. E. MARBURGER: Phys. Rev. **96**, 533 (1954).

from the barrier. To explain this peak in the response REYNOLDS and CZYZAK[1] suggest that impurity levels are not discrete but form a subsidiary energy band in the normally forbidden gap, a model suggested originally by BROSER, KALLMANN and WARMINSKY[2] to explain photoconduction characteristics. To obtain a steady photovoltaic effect both holes and electrons must be able to move and so discrete impurity levels would not explain the effect.

WLERICK[3] has studied the variation of the contact potential at a CdS layer with illumination (cf. BRATTAIN and BARDEEN[4]). He mounted the layer in vacuo, arranged a beam of slow electrons to be incident on the surface and measured the voltage current relations for the system. He concludes that the shift in contact potential with light intensity can be explained by a shift in the FERMI level in the bulk crystal and invocation of surface state changes is not necessary.

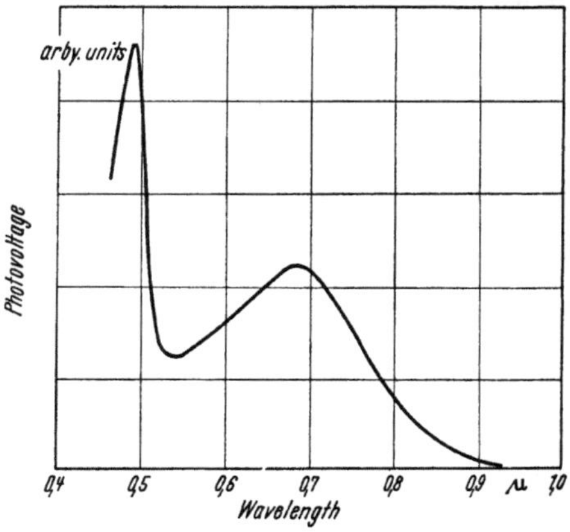


Fig. 45. Spectral response of the photovoltaic effect in a cadmium sulphide crystal with indium base electrode and various metals as rectifier contact (after REYNOLDS *et coll.* ref. 1).

**41. Zinc and cadmium selenide, cadmium telluride.** The spectral response of photoconductivity in cadmium selenide has been measured by SCHWARZ[5], and more recently by BUBE[6] together with that of ZnSe and CdTe. The latter's results are shown in Fig. 46. Two peaks appear in the response curves near the absorption edge for CdS and CdSe with separations of 35 and 65 Å respectively, these values being independent of temperature. We have already noted the results of GOBRECHT and BARTSCHAT[7] (Fig. 38) showing peaks in the response results of CdS crystals which depend for position on the incidence direction of the electric vector of the exciting light. The maximum shift shown in Fig. 38 is ~40 Å but the actual peak positions are 100 Å apart and at different wavelength values from those observed by BUBE. However, the anisotropic effect shown by the polarisation experiments is of the same order as the value 35 Å above. Cadmium selenide has more anisotropy than the sulphide, using the axial spacings as an indication, and so the larger peak separation might be expected.

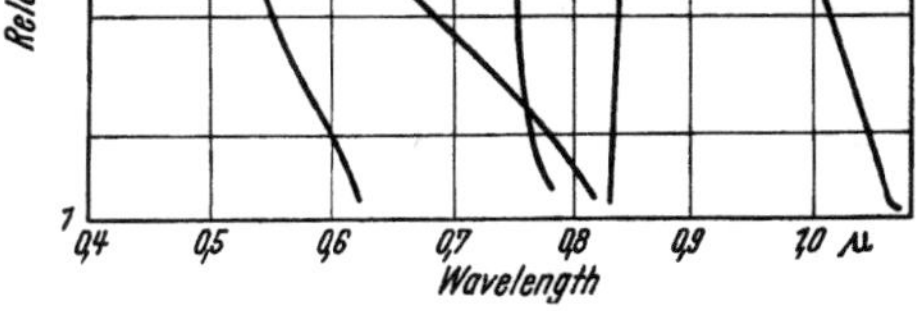


Fig. 46. Spectral response of photoconduction in various crystals. *a* CdS, *b* ZnSe, *c* CdSe, *d* CdTe (after BUBE ref. 6).

[1] D. C. REYNOLDS and S. J. CZYZAK: Phys. Rev. **96**, 1705 (1954).
[2] I. BROSER, H. KALLMANN and R. WARMINSKY: Z. Naturforsch. **4**a, 631 (1949).
[3] G. WLERICK: Physica, Haag **20**, 1099 (1954).
[4] W. H. BRATTAIN and J. BARDEEN: Bell. Syst. Techn. J. **32**, 1 (1953).
[5] E. SCHWARZ: Proc. Phys. Soc. Lond. B **63**, 624 (1950); **64**, 821 (1951).
[6] R. H. BUBE: Phys. Rev. **98**, 431 (1955).
[7] H. GOBRECHT and A. BARTSCHAT: Z. Physik **136**, 224 (1953).

Cadmium telluride forms a cubic lattice and is not a true member of this group. As shown in Fig. 46 its photoconduction response falls off rapidly at shorter wavelengths. The threshold response for CdTe obtained by BUBE (0.92 μ or 1.35 eV) shows reasonable agreement with that found by MIYASAWA and SUGAIKE[1] of about 1.42 eV.

Evidence of trapping states in cadmium selenide has been found by JENSEN and CASHMAN[2] using the dark current glow curve techniques. Peaks of current occur at 120 and 200° K and the effective charge released is as high as $10^{-3}$ coulombs.

Mercury sulphide (natural cinnabar crystal) was investigated in the classical experiments of GUDDEN and POHL and coll. (see Sect. 38), its threshold for intrinsic photoconduction being at about 0.64 μ. We give below the threshold characteristics of tellurides investigated by BRAITHWAITE[3] which include mercury telluride.

Table 2. *Telluride photoconductors.*

| Material | Peak response wavelength | Threshold wavelength ($\lambda_{\frac{1}{2}}$) |
|---|---|---|
| $Cu_2Te$ . . . | — | 1.2 μ |
| $Ag_2Te$ . . . | — | 1.3 μ |
| ZnTe . . . | 0.8 μ | 1.4 μ |
| HgTe . . . | 2.6 μ | 3.1 μ |
| $Tl_2Te$ . . . | 12. μ | 1.6 μ |
| $Sb_2Te$ . . . | — | 1.6 μ |
| $Mo_2Te$ . . . | 0.9 μ | 1.6 μ |
| $W_2Te$ . . . | 1.0 μ | 1.35 μ |
| $U_2Te$ . . . | 1.0 μ | 1.1 μ |

## d) Photoconduction in some oxides.

**42. Magnesium oxide.** Magnesium oxide crystals are of ionic type and have the rock-salt cubic structure. They can be grown in single crystal form to large dimensions and with a high degree of purity. Excess of either lattice constituent can be obtained by oxidation or reduction[4]. The refractive index is 1.74 and so the GUDDEN and POHL[5] criterion for intrinsic photoconduction is not fulfilled. The observed photoconduction, described below, is characteristic of crystal defect levels as for F centres in the alkali halides. DAY[6] has investigated the photoconductivity in ultra violet and neutron irradiated specimens. The photocurrents are proportional to applied electric field and to light intensity. The response spectra of specimens, untreated and after ultra violet and neutron irradiation are shown in Fig. 47. It is found that "fresh" crystals have response peaks at 2.1, 3.7 and 4.8 eV while the treated crystals have much enhanced photoconduction for these regions and a new response peak at 1.2 μ. The enhancement by

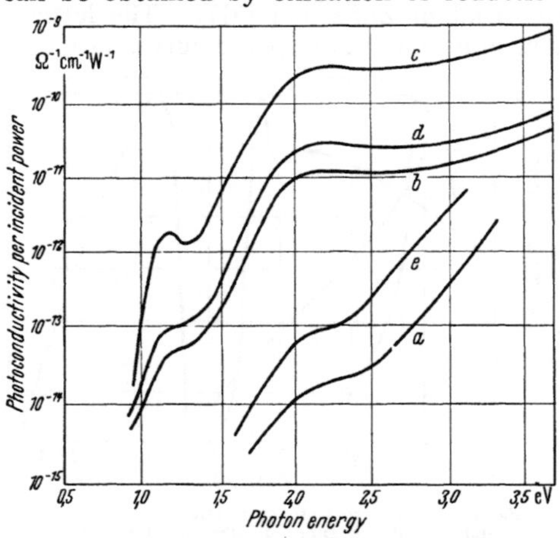


Fig. 47. Spectral response of photoconduction in a MgO crystal after ultra violet (0.31 μ) and neutron irradiation. *a* Unirradiated crystal. *b* After prolonged u.v. irradiation. *c* Immediately after neutron irradiation. *d* After neutron irradiation and saturation irradiation by u.v. *e* Crystal finally bleached by visible light irradiation (after DAY ref. 6).

[1] H. MIYASAWA and S. SUGAIKE: J. Phys. Soc. Japan **9**, 648 (1954).
[2] H. C. JENSEN and R. J. CASHMAN: Phys. Rev. **96**, 798 (1954).
[3] J. G. N. BRAITHWAITE: Proc. Phys. Soc. Lond. B **64**, 274 (1951).
[4] H. WEBER: Z. Physik **130**, 392 (1951).
[5] B. GUDDEN: Lichtelektrische Erscheinungen. Leipzig 1928.
[6] H. R. DAY: Phys. Rev. **91**, 822 (1953).

ultraviolet radiation (0.312 μ) reaches a saturation value independent of ultraviolet intensity which indicates the number of crystal defects. The latter and the response spectrum are increased by neutron irradiation. An interesting fact is that illumination of a narrow region of the specimen produces the same photoconduction as uniform illumination for equal absorbed power. The enhancing due to 0.312 μ irradiation can be bleached or removed by irradiation at 0.580 μ. It also decreases in the dark becoming rapid with increase in temperature. Neutron irradiation enhances the photoresponse as shown but the retention of the crystal in the dark of several days remove the effect. However, subsequent ultra violet irradiation produces a higher saturation level than before which suggests that new and stable lattice defects are formed by the neutron bombardment.

Photoconduction is apparently of *P*-type and decreases markedly with fall in temperature (a factor of $10^5$ between 290 and 90° K). This indicates either a two stage process involving thermal activation as for F centres in alkali halides or trapping effects which decrease the effective mobility.

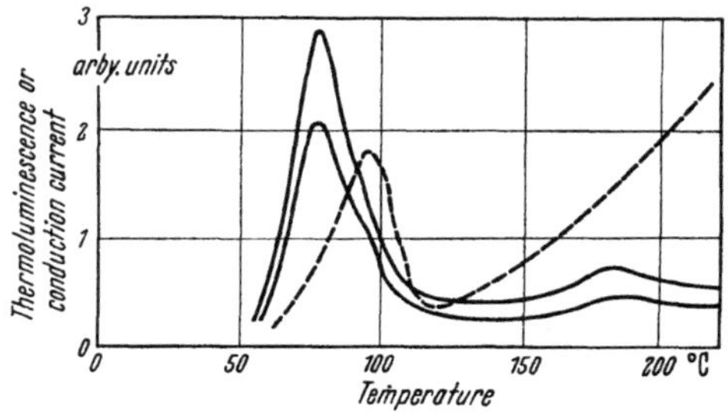


a

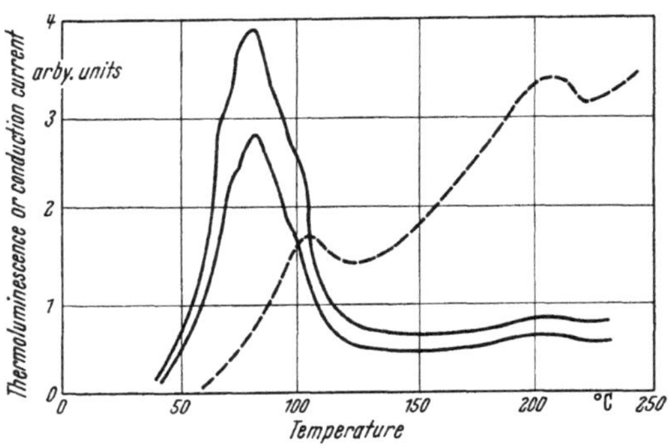


b

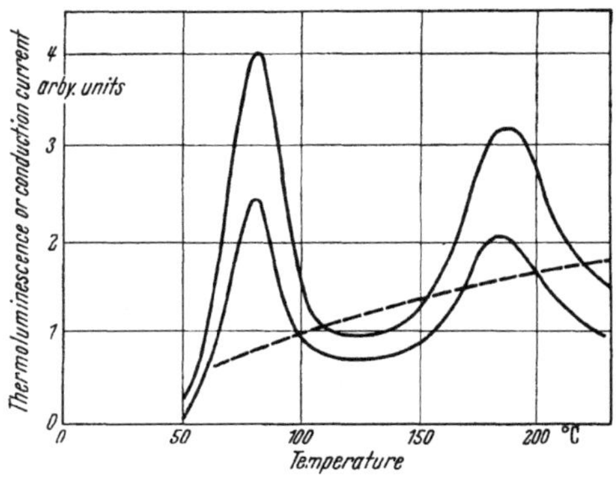


c

Fig. 48a–c. Thermoluminescence (full lines) and corresponding conductivity (broken line) curves for various MgO crystals. The higher thermoluminescence curve is for unfiltered emission the lower for $\lambda_{\text{emis}} > 0.56\,\mu$. a) Uncoloured crystal. b) Crystal with excess magnesium. c) Crystal with excess oxygen (after Yamaka ref. 3).

From absorption measurements on additively coloured crystals Weber[1] identifies the absorption bands at 2.1, 3.7 and 4.8 eV with excess magnesium and bands at 4.3 and 5.6 eV with excess oxygen. Neutron bombardment should increase both magnesium and oxygen vacancies but only photoresponse effects for excess magnesium were detected. Eisenstein[2] observed slow phosphorescence only for the 3.6 eV band of the luminescence spectrum which followed a power law or hyperbolic decay. X-ray excited photoconductivity also showed a slow decay which could be stimulated by 2.3 or 3.4 eV irradiation.

Yamaka[3] has measured for magnesium oxide crystals, with and without additive colouration, the thermoluminescence glow curves after excitation and the corresponding electrical dark current curves as shown in Fig. 48a, b and c.

[1] H. Weber: Z. Physik **130**, 392 (1951).
[2] A. S. Eisenstein: Phys. Rev. **93**, 1017 (1954).
[3] E. Yamaka: Phys. Rev. **96**, 293 (1954).

Several peaks occur with relation to the presence of excess oxygen or excess magnesium but results are not sufficiently adequate to determine the thermal activation energies, the change in emission spectra with temperature or the relation of glow curves to thermal bleaching of the absorption bands. However, the experiments are preliminary and worthy of extension. An important result is the determination of activation energies for semiconduction which are 2.1 eV at high temperatures and 0.58 eV at lower temperatures. If we take into account the ratio of dielectric constants for low and for optical frequencies (9.7/2.95) and assume the value of 2.1 eV for the intrinsic semiconduction then the optical absorption edge will be at ~7 eV as suggested by optical studies. For this reason the attempts of SAKSENA and PANT[1] to correlate luminescence spectra with the optical absorption and photoconduction response spectra must be considered doubtful. FRANCK-CONDON shifts are ignored in their treatment[2].

With respect to future studies of photoconduction and optical absorption in magnesium oxide it would be better to use high energy electrons (~2 MeV) for controlled defect production as the defects would be more widely separated from each other compared with the distribution from neutron irradiation.

**43. Zinc oxide.** Zinc oxide received much attention in the past as a semiconductor which could be obtained as a compact sintered mass and whose properties could be altered by heat treatment in oxidising or reducing atmospheres. A review of literature on the material up to 1952 has been given by HEILAND[3]. In the past few years more precise studies of the photoconductivity in zinc oxide relative to semiconduction and optical properties have been made for sintered samples, and very thin films. Some optical and semiconduction measurements have been made on single crystals.

MOLLWO and STÖCKMANN[4] made very thin films of zinc oxide by evaporation of zinc metal on to glass plates in a blow pipe flame, the departure from stoichiometry being controlled by the amount of oxygen in the flame. Fig. 49a shows their measurements of optical absorption spectra for films with near stoichiometric constitution and with excess zinc respectively. Fig. 49b shows the measured photoconduction response for such specimens. The yellow or reduced films show increased photosensitivity and also increased semiconduction. A relation was found between the photocurrent ($i_{ph}$) the semiconduction current ($i_s$) and the excitation intensity $J$ as follows:

$$i_{ph}^3 = \text{const}\, i_s^2 \cdot J. \tag{43.1}$$

A theoretical treatment of the results is given which assumes that the excess of zinc gives rise to $Zn^+$ ions in the lattice by thermal ionisation of the excess zinc atoms. However, their results are probably more difficult to interpret than they suggest since other work shows that in sintered zinc oxide the main effects are controlled by the properties of the grain boundaries. A study of the temperature dependence of the film characteristics would have given useful data with respect to the theory since the latter included a thermal activation process. MILLER[5] points out that sintered zinc oxide specimens give complex results, for example, the illumination of samples with white light causes slow changes over several hours. In experiments by HAHN, RUSSEL and MILLER[6] the

[1] B. D. SAKSENA and L. M. PANT: J. Chem. Phys. **23**, 989 (1955).

[2] For a further treatment of MgO see the present writer' scontribution to vol. XXVI of this Encyclopedia.

[3] G. HEILAND: Z. Physik **132**, 354 (1952).

[4] E. MOLLWO and F. STOCKMANN: Ann. Phys., Lpz. **3**, 223 (1948).

[5] P. H. MILLER: Semi-Conducting Materials, p. 172. London: Butterworths 1951.

[6] E. E. HAHN, B. R. RUSSEL and P. H. MILLER: Phys. Rev. **75**, 1631 (1949); **78**, 348 (1950).

measurement of intrinsic semiconduction gives an energy gap of 2 eV while the optical absorption edge is found at 3.2 eV. Photoconduction quantum yields can be very much greater than unity for sintered zinc oxide samples[1] and the photocurrent is very much dependent on adsorbed oxygen. MELNICK and NEURINGEN[2] suggest that adsorbed oxygen at the grain boundaries can capture conduction electrons. In experiments on the photovoltaic effect for zinc oxide PUTSEIKO[3] found that reduction of oxygen pressure on the specimens lowered the photovoltage reversibly except in the case of prolonged treatment at low pressures. He has also studied with TERENIN[4] the effect of adsorbed dyes on the photoconduction of zinc oxide. New spectral response peaks appear (at 0.43 and 0.65 μ for chlorophyll) outside the lattice edge peak at 0.37 μ. The suggestion is

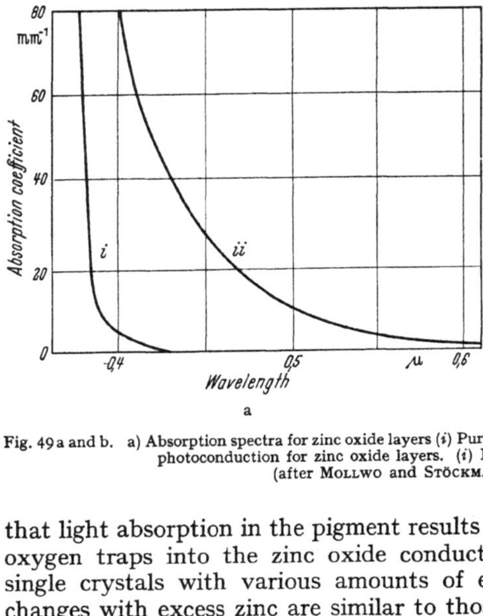


a

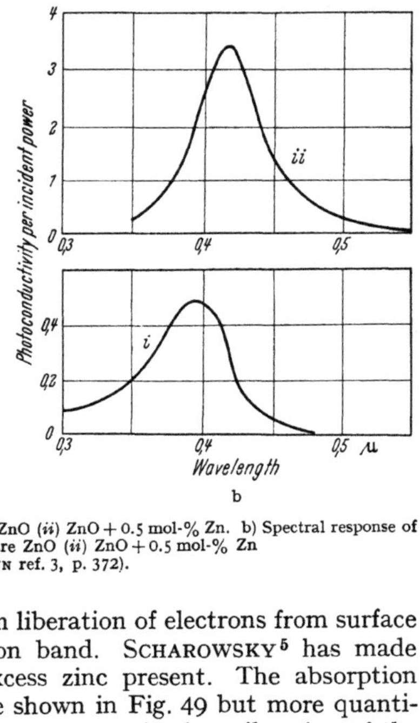


b

Fig. 49a and b. a) Absorption spectra for zinc oxide layers (*i*) Pure ZnO (*ii*) ZnO + 0.5 mol-% Zn. b) Spectral response of photoconduction for zinc oxide layers. (*i*) Pure ZnO (*ii*) ZnO + 0.5 mol-% Zn (after MOLLWO and STÖCKMANN ref. 3, p. 372).

that light absorption in the pigment results in liberation of electrons from surface oxygen traps into the zinc oxide conduction band. SCHAROWSKY[5] has made single crystals with various amounts of excess zinc present. The absorption changes with excess zinc are similar to those shown in Fig. 49 but more quantitative measurements show that the absorption constant in the tail region of the lattice absorption increases linearly with the concentration of zinc atoms in the vapour used in preparation of the crystals.

There are at present serious discrepancies between the thermally measured energy gap for zinc oxide (~2.0 eV) and the optical absorption edge (3.2 eV) However, as discussed for other oxide photoconductors, there should be a difference between the two determined by the values of the dielectric constant at optical and at low frequencies. For such oxides the optical value is a factor of three or more less than the static value. In this case the absorption edge for zinc oxide should be at an energy greater than 6 eV. Now that single crystals are available it should be possible to determine whether the absorption edge at 3.2 eV is due to band to band transitions or is an "exciton" absorption.

---

[1] See footnote 3, p. 372.
[2] D. A. MELNICK and L. J. NEURINGEN: Phys. Rev. **91**, 229 (1953).
[3] E. K. PUTSEIKO: Dokl. Akad. Nauk SSSR. **91**, 1071 (1953).
[4] E. K. PUTSEIKO and A. N. TERENIN: Dokl. Akad. Nauk SSSR. **90**, 1005 (1953).
[5] E. SCHAROWSKY: Z. Physik **135**, 318 (1953).

**44. Barium oxide.** Barium oxide is well known as the basic material of the oxide coated thermionic cathode. Studies of photoconduction and optical properties have been made as an aid to an understanding of the thermionic processes. In order to avoid the rather complex conditions in cathode coatings single crystals were prepared by SPROULL and TYLER[1]. Absorption and photoconduction response spectra for a typical specimen are shown in Fig. 50. The photoconduction shows a marked increase with temperature over the range $-70°$ to $+60°$ C but since modulated light was used and the corresponding a.c. component of current measured, the increase may be associated with time constant changes.

More recent and more extensive measurements on barium oxide crystals by DASH[2] have shown that photoconduction response occurs at longer wavelengths

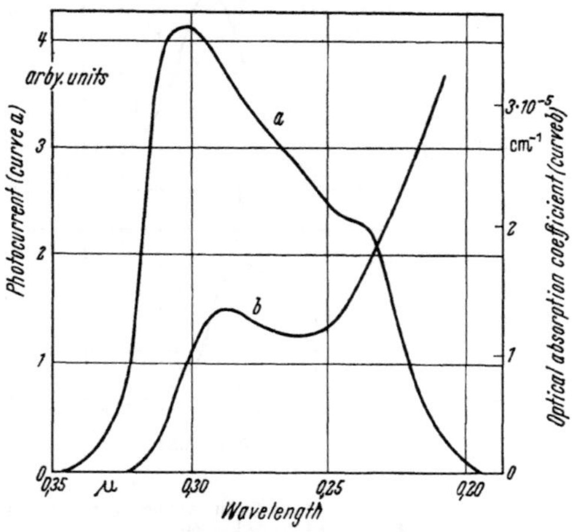


Fig. 50. *a* Spectral response of photoconduction. *b* Optical absorption spectrum for a single crystal of barium oxide (after SPROULL and TYLER ref. 1).

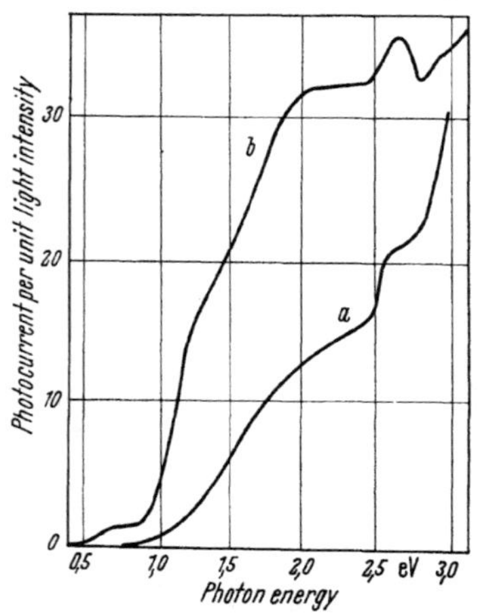


Fig. 51. Spectral response of photoconduction for a barium oxide crystal. *a* Before, *b* during u.v. irradiation (after DASH ref. 2).

than those shown in Fig. 50 and is dependent on previous ultraviolet or X-ray irradiation of the crystals. The photoconduction response for a crystal before and after ultra violet irradiation is shown in Fig. 51. Absorption peaks found at ~0.7 eV and 1.4 eV agree with the structure in the above curves and analysis of the response increment for ultra violet and for X-ray irradiation suggests peaks at 2.0 and 2.6 eV also.

By analogy with colouration in the alkali halides studies of additively coloured barium oxide crystals were made. Heating in barium vapour produces blue coloured specimens with an absorption peak at 2.0 eV and a suggestion of a weak one at 1.4 eV, while crystals grown from solution in molten barium are red and show no peaks in this infra red region but a gradual rise of absorption towards the visible and towards longer wavelengths. Photoconduction response spectra in blue, red and uncoloured crystals are shown in Fig. 52. The additively coloured crystals were not easy to use as considerable current noise was encountered. The blue colouration (2.0 eV absorption band) in additively coloured crystals is attributed to oxygen vacancies[3]. The general evidence from the above studies

[1] R. L. SPROULL and W. W. TYLER: Semi-Conducting Materials, p. 122. London Butterworths 1951. — Phys. Rev. **83**, 548 (1951).
[2] W. C. DASH: Phys. Rev. **92**, 68 (1953).
[3] R. L. SPROULL, R. S. BEVER and G. LIBOWITZ: Phys. Rev. **92**, 77 (1955).

is that the peaks at 0.8 and 1.4 eV are due to separate centres of unknown structure. In studies of photoconductivity for different states of activation of a barium oxide cathode DE VORE[1] found that a 2.3 eV response developed with a tail to longer wavelengths indicating a peak at ~1.5 eV. ORTUSI[2] gives a theoretical treatment in which the 2.3 eV response is associated with an oxygen vacancy (F centre equivalent). His calculations suggest that the 1.5 eV band is due to colloidal barium levels. The production of the 1.4 eV peak has been observed for barium oxide cathodes by SAMAKATO[3] as shown in the curves of Fig. 53. In other studies of the effect of oxygen on the photoconduction of the cathode SAMAKATO *et coll.*[4] suggested that the 3.7 eV response peal is not found in single crystals but this is not true as shown by Fig. 50 above for a single crystal. They find that the 4.2 eV response falls slowly with oxygenation while the 3.7 eV peak is very rapidly reduced. The slow fall of the 4.2 eV peak is attributed to increase in scattering of conduction electrons as the oxygen diffuses into the cathode pores. It is clear that the results for cathode layers are not easy to interpret compared with those for single crystals.

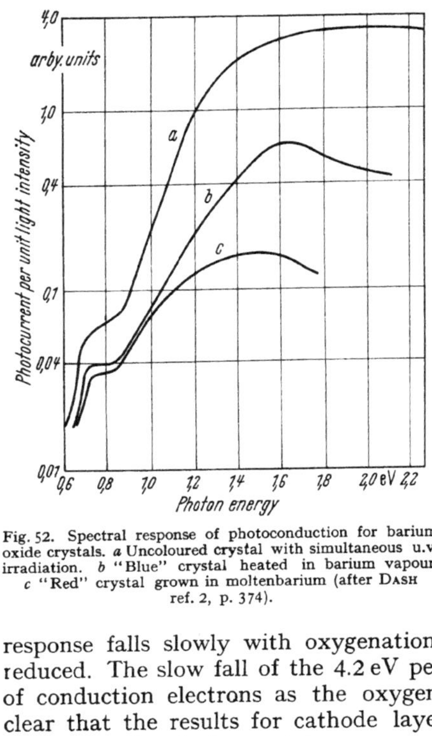


Fig. 52. Spectral response of photoconduction for barium oxide crystals. *a* Uncoloured crystal with simultaneous u.v. irradiation. *b* "Blue" crystal heated in barium vapour. *c* "Red" crystal grown in moltenbarium (after DASH ref. 2, p. 374).

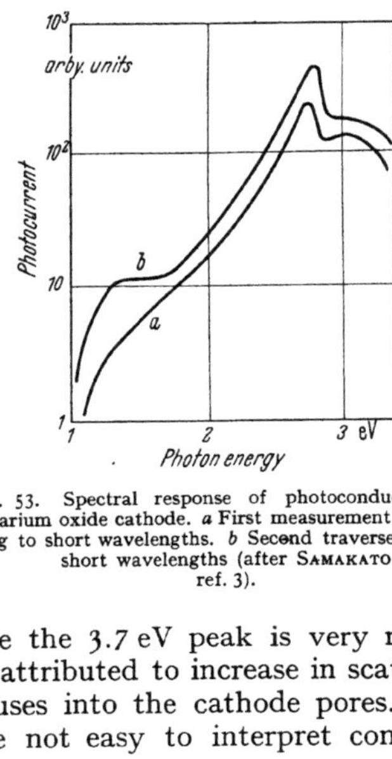


Fig. 53. Spectral response of photoconduction for a barium oxide cathode. *a* First measurement traverse, long to short wavelengths. *b* Second traverse, long to short wavelengths (after SAMAKATO ref. 3).

**45. Titanium dioxide-rutile.** Titanium dioxide forms the basic material in the titanate groups which have become widely known for their ferro-electric properties. Rutile is the most stable form of titanium oxide and has a sixfold oxygen coordination. In recent years synthetic rutile crystals have been produced and their optical and photoconduction properties have been investigated by CRONEMEYER[5]. The crystal is anisotropic with refractive indices of 2.4 and 2.57. From semiconduction measurements the forbidden energy gap activation energy is not single valued. Measurements below 900° C give 3.05 eV while those above

[1] H. B. DE VORE: RCA-Review **13**, 453 (1952).
[2] J. ORTUSI: Vide **9**, 100 (1954).
[3] M. SAMAKATO: Vide **9**, 109 (1954).
[4] M. SAMAKATO, S. KOBAYASHI and S. ISHII: Phys. Rev. **98**, 552 (1955).
[5] D. C. CRONEMEYER: Tech. Rep. 46, Lab. Insulation Res. M.I.T. 1951. — Phys. Rev. **87**, 876 (1952).

900° C give 3,67 eV. The photoconduction response and its variation with temperature are shown in Fig. 54. The threshold moves to shorter wavelengths as the temperature falls. A similar shift is shown by the ansorption edge with which the photoconduction ıesponse shows correlation, the temperature coefficient of the shift for absorption and photoconduction response being $\sim 10^{-3}$ eV/°C. Pure crystals aıe transparent between 0.4 and 7 μ the latter wavelength being determined by the Reststrahlen frequencies of the lattice. A long tail of photoconduction response extends to 1.2 μ and perhaps as far as 2 μ. When the crystals are reduced, which colours them blue, a new optical absorption band appears at 1.2 μ but no corresponding band is produced in the photoconduction response.

In other studies BRECKENBRIDGE and HOSLER[1] have made semiconduction measurements on titanium oxide crystals and found an intrinsic activation energy of 3.67 eV in agreement with CRONEMEYER's higher value. However, the thermal value suggests that the optical absorption edge and the photoconduction threshold at $\sim 0.41$ μ do not correspond to band to band transitions. BRECKENBRIDGE and HOSLER have given the following interpretation.

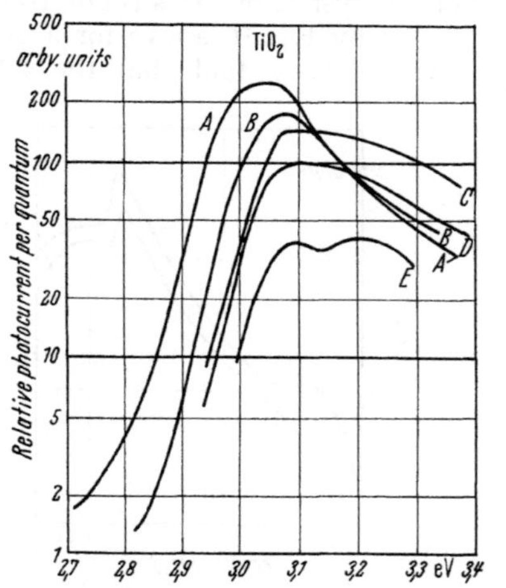


Fig. 54. Spectral response of photoconduction for a titanium dioxide crystal at various temperatures *A* 295° K, *B* 228° K, *C* 190° K, *D* 166° K, *E* 103° K (after CRONEMEYER ref. 5, p. 375).

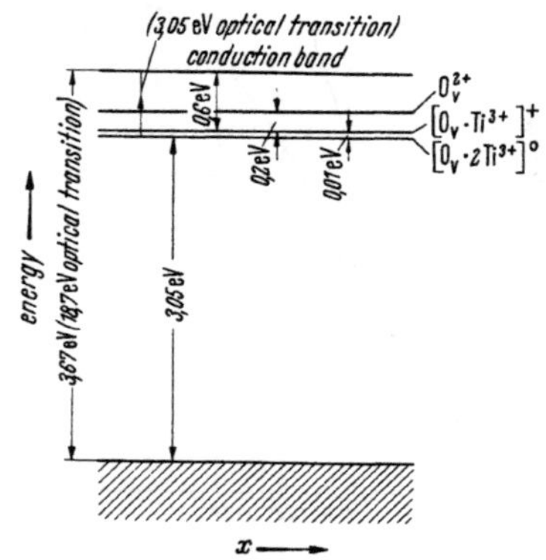


Fig. 55. Energy level diagram for titanium dioxide (after BRECKENBRIDGE and HOSLER ref. 1).

The energy level diagram for thermal equilibrium in a slightly reduced crystal is shown in Fig. 55. Semiconduction shows a donor activation energy of 0.2 eV for thermal bleaching of the 1.2 μ band and another activation energy of 0.01 eV. Reduction is assumed to produce two types of oxygen vacancy defects, namely a vacancy with two associated $Ti^{3+}$ ions having neutral properties $[O_v \cdot 2Ti^{3+}]^0$ and a vacancy with one associated $Ti^{3+}$ having a net positive charge $[O_v \cdot Ti^{3+}]^+$. The former has the 0.01 eV activation energy to the latter centre as shown in Fig. 55. The 0.2 eV energy represents the transformation $[O_v \cdot Ti^{3+}]^+ \rightarrow O_v^{++}$. These authors consider that the large ratio between thermal and optical dielectric constants will, according to the theory of MOTT and GURNEY[1], give a similar ratio of optical and thermal activation energies. They thus identify the 0.2 eV thermal activation energy for bleaching of the 1.2 μ band with energy of 1.0 eV for optical transitions in this band, and the absorption edge and photoconduction threshold at 3.05 eV with the excitation of electrons from the centres $[O_v \cdot 2Ti^{3+}]^0$ and $[O_v \cdot Ti^{3+}]^+$ to the conduction band (thermal jump 0.62 eV). With 1.2 μ

[1] R. G. BRECKENBRIDGE and W. R. HOSLER: Phys. Rev. 91, 793 (1953).

irradiation bleaching of the band is expected but no photoconductivity since the electrons do not reach conduction levels. The two peaks in the low temperature curve of Fig. 54 would then correspond to the 0.01 eV separation of the above vacancy centres. This model implies that the forbidden energy gap between the bands corresponds to an optical transition of energy 18.7 eV (0.067 μ) which is consistent with results for other oxides[1]. The above model should be capable of verification by photoconduction measurements. Reduction of the crystals should provide more vacancy centres, more carriers from the optical transition at 3.05 eV and hence increased photosensitivity.

**46. Cuprous oxide.** Cuprous oxide ($Cu_2O$) layers on copper have been known for many years to be photoconducting and to show the photovoltaic effect. The first investigations were made by PFUND[2] but the more precise study of photoconductivity begins with the work of SCHÖNWALD[3] and photovoltaic studies by LANGE[4]. A critical discussion of the knowledge of the semiconduction and photoeffects in cuprous oxide was given in 1940 by MOTT and GURNEY[1]. Since then some notable contributions to the investigations of cuprous oxide have been made by Russian workers, to be described below. We first collect the information obtained in the earlier work and the discussion provided by MOTT and GURNEY.

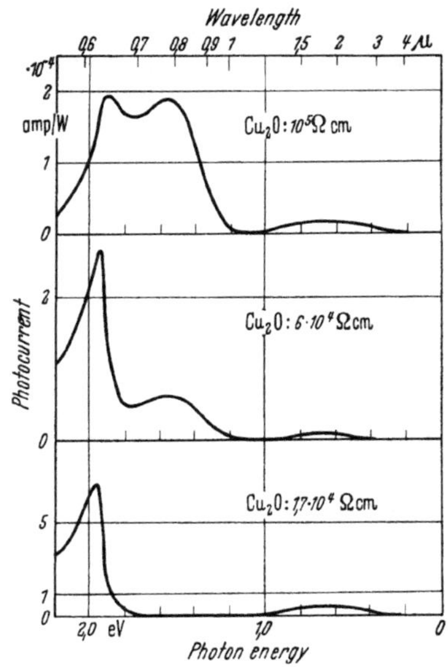


Fig. 56. Spectral response of photoconduction in cuprous oxide crystals (after SCHÖNWALD ref. 3).

Fig. 56 gives the spectral response of photoconduction obtained by SCHÖNWALD[3] and still generally accepted. He used polycrystalline plates of the oxide as large as $11 \times 6.5 \times 1$ mm. with evaporated silver electrodes. Polycrystalline samples were easier to prepare than single crystals but similar properties were observed for both. The measurements were made at room temperature. The observed photocurrents were proportional to the applied field and to exciting radiation intensity and time constants were of the order of $10^{-5}$ sec. The photocurrents are incremental on the dark current background. As shown, response peaks were obtained at 0.63 μ (1.96 eV), 0.8 μ (1.6 eV) and ~2 μ (0.65 eV) the first coinciding with the optical absorption edge of $Cu_2O$. In studies of the photovoltaic effect LANGE[4] found a spectral response which included the 0.63 μ and the 0.8 μ peaks but having a limit of detection at 1.4 μ. The 0.8 μ peak disappears at low temperature (−150° C) leaving only that associated with the lattice absorption. MOTT and GURNEY suggest that the absorption up to 1.4 μ produces carriers of both signs but it is clear that this is not effective at low temperatures. From HALL measurements[5] and from mode of preparation it is concluded that $Cu_2O$ is a *P*-type semiconductor. MOTT and GURNEY suggest that the highest occupied band in the energy band model is due to $Cu^+$ levels the $O^{2-}$ levels

[1] N. F. MOTT and R. W. GURNEY: Electronic Processes in Ionic Crystals. Oxford 1940.
[2] A. H. PFUND: Phys. Rev. **7**, 289 (1916).
[3] B. SCHÖNWALD: Ann. Phys., Lpz. **15**, 395 (1932).
[4] B. LANGE: Phys. Z. **32**, 850 (1931).
[5] E. ENGELHARD: Ann. Phys., Lpz. **17**, 501 (1933).

forming a band at a lower energy. The *P*-type conduction results from copper ion vacancies in the crystal lattice which act as acceptor levels above the highest filled band. To these conclusions may be added those based on more recent investigations. In 1946 LASHKAREV and KOSSONOGOVA[1] reported an infra red luminescence emission at 1 μ excited by shorter wavelength irradiation of $Cu_2O$ specimens which occurred at room temperature and below and showed the usual thermal quenching at higher temperatures. The thermal activation process involved in the quenching had an activation energy of ~0.67 eV. The emission was also better in specimens which were relatively copper rich. These observations have been confirmed by M. J. DUMBLETON and the writer together

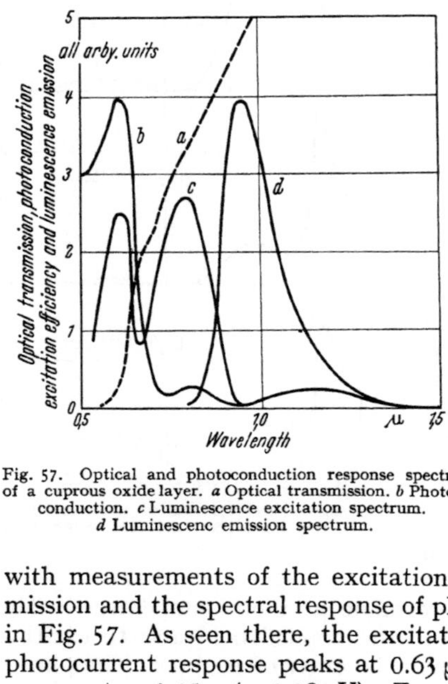


Fig. 57. Optical and photoconduction response spectra of a cuprous oxide layer. *a* Optical transmission. *b* Photoconduction. *c* Luminescence excitation spectrum. *d* Luminescenc emission spectrum.

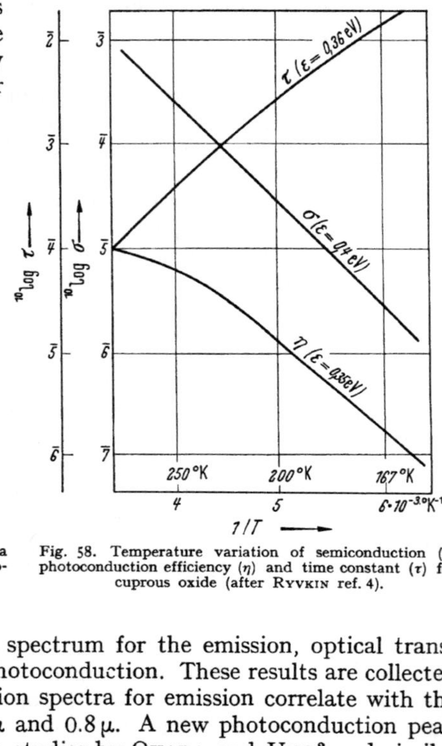


Fig. 58. Temperature variation of semiconduction (σ) photoconduction efficiency (η) and time constant (τ) for cuprous oxide (after RYVKIN ref. 4).

with measurements of the excitation spectrum for the emission, optical transmission and the spectral response of photoconduction. These results are collected in Fig. 57. As seen there, the excitation spectra for emission correlate with the photocurrent response peaks at 0.63 μ and 0.8 μ. A new photoconduction peak occurs at ~1.15 μ (~1.08 eV). From studies by OKADA and UNO[2] and similar ones by ANDRIEVSKII and RZACHEW[3] it is known that reduction of the surface layer of the oxide (copper enrichment) improves the photoconduction response at 0.8μ which is also a peak in the excitation spectrum for the 1 μ emission, itself enhanced by copper enrichment. Studies of the dependence of photoconduction on temperature by RYVKIN[4] using a modulated exciting radiation method developed by ZHUZE and RYVKIN[5] have shown that the time constant $\tau$ and quantum efficiency $\eta$ for the photocurrent alter exponentially with temperature as follows:

$$\tau \propto \exp(\mathcal{E}/2kT), \tag{46.1}$$

$$\eta \propto \exp(-\mathcal{E}/2kT) \tag{46.2}$$

[1] V. E. LASHKAREV and K. M. KOSSONOGOVA: C. R. Acad. Sci. URSS. **54**, 125 (1946).
[2] T. OKADA and R. UNO: J. Phys. Soc. Japan **4**, 351 (1949).
[3] A. I. ANDRIEVISKII and A. L. RVACHEV: Dokl. Akad. Nauk SSSR. **89**, 245 (1953).
[4] S. M. RYVKIN: Dokl. Akad. Nauk SSSR. **68**, 487 (1949).
[5] V. P. ZHUZE and S. M. RYVKIN: Dokl. Akad. Nauk USSSR. **58**, 1629 (1947).

where $\mathcal{E}$ is $\approx 0.35$ eV and agrees reasonably well with the activation energy for semiconduction over this range. RYVKIN's results are given in Fig. 58. From earlier studies of semiconduction JUSÉ and KURTSCHATOV[1] obtained an intrinsic activation energy of 1.4 eV and a value at lower temperatures of $\sim 0.26$ eV. ENGELHARD's[1] measurements of HALL coefficients and semiconduction gave $\mathcal{E}$ values of 0.7 eV and 0.36 eV.

From the above data we are now in a position to construct an energy band diagram for $Cu_2O$ as shown in Fig. 59. It is similar to that proposed by ZHUZE and RYVKIN[2] in their quantitative treatment of RYVKIN's results[3], except that we have attempted to coordinate all the present available evidence from photo and semiconduction, photovoltaic effect and from luminescence excitation and emission spectra. Fig. 59a is for optical transitions while Fig. 59b is for thermal equilibrium transitions (semiconduction) only one of each kind of defect level being inserted for the sake of clarity. The transitions for the 1.96 eV absorption are shown as transitions to exciton levels for which there is now good evidence in $Cu_2O$[4]. This means that the real energy gap for $Cu_2O$ is somewhat greater. On the basis of the evidence below and collected in Fig. 59 we suggest that it is about 2.3 eV. The luminescence centre is identified with the level at 0.65 eV (thermal) above the filled band. In Fig. 59a we thus have all the known optical transitions for photoconduction and luminescence. In Fig. 59b we have the known transitions for semiconduction and for the thermal quenching of luminescence. To correlate the two diagrams we have to consider the relation between optical and thermal activation energies, as for the other oxides above. The square of optical dielectric constant for $Cu_2O$ is $2.75^2 = 7.56$ and if we assume that the static dielectric constant is $\sim 12$ the ratio of the two justifies the identity between the corresponding levels in each diagram. The activation energy for luminescence quenching (0.67 eV) represents the thermal filling of the ground state $B$ of the centres by electrons from the filled band thus preventing the luminescence transitions. The exponential dependence of the efficiency on temperature is determined by the number of electrons in the defect levels $A$. These must reach these levels by thermal activation ($\mathcal{E} \sim 0.35$) and so the identity with the lower temperature semiconduction process is justified. The experiments of ENGELHARD[5] revealed an activation energy of $\sim 0.7$ eV. This is probably due to thermal transitions of electrons into the $B$ levels as in thermal quenching.

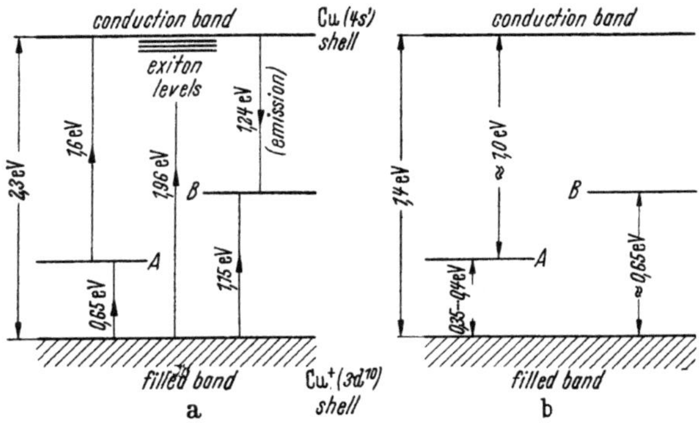


Fig. 59a and b. Energy band scheme for cuprous oxide. a) Optical transition energies. b) Thermal transition energies.

---

[1] W. P. JUSÉ and B. W. KURTSCHATOV: Phys. Z. Sowjet. **2**, 453 (1933).
[2] V. P. ZHUZE and S. M. RYVKIN: Dokl. Akad. Nauk SSSR. **68**, 673 (1949).
[3] S. M. RYVKIN: Dokl. Akad. Nauk SSSR. **68**, 487 (1949).
[4] E. F. GROSS and N. A. KARRYEV: Dokl. Akad. Nauk SSSR. **84**, 261 (1952). — E. F. GROSS and B. P. ZAKHACHENYA: Dokl. Akad. Nauk SSSR. **90**, 745 (1953).
[5] E. ENGELHARD: Ann. Phys., Lpz. **17**, 501 (1933).

The exponential increase of the time constant $\tau$ with decrease in temperature must be due to the thermal repletion rate for electrons in the $A$ levels which will have the activation energy (0.35 eV) as for semiconduction.

With respect to the photovoltaic effect the spectral response at $\sim 0.8\,\mu$ (1.6 eV) should disappear at low temperatures as observed since free holes will not be produced by activation of electrons into the $A$ levels. Reverting to photoconduction, there appears to be an experiment possible which would test the validity of the above model. Irradiation in the $2\,\mu$ region should raise electrons into $A$ levels by optical transitions. These would then be available for transitions due to $0.8\,\mu$ irradiation. Thus at low temperatures the photocurrent and the photovoltaic effect in the $0.8\,\mu$ region should be enhanced by simultaneous irradiation in the $2\,\mu$ region. A further experiment also suggests itself. At low temperatures where thermal transitions from the filled band to the $A$ levels are not possible lattice absorption should be followed by optical transitions of conduction electrons into the $A$ levels giving luminescence to shorter wavelengths than the $1\,\mu$ emission from transitions into $B$ levels. Possibilities are also indicated by this model of emission following minority carrier injection (electrons) as in germanium[1].

### e) Photoconduction in silver and thallous halides.

**47. General characteristics of the halides and their photoconductivity.** The silver halides are of considerable importance since they are the essential constituents of photographic emulsions. As seen in Sect. 1 above their properties and those of the thallous halides were studied by the Göttingen school using large single crystals. Reviews of this work were given by HILSCH and POHL[2] and by MOTT and GURNEY[3]. A more recent survey with a critical discussion based on modern concepts has been given by SEITZ[4] while the general physical properties of silver halides and their relation to the photographic process have been the subject of a recent conference[5].

The absorption spectra of the silver and thallous halides were measured by FESEFELDT[6] and some are shown in Fig. 60a, b and c. The fundamental lattice absorption processes an extensive "tail" which stretches to longer wavelengths and is usually associated with exciton production. Absorption in the lattice region or in the tail leads to photoconduction by electrons but in the earlier measurements no hole motion was detected[7,8]. In recent observations referred to by SEITZ[4] hole motion has been observed in silver chloride crystals and hole mobilities of up to 40 $cm^2$/volt sec. obtained. The photoelectrons in these halides have relatively long ranges and especially when the crystals are well annealed. It appears from the early measurements of the Göttingen group and from recent work of HAYNES and SHOCKLEY[9] that the trap concentration in annealed crystals

[1] For extensive semiconduction studies (excluding photoconduction) of $Cu_2O$ crystals made under controlled conditions see a series of papers by BLANKENBURG *et coll.* appearing in Ann. Phys., Lpz. **10**, 201, 211, 217, 232, 241 (1952); **12**, 281, 297 (1953); **14**, 135, 265, 290, 308 (1954).

[2] R. HILSCH and R. W. POHL: Z. Physik **64**, 606 (1930). — Trans. Faraday Soc. **54**, 883 (1938).

[3] N. F. MOTT and R. W. GURNEY: Electronic Processes in Ionic Crystals. Oxford 1940.

[4] F. SEITZ: Rev. Mod. Phys. **23**, 328 (1951).

[5] Mechanisms of Photographic Sensitivity, edit. J. W. MITCHELL. London: Butterworths 1951.

[6] H. FESEFELDT: Z. Physik **64**, 628 (1930).

[7] W. LEHFELDT: Göttinger Nachr., Fachgr. II **1**, 171 (1935). See also ref. 2, Chap. IV.

[8] K. HECHT: Z. Physik **77**, 234 (1932).

[9] J. R. HAYNES and W. SHOCKLEY: Phys. Rev. **82**, 935 (1951).

is less than $10^{16}$ per cm.[3] (cf. cadmium sulphide or alkali halides where minimum concentrations are $\approx 10^{16}$ per cm.[3]). Carrier life times are of the order of a microsecond.

The spectral response for photoconduction correlates well with the absorption spectrum and with the photographic sensitivity spectrum. At peak response the quantum efficiencies are 0.53, 0.61 and 0.45 for AgCl, AgBr and TlBr respectively falling to values of 0.22, 0.17 and 0.06 well within the lattice absorption region. SEITZ[1] has discussed the interpretation of the maximum yield values and suggests that the most likely hypothesis is that absorption at the optimum spectral region produces excitons which do not dissociate and free carriers in about equal numbers.

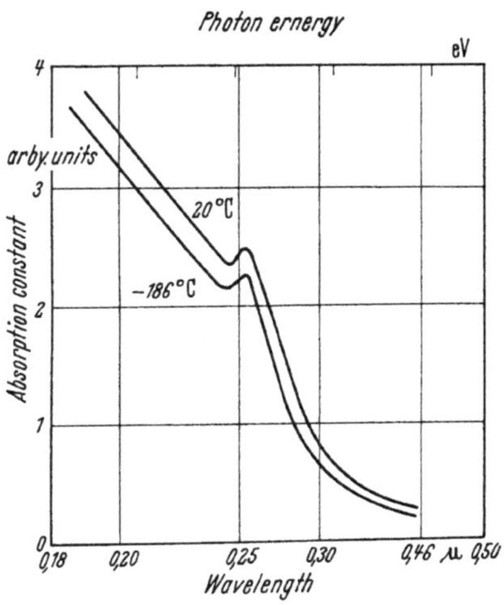


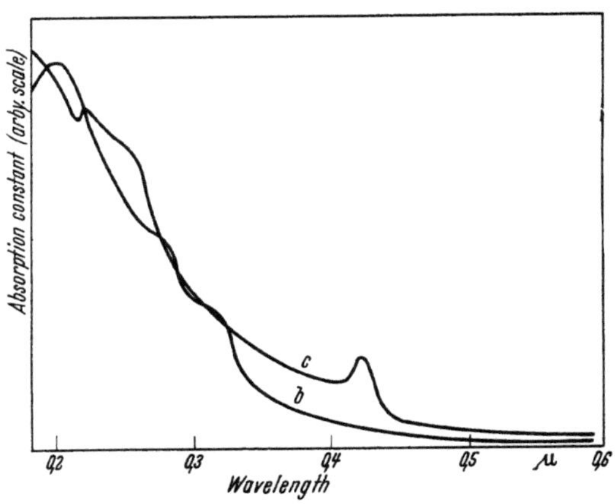


Fig. 60. Absorption spectra for silver halides. a AgCl, *b* AgBr, *c* AgI (after FESEFELDT ref. 6, p. 380).

**48. Crystal defects and trapping states and their effect on photoconduction.** In early experiments by LEHFELDT[2] it was found that at low temperatures irradiation increased the carrier electron range. This is explained[1] by the filling and neutralising of traps with carrier electrons so that subsequent carriers are not trapped or impeded. At ordinary temperatures irradiation produces a latent image and on prolongation the familiar print-out or darkening effect in the crystals. The colloidal silver produced in the crystal causes the new absorption at longer wavelenths and enhances the photoconduction response at these wavelengths. The range of photoelectrons produced by colloidal band absorption is also enhanced at low temperatures by ultra violet irradiation which, as stated above, appears to neutralise electron traps. Cold working of silver chloride reduces the electron range, presumably due to the creation of trapping defects. Prolonged irradiation at low temperatures leads to the production of so called "secondary" currents which continue after irradiation is cut off. These were investigated by LEHFELDT[3] and a theoretical explanation proposed by MOTT and GURNEY[4]. Electron carriers produced will leave behind a positive space charge and for illumination near electrodes this will produce a lowering of the conduction levels so that electrons can pass easily into the crystal and through

[1] See footnote 4, p. 380.
[2] See footnote 7, p. 380.
[3] W. LEHFELDT: Göttinger Nachr., Fachgr. II **1**, 171 (1935). See also ref. 3, Chap. IV.
[4] N. F. MOTT and R. W. GURNEY: Electronic Processes in Ionic Crystals. Oxford 1940.

it from the electrodes. At low temperatures where ionic mobility is negligible the effects persist but are disturbed by warming the crystal to room temperature.

SEITZ[1] in his review has given much attention to the problem of carrier traps and defect production in the silver halides and we give a brief outline of his ideas here. It is apparent that even in well annealed crystals holes are easily trapped while electrons are not. In such crystals dislocations exist and jogs in such dislocations (TAYLOR type) will possess sites which form loci for incipient vacancies of positive or negative ion types. The incipient vacancy should be able to trap a hole or electron at sufficiently low temperatures. The experimental evidence for the silver halides suggests that the electrons are not trapped in the average unbroken region of the dislocations and hence do not have a large chance of finding an incipient negative ion vacancy. The holes on the other hand are trapped by the dislocation and wander along it till they find an incipient vacancy. In plastic flow large numbers of vacant lattice sites are produced which can trap holes and electrons and so, as observed[2], the mean range or electrons is reduced. Vacancies of one type may be deliberately introduced by additive colouration as in the case of the alkali halides. In LEHFELDT'S[3] studies heating of silver chloride in excess halogen reduced the photoconduction response to a low value. Presumably excess of the halogen causes migration of silver ions out of the crystal leaving positive ion vacancies while positive holes move in and combine with the vacancies to form $V$ type centres in which electrons are readily trapped thus decreasing their mean free life time. No corresponding studies of photoconduction for silver addition have been made.

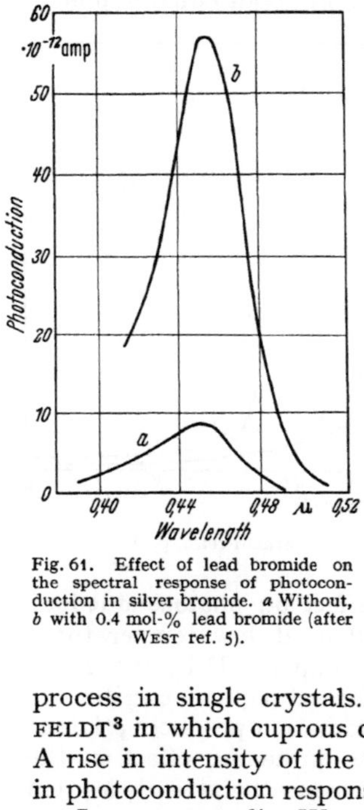


Fig. 61. Effect of lead bromide on the spectral response of photoconduction in silver bromide. *a* Without, *b* with 0.4 mol-% lead bromide (after WEST ref. 5).

The effect of foreign ions on photographic sensitivity has been studied but relatively few studies have been made of the effect on the photoconduction process in single crystals. SEITZ[1] mentions studies by HECHT[4] and by LEHFELDT[3] in which cuprous chloride additions are made to silver chloride crystals. A rise in intensity of the tail of the lattice absorption is paralleled by increase in photoconduction response.

In recent studies WEST[5] has made inclusions of divalent ions in fused silver halide films. Additions of small quantities of silver sulphide accelerate the darkening of the crystals but the photoconductivity is two orders of magnitude less than in the pure crystal. This decrease in photoconduction for an increase in photochemical activity is explained by carrier electron trapping at the sites introduced by the sulphide the trapping thus increasing silver build up but decreasing the carrier range. An opposite effect is produced by cadmium or lead ions which inhibit darkening[6]. WEST finds that these additions enhance the photosensitivity as shown in Fig. 61 although e.g. lead inclusions produce only

[1] F. SEITZ: Rev. Mod. Phys. **23**, 328 (1951).
[2] J. R. HAYNES and W. SHOCKLEY: Phys. Rev. **82**, 935 (1951).
[3] See footnote 3, p. 381.
[4] K. HECHT: Z. Physik **77**, 234 (1932).
[5] W. WEST: Mechanisms of Photographic Sensitivity, edit. J. W. MITCHELL, p. 99. London: Butterworths 1951.
[6] W. G. BURGERS and J. N. KOOY: Rec. Trav. chim. Pays-Bas **67**, 16 (1948).

small changes in the absorption spectrum. These inclusions also remove the "negative photoconductivity" (reduction in steady dark current by irradiation) in well annealed crystals. The explanation of the enhancing of photoconduction by divalent metal ions is based on MITCHELL'S model for silver halides[1]. Vacant halide ion sites will trap electrons but introduction of lead ions, for example, will increase the number of silver ion vacancies. To preserve equilibrium the number of halide ion vacancies must be reduced and hence their electron trapping effects removed with enhancing of carrier life time. The negative ion vacancies also promote the negative photoconduction effect (for details see WEST[2]) and so the latter is also reduced by the lead inclusion.

**49. Photoconductivity in photographic emulsions.** Photographic emulsion have large electrolytic dark currents at ordinary temperatures and the photocurrents are very much smaller than these. However, interest in emulsions has been stimulated by the fact that factors affecting photographic sensitivity also influence photoconduction. In 1929 SHEPPARD *et coll.*[3] observed that the spectral response of the photovoltaic effect in Ag: AgBr electrodes in solution was extended to longer wavelengths by dyes. In more recent studies WEST and CARROLL[4] have shown that the dye sensitisation of the photographic emulsion also alters the spectral response of its photoconduction in a similar way. Further studies have since been reported by WEST[2], who has also reported the effects of divalent ions on the photoconduction, and independently by AMSLER[5].

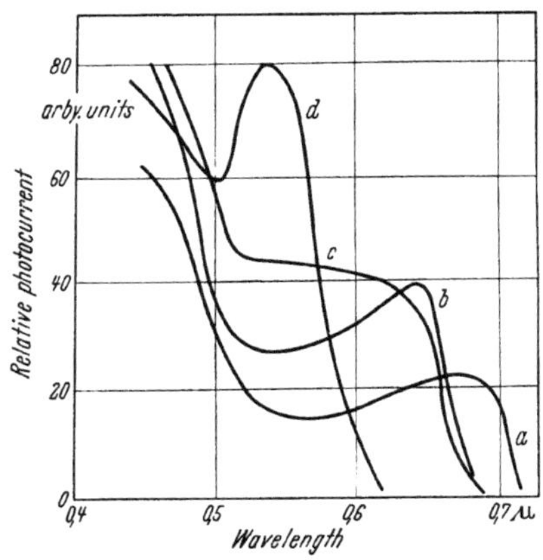

Fig. 62. Spectral response of photoconduction for various dye sensitised photographic emulsions. *a* α-dibenzothiocarbocyanin bromide. *b* α-thiocarbocyanin iodide. *c* α-selenacarbocyanin bromide. *d* α-merocyanin (after WEST ref. 2).

In unsensitised emulsions the photocurrent rises and falls slowly with onset and cessation of irradiation. It is proportional to the applied field and for intensities in the latent image exposure region is proportional to light intensity. At higher intensities a square root relation holds as in the simple recombination theory. These facts indicate that trapping effects are marked. Since the latter can form nuclei for latent image build up some connection between photoconduction and photographic sensitivity would be expected. One of the important contributions of WEST and CARROLL[4] and of WEST[2] has been on dye sensitised emulsions. Adsorbed dyes which extend the photographic sensitivity to longer wavelengths also increase the long wavelength response of photoconduction. Fig. 62 shows photoconduction response curves obtained by WEST[2] for various dye sensitised emulsions. These are identical with the spectral distribution of photographic sensitivity and correlate with the dye absorption bands. A further correlation occurs in the case of super-sensitisation by a second dye, not itself an effective sensitiser and not absorbing in the region for the first and sensitising

[1] J. W. MITCHELL: Phil. Mag. **40**, 244 (1949).
[2] W. WEST: Mechanisms of Photographic Sensitivity, edit. J. W. MITCHELL, p. 99. London: Butterworths 1951.
[3] S. E. SHEPPARD *et coll.*: J. phys. Chem. **33**, 331, 1403 (1949).
[4] W. WEST and B. H. CARROLL: J. Chem. Phys. **15**, 529, 539 (1947).
[5] H. AMSLER: Diss. Zürich (1953).

dye. Fig. 63 shows West's results for a typical dye sensitised emulsion with and without the supersensitiser. The results suggest that the absorption in the sensitiser results in a liberation of an electron with subsequent transfer to the halide grain conduction band. The action of the supersensitiser would appear to be that of improving the efficiency of electron transfer from dye to halide grain.

Desensitising dyes for the photographic process do not usually affect the fundamental photoconduction response of the halide and in certain cases may cause a small increase in the spectral response of the latter at longer wavelengths. These dyes probably provide electron trapping states alternative to those of the halide sensitivity centres (for latent image formation). Absorption in the desensitiser bands produces only small enhancement of photoconduction presumably because the transfer efficiency to the halide is low. A further effect of certain dyes may be mentioned. The cyanin dyes not only produce the observed long wavelength response but also increase the photosensitivity for absorption in the lattice band tail. It is thought probable that the dyes act as halogen acceptors and diminish the trapping of electrons at halogen atoms appearing in the grain surfaces due to the photolytic process. Well known halogen acceptors in emulsions produce a similar effect if they contain ring structures but simple acceptors such as potassium nitrite have little effect.

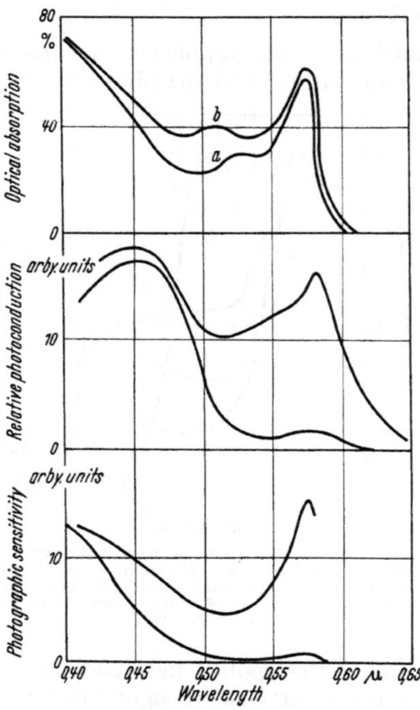


Fig. 63. Comparison of absorption, photoconduction and photographic sensitivity spectra for dye sensitised and supersensitised emulsions. *a* 1.1. diethyl 2.2. cyanine iodide dye sensitised. *b* Dye plus super-sensitiser (after West ref. 2, p. 383).

The above account shows that photoconductivity measurements in silver halide crystals and in photographic emulsions provide important evidence for the establishment of the theory of the latent image formation and such processes as sensitisation. Future studies might include comprehensive measurements on the same specimen or film of the photographic photoconduction, luminescence and absorption characteristics. Application of light and electrical "glow curve" and photodielectric techniques to the systems should prove very fruitful especially for single crystals of controlled constitution.

## f) Photoconductivity in intermetallic compounds.

**50. The III to V compounds.** Among compounds classed as intermetallic some of those formed by combinations of elements of groups III and V of ther periodic table have attracted considerable interest in recent years because of their close relation in characteristics to the group IV elements discussed in section III above. A useful survey of the general position and properties of these compounds relative to other intermetallic systems has recently been provided by Welker[1].

[1] H. Welker: Physica, Haag **20**, 893 (1954).

As yet little information is available, except in a few cases, on photoconduction in these compounds. We give below an account of published studies on the latter. As in the known photoconductors, a close correlation is found between optical absorption and photoconduction thresholds. We give in Fig. 64 the absorption spectra of the semiconducting III-V compounds measured by OSWALD and SCHADE and included in WELKER's survey[1].

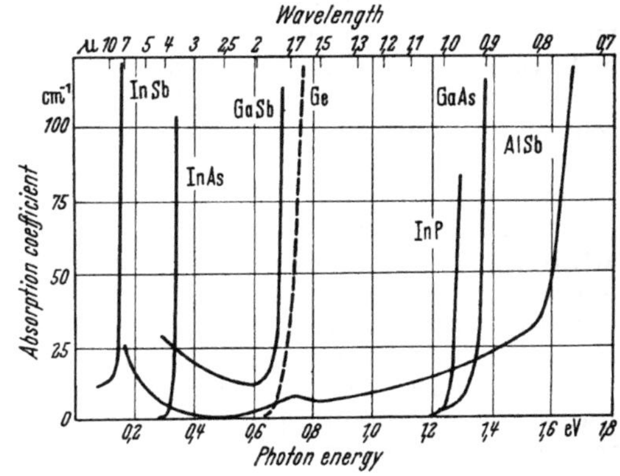


Fig. 64. Absorption spectra for various intermetallic compounds with that of germanium as a comparison (after OSWALD and SCHADE ref. 1, p. 384).

**51. Indium antimonide.** The photoconduction and photovoltaic effects for indium antimonide have been recently measured by AVERY, GOODWIN, LAWSON and MOSS[2]. Independent measurements for photoconduction have been reported by TAUC and ABRAHAM[3] and for the photovoltaic effect at a *P-N* junction by MITCHELL, GOLDBERG and KURNICK[4]. The spectral response of the photoconduction obtained by AVERY *et coll*[2] is shown in Fig. 65. The results of TAUC and ABRAHAM give a similar threshold value at room temperature. Those of MITCHELL *et coll.* give a threshold for the photovoltaic effect of 5.7 μ at 77° K. They studied the photovoltage at a *P-N* junction in the crystal. The photoconduction threshold corresponds to the absorption limit and moves in the same way to shorter wavelengths at lower temperatures. Independent measurements of absorption spectra have been made by TANNENBAUM and BRIGGS[5]. A high level of purity of the antimonide is essential to obtain the long wave length absorption and photoresponse.

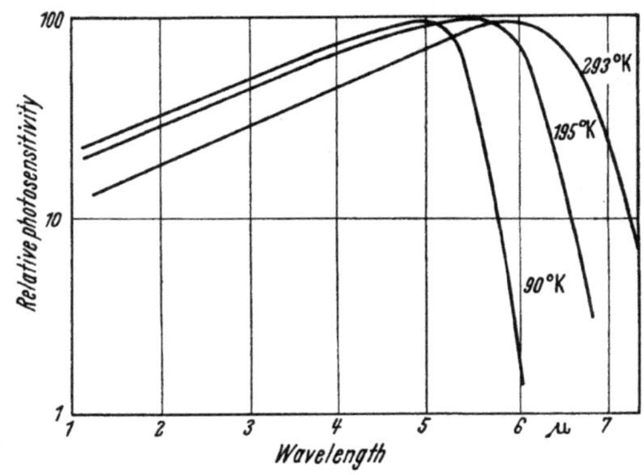


Fig. 65. Response spectra of photoconduction in indium antimonide (after AVERY *et coll.*[2]).

MOSS[6] has given an extensive theoretical discussion of the optical and electrical characteristics of indium antimonide. The shorter wavelength threshold for the photovoltaic effect (7.15 μ at 291° K) compared with that for photoconduction (7.7 μ; 0.18 eV) is explained as follows. Photovoltaic effects will decrease as soon as the optical absorption coefficient is low enough to produce carrier pairs at some distance from the photovoltaic barrier. The critical

[1] See footnote 1, p. 384.

[2] D. G. AVERY, D. W. GOODWIN, W. D. LAWSON and T. S. MOSS: Proc. Phys. Soc. Lond. B **67**, 761 (1954).

[3] J. TAUC and A. ABRAHAM: Czech. J. Phys. **4**, 478 (1954).

[4] G. R. MITCHELL, A. E. GOLDBERG and S. W. KURNICK: Phys. Rev. **97**, 239 (1955).

[5] M. TANENBAUM and H. B. BRIGGS: Phys. Rev. **91**, 1561 (1953).

[6] T. S. MOSS: Proc. Phys. Lond. B **67**, 775 (1954).

absorption coefficient will be that for which the effective penetration depth is greater than the diffusion length for the carrier. This length is given by $(D\tau)^{\frac{1}{2}}$ as in Sect. 16 above. For the $\lambda_{\frac{1}{2}}$ value at 7.15 μ the reciprocal of the wavelength will be of the order of this length. The absorption coefficient at this value is 200 cm.$^{-1}$ at room temperature which gives $(D\tau)^{\frac{1}{2}} \sim 50$ μ. From other measurements the electron carrier mobility is found to be 42000 cm.$^2$/voltsec., a very high value, and since $D = \mu k T/e$ (Sect. 76) the carrier life time at room temperature is ~0.025 sec. The value at 90° K measured from the rise of photoconduction is ~0.8 μ sec.[1]. From the ratio of carrier concentrations at the two temperatures (20:1) the value of 0.025 sec. is considered by MOSS to be reasonable.

Indium antimonide has, as seen, a very high electron mobility and this is attributed to the low effective electron mass which is estimated to be 0.03 that of the normal value[2]. There is also a low density of energy states in the conduction band so the absorption edge is affected markedly by a relatively low density of conduction electrons which will also determine the temperature shift of the absorption edge. For impure specimens giving high conduction electron concentration the absorption edge moves to shorter wavelengths ~3 μ (carrier concentration ~$10^{18}$ cm.$^3$)[1,3].

**52. Aluminium and gallium antimonides.** So far no measurements of photoconduction have been reported for these materials but optical characteristics have been reported[4,5,6] which might serve as a guide for future investigations. From electrical measurements (HALL coefficient and resistivity variations with temperature) the intrinsic energy gap for aluminium antimonide is 1.6 eV[5] while for gallium antimonide it is 0.77 eV[6] (see also Fig. 64).

RITTNER[7] has pointed out that aluminium antimonide should be a very promising material for use as a photovoltaic solar energy converter since its energy gap is near the ideal value. Its ultimate performance would depend on obtaining as large a value as possible for the minority carrier life times.

**53. Gallium arsenide.** Photoconduction in gallium arsenide has been reported by GANS, LAGRENAUDIE and SEGUIN[8] and recently by BARRIE *et coll.*[9]. Their response curve is included in Fig. 66. They obtain a value for $\lambda_{\frac{1}{2}}$ of 1.25 eV. GANS *et coll.*[8] give a threshold of ~1.13 eV with a thermal energy gap of 1.1 eV showing reasonable agreement. BARRIE *et coll.* find much higher photosensitivity in polycrystalline specimens than in single crystals and by light spot measurements determine that the higher sensitivity is associated with grain boundaries. The time constant measurements for the polycrystalline specimens used are large ~0.01 sec. The effective electron mass obtained by thermoelectric data is about 0.03 of the normal value and so a high mobility might be expected (as in indium antimonide) for pure samples, though not in the specimens used in the above measurements.

---

[1] D. G. AVERY, D. W. GOODWIN, W. D. LAWSON and T. S. MOSS: Proc. Phys. Soc. Lond. B **67**, 761 (1954).

[2] T. S. MOSS: Proc. Phys. Soc. Lond. B **67**, 775 (1954).

[3] M. TANENBAUM and H. B. BRIGGS: Phys. Rev. **91**, 1561 (1953).

[4] H. WELKER: Physica, Haag **20**, 893 (1954).

[5] R. F. BLUNT, H. P. R. FREDERIKSE, J. H. BECKER and W. R. HOSLER: Phys. Rev. **96**, 578 (1954).

[6] R. F. BLUNT, W. R. HOSLER and H. P. R. FREDERIKSE: Phys. Rev. **96**, 576 (1954).

[7] E. S. RITTNER: Phys. Rev. **96**, 1708 (1954).

[8] F. GANS, J. LAGRENAUDIE and P. SEGUIN: C. R. Acad. Sci. Paris **237**, 310 (1953).

[9] R. BARRIE, F. A. CUNNELL, J. T. EDMOND and I. M. ROSS: Physica, Haag **20**, 1087 (1954).

Photovoltaic effects in gallium arsenide have been measured by GREMMELMAIER[1] with some interest in its behaviour as a solar energy converter. The photovoltage shows the usual logarithmic relation to light intensity over a wide range but gives a conversion efficiency for sunlight of only 1%. However, with an energy gap of ~1.1 eV it is of interest with respect to the theoretical calculations for an optimum material (see Sect. 20 above).

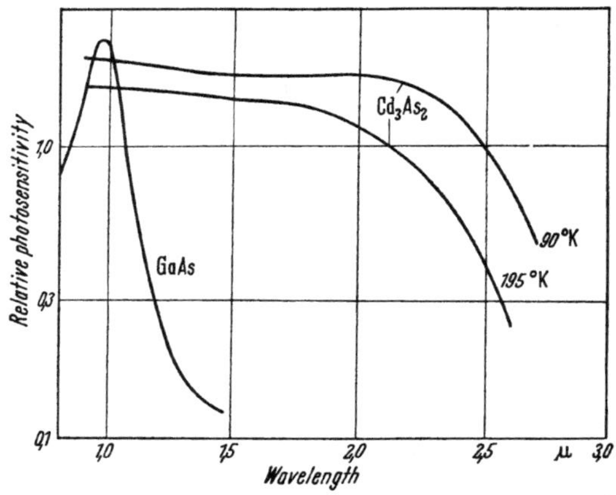


Fig. 66. Response spectra of photoconduction in gallium and cadmium arsenides (after BARRIE *et coll*, for GaAs ref. 9, p. 386; after Moss for $Cd_3As_2$ ref. 1, p. 388).

**54. Indium selenide, telluride and oxide.** The photoconduction of indium selenide was first reported by KOLOMIYETS and RYVKIN[2] in 1947 and since then has been studied by BODE and LEVINSTEIN[3] using evaporated layers and by DAMON and REDINGTON[4] using single crystals. The optical absorption edge is at about 1.2 eV, the photoconduction decreasing rapidly at this edge towards longer wavelengths. Layer type specimens[3] show a fall in response at the same region but a longer wavelength peak is also observed, either due to defects in the layers. The mode of preparation of the layer suggests this. For layers the response time in the intrinsic region is larger than at longer wavelengths.

Photoconduction in a single crystal of indium telluride has been measured by MIYASAWA and SUGAIKE[5]. The response is reasonably constant for wavelengths greater than 1 μ but at 1 μ falls rapidly with wavelength giving a threshold of about 1.2 eV.

Thin films of indium oxide have been prepared by RUPRECHT[6]. The photoconduction is *N*-type and is sensitive to adsorbed oxygen layers showing a permanent rise when the latter are removed.

**55. Miscellaneous compounds.** *α) Zinc phosphide.* The photoconduction of evaporated layers of zinc phosphide ($Zn_3P_2$) has been investigated by LAGRENAUDIE[7] who has since published work on small single crystals of the material[8] ~1 mm. size. The threshold response is at 1.15 eV and corresponds to the intrinsic semiconduction activation energy. Time constants are of the order of 50 μsec at room temperature; the optimum quantum efficiency is only of the order of 1%.

*β) Cadmium arsenide.* The photoconduction of cadmium arsenide has been investigated by MOSS[9] and spectral response curves obtained as shown in Fig. 66.

[1] R. GREMMELMAIER: Z. Naturforsch. **10**a, 501 (1955).
[2] B. T. KOLOMIYETS and S. M. RYVKIN:. Ž. Tekh. Fyz. **9**, 987 (1947).
[3] D. E. BODE and H. LEVINSTEIN: J. Opt. Soc. Amer. **43**, 1209 (1953).
[4] R. W. DAMON and R. W. REDINGTON: Phys. Rev. **96**, 1498 (1954).
[5] H. MIYASAWA and S. SUGAIKE: J. Phys. Soc. Japan **9**, 648 (1954).
[6] G. RUPRECHT: Z. Physik **139**, 5041 (1954).
[7] J. LAGRENAUDIE: J. Chim. Phys. **50**, 545 (1953).
[8] J. LAGRENAUDIE: J. Phys. Radium **16**, 234 (1955).
[9] T. S. MOSS: Proc. Phys. Soc. Lond. B **63**, 172 (1950).

*γ) Arsenic trisulphide.* Using layers of arsenic trisulphide under electron excitation ANSBACHER and EHRENBERG[1] found a current multiplication effect of up to $4 \times 10^4$ of the primary beam current. Multiplication of photoconduction currents in single crystals of $As_2S_3$ in high fields ($\geq 10^4$ V/cm.) has been observed by REKALOVA[2] who has also measured the spectral response at 90° K.

*δ) Antimony trisulphide.* Layers of antimony trisulphide prepared as targets for television pick-up tubes have been investigated by FORGUE, GOODRICH and COPE[3]. The layers show a photoconductivity which peaks at 0.71 μ and has a $\lambda_{\frac{1}{2}}$ value of 0.75 μ. The photocurrent increases as the 0.7 power of light intensity over a wide range, this relation being explained by ROSE's[4] theory of the effects of trapping state distribution (see Sect. 7 above). The photosensitivity shows an approximately linear increase with temperature over the range up to 180° C, irreversible changes taking place above this temperature.

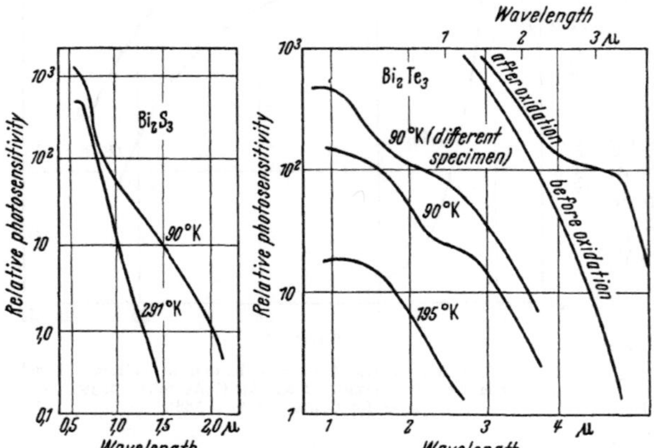


Fig. 67. Response spectra of photoconduction in bismuth sulphide and telluride layers (after GIBSON and MOSS[5]).

*ε) Bismuth sulphide, selenide and telluride.* Layers of $Bi_2S_3$, $Bi_2Se$ and $Bi_2Te_3$ investigated by GIBSON and MOSS[5] show a spectral response of photoconduction which extends to longer wavelengths as the temperature falls. The response spectra are shown in Fig. 67 for $Bi_2S_3$ and $Bi_2Te_3$ layers. For bismuth telluride layers oxidation produces a new response peak at longer wavelengths than that of the intıinsic threshold.

*ζ) Magnesium antimonide.* Evaporated layers of magnesium antimonide ($Mg_3Sb_2$) have been made by MOSS[6] which exhibit photoconduction, reported earlier by ZHUZE, MOCHAN and RYVKIN[7]. The long wavelength threshold of the intrinsic effect is at 1.5 μ with impurity levels giving response threshold at 2.6 and 3.5 μ. These threshold correlate with the activation energies obtained from semiconduction characteristics[8].

## g) Miscellaneous inorganic photoconductors.

**56. Thallous sulphide.** Photoconductive cells made with thallous sulphide layers have been known for some time but during the late war considerable

[1] F. ANSBACHER and W. EHRENBERG: Proc. Phys. Soc. Lond. A **64**, 362 (1951).
[2] G. I. REKALOVA: Ž. tekh. Fyz. **21**, 1041 (1951).
[3] S. V. FORGUE, R. R. GOODRICH and A. D. COPE: RCA-Review **12**, 335 (1951).
[4] A. ROSE: RCA-Review **12**, 362 (1951).
[5] A. F. GIBSON and T. S. MOSS: Proc. Phys. Soc. Lond. A **63**, 176 (1950).
[6] T. S. MOSS: Proc. Phys. Soc. Lond. B **63**, 982 (1950).
[7] Z. P. ZHUZE, I. V. MOCHAN and S. M. RYVKIN: Ž. tekh. Fyz. **18**, 1494 (1948).
[8] Further evidence on photoconductivity in intermetallic compounds has been given since this article was written by H. P. R. FREDERIKSE and R. F. BLUNT, Proc. Amer. I. R. E. **43**, 1828 (1955).

effort was devoted to their development. Extensive experimental and theoretical studies were made by HIPPEL *et coll.*[1] in the U.S.A. which showed that, as for lead sulphide above, the maximum photosensitivity was obtained by oxygen treatment of the evaporated sulphide layer making it slightly *P*-type. The photocurrent is observed as an increment to the standing dark current (layer resistances $\sim 5$ MΩ) and both currents show linear behaviour with applied voltage. The dependence of photocurrent on light intensity shows the form expected from recombination theory (Sect. 6) i.e. linearity at low intensities and a square root relation at higher levels. Time constants are of the order of milliseconds. A typical spectral response curve is shown in Fig. 68.

A simple bimolecular recombination theory has been shown by EWALD[2] be inadequate to explain the layer characteristics and by analogy with lead sulphide a barrier mechanism would appear to be involved. EWALD shows by analysis of previous results that the initial process is a liberation of carriers in the volume of the $Tl_2S$ but that the temperature dependence of photocurrent particularly that of its rise and decay is determined by the variation of the heights of the barriers with temperature. Other studies by OSTROWSKI and SOSNOWSKI[3] suggest that barrier modulation effects are of second order and that recombination processes largely determine the characteristics.

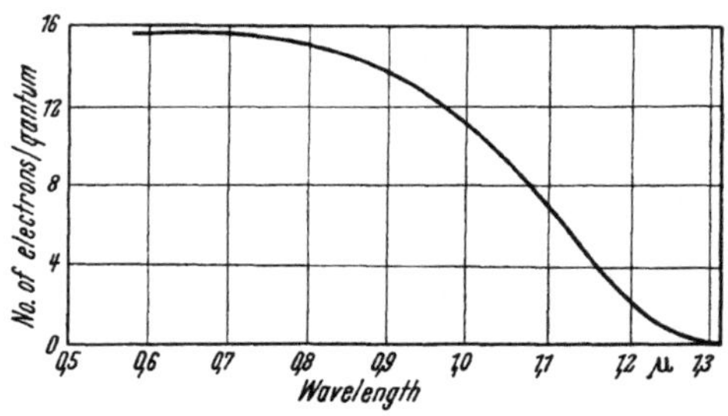


Fig. 68. Response spectrum of photoconduction in thallous sulphide layers (after HIPPEL *et coll.*[1]).

**57. The molybdenite group.** The photoconductivity of molybdenite ($MoS_2$) was listed by CASE[4]. In a recent publication LAGRENAUDIE[5] has given a tabulated comparison of molybdenum, tungsten and rhenium sulphides. As photovoltaic cells all three sulphides exhibit the same threshold wavelength of 1.1 eV and at shorter wavelengths there is a sharp fall in efficiency, probably due to increased recombination in surface states. MOSS[6] gives the threshold for photoconductivity in molybdenum sulphide as 2.0 μ.

## V. Photoconductivity in organic systems.

**58. Organic crystals.** It has already been shown above (Sect. 49) that dye sensitisation of photographic emulsions produces an extension of the photoconduction response to longer wavelengths the response showing close agreement with the absorption bands of the dye. It was also mentioned (Sect. 43) that the photoconduction in zinc oxide could be sensitised by adsorbed dyes. The same effect has been observed for the alkali halides[7] absorption being in the dye bands.

Photoconduction in anthracene was investigated by VOLMER[8] as early as 1913 and found to correlate in its spectral response with molecular absorption. This has since been confirmed by VARTANIAN[9] and more recently by other wor-

[1] A. HIPPEL et coll.: J. Chem. Phys. **14**, 355, 370 (1946).
[2] A. W. EWALD: Phys. Rev. **81**, 607 (1951).
[3] J. W. OSTROWSKI and L. SOSNOWSKI: Bull. Acad. polon. Sci., Cl. 3 **2**, 385 (1954).
[4] T. W. CASE: Phys. Rev. **9**, 305 (1917).
[5] J. LAGRENAUDIE: Phys. Radium **15**, 299 (1954).
[6] T. S. MOSS: Proc. Phys. Soc. Lond. A **63**, 167 (1950).
[7] H. YAGI: Mem. Coll. Sci., Kyoto A **26**, 75 (1950).
[8] M. VOLMER: Ann. Phys., Lpz. **40**, 775 (1913).
[9] A. T. VARTANIAN: Dokl. Akad. Nauk SSSR. **17**, 641 (1950).

kers[1–4]. Fig. 69 shows the photoconduction response and absorption spectra for solid and solution as given by Carswell[1].

Chynoweth and Schneider[2] have made extensive investigations of the photocurrents in single crystals. The initial photocurrent on illumination is subsequently reduced byspace charge effects associated with the trapping of charge carriers. The peak and space charge limited currents are both proportional to the light intensity. The space charge can be removed by light within the molecular absorption bands if the applied field is first removed. Variation of the photocurrent with temperature follows the form predicted by Rose for an exponential distribution of trap depths (see Sect. 7). The current in the crystals is apparently carried by positive holes which agrees with Putseiko's[5] observations on a number of substances.

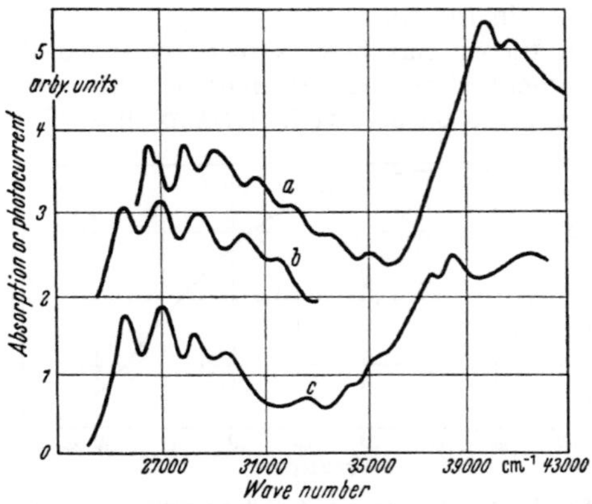


Fig. 69. Optical absorption and photoconduction response spectra for anthracene. *a* Absorption in solution. *b* Crystal absorption. *c* Crystal photoconduction (after Carswell[1]).

In the presence of oxygen the anthracene crystal surfaces can be photooxidised by the exciting radiation which causes a large increase in the crystal photoconductivity. This is probably due to the effect of oxygen in prolonging carrier life time. This is likely as in crystals where the electrodes are on opposite faces of a parallel sided specimen the effect is less than where the two electrodes are on the same crystal surface. The fundamental process of release of holes during excitation is still not decided but the temperature dependence of photocurrent suggests an initial exciton formation in a molecule which then loses a hole by thermal activation.

**59. Photoconductivity in dye films.** Photoconductivity in certain classes of dyes has been known for a long time[6] and as seen above it can provide an additional phenomenon for study of dye sensitisation of emulsions. Recently studies of dye films deposited on platinum electrodes on glass have been made by Nelson[7,8]. In his first investigations he shows that photoconducting dyes are basic and have coloured cations. Chlorides, sulphates and oxalates are more photosensitive than e.g. bisulphates the criterion being that the cation is not more than singly charged. Even negative photoconduction effects can be observed for doubly charged cations e.g. methyl green.

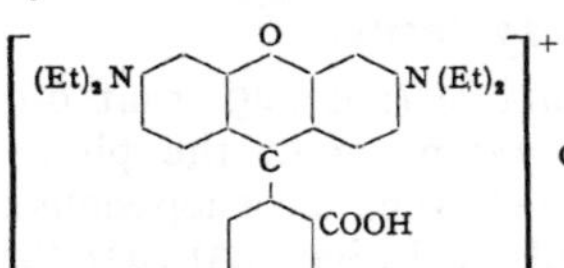


In later work Nelson uses deposited films of Rhodamine B, crystal violet and basic fuschin. Rhodamine B has the structure as shown and forms glossy

[1] D. J. Carswell: J. Chem. Phys. **21**, 1891 (1953).
[2] A. G. Chynoweth and W. G. Schneider: J. Chem. Phys. **22**, 1021 (1954).
[3] A. G. Chynoweth: J. Chem. Phys. **22**, 1029 (1954).
[4] A. Bree and L. E. Lyons: J. Chem. Phys. **22**, 1630 (1954).
[5] E. K. Putseiko: Dokl. Akad. Nauk. SSSR. **67**, 1009 (1949).
[6] See e.g. Z. Petrikalu: Z. phys. Chem. B **10**, 9 (1930).
[7] R. C. Nelson: J. Chem. Phys. **19**, 798 (1951).
[8] R. C. Nelson: J. Chem. Phys. **20**, 1327 (1952); **22**, 885 (1954).

films with a golden lustre there being a strong optical absorption at 5500 Å. The induced photoconduction decays relatively slowly[1] and the decay is of bimolecular form after a rapid initial fall. The initial rate of rise on excitation is proportional to the light intensity and the steady state photocurrent to the square root of the intensity as expected for a straightforward bimolecular mechanism. Temperature dependence of the photoconduction indicates that the slow decay is governed by a trapping process with a single activation energy of 0.57 eV. The bimolecular form and the single activation energy suggest that there is retrapping in the decay process. In view of the layer form of the dye it could well be that the "trapping effects" are in fact potential barrier effects as in inorganic layers like thallous sulphide.

Investigations of crystal violet films (4,4′,4″-tri(dimethylamino)triphenylmethylchloride) show that the photoconduction is a volume effect since illumination of a small area away from the electrodes produces a current with no polarisation effects. Films of crystal violet which have been partally oxidised show a distribution of "trap depths" from 0.36 to 0.5 eV but in unoxidised layers the depth is almost single valued at ~0.37 eV. Oxygen is assumed to provide trapping states but these may well be at surfaces and layer barriers. NELSON observed what he calls a memory effect in crystal violet. The rate of decay depends on the initial level of current before cut off but for several previous illumination it depends on the highest previous intensity of excitation. It is thought that the current carriers are confined in localised regions and that the number of these regions increases with light intensity. However, it seems more likely that the effect involves light perturbation of potential barriers around these regions with some kind of carrier trapping since the "memory" effects are removable by heating the film at 100° C for several hours.

Similar memory effects are found for basic fusch in films (3-methyl-4,4′, 4″-triaminotriphenyl-methylchloride). The rate of rise of photocurrent on illumination is faster if previous illumination occurs. This suggests a trap filling process. It is interesting to note that an exponential increase of dark current occurs in the film over the range 40 to 80° C but the photocurrent decay rate is unchanged. The quantum efficiency is markedly temperature dependent and suggests that photon absorption produces an excited molecular state followed by thermal ionisation of the carriers.

## VI. A note on the photodielectric effect in solids.

**60.** Changes in dielectric constant on illumination have been observed in many solids and over many years. A review of studies on phosphors up to 1949 has been given by the writer[2]. It was suggested by GARLICK and GIBSON[3] on the basis of quantitative evidence that increases in dielectric constant and loss in zinc and cadmium sulphides on illumination could be explained by the dielectric polarisation due to trapped electrons. It now seems certain that this is not the correct explanation although traps play a large part in the effect. The explanation can be given in terms of a dielectric composed of photoconducting microcrystals with relatively insulating intercrystalline boundaries at which carriers are trapped. Experimental and theoretical analyses of such a system have been made by MAC DONALD[4], by KALLMANN, KRAMER and PERLMUTTER[5]

[1] See footnote 8, p. 390.
[2] G. F. J. GARLICK: Luminescent Materials. Oxford 1949.
[3] G. F. J. GARLICK and A. F. GIBSON: Proc. Roy. Soc. Lond., Ser A **188**, 485 (1947)
[4] J. R. MACDONALD: Phys. Rev. **85**, 381 (1952). — J. Chem. Phys. **23**, 275 (1955).
[5] H. KALLMANN, B. KRAMER and A. PERLMUTTER: Phys. Rev. **89**, 700 (1953).

and by Dropkin *et coll.*[1]. As shown recently[2] large capacity changes occur for single crystals ($\sim 10^6 \times$ dark value) on illumination if relatively insulating barriers occur at the electrodes. This is because the crystal is effectively a conductor on illumination leaving only the boundaries still insulating.

The photodielectric effect is thus useful as a means of measuring photoconduction in microcrystalline powders without the space charge difficulties of d.c. measurements. Extensive application to phosphors in powder form has recently been made by the above workers.

## VII. Some experimental techniques in photoconduction measurements.

**61. Types of measurement.** In the preceeding sections it has been seen that the special characteristics of interest in photoconduction are spectral response, forms of rise and decay, quantum efficiency, distribution of trapping states and surface states carrier sign mobility, and diffusion and recombination coefficients and the variation of each of these quantities with temperature and crystal constitution and purity. In addition other physical characteristics must be determined inclusive of optical absorption spectra and semiconduction properties.

Many of the experimental techniques are now well known but one or two of recent origin merit more detailed description as follows:

**62. Sign and mobility of liberated carriers by Hall coefficient measurement.** Measurements of the Hall effect in insulating photoconductors are difficult by using the conventional techniques since small space charge effects, particularly near the current electrodes, can upset the observations. In recent years Redfield[3] has developed a method for Hall coefficient measurement in which no electrical contact need be made to the crystal at all. The schematic form of the method is shown in Fig. 70. The crystal is held between glass plates of larger dimensions than itself with semiconducting surface coatings each coating having a potential gradient along its length provided by batteries. One of the plates is connected via a battery centre tap to one terminal of a vibrating reed electrometer, the other electrometer terminal being earthed. The centre tap of the lower plate system goes to a variable potentiometer tapping on another battery supply of which the centre tap is earthed. Potentials of the order of 100 V are available in each battery system. The field in the crystal due to the potential ($V_L$) across each plate is in the $x$ direction. The potentiometer tapping gives a potential ($V_T$) which gives a small crystal field in the $y$ direction. With the magnetic field on in the $z$ direction the crystal is excited and the potentiometer adjusted to give a null deflection of the electrometer. The magnetic induction is then reversed and $V_T$ again adjusted for zero deflection. This means that the resultant electric field in the crystal is rotated through twice the Hall angle between the two measurements (see Fig. 70b). This angle can be obtained approximately from the voltage change and the electrode geometry ($d$ = crystal thickness, $l$ = length of resistance film on each electrode) as follows:

$$\frac{2\mu_H B}{c} \approx 2\vartheta \approx \frac{\Delta E_y}{E_x} \approx \frac{\Delta V_T/d}{V_L/l} \tag{62.1}$$

where $\mu_H$ is the Hall mobility, $B$ the magnetic induction, $c$ the velocity of light). This relation holds for small $\vartheta$. In general it must be corrected for edge effects etc. Such corrections are discussed by Redfield[3].

[1] J. J. Dropkin: ONR-U.S.A. report N 6 onr 26312, 1952; 26313, 1953.
[2] H. Kallmann, B. Kramer and P. Mark: Phys. Rev. **99**, 1328 (1955).
[3] A. Redfield: Phys. Rev. **91**, 244, 753 (1953); **94**, 526, 537 (1954).

In a later modification the top glass plate is connected direct to the electrometer without a potential $V_L$ as charge leakage from the batteries proved a nuisance. This distorts the crystal field but analysis of results is still possible and the actual data are more reproducible and free from leakage noise. A mean value $\overline{E}_x$ must now be taken for the longitudinal crystal field. $E_x$ increases towards the lower plate and so if the sample is arranged to have a non photoconducting spacer above it the mean value of can be improved and can even be $\approx V_L/l$. Now a larger applied potential $V_T$ is required to give the transverse fields which improves the experimental accuracy. The spacer separation reduces the electrometer sensitivity to charge motion between the electrodes but this is offset by a reduction in noise due to fluctuations in the resistance film on the lower electrode. Even with crystals of poor geometry it is possible to use this method for a rough measurement of carrier sign or of temperature dependence of the HALL effect.

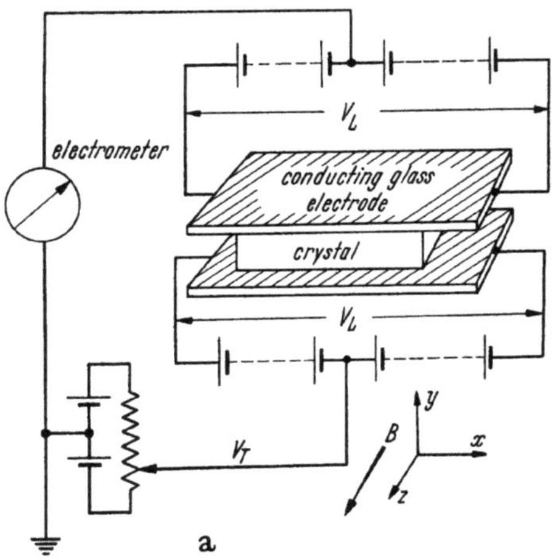


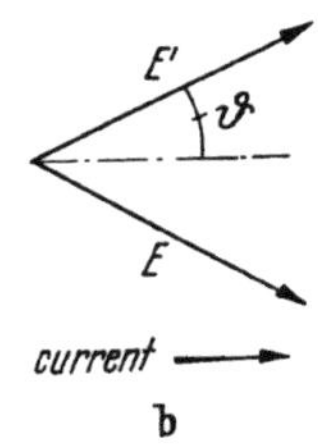


Fig. 70. Schematic diagram of REDFIELD's method for measurement of HALL effect in photoconductors (after REDFIELD ref. 3, p. 392).

**63. Diffusion, mobility and life-time of carriers by the flying spot technique.** The carrier mobility and life-time can be determined for a photoconductor by observing the photovoltaic effect at a probe on its surface (or at a *p-n* junction) as a light spot is swept across the probe at a known velocity. We have already noted the measurements of GOUCHER[1] using a stationary light spot at various distances from a *P-N* junction in germanium. In a recent paper ADAM[2] has described the advantages of the flying spot method which is now widely employed. Fig. 71 taken from his results shows a suitable experimental arrangement. The image of the illuminated slit *A* is, by means of the lens *B* and rotating mirror *M*, scanned over a probe on the photoconductor surface at *B*. The light spot also scans over a grid *E* which provides a photoelectric cell with light pulses and thus gives a time scale calibration on the recording oscilloscope (C.R.O.). Triggering of the oscilloscope time base is achieved by the photocell *F* which receives pulses from a light source through holes in the rotating disc *C* on the mirror shaft at appropriate times in the mirror revolution. With the light spot stationary at various distances from the probe the photovoltage (shown logarithmically) varies with distance as shown in curve (a) of Fig. 72 (cf. Fig. 20 for a *P-N* junction) and the slopes of the sides of the curve are equal and give the inverse of the carrier diffusion length $L=(D\tau)^{\frac{1}{2}}$ (see Sect. 16). This static method

[1] F. S. GOUCHER: Phys. Rev. **81**, 475 (1951).
[2] G. ADAM: Physica, Haag **20**, 1037 (1954). — Z. Naturforsch. **9a**, 607 (1954).

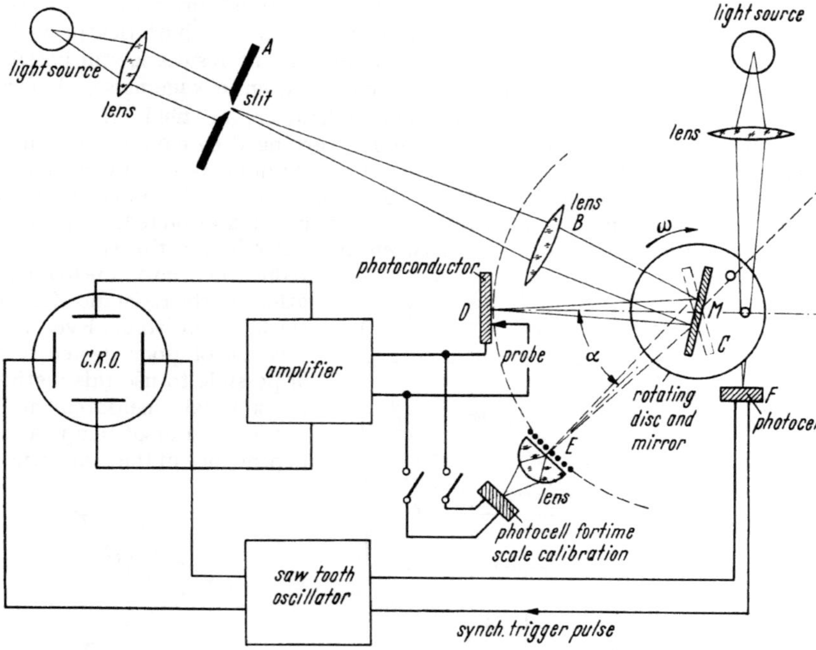


Fig. 71. Schematic diagram of experimental arrangement for the flying light spot method of measuring carrier mobility and life time in photoconductors (after ADAM ref. 2, p. 393).

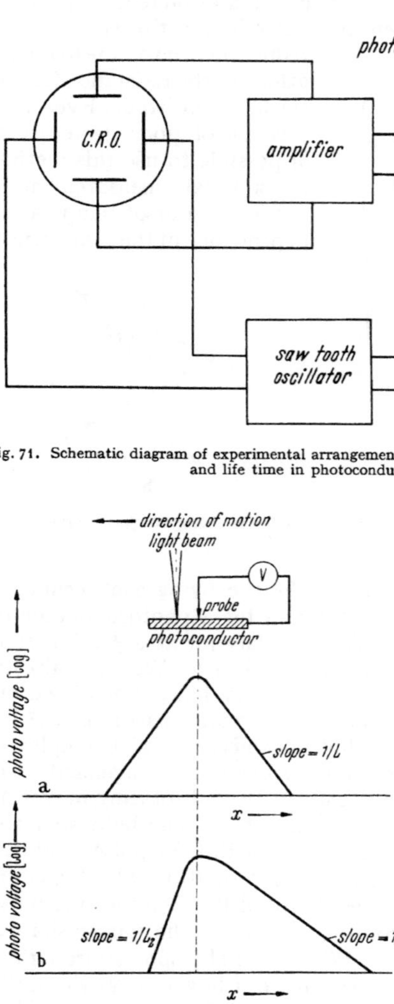


Fig. 72a and b. Schematic diagram showing effect of light spot velocity on photovoltage at photovoltaic barrier or junction. a) Velocity zero, b) finite velocity $c$ (after ADAM ref. 2, p. 393).

only gives the product $D\tau$. By scanning the light spot at a velocity $c$ across the probe the diffusion velocity $v_D$ of the liberated carriers now has to compete with the velocity $c$. As a result the approach slope is different from the retreating slope as shown in curve (b) of Fig. 72. If these slopes give $1/L_1$ and $1/L_2$ respectively then it is easy to show that for the case of a very small spot moving in one dimension:

$$\left.\begin{aligned}\tau &= (L_1 - L_2)/c\\ &\text{and}\\ D &= c/(1/L_2 - 1/L_1) = \mu\, kT/e.\end{aligned}\right\}\quad (63.1)$$

When a line image is used instead of a spot the analysis is more complex for the curve at small distances from the probe but the above relations are applicable at larger distances.

# Bibliography of important literature.

FASSBENDER, J., and H. LEHMANN: Ann. Phys., Lpz. **6**, 215 (1949).
GUDDEN, B.: Lichtelektrische Erscheinungen. Leipzig 1928. A useful survey of the Göttingen work is given in this volume.
HOFSTADTER, R.: Nucleonics **4**, 2, 29 (1949). Gives a review of the development of conductivity or crystal counters for nuclear particles.
HUGHES, A. L., and L. A. DUBRIDGE: Photoelectric Phenomena. New York 1932. Gives an exhaustive review of photoconductivity up to date of publication.
MAC DONALD, J. R.: Phys. Rev. **85**, 381 (1952); **92**, 4 (1953). J. Chem. Phys. **23**, 275 (1955). Gives a treatment of the photodielectric effect in solids which is important for analysis of dielectric constant changes in polycrystalline photoconductors.
MITCHELL, J. W. (Editor): Mechanisms of Photographic Sensitivity. London 1951. Contains the proceedings of a Conference at Bristol in 1950 on photographic materials inclusive of fundamental research. The paper of WEST is in this volume.
MOSS, T. S.: Photoconductivity in the Elements. London 1952. An exhaustive account and critical discussion of photoconduction in monatomic solids with initial chapters giving the theories of photoconduction, the photovoltaic effect and spectral response.
MOTT, N. F., and R. W. GURNEY: Electronic Processes in Ionic Crystals. Oxford 1940. An important source book in solid state physics-useful for relation of photoconduction to other phenomena such as semiconduction and colour centres in alkali halides.
MUSCHEID, W.: Ann. Phys., Lpz. **13**, 305 (1953).
NIEKISCH, E. A.: Ann. Phys., Lpz. **15**, 279, 288 (1955). These papers describe applications of a light modulation system to measurement of carrier mobility and trapping state distributions in photoconducting crystals.
REDFIELD, A.: Phys. Rev. **91**, 244, 753 (1953); **94**, 526, 537 (1954). These papers describe a novel method of measuring the HALL effect in insulating photoconductors where conventional techniques are inadequate.
SEITZ, F.: Modern Theory of Solids. New York 1940. A similar coverage to that of MOTT and GURNEY.
— Rev. Mod. Phys. **23**, 328 (1951). An extensive review of photographic and photoconduction characteristics of silver halides with a critical theoretical discussion.
— Rev. Mod. Phys. **26**, 7 (1954). An exhaustive review of the knowledge of colour centres in the alkali halides with a section on photoconductivity in these crystals.
SHOCKLEY, W.: Electrons and Holes in Semiconductors. New York 1950. The first part concerns the physical properties of germanium and the germanium transistor. The last half gives an extensive treatment of solid state theory.
SMITH, R. A.: Adv. Physics **2**, 321 (1953). Physica, Haag **20**, 910 (1954). Gives exhaustive review and critical discussions of work on lead sulphide, selenide and telluride both as infra red detectors and from a fundamental aspect.
WILSON, A. H.: Theory of Metals, 2nd edit. Cambridge 1953. Provides a detailed, up to date treatment of energy levels in solids with treatment of semiconduction theory.
Several authors: Physica, Haag **20**, 801—1138 (1954). Contains the papers of a Conference at Amsterdam in 1954 on semi and photoconduction. Several references to these papers appear in the above text. Important materials dealt with are germanium, lead sulphide etc. and the intermetallic compounds.

Also available since this article was written:

Photoconductivity. New York 1955. Proceedings of recent American Conference on the subject.

*Conductivity Counters.*

Information on conductivity counters for nuclear radiation detection may be found in the review by HOFSTADTER and in a review by CHYNOWETH: A. G. CHYNOWETH: Amer. J. Phys. **20**, 218 (1952). For some fundamental studies on conductivity induced by particles see CHYNOWETH (ref. 3, p. 341), TROTT (ref. 4, p. 341), CHAMPION (ref. 6, p. 341).

*Energy states and transfer in organic systems.*

For information on the theoretical treatment of fundamental processes of excitation in organic systems relevant to section V above see TH. FÖRSTER: Fluoreszenz organischer Verbindungen. Göttingen 1951.

# Sachverzeichnis.

## (Deutsch-Englisch.)

Bei gleicher Schreibweise in beiden Sprachen sind die Stichwörter nur einmal aufgeführt.

# Subject Index.

## (English-German.)

Where English and German spelling of a word is identical the German version is omitted.

MIX
Papier aus verantwortungsvollen Quellen
Paper from responsible sources
FSC® C105338

If you have any concerns about our products,
you can contact us on
**ProductSafety@springernature.com**

In case Publisher is established outside the EU,
the EU authorized representative is:
**Springer Nature Customer Service Center GmbH**
**Europaplatz 3, 69115 Heidelberg, Germany**

Printed by Libri Plureos GmbH
in Hamburg, Germany